Introduction to Electrodynamics

David J. Griffiths
Reed College

Prentice Hall
Upper Saddle River, New Jersey 07458

Original English language title from Proprietor's edition of the Work..

Original English language title: Introduction to Electrodynamics 3rd ed., David J. Griffiths, copyright © 1999，1989,1981

Published by arrangement with the original publisher, Pearson Education, Inc., publishing as Prentice Hall, Inc.

本书影印版由 Pearson Education, Inc.和世界图书出版公司北京公司合作出版发行。

ISBN 0-13-805326-X

影印版前言

电动力学是大学本科物理系的一门支柱课程。它和理论力学、热力学和统计物理学以及量子力学一起通常称为“四大力学”，构成对学生进行基础物理学理论知识训练的核心，也是进一步学习更高等的基础课和各类专业课必不可少的准备知识。

在自然界已知的 4 种相互作用力（引力、电磁力、强相互作用力和弱相互作用力）中，电磁力是我们了解得最清楚的。它在我们日常的生活与生产活动以及各类技术的进步中起着无可置疑的支配作用。而电动力学正是介绍电磁规律的理论课程。因此，它不仅是物理系本科生的必修课，也是范围越来越广的理工科学生选修的课程。

从事电动力学教和学的师生，多年来一直有一个困惑，或者有一点遗憾：即除了国内一些专家撰写的教材以外，国外引进的适合本科使用的教材难得一见。大家熟悉的 Jackson 的名著《经典电动力学》（Classical Electrodynamics）是以研究生为对象的，其内容的深度与广度都不适宜本科生使用。而量子力学的情况则截然不同，随便就可以找到适合各种程度，风格各异的十几种外文教材。它们让师生们大开眼界，使他们不仅可以取长补短，改进教学，还可以在比较中鉴别，深化对于该课程的理解，进而发展自己具有独特风格的教材。

我们现在终于有机会向大家推荐由 Prentice Hall 出版社出版的一本面向本科生的优秀教材，它是由格里菲斯（David J. Griffiths）所著的《电动力学导论》（Introduction to Electrodynamics）。该书的第 1 版出版于 1981 年，由于其丰富的内容，新颖的叙述风格以及大量完整解出的例题而得到好评，被美国多所大学选为教材。这里影印的是 1999 年出版的第 3 版。该版改动了不少细节，把第 1 版第 7 章的两节扩充为新的两章，充实了对麦克斯韦方程物理意义的讨论。

本书是一部很有实际应用价值的教科书。清晰和完美是作者努力追求的目标。为了使其能自成完整的体系，本书的第 1 章用了很大篇幅讲述矢量分析，为电动力学的主要数学工具作了充分的准备。作者从一个比较低的起点开始正文的阐述，纳入了一些普通物理学中的电磁学知识，由浅入深地展开全文。全书的公式推导很详尽，这使本书易读易懂，适应更广泛的读者对象。本书设计为两个学期教完，但对于基础比较好的学生，很多部分可以跳过去，一个学期完成全书的讲授是完全可能的。

本书选编了104个例题和530个习题，它们穿插安排在各个章节，成为本书重要的组成部分。它们对于理解课程内容，掌握解决实际问题的方法，扩大知识面至关重要。不少题目难度比较大，有的直接取自《美国物理学杂志》(American Journal of Physics)。因此有些读者曾经建议把习题解答附在书后，但遭到作者的强烈反对。他只同意对某些特别困难的题目提供答案或简单的提示。作为例外的是，对于集体购买超过一定数量，而且可以证明把本书选作教材的任课教师，可以向出版商索取带有全部解答的教学指南。此外还需要特别指出，本书的很多插图制作非常精美，这些清晰而直观的图像，对于理解物理结果有很大的帮助。

我们相信，本书的影印出版不仅会受到物理类理工科本科师生的欢迎，相关专业的研究生和科研人员也会从中受益。

中国科学院研究生院物理科学学院教授 丁亦兵

2005年12月

Contents

Preface ix

Advertisement xi

1 Vector Analysis 1
1.1 Vector Algebra 1
1.1.1 Vector Operations 1
1.1.2 Vector Algebra: Component Form 4
1.1.3 Triple Products 7
1.1.4 Position, Displacement, and Separation Vectors 8
1.1.5 How Vectors Transform 10
1.2 Differential Calculus 13
1.2.1 "Ordinary" Derivatives 13
1.2.2 Gradient 13
1.2.3 The Operator ∇ 16
1.2.4 The Divergence 17
1.2.5 The Curl 19
1.2.6 Product Rules 20
1.2.7 Second Derivatives 22
1.3 Integral Calculus 24
1.3.1 Line, Surface, and Volume Integrals 24
1.3.2 The Fundamental Theorem of Calculus 28
1.3.3 The Fundamental Theorem for Gradients 29
1.3.4 The Fundamental Theorem for Divergences 31
1.3.5 The Fundamental Theorem for Curls 34
1.3.6 Integration by Parts 37
1.4 Curvilinear Coordinates 38
1.4.1 Spherical Polar Coordinates 38
1.4.2 Cylindrical Coordinates 43
1.5 The Dirac Delta Function 45
1.5.1 The Divergence of $\hat{\mathbf{r}}/r^2$ 45
1.5.2 The One-Dimensional Dirac Delta Function 46

1.5.3 The Three-Dimensional Delta Function 50
1.6 The Theory of Vector Fields 52
1.6.1 The Helmholtz Theorem 52
1.6.2 Potentials 53

2 Electrostatics **58**
2.1 The Electric Field 58
2.1.1 Introduction 58
2.1.2 Coulomb's Law 59
2.1.3 The Electric Field 60
2.1.4 Continuous Charge Distributions 61
2.2 Divergence and Curl of Electrostatic Fields 65
2.2.1 Field Lines, Flux, and Gauss's Law 65
2.2.2 The Divergence of E 69
2.2.3 Applications of Gauss's Law 70
2.2.4 The Curl of E 76
2.3 Electric Potential 77
2.3.1 Introduction to Potential 77
2.3.2 Comments on Potential 79
2.3.3 Poisson's Equation and Laplace's Equation 83
2.3.4 The Potential of a Localized Charge Distribution 83
2.3.5 Summary; Electrostatic Boundary Conditions 87
2.4 Work and Energy in Electrostatics 90
2.4.1 The Work Done to Move a Charge 90
2.4.2 The Energy of a Point Charge Distribution 91
2.4.3 The Energy of a Continuous Charge Distribution 93
2.4.4 Comments on Electrostatic Energy 95
2.5 Conductors 96
2.5.1 Basic Properties 96
2.5.2 Induced Charges 98
2.5.3 Surface Charge and the Force on a Conductor 102
2.5.4 Capacitors 103

3 Special Techniques **110**
3.1 Laplace's Equation 110
3.1.1 Introduction 110
3.1.2 Laplace's Equation in One Dimension 111
3.1.3 Laplace's Equation in Two Dimensions 112
3.1.4 Laplace's Equation in Three Dimensions 114
3.1.5 Boundary Conditions and Uniqueness Theorems 116
3.1.6 Conductors and the Second Uniqueness Theorem 118
3.2 The Method of Images 121
3.2.1 The Classic Image Problem 121
3.2.2 Induced Surface Charge 123

3.2.3 Force and Energy 123
3.2.4 Other Image Problems 124
3.3 Separation of Variables 127
3.3.1 Cartesian Coordinates 127
3.3.2 Spherical Coordinates 137
3.4 Multipole Expansion 146
3.4.1 Approximate Potentials at Large Distances 146
3.4.2 The Monopole and Dipole Terms 149
3.4.3 Origin of Coordinates in Multipole Expansions 151
3.4.4 The Electric Field of a Dipole 153

4 Electric Fields in Matter **160**
4.1 Polarization 160
4.1.1 Dielectrics 160
4.1.2 Induced Dipoles 160
4.1.3 Alignment of Polar Molecules 163
4.1.4 Polarization 166
4.2 The Field of a Polarized Object 166
4.2.1 Bound Charges 166
4.2.2 Physical Interpretation of Bound Charges 170
4.2.3 The Field Inside a Dielectric 173
4.3 The Electric Displacement 175
4.3.1 Gauss's Law in the Presence of Dielectrics 175
4.3.2 A Deceptive Parallel 178
4.3.3 Boundary Conditions 178
4.4 Linear Dielectrics 179
4.4.1 Susceptibility, Permittivity, Dielectric Constant 179
4.4.2 Boundary Value Problems with Linear Dielectrics 186
4.4.3 Energy in Dielectric Systems 191
4.4.4 Forces on Dielectrics 193

5 Magnetostatics **202**
5.1 The Lorentz Force Law 202
5.1.1 Magnetic Fields 202
5.1.2 Magnetic Forces 204
5.1.3 Currents 208
5.2 The Biot-Savart Law 215
5.2.1 Steady Currents 215
5.2.2 The Magnetic Field of a Steady Current 215
5.3 The Divergence and Curl of B 221
5.3.1 Straight-Line Currents 221
5.3.2 The Divergence and Curl of B 222
5.3.3 Applications of Ampère's Law 225
5.3.4 Comparison of Magnetostatics and Electrostatics 232

5.4 Magnetic Vector Potential 234
5.4.1 The Vector Potential 234
5.4.2 Summary; Magnetostatic Boundary Conditions 240
5.4.3 Multipole Expansion of the Vector Potential 242

6 Magnetic Fields in Matter **255**
6.1 Magnetization 255
6.1.1 Diamagnets, Paramagnets, Ferromagnets 255
6.1.2 Torques and Forces on Magnetic Dipoles 255
6.1.3 Effect of a Magnetic Field on Atomic Orbits 260
6.1.4 Magnetization 262
6.2 The Field of a Magnetized Object 263
6.2.1 Bound Currents 263
6.2.2 Physical Interpretation of Bound Currents 266
6.2.3 The Magnetic Field Inside Matter 268
6.3 The Auxiliary Field H 269
6.3.1 Ampère's law in Magnetized Materials 269
6.3.2 A Deceptive Parallel 273
6.3.3 Boundary Conditions 273
6.4 Linear and Nonlinear Media 274
6.4.1 Magnetic Susceptibility and Permeability 274
6.4.2 Ferromagnetism 278

7 Electrodynamics **285**
7.1 Electromotive Force 285
7.1.1 Ohm's Law 285
7.1.2 Electromotive Force 292
7.1.3 Motional emf 294
7.2 Electromagnetic Induction 301
7.2.1 Faraday's Law 301
7.2.2 The Induced Electric Field 305
7.2.3 Inductance 310
7.2.4 Energy in Magnetic Fields 317
7.3 Maxwell's Equations 321
7.3.1 Electrodynamics Before Maxwell 321
7.3.2 How Maxwell Fixed Ampère's Law 323
7.3.3 Maxwell's Equations 326
7.3.4 Magnetic Charge 327
7.3.5 Maxwell's Equations in Matter 328
7.3.6 Boundary Conditions 331

8 Conservation Laws **345**
8.1 Charge and Energy 345
8.1.1 The Continuity Equation 345
8.1.2 Poynting's Theorem 346
8.2 Momentum 349
8.2.1 Newton's Third Law in Electrodynamics 349
8.2.2 Maxwell's Stress Tensor 351
8.2.3 Conservation of Momentum 355
8.2.4 Angular Momentum 358

9 Electromagnetic Waves **364**
9.1 Waves in One Dimension 364
9.1.1 The Wave Equation 364
9.1.2 Sinusoidal Waves 367
9.1.3 Boundary Conditions: Reflection and Transmission 370
9.1.4 Polarization 373
9.2 Electromagnetic Waves in Vacuum 375
9.2.1 The Wave Equation for **E** and **B** 375
9.2.2 Monochromatic Plane Waves 376
9.2.3 Energy and Momentum in Electromagnetic Waves 380
9.3 Electromagnetic Waves in Matter 382
9.3.1 Propagation in Linear Media 382
9.3.2 Reflection and Transmission at Normal Incidence 384
9.3.3 Reflection and Transmission at Oblique Incidence 386
9.4 Absorption and Dispersion 392
9.4.1 Electromagnetic Waves in Conductors 392
9.4.2 Reflection at a Conducting Surface 396
9.4.3 The Frequency Dependence of Permittivity 398
9.5 Guided Waves 405
9.5.1 Wave Guides 405
9.5.2 TE Waves in a Rectangular Wave Guide 408
9.5.3 The Coaxial Transmission Line 411

10 Potentials and Fields **416**
10.1 The Potential Formulation 416
10.1.1 Scalar and Vector Potentials 416
10.1.2 Gauge Transformations 419
10.1.3 Coulomb Gauge and Lorentz* Gauge 421
10.2 Continuous Distributions 422
10.2.1 Retarded Potentials 422
10.2.2 Jefimenko's Equations 427
10.3 Point Charges 429
10.3.1 Liénard-Wiechert Potentials 429
10.3.2 The Fields of a Moving Point Charge 435

11 Radiation **443**
11.1 Dipole Radiation . . . 443
11.1.1 What is Radiation? . . . 443
11.1.2 Electric Dipole Radiation . . . 444
11.1.3 Magnetic Dipole Radiation . . . 451
11.1.4 Radiation from an Arbitrary Source . . . 454
11.2 Point Charges . . . 460
11.2.1 Power Radiated by a Point Charge . . . 460
11.2.2 Radiation Reaction . . . 465
11.2.3 The Physical Basis of the Radiation Reaction . . . 469

12 Electrodynamics and Relativity **477**
12.1 The Special Theory of Relativity . . . 477
12.1.1 Einstein's Postulates . . . 477
12.1.2 The Geometry of Relativity . . . 483
12.1.3 The Lorentz Transformations . . . 493
12.1.4 The Structure of Spacetime . . . 500
12.2 Relativistic Mechanics . . . 507
12.2.1 Proper Time and Proper Velocity . . . 507
12.2.2 Relativistic Energy and Momentum . . . 509
12.2.3 Relativistic Kinematics . . . 511
12.2.4 Relativistic Dynamics . . . 516
12.3 Relativistic Electrodynamics . . . 522
12.3.1 Magnetism as a Relativistic Phenomenon . . . 522
12.3.2 How the Fields Transform . . . 525
12.3.3 The Field Tensor . . . 535
12.3.4 Electrodynamics in Tensor Notation . . . 537
12.3.5 Relativistic Potentials . . . 541

A Vector Calculus in Curvilinear Coordinates **547**
A.1 Introduction . . . 547
A.2 Notation . . . 547
A.3 Gradient . . . 548
A.4 Divergence . . . 549
A.5 Curl . . . 552
A.6 Laplacian . . . 554

B The Helmholtz Theorem **555**

C Units **558**

Index **562**

Preface

This is a textbook on electricity and magnetism, designed for an undergraduate course at the junior or senior level. It can be covered comfortably in two semesters, maybe even with room to spare for special topics (AC circuits, numerical methods, plasma physics, transmission lines, antenna theory, etc.) A one-semester course could reasonably stop after Chapter 7. Unlike quantum mechanics or thermal physics (for example), there is a fairly general consensus with respect to the teaching of electrodynamics; the subjects to be included, and even their order of presentation, are not particularly controversial, and textbooks differ mainly in style and tone. My approach is perhaps less formal than most; I think this makes difficult ideas more interesting and accessible.

For the third edition I have made a large number of small changes, in the interests of clarity and grace. I have also modified some notation to avoid inconsistencies or ambiguities. Thus the Cartesian unit vectors $\hat{\imath}$, $\hat{\jmath}$, and $\hat{k}$ have been replaced with $\hat{\mathbf{x}}$, $\hat{\mathbf{y}}$, and $\hat{\mathbf{z}}$, so that all vectors are bold, and all unit vectors inherit the letter of the corresponding coordinate. (This also frees up $\mathbf{k}$ to be the propagation vector for electromagnetic waves.) It has always bothered me to use the same letter r for the spherical coordinate (distance from the origin) and the cylindrical coordinate (distance from the z axis). A common alternative for the latter is ρ, but that has more important business in electrodynamics, and after an exhaustive search I settled on the underemployed letter s; I hope this unorthodox usage will not be confusing.

Some readers have urged me to abandon the script letter $\boldsymbol{\mathscr{r}}$ (the vector from a source point $\mathbf{r}'$ to the field point $\mathbf{r}$) in favor of the more explicit $\mathbf{r} - \mathbf{r}'$. But this makes many equations distractingly cumbersome, especially when the unit vector $\hat{\boldsymbol{\mathscr{r}}}$ is involved. I know from my own teaching experience that unwary students are tempted to read $\boldsymbol{\mathscr{r}}$ as $\mathbf{r}$—it certainly makes the integrals easier! I have inserted a section in Chapter 1 explaining this notation, and I hope that will help. If you are a student, please take note: $\boldsymbol{\mathscr{r}} \equiv \mathbf{r} - \mathbf{r}'$, which is *not* the same as $\mathbf{r}$. If you're a teacher, please warn your students to pay close attention to the meaning of $\boldsymbol{\mathscr{r}}$. I think it's *good* notation, but it does have to be handled with care.

The main structural change is that I have removed the conservation laws and potentials from Chapter 7, creating two new short chapters (8 and 10). This should more smoothly accommodate one-semester courses, and it gives a tighter focus to Chapter 7.

I have added some problems and examples (and removed a few that were not effective). And I have included more references to the accessible literature (particularly the *American Journal of Physics*). I realize, of course, that most readers will not have the time or inclination to consult these resources, but I think it is worthwhile anyway, if only to emphasize that electrodynamics, notwithstanding its venerable age, is very much alive, and intriguing new discoveries are being made all the time. I hope that occasionally a problem will pique your curiosity, and you will be inspired to look up the reference—some of them are real gems.

As in the previous editions, I distinguish two kinds of problems. Some have a specific pedagogical purpose, and should be worked immediately after reading the section to which they pertain; these I have placed at the pertinent point within the chapter. (In a few cases the solution to a problem is used later in the text; these are indicated by a bullet (•) in the left margin.) Longer problems, or those of a more general nature, will be found at the end of each chapter. When I teach the subject I assign some of these, and work a few of them in class. Unusually challenging problems are flagged by an exclamation point (!) in the margin. Many readers have asked that the answers to problems be provided at the back of the book; unfortunately, just as many are strenuously opposed. I have compromised, supplying answers when this seems particularly appropriate. A complete solution manual is available (to instructors) from the publisher.

I have benefitted from the comments of many colleagues—I cannot list them all here. But I would like to thank the following people for suggestions that contributed specifically to the third edition: Burton Brody (Bard), Steven Grimes (Ohio), Mark Heald (Swarthmore), Jim McTavish (Liverpool), Matthew Moelter (Puget Sound), Paul Nachman (New Mexico State), Gigi Quartapelle (Milan), Carl A. Rotter (West Virginia), Daniel Schroeder (Weber State), Juri Silmberg (Ryerson Polytechnic), Walther N. Spjeldvik (Weber State), Larry Tankersley (Naval Academy), and Dudley Towne (Amherst). Practically everything I know about electrodynamics—certainly about teaching electrodynamics—I owe to Edward Purcell.

David J. Griffiths

Advertisement

What is electrodynamics, and how does it fit into the general scheme of physics?

Four Realms of Mechanics

In the diagram below I have sketched out the four great realms of mechanics:

Classical Mechanics (Newton)	**Quantum Mechanics** (Bohr, Heisenberg, Schrödinger, *et al.*)
Special Relativity (Einstein)	**Quantum Field Theory** (Dirac, Pauli, Feynman, Schwinger, *et al.*)

Newtonian mechanics was found to be inadequate in the early years of this century—it's all right in "everyday life," but for objects moving at high speeds (near the speed of light) it is incorrect, and must be replaced by special relativity (introduced by Einstein in 1905); for objects that are extremely small (near the size of atoms) it fails for different reasons, and is superseded by quantum mechanics (developed by Bohr, Schrödinger, Heisenberg, and many others, in the twenties, mostly). For objects that are both very fast *and* very small (as is common in modern particle physics), a mechanics that combines relativity and quantum principles is in order: this relativistic quantum mechanics is known as quantum field theory—it was worked out in the thirties and forties, but even today it cannot claim to be a completely satisfactory system. In this book, save for the last chapter, we shall work exclusively in the domain of classical mechanics, although electrodynamics extends with unique simplicity to the other three realms. (In fact, the theory is in most respects *automatically* consistent with special relativity, for which it was, historically, the main stimulus.)

Four Kinds of Forces

Mechanics tells us how a system will behave when subjected to a given *force*. There are just *four* basic forces known (presently) to physics: I list them in the order of decreasing strength:

1. Strong
2. Electromagnetic
3. Weak
4. Gravitational

The brevity of this list may surprise you. Where is friction? Where is the "normal" force that keeps you from falling through the floor? Where are the chemical forces that bind molecules together? Where is the force of impact between two colliding billiard balls? The answer is that *all* these forces are *electromagnetic*. Indeed, it is scarcely an exaggeration to say that we live in an electromagnetic world—for virtually every force we experience in everyday life, with the exception of gravity, is electromagnetic in origin.

The **strong forces**, which hold protons and neutrons together in the atomic nucleus, have extremely short range, so we do not "feel" them, in spite of the fact that they are a hundred times more powerful than electrical forces. The **weak forces**, which account for certain kinds of radioactive decay, are not only of short range; they are far weaker than electromagnetic ones to begin with. As for gravity, it is so pitifully feeble (compared to all of the others) that it is only by virtue of huge mass concentrations (like the earth and the sun) that we ever notice it at all. The electrical repulsion between two electrons is 10^{42} times as large as their gravitational attraction, and if atoms were held together by gravitational (instead of electrical) forces, a single hydrogen atom would be much larger than the known universe.

Not only are electromagnetic forces overwhelmingly the dominant ones in everyday life, they are also, at present, the *only* ones that are completely understood. There is, of course, a classical theory of gravity (Newton's law of universal gravitation) and a relativistic one (Einstein's general relativity), but no entirely satisfactory quantum mechanical theory of gravity has been constructed (though many people are working on it). At the present time there is a very successful (if cumbersome) theory for the weak interactions, and a strikingly attractive candidate (called **chromodynamics**) for the strong interactions. All these theories draw their inspiration from electrodynamics; none can claim conclusive experimental verification at this stage. So electrodynamics, a beautifully complete and successful theory, has become a kind of paradigm for physicists: an ideal model that other theories strive to emulate.

The laws of classical electrodynamics were discovered in bits and pieces by Franklin, Coulomb, Ampère, Faraday, and others, but the person who completed the job, and packaged it all in the compact and consistent form it has today, was James Clerk Maxwell. The theory is now a little over a hundred years old.

The Unification of Physical Theories

In the beginning, **electricity** and **magnetism** were entirely separate subjects. The one dealt with glass rods and cat's fur, pith balls, batteries, currents, electrolysis, and lightning; the other with bar magnets, iron filings, compass needles, and the North Pole. But in 1820 Oersted noticed that an *electric* current could deflect a *magnetic* compass needle. Soon afterward, Ampère correctly postulated that *all* magnetic phenomena are due to electric charges in motion. Then, in 1831, Faraday discovered that a moving *magnet* generates an *electric* current. By the time Maxwell and Lorentz put the finishing touches on the theory, electricity and magnetism were inextricably intertwined. They could no longer be regarded as separate subjects, but rather as two *aspects* of a *single* subject: **electromagnetism**.

Faraday had speculated that light, too, is electrical in nature. Maxwell's theory provided spectacular justification for this hypothesis, and soon **optics**—the study of lenses, mirrors, prisms, interference, and diffraction—was incorporated into electromagnetism. Hertz, who presented the decisive experimental confirmation for Maxwell's theory in 1888, put it this way: "The connection between light and electricity is now established ... In every flame, in every luminous particle, we see an electrical process ... Thus, the domain of electricity extends over the whole of nature. It even affects ourselves intimately: we perceive that we possess ... an electrical organ—the eye." By 1900, then, three great branches of physics, electricity, magnetism, and optics, had merged into a single unified theory. (And it was soon apparent that visible light represents only a tiny "window" in the vast spectrum of electromagnetic radiation, from radio through microwaves, infrared and ultraviolet, to x-rays and gamma rays.)

Einstein dreamed of a further unification, which would combine gravity and electrodynamics, in much the same way as electricity and magnetism had been combined a century earlier. His **unified field theory** was not particularly successful, but in recent years the same impulse has spawned a hierarchy of increasingly ambitious (and speculative) unification schemes, beginning in the 1960s with the **electroweak** theory of Glashow, Weinberg, and Salam (which joins the weak and electromagnetic forces), and culminating in the 1980s with the **superstring** theory (which, according to its proponents, incorporates all four forces in a single "theory of everything"). At each step in this hierarchy the mathematical difficulties mount, and the gap between inspired conjecture and experimental test widens; nevertheless, it is clear that the unification of forces initiated by electrodynamics has become a major theme in the progress of physics.

The Field Formulation of Electrodynamics

The fundamental problem a theory of electromagnetism hopes to solve is this: I hold up a bunch of electric charges *here* (and maybe shake them around)—what happens to some other charge, over *there?* The classical solution takes the form of a **field theory**: We say that the space around an electric charge is permeated by electric and magnetic **fields** (the electromagnetic "odor," as it were, of the charge). A second charge, in the presence of these fields, experiences a force; the fields, then, transmit the influence from one charge to the other—they mediate the interaction.

When a charge undergoes *acceleration,* a portion of the field "detaches" itself, in a sense, and travels off at the speed of light, carrying with it energy, momentum, and angular momentum. We call this **electromagnetic radiation**. Its existence invites (if not *compels*) us to regard the fields as independent dynamical entities in their own right, every bit as "real" as atoms or baseballs. Our interest accordingly shifts from the study of forces between charges to the theory of the fields themselves. But it takes a charge to *produce* an electromagnetic field, and it takes another charge to *detect* one, so we had best begin by reviewing the essential properties of electric charge.

Electric Charge

1. *Charge comes in two varieties,* which we call "plus" and "minus," because their effects tend to *cancel* (if you have $+q$ and $-q$ at the same point, electrically it is the same as having no charge there at all). This may seem too obvious to warrant comment, but I encourage you to contemplate other possibilities: what if there were 8 or 10 different species of charge? (In chromodynamics there are, in fact, *three* quantities analogous to electric charge, each of which may be positive or negative.) Or what if the two kinds did not tend to cancel? The extraordinary fact is that plus and minus charges occur in *exactly* equal amounts, to fantastic precision, in bulk matter, so that their effects are almost completely neutralized. Were it not for this, we would be subjected to enormous forces: a potato would explode violently if the cancellation were imperfect by as little as one part in 10^{10}.

2. *Charge is conserved:* it cannot be created or destroyed—what there is now has always been. (A plus charge can "annihilate" an equal minus charge, but a plus charge cannot simply disappear by itself—*something* must account for that electric charge.) So the total charge of the universe is fixed for all time. This is called **global** conservation of charge. Actually, I can say something much stronger: Global conservation would allow for a charge to disappear in New York and instantly reappear in San Francisco (that wouldn't affect the *total*), and yet we know this doesn't happen. If the charge *was* in New York and it *went* to San Francisco, then it must have passed along some continuous path from one to the other. This is called **local** conservation of charge. Later on we'll see how to formulate a precise mathematical law expressing local conservation of charge—it's called the **continuity equation**.

3. *Charge is quantized.* Although nothing in classical electrodynamics requires that it be so, the *fact* is that electric charge comes only in discrete lumps—integer multiples of the basic unit of charge. If we call the charge on the proton $+e$, then the electron carries charge $-e$, the neutron charge zero, the pi mesons $+e$, 0, and $-e$, the carbon nucleus $+6e$, and so on (never $7.392e$, or even $1/2e$).[1] This fundamental unit of charge is extremely small, so for practical purposes it is usually appropriate to ignore quantization altogether. Water, too, "really" consists of discrete lumps (molecules); yet, if we are dealing with reasonably large large quantities of it we can treat it as a continuous fluid. This is in fact much closer to Maxwell's own view; he knew nothing of electrons and protons—he must have pictured

[1] Actually, protons and neutrons are composed of three **quarks**, which carry fractional charges ($\pm\frac{2}{3}e$ and $\pm\frac{1}{3}e$). However, *free* quarks do not appear to exist in nature, and in any event this does not alter the fact that charge is quantized; it merely reduces the size of the basic unit.

charge as a kind of "jelly" that could be divided up into portions of any size and smeared out at will.

These, then, are the basic properties of charge. Before we discuss the forces *between* charges, some mathematical tools are necessary; their introduction will occupy us in Chapter 1.

Units

The subject of electrodynamics is plagued by competing systems of units, which sometimes render it difficult for physicists to communicate with one another. The problem is far worse than in mechanics, where Neanderthals still speak of pounds and feet; for in mechanics at least all equations *look* the same, regardless of the units used to measure quantities. Newton's second law remains $\mathbf{F} = m\mathbf{a}$, whether it is feet-pounds-seconds, kilograms-meters-seconds, or whatever. But this is not so in electromagnetism, where Coulomb's law may appear variously as

$$\frac{q_1 q_2}{\mathscr{r}^2}\hat{\mathscr{r}} \text{ (Gaussian), or } \frac{1}{4\pi\epsilon_0}\frac{q_1 q_2}{\mathscr{r}^2}\hat{\mathscr{r}} \text{ (SI), or } \frac{1}{4\pi}\frac{q_1 q_2}{\mathscr{r}^2}\hat{\mathscr{r}} \text{ (HL).}$$

Of the systems in common use, the two most popular are **Gaussian** (cgs) and **SI** (mks). Elementary particle theorists favor yet a third system: **Heaviside-Lorentz**. Although Gaussian units offer distinct theoretical advantages, most undergraduate instructors seem to prefer SI, I suppose because they incorporate the familiar household units (volts, amperes, and watts). In this book, therefore, I have used SI units. Appendix C provides a "dictionary" for converting the main results into Gaussian units.

Chapter 1

Vector Analysis

1.1 Vector Algebra

1.1.1 Vector Operations

If you walk 4 miles due north and then 3 miles due east (Fig. 1.1), you will have gone a total of 7 miles, but you're *not* 7 miles from where you set out—you're only 5. We need an arithmetic to describe quantities like this, which evidently do not add in the ordinary way. The reason they don't, of course, is that **displacements** (straight line segments going from one point to another) have *direction* as well as *magnitude* (length), and it is essential to take both into account when you combine them. Such objects are called **vectors**: velocity, acceleration, force and momentum are other examples. By contrast, quantities that have magnitude but no direction are called **scalars**: examples include mass, charge, density, and temperature. I shall use **boldface** (**A**, **B**, and so on) for vectors and ordinary type for scalars. The magnitude of a vector **A** is written |**A**| or, more simply, A. In diagrams, vectors are denoted by arrows: the length of the arrow is proportional to the magnitude of the vector, and the arrowhead indicates its direction. *Minus* **A** (−**A**) is a vector with the

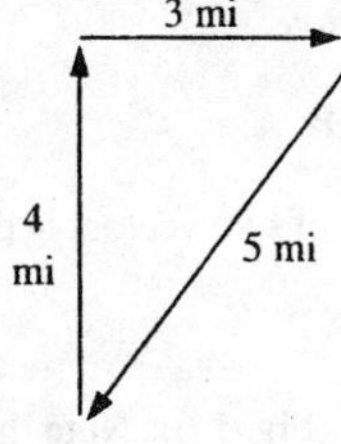

Figure 1.1

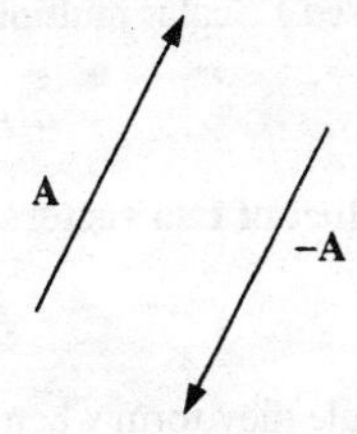

Figure 1.2

same magnitude as **A** but of opposite direction (Fig. 1.2). Note that vectors have magnitude and direction but *not location:* a displacement of 4 miles due north from Washington is represented by the same vector as a displacement 4 miles north from Baltimore (neglecting, of course, the curvature of the earth). On a diagram, therefore, you can slide the arrow around at will, as long as you don't change its length or direction.

We define four vector operations: addition and three kinds of multiplication.

(i) Addition of two vectors. Place the tail of **B** at the head of **A**; the sum, $\mathbf{A}+\mathbf{B}$, is the vector from the tail of **A** to the head of **B** (Fig. 1.3). (This rule generalizes the obvious procedure for combining two displacements.) Addition is *commutative:*

$$\mathbf{A}+\mathbf{B}=\mathbf{B}+\mathbf{A};$$

3 miles east followed by 4 miles north gets you to the same place as 4 miles north followed by 3 miles east. Addition is also *associative:*

$$(\mathbf{A}+\mathbf{B})+\mathbf{C}=\mathbf{A}+(\mathbf{B}+\mathbf{C}).$$

To subtract a vector (Fig. 1.4), add its opposite:

$$\mathbf{A}-\mathbf{B}=\mathbf{A}+(-\mathbf{B}).$$

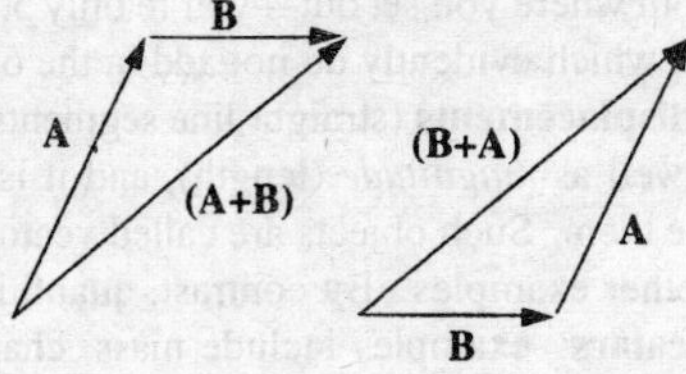

Figure 1.3

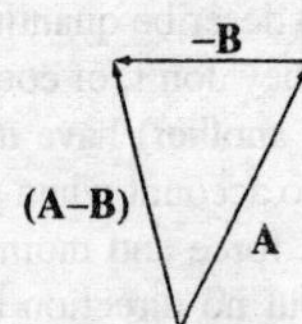

Figure 1.4

(ii) Multiplication by a scalar. Multiplication of a vector by a positive scalar a multiplies the *magnitude* but leaves the direction unchanged (Fig. 1.5). (If a is negative, the direction is reversed.) Scalar multiplication is *distributive:*

$$a(\mathbf{A}+\mathbf{B})=a\mathbf{A}+a\mathbf{B}.$$

(iii) Dot product of two vectors. The dot product of two vectors is defined by

$$\mathbf{A}\cdot\mathbf{B}\equiv AB\cos\theta, \tag{1.1}$$

where θ is the angle they form when placed tail-to-tail (Fig. 1.6). Note that $\mathbf{A}\cdot\mathbf{B}$ is itself a *scalar* (hence the alternative name **scalar product**). The dot product is *commutative,*

$$\mathbf{A}\cdot\mathbf{B}=\mathbf{B}\cdot\mathbf{A},$$

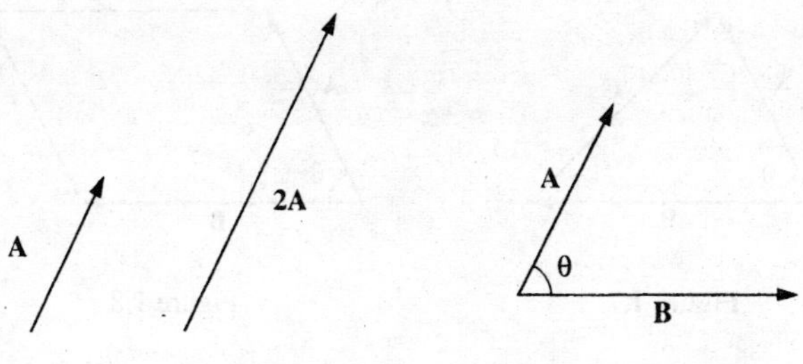

Figure 1.5

Figure 1.6

and *distributive*,

$$\mathbf{A} \cdot (\mathbf{B} + \mathbf{C}) = \mathbf{A} \cdot \mathbf{B} + \mathbf{A} \cdot \mathbf{C}. \tag{1.2}$$

Geometrically, $\mathbf{A} \cdot \mathbf{B}$ is the product of A times the projection of $\mathbf{B}$ along $\mathbf{A}$ (or the product of B times the projection of $\mathbf{A}$ along $\mathbf{B}$). If the two vectors are parallel, then $\mathbf{A} \cdot \mathbf{B} = AB$. In particular, for any vector $\mathbf{A}$,

$$\mathbf{A} \cdot \mathbf{A} = A^2. \tag{1.3}$$

If $\mathbf{A}$ and $\mathbf{B}$ are perpendicular, then $\mathbf{A} \cdot \mathbf{B} = 0$.

Example 1.1

Let $\mathbf{C} = \mathbf{A} - \mathbf{B}$ (Fig. 1.7), and calculate the dot product of $\mathbf{C}$ with itself.

Solution:

$$\mathbf{C} \cdot \mathbf{C} = (\mathbf{A} - \mathbf{B}) \cdot (\mathbf{A} - \mathbf{B}) = \mathbf{A} \cdot \mathbf{A} - \mathbf{A} \cdot \mathbf{B} - \mathbf{B} \cdot \mathbf{A} + \mathbf{B} \cdot \mathbf{B},$$

or

$$C^2 = A^2 + B^2 - 2AB\cos\theta.$$

This is the **law of cosines.**

(iv) Cross product of two vectors. The cross product of two vectors is defined by

$$\mathbf{A} \times \mathbf{B} \equiv AB \sin\theta \, \hat{\mathbf{n}}, \tag{1.4}$$

where $\hat{\mathbf{n}}$ is a **unit vector** (vector of length 1) pointing perpendicular to the plane of $\mathbf{A}$ and $\mathbf{B}$. (I shall use a hat (ˆ) to designate unit vectors.) Of course, there are *two* directions perpendicular to any plane: "in" and "out." The ambiguity is resolved by the **right-hand rule**: let your fingers point in the direction of the first vector and curl around (via the smaller angle) toward the second; then your thumb indicates the direction of $\hat{\mathbf{n}}$. (In Fig. 1.8 $\mathbf{A} \times \mathbf{B}$ points *into* the page; $\mathbf{B} \times \mathbf{A}$ points *out* of the page.) Note that $\mathbf{A} \times \mathbf{B}$ is itself a *vector* (hence the alternative name **vector product**). The cross product is *distributive*,

$$\mathbf{A} \times (\mathbf{B} + \mathbf{C}) = (\mathbf{A} \times \mathbf{B}) + (\mathbf{A} \times \mathbf{C}), \tag{1.5}$$

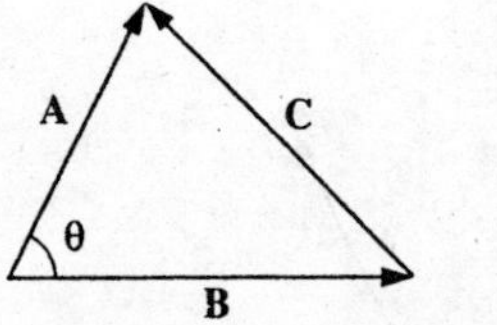

Figure 1.7

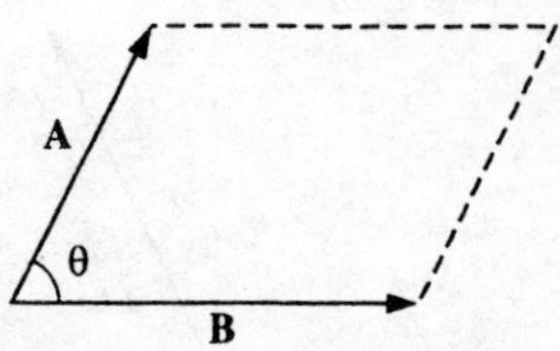

Figure 1.8

but *not commutative.* In fact,

$$(\mathbf{B} \times \mathbf{A}) = -(\mathbf{A} \times \mathbf{B}). \tag{1.6}$$

Geometrically, $|\mathbf{A} \times \mathbf{B}|$ is the area of the parallelogram generated by **A** and **B** (Fig. 1.8). If two vectors are parallel, their cross product is zero. In particular,

$$\mathbf{A} \times \mathbf{A} = \mathbf{0}$$

for any vector **A**.

Problem 1.1 Using the definitions in Eqs. 1.1 and 1.4, and appropriate diagrams, show that the dot product and cross product are distributive,

a) when the three vectors are coplanar;

! b) in the general case.

Problem 1.2 Is the cross product associative?

$$(\mathbf{A} \times \mathbf{B}) \times \mathbf{C} \stackrel{?}{=} \mathbf{A} \times (\mathbf{B} \times \mathbf{C}).$$

If so, *prove* it; if not, provide a counterexample.

1.1.2 Vector Algebra: Component Form

In the previous section I defined the four vector operations (addition, scalar multiplication, dot product, and cross product) in "abstract" form—that is, without reference to any particular coordinate system. In practice, it is often easier to set up Cartesian coordinates x, y, z and work with vector "components." Let $\hat{\mathbf{x}}$, $\hat{\mathbf{y}}$, and $\hat{\mathbf{z}}$ be unit vectors parallel to the x, y, and z axes, respectively (Fig. 1.9(a)). An arbitrary vector **A** can be expanded in terms of these **basis vectors** (Fig. 1.9(b)):

$$\mathbf{A} = A_x\hat{\mathbf{x}} + A_y\hat{\mathbf{y}} + A_z\hat{\mathbf{z}}.$$

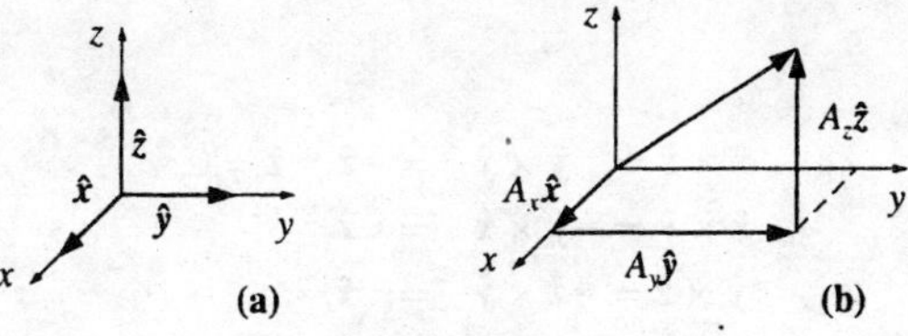

Figure 1.9

The numbers A_x, A_y, and A_z, are called **components** of **A**; geometrically, they are the projections of **A** along the three coordinate axes. We can now reformulate each of the four vector operations as a rule for manipulating components:

$$\begin{aligned}\mathbf{A}+\mathbf{B} &= (A_x\hat{\mathbf{x}}+A_y\hat{\mathbf{y}}+A_z\hat{\mathbf{z}})+(B_x\hat{\mathbf{x}}+B_y\hat{\mathbf{y}}+B_z\hat{\mathbf{z}})\\ &= (A_x+B_x)\hat{\mathbf{x}}+(A_y+B_y)\hat{\mathbf{y}}+(A_z+B_z)\hat{\mathbf{z}}.\end{aligned} \tag{1.7}$$

(i) Rule: *To add vectors, add like components.*

$$a\mathbf{A} = (aA_x)\hat{\mathbf{x}}+(aA_y)\hat{\mathbf{y}}+(aA_z)\hat{\mathbf{z}}. \tag{1.8}$$

(ii) Rule: *To multiply by a scalar, multiply each component.*

Because $\hat{\mathbf{x}}$, $\hat{\mathbf{y}}$, and $\hat{\mathbf{z}}$ are mutually perpendicular unit vectors,

$$\hat{\mathbf{x}}\cdot\hat{\mathbf{x}}=\hat{\mathbf{y}}\cdot\hat{\mathbf{y}}=\hat{\mathbf{z}}\cdot\hat{\mathbf{z}}=1; \quad \hat{\mathbf{x}}\cdot\hat{\mathbf{y}}=\hat{\mathbf{x}}\cdot\hat{\mathbf{z}}=\hat{\mathbf{y}}\cdot\hat{\mathbf{z}}=0. \tag{1.9}$$

Accordingly,

$$\begin{aligned}\mathbf{A}\cdot\mathbf{B} &= (A_x\hat{\mathbf{x}}+A_y\hat{\mathbf{y}}+A_z\hat{\mathbf{z}})\cdot(B_x\hat{\mathbf{x}}+B_y\hat{\mathbf{y}}+B_z\hat{\mathbf{z}})\\ &= A_xB_x+A_yB_y+A_zB_z.\end{aligned} \tag{1.10}$$

(iii) Rule: *To calculate the dot product, multiply like components, and add.*

In particular,

$$\mathbf{A}\cdot\mathbf{A}=A_x^2+A_y^2+A_z^2,$$

so

$$A=\sqrt{A_x^2+A_y^2+A_z^2}. \tag{1.11}$$

(This is, if you like, the three-dimensional generalization of the Pythagorean theorem.) Note that the dot product of **A** with any *unit* vector is the component of **A** along that direction (thus $\mathbf{A}\cdot\hat{\mathbf{x}}=A_x$, $\mathbf{A}\cdot\hat{\mathbf{y}}=A_y$, and $\mathbf{A}\cdot\hat{\mathbf{z}}=A_z$).

Similarly,[1]

$$\begin{aligned} \hat{\mathbf{x}}\times\hat{\mathbf{x}} = \hat{\mathbf{y}}\times\hat{\mathbf{y}} &= \hat{\mathbf{z}}\times\hat{\mathbf{z}} = 0, \\ \hat{\mathbf{x}}\times\hat{\mathbf{y}} = -\hat{\mathbf{y}}\times\hat{\mathbf{x}} &= \hat{\mathbf{z}}, \\ \hat{\mathbf{y}}\times\hat{\mathbf{z}} = -\hat{\mathbf{z}}\times\hat{\mathbf{y}} &= \hat{\mathbf{x}}, \\ \hat{\mathbf{z}}\times\hat{\mathbf{x}} = -\hat{\mathbf{x}}\times\hat{\mathbf{z}} &= \hat{\mathbf{y}}. \end{aligned} \tag{1.12}$$

Therefore,

$$\begin{aligned} \mathbf{A}\times\mathbf{B} &= (A_x\hat{\mathbf{x}} + A_y\hat{\mathbf{y}} + A_z\hat{\mathbf{z}}) \times (B_x\hat{\mathbf{x}} + B_y\hat{\mathbf{y}} + B_z\hat{\mathbf{z}}) \\ &= (A_yB_z - A_zB_y)\hat{\mathbf{x}} + (A_zB_x - A_xB_z)\hat{\mathbf{y}} + (A_xB_y - A_yB_x)\hat{\mathbf{z}}. \end{aligned} \tag{1.13}$$

This cumbersome expression can be written more neatly as a determinant:

$$\mathbf{A}\times\mathbf{B} = \begin{vmatrix} \hat{\mathbf{x}} & \hat{\mathbf{y}} & \hat{\mathbf{z}} \\ A_x & A_y & A_z \\ B_x & B_y & B_z \end{vmatrix}. \tag{1.14}$$

(iv) Rule: *To calculate the cross product, form the determinant whose first row is* $\hat{\mathbf{x}}$, $\hat{\mathbf{y}}$, $\hat{\mathbf{z}}$, *whose second row is* **A** *(in component form), and whose third row is* **B**.

Example 1.2

Find the angle between the face diagonals of a cube.

Solution: We might as well use a cube of side 1, and place it as shown in Fig. 1.10, with one corner at the origin. The face diagonals **A** and **B** are

$$\mathbf{A} = 1\,\hat{\mathbf{x}} + 0\,\hat{\mathbf{y}} + 1\,\hat{\mathbf{z}}; \qquad \mathbf{B} = 0\,\hat{\mathbf{x}} + 1\,\hat{\mathbf{y}} + 1\,\hat{\mathbf{z}}.$$

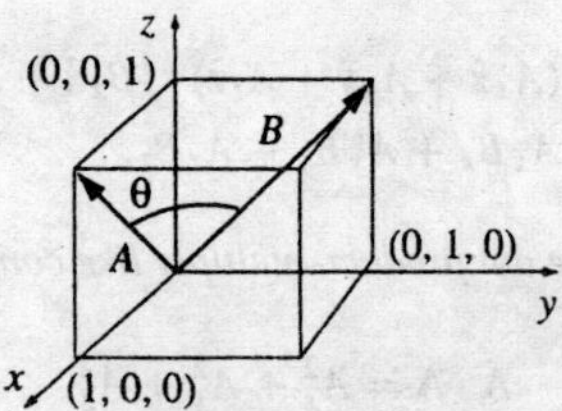

Figure 1.10

---

[1]These signs pertain to a *right-handed* coordinate system (x-axis out of the page, y-axis to the right, z-axis up, or any rotated version thereof). In a *left-handed* system (z-axis down) the signs are reversed: $\hat{\mathbf{x}}\times\hat{\mathbf{y}} = -\hat{\mathbf{z}}$, and so on. We shall use right-handed systems exclusively.

So, in component form,

$$\mathbf{A}\cdot\mathbf{B} = 1\cdot 0 + 0\cdot 1 + 1\cdot 1 = 1.$$

On the other hand, in "abstract" form,

$$\mathbf{A}\cdot\mathbf{B} = AB\cos\theta = \sqrt{2}\sqrt{2}\cos\theta = 2\cos\theta.$$

Therefore,

$$\cos\theta = 1/2, \quad \text{or} \quad \theta = 60^\circ.$$

Of course, you can get the answer more easily by drawing in a diagonal across the top of the cube, completing the equilateral triangle. But in cases where the geometry is not so simple, this device of comparing the abstract and component forms of the dot product can be a very efficient means of finding angles.

Problem 1.3 Find the angle between the body diagonals of a cube.

Problem 1.4 Use the cross product to find the components of the unit vector $\hat{\mathbf{n}}$ perpendicular to the plane shown in Fig. 1.11.

1.1.3 Triple Products

Since the cross product of two vectors is itself a vector, it can be dotted or crossed with a third vector to form a *triple* product.

(i) Scalar triple product: $\mathbf{A}\cdot(\mathbf{B}\times\mathbf{C})$. Geometrically, $|\mathbf{A}\cdot(\mathbf{B}\times\mathbf{C})|$ is the volume of the parallelepiped generated by **A**, **B**, and **C**, since $|\mathbf{B}\times\mathbf{C}|$ is the area of the base, and $|\mathbf{A}\cos\theta|$ is the altitude (Fig. 1.12). Evidently,

$$\mathbf{A}\cdot(\mathbf{B}\times\mathbf{C}) = \mathbf{B}\cdot(\mathbf{C}\times\mathbf{A}) = \mathbf{C}\cdot(\mathbf{A}\times\mathbf{B}), \tag{1.15}$$

for they all correspond to the same figure. Note that "alphabetical" order is preserved—in view of Eq. 1.6, the "nonalphabetical" triple products,

$$\mathbf{A}\cdot(\mathbf{C}\times\mathbf{B}) = \mathbf{B}\cdot(\mathbf{A}\times\mathbf{C}) = \mathbf{C}\cdot(\mathbf{B}\times\mathbf{A}),$$

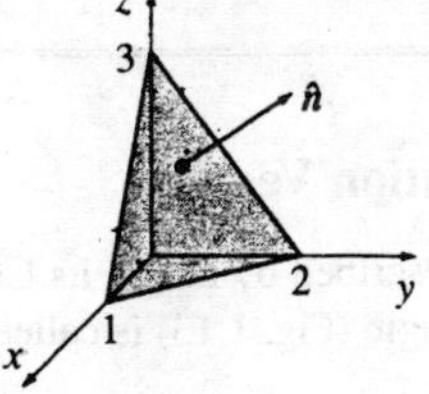

Figure 1.11

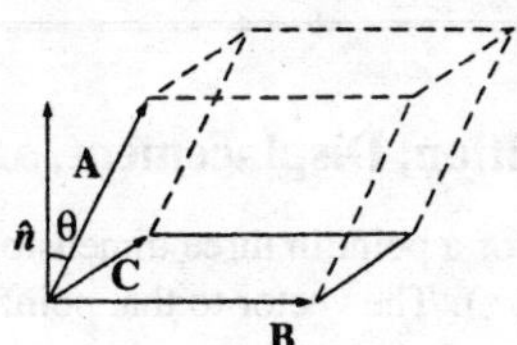

Figure 1.12

have the opposite sign. In component form,

$$\mathbf{A}\cdot(\mathbf{B}\times\mathbf{C}) = \begin{vmatrix} A_x & A_y & A_z \\ B_x & B_y & B_z \\ C_x & C_y & C_z \end{vmatrix}. \tag{1.16}$$

Note that the dot and cross can be interchanged:

$$\mathbf{A}\cdot(\mathbf{B}\times\mathbf{C}) = (\mathbf{A}\times\mathbf{B})\cdot\mathbf{C}$$

(this follows immediately from Eq. 1.15); however, the placement of the parentheses is critical: $(\mathbf{A}\cdot\mathbf{B})\times\mathbf{C}$ is a meaningless expression—you can't make a cross product from a *scalar* and a vector.

(ii) Vector triple product: $\mathbf{A}\times(\mathbf{B}\times\mathbf{C})$. The vector triple product can be simplified by the so-called **BAC-CAB** rule:

$$\mathbf{A}\times(\mathbf{B}\times\mathbf{C}) = \mathbf{B}(\mathbf{A}\cdot\mathbf{C}) - \mathbf{C}(\mathbf{A}\cdot\mathbf{B}). \tag{1.17}$$

Notice that

$$(\mathbf{A}\times\mathbf{B})\times\mathbf{C} = -\mathbf{C}\times(\mathbf{A}\times\mathbf{B}) = -\mathbf{A}(\mathbf{B}\cdot\mathbf{C}) + \mathbf{B}(\mathbf{A}\cdot\mathbf{C})$$

is an entirely different vector. Incidentally, all *higher* vector products can be similarly reduced, often by repeated application of Eq. 1.17, so it is never necessary for an expression to contain more than one cross product in any term. For instance,

$$\begin{aligned} (\mathbf{A}\times\mathbf{B})\cdot(\mathbf{C}\times\mathbf{D}) &= (\mathbf{A}\cdot\mathbf{C})(\mathbf{B}\cdot\mathbf{D}) - (\mathbf{A}\cdot\mathbf{D})(\mathbf{B}\cdot\mathbf{C}); \\ \mathbf{A}\times(\mathbf{B}\times(\mathbf{C}\times\mathbf{D})) &= \mathbf{B}(\mathbf{A}\cdot(\mathbf{C}\times\mathbf{D})) - (\mathbf{A}\cdot\mathbf{B})(\mathbf{C}\times\mathbf{D}). \end{aligned} \tag{1.18}$$

Problem 1.5 Prove the **BAC-CAB** rule by writing out both sides in component form.

Problem 1.6 Prove that

$$[\mathbf{A}\times(\mathbf{B}\times\mathbf{C})] + [\mathbf{B}\times(\mathbf{C}\times\mathbf{A})] + [\mathbf{C}\times(\mathbf{A}\times\mathbf{B})] = 0.$$

Under what conditions does $\mathbf{A}\times(\mathbf{B}\times\mathbf{C}) = (\mathbf{A}\times\mathbf{B})\times\mathbf{C}$?

1.1.4 Position, Displacement, and Separation Vectors

The location of a point in three dimensions can be described by listing its Cartesian coordinates (x, y, z). The vector to that point from the origin (Fig. 1.13) is called the **position vector**:

$$\mathbf{r} \equiv x\,\hat{\mathbf{x}} + y\,\hat{\mathbf{y}} + z\,\hat{\mathbf{z}}. \tag{1.19}$$

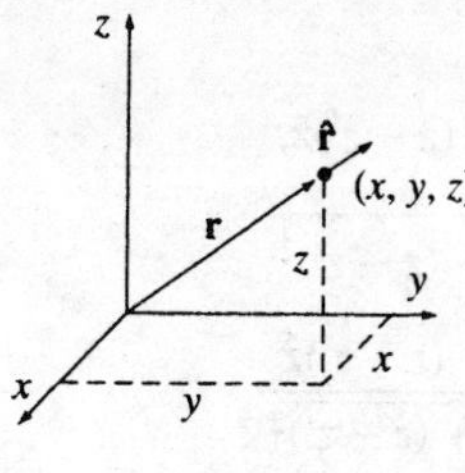

Figure 1.13

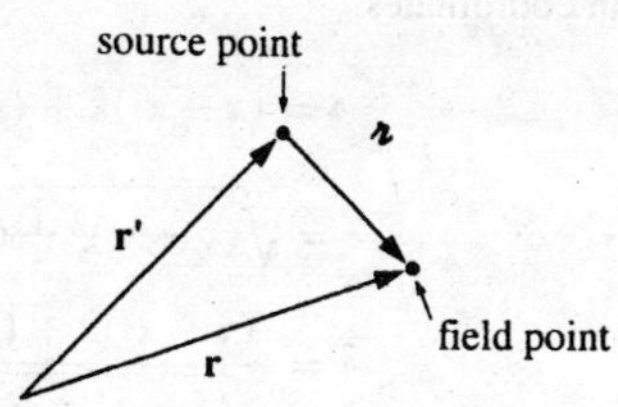

Figure 1.14

I will reserve the letter **r** for this purpose, throughout the book. Its magnitude,

$$r = \sqrt{x^2 + y^2 + z^2}, \tag{1.20}$$

is the distance from the origin, and

$$\hat{\mathbf{r}} = \frac{\mathbf{r}}{r} = \frac{x\,\hat{\mathbf{x}} + y\,\hat{\mathbf{y}} + z\,\hat{\mathbf{z}}}{\sqrt{x^2 + y^2 + z^2}} \tag{1.21}$$

is a unit vector pointing radially outward. The **infinitesimal displacement vector**, from (x, y, z) to $(x + dx, y + dy, z + dz)$, is

$$d\mathbf{l} = dx\,\hat{\mathbf{x}} + dy\,\hat{\mathbf{y}} + dz\,\hat{\mathbf{z}}. \tag{1.22}$$

(We could call this $d\mathbf{r}$, since that's what it *is*, but it is useful to reserve a special letter for infinitesimal displacements.)

In electrodynamics one frequently encounters problems involving *two* points—typically, a **source point**, $\mathbf{r}'$, where an electric charge is located, and a **field point**, $\mathbf{r}$, at which you are calculating the electric or magnetic field (Fig. 1.14). It pays to adopt right from the start some short-hand notation for the **separation vector** from the source point to the field point. I shall use for this purpose the script letter $\boldsymbol{\mathscr{r}}$:

$$\boldsymbol{\mathscr{r}} \equiv \mathbf{r} - \mathbf{r}'. \tag{1.23}$$

Its magnitude is

$$\mathscr{r} = |\mathbf{r} - \mathbf{r}'|, \tag{1.24}$$

and a unit vector in the direction from $\mathbf{r}'$ to $\mathbf{r}$ is

$$\hat{\boldsymbol{\mathscr{r}}} = \frac{\boldsymbol{\mathscr{r}}}{\mathscr{r}} = \frac{\mathbf{r} - \mathbf{r}'}{|\mathbf{r} - \mathbf{r}'|}. \tag{1.25}$$

In Cartesian coordinates,

$$\boldsymbol{\mathscr{r}} = (x - x')\hat{\mathbf{x}} + (y - y')\hat{\mathbf{y}} + (z - z')\hat{\mathbf{z}}, \tag{1.26}$$

$$\mathscr{r} = \sqrt{(x - x')^2 + (y - y')^2 + (z - z')^2}, \tag{1.27}$$

$$\hat{\boldsymbol{\mathscr{r}}} = \frac{(x - x')\hat{\mathbf{x}} + (y - y')\hat{\mathbf{y}} + (z - z')\hat{\mathbf{z}}}{\sqrt{(x - x')^2 + (y - y')^2 + (z - z')^2}} \tag{1.28}$$

(from which you can begin to appreciate the advantage of the script-$\mathscr{r}$ notation).

Problem 1.7 Find the separation vector $\boldsymbol{\mathscr{r}}$ from the source point (2,8,7) to the field point (4,6,8). Determine its magnitude ($\mathscr{r}$), and construct the unit vector $\hat{\boldsymbol{\mathscr{r}}}$.

1.1.5 How Vectors Transform

The definition of a vector as "a quantity with a magnitude and direction" is not altogether satisfactory: What precisely does "direction" *mean*?[2] This may seem a pedantic question, but we shall shortly encounter a species of derivative that *looks* rather like a vector, and we'll want to know for sure whether it *is* one. You might be inclined to say that a vector is anything that has three components that combine properly under addition. Well, how about this: We have a barrel of fruit that contains N_x pears, N_y apples, and N_z bananas. Is $\mathbf{N} = N_x\hat{\mathbf{x}} + N_y\hat{\mathbf{y}} + N_z\hat{\mathbf{z}}$ a vector? It has three components, and when you add another barrel with M_x pears, M_y apples, and M_z bananas the result is $(N_x + M_x)$ pears, $(N_y + M_y)$ apples, $(N_z + M_z)$ bananas. So it does *add* like a vector. Yet it's obviously *not* a vector, in the physicist's sense of the word, because it doesn't really have a direction. What exactly is wrong with it?

The answer is that **N** *does not transform properly when you change coordinates.* The coordinate frame we use to describe positions in space is of course entirely arbitrary, but there is a specific geometrical transformation law for converting vector components from one frame to another. Suppose, for instance, the $\overline{x}, \overline{y}, \overline{z}$ system is rotated by angle ϕ, relative to x, y, z, about the common $x = \overline{x}$ axes. From Fig. 1.15,

$$A_y = A\cos\theta, \qquad A_z = A\sin\theta,$$

while

$$\begin{aligned}
\overline{A}_y &= A\cos\overline{\theta} = A\cos(\theta - \phi) = A(\cos\theta\cos\phi + \sin\theta\sin\phi) \\
&= \cos\phi A_y + \sin\phi A_z, \\
\overline{A}_z &= A\sin\overline{\theta} = A\sin(\theta - \phi) = A(\sin\theta\cos\phi - \cos\theta\sin\phi) \\
&= -\sin\phi A_y + \cos\phi A_z.
\end{aligned}$$

[2] This section can be skipped without loss of continuity.

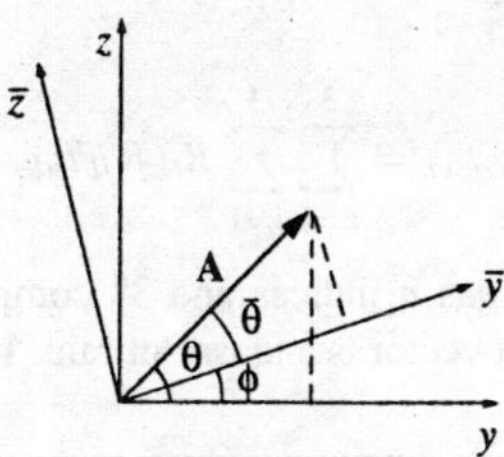

Figure 1.15

We might express this conclusion in matrix notation:

$$\begin{pmatrix} \overline{A}_y \\ \overline{A}_z \end{pmatrix} = \begin{pmatrix} \cos\phi & \sin\phi \\ -\sin\phi & \cos\phi \end{pmatrix} \begin{pmatrix} A_y \\ A_z \end{pmatrix}. \tag{1.29}$$

More generally, for rotation about an *arbitrary* axis in three dimensions, the transformation law takes the form

$$\begin{pmatrix} \overline{A}_x \\ \overline{A}_y \\ \overline{A}_z \end{pmatrix} = \begin{pmatrix} R_{xx} & R_{xy} & R_{xz} \\ R_{yx} & R_{yy} & R_{yz} \\ R_{zx} & R_{zy} & R_{zz} \end{pmatrix} \begin{pmatrix} A_x \\ A_y \\ A_z \end{pmatrix}, \tag{1.30}$$

or, more compactly,

$$\overline{A}_i = \sum_{j=1}^{3} R_{ij} A_j, \tag{1.31}$$

where the index 1 stands for x, 2 for y, and 3 for z. The elements of the matrix R can be ascertained, for a given rotation, by the same sort of geometrical arguments as we used for a rotation about the x axis.

Now: *Do* the components of **N** transform in this way? Of *course* not—it doesn't matter what coordinates you use to represent positions in space, there is still the same number of apples in the barrel. You can't convert a pear into a banana by choosing a different set of axes, but you *can* turn A_x into $\overline{A}_y$. Formally, then, a *vector is any set of three components that transforms in the same manner as a displacement when you change coordinates.* As always, displacement is the *model* for the behavior of all vectors.

By the way, a (second-rank) **tensor** is a quantity with *nine* components, T_{xx}, T_{xy}, T_{xz}, $T_{yx}, \ldots, T_{zz}$, which transforms with *two* factors of R:

$$\begin{aligned} \overline{T}_{xx} = R_{xx}(&R_{xx}T_{xx} + R_{xy}T_{xy} + R_{xz}T_{xz}) \\ +R_{xy}(&R_{xx}T_{yx} + R_{xy}T_{yy} + R_{xz}T_{yz}) \\ +R_{xz}(&R_{xx}T_{zx} + R_{xy}T_{zy} + R_{xz}T_{zz}), \ldots \end{aligned}$$

or, more compactly,

$$\overline{T}_{ij} = \sum_{k=1}^{3}\sum_{l=1}^{3} R_{ik}R_{jl}T_{kl}. \tag{1.32}$$

In general, an nth-rank tensor has n indices and 3^n components, and transforms with n factors of R. In this hierarchy, a vector is a tensor of rank 1, and a scalar is a tensor of rank zero.

Problem 1.8

(a) Prove that the two-dimensional rotation matrix (1.29) preserves dot products. (That is, show that $\overline{A}_y\overline{B}_y + \overline{A}_z\overline{B}_z = A_yB_y + A_zB_z$.)

(b) What constraints must the elements (R_{ij}) of the three-dimensional rotation matrix (1.30) satisfy in order to preserve the length of **A** (for all vectors **A**)?

Problem 1.9 Find the transformation matrix R that describes a rotation by 120° about an axis from the origin through the point (1, 1, 1). The rotation is clockwise as you look down the axis toward the origin.

Problem 1.10

(a) How do the components of a vector transform under a **translation** of coordinates ($\overline{x} = x$, $\overline{y} = y - a$, $\overline{z} = z$, Fig. 1.16a)?

(b) How do the components of a vector transform under an **inversion** of coordinates ($\overline{x} = -x$, $\overline{y} = -y$, $\overline{z} = -z$, Fig. 1.16b)?

(c) How does the cross product (1.13) of two vectors transform under inversion? [The cross-product of two vectors is properly called a **pseudovector** because of this "anomalous" behavior.] Is the cross product of two pseudovectors a vector, or a pseudovector? Name two pseudovector quantities in classical mechanics.

(d) How does the scalar triple product of three vectors transform under inversions? (Such an object is called a **pseudoscalar**.)

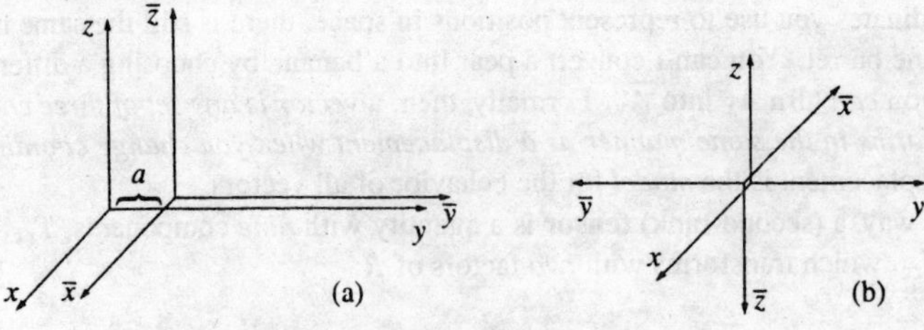

Figure 1.16

1.2 Differential Calculus

1.2.1 "Ordinary" Derivatives

Question: Suppose we have a function of one variable: $f(x)$. What does the derivative, df/dx, do for us? *Answer:* It tells us how rapidly the function $f(x)$ varies when we change the argument x by a tiny amount, dx:

$$df = \left(\frac{df}{dx}\right) dx. \tag{1.33}$$

In words: If we change x by an amount dx, then f changes by an amount df; the derivative is the proportionality factor. For example, in Fig. 1.17(a), the function varies slowly with x, and the derivative is correspondingly small. In Fig. 1.17(b), f increases rapidly with x, and the derivative is large, as you move away from $x = 0$.

Geometrical Interpretation: The derivative df/dx is the *slope* of the graph of f versus x.

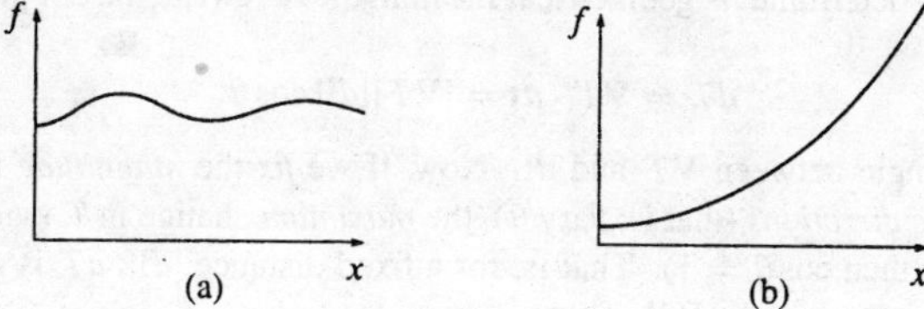

Figure 1.17

1.2.2 Gradient

Suppose, now, that we have a function of *three* variables—say, the temperature $T(x, y, z)$ in a room. (Start out in one corner, and set up a system of axes; then for each point (x, y, z) in the room, T gives the temperature at that spot.) We want to generalize the notion of "derivative" to functions like T, which depend not on *one* but on *three* variables.

Now a derivative is supposed to tell us how fast the function varies, if we move a little distance. But this time the situation is more complicated, because it depends on what *direction* we move: If we go straight up, then the temperature will probably increase fairly rapidly, but if we move horizontally, it may not change much at all. In fact, the question "How fast does T vary?" has an infinite number of answers, one for each direction we might choose to explore.

Fortunately, the problem is not as bad as it looks. A theorem on partial derivatives states that

$$dT = \left(\frac{\partial T}{\partial x}\right) dx + \left(\frac{\partial T}{\partial y}\right) dy + \left(\frac{\partial T}{\partial z}\right) dz. \tag{1.34}$$

This tells us how T changes when we alter all three variables by the infinitesimal amounts dx, dy, dz. Notice that we do *not* require an infinite number of derivatives—*three* will suffice: the *partial* derivatives along each of the three coordinate directions.

Equation 1.34 is reminiscent of a dot product:

$$\begin{aligned} dT &= \left(\frac{\partial T}{\partial x}\hat{\mathbf{x}} + \frac{\partial T}{\partial y}\hat{\mathbf{y}} + \frac{\partial T}{\partial z}\hat{\mathbf{z}}\right) \cdot (dx\,\hat{\mathbf{x}} + dy\,\hat{\mathbf{y}} + dz\,\hat{\mathbf{z}}) \\ &= (\nabla T) \cdot (d\mathbf{l}), \end{aligned} \tag{1.35}$$

where

$$\nabla T \equiv \frac{\partial T}{\partial x}\hat{\mathbf{x}} + \frac{\partial T}{\partial y}\hat{\mathbf{y}} + \frac{\partial T}{\partial z}\hat{\mathbf{z}} \tag{1.36}$$

is the **gradient** of T. ∇T is a *vector* quantity, with three components; it is the generalized derivative we have been looking for. Equation 1.35 is the three-dimensional version of Eq. 1.33.

Geometrical Interpretation of the Gradient: Like any vector, the gradient has *magnitude* and *direction*. To determine its geometrical meaning, let's rewrite the dot product (1.35) in abstract form:

$$dT = \nabla T \cdot d\mathbf{l} = |\nabla T||d\mathbf{l}|\cos\theta, \tag{1.37}$$

where θ is the angle between ∇T and $d\mathbf{l}$. Now, if we *fix* the *magnitude* $|d\mathbf{l}|$ and search around in various *directions* (that is, vary θ), the *maximum* change in T evidentally occurs when $\theta = 0$ (for then $\cos\theta = 1$). That is, for a fixed distance $|d\mathbf{l}|$, dT is greatest when I move in the *same direction* as ∇T. Thus:

> *The gradient ∇T points in the direction of maximum increase of the function T.*

Moreover:

> *The magnitude $|\nabla T|$ gives the slope (rate of increase) along this maximal direction.*

Imagine you are standing on a hillside. Look all around you, and find the direction of steepest ascent. That is the *direction* of the gradient. Now measure the *slope* in that direction (rise over run). That is the *magnitude* of the gradient. (Here the function we're talking about is the height of the hill, and the coordinates it depends on are positions—latitude and longitude, say. This function depends on only *two* variables, not *three*, but the geometrical meaning of the gradient is easier to grasp in two dimensions.) Notice from Eq. 1.37 that the direction of maximum *descent* is opposite to the direction of maximum *ascent*, while at right angles ($\theta = 90°$) the slope is zero (the gradient is perpendicular to the contour lines). You can conceive of surfaces that do not have these properties, but they always have "kinks" in them and correspond to nondifferentiable functions.

What would it mean for the gradient to vanish? If $\nabla T = 0$ at (x, y, z), then $dT = 0$ for small displacements about the point (x, y, z). This is, then, a **stationary point** of the function $T(x, y, z)$. It could be a maximum (a summit), a minimum (a valley), a saddle point (a pass), or a "shoulder." This is analogous to the situation for functions of *one* variable, where a vanishing derivative signals a maximum, a minimum, or an inflection. In particular, if you want to locate the extrema of a function of three variables, set its gradient equal to zero.

Example 1.3

Find the gradient of $r = \sqrt{x^2 + y^2 + z^2}$ (the magnitude of the position vector).

Solution:

$$\begin{aligned}\nabla r &= \frac{\partial r}{\partial x}\hat{\mathbf{x}} + \frac{\partial r}{\partial y}\hat{\mathbf{y}} + \frac{\partial r}{\partial z}\hat{\mathbf{z}} \\ &= \frac{1}{2}\frac{2x}{\sqrt{x^2+y^2+z^2}}\hat{\mathbf{x}} + \frac{1}{2}\frac{2y}{\sqrt{x^2+y^2+z^2}}\hat{\mathbf{y}} + \frac{1}{2}\frac{2z}{\sqrt{x^2+y^2+z^2}}\hat{\mathbf{z}} \\ &= \frac{x\,\hat{\mathbf{x}} + y\,\hat{\mathbf{y}} + z\,\hat{\mathbf{z}}}{\sqrt{x^2+y^2+z^2}} = \frac{\mathbf{r}}{r} = \hat{\mathbf{r}}.\end{aligned}$$

Does this make sense? Well, it says that the distance from the origin increases most rapidly in the radial direction, and that its *rate* of increase in that direction is 1...just what you'd expect.

Problem 1.11 Find the gradients of the following functions:

(a) $f(x, y, z) = x^2 + y^3 + z^4$.

(b) $f(x, y, z) = x^2y^3z^4$.

(c) $f(x, y, z) = e^x \sin(y) \ln(z)$.

Problem 1.12 The height of a certain hill (in feet) is given by

$$h(x, y) = 10(2xy - 3x^2 - 4y^2 - 18x + 28y + 12),$$

where y is the distance (in miles) north, x the distance east of South Hadley.

(a) Where is the top of the hill located?

(b) How high is the hill?

(c) How steep is the slope (in feet per mile) at a point 1 mile north and one mile east of South Hadley? In what direction is the slope steepest, at that point?

Problem 1.13 Let $\boldsymbol{\mathscr{r}}$ be the separation vector from a fixed point (x', y', z') to the point (x, y, z), and let $\mathscr{r}$ be its length. Show that

(a) $\nabla(\mathscr{r}^2) = 2\boldsymbol{\mathscr{r}}$.

(b) $\nabla(1/\mathscr{r}) = -\hat{\boldsymbol{\mathscr{r}}}/\mathscr{r}^2$.

(c) What is the *general* formula for $\nabla(\mathscr{r}^n)$?

! **Problem 1.14** Suppose that f is a function of two variables (y and z) only. Show that the gradient $\nabla f = (\partial f/\partial y)\hat{\mathbf{y}} + (\partial f/\partial z)\hat{\mathbf{z}}$ transforms as a vector under rotations, Eq. 1.29. [*Hint:* $(\partial f/\partial \overline{y}) = (\partial f/\partial y)(\partial y/\partial \overline{y}) + (\partial f/\partial z)(\partial z/\partial \overline{y})$, and the analogous formula for $\partial f/\partial \overline{z}$. We know that $\overline{y} = y\cos\phi + z\sin\phi$ and $\overline{z} = -y\sin\phi + z\cos\phi$; "solve" these equations for y and z (as functions of $\overline{y}$ and $\overline{z}$), and compute the needed derivatives $\partial y/\partial \overline{y}$, $\partial z/\partial \overline{y}$, etc.]

1.2.3 The Operator ∇

The gradient has the formal appearance of a vector, ∇, "multiplying" a scalar T:

$$\nabla T = \left(\hat{\mathbf{x}}\frac{\partial}{\partial x} + \hat{\mathbf{y}}\frac{\partial}{\partial y} + \hat{\mathbf{z}}\frac{\partial}{\partial z}\right) T. \tag{1.38}$$

(For once I write the unit vectors to the *left*, just so no one will think this means $\partial\hat{\mathbf{x}}/\partial x$, and so on—which would be zero, since $\hat{\mathbf{x}}$ is constant.) The term in parentheses is called **"del"**:

$$\boxed{\nabla = \hat{\mathbf{x}}\frac{\partial}{\partial x} + \hat{\mathbf{y}}\frac{\partial}{\partial y} + \hat{\mathbf{z}}\frac{\partial}{\partial z}.} \tag{1.39}$$

Of course, del is *not* a vector, in the usual sense. Indeed, it is without specific meaning until we provide it with a function to act upon. Furthermore, it does not "multiply" T; rather, it is an instruction to *differentiate* what follows. To be precise, then, we should say that ∇ is a **vector operator** that *acts upon* T, not a vector that multiplies T.

With this qualification, though, ∇ mimics the behavior of an ordinary vector in virtually every way; almost anything that can be done with other vectors can also be done with ∇, if we merely translate "multiply" by "act upon." So by all means take the vector appearance of ∇ seriously: it is a marvelous piece of notational simplification, as you will appreciate if you ever consult Maxwell's original work on electromagnetism, written without the benefit of ∇.

Now an ordinary vector $\mathbf{A}$ can multiply in three ways:

1. Multiply a scalar a : $\mathbf{A}a$;
2. Multiply another vector $\mathbf{B}$, via the dot product: $\mathbf{A}\cdot\mathbf{B}$;
3. Multiply another vector via the cross product: $\mathbf{A}\times\mathbf{B}$.

Correspondingly, there are three ways the operator ∇ can act:

1. On a scalar function T : ∇T (the gradient);
2. On a vector function $\mathbf{v}$, via the dot product: $\nabla\cdot\mathbf{v}$ (the **divergence**);
3. On a vector function $\mathbf{v}$, via the cross product: $\nabla\times\mathbf{v}$ (the **curl**).

We have already discussed the gradient. In the following sections we examine the other two vector derivatives: divergence and curl.

1.2.4 The Divergence

From the definition of ∇ we construct the divergence:

$$\begin{aligned}\nabla \cdot \mathbf{v} &= \left(\hat{\mathbf{x}}\frac{\partial}{\partial x} + \hat{\mathbf{y}}\frac{\partial}{\partial y} + \hat{\mathbf{z}}\frac{\partial}{\partial z}\right) \cdot (v_x \hat{\mathbf{x}} + v_y \hat{\mathbf{y}} + v_z \hat{\mathbf{z}}) \\ &= \frac{\partial v_x}{\partial x} + \frac{\partial v_y}{\partial y} + \frac{\partial v_z}{\partial z}.\end{aligned} \tag{1.40}$$

Observe that the divergence of a vector function $\mathbf{v}$ is itself a *scalar* $\nabla \cdot \mathbf{v}$. (You can't have the divergence of a scalar: that's meaningless.)

Geometrical Interpretation: The name **divergence** is well chosen, for $\nabla \cdot \mathbf{v}$ is a measure of how much the vector $\mathbf{v}$ spreads out (diverges) from the point in question. For example, the vector function in Fig. 1.18a has a large (positive) divergence (if the arrows pointed *in*, it would be a large *negative* divergence), the function in Fig. 1.18b has zero divergence, and the function in Fig. 1.18c again has a positive divergence. (Please understand that $\mathbf{v}$ here is a *function*—there's a different vector associated with every point in space. In the diagrams,

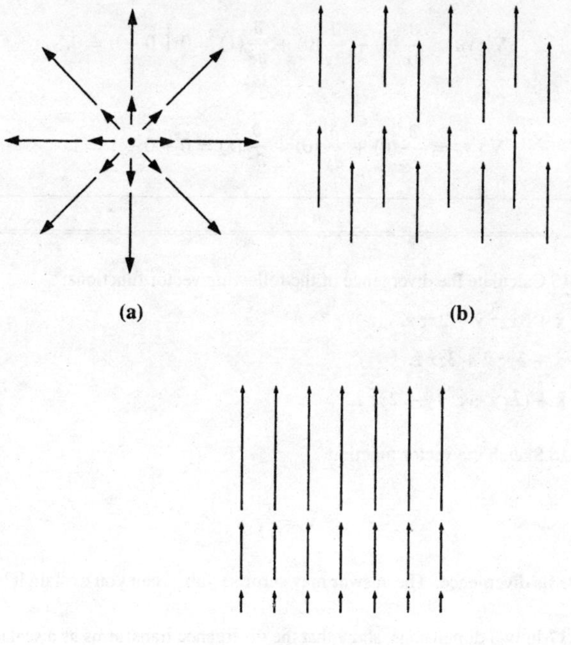

Figure 1.18

of course, I can only draw the arrows at a few representative locations.) Imagine standing at the edge of a pond. Sprinkle some sawdust or pine needles on the surface. If the material spreads out, then you dropped it at a point of positive divergence; if it collects together, you dropped it at a point of negative divergence. (The vector function **v** in this model is the velocity of the water—this is a *two*-dimensional example, but it helps give one a "feel" for what the divergence means. A point of positive divergence is a source, or "faucet"; a point of negative divergence is a sink, or "drain.")

Example 1.4

Suppose the functions in Fig. 1.18 are $\mathbf{v}_a = \mathbf{r} = x\,\hat{\mathbf{x}} + y\,\hat{\mathbf{y}} + z\,\hat{\mathbf{z}}$, $\mathbf{v}_b = \hat{\mathbf{z}}$, and $\mathbf{v}_c = z\,\hat{\mathbf{z}}$. Calculate their divergences.

Solution:

$$\nabla \cdot \mathbf{v}_a = \frac{\partial}{\partial x}(x) + \frac{\partial}{\partial y}(y) + \frac{\partial}{\partial z}(z) = 1 + 1 + 1 = 3.$$

As anticipated, this function has a positive divergence.

$$\nabla \cdot \mathbf{v}_b = \frac{\partial}{\partial x}(0) + \frac{\partial}{\partial y}(0) + \frac{\partial}{\partial z}(1) = 0 + 0 + 0 = 0,$$

as expected.

$$\nabla \cdot \mathbf{v}_c = \frac{\partial}{\partial x}(0) + \frac{\partial}{\partial y}(0) + \frac{\partial}{\partial z}(z) = 0 + 0 + 1 = 1.$$

Problem 1.15 Calculate the divergence of the following vector functions:

(a) $\mathbf{v}_a = x^2\,\hat{\mathbf{x}} + 3xz^2\,\hat{\mathbf{y}} - 2xz\,\hat{\mathbf{z}}$.

(b) $\mathbf{v}_b = xy\,\hat{\mathbf{x}} + 2yz\,\hat{\mathbf{y}} + 3zx\,\hat{\mathbf{z}}$.

(c) $\mathbf{v}_c = y^2\,\hat{\mathbf{x}} + (2xy + z^2)\,\hat{\mathbf{y}} + 2yz\,\hat{\mathbf{z}}$.

Problem 1.16 Sketch the vector function

$$\mathbf{v} = \frac{\hat{\mathbf{r}}}{r^2},$$

and compute its divergence. The answer may surprise you... can you explain it?

Problem 1.17 In two dimensions, show that the divergence transforms as a scalar under rotations. [*Hint:* Use Eq. 1.29 to determine $\bar{v}_y$ and $\bar{v}_z$, and the method of Prob. 1.14 to calculate the derivatives. Your aim is to show that $\partial\bar{v}_y/\partial\bar{y} + \partial\bar{v}_z/\partial\bar{z} = \partial v_y/\partial y + \partial v_z/\partial z$.]

1.2.5 The Curl

From the definition of ∇ we construct the curl:

$$\nabla \times \mathbf{v} = \begin{vmatrix} \hat{\mathbf{x}} & \hat{\mathbf{y}} & \hat{\mathbf{z}} \\ \partial/\partial x & \partial/\partial y & \partial/\partial z \\ v_x & v_y & v_z \end{vmatrix}$$

$$= \hat{\mathbf{x}}\left(\frac{\partial v_z}{\partial y} - \frac{\partial v_y}{\partial z}\right) + \hat{\mathbf{y}}\left(\frac{\partial v_x}{\partial z} - \frac{\partial v_z}{\partial x}\right) + \hat{\mathbf{z}}\left(\frac{\partial v_y}{\partial x} - \frac{\partial v_x}{\partial y}\right). \quad (1.41)$$

Notice that the curl of a vector function **v** is, like any cross product, a *vector*. (You cannot have the curl of a scalar; that's meaningless.)

Geometrical Interpretation: The name **curl** is also well chosen, for $\nabla \times \mathbf{v}$ is a measure of how much the vector **v** "curls around" the point in question. Thus the three functions in Fig. 1.18 all have zero curl (as you can easily check for yourself), whereas the functions in Fig. 1.19 have a substantial curl, pointing in the z-direction, as the natural right-hand rule would suggest. Imagine (again) you are standing at the edge of a pond. Float a small paddlewheel (a cork with toothpicks pointing out radially would do); if it starts to rotate, then you placed it at a point of nonzero *curl*. A whirlpool would be a region of large curl.

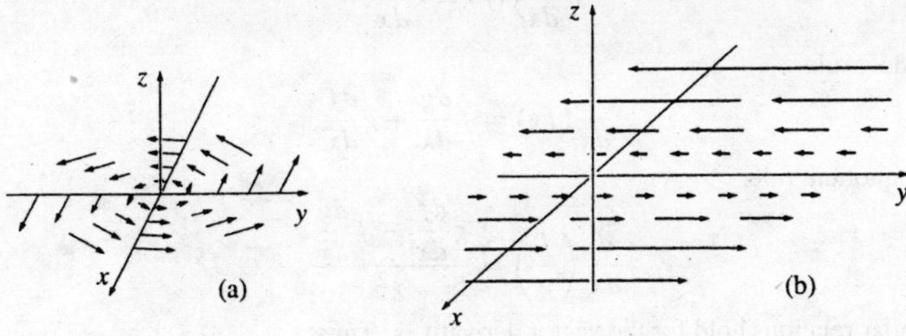

Figure 1.19

Example 1.5

Suppose the function sketched in Fig. 1.19a is $\mathbf{v}_a = -y\hat{\mathbf{x}} + x\hat{\mathbf{y}}$, and that in Fig. 1.19b is $\mathbf{v}_b = x\hat{\mathbf{y}}$. Calculate their curls.

Solution:

$$\nabla \times \mathbf{v}_a = \begin{vmatrix} \hat{\mathbf{x}} & \hat{\mathbf{y}} & \hat{\mathbf{z}} \\ \partial/\partial x & \partial/\partial y & \partial/\partial z \\ -y & x & 0 \end{vmatrix} = 2\hat{\mathbf{z}},$$

and

$$\nabla \times \mathbf{v}_b = \begin{vmatrix} \hat{\mathbf{x}} & \hat{\mathbf{y}} & \hat{\mathbf{z}} \\ \partial/\partial x & \partial/\partial y & \partial/\partial z \\ 0 & x & 0 \end{vmatrix} = \hat{\mathbf{z}}.$$

As expected, these curls point in the $+z$ direction. (Incidentally, they both have zero divergence, as you might guess from the pictures: nothing is "spreading out"... it just "curls around.")

Problem 1.18 Calculate the curls of the vector functions in Prob. 1.15.

Problem 1.19 Construct a vector function that has zero divergence and zero curl everywhere. (A *constant* will do the job, of course, but make it something a little more interesting than that!)

1.2.6 Product Rules

The calculation of ordinary derivatives is facilitated by a number of general rules, such as the sum rule:

$$\frac{d}{dx}(f+g) = \frac{df}{dx} + \frac{dg}{dx},$$

the rule for multiplying by a constant:

$$\frac{d}{dx}(kf) = k\frac{df}{dx},$$

the product rule:

$$\frac{d}{dx}(fg) = f\frac{dg}{dx} + g\frac{df}{dx},$$

and the quotient rule:

$$\frac{d}{dx}\left(\frac{f}{g}\right) = \frac{g\dfrac{df}{dx} - f\dfrac{dg}{dx}}{g^2}.$$

Similar relations hold for the vector derivatives. Thus,

$$\nabla(f+g) = \nabla f + \nabla g, \qquad \nabla\cdot(\mathbf{A}+\mathbf{B}) = (\nabla\cdot\mathbf{A}) + (\nabla\cdot\mathbf{B}),$$

$$\nabla\times(\mathbf{A}+\mathbf{B}) = (\nabla\times\mathbf{A}) + (\nabla\times\mathbf{B}),$$

and

$$\nabla(kf) = k\nabla f, \qquad \nabla\cdot(k\mathbf{A}) = k(\nabla\cdot\mathbf{A}), \qquad \nabla\times(k\mathbf{A}) = k(\nabla\times\mathbf{A}),$$

as you can check for yourself. The product rules are not quite so simple. There are two ways to construct a scalar as the product of two functions:

fg (product of two scalar functions),
$\mathbf{A}\cdot\mathbf{B}$ (dot product of two vector functions),

and two ways to make a vector:

$f\mathbf{A}$ (scalar times vector),
$\mathbf{A}\times\mathbf{B}$ (cross product of two vectors).

Accordingly, there are *six* product rules, two for gradients:

(i) $$\nabla(fg) = f\nabla g + g\nabla f,$$

(ii) $$\nabla(\mathbf{A}\cdot\mathbf{B}) = \mathbf{A}\times(\nabla\times\mathbf{B}) + \mathbf{B}\times(\nabla\times\mathbf{A}) + (\mathbf{A}\cdot\nabla)\mathbf{B} + (\mathbf{B}\cdot\nabla)\mathbf{A},$$

two for divergences:

(iii) $$\nabla\cdot(f\mathbf{A}) = f(\nabla\cdot\mathbf{A}) + \mathbf{A}\cdot(\nabla f),$$

(iv) $$\nabla\cdot(\mathbf{A}\times\mathbf{B}) = \mathbf{B}\cdot(\nabla\times\mathbf{A}) - \mathbf{A}\cdot(\nabla\times\mathbf{B}),$$

and two for curls:

(v) $$\nabla\times(f\mathbf{A}) = f(\nabla\times\mathbf{A}) - \mathbf{A}\times(\nabla f),$$

(vi) $$\nabla\times(\mathbf{A}\times\mathbf{B}) = (\mathbf{B}\cdot\nabla)\mathbf{A} - (\mathbf{A}\cdot\nabla)\mathbf{B} + \mathbf{A}(\nabla\cdot\mathbf{B}) - \mathbf{B}(\nabla\cdot\mathbf{A}).$$

You will be using these product rules so frequently that I have put them on the inside front cover for easy reference. The proofs come straight from the product rule for ordinary derivatives. For instance,

$$\begin{aligned}
\nabla\cdot(f\mathbf{A}) &= \frac{\partial}{\partial x}(fA_x) + \frac{\partial}{\partial y}(fA_y) + \frac{\partial}{\partial z}(fA_z) \\
&= \left(\frac{\partial f}{\partial x}A_x + f\frac{\partial A_x}{\partial x}\right) + \left(\frac{\partial f}{\partial y}A_y + f\frac{\partial A_y}{\partial y}\right) + \left(\frac{\partial f}{\partial z}A_z + f\frac{\partial A_z}{\partial z}\right) \\
&= (\nabla f)\cdot\mathbf{A} + f(\nabla\cdot\mathbf{A}).
\end{aligned}$$

It is also possible to formulate three quotient rules:

$$\begin{aligned}
\nabla\left(\frac{f}{g}\right) &= \frac{g\nabla f - f\nabla g}{g^2}, \\
\nabla\cdot\left(\frac{\mathbf{A}}{g}\right) &= \frac{g(\nabla\cdot\mathbf{A}) - \mathbf{A}\cdot(\nabla g)}{g^2}, \\
\nabla\times\left(\frac{\mathbf{A}}{g}\right) &= \frac{g(\nabla\times\mathbf{A}) + \mathbf{A}\times(\nabla g)}{g^2}.
\end{aligned}$$

However, since these can be obtained quickly from the corresponding product rules, I haven't bothered to put them on the inside front cover.

Problem 1.20 Prove product rules (i), (iv), and (v).

Problem 1.21

(a) If $\mathbf{A}$ and $\mathbf{B}$ are two vector functions, what does the expression $(\mathbf{A}\cdot\nabla)\mathbf{B}$ mean? (That is, what are its x, y, and z components in terms of the Cartesian components of $\mathbf{A}$, $\mathbf{B}$, and ∇?)

(b) Compute $(\hat{\mathbf{r}}\cdot\nabla)\hat{\mathbf{r}}$, where $\hat{\mathbf{r}}$ is the unit vector defined in Eq. 1.21.

(c) For the functions in Prob. 1.15, evaluate $(\mathbf{v}_a\cdot\nabla)\mathbf{v}_b$.

Problem 1.22 (For masochists only.) Prove product rules (ii) and (vi). Refer to Prob. 1.21 for the definition of $(\mathbf{A}\cdot\nabla)\mathbf{B}$.

Problem 1.23 Derive the three quotient rules.

Problem 1.24

(a) Check product rule (iv) (by calculating each term separately) for the functions

$$\mathbf{A} = x\,\hat{\mathbf{x}} + 2y\,\hat{\mathbf{y}} + 3z\,\hat{\mathbf{z}};\qquad \mathbf{B} = 3y\,\hat{\mathbf{x}} - 2x\,\hat{\mathbf{y}}.$$

(b) Do the same for product rule (ii).

(c) The same for rule (vi).

1.2.7 Second Derivatives

The gradient, the divergence, and the curl are the only first derivatives we can make with ∇; by applying ∇ *twice* we can construct five species of *second* derivatives. The gradient ∇T is a *vector*, so we can take the *divergence* and *curl* of it:

(1) Divergence of gradient: $\nabla\cdot(\nabla T)$.

(2) Curl of gradient: $\nabla\times(\nabla T)$.

The divergence $\nabla\cdot\mathbf{v}$ is a *scalar*—all we can do is take its *gradient:*

(3) Gradient of divergence: $\nabla(\nabla\cdot\mathbf{v})$.

The curl $\nabla\times\mathbf{v}$ is a *vector*, so we can take its *divergence* and *curl:*

(4) Divergence of curl: $\nabla\cdot(\nabla\times\mathbf{v})$.

(5) Curl of curl: $\nabla\times(\nabla\times\mathbf{v})$.

This exhausts the possibilities, and in fact not all of them give anything new. Let's consider them one at a time:

$$\text{(1)}\quad \nabla\cdot(\nabla T) = \left(\hat{\mathbf{x}}\frac{\partial}{\partial x}+\hat{\mathbf{y}}\frac{\partial}{\partial y}+\hat{\mathbf{z}}\frac{\partial}{\partial z}\right)\cdot\left(\frac{\partial T}{\partial x}\hat{\mathbf{x}}+\frac{\partial T}{\partial y}\hat{\mathbf{y}}+\frac{\partial T}{\partial z}\hat{\mathbf{z}}\right)$$
$$= \frac{\partial^2 T}{\partial x^2}+\frac{\partial^2 T}{\partial y^2}+\frac{\partial^2 T}{\partial z^2}. \tag{1.42}$$

This object, which we write $\nabla^2 T$ for short, is called the **Laplacian** of T; we shall be studying it in great detail later on. Notice that the Laplacian of a *scalar* T is a *scalar*. Occasionally, we shall speak of the Laplacian of a *vector*, $\nabla^2\mathbf{v}$. By this we mean a *vector* quantity whose x-component is the Laplacian of v_x, and so on:[3]

$$\nabla^2\mathbf{v} \equiv (\nabla^2 v_x)\hat{\mathbf{x}} + (\nabla^2 v_y)\hat{\mathbf{y}} + (\nabla^2 v_z)\hat{\mathbf{z}}. \tag{1.43}$$

This is nothing more than a convenient *extension* of the meaning of ∇^2.

(2) The curl of a gradient is always *zero:*

$$\nabla \times (\nabla T) = 0. \tag{1.44}$$

This is an important fact, which we shall use repeatedly; you can easily prove it from the definition of ∇, Eq. 1.39. *Beware*: You might think Eq. 1.44 is "obviously" true—isn't it just $(\nabla \times \nabla)T$, and isn't the cross product of *any* vector (in this case, ∇) with itself always zero? This reasoning is *suggestive* but not quite *conclusive,* since ∇ is an *operator* and does not "multiply" in the usual way. The proof of Eq. 1.44, in fact, hinges on the equality of cross derivatives:

$$\frac{\partial}{\partial x}\left(\frac{\partial T}{\partial y}\right) = \frac{\partial}{\partial y}\left(\frac{\partial T}{\partial x}\right). \tag{1.45}$$

If you think I'm being fussy, test your intuition on this one:

$$(\nabla T) \times (\nabla S).$$

Is *that* always zero? (It *would* be, of course, if you replaced the ∇'s by an ordinary vector.)

(3) $\nabla(\nabla \cdot \mathbf{v})$ for some reason seldom occurs in physical applications, and it has not been given any special name of its own—it's just **the gradient of the divergence.** Notice that $\nabla(\nabla \cdot \mathbf{v})$ is *not* the same as the Laplacian of a vector: $\nabla^2\mathbf{v} = (\nabla \cdot \nabla)\mathbf{v} \neq \nabla(\nabla \cdot \mathbf{v})$.

(4) The divergence of a curl, like the curl of a gradient, is *always zero:*

$$\nabla \cdot (\nabla \times \mathbf{v}) = 0. \tag{1.46}$$

You can prove this for yourself. (Again, there is a fraudulent short-cut proof, using the vector identity $\mathbf{A} \cdot (\mathbf{B} \times \mathbf{C}) = (\mathbf{A} \times \mathbf{B}) \cdot \mathbf{C}$.)

(5) As you can check from the definition of ∇:

$$\nabla \times (\nabla \times \mathbf{v}) = \nabla(\nabla \cdot \mathbf{v}) - \nabla^2\mathbf{v}. \tag{1.47}$$

So curl-of-curl gives nothing new; the first term is just number (3) and the second is the Laplacian (of a vector). (In fact, Eq. 1.47 is often used to *define* the Laplacian of a vector, in preference to Eq. 1.43, which makes specific reference to Cartesian coordinates.)

Really, then, there are just two kinds of second derivatives: the Laplacian (which is of fundamental importance) and the gradient-of-divergence (which we seldom encounter).

---

[3] In curvilinear coordinates, where the unit vectors themselves depend on position, they too must be differentiated (see Sect. 1.4.1).

We could go through a similar ritual to work out *third* derivatives, but fortunately second derivatives suffice for practically all physical applications.

A final word on vector differential calculus: It *all* flows from the operator $\boldsymbol{\nabla}$, and from taking seriously its vector character. Even if you remembered *only* the definition of $\boldsymbol{\nabla}$, you should be able, in principle, to reconstruct all the rest.

Problem 1.25 Calculate the Laplacian of the following functions:

(a) $T_a = x^2 + 2xy + 3z + 4$.

(b) $T_b = \sin x \sin y \sin z$.

(c) $T_c = e^{-5x} \sin 4y \cos 3z$.

(d) $\mathbf{v} = x^2\,\hat{\mathbf{x}} + 3xz^2\,\hat{\mathbf{y}} - 2xz\,\hat{\mathbf{z}}$.

Problem 1.26 Prove that the divergence of a curl is always zero. *Check* it for function $\mathbf{v}_a$ in Prob. 1.15.

Problem 1.27 Prove that the curl of a gradient is always zero. *Check* it for function (b) in Prob. 1.11.

1.3 Integral Calculus

1.3.1 Line, Surface, and Volume Integrals

In electrodynamics we encounter several different kinds of integrals, among which the most important are **line** (or **path**) **integrals**, **surface integrals** (or **flux**), and **volume integrals**.

(a) **Line Integrals.** A line integral is an expression of the form

$$\int_{\mathbf{a}\,\mathcal{P}}^{\mathbf{b}} \mathbf{v} \cdot d\mathbf{l}, \tag{1.48}$$

where $\mathbf{v}$ is a vector function, $d\mathbf{l}$ is the infinitesimal displacement vector (Eq. 1.22), and the integral is to be carried out along a prescribed path $\mathcal{P}$ from point $\mathbf{a}$ to point $\mathbf{b}$ (Fig. 1.20). If the path in question forms a closed loop (that is, if $\mathbf{b} = \mathbf{a}$), I shall put a circle on the integral sign:

$$\oint \mathbf{v} \cdot d\mathbf{l}. \tag{1.49}$$

At each point on the path we take the dot product of $\mathbf{v}$ (evaluated at that point) with the displacement $d\mathbf{l}$ to the next point on the path. To a physicist, the most familiar example of a line integral is the work done by a force $\mathbf{F}$: $W = \int \mathbf{F} \cdot d\mathbf{l}$.

Ordinarily, the value of a line integral depends critically on the particular path taken from $\mathbf{a}$ to $\mathbf{b}$, but there is an important special class of vector functions for which the line integral is *independent* of the path, and is determined entirely by the end points. It will be our business in due course to characterize this special class of vectors. (A *force* that has this property is called **conservative**.)

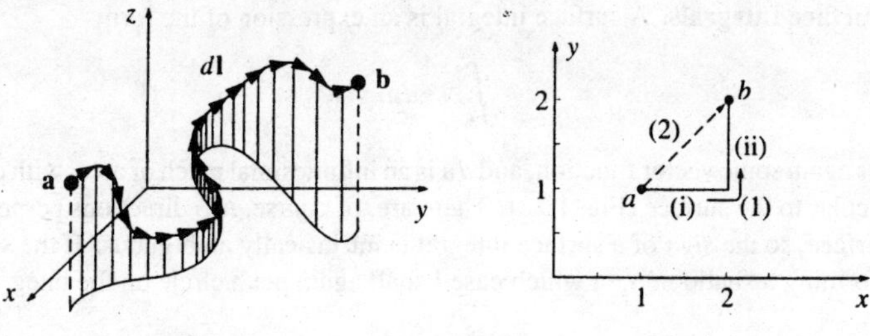

Figure 1.20

Figure 1.21

Example 1.6

Calculate the line integral of the function $\mathbf{v} = y^2\,\hat{\mathbf{x}} + 2x(y+1)\,\hat{\mathbf{y}}$ from the point $\mathbf{a} = (1, 1, 0)$ to the point $\mathbf{b} = (2, 2, 0)$, along the paths (1) and (2) in Fig. 1.21. What is $\oint \mathbf{v}\cdot d\mathbf{l}$ for the loop that goes from $\mathbf{a}$ to $\mathbf{b}$ along (1) and returns to $\mathbf{a}$ along (2)?

Solution: As always, $d\mathbf{l} = dx\,\hat{\mathbf{x}} + dy\,\hat{\mathbf{y}} + dz\,\hat{\mathbf{z}}$. Path (1) consists of two parts. Along the "horizontal" segment $dy = dz = 0$, so

(i) $d\mathbf{l} = dx\,\hat{\mathbf{x}},\ y = 1,\ \mathbf{v}\cdot d\mathbf{l} = y^2\,dx = dx$, so $\int \mathbf{v}\cdot d\mathbf{l} = \int_1^2 dx = 1$.

On the "vertical" stretch $dx = dz = 0$, so

(ii) $d\mathbf{l} = dy\,\hat{\mathbf{y}},\ x = 2,\ \mathbf{v}\cdot d\mathbf{l} = 2x(y+1)\,dy = 4(y+1)\,dy$, so

$$\int \mathbf{v}\cdot d\mathbf{l} = 4\int_1^2 (y+1)\,dy = 10.$$

By path (1), then,

$$\int_{\mathbf{a}}^{\mathbf{b}} \mathbf{v}\cdot d\mathbf{l} = 1 + 10 = 11.$$

Meanwhile, on path (2) $x = y,\ dx = dy$, and $dz = 0$, so

$$d\mathbf{l} = dx\,\hat{\mathbf{x}} + dy\,\hat{\mathbf{y}},\ \mathbf{v}\cdot d\mathbf{l} = x^2\,dx + 2x(x+1)\,dx = (3x^2 + 2x)\,dx,$$

so

$$\int_{\mathbf{a}}^{\mathbf{b}} \mathbf{v}\cdot d\mathbf{l} = \int_1^2 (3x^2 + 2x)\,dx = (x^3 + x^2)\Big|_1^2 = 10.$$

(The strategy here is to get everything in terms of one variable; I could just as well have eliminated x in favor of y.)

For the loop that goes *out* (1) and *back* (2), then,

$$\oint \mathbf{v}\cdot d\mathbf{l} = 11 - 10 = 1.$$

(b) **Surface Integrals.** A surface integral is an expression of the form

$$\int_{\mathcal{S}} \mathbf{v} \cdot d\mathbf{a}, \tag{1.50}$$

where $\mathbf{v}$ is again some vector function, and $d\mathbf{a}$ is an infinitesimal patch of area, with direction perpendicular to the surface (Fig. 1.22). There are, of course, *two* directions perpendicular to any surface, so the *sign* of a surface integral is intrinsically ambiguous. If the surface is *closed* (forming a "balloon"), in which case I shall again put a circle on the integral sign

$$\oint \mathbf{v} \cdot d\mathbf{a},$$

then tradition dictates that "outward" is positive, but for open surfaces it's arbitrary. If $\mathbf{v}$ describes the flow of a fluid (mass per unit area per unit time), then $\int \mathbf{v} \cdot d\mathbf{a}$ represents the total mass per unit time passing through the surface—hence the alternative name, "flux."

Ordinarily, the value of a surface integral depends on the particular surface chosen, but there is a special class of vector functions for which it is *independent* of the surface, and is determined entirely by the boundary line. We shall soon be in a position to characterize this special class.

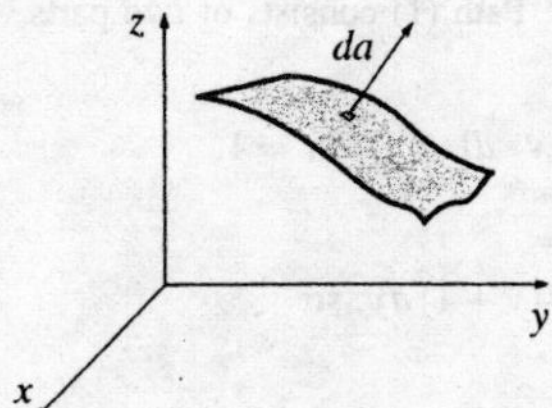

Figure 1.22

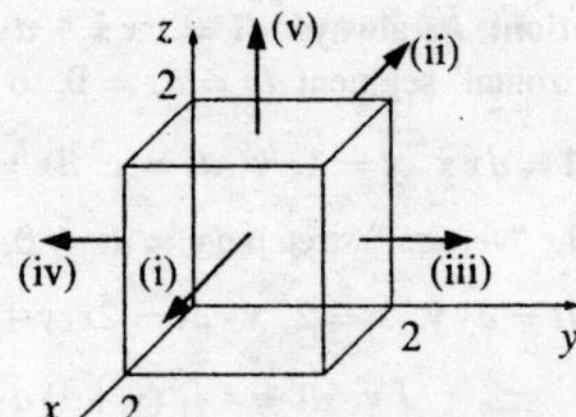

Figure 1.23

Example 1.7

Calculate the surface integral of $\mathbf{v} = 2xz\,\hat{\mathbf{x}} + (x+2)\,\hat{\mathbf{y}} + y(z^2-3)\,\hat{\mathbf{z}}$ over five sides (excluding the bottom) of the cubical box (side 2) in Fig. 1.23. Let "upward and outward" be the positive direction, as indicated by the arrows.

Solution: Taking the sides one at a time:

(i) $x = 2,\ d\mathbf{a} = dy\,dz\,\hat{\mathbf{x}},\ \mathbf{v} \cdot d\mathbf{a} = 2xz\,dy\,dz = 4z\,dy\,dz$, so

$$\int \mathbf{v} \cdot d\mathbf{a} = 4\int_0^2 dy \int_0^2 z\,dz = 16.$$

(ii) $x = 0,\ d\mathbf{a} = -dy\,dz\,\hat{\mathbf{x}},\ \mathbf{v} \cdot d\mathbf{a} = -2xz\,dy\,dz = 0$, so

$$\int \mathbf{v} \cdot d\mathbf{a} = 0.$$

(iii) $y = 2$, $d\mathbf{a} = dx\,dz\,\hat{\mathbf{y}}$, $\mathbf{v}\cdot d\mathbf{a} = (x+2)\,dx\,dz$, so

$$\int \mathbf{v}\cdot d\mathbf{a} = \int_0^2 (x+2)\,dx \int_0^2 dz = 12.$$

(iv) $y = 0$, $d\mathbf{a} = -dx\,dz\,\hat{\mathbf{y}}$, $\mathbf{v}\cdot d\mathbf{a} = -(x+2)\,dx\,dz$, so

$$\int \mathbf{v}\cdot d\mathbf{a} = -\int_0^2 (x+2)\,dx \int_0^2 dz = -12.$$

(v) $z = 2$, $d\mathbf{a} = dx\,dy\,\hat{\mathbf{z}}$, $\mathbf{v}\cdot d\mathbf{a} = y(z^2-3)\,dx\,dy = y\,dx\,dy$, so

$$\int \mathbf{v}\cdot d\mathbf{a} = \int_0^2 dx \int_0^2 y\,dy = 4.$$

Evidently the *total* flux is

$$\int_{\text{surface}} \mathbf{v}\cdot d\mathbf{a} = 16 + 0 + 12 - 12 + 4 = 20.$$

(c) **Volume Integrals.** A volume integral is an expression of the form

$$\int_{\mathcal{V}} T\,d\tau, \tag{1.51}$$

where T is a scalar function and $d\tau$ is an infinitesimal volume element. In Cartesian coordinates,

$$d\tau = dx\,dy\,dz. \tag{1.52}$$

For example, if T is the density of a substance (which might vary from point to point), then the volume integral would give the total mass. Occasionally we shall encounter volume integrals of *vector* functions:

$$\int \mathbf{v}\,d\tau = \int (v_x\,\hat{\mathbf{x}} + v_y\,\hat{\mathbf{y}} + v_z\,\hat{\mathbf{z}})d\tau = \hat{\mathbf{x}}\int v_x d\tau + \hat{\mathbf{y}}\int v_y d\tau + \hat{\mathbf{z}}\int v_z d\tau; \tag{1.53}$$

because the unit vectors are constants, they come outside the integral.

Example 1.8

Calculate the volume integral of $T = xyz^2$ over the prism in Fig. 1.24.

Solution: You can do the three integrals in any order. Let's do x first: it runs from 0 to $(1-y)$; then y (it goes from 0 to 1); and finally z (0 to 3):

$$\int T\,d\tau = \int_0^3 z^2 \left\{\int_0^1 y\left[\int_0^{1-y} x\,dx\right]dy\right\}dz =$$

$$\frac{1}{2}\int_0^3 z^2\,dz \int_0^1 (1-y)^2 y\,dy = \tfrac{1}{2}(9)(\tfrac{1}{12}) = \tfrac{3}{8}.$$

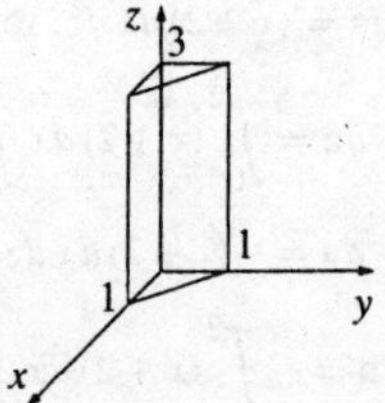

Figure 1.24

Problem 1.28 Calculate the line integral of the function $\mathbf{v} = x^2\,\hat{\mathbf{x}} + 2yz\,\hat{\mathbf{y}} + y^2\,\hat{\mathbf{z}}$ from the origin to the point (1,1,1) by three different routes:

(a) $(0,0,0) \to (1,0,0) \to (1,1,0) \to (1,1,1)$;

(b) $(0,0,0) \to (0,0,1) \to (0,1,1) \to (1,1,1)$;

(c) The direct straight line.

(d) What is the line integral around the closed loop that goes *out* along path (a) and *back* along path (b)?

Problem 1.29 Calculate the surface integral of the function in Ex. 1.7, over the *bottom* of the box. For consistency, let "upward" be the positive direction. Does the surface integral depend only on the boundary line for this function? What is the total flux over the *closed* surface of the box (*including* the bottom)? [*Note:* For the *closed* surface the positive direction is "outward," and hence "down," for the bottom face.]

Problem 1.30 Calculate the volume integral of the function $T = z^2$ over the tetrahedron with corners at (0,0,0), (1,0,0), (0,1,0), and (0,0,1).

1.3.2 The Fundamental Theorem of Calculus

Suppose $f(x)$ is a function of one variable. The **fundamental theorem of calculus** states:

$$\int_a^b \frac{df}{dx}\,dx = f(b) - f(a). \tag{1.54}$$

In case this doesn't look familiar, let's write it another way:

$$\int_a^b F(x)\,dx = f(b) - f(a),$$

where $df/dx = F(x)$. The fundamental theorem tells you how to integrate $F(x)$: you think up a function $f(x)$ whose *derivative* is equal to F.

Geometrical Interpretation: According to Eq. 1.33, $df = (df/dx)dx$ is the infinitesimal change in f when you go from (x) to $(x + dx)$. The fundamental theorem (1.54) says that if you chop the interval from a to b (Fig. 1.25) into many tiny pieces, dx, and add up the increments df from each little piece, the result is (not surprisingly) equal to the total change in f: $f(b) - f(a)$. In other words, there are two ways to determine the total change in the function: *either* subtract the values at the ends *or* go step-by-step, adding up all the tiny increments as you go. You'll get the same answer either way.

Notice the basic format of the fundamental theorem: *the integral of a derivative over an interval is given by the value of the function at the end points (boundaries).* In vector calculus there are three species of derivative (gradient, divergence, and curl), and each has its own "fundamental theorem," with essentially the same format. I don't plan to prove these theorems here; rather, I shall explain what they *mean*, and try to make them *plausible*. Proofs are given in Appendix A.

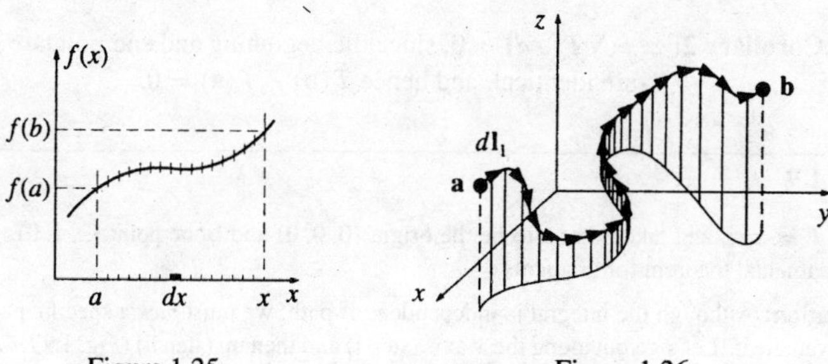

Figure 1.25

Figure 1.26

1.3.3 The Fundamental Theorem for Gradients

Suppose we have a scalar function of three variables $T(x, y, z)$. Starting at point **a**, we move a small distance $d\mathbf{l}_1$ (Fig. 1.26). According to Eq. 1.37, the function T will change by an amount

$$dT = (\nabla T) \cdot d\mathbf{l}_1.$$

Now we move a little further, by an additional small displacement $d\mathbf{l}_2$; the incremental change in T will be $(\nabla T) \cdot d\mathbf{l}_2$. In this manner, proceeding by infinitesimal steps, we make the journey to point **b**. At each step we compute the gradient of T (at that point) and dot it into the displacement $d\mathbf{l}$... this gives us the change in T. Evidently the *total* change in T in going from **a** to **b** *along the path selected* is

$$\boxed{\int_{\mathbf{a}\atop\mathcal{P}}^{\mathbf{b}} (\nabla T) \cdot d\mathbf{l} = T(\mathbf{b}) - T(\mathbf{a}).} \tag{1.55}$$

This is called the **fundamental theorem for gradients**; like the "ordinary" fundamental theorem, it says that the integral (here a *line* integral) of a derivative (here the *gradient*) is given by the value of the function at the boundaries (**a** and **b**).

Geometrical Interpretation: Suppose you wanted to determine the height of the Eiffel Tower. You could climb the stairs, using a ruler to measure the rise at each step, and adding them all up (that's the left side of Eq. 1.55), or you could place altimeters at the top and the bottom, and subtract the two readings (that's the right side); you should get the same answer either way (that's the fundamental theorem).

Incidentally, as we found in Ex. 1.6, line integrals ordinarily depend on the *path* taken from **a** to **b**. But the *right* side of Eq. 1.55 makes no reference to the path—only to the end points. Evidently, *gradients* have the special property that their line integrals are path independent:

Corollary 1: $\int_{\mathbf{a}}^{\mathbf{b}}(\nabla T)\cdot d\mathbf{l}$ is independent of path taken from **a** to **b**.

Corollary 2: $\oint(\nabla T)\cdot d\mathbf{l}=0$, since the beginning and end points are identical, and hence $T(\mathbf{b})-T(\mathbf{a})=0$.

Example 1.9

Let $T = xy^2$, and take point **a** to be the origin (0, 0, 0) and **b** the point (2, 1, 0). Check the fundamental theorem for gradients.

Solution: Although the integral is independent of path, we must *pick* a specific path in order to evaluate it. Let's go out along the x axis (step i) and then up (step ii) (Fig. 1.27). As always, $d\mathbf{l} = dx\,\hat{\mathbf{x}} + dy\,\hat{\mathbf{y}} + dz\,\hat{\mathbf{z}}$; $\nabla T = y^2\,\hat{\mathbf{x}} + 2xy\,\hat{\mathbf{y}}$.

(i) $y = 0$; $d\mathbf{l} = dx\,\hat{\mathbf{x}}$, $\nabla T\cdot d\mathbf{l} = y^2\,dx = 0$, so

$$\int_{\mathrm{i}} \nabla T\cdot d\mathbf{l} = 0.$$

(ii) $x = 2$; $d\mathbf{l} = dy\,\hat{\mathbf{y}}$, $\nabla T\cdot d\mathbf{l} = 2xy\,dy = 4y\,dy$, so

$$\int_{\mathrm{ii}} \nabla T\cdot d\mathbf{l} = \int_0^1 4y\,dy = 2y^2\Big|_0^1 = 2.$$

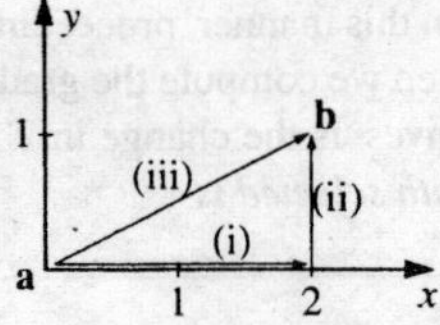

Figure 1.27

Evidently the total line integral is 2. Is this consistent with the fundamental theorem? Yes: $T(\mathbf{b}) - T(\mathbf{a}) = 2 - 0 = 2$.

Now, just to convince you that the answer is independent of path, let me calculate the same integral along path iii (the straight line from **a** to **b**):

(iii) $y = \frac{1}{2}x,\ dy = \frac{1}{2}\,dx,\ \nabla T \cdot d\mathbf{l} = y^2\,dx + 2xy\,dy = \frac{3}{4}x^2\,dx$, so

$$\int_{\text{iii}} \nabla T \cdot d\mathbf{l} = \int_0^2 \tfrac{3}{4}x^2\,dx = \tfrac{1}{4}x^3\Big|_0^2 = 2.$$

Problem 1.31 Check the fundamental theorem for gradients, using $T = x^2 + 4xy + 2yz^3$, the points $\mathbf{a} = (0, 0, 0)$, $\mathbf{b} = (1, 1, 1)$, and the three paths in Fig. 1.28:

(a) $(0, 0, 0) \to (1, 0, 0) \to (1, 1, 0) \to (1, 1, 1)$;

(b) $(0, 0, 0) \to (0, 0, 1) \to (0, 1, 1) \to (1, 1, 1)$;

(c) the parabolic path $z = x^2$; $y = x$.

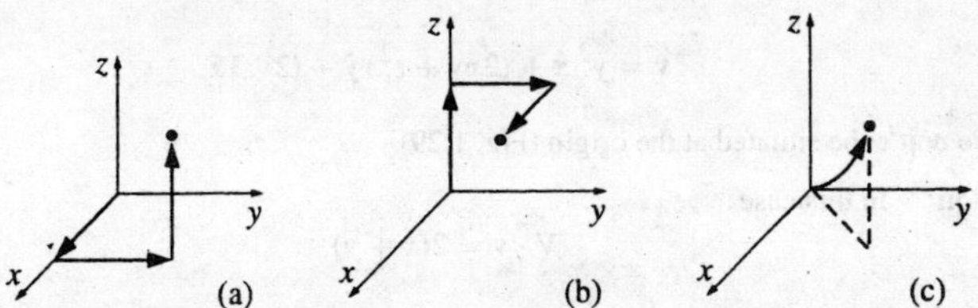

Figure 1.28

1.3.4 The Fundamental Theorem for Divergences

The fundamental theorem for divergences states that:

$$\boxed{\int_{\mathcal{V}} (\nabla \cdot \mathbf{v})\,d\tau = \oint_{\mathcal{S}} \mathbf{v} \cdot d\mathbf{a}.} \tag{1.56}$$

In honor, I suppose of its great importance, this theorem has at least three special names: **Gauss's theorem, Green's theorem**, or, simply, the **divergence theorem**. Like the other "fundamental theorems," it says that the *integral* of a *derivative* (in this case the *divergence*) over a *region* (in this case a *volume*) is equal to the value of the function at the *boundary* (in this case the *surface* that bounds the volume). Notice that the boundary term is itself an integral (specifically, a surface integral). This is reasonable: the "boundary" of a *line* is just two end points, but the boundary of a *volume* is a (closed) surface.

Geometrical Interpretation: If $\mathbf{v}$ represents the flow of an incompressible fluid, then the *flux* of $\mathbf{v}$ (the right side of Eq. 1.56) is the total amount of fluid passing out through the surface, per unit time. Now, the divergence measures the "spreading out" of the vectors from a point—a place of high divergence is like a "faucet," pouring out liquid. If we have lots of faucets in a region filled with incompressible fluid, an equal amount of liquid will be forced out through the boundaries of the region. In fact, there are *two* ways we could determine how much is being produced: (a) we could count up all the faucets, recording how much each puts out, or (b) we could go around the boundary, measuring the flow at each point, and add it all up. You get the same answer either way:

$$\int (\text{faucets within the volume}) = \oint (\text{flow out through the surface}).$$

This, in essence, is what the divergence theorem says.

Example 1.10

Check the divergence theorem using the function

$$\mathbf{v} = y^2\,\hat{\mathbf{x}} + (2xy + z^2)\,\hat{\mathbf{y}} + (2yz)\,\hat{\mathbf{z}}$$

and the unit cube situated at the origin (Fig. 1.29).

Solution: In this case

$$\nabla \cdot \mathbf{v} = 2(x + y),$$

and

$$\int_{\mathcal{V}} 2(x + y)\,d\tau = 2\int_0^1\int_0^1\int_0^1 (x + y)\,dx\,dy\,dz,$$

$$\int_0^1 (x + y)\,dx = \tfrac{1}{2} + y, \quad \int_0^1 (\tfrac{1}{2} + y)\,dy = 1, \quad \int_0^1 1\,dz = 1.$$

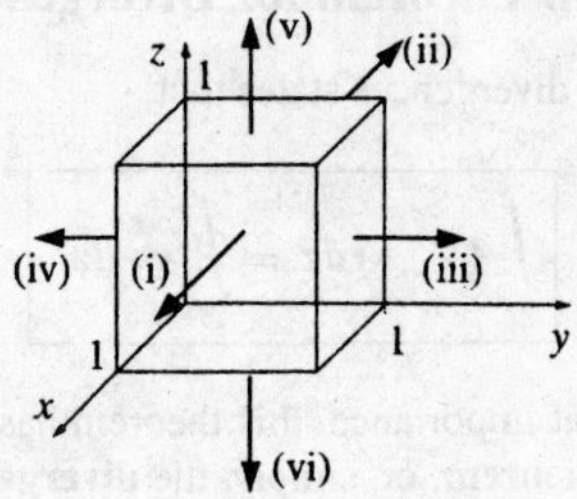

Figure 1.29

Evidently,

$$\int_{\mathcal{V}} \nabla \cdot \mathbf{v}\, d\tau = 2.$$

So much for the left side of the divergence theorem. To evaluate the surface integral we must consider separately the six sides of the cube:

(i) $$\int \mathbf{v} \cdot d\mathbf{a} = \int_0^1 \int_0^1 y^2 dy\, dz = \tfrac{1}{3}.$$

(ii) $$\int \mathbf{v} \cdot d\mathbf{a} = -\int_0^1 \int_0^1 y^2\, dy\, dz = -\tfrac{1}{3}.$$

(iii) $$\int \mathbf{v} \cdot d\mathbf{a} = \int_0^1 \int_0^1 (2x + z^2)\, dx\, dz = \tfrac{4}{3}.$$

(iv) $$\int \mathbf{v} \cdot d\mathbf{a} = -\int_0^1 \int_0^1 z^2\, dx\, dz = -\tfrac{1}{3}.$$

(v) $$\int \mathbf{v} \cdot d\mathbf{a} = \int_0^1 \int_0^1 2y\, dx\, dy = 1.$$

(vi) $$\int \mathbf{v} \cdot d\mathbf{a} = -\int_0^1 \int_0^1 0\, dx\, dy = 0.$$

So the total flux is:

$$\oint_{\mathcal{S}} \mathbf{v} \cdot d\mathbf{a} = \tfrac{1}{3} - \tfrac{1}{3} + \tfrac{4}{3} - \tfrac{1}{3} + 1 + 0 = 2,$$

as expected.

Problem 1.32 Test the divergence theorem for the function $\mathbf{v} = (xy)\,\hat{\mathbf{x}} + (2yz)\,\hat{\mathbf{y}} + (3zx)\,\hat{\mathbf{z}}$. Take as your volume the cube shown in Fig. 1.30, with sides of length 2.

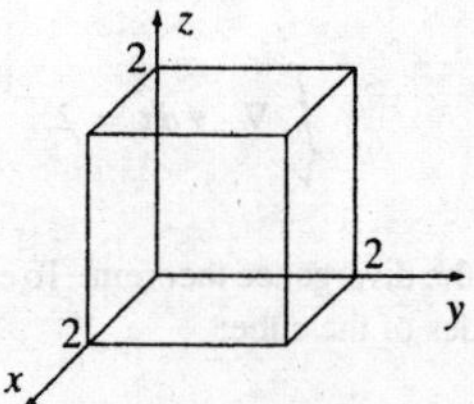

Figure 1.30

1.3.5 The Fundamental Theorem for Curls

The fundamental theorem for curls, which goes by the special name of **Stokes' theorem**, states that

$$\boxed{\int_{\mathcal{S}} (\nabla \times \mathbf{v}) \cdot d\mathbf{a} = \oint_{\mathcal{P}} \mathbf{v} \cdot d\mathbf{l}.} \tag{1.57}$$

As always, the *integral* of a *derivative* (here, the *curl*) over a *region* (here, a patch of *surface*) is equal to the value of the function at the *boundary* (here, the perimeter of the patch). As in the case of the divergence theorem, the boundary term is itself an integral—specifically, a closed line integral.

Geometrical Interpretation: Recall that the curl measures the "twist" of the vectors **v**; a region of high curl is a whirlpool—if you put a tiny paddle wheel there, it will rotate. Now, the *integral* of the curl over some surface (or, more precisely, the *flux* of the curl *through* that surface) represents the "total amount of swirl," and we can determine that swirl just as well by going around the *edge* and finding how much the flow is following the boundary (Fig. 1.31). You may find this a rather forced interpretation of Stokes' theorem, but it's a helpful mnemonic, if nothing else.

You might have noticed an apparent ambiguity in Stokes' theorem: concerning the boundary line integral, which *way* are we supposed to go around (clockwise or counter-clockwise)? If we go the "wrong" way we'll pick up an overall sign error. The answer is that it doesn't *matter* which way you go as long as you are *consistent,* for there is a compensating sign ambiguity in the surface integral: Which way does $d\mathbf{a}$ point? For a *closed* surface (as in the divergence theorem) $d\mathbf{a}$ points in the direction of the *outward* normal; but for an *open* surface, which way is "out?" Consistency in Stokes' theorem (as in all such matters) is given by the right-hand rule: If your fingers point in the direction of the line integral, then your thumb fixes the direction of $d\mathbf{a}$ (Fig. 1.32).

Now, there are plenty of surfaces (infinitely many) that share any given boundary line. Twist a paper clip into a loop and dip it in soapy water. The soap film constitutes a surface, with the wire loop as its boundary. If you blow on it, the soap film will expand, making a larger surface, with the same boundary. Ordinarily, a flux integral depends critically on what surface you integrate over, but evidently this is *not* the case with curls. For Stokes'

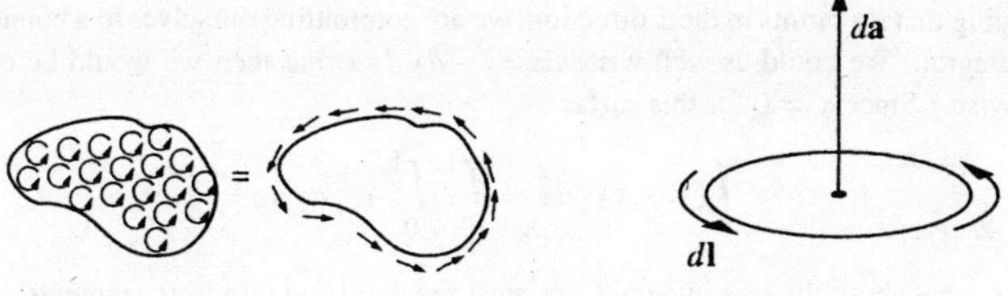

Figure 1.31 Figure 1.32

theorem says that $\int(\nabla \times \mathbf{v}) \cdot d\mathbf{a}$ is equal to the line integral of $\mathbf{v}$ around the boundary, and the latter makes no reference to the specific surface you choose.

Corollary 1: $\int(\nabla \times \mathbf{v}) \cdot d\mathbf{a}$ depends only on the boundary line, not on the particular surface used.

Corollary 2: $\oint(\nabla \times \mathbf{v}) \cdot d\mathbf{a} = 0$ for any closed surface, since the boundary line, like the mouth of a balloon, shrinks down to a point, and hence the right side of Eq. 1.57 vanishes.

These corollaries are analogous to those for the gradient theorem. We shall develop the parallel further in due course.

Example 1.11

Suppose $\mathbf{v} = (2xz + 3y^2)\hat{\mathbf{y}} + (4yz^2)\hat{\mathbf{z}}$. Check Stokes' theorem for the square surface shown in Fig. 1.33.

Solution: Here

$$\nabla \times \mathbf{v} = (4z^2 - 2x)\,\hat{\mathbf{x}} + 2z\,\hat{\mathbf{z}} \quad \text{and} \quad d\mathbf{a} = dy\,dz\,\hat{\mathbf{x}}.$$

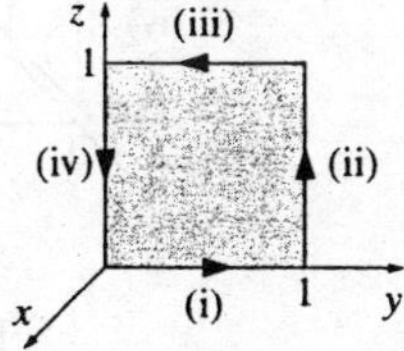

Figure 1.33

(In saying that $d\mathbf{a}$ points in the x direction, we are committing ourselves to a counterclockwise line integral. We could as well write $d\mathbf{a} = -dy\,dz\,\hat{\mathbf{x}}$, but then we would be obliged to go clockwise.) Since $x = 0$ for this surface,

$$\int (\nabla \times \mathbf{v}) \cdot d\mathbf{a} = \int_0^1 \int_0^1 4z^2\,dy\,dz = \tfrac{4}{3}.$$

Now, what about the line integral? We must break this up into four segments:

$$\text{(i)} \quad x = 0, \quad z = 0, \quad \mathbf{v} \cdot d\mathbf{l} = 3y^2\,dy, \quad \int \mathbf{v} \cdot d\mathbf{l} = \int_0^1 3y^2\,dy = 1,$$

$$\text{(ii)} \quad x = 0, \quad y = 1, \quad \mathbf{v} \cdot d\mathbf{l} = 4z^2\,dz, \quad \int \mathbf{v} \cdot d\mathbf{l} = \int_0^1 4z^2\,dz = \tfrac{4}{3},$$

$$\text{(iii)} \quad x = 0, \quad z = 1, \quad \mathbf{v} \cdot d\mathbf{l} = 3y^2\,dy, \quad \int \mathbf{v} \cdot d\mathbf{l} = \int_1^0 3y^2\,dy = -1,$$

$$\text{(iv)} \quad x = 0, \quad y = 0, \quad \mathbf{v} \cdot d\mathbf{l} = 0, \quad \int \mathbf{v} \cdot d\mathbf{l} = \int_1^0 0\,dz = 0.$$

So

$$\oint \mathbf{v} \cdot d\mathbf{l} = 1 + \tfrac{4}{3} - 1 + 0 = \tfrac{4}{3}.$$

It checks.

A point of strategy: notice how I handled step (iii). There is a temptation to write $d\mathbf{l} = -dy\,\hat{\mathbf{y}}$ here, since the path goes to the left. You can get away with this, if you insist, by running the integral from $0 \rightarrow 1$. Personally, I prefer to say $d\mathbf{l} = dx\,\hat{\mathbf{x}} + dy\,\hat{\mathbf{y}} + dz\,\hat{\mathbf{z}}$ *always* (never any minus signs) and let the limits of the integral take care of the direction.

Problem 1.33 Test Stokes' theorem for the function $\mathbf{v} = (xy)\,\hat{\mathbf{x}} + (2yz)\,\hat{\mathbf{y}} + (3zx)\,\hat{\mathbf{z}}$, using the triangular shaded area of Fig. 1.34.

Problem 1.34 Check Corollary 1 by using the same function and boundary line as in Ex. 1.11, but integrating over the five sides of the cube in Fig. 1.35. The back of the cube is open.

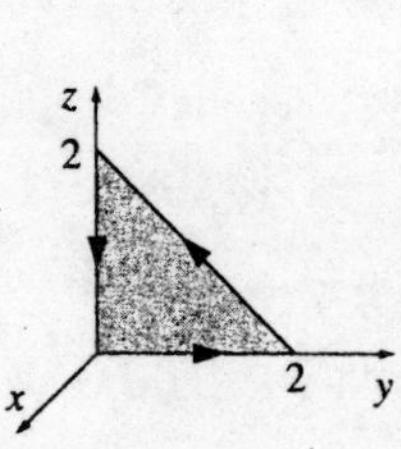

Figure 1.34

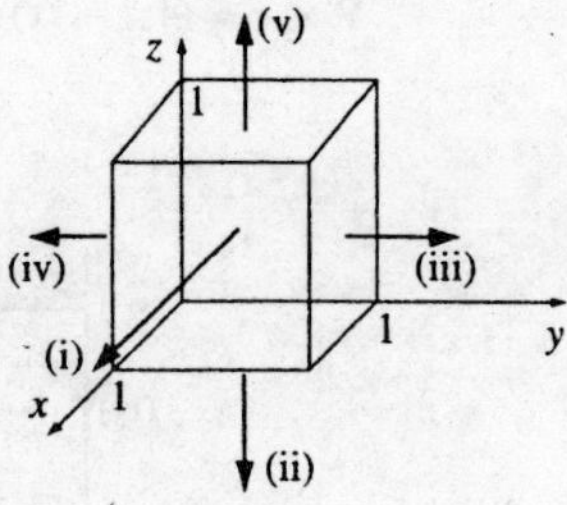

Figure 1.35

1.3.6 Integration by Parts

The technique known (awkwardly) as **integration by parts** exploits the product rule for derivatives:

$$\frac{d}{dx}(fg) = f\left(\frac{dg}{dx}\right) + g\left(\frac{df}{dx}\right).$$

Integrating both sides, and invoking the fundamental theorem:

$$\int_a^b \frac{d}{dx}(fg)\,dx = fg\Big|_a^b = \int_a^b f\left(\frac{dg}{dx}\right)dx + \int_a^b g\left(\frac{df}{dx}\right)dx,$$

or

$$\int_a^b f\left(\frac{dg}{dx}\right)dx = -\int_a^b g\left(\frac{df}{dx}\right)dx + fg\Big|_a^b. \tag{1.58}$$

That's integration by parts. It pertains to the situation in which you are called upon to integrate the product of one function (f) and the *derivative* of another (g); it says you can *transfer the derivative from g to f*, at the cost of a minus sign and a boundary term.

Example 1.12

Evaluate the integral

$$\int_0^\infty xe^{-x}\,dx.$$

Solution: The exponential can be expressed as a derivative:

$$e^{-x} = \frac{d}{dx}\left(-e^{-x}\right);$$

in this case, then, $f(x) = x$, $g(x) = -e^{-x}$, and $df/dx = 1$, so

$$\int_0^\infty xe^{-x}\,dx = \int_0^\infty e^{-x}\,dx - xe^{-x}\Big|_0^\infty = -e^{-x}\Big|_0^\infty = 1.$$

We can exploit the product rules of vector calculus, together with the appropriate fundamental theorems, in exactly the same way. For example, integrating

$$\nabla\cdot(f\mathbf{A}) = f(\nabla\cdot\mathbf{A}) + \mathbf{A}\cdot(\nabla f)$$

over a volume, and invoking the divergence theorem, yields

$$\int \nabla\cdot(f\mathbf{A})\,d\tau = \int f(\nabla\cdot\mathbf{A})\,d\tau + \int \mathbf{A}\cdot(\nabla f)\,d\tau = \oint f\mathbf{A}\cdot d\mathbf{a},$$

or

$$\int_{\mathcal{V}} f(\nabla\cdot\mathbf{A})\,d\tau = -\int_{\mathcal{V}} \mathbf{A}\cdot(\nabla f)\,d\tau + \oint_{\mathcal{S}} f\mathbf{A}\cdot d\mathbf{a}. \tag{1.59}$$

Here again the integrand is the product of one function (f) and the derivative (in this case the *divergence*) of another (**A**), and integration by parts licenses us to transfer the derivative from **A** to f (where it becomes a *gradient*), at the cost of a minus sign and a boundary term (in this case a surface integral).

You might wonder how often one is likely to encounter an integral involving the product of one function and the derivative of another; the answer is *surprisingly* often, and integration by parts turns out to be one of the most powerful tools in vector calculus.

Problem 1.35

(a) Show that

$$\int_{\mathcal{S}} f(\nabla \times \mathbf{A}) \cdot d\mathbf{a} = \int_{\mathcal{S}} [\mathbf{A} \times (\nabla f)] \cdot d\mathbf{a} + \oint_{\mathcal{P}} f\mathbf{A} \cdot d\mathbf{l}. \tag{1.60}$$

(b) Show that

$$\int_{\mathcal{V}} \mathbf{B} \cdot (\nabla \times \mathbf{A})\, d\tau = \int_{\mathcal{V}} \mathbf{A} \cdot (\nabla \times \mathbf{B})\, d\tau + \oint_{\mathcal{S}} (\mathbf{A} \times \mathbf{B}) \cdot d\mathbf{a}. \tag{1.61}$$

1.4 Curvilinear Coordinates

1.4.1 Spherical Polar Coordinates

The spherical polar coordinates (r, θ, ϕ) of a point P are defined in Fig. 1.36; r is the distance from the origin (the magnitude of the position vector), θ (the angle down from the z axis) is called the **polar angle**, and ϕ (the angle around from the x axis) is the **azimuthal angle**. Their relation to Cartesian coordinates (x, y, z) can be read from the figure:

$$x = r \sin\theta \cos\phi, \qquad y = r \sin\theta \sin\phi, \qquad z = r\cos\theta. \tag{1.62}$$

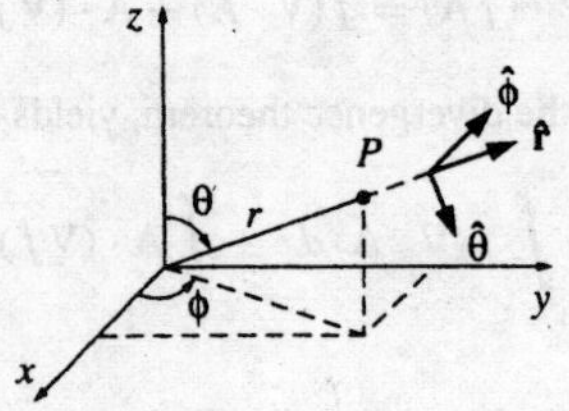

Figure 1.36

Figure 1.36 also shows three unit vectors, $\hat{\mathbf{r}}$, $\hat{\boldsymbol{\theta}}$, $\hat{\boldsymbol{\phi}}$, pointing in the direction of increase of the corresponding coordinates. They constitute an orthogonal (mutually perpendicular) basis set (just like $\hat{\mathbf{x}}$, $\hat{\mathbf{y}}$, $\hat{\mathbf{z}}$), and any vector $\mathbf{A}$ can be expressed in terms of them in the usual way:

$$\mathbf{A} = A_r\,\hat{\mathbf{r}} + A_\theta\,\hat{\boldsymbol{\theta}} + A_\phi\,\hat{\boldsymbol{\phi}}. \tag{1.63}$$

A_r, A_θ, and A_ϕ are the radial, polar, and azimuthal components of $\mathbf{A}$. In terms of the Cartesian unit vectors,

$$\left.\begin{aligned} \hat{\mathbf{r}} &= \sin\theta\cos\phi\,\hat{\mathbf{x}} + \sin\theta\sin\phi\,\hat{\mathbf{y}} + \cos\theta\,\hat{\mathbf{z}}, \\ \hat{\boldsymbol{\theta}} &= \cos\theta\cos\phi\,\hat{\mathbf{x}} + \cos\theta\sin\phi\,\hat{\mathbf{y}} - \sin\theta\,\hat{\mathbf{z}}, \\ \hat{\boldsymbol{\phi}} &= -\sin\phi\,\hat{\mathbf{x}} + \cos\phi\,\hat{\mathbf{y}}, \end{aligned}\right\} \tag{1.64}$$

as you can easily check for yourself (Prob. 1.37). I have put these formulas inside the back cover, for easy reference.

But there is a poisonous snake lurking here that I'd better warn you about: $\hat{\mathbf{r}}$, $\hat{\boldsymbol{\theta}}$, and $\hat{\boldsymbol{\phi}}$ are associated with a *particular point P*, and they *change direction* as *P* moves around. For example, $\hat{\mathbf{r}}$ always points radially outward, but "radially outward" can be the *x* direction, the *y* direction, or any other direction, depending on where you are. In Fig. 1.37, $\mathbf{A} = \hat{\mathbf{y}}$ and $\mathbf{B} = -\hat{\mathbf{y}}$, and yet *both* of them would be written as $\hat{\mathbf{r}}$ in spherical coordinates. One could take account of this by explicitly indicating the point of reference: $\hat{\mathbf{r}}(\theta, \phi)$, $\hat{\boldsymbol{\theta}}(\theta, \phi)$, $\hat{\boldsymbol{\phi}}(\theta, \phi)$, but this would be cumbersome, and as long as you are alert to the problem I don't think it will cause difficulties.[4] In particular, do not naïvely combine the spherical components of vectors associated with different points (in Fig. 1.37, $\mathbf{A} + \mathbf{B} = 0$, not $2\hat{\mathbf{r}}$, and $\mathbf{A} \cdot \mathbf{B} = -1$, not $+1$). Beware of differentiating a vector that is expressed in spherical coordinates, since the unit vectors themselves are functions of position ($\partial\hat{\mathbf{r}}/\partial\theta = \hat{\boldsymbol{\theta}}$, for example). And do not take $\hat{\mathbf{r}}$, $\hat{\boldsymbol{\theta}}$, and $\hat{\boldsymbol{\phi}}$ outside an integral, as we did with $\hat{\mathbf{x}}$, $\hat{\mathbf{y}}$, and $\hat{\mathbf{z}}$ in Eq. 1.53. In general, if you're uncertain about the validity of an operation, reexpress the problem in Cartesian coordinates, where this difficulty does not arise.

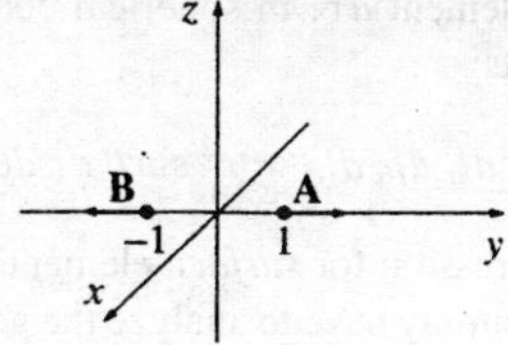

Figure 1.37

[4] I claimed on the very first page that vectors have no location, and I'll stand by that. The vectors themselves live "out there," completely independent of our choice of coordinates. But the *notation* we use to represent them *does* depend on the point in question, in curvilinear coordinates.

An infinitesimal displacement in the $\hat{\mathbf{r}}$ direction is simply dr (Fig. 1.38a), just as an infinitesimal element of length in the x direction is dx:

$$dl_r = dr. \tag{1.65}$$

On the other hand, an infinitesimal element of length in the $\hat{\boldsymbol{\theta}}$ direction (Fig. 1.38b) is *not* just $d\theta$ (that's an *angle*—it doesn't even have the right *units* for a length), but rather $r\,d\theta$:

$$dl_\theta = r\,d\theta. \tag{1.66}$$

Similarly, an infinitesimal element of length in the $\hat{\boldsymbol{\phi}}$ direction (Fig. 1.38c) is $r\sin\theta\,d\phi$:

$$dl_\phi = r\sin\theta\,d\phi. \tag{1.67}$$

Thus, the general infinitesimal displacement $d\mathbf{l}$ is

$$d\mathbf{l} = dr\,\hat{\mathbf{r}} + r\,d\theta\,\hat{\boldsymbol{\theta}} + r\sin\theta\,d\phi\,\hat{\boldsymbol{\phi}}. \tag{1.68}$$

This plays the role (in line integrals, for example) that $d\mathbf{l} = dx\,\hat{\mathbf{x}} + dy\,\hat{\mathbf{y}} + dz\,\hat{\mathbf{z}}$ played in Cartesian coordinates.

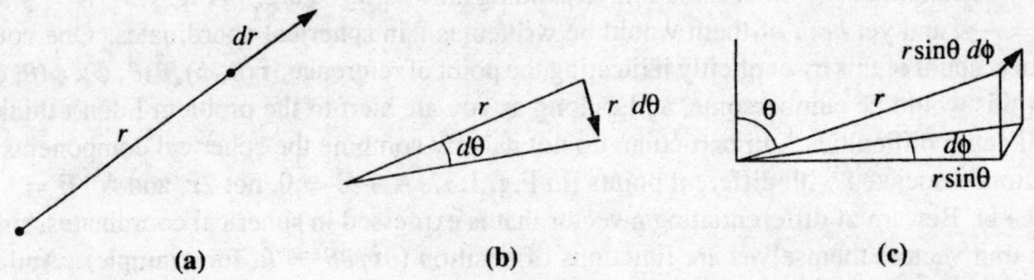

Figure 1.38

The infinitesimal volume element $d\tau$, in spherical coordinates, is the product of the three infinitesimal displacements:

$$d\tau = dl_r\,dl_\theta\,dl_\phi = r^2\sin\theta\,dr\,d\theta\,d\phi. \tag{1.69}$$

I cannot give you a general expression for *surface* elements $d\mathbf{a}$, since these depend on the orientation of the surface. You simply have to analyze the geometry for any given case (this goes for Cartesian and curvilinear coordinates alike). If you are integrating over the surface of a sphere, for instance, then r is constant, whereas θ and ϕ change (Fig. 1.39), so

$$d\mathbf{a}_1 = dl_\theta\,dl_\phi\,\hat{\mathbf{r}} = r^2\sin\theta\,d\theta\,d\phi\,\hat{\mathbf{r}}.$$

On the other hand, if the surface lies in the xy plane, say, so that θ is constant (to wit: $\pi/2$) while r and ϕ vary, then

$$d\mathbf{a}_2 = dl_r\,dl_\phi\,\hat{\boldsymbol{\theta}} = r\,dr\,d\phi\,\hat{\boldsymbol{\theta}}.$$

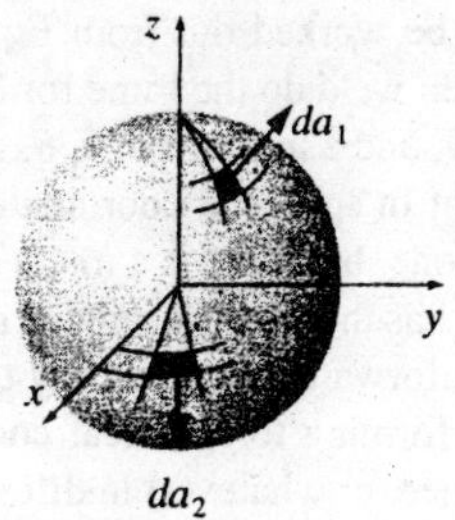

Figure 1.39

Notice, finally, that r ranges from 0 to ∞, ϕ from 0 to 2π, and θ from 0 to π (*not* 2π—that would count every point twice).[5]

Example 1.13

Find the volume of a sphere of radius R.

Solution:

$$\begin{aligned} V &= \int d\tau = \int_{r=0}^{R}\int_{\theta=0}^{\pi}\int_{\phi=0}^{2\pi} r^2 \sin\theta \, dr \, d\theta \, d\phi \\ &= \left(\int_0^R r^2 \, dr\right)\left(\int_0^{\pi} \sin\theta \, d\theta\right)\left(\int_0^{2\pi} d\phi\right) \\ &= \left(\frac{R^3}{3}\right)(2)(2\pi) = \tfrac{4}{3}\pi R^3. \end{aligned}$$

(Not a big surprise.)

So far we have talked only about the *geometry* of spherical coordinates. Now I would like to "translate" the vector derivatives (gradient, divergence, curl, and Laplacian) into r, θ, ϕ notation. In principle this is entirely straightforward: in the case of the gradient,

$$\nabla T = \frac{\partial T}{\partial x}\hat{\mathbf{x}} + \frac{\partial T}{\partial y}\hat{\mathbf{y}} + \frac{\partial T}{\partial z}\hat{\mathbf{z}},$$

for instance, we would first use the chain rule to reexpress the partials:

$$\frac{\partial T}{\partial x} = \frac{\partial T}{\partial r}\left(\frac{\partial r}{\partial x}\right) + \frac{\partial T}{\partial \theta}\left(\frac{\partial \theta}{\partial x}\right) + \frac{\partial T}{\partial \phi}\left(\frac{\partial \phi}{\partial x}\right).$$

[5] Alternatively, you could run ϕ from 0 to π (the "eastern hemisphere") and cover the "western hemisphere" by extending θ from π up to 2π. But this is very bad notation, since, among other things, $\sin\theta$ will then run negative, and you'll have to put absolute value signs around that term in volume and surface elements (area and volume being intrinsically positive quantities).

The terms in parentheses could be worked out from Eq. 1.62—or rather, the *inverse* of those equations (Prob. 1.36). Then we'd do the same for $\partial T/\partial y$ and $\partial T/\partial z$. Finally, we'd substitute in the formulas for $\hat{\mathbf{x}}$, $\hat{\mathbf{y}}$, and $\hat{\mathbf{z}}$ in terms of $\hat{\mathbf{r}}$, $\hat{\boldsymbol{\theta}}$, and $\hat{\boldsymbol{\phi}}$ (Prob. 1.37). It would take an hour to figure out the gradient in spherical coordinates by this brute-force method. I suppose this is how it was first done, but there is a much more efficient indirect approach, explained in Appendix A, which has the extra advantage of treating all coordinate systems at once. I described the "straightforward" method only to show you that there is nothing subtle or mysterious about transforming to spherical coordinates: you're expressing the *same quantity* (gradient, divergence, or whatever) in different notation, that's all.

Here, then, are the vector derivatives in spherical coordinates:

Gradient:

$$\nabla T = \frac{\partial T}{\partial r}\hat{\mathbf{r}} + \frac{1}{r}\frac{\partial T}{\partial \theta}\hat{\boldsymbol{\theta}} + \frac{1}{r\sin\theta}\frac{\partial T}{\partial \phi}\hat{\boldsymbol{\phi}}. \tag{1.70}$$

Divergence:

$$\nabla \cdot \mathbf{v} = \frac{1}{r^2}\frac{\partial}{\partial r}(r^2 v_r) + \frac{1}{r\sin\theta}\frac{\partial}{\partial\theta}(\sin\theta\, v_\theta) + \frac{1}{r\sin\theta}\frac{\partial v_\phi}{\partial\phi}. \tag{1.71}$$

Curl:

$$\begin{aligned}\nabla\times\mathbf{v} &= \frac{1}{r\sin\theta}\left[\frac{\partial}{\partial\theta}(\sin\theta\, v_\phi) - \frac{\partial v_\theta}{\partial\phi}\right]\hat{\mathbf{r}} + \frac{1}{r}\left[\frac{1}{\sin\theta}\frac{\partial v_r}{\partial\phi} - \frac{\partial}{\partial r}(r v_\phi)\right]\hat{\boldsymbol{\theta}} \\ &+ \frac{1}{r}\left[\frac{\partial}{\partial r}(r v_\theta) - \frac{\partial v_r}{\partial\theta}\right]\hat{\boldsymbol{\phi}}.\end{aligned} \tag{1.72}$$

Laplacian:

$$\nabla^2 T = \frac{1}{r^2}\frac{\partial}{\partial r}\left(r^2\frac{\partial T}{\partial r}\right) + \frac{1}{r^2\sin\theta}\frac{\partial}{\partial\theta}\left(\sin\theta\frac{\partial T}{\partial\theta}\right) + \frac{1}{r^2\sin^2\theta}\frac{\partial^2 T}{\partial\phi^2}. \tag{1.73}$$

For reference, these formulas are listed inside the front cover.

Problem 1.36 Find formulas for r, θ, ϕ in terms of x, y, z (the inverse, in other words, of Eq. 1.62).

• **Problem 1.37** Express the unit vectors $\hat{\mathbf{r}}, \hat{\boldsymbol{\theta}}, \hat{\boldsymbol{\phi}}$ in terms of $\hat{\mathbf{x}}, \hat{\mathbf{y}}, \hat{\mathbf{z}}$ (that is, derive Eq. 1.64). Check your answers several ways ($\hat{\mathbf{r}}\cdot\hat{\mathbf{r}} \stackrel{?}{=} 1$, $\hat{\boldsymbol{\theta}}\cdot\hat{\boldsymbol{\phi}} \stackrel{?}{=} 0$, $\hat{\mathbf{r}}\times\hat{\boldsymbol{\theta}} \stackrel{?}{=} \hat{\boldsymbol{\phi}}, \ldots$). Also work out the inverse formulas, giving $\hat{\mathbf{x}}, \hat{\mathbf{y}}, \hat{\mathbf{z}}$ in terms of $\hat{\mathbf{r}}, \hat{\boldsymbol{\theta}}, \hat{\boldsymbol{\phi}}$ (and θ, ϕ).

• **Problem 1.38**

(a) Check the divergence theorem for the function $\mathbf{v}_1 = r^2\hat{\mathbf{r}}$, using as your volume the sphere of radius R, centered at the origin.

b) Do the same for $\mathbf{v}_2 = (1/r^2)\hat{\mathbf{r}}$. (If the answer surprises you, look back at Prob. 1.16.)

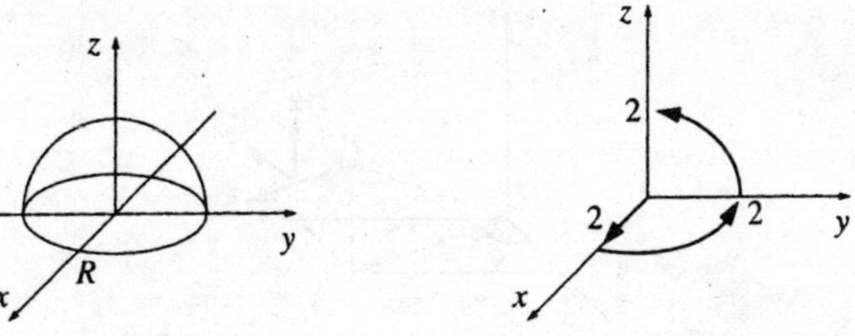

Figure 1.40

Figure 1.41

Problem 1.39 Compute the divergence of the function

$$\mathbf{v} = (r\cos\theta)\,\hat{\mathbf{r}} + (r\sin\theta)\,\hat{\boldsymbol{\theta}} + (r\sin\theta\cos\phi)\,\hat{\boldsymbol{\phi}}.$$

Check the divergence theorem for this function, using as your volume the inverted hemispherical bowl of radius R, resting on the xy plane and centered at the origin (Fig. 1.40).

Problem 1.40 Compute the gradient and Laplacian of the function $T = r(\cos\theta + \sin\theta\cos\phi)$. Check the Laplacian by converting T to Cartesian coordinates and using Eq. 1.42. Test the gradient theorem for this function, using the path shown in Fig. 1.41, from (0, 0, 0) to (0, 0, 2).

1.4.2 Cylindrical Coordinates

The cylindrical coordinates (s, ϕ, z) of a point P are defined in Fig. 1.42. Notice that ϕ has the same meaning as in spherical coordinates, and z is the same as Cartesian; s is the distance to P *from the z axis,* whereas the spherical coordinate r is the distance from the *origin.* The relation to Cartesian coordinates is

$$x = s\cos\phi, \qquad y = s\sin\phi, \qquad z = z. \tag{1.74}$$

The unit vectors (Prob. 1.41) are

$$\left.\begin{array}{rcl} \hat{\mathbf{s}} & = & \cos\phi\,\hat{\mathbf{x}} + \sin\phi\,\hat{\mathbf{y}}, \\ \hat{\boldsymbol{\phi}} & = & -\sin\phi\,\hat{\mathbf{x}} + \cos\phi\,\hat{\mathbf{y}}, \\ \hat{\mathbf{z}} & = & \hat{\mathbf{z}}. \end{array}\right\} \tag{1.75}$$

The infinitesimal displacements are

$$dl_s = ds, \qquad dl_\phi = s\,d\phi, \qquad dl_z = dz, \tag{1.76}$$

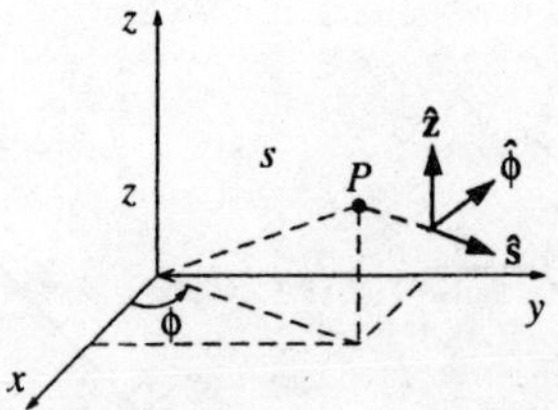

Figure 1.42

so

$$d\mathbf{l} = ds\,\hat{\mathbf{s}} + s\,d\phi\,\hat{\boldsymbol{\phi}} + dz\,\hat{\mathbf{z}}, \tag{1.77}$$

and the volume element is

$$d\tau = s\,ds\,d\phi\,dz. \tag{1.78}$$

The range of s is $0 \to \infty$, ϕ goes from $0 \to 2\pi$, and z from $-\infty$ to ∞.

The vector derivatives in cylindrical coordinates are:

Gradient:

$$\nabla T = \frac{\partial T}{\partial s}\,\hat{\mathbf{s}} + \frac{1}{s}\frac{\partial T}{\partial \phi}\,\hat{\boldsymbol{\phi}} + \frac{\partial T}{\partial z}\,\hat{\mathbf{z}}. \tag{1.79}$$

Divergence:

$$\nabla \cdot \mathbf{v} = \frac{1}{s}\frac{\partial}{\partial s}(s v_s) + \frac{1}{s}\frac{\partial v_\phi}{\partial \phi} + \frac{\partial v_z}{\partial z}. \tag{1.80}$$

Curl:

$$\nabla \times \mathbf{v} = \left(\frac{1}{s}\frac{\partial v_z}{\partial \phi} - \frac{\partial v_\phi}{\partial z}\right)\hat{\mathbf{s}} + \left(\frac{\partial v_s}{\partial z} - \frac{\partial v_z}{\partial s}\right)\hat{\boldsymbol{\phi}} + \frac{1}{s}\left[\frac{\partial}{\partial s}(s v_\phi) - \frac{\partial v_s}{\partial \phi}\right]\hat{\mathbf{z}}. \tag{1.81}$$

Laplacian:

$$\nabla^2 T = \frac{1}{s}\frac{\partial}{\partial s}\left(s\frac{\partial T}{\partial s}\right) + \frac{1}{s^2}\frac{\partial^2 T}{\partial \phi^2} + \frac{\partial^2 T}{\partial z^2}. \tag{1.82}$$

These formulas are also listed inside the front cover.

Problem 1.41 Express the cylindrical unit vectors $\hat{\mathbf{s}}$, $\hat{\boldsymbol{\phi}}$, $\hat{\mathbf{z}}$ in terms of $\hat{\mathbf{x}}$, $\hat{\mathbf{y}}$, $\hat{\mathbf{z}}$ (that is, derive Eq. 1.75). "Invert" your formulas to get $\hat{\mathbf{x}}$, $\hat{\mathbf{y}}$, $\hat{\mathbf{z}}$ in terms of $\hat{\mathbf{s}}$, $\hat{\boldsymbol{\phi}}$, $\hat{\mathbf{z}}$ (and ϕ).

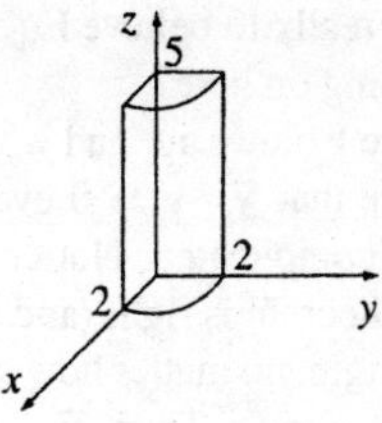

Figure 1.43

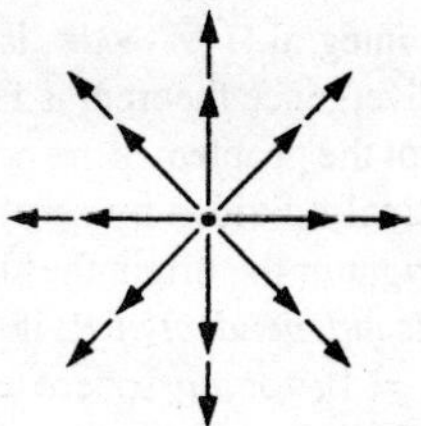

Figure 1.44

Problem 1.42

(a) Find the divergence of the function

$$\mathbf{v} = s(2 + \sin^2 \phi)\,\hat{\mathbf{s}} + s \sin\phi \cos\phi\,\hat{\boldsymbol{\phi}} + 3z\,\hat{\mathbf{z}}.$$

(b) Test the divergence theorem for this function, using the quarter-cylinder (radius 2, height 5) shown in Fig. 1.43.

(c) Find the curl of $\mathbf{v}$.

1.5 The Dirac Delta Function

1.5.1 The Divergence of $\hat{\mathbf{r}}/r^2$

Consider the vector function

$$\mathbf{v} = \frac{1}{r^2}\hat{\mathbf{r}}. \tag{1.83}$$

At every location, $\mathbf{v}$ is directed radially outward (Fig. 1.44); if ever there was a function that ought to have a large positive divergence, this is it. And yet, when you actually *calculate* the divergence (using Eq. 1.71), you get precisely *zero*:

$$\nabla \cdot \mathbf{v} = \frac{1}{r^2}\frac{\partial}{\partial r}\left(r^2 \frac{1}{r^2}\right) = \frac{1}{r^2}\frac{\partial}{\partial r}(1) = 0. \tag{1.84}$$

(You will have encountered this paradox already, if you worked Prob. 1.16.) The plot thickens if you apply the divergence theorem to this function. Suppose we integrate over a sphere of radius R, centered at the origin (Prob. 1.38b); the surface integral is

$$\begin{aligned}\oint \mathbf{v} \cdot d\mathbf{a} &= \int \left(\frac{1}{R^2}\hat{\mathbf{r}}\right) \cdot (R^2 \sin\theta\, d\theta\, d\phi\, \hat{\mathbf{r}}) \\ &= \left(\int_0^{\pi} \sin\theta\, d\theta\right)\left(\int_0^{2\pi} d\phi\right) = 4\pi. \end{aligned} \tag{1.85}$$

But the *volume* integral, $\int \nabla \cdot \mathbf{v}\, d\tau$, is *zero*, if we are really to believe Eq. 1.84. Does this mean that the divergence theorem is false? What's going on here?

The source of the problem is the point $r = 0$, where $\mathbf{v}$ blows up (and where, in Eq. 1.84, we have unwittingly divided by zero). It is quite true that $\nabla \cdot \mathbf{v} = 0$ everywhere *except* the origin, but right *at* the origin the situation is more complicated. Notice that the surface integral (1.85) is *independent of R*; if the divergence theorem is right (and it *is*), we should get $\int (\nabla \cdot \mathbf{v})\, d\tau = 4\pi$ for *any* sphere centered at the origin, no matter how small. Evidently the entire contribution must be coming from the point $r = 0$! Thus, $\nabla \cdot \mathbf{v}$ has the bizarre property that it vanishes everywhere except at one point, and yet its *integral* (over any volume containing that point) is 4π. No ordinary function behaves like that. (On the other hand, a *physical* example *does* come to mind: the density (mass per unit volume) of a point particle. It's zero except at the exact location of the particle, and yet its *integral* is finite—namely, the mass of the particle.) What we have stumbled on is a mathematical object known to physicists as the **Dirac delta function**. It arises in many branches of theoretical physics. Moreover, the specific problem at hand (the divergence of the function $\hat{\mathbf{r}}/r^2$) is not just some arcane curiosity—it is, in fact, central to the whole theory of electrodynamics. So it is worthwhile to pause here and study the Dirac delta function with some care.

1.5.2 The One-Dimensional Dirac Delta Function

The one dimensional Dirac delta function, $\delta(x)$, can be pictured as an infinitely high, infinitesimally narrow "spike," with area 1 (Fig. 1.45). That is to say:

$$\delta(x) = \left\{ \begin{array}{ll} 0, & \text{if } x \neq 0 \\ \infty, & \text{if } x = 0 \end{array} \right\} \tag{1.86}$$

and

$$\int_{-\infty}^{\infty} \delta(x)\, dx = 1. \tag{1.87}$$

Technically, $\delta(x)$ is not a function at all, since its value is not finite at $x = 0$. In the mathematical literature it is known as a **generalized function**, or **distribution**. It is, if you

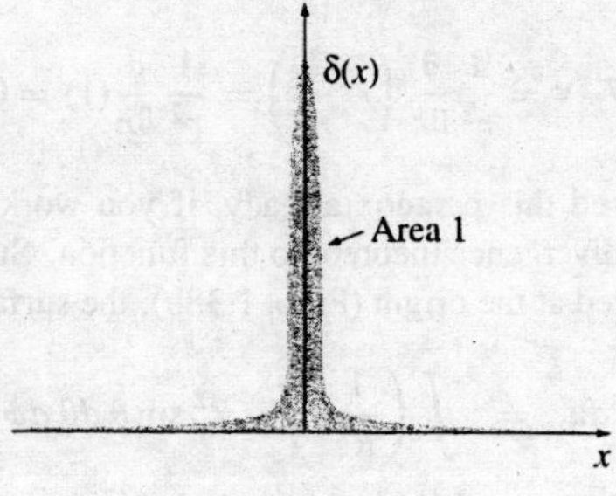

Figure 1.45

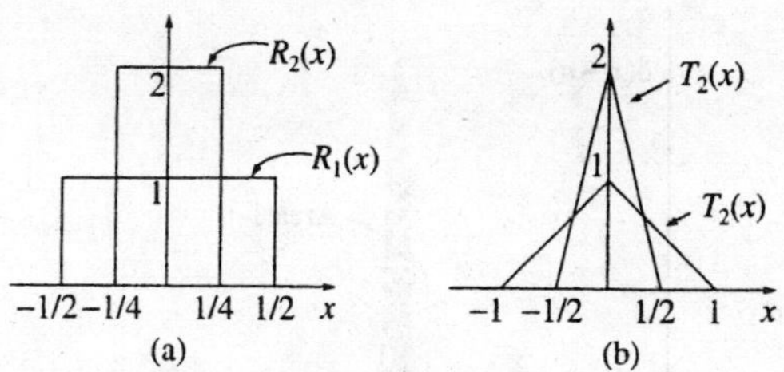

Figure 1.46

like, the *limit* of a *sequence* of functions, such as rectangles $R_n(x)$, of height n and width $1/n$, or isosceles triangles $T_n(x)$, of height n and base $2/n$ (Fig. 1.46).

If $f(x)$ is some "ordinary" function (that is, *not* another delta function—in fact, just to be on the safe side let's say that $f(x)$ is *continuous*), then the *product* $f(x)\delta(x)$ is zero everywhere except at $x = 0$. It follows that

$$f(x)\delta(x) = f(0)\delta(x). \tag{1.88}$$

(This is the most important fact about the delta function, so make sure you understand why it is true: since the product is zero anyway *except* at $x = 0$, we may as well replace $f(x)$ by the value it assumes at the origin.) In particular

$$\int_{-\infty}^{\infty} f(x)\delta(x)\,dx = f(0)\int_{-\infty}^{\infty} \delta(x)\,dx = f(0). \tag{1.89}$$

Under an integral, then, the delta function "picks out" the value of $f(x)$ at $x = 0$. (Here and below, the integral need not run from $-\infty$ to $+\infty$; it is sufficient that the domain extend across the delta function, and $-\epsilon$ to $+\epsilon$ would do as well.)

Of course, we can shift the spike from $x = 0$ to some other point, $x = a$ (Fig. 1.47):

$$\delta(x-a) = \left\{ \begin{array}{ll} 0, & \text{if } x \neq a \\ \infty, & \text{if } x = a \end{array} \right\} \text{with} \int_{-\infty}^{\infty} \delta(x-a)\,dx = 1. \tag{1.90}$$

Equation 1.88 becomes

$$f(x)\delta(x-a) = f(a)\delta(x-a), \tag{1.91}$$

and Eq. 1.89 generalizes to

$$\int_{-\infty}^{\infty} f(x)\delta(x-a)\,dx = f(a). \tag{1.92}$$

Example 1.14

Evaluate the integral

$$\int_0^3 x^3\delta(x-2)\,dx.$$

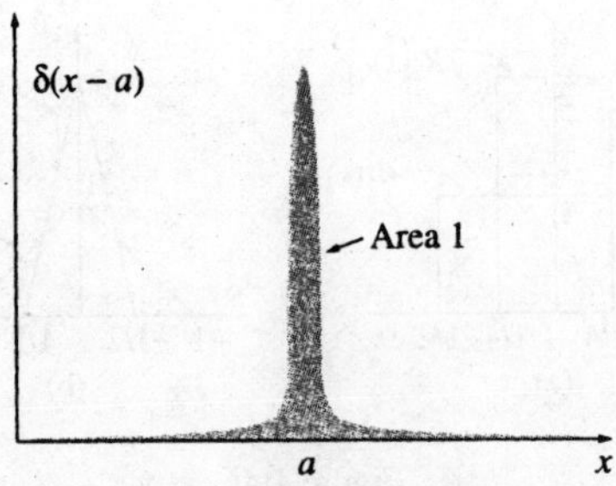

Figure 1.47

Solution: The delta function picks out the value of x^3 at the point $x = 2$, so the integral is $2^3 = 8$. Notice, however, that if the upper limit had been 1 (instead of 3) the answer would be 0, because the spike would then be outside the domain of integration.

Although δ itself is not a legitimate function, *integrals* over δ are perfectly acceptable. In fact, it's best to think of the delta function as something that is *always intended for use under an integral sign.* In particular, two expressions involving delta functions (say, $D_1(x)$ and $D_2(x)$) are considered equal if [6]

$$\int_{-\infty}^{\infty} f(x)D_1(x)\,dx = \int_{-\infty}^{\infty} f(x)D_2(x)\,dx, \tag{1.93}$$

for all ("ordinary") functions $f(x)$.

Example 1.15

Show that

$$\delta(kx) = \frac{1}{|k|}\delta(x), \tag{1.94}$$

where k is any (nonzero) constant. (In particular, $\delta(-x) = \delta(x)$.)

Solution: For an arbitrary test function $f(x)$, consider the integral

$$\int_{-\infty}^{\infty} f(x)\delta(kx)\,dx.$$

Changing variables, we let $y \equiv kx$, so that $x = y/k$, and $dx = 1/k\,dy$. If k is positive, the integration still runs from $-\infty$ to $+\infty$, but if k is *negative*, then $x = \infty$ implies $y = -\infty$, and

[6] This is not as arbitrary as it may sound. The crucial point is that the integrals must be equal for *any* $f(x)$. Suppose $D_1(x)$ and $D_2(x)$ actually *differed,* say, in the neighborhood of the point $x = 17$. Then we could pick a function $f(x)$ that was sharply peaked about $x = 17$, and the integrals would not be equal.

vice versa, so the order of the limits is reversed. Restoring the "proper" order costs a minus sign. Thus

$$\int_{-\infty}^{\infty} f(x)\delta(kx)\,dx = \pm\int_{-\infty}^{\infty} f(y/k)\delta(y)\frac{dy}{k} = \pm\frac{1}{k}f(0) = \frac{1}{|k|}f(0).$$

(The lower signs apply when k is negative, and we account for this neatly by putting absolute value bars around the final k, as indicated.) Under the integral sign, then, $\delta(kx)$ serves the same purpose as $(1/|k|)\delta(x)$:

$$\int_{-\infty}^{\infty} f(x)\delta(kx)\,dx = \int_{-\infty}^{\infty} f(x)\left[\frac{1}{|k|}\delta(x)\right]dx.$$

According to criterion 1.93, therefore, $\delta(kx)$ and $(1/|k|)\delta(x)$ are equal.

Problem 1.43 Evaluate the following integrals:

(a) $\int_2^6 (3x^2 - 2x - 1)\,\delta(x-3)\,dx$.

(b) $\int_0^5 \cos x\,\delta(x-\pi)\,dx$.

(c) $\int_0^3 x^3\delta(x+1)\,dx$.

(d) $\int_{-\infty}^{\infty} \ln(x+3)\,\delta(x+2)\,dx$.

Problem 1.44 Evaluate the following integrals:

(a) $\int_{-2}^{2} (2x+3)\,\delta(3x)\,dx$.

(b) $\int_0^2 (x^3+3x+2)\,\delta(1-x)\,dx$.

(c) $\int_{-1}^{1} 9x^2\delta(3x+1)\,dx$.

(d) $\int_{-\infty}^{a} \delta(x-b)\,dx$.

Problem 1.45

(a) Show that

$$x\frac{d}{dx}(\delta(x)) = -\delta(x).$$

[*Hint:* Use integration by parts.]

(b) Let $\theta(x)$ be the **step function**:

$$\theta(x) \equiv \left\{\begin{array}{ll} 1, & \text{if } x > 0 \\ 0, & \text{if } x \le 0 \end{array}\right\}. \tag{1.95}$$

Show that $d\theta/dx = \delta(x)$.

1.5.3 The Three-Dimensional Delta Function

It is an easy matter to generalize the delta function to three dimensions:

$$\delta^3(\mathbf{r}) = \delta(x)\,\delta(y)\,\delta(z). \tag{1.96}$$

(As always, $\mathbf{r} \equiv x\,\hat{\mathbf{x}} + y\,\hat{\mathbf{y}} + z\,\hat{\mathbf{z}}$ is the position vector, extending from the origin to the point (x, y, z)). This three-dimensional delta function is zero everywhere except at $(0, 0, 0)$, where it blows up. Its volume integral is 1:

$$\int_{\text{all space}} \delta^3(\mathbf{r})\,d\tau = \int_{-\infty}^{\infty}\int_{-\infty}^{\infty}\int_{-\infty}^{\infty} \delta(x)\,\delta(y)\,\delta(z)\,dx\,dy\,dz = 1. \tag{1.97}$$

And, generalizing Eq. 1.92,

$$\int_{\text{all space}} f(\mathbf{r})\delta^3(\mathbf{r}-\mathbf{a})\,d\tau = f(\mathbf{a}). \tag{1.98}$$

As in the one-dimensional case, integration with δ picks out the value of the function f at the location of the spike.

We are now in a position to resolve the paradox introduced in Sect. 1.5.1. As you will recall, we found that the divergence of $\hat{\mathbf{r}}/r^2$ is zero everywhere except at the origin, and yet its *integral* over any volume containing the origin is a constant (to wit: 4π). These are precisely the defining conditions for the Dirac delta function; evidently

$$\nabla \cdot \left(\frac{\hat{\mathbf{r}}}{r^2}\right) = 4\pi\delta^3(\mathbf{r}). \tag{1.99}$$

More generally,

$$\boxed{\nabla \cdot \left(\frac{\hat{\boldsymbol{\mathscr{r}}}}{\mathscr{r}^2}\right) = 4\pi\delta^3(\boldsymbol{\mathscr{r}}),} \tag{1.100}$$

where, as always, $\boldsymbol{\mathscr{r}}$ is the separation vector: $\boldsymbol{\mathscr{r}} \equiv \mathbf{r} - \mathbf{r}'$. Note that differentiation here is with respect to $\mathbf{r}$, while $\mathbf{r}'$ is held constant. Incidentally, since

$$\nabla\left(\frac{1}{\mathscr{r}}\right) = -\frac{\hat{\boldsymbol{\mathscr{r}}}}{\mathscr{r}^2} \tag{1.101}$$

(Prob. 1.13), it follows that

$$\nabla^2\frac{1}{\mathscr{r}} = -4\pi\delta^3(\boldsymbol{\mathscr{r}}). \tag{1.102}$$

Example 1.16

Evaluate the integral

$$J = \int_{\mathcal{V}} (r^2 + 2)\, \nabla \cdot \left(\frac{\hat{\mathbf{r}}}{r^2} \right) d\tau,$$

where $\mathcal{V}$ is a sphere of radius R centered at the origin.

Solution 1: Use Eq. 1.99 to rewrite the divergence, and Eq. 1.98 to do the integral:

$$J = \int_{\mathcal{V}} (r^2 + 2) 4\pi \delta^3(\mathbf{r})\, d\tau = 4\pi(0 + 2) = 8\pi.$$

This one-line solution demonstrates something of the power and beauty of the delta function, but I would like to show you a second method, which is much more cumbersome but serves to illustrate the method of integration by parts, Sect. 1.3.6.

Solution 2: Using Eq. 1.59, we transfer the derivative from $\hat{\mathbf{r}}/r^2$ to $(r^2 + 2)$:

$$J = -\int \frac{\hat{\mathbf{r}}}{r^2} \cdot [\nabla(r^2 + 2)]\, d\tau + \oint (r^2 + 2) \frac{\hat{\mathbf{r}}}{r^2} \cdot d\mathbf{a}.$$

The gradient is

$$\nabla(r^2 + 2) = 2r\hat{\mathbf{r}},$$

so the volume integral becomes

$$\int \frac{2}{r}\, d\tau = \int \frac{2}{r} r^2 \sin\theta\, dr\, d\theta\, d\phi = 8\pi \int_0^R r\, dr = 4\pi R^2.$$

Meanwhile, on the boundary of the sphere (where $r = R$),

$$d\mathbf{a} = R^2 \sin\theta\, d\theta\, d\phi\, \hat{\mathbf{r}},$$

so the surface integral becomes

$$\int (R^2 + 2) \sin\theta\, d\theta\, d\phi = 4\pi(R^2 + 2).$$

Putting it all together, then,

$$J = -4\pi R^2 + 4\pi(R^2 + 2) = 8\pi,$$

as before.

Problem 1.46

(a) Write an expression for the volume charge density $\rho(\mathbf{r})$ of a point charge q at $\mathbf{r}'$. Make sure that the volume integral of ρ equals q.

(b) What is the volume charge density of an electric dipole, consisting of a point charge $-q$ at the origin and a point charge $+q$ at $\mathbf{a}$?

(c) What is the volume charge density of a uniform, infinitesimally thin spherical shell of radius R and total charge Q, centered at the origin? [*Beware:* the integral over all space must equal Q.]

Problem 1.47 Evaluate the following integrals:

(a) $\int_{\text{all space}} (r^2 + \mathbf{r} \cdot \mathbf{a} + a^2)\delta^3(\mathbf{r} - \mathbf{a})\, d\tau$, where $\mathbf{a}$ is a fixed vector and a is its magnitude.

(b) $\int_{\mathcal{V}} |\mathbf{r} - \mathbf{b}|^2 \delta^3(5\mathbf{r})\, d\tau$, where $\mathcal{V}$ is a cube of side 2, centered on the origin, and $\mathbf{b} = 4\,\hat{\mathbf{y}} + 3\,\hat{\mathbf{z}}$.

(c) $\int_{\mathcal{V}} (r^4 + r^2(\mathbf{r} \cdot \mathbf{c}) + c^4)\delta^3(\mathbf{r} - \mathbf{c})\, d\tau$, where $\mathcal{V}$ is a sphere of radius 6 about the origin, $\mathbf{c} = 5\,\hat{\mathbf{x}} + 3\,\hat{\mathbf{y}} + 2\,\hat{\mathbf{z}}$, and c is its magnitude.

(d) $\int_{\mathcal{V}} \mathbf{r} \cdot (\mathbf{d} - \mathbf{r})\delta^3(\mathbf{e} - \mathbf{r})\, d\tau$, where $\mathbf{d} = (1, 2, 3)$, $\mathbf{e} = (3, 2, 1)$, and $\mathcal{V}$ is a sphere of radius 1.5 centered at (2, 2, 2).

Problem 1.48 Evaluate the integral

$$J = \int_{\mathcal{V}} e^{-r} \left(\nabla \cdot \frac{\hat{\mathbf{r}}}{r^2} \right) d\tau$$

(where $\mathcal{V}$ is a sphere of radius R, centered at the origin) by two different methods, as in Ex. 1.16.

1.6 The Theory of Vector Fields

1.6.1 The Helmholtz Theorem

Ever since Faraday, the laws of electricity and magnetism have been expressed in terms of **electric** and **magnetic fields**, **E** and **B**. Like many physical laws, these are most compactly expressed as differential equations. Since **E** and **B** are *vectors*, the differential equations naturally involve vector derivatives: divergence and curl. Indeed, Maxwell reduced the entire theory to four equations, specifying respectively the divergence and the curl of **E** and **B**.[7]

[7] Strictly speaking, this is only true in the static case; in general, the divergence and curl are given in terms of time derivatives of the fields themselves.

Maxwell's formulation raises an important mathematical question: To what extent is a vector function determined by its divergence and curl? In other words, if I tell you that the *divergence* of **F** (which stands for **E** or **B**, as the case may be) is a specified (scalar) function D,

$$\nabla \cdot \mathbf{F} = D,$$

and the curl of **F** is a specified (vector) function **C**,

$$\nabla \times \mathbf{F} = \mathbf{C},$$

(for consistency, **C** must be divergenceless,

$$\nabla \cdot \mathbf{C} = 0,$$

because the divergence of a curl is always zero), can you then determine the function **F**?

Well... not quite. For example, as you may have discovered in Prob. 1.19, there are many functions whose divergence and curl are both zero everywhere—the trivial case $\mathbf{F} = 0$, of course, but also $\mathbf{F} = yz\,\hat{\mathbf{x}} + zx\,\hat{\mathbf{y}} + xy\,\hat{\mathbf{z}}$, $\mathbf{F} = \sin x \cosh y\,\hat{\mathbf{x}} - \cos x \sinh y\,\hat{\mathbf{y}}$, etc. To solve a differential equation you must also be supplied with appropriate **boundary conditions**. In electrodynamics we typically require that the fields go to zero "at infinity" (far away from all charges).[8] With that extra information the **Helmholtz theorem** guarantees that the field is uniquely determined by its divergence and curl. (A proof of the Helmholtz theorem is given in Appendix B.)

1.6.2 Potentials

If the curl of a vector field (**F**) vanishes (everywhere), then **F** can be written as the gradient of a **scalar potential** (V):

$$\nabla \times \mathbf{F} = 0 \Longleftrightarrow \mathbf{F} = -\nabla V. \qquad (1.103)$$

(The minus sign is purely conventional.) That's the essential burden of the following theorem:

Theorem 1: **Curl-less** (or "**irrotational**") **fields**. The following conditions are equivalent (that is, **F** satisfies one if and only if it satisfies all the others):
(a) $\nabla \times \mathbf{F} = 0$ everywhere.
(b) $\int_{\mathbf{a}}^{\mathbf{b}} \mathbf{F} \cdot d\mathbf{l}$ is independent of path, for any given end points.
(c) $\oint \mathbf{F} \cdot d\mathbf{l} = 0$ for any closed loop.
(d) **F** is the gradient of some scalar, $\mathbf{F} = -\nabla V$.

[8] In some textbook problems the charge itself extends to infinity (we speak, for instance, of the electric field of an infinite plane, or the magnetic field of an infinite wire). In such cases the normal boundary conditions do not apply, and one must invoke symmetry arguments to determine the fields uniquely.

The scalar potential is not unique—any constant can be added to V with impunity, since this will not affect its gradient.

If the divergence of a vector field (**F**) vanishes (everywhere), then **F** can be expressed as the curl of a **vector potential** (**A**):

$$\nabla \cdot \mathbf{F} = 0 \Longleftrightarrow \mathbf{F} = \nabla \times \mathbf{A}. \qquad (1.104)$$

That's the main conclusion of the following theorem:

Theorem 2: **Divergence-less** (or "**solenoidal**") **fields.** The following conditions are equivalent:
(a) $\nabla \cdot \mathbf{F} = 0$ everywhere.
(b) $\int \mathbf{F} \cdot d\mathbf{a}$ is independent of surface, for any given boundary line.
(c) $\oint \mathbf{F} \cdot d\mathbf{a} = 0$ for any closed surface.
(d) **F** is the curl of some vector, $\mathbf{F} = \nabla \times \mathbf{A}$.

The vector potential is not unique—the gradient of any scalar function can be added to **A** without affecting the curl, since the curl of a gradient is zero.

You should by now be able to prove all the connections in these theorems, save for the ones that say (a), (b), or (c) implies (d). Those are more subtle, and will come later. Incidentally, in *all* cases (*whatever* its curl and divergence may be) a vector field **F** can be written as the gradient of a scalar plus the curl of a vector:

$$\mathbf{F} = -\nabla V + \nabla \times \mathbf{A} \quad \text{(always)}. \qquad (1.105)$$

Problem 1.49

(a) Let $\mathbf{F}_1 = x^2\,\hat{\mathbf{z}}$ and $\mathbf{F}_2 = x\,\hat{\mathbf{x}} + y\,\hat{\mathbf{y}} + z\,\hat{\mathbf{z}}$. Calculate the divergence and curl of $\mathbf{F}_1$ and $\mathbf{F}_2$. Which one can be written as the gradient of a scalar? Find a scalar potential that does the job. Which one can be written as the curl of a vector? Find a suitable vector potential.

(b) Show that $\mathbf{F}_3 = yz\,\hat{\mathbf{x}} + zx\,\hat{\mathbf{y}} + xy\,\hat{\mathbf{z}}$ can be written both as the gradient of a scalar and as the curl of a vector. Find scalar and vector potentials for this function.

Problem 1.50 For Theorem 1 show that (d) ⇒ (a), (a) ⇒ (c), (c) ⇒ (b), (b) ⇒ (c), and (c) ⇒ (a).

Problem 1.51 For Theorem 2 show that (d) ⇒ (a), (a) ⇒ (c), (c) ⇒ (b), (b) ⇒ (c), and (c) ⇒ (a).

Problem 1.52

(a) Which of the vectors in Problem 1.15 can be expressed as the gradient of a scalar? Find a scalar function that does the job.

(b) Which can be expressed as the curl of a vector? Find such a vector.

More Problems on Chapter 1

Problem 1.53 Check the divergence theorem for the function

$$\mathbf{v} = r^2 \cos\theta\,\hat{\mathbf{r}} + r^2 \cos\phi\,\hat{\boldsymbol{\theta}} - r^2 \cos\theta \sin\phi\,\hat{\boldsymbol{\phi}},$$

using as your volume one octant of the sphere of radius R (Fig. 1.48). Make sure you include the *entire* surface. [*Answer:* $\pi R^4/4$]

Problem 1.54 Check Stokes' theorem using the function $\mathbf{v} = ay\,\hat{\mathbf{x}} + bx\,\hat{\mathbf{y}}$ (a and b are constants) and the circular path of radius R, centered at the origin in the xy plane. [*Answer:* $\pi R^2(b-a)$]

Problem 1.55 Compute the line integral of

$$\mathbf{v} = 6\,\hat{\mathbf{x}} + yz^2\,\hat{\mathbf{y}} + (3y+z)\,\hat{\mathbf{z}}$$

along the triangular path shown in Fig. 1.49. Check your answer using Stokes' theorem. [*Answer:* 8/3]

Problem 1.56 Compute the line integral of

$$\mathbf{v} = (r\cos^2\theta)\,\hat{\mathbf{r}} - (r\cos\theta\sin\theta)\,\hat{\boldsymbol{\theta}} + 3r\,\hat{\boldsymbol{\phi}}$$

around the path shown in Fig. 1.50 (the points are labeled by their Cartesian coordinates). Do it either in cylindrical or in spherical coordinates. Check your answer, using Stokes' theorem. [*Answer:* $3\pi/2$]

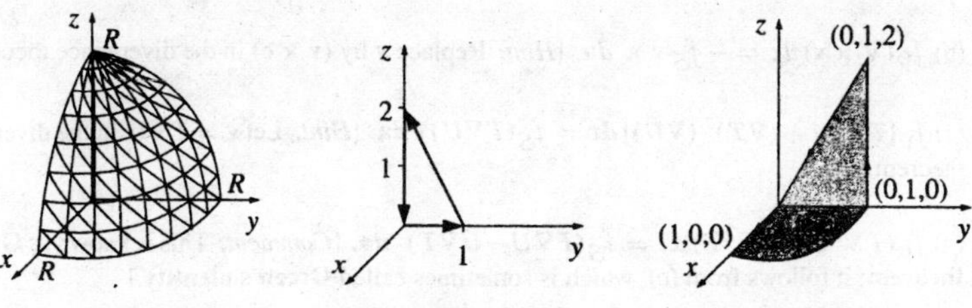

Figure 1.48 Figure 1.49 Figure 1.50

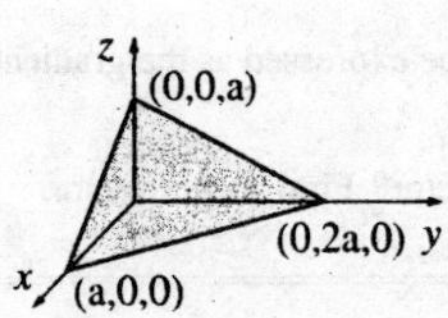

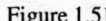

Figure 1.51

Figure 1.52

Problem 1.57 Check Stokes' theorem for the function $\mathbf{v} = y\,\hat{\mathbf{z}}$, using the triangular surface shown in Fig. 1.51. [*Answer:* a^2]

Problem 1.58 Check the divergence theorem for the function

$$\mathbf{v} = r^2 \sin\theta\,\hat{\mathbf{r}} + 4r^2 \cos\theta\,\hat{\boldsymbol{\theta}} + r^2 \tan\theta\,\hat{\boldsymbol{\phi}},$$

using the volume of the "ice-cream cone" shown in Fig. 1.52 (the top surface is spherical, with radius R and centered at the origin). [*Answer:* $(\pi R^4/12)(2\pi + 3\sqrt{3})$]

Problem 1.59 Here are two cute checks of the fundamental theorems:

(a) Combine Corollary 2 to the gradient theorem with Stokes' theorem ($\mathbf{v} = \nabla T$, in this case). Show that the result is consistent with what you already knew about second derivatives.

(b) Combine Corollary 2 to Stokes' theorem with the divergence theorem. Show that the result is consistent with what you already knew.

Problem 1.60 Although the gradient, divergence, and curl theorems are the fundamental integral theorems of vector calculus, it is possible to derive a number of corollaries from them. Show that:

(a) $\int_{\mathcal{V}} (\nabla T)\,d\tau = \oint_{\mathcal{S}} T\,d\mathbf{a}$. [*Hint:* Let $\mathbf{v} = \mathbf{c}T$, where $\mathbf{c}$ is a constant, in the divergence theorem; use the product rules.]

(b) $\int_{\mathcal{V}} (\nabla \times \mathbf{v})\,d\tau = -\oint_{\mathcal{S}} \mathbf{v} \times d\mathbf{a}$. [*Hint:* Replace $\mathbf{v}$ by $(\mathbf{v} \times \mathbf{c})$ in the divergence theorem.]

(c) $\int_{\mathcal{V}} [T\nabla^2 U + (\nabla T)\cdot(\nabla U)]\,d\tau = \oint_{\mathcal{S}} (T\nabla U)\cdot d\mathbf{a}$. [*Hint:* Let $\mathbf{v} = T\nabla U$ in the divergence theorem.]

(d) $\int_{\mathcal{V}} (T\nabla^2 U - U\nabla^2 T)\,d\tau = \oint_{\mathcal{S}} (T\nabla U - U\nabla T)\cdot d\mathbf{a}$. [*Comment:* This is known as **Green's theorem**; it follows from (c), which is sometimes called **Green's identity**.]

(e) $\int_{\mathcal{S}} \nabla T \times d\mathbf{a} = -\oint_{\mathcal{P}} T\,d\mathbf{l}$. [*Hint:* Let $\mathbf{v} = \mathbf{c}T$ in Stokes' theorem.]

• **Problem 1.61** The integral

$$\mathbf{a} \equiv \int_{\mathcal{S}} d\mathbf{a} \tag{1.106}$$

is sometimes called the **vector area** of the surface $\mathcal{S}$. If $\mathcal{S}$ happens to be *flat*, then $|\mathbf{a}|$ is the *ordinary* (scalar) area, obviously.

(a) Find the vector area of a hemispherical bowl of radius R.

(b) Show that $\mathbf{a} = 0$ for any *closed* surface. [*Hint:* Use Prob. 1.60a.]

(c) Show that $\mathbf{a}$ is the same for all surfaces sharing the same boundary.

(d) Show that

$$\mathbf{a} = \frac{1}{2}\oint \mathbf{r} \times d\mathbf{l}, \tag{1.107}$$

where the integral is around the boundary line. [*Hint:* One way to do it is to draw the cone subtended by the loop at the origin. Divide the conical surface up into infinitesimal triangular wedges, each with vertex at the origin and opposite side $d\mathbf{l}$, and exploit the geometrical interpretation of the cross product (Fig. 1.8).]

(e) Show that

$$\oint (\mathbf{c} \cdot \mathbf{r})\, d\mathbf{l} = \mathbf{a} \times \mathbf{c}, \tag{1.108}$$

for any constant vector $\mathbf{c}$. [*Hint:* let $T = \mathbf{c} \cdot \mathbf{r}$ in Prob. 1.60e.]

• **Problem 1.62**

(a) Find the divergence of the function

$$\mathbf{v} = \frac{\hat{\mathbf{r}}}{r}.$$

First compute it directly, as in Eq. 1.84. Test your result using the divergence theorem, as in Eq. 1.85. Is there a delta function at the origin, as there was for $\hat{\mathbf{r}}/r^2$? What is the general formula for the divergence of $r^n\hat{\mathbf{r}}$? [*Answer:* $\nabla \cdot (r^n\hat{\mathbf{r}}) = (n+2)r^{n-1}$, unless $n = -2$, in which case it is $4\pi\delta^3(\mathbf{r})$]

(b) Find the *curl* of $r^n\hat{\mathbf{r}}$. Test your conclusion using Prob. 1.60b. [*Answer:* $\nabla \times (r^n\hat{\mathbf{r}}) = 0$]

Chapter 2

Electrostatics

2.1 The Electric Field

2.1.1 Introduction

The fundamental problem electromagnetic theory hopes to solve is this (Fig. 2.1): We have some electric charges, $q_1, q_2, q_3, \ldots$ (call them **source charges**); what force do they exert on another charge, Q (call it the **test charge**)? The positions of the source charges are *given* (as functions of time); the trajectory of the test particle is *to be calculated.* In general, both the source charges and the test charge are in motion.

The solution to this problem is facilitated by the **principle of superposition**, which states that the interaction between any two charges is completely unaffected by the presence of others. This means that to determine the force on Q, we can first compute the force $\mathbf{F}_1$, due to q_1 alone (ignoring all the others); then we compute the force $\mathbf{F}_2$, due to q_2 alone; and so on. Finally, we take the vector sum of all these individual forces: $\mathbf{F} = \mathbf{F}_1 + \mathbf{F}_2 + \mathbf{F}_3 + \ldots$ Thus, if we can find the force on Q due to a *single* source charge q, we are, in principle, done (the rest is just a question of repeating the same operation over and over, and adding it all up).[1]

Well, at first sight this sounds very easy: Why don't I just write down the formula for the force on Q due to q, and be done with it? I *could,* and in Chapter 10 I shall, but you would be shocked to see it at this stage, for not only does the force on Q depend on the separation distance $\mathscr{r}$ between the charges (Fig. 2.2), it also depends on *both* their velocities and on the *acceleration* of q. Moreover, it is not the position, velocity, and acceleration of q *right now* that matter: Electromagnetic "news" travels at the speed of light, so what concerns Q is the position, velocity, and acceleration q *had* at some earlier time, when the message left.

[1] The principle of superposition may seem "obvious" to you, but it did not *have* to be so simple: if the electromagnetic force were proportional to the *square* of the total source charge, for instance, the principle of superposition would not hold, since $(q_1 + q_2)^2 \neq q_1^2 + q_2^2$ (there would be "cross terms" to consider). Superposition is not a *logical necessity*, but an *experimental fact.*

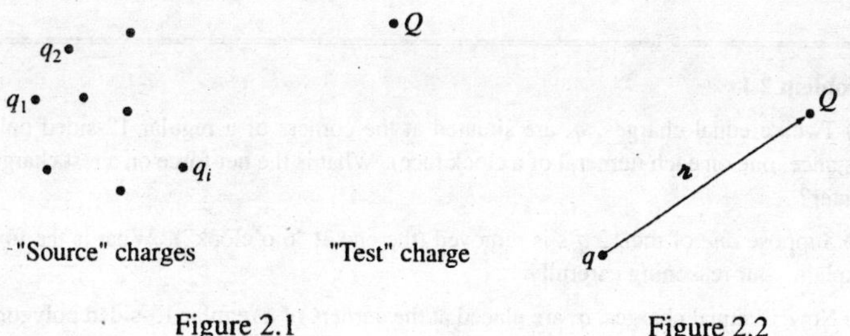

Figure 2.1

Figure 2.2

Therefore, in spite of the fact that the basic question ("What is the force on Q due to q?") is easy to state, it does not pay to confront it head on; rather, we shall go at it by stages. In the meantime, the theory we develop will permit the solution of more subtle electromagnetic problems that do not present themselves in quite this simple format. To begin with, we shall consider the special case of **electrostatics** in which all the *source charges are stationary* (though the test charge may be moving).

2.1.2 Coulomb's Law

What is the force on a test charge Q due to a single point charge q which is at *rest* a distance $\mathscr{r}$ away? The answer (based on experiments) is given by **Coulomb's law**:

$$\boxed{\mathbf{F} = \frac{1}{4\pi\epsilon_0}\frac{qQ}{\mathscr{r}^2}\hat{\boldsymbol{\mathscr{r}}}.} \tag{2.1}$$

The constant ϵ_0 is called the **permitivity of free space.** In SI units, where force is in Newtons (N), distance in meters (m), and charge in coulombs (C),

$$\epsilon_0 = 8.85 \times 10^{-12} \frac{\mathrm{C}^2}{\mathrm{N}\cdot\mathrm{m}^2}.$$

In words, the force is proportional to the product of the charges and inversely proportional to the square of the separation distance. As always (Sect. 1.1.4), $\boldsymbol{\mathscr{r}}$ is the separation vector from $\mathbf{r}'$ (the location of q) to $\mathbf{r}$ (the location of Q):

$$\boldsymbol{\mathscr{r}} = \mathbf{r} - \mathbf{r}'; \tag{2.2}$$

$\mathscr{r}$ is its magnitude, and $\hat{\boldsymbol{\mathscr{r}}}$ is its direction. The force points along the line from q to Q; it is repulsive if q and Q have the same sign, and attractive if their signs are opposite.

Coulomb's law and the principle of superposition constitute the physical input for electrostatics—the rest, except for some special properties of matter, is mathematical elaboration of these fundamental rules.

Problem 2.1

(a) Twelve equal charges, q, are situated at the corners of a regular 12-sided polygon (for instance, one on each numeral of a clock face). What is the net force on a test charge Q at the center?

(b) Suppose *one* of the 12 q's is removed (the one at "6 o'clock"). What is the force on Q? Explain your reasoning carefully.

(c) Now 13 equal charges, q, are placed at the corners of a regular 13-sided polygon. What is the force on a test charge Q at the center?

(d) If one of the 13 q's is removed, what is the force on Q? Explain your reasoning.

2.1.3 The Electric Field

If we have *several* point charges $q_1, q_2, \ldots, q_n$, at distances $\mathscr{r}_1, \mathscr{r}_2, \ldots, \mathscr{r}_n$ from Q, the total force on Q is evidently

$$\begin{aligned} \mathbf{F} &= \mathbf{F}_1 + \mathbf{F}_2 + \ldots = \frac{1}{4\pi\epsilon_0}\left(\frac{q_1 Q}{\mathscr{r}_1^2}\hat{\boldsymbol{\mathscr{r}}}_1 + \frac{q_2 Q}{\mathscr{r}_2^2}\hat{\boldsymbol{\mathscr{r}}}_2 + \ldots\right) \\ &= \frac{Q}{4\pi\epsilon_0}\left(\frac{q_1\hat{\boldsymbol{\mathscr{r}}}_1}{\mathscr{r}_1^2} + \frac{q_2\hat{\boldsymbol{\mathscr{r}}}_2}{\mathscr{r}_2^2} + \frac{q_3\hat{\boldsymbol{\mathscr{r}}}_3}{\mathscr{r}_3^2} + \ldots\right), \end{aligned}$$

or

$$\boxed{\mathbf{F} = Q\mathbf{E},} \tag{2.3}$$

where

$$\mathbf{E}(\mathbf{r}) \equiv \frac{1}{4\pi\epsilon_0}\sum_{i=1}^{n}\frac{q_i}{\mathscr{r}_i^2}\hat{\boldsymbol{\mathscr{r}}}_i. \tag{2.4}$$

$\mathbf{E}$ is called the **electric field** of the source charges. Notice that it is a function of position ($\mathbf{r}$), because the separation vectors $\boldsymbol{\mathscr{r}}_i$ depend on the location of the field point P (Fig. 2.3). But it makes no reference to the test charge Q. The electric field is a vector quantity that varies

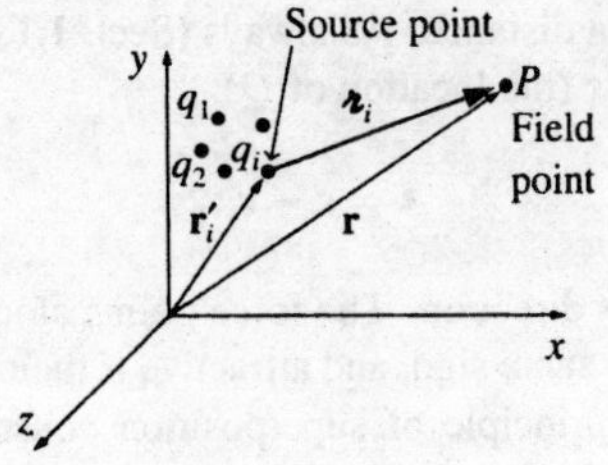

Figure 2.3

from point to point and is determined by the configuration of source charges; physically, $\mathbf{E}(\mathbf{r})$ is the force per unit charge that would be exerted on a test charge, if you were to place one at P.

What exactly *is* an electric field? I have deliberately begun with what you might call the "minimal" interpretation of $\mathbf{E}$, as an intermediate step in the calculation of electric forces. But I encourage you to think of the field as a "real" physical entity, filling the space in the neighborhood of any electric charge. Maxwell himself came to believe that electric and magnetic fields represented actual stresses and strains in an invisible primordial jellylike "ether." Special relativity has forced us to abandon the notion of ether, and with it Maxwell's mechanical interpretation of electromagnetic fields. (It is even possible, though cumbersome, to formulate classical electrodynamics as an "action-at-a-distance" theory, and dispense with the field concept altogether.) I can't tell you, then, what a field *is*—only how to calculate it and what it can do for you once you've got it.

Problem 2.2

(a) Find the electric field (magnitude and direction) a distance z above the midpoint between two equal charges, q, a distance d apart (Fig. 2.4). Check that your result is consistent with what you'd expect when $z \gg d$.

(b) Repeat part (a), only this time make the right-hand charge $-q$ instead of $+q$.

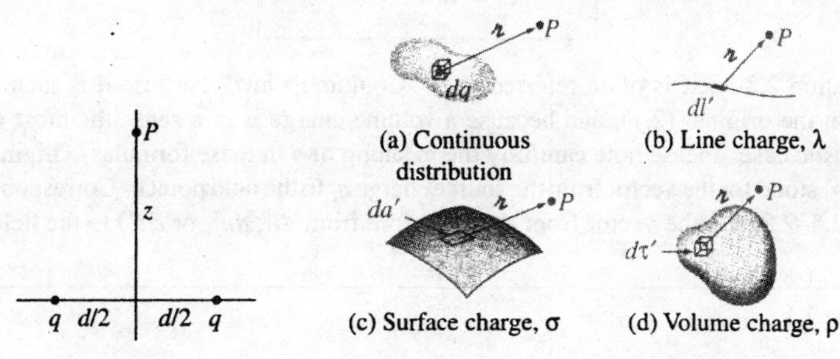

Figure 2.4

Figure 2.5

2.1.4 Continuous Charge Distributions

Our definition of the electric field (Eq. 2.4), assumes that the source of the field is a collection of discrete point charges q_i. If, instead, the charge is distributed continuously over some region, the sum becomes an integral (Fig. 2.5a):

$$\mathbf{E}(\mathbf{r}) = \frac{1}{4\pi\epsilon_0}\int \frac{1}{\mathscr{r}^2}\hat{\mathscr{r}}\, dq. \tag{2.5}$$

If the charge is spread out along a *line* (Fig. 2.5b), with charge-per-unit-length λ, then $dq = \lambda\, dl'$ (where dl' is an element of length along the line); if the charge is smeared out over a *surface* (Fig. 2.5c), with charge-per-unit-area σ, then $dq = \sigma\, da'$ (where da' is an element of area on the surface); and if the charge fills a *volume* (Fig. 2.5d), with charge-per-unit-volume ρ, then $dq = \rho\, d\tau'$ (where $d\tau'$ is an element of volume):

$$dq \to \lambda\, dl' \sim \sigma\, da' \sim \rho\, d\tau'.$$

Thus the electric field of a line charge is

$$\mathbf{E}(\mathbf{r}) = \frac{1}{4\pi\epsilon_0}\int_{\mathcal{P}} \frac{\lambda(\mathbf{r}')}{\boldsymbol{\imath}^2}\hat{\boldsymbol{\imath}}\, dl'; \tag{2.6}$$

for a surface charge,

$$\mathbf{E}(\mathbf{r}) = \frac{1}{4\pi\epsilon_0}\int_{\mathcal{S}} \frac{\sigma(\mathbf{r}')}{\boldsymbol{\imath}^2}\hat{\boldsymbol{\imath}}\, da'; \tag{2.7}$$

and for a volume charge,

$$\boxed{\mathbf{E}(\mathbf{r}) = \frac{1}{4\pi\epsilon_0}\int_{\mathcal{V}} \frac{\rho(\mathbf{r}')}{\boldsymbol{\imath}^2}\hat{\boldsymbol{\imath}}\, d\tau'.} \tag{2.8}$$

Equation 2.8 itself is often referred to as "Coulomb's law," because it is such a short step from the original (2.1), and because a volume charge is in a sense the most general and realistic case. Please note carefully the meaning of $\boldsymbol{\imath}$ in these formulas. Originally, in Eq. 2.4, $\boldsymbol{\imath}_i$ stood for the vector from the source charge q_i to the field point $\mathbf{r}$. Correspondingly, in Eqs. 2.5–2.8, $\boldsymbol{\imath}$ is the vector from dq (therefore from dl', da', or $d\tau'$) to the field point $\mathbf{r}$.[2]

Example 2.1

Find the electric field a distance z above the midpoint of a straight line segment of length $2L$, which carries a uniform line charge λ (Fig. 2.6).

Solution: It is advantageous to chop the line up into symmetrically placed pairs (at $\pm x$), for then the horizontal components of the two fields cancel, and the net field of the pair is

$$d\mathbf{E} = 2\frac{1}{4\pi\epsilon_0}\left(\frac{\lambda\, dx}{\boldsymbol{\imath}^2}\right)\cos\theta\,\hat{\mathbf{z}}.$$

---

[2] *Warning:* The unit vector $\hat{\boldsymbol{\imath}}$ is *not* constant; its *direction* depends on the source point $\mathbf{r}'$, and hence it *cannot* be taken outside the integrals 2.5–2.8. In practice, you must work with *Cartesian* components ($\hat{\mathbf{x}}, \hat{\mathbf{y}}, \hat{\mathbf{z}}$ *are* constant, and *do* come out), even if you use curvilinear coordinates to perform the integration.

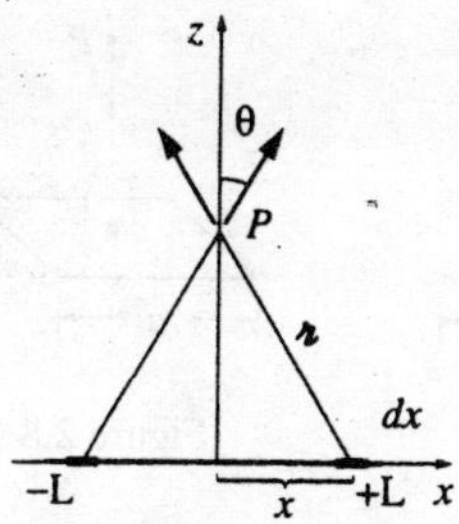

Figure 2.6

Here $\cos\theta = z/\mathscr{r}$, $\mathscr{r} = \sqrt{z^2+x^2}$, and x runs from 0 to L:

$$\begin{aligned} E &= \frac{1}{4\pi\epsilon_0}\int_0^L \frac{2\lambda z}{(z^2+x^2)^{3/2}}\,dx \\ &= \frac{2\lambda z}{4\pi\epsilon_0}\left[\frac{x}{z^2\sqrt{z^2+x^2}}\right]_0^L \\ &= \frac{1}{4\pi\epsilon_0}\frac{2\lambda L}{z\sqrt{z^2+L^2}}, \end{aligned}$$

and it aims in the z-direction.

For points far from the line ($z \gg L$), this result simplifies:

$$E \cong \frac{1}{4\pi\epsilon_0}\frac{2\lambda L}{z^2},$$

which makes sense: From far away the line "looks" like a point charge $q = 2\lambda L$, so the field reduces to that of point charge $q/(4\pi\epsilon_0 z^2)$. In the limit $L \to \infty$, on the other hand, we obtain the field of an infinite straight wire:

$$E = \frac{1}{4\pi\epsilon_0}\frac{2\lambda}{z};$$

or, more generally,

$$E = \frac{1}{4\pi\epsilon_0}\frac{2\lambda}{s}, \tag{2.9}$$

where s is the distance from the wire.

Problem 2.3 Find the electric field a distance z above one end of a straight line segment of length L (Fig. 2.7), which carries a uniform line charge λ. Check that your formula is consistent with what you would expect for the case $z \gg L$.

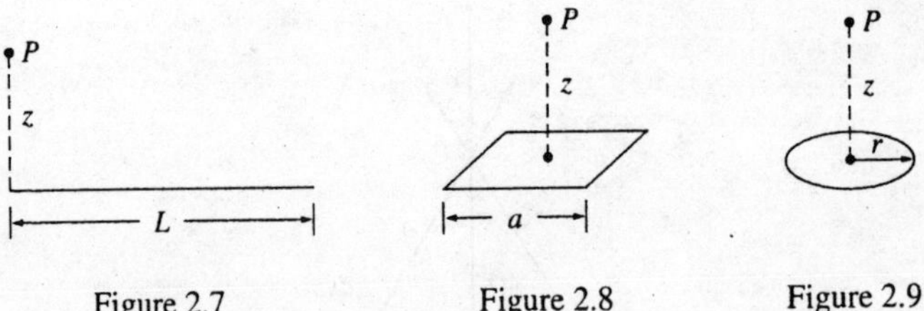

Figure 2.7 Figure 2.8 Figure 2.9

Problem 2.4 Find the electric field a distance z above the center of a square loop (side a) carrying uniform line charge λ (Fig. 2.8). [*Hint:* Use the result of Ex. 2.1.]

Problem 2.5 Find the electric field a distance z above the center of a circular loop of radius r (Fig. 2.9), which carries a uniform line charge λ.

Problem 2.6 Find the electric field a distance z above the center of a flat circular disk of radius R (Fig. 2.10), which carries a uniform surface charge σ. What does your formula give in the limit $R \to \infty$? Also check the case $z \gg R$.

Problem 2.7 Find the electric field a distance z from the center of a spherical surface of radius R (Fig. 2.11), which carries a uniform charge density σ. Treat the case $z < R$ (inside) as well as $z > R$ (outside). Express your answers in terms of the total charge q on the sphere. [*Hint:* Use the law of cosines to write $\mathscr{r}$ in terms of R and θ. Be sure to take the *positive* square root: $\sqrt{R^2 + z^2 - 2Rz} = (R - z)$ if $R > z$, but it's $(z - R)$ if $R < z$.]

Problem 2.8 Use your result in Prob. 2.7 to find the field inside and outside a sphere of radius R, which carries a uniform volume charge density ρ. Express your answers in terms of the total charge of the sphere, q. Draw a graph of $|\mathbf{E}|$ as a function of the distance from the center.

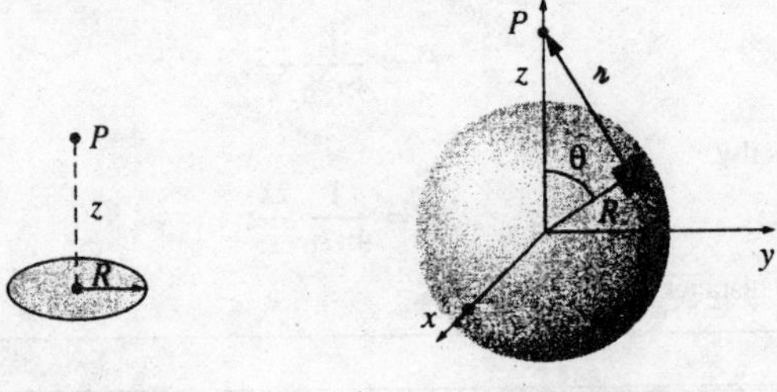

Figure 2.10 Figure 2.11

2.2 Divergence and Curl of Electrostatic Fields

2.2.1 Field Lines, Flux, and Gauss's Law

In principle, we are *done* with the subject of electrostatics. Equation 2.8 tells us how to compute the field of a charge distribution, and Eq. 2.3 tells us what the force on a charge Q placed in this field will be. Unfortunately, as you may have discovered in working Prob. 2.7, the integrals involved in computing **E** can be formidable, even for reasonably simple charge distributions. Much of the rest of electrostatics is devoted to assembling a bag of tools and tricks for *avoiding* these integrals. It all begins with the divergence and curl of **E**. I shall calculate the divergence of **E** directly from Eq. 2.8, in Sect. 2.2.2, but first I want to show you a more qualitative, and perhaps more illuminating, intuitive approach.

Let's begin with the simplest possible case: a single point charge q, situated at the origin:

$$\mathbf{E}(\mathbf{r}) = \frac{1}{4\pi\epsilon_0}\frac{q}{r^2}\hat{\mathbf{r}}. \tag{2.10}$$

To get a "feel" for this field, I might sketch a few representative vectors, as in Fig. 2.12a. Because the field falls off like $1/r^2$, the vectors get shorter as you go farther away from the origin; they always point radially outward. But there is a nicer way to represent this field, and that's to connect up the arrows, to form **field lines** (Fig. 2.12b). You might think that I have thereby thrown away information about the *strength* of the field, which was contained in the length of the arrows. But actually I have not. The magnitude of the field is indicated by the *density* of the field lines: it's strong near the center where the field lines are close together, and weak farther out, where they are relatively far apart.

In truth, the field-line diagram is deceptive, when I draw it on a two-dimensional surface, for the density of lines passing through a circle of radius r is the total number divided by the circumference $(n/2\pi r)$, which goes like $(1/r)$, not $(1/r^2)$. But if you imagine the model in three dimensions (a pincushion with needles sticking out in all directions), then the density of lines is the total number divided by the area of the sphere $(n/4\pi r^2)$, which *does* go like $(1/r^2)$.

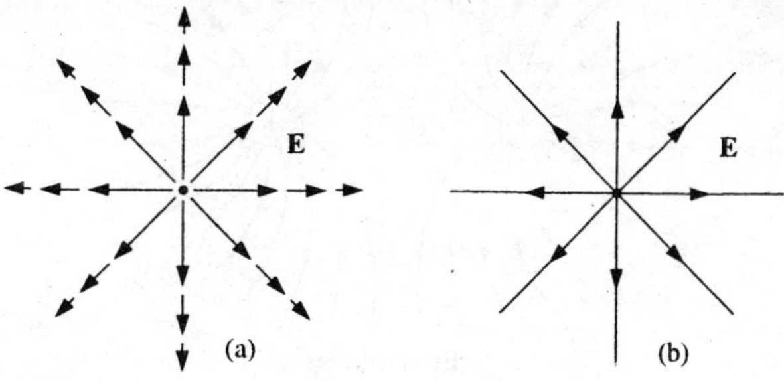

Figure 2.12

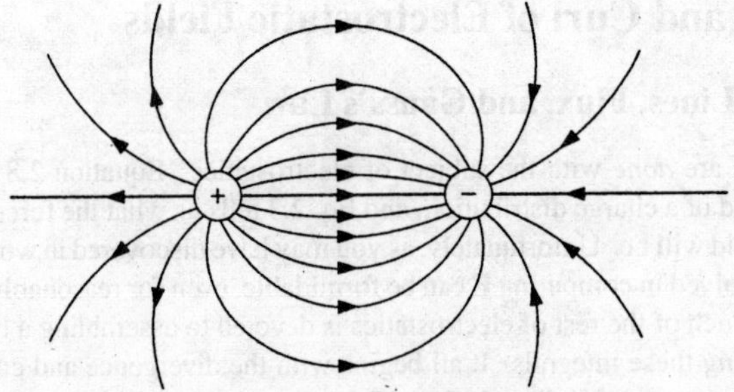

Equal but opposite charges

Figure 2.13

Such diagrams are also convenient for representing more complicated fields. Of course, the number of lines you draw depends on how energetic you are (and how sharp your pencil is), though you ought to include enough to get an accurate sense of the field, and you must be consistent: If charge q gets 8 lines, then $2q$ deserves 16. And you must space them fairly—they emanate from a point charge symmetrically in all directions. Field lines begin on positive charges and end on negative ones; they cannot simply terminate in midair, though they may extend out to infinity. Moreover, field lines can never cross—at the intersection, the field would have two different directions at once! With all this in mind, it is easy to sketch the field of any simple configuration of point charges: Begin by drawing the lines in the neighborhood of each charge, and then connect them up or extend them to infinity (Figs. 2.13 and 2.14).

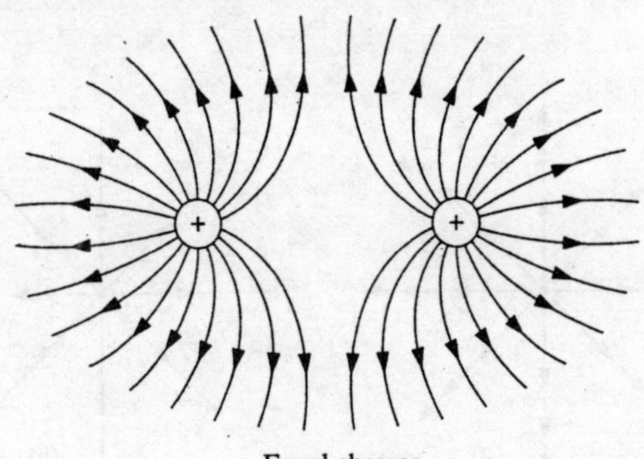

Equal charges

Figure 2.14

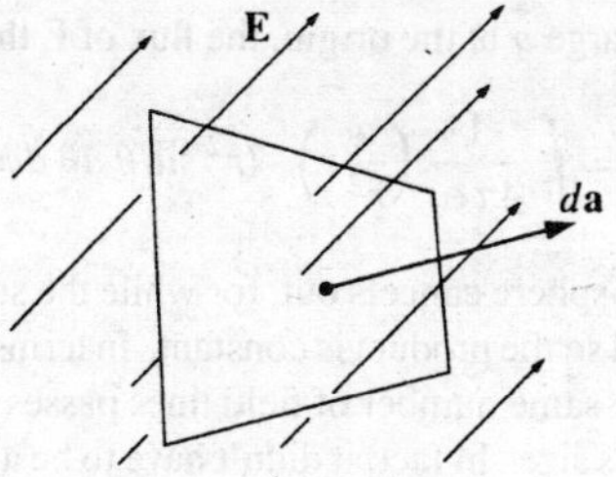

Figure 2.15

In this model the *flux* of **E** through a surface $\mathcal{S}$,

$$\Phi_E \equiv \int_{\mathcal{S}} \mathbf{E} \cdot d\mathbf{a}, \tag{2.11}$$

is a measure of the "number of field lines" passing through $\mathcal{S}$. I put this in quotes because of course we can only draw a representative *sample* of the field lines—the *total* number would be infinite. But *for a given sampling rate* the flux is *proportional* to the number of lines drawn, because the field strength, remember, is proportional to the density of field lines (the number per unit area), and hence $\mathbf{E} \cdot d\mathbf{a}$ is proportional to the number of lines passing through the infinitesimal area $d\mathbf{a}$. (The dot product picks out the component of $d\mathbf{a}$ along the direction of **E**, as indicated in Fig. 2.15. It is only the area *in the plane perpendicular to* **E** that we have in mind when we say that the density of field lines is the number per unit area.)

This suggests that the flux through any *closed* surface is a measure of the total charge inside. For the field lines that originate on a positive charge must either pass out through the surface or else terminate on a negative charge inside (Fig. 2.16a). On the other hand, a charge *outside* the surface will contribute nothing to the total flux, since its field lines pass in one side and out the other (Fig. 2.16b). This is the *essence* of **Gauss's law**. Now let's make it quantitative.

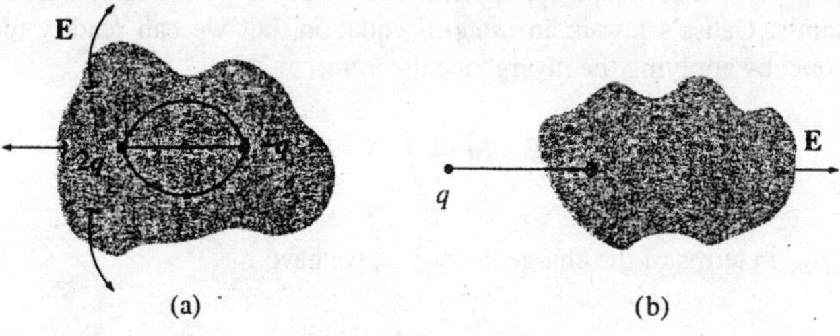

Figure 2.16

In the case of a point charge q at the origin, the flux of $\mathbf{E}$ through a sphere of radius r is

$$\oint \mathbf{E} \cdot d\mathbf{a} = \int \frac{1}{4\pi\epsilon_0}\left(\frac{q}{r^2}\hat{\mathbf{r}}\right) \cdot (r^2 \sin\theta \, d\theta \, d\phi \, \hat{\mathbf{r}}) = \frac{1}{\epsilon_0}q. \tag{2.12}$$

Notice that the radius of the sphere cancels out, for while the surface area goes *up* as r^2, the field goes *down* as $1/r^2$, and so the product is constant. In terms of the field-line picture, this makes good sense, since the same number of field lines passes through any sphere centered at the origin, regardless of its size. In fact, it didn't have to be a sphere—any closed surface, whatever its shape, would trap the same number of field lines. Evidently *the flux through any surface enclosing the charge is* q/ϵ_0.

Now suppose that instead of a single charge at the origin, we have a bunch of charges scattered about. According to the principle of superposition, the total field is the (vector) sum of all the individual fields:

$$\mathbf{E} = \sum_{i=1}^{n} \mathbf{E}_i.$$

The flux through a surface that encloses them all, then, is

$$\oint \mathbf{E} \cdot d\mathbf{a} = \sum_{i=1}^{n}\left(\oint \mathbf{E}_i \cdot d\mathbf{a}\right) = \sum_{i=1}^{n}\left(\frac{1}{\epsilon_0}q_i\right).$$

For any closed surface, then,

$$\boxed{\oint_{\mathcal{S}} \mathbf{E} \cdot d\mathbf{a} = \frac{1}{\epsilon_0}Q_{\text{enc}},} \tag{2.13}$$

where Q_{enc} is the total charge enclosed within the surface. This is the quantitative statement of Gauss's law. Although it contains no information that was not already present in Coulomb's law and the principle of superposition, it is of almost magical power, as you will see in Sect. 2.2.3. Notice that it all hinges on the $1/r^2$ character of Coulomb's law; without that the crucial cancellation of the r's in Eq. 2.12 would not take place, and the total flux of $\mathbf{E}$ would depend on the surface chosen, not merely on the total charge enclosed. Other $1/r^2$ forces (I am thinking particularly of Newton's law of universal gravitation) will obey "Gauss's laws" of their own, and the applications we develop here carry over directly.

As it stands, Gauss's law is an *integral* equation, but we can readily turn it into a *differential* one, by applying the divergence theorem:

$$\oint_{\mathcal{S}} \mathbf{E} \cdot d\mathbf{a} = \int_{\mathcal{V}} (\nabla \cdot \mathbf{E}) \, d\tau.$$

Rewriting Q_{enc} in terms of the charge density ρ, we have

$$Q_{\text{enc}} = \int_{\mathcal{V}} \rho \, d\tau.$$

So Gauss's law becomes

$$\int_{\mathcal{V}} (\nabla \cdot \mathbf{E})\, d\tau = \int_{\mathcal{V}} \left(\frac{\rho}{\epsilon_0}\right) d\tau.$$

And since this holds for *any* volume, the integrands must be equal:

$$\boxed{\nabla \cdot \mathbf{E} = \frac{1}{\epsilon_0}\rho.} \tag{2.14}$$

Equation 2.14 carries the same message as Eq. 2.13; it is **Gauss's law in differential form.** The differential version is tidier, but the integral form has the advantage in that it accommodates point, line, and surface charges more naturally.

Problem 2.9 Suppose the electric field in some region is found to be $\mathbf{E} = kr^3\hat{\mathbf{r}}$, in spherical coordinates (k is some constant).

(a) Find the charge density ρ.

(b) Find the total charge contained in a sphere of radius R, centered at the origin. (Do it two different ways.)

Problem 2.10 A charge q sits at the back corner of a cube, as shown in Fig. 2.17. What is the flux of $\mathbf{E}$ through the shaded side?

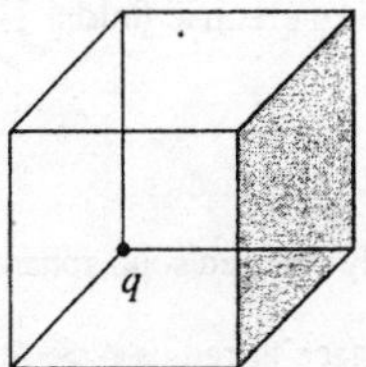

Figure 2.17

2.2.2 The Divergence of E

Let's go back, now, and calculate the divergence of **E** directly from Eq. 2.8:

$$\mathbf{E}(\mathbf{r}) = \frac{1}{4\pi\epsilon_0} \int_{\text{all space}} \frac{\hat{\boldsymbol{\imath}}}{\imath^2} \rho(\mathbf{r}')\, d\tau'. \tag{2.15}$$

(Originally the integration was over the volume occupied by the charge, but I may as well extend it to all space, since $\rho = 0$ in the exterior region anyway.) Noting that the

r-dependence is contained in $\boldsymbol{\mathscr{r}} = \mathbf{r} - \mathbf{r}'$, we have

$$\nabla \cdot \mathbf{E} = \frac{1}{4\pi\epsilon_0} \int \nabla \cdot \left(\frac{\hat{\boldsymbol{\mathscr{r}}}}{\mathscr{r}^2} \right) \rho(\mathbf{r}')\, d\tau'.$$

This is precisely the divergence we calculated in Eq. 1.100:

$$\nabla \cdot \left(\frac{\hat{\boldsymbol{\mathscr{r}}}}{\mathscr{r}^2} \right) = 4\pi \delta^3(\boldsymbol{\mathscr{r}}).$$

Thus

$$\nabla \cdot \mathbf{E} = \frac{1}{4\pi\epsilon_0} \int 4\pi \delta^3(\mathbf{r} - \mathbf{r}') \rho(\mathbf{r}')\, d\tau' = \frac{1}{\epsilon_0} \rho(\mathbf{r}), \tag{2.16}$$

which is Gauss's law in differential form (2.14). To recover the integral form (2.13), we run the previous argument in reverse—integrate over a volume and apply the divergence theorem:

$$\int_{\mathcal{V}} \nabla \cdot \mathbf{E}\, d\tau = \oint_{\mathcal{S}} \mathbf{E} \cdot d\mathbf{a} = \frac{1}{\epsilon_0} \int_{\mathcal{V}} \rho\, d\tau = \frac{1}{\epsilon_0} Q_{\text{enc}}.$$

2.2.3 Applications of Gauss's Law

I must interrupt the theoretical development at this point to show you the extraordinary power of Gauss's law, in integral form. When symmetry permits, it affords *by far* the quickest and easiest way of computing electric fields. I'll illustrate the method with a series of examples.

Example 2.2

Find the field outside a uniformly charged solid sphere of radius R and total charge q.

Solution: Draw a spherical surface at radius $r > R$ (Fig. 2.18); this is called a "Gaussian surface" in the trade. Gauss's law says that for this surface (as for any other)

$$\oint_{\mathcal{S}} \mathbf{E} \cdot d\mathbf{a} = \frac{1}{\epsilon_0} Q_{\text{enc}},$$

and $Q_{\text{enc}} = q$. At first glance this doesn't seem to get us very far, because the quantity we want (**E**) is buried inside the surface integral. Luckily, symmetry allows us to extract **E** from under the integral sign: **E** certainly points radially outward,[3] as does $d\mathbf{a}$, so we can drop the dot product,

$$\int_{\mathcal{S}} \mathbf{E} \cdot d\mathbf{a} = \int_{\mathcal{S}} |\mathbf{E}|\, da,$$

---

[3]If you doubt that **E** is radial, consider the alternative. Suppose, say, that it points due east, at the "equator." But the orientation of the equator is perfectly arbitrary—nothing is spinning here, so there is no natural "north-south" axis—any argument purporting to show that **E** points east could just as well be used to show it points west, or north, or any other direction. The only *unique* direction on a sphere is *radial*.

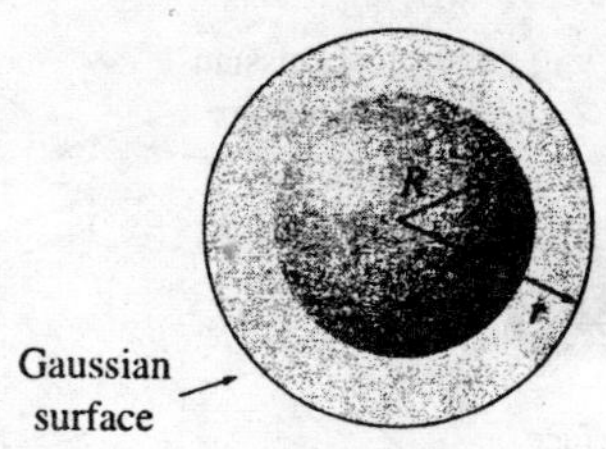

Figure 2.18

and the *magnitude* of **E** is constant over the Gaussian surface, so it comes outside the integral:

$$\int_S |\mathbf{E}|\,da = |\mathbf{E}| \int_S da = |\mathbf{E}|\,4\pi r^2.$$

Thus

$$|\mathbf{E}|\,4\pi r^2 = \frac{1}{\epsilon_0} q,$$

or

$$\mathbf{E} = \frac{1}{4\pi\epsilon_0}\frac{q}{r^2}\hat{\mathbf{r}}.$$

Notice a remarkable feature of this result: The field outside the sphere is exactly the *same as it would have been if all the charge had been concentrated at the center.*

Gauss's law is always *true*, but it is not always *useful.* If ρ had not been uniform (or, at any rate, not spherically symmetrical), or if I had chosen some other shape for my Gaussian surface, it would still have been true that the flux of **E** is $(1/\epsilon_0)q$, but I would not have been certain that **E** was in the same direction as $d\mathbf{a}$ and constant in magnitude over the surface, and without that I could not pull $|\mathbf{E}|$ out of the integral. *Symmetry is crucial* to this application of Gauss's law. As far as I know, there are only three kinds of symmetry that work:

1. *Spherical symmetry.* Make your Gaussian surface a concentric sphere.
2. *Cylindrical symmetry.* Make your Gaussian surface a coaxial cylinder (Fig. 2.19).
3. *Plane symmetry.* Use a Gaussian "pillbox," which straddles the surface (Fig. 2.20).

Although (2) and (3) technically require infinitely long cylinders, and planes extending to infinity in all directions, we shall often use them to get approximate answers for "long" cylinders or "large" plane surfaces, at points far from the edges.

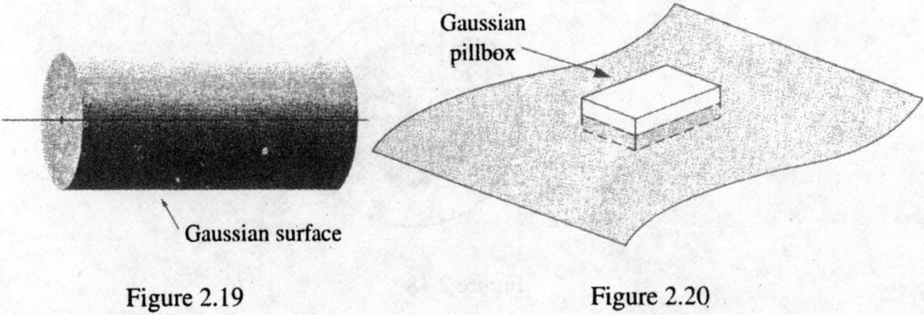

Figure 2.19

Figure 2.20

Example 2.3

A long cylinder (Fig. 2.21) carries a charge density that is proportional to the distance from the axis: $\rho = ks$, for some constant k. Find the electric field inside this cylinder.

Solution: Draw a Gaussian cylinder of length l and radius s. For this surface, Gauss's law states:

$$\oint_{\mathcal{S}} \mathbf{E} \cdot d\mathbf{a} = \frac{1}{\epsilon_0} Q_{\text{enc}}.$$

The enclosed charge is

$$Q_{\text{enc}} = \int \rho \, d\tau = \int (ks')(s' \, ds' \, d\phi \, dz) = 2\pi k l \int_0^s s'^2 \, ds' = \tfrac{2}{3}\pi k l s^3.$$

(I used the volume element appropriate to cylindrical coordinates, Eq. 1.78, and integrated ϕ from 0 to 2π, dz from 0 to l. I put a prime on the integration variable s', to distinguish it from the radius s of the Gaussian surface.)

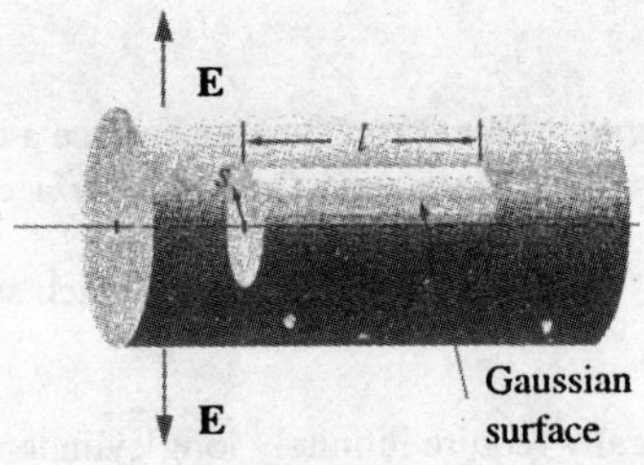

Figure 2.21

Now, symmetry dictates that **E** must point radially outward, so for the curved portion of the Gaussian cylinder we have:

$$\int \mathbf{E}\cdot d\mathbf{a} = \int |\mathbf{E}|\, da = |\mathbf{E}| \int da = |\mathbf{E}|\, 2\pi s l,$$

while the two ends contribute nothing (here **E** is perpendicular to $d\mathbf{a}$). Thus,

$$|\mathbf{E}|\, 2\pi s l = \frac{1}{\epsilon_0}\frac{2}{3}\pi k l s^3,$$

or, finally,

$$\mathbf{E} = \frac{1}{3\epsilon_0} k s^2 \hat{\mathbf{s}}.$$

Example 2.4

An infinite plane carries a uniform surface charge σ. Find its electric field.

Solution: Draw a "Gaussian pillbox," extending equal distances above and below the plane (Fig. 2.22). Apply Gauss's law to this surface:

$$\oint \mathbf{E}\cdot d\mathbf{a} = \frac{1}{\epsilon_0} Q_{\text{enc}}.$$

In this case, $Q_{\text{enc}} = \sigma A$, where A is the area of the lid of the pillbox. By symmetry, **E** points away from the plane (upward for points above, downward for points below). Thus, the top and bottom surfaces yield

$$\int \mathbf{E}\cdot d\mathbf{a} = 2A|\mathbf{E}|,$$

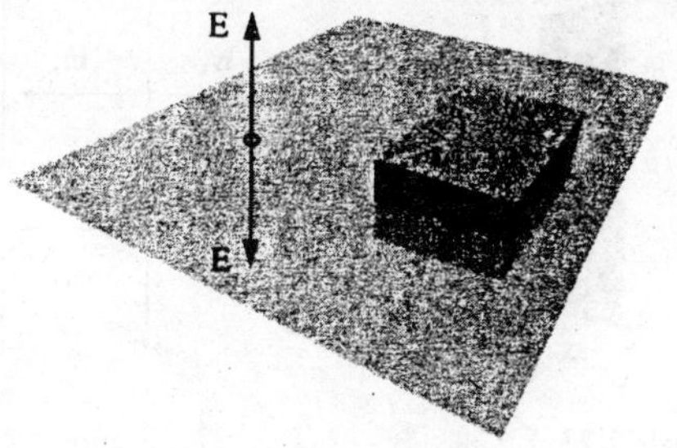

Figure 2.22

whereas the sides contribute nothing. Thus

$$2A\,|\mathbf{E}| = \frac{1}{\epsilon_0}\sigma A,$$

or

$$\mathbf{E} = \frac{\sigma}{2\epsilon_0}\hat{\mathbf{n}} \tag{2.17}$$

where $\hat{\mathbf{n}}$ is a unit vector pointing away from the surface. In Prob. 2.6, you obtained this same result by a much more laborious method.

It seems surprising, at first, that the field of an infinite plane is *independent of how far away you are.* What about the $1/r^2$ in Coulomb's law? Well, the point is that as you move farther and farther away from the plane, more and more charge comes into your "field of view" (a cone shape extending out from your eye), and this compensates for the diminishing influence of any particular piece. The electric field of a sphere falls off like $1/r^2$; the electric field of an infinite line falls off like $1/r$; and the electric field of an infinite plane does not fall off at all.

Although the direct use of Gauss's law to compute electric fields is limited to cases of spherical, cylindrical, and planar symmetry, we can put together *combinations* of objects possessing such symmetry, even though the arrangement as a whole is not symmetrical. For example, invoking the principle of superposition, we could find the field in the vicinity of two uniformly charged parallel cylinders, or a sphere near an infinite charged plane.

Example 2.5

Two infinite parallel planes carry equal but opposite uniform charge densities $\pm\sigma$ (Fig. 2.23). Find the field in each of the three regions: (i) to the left of both, (ii) between them, (iii) to the right of both.

Solution: The left plate produces a field $(1/2\epsilon_0)\sigma$ which points away from it (Fig. 2.24)—to the left in region (i) and to the right in regions (ii) and (iii). The right plate, being negatively charged, produces a field $(1/2\epsilon_0)\sigma$, which points *toward* it—to the right in regions (i) and (ii) and to the left in region (iii). The two fields cancel in regions (i) and (iii); they conspire in region (ii). *Conclusion:* The field is $(1/\epsilon_0)\sigma$, and points to the right, between the planes; elsewhere it is zero.

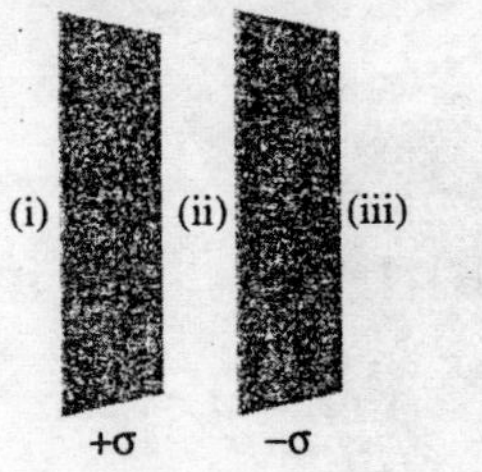

Figure 2.23

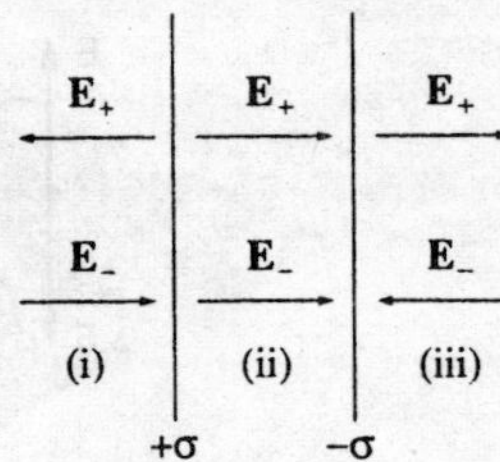

Figure 2.24

Problem 2.11 Use Gauss's law to find the electric field inside and outside a spherical shell of radius R, which carries a uniform surface charge density σ. Compare your answer to Prob. 2.7.

Problem 2.12 Use Gauss's law to find the electric field inside a uniformly charged sphere (charge density ρ). Compare your answer to Prob. 2.8.

Problem 2.13 Find the electric field a distance s from an infinitely long straight wire, which carries a uniform line charge λ. Compare Eq. 2.9.

Problem 2.14 Find the electric field inside a sphere which carries a charge density proportional to the distance from the origin, $\rho = kr$, for some constant k. [*Hint:* This charge density is *not* uniform, and you must *integrate* to get the enclosed charge.]

Problem 2.15 A hollow spherical shell carries charge density

$$\rho = \frac{k}{r^2}$$

in the region $a \le r \le b$ (Fig. 2.25). Find the electric field in the three regions: (i) $r < a$, (ii) $a < r < b$, (iii) $r > b$. Plot $|\mathbf{E}|$ as a function of r.

Problem 2.16 A long coaxial cable (Fig. 2.26) carries a uniform *volume* charge density ρ on the inner cylinder (radius a), and a uniform *surface* charge density on the outer cylindrical shell (radius b). This surface charge is negative and of just the right magnitude so that the cable as a whole is electrically neutral. Find the electric field in each of the three regions: (i) inside the inner cylinder ($s < a$), (ii) between the cylinders ($a < s < b$), (iii) outside the cable ($s > b$). Plot $|\mathbf{E}|$ as a function of s.

Problem 2.17 An infinite plane slab, of thickness $2d$, carries a uniform volume charge density ρ (Fig. 2.27). Find the electric field, as a function of y, where $y = 0$ at the center. Plot E versus y, calling E positive when it points in the $+y$ direction and negative when it points in the $-y$ direction.

Problem 2.18 Two spheres, each of radius R and carrying uniform charge densities $+\rho$ and $-\rho$, respectively, are placed so that they partially overlap (Fig. 2.28). Call the vector from the positive center to the negative center $\mathbf{d}$. Show that the field in the region of overlap is constant, and find its value. [*Hint:* Use the answer to Prob. 2.12.]

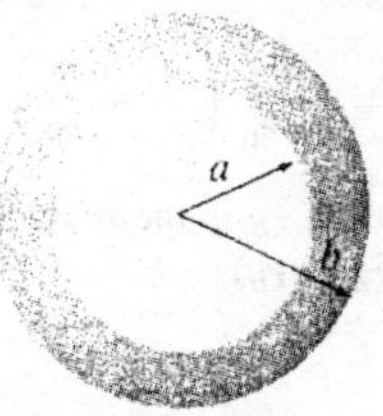

Figure 2.25

Figure 2.26

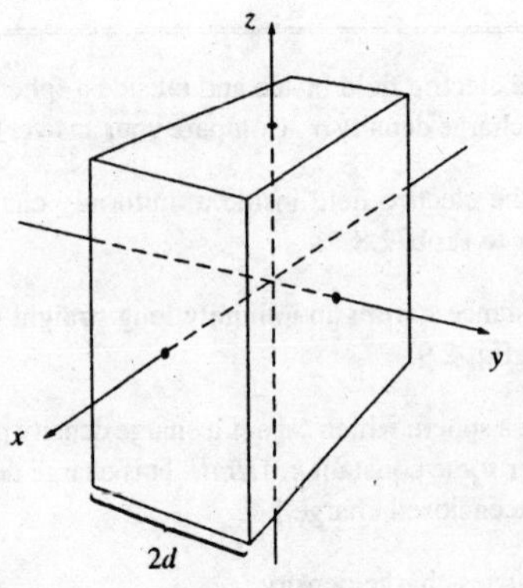

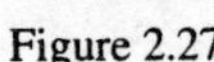

Figure 2.27

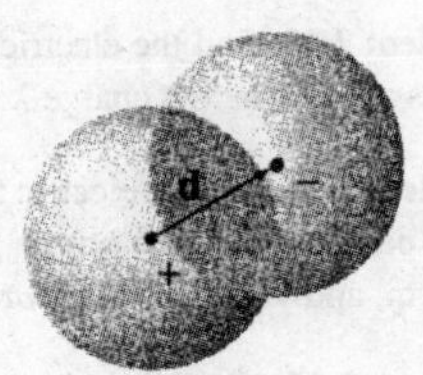

Figure 2.28

2.2.4 The Curl of E

I'll calculate the curl of **E**, as I did the divergence in Sect. 2.2.1, by studying first the simplest possible configuration: a point charge at the origin. In this case

$$\mathbf{E} = \frac{1}{4\pi\epsilon_0}\frac{q}{r^2}\hat{\mathbf{r}}.$$

Now, a glance at Fig. 2.12 should convince you that the curl of this field has to be zero, but I suppose we ought to come up with something a little more rigorous than that. What if we calculate the line integral of this field from some point **a** to some other point **b** (Fig. 2.29):

$$\int_{\mathbf{a}}^{\mathbf{b}} \mathbf{E}\cdot d\mathbf{l}.$$

In spherical coordinates, $d\mathbf{l} = dr\,\hat{\mathbf{r}} + r\,d\theta\,\hat{\boldsymbol{\theta}} + r\sin\theta\,d\phi\,\hat{\boldsymbol{\phi}}$, so

$$\mathbf{E}\cdot d\mathbf{l} = \frac{1}{4\pi\epsilon_0}\frac{q}{r^2}dr.$$

Therefore,

$$\int_{\mathbf{a}}^{\mathbf{b}} \mathbf{E}\cdot d\mathbf{l} = \frac{1}{4\pi\epsilon_0}\int_{\mathbf{a}}^{\mathbf{b}}\frac{q}{r^2}dr = \left.\frac{-1}{4\pi\epsilon_0}\frac{q}{r}\right|_{r_a}^{r_b} = \frac{1}{4\pi\epsilon_0}\left(\frac{q}{r_a}-\frac{q}{r_b}\right), \tag{2.18}$$

where r_a is the distance from the origin to the point **a** and r_b is the distance to **b**. The integral around a *closed* path is evidently zero (for then $r_a = r_b$):

$$\boxed{\oint \mathbf{E}\cdot d\mathbf{l} = 0,} \tag{2.19}$$

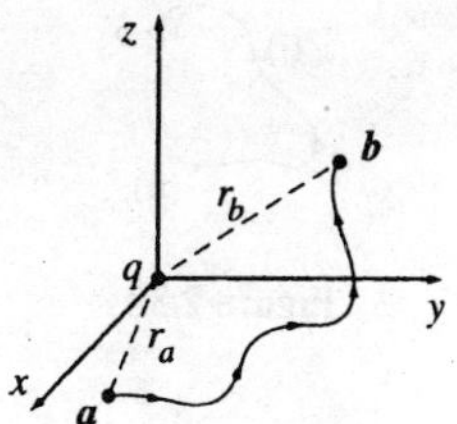

Figure 2.29

and hence, applying Stokes' theorem,

$$\boxed{\nabla \times \mathbf{E} = 0.} \tag{2.20}$$

Now, I proved Eqs. 2.19 and 2.20 only for the field of a single point charge at the *origin*, but these results make no reference to what is, after all, a perfectly arbitrary choice of coordinates; they also hold no matter *where* the charge is located. Moreover, if we have many charges, the principle of superposition states that the total field is a vector sum of their individual fields:

$$\mathbf{E} = \mathbf{E}_1 + \mathbf{E}_2 + \ldots,$$

so

$$\nabla \times \mathbf{E} = \nabla \times (\mathbf{E}_1 + \mathbf{E}_2 + \ldots) = (\nabla \times \mathbf{E}_1) + (\nabla \times \mathbf{E}_2) + \ldots = 0.$$

Thus, Eqs. 2.19 and 2.20 hold for *any static charge distribution whatever.*

Problem 2.19 Calculate $\nabla \times \mathbf{E}$ directly from Eq. 2.8, by the method of Sect. 2.2.2. Refer to Prob. 1.62 if you get stuck.

2.3 Electric Potential

2.3.1 Introduction to Potential

The electric field **E** is not just *any* old vector function; it is a very special *kind* of vector function, one whose curl is always zero. $\mathbf{E} = y\hat{\mathbf{x}}$, for example, could not possibly be an electrostatic field; *no* set of charges, regardless of their sizes and positions, could ever produce such a field. In this section we're going to exploit this special property of electric fields to reduce a vector problem (finding **E**) down to a much simpler scalar problem. The first theorem in Sect. 1.6.2 asserts that any vector whose curl is zero is equal to the gradient of some scalar. What I'm going to do now amounts to a proof of that claim, in the context of electrostatics.

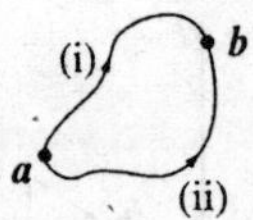

Figure 2.30

Because $\nabla \times \mathbf{E} = 0$, the line integral of $\mathbf{E}$ around any closed loop is zero (that follows from Stokes' theorem). Because $\oint \mathbf{E} \cdot d\mathbf{l} = 0$, the line integral of $\mathbf{E}$ from point $\mathbf{a}$ to point $\mathbf{b}$ is the same for all paths (otherwise you could go *out* along path (i) and return along path (ii)—Fig. 2.30—and obtain $\oint \mathbf{E} \cdot d\mathbf{l} \neq 0$). Because the line integral is independent of path, we can define a function[4]

$$\boxed{V(\mathbf{r}) \equiv -\int_{\mathcal{O}}^{\mathbf{r}} \mathbf{E} \cdot d\mathbf{l}.} \tag{2.21}$$

Here $\mathcal{O}$ is some standard reference point on which we have agreed beforehand; V then depends only on the point $\mathbf{r}$. It is called the **electric potential.**

Evidently, the potential *difference* between two points $\mathbf{a}$ and $\mathbf{b}$ is

$$\begin{aligned} V(\mathbf{b}) - V(\mathbf{a}) &= -\int_{\mathcal{O}}^{\mathbf{b}} \mathbf{E} \cdot d\mathbf{l} + \int_{\mathcal{O}}^{\mathbf{a}} \mathbf{E} \cdot d\mathbf{l} \\ &= -\int_{\mathcal{O}}^{\mathbf{b}} \mathbf{E} \cdot d\mathbf{l} - \int_{\mathbf{a}}^{\mathcal{O}} \mathbf{E} \cdot d\mathbf{l} = -\int_{\mathbf{a}}^{\mathbf{b}} \mathbf{E} \cdot d\mathbf{l}. \end{aligned} \tag{2.22}$$

Now, the fundamental theorem for gradients states that

$$V(\mathbf{b}) - V(\mathbf{a}) = \int_{\mathbf{a}}^{\mathbf{b}} (\nabla V) \cdot d\mathbf{l},$$

so

$$\int_{\mathbf{a}}^{\mathbf{b}} (\nabla V) \cdot d\mathbf{l} = -\int_{\mathbf{a}}^{\mathbf{b}} \mathbf{E} \cdot d\mathbf{l}.$$

Since, finally, this is true for *any* points $\mathbf{a}$ and $\mathbf{b}$, the integrands must be equal:

$$\boxed{\mathbf{E} = -\nabla V.} \tag{2.23}$$

Equation 2.23 is the differential version of Eq. 2.21; it says that the electric field is the gradient of a scalar potential, which is what we set out to prove.

[4]To avoid any possible ambiguity I should perhaps put a prime on the integration variable:

$$V(\mathbf{r}) = -\int_{\mathcal{O}}^{\mathbf{r}} \mathbf{E}(\mathbf{r}') \cdot d\mathbf{l}'.$$

But this makes for cumbersome notation, and I prefer whenever possible to reserve the primes for source points. However, when (as in Ex. 2.6) we calculate such integrals explicitly, I shall put in the primes.

Notice the subtle but crucial role played by path independence (or, equivalently, the fact that $\nabla \times \mathbf{E} = 0$) in this argument. If the line integral of $\mathbf{E}$ depended on the path taken, then the "definition" of V, Eq. 2.21, would be nonsense. It simply would not define a function, since changing the path would alter the value of $V(\mathbf{r})$. By the way, don't let the minus sign in Eq. 2.23 distract you; it carries over from 2.21 and is largely a matter of convention.

Problem 2.20 One of these is an impossible electrostatic field. Which one?

(a) $\mathbf{E} = k[xy\,\hat{\mathbf{x}} + 2yz\,\hat{\mathbf{y}} + 3xz\,\hat{\mathbf{z}}]$;

(b) $\mathbf{E} = k[y^2\,\hat{\mathbf{x}} + (2xy + z^2)\,\hat{\mathbf{y}} + 2yz\,\hat{\mathbf{z}}]$.

Here k is a constant with the appropriate units. For the *possible* one, find the potential, using the *origin* as your reference point. Check your answer by computing ∇V. [*Hint:* You must select a specific path to integrate along. It doesn't matter *what* path you choose, since the answer is path-independent, but you simply cannot integrate unless you have a particular path in mind.]

2.3.2 Comments on Potential

(i) The name. The word "potential" is a hideous misnomer because it inevitably reminds you of potential *energy*. This is particularly confusing, because there *is* a connection between "potential" and "potential energy," as you will see in Sect. 2.4. I'm sorry that it is impossible to escape this word. The best I can do is to insist once and for all that "potential" and "potential energy" are completely different terms and should, by all rights, have different names. Incidentally, a surface over which the potential is constant is called an **equipotential**.

(ii) Advantage of the potential formulation. If you know V, you can easily get $\mathbf{E}$—just take the gradient: $\mathbf{E} = -\nabla V$. This is quite extraordinary when you stop to think about it, for $\mathbf{E}$ is a *vector* quantity (three components), but V is a *scalar* (one component). How can *one* function possibly contain all the information that *three* independent functions carry? The answer is that the three components of $\mathbf{E}$ are not really as independent as they look; in fact, they are explicitly interrelated by the very condition we started with, $\nabla \times \mathbf{E} = 0$. In terms of components,

$$\frac{\partial E_x}{\partial y} = \frac{\partial E_y}{\partial x}, \qquad \frac{\partial E_z}{\partial y} = \frac{\partial E_y}{\partial z}, \qquad \frac{\partial E_x}{\partial z} = \frac{\partial E_z}{\partial x}.$$

This brings us back to my observation at the beginning of Sect. 2.3.1: **E** *is a very special kind of vector.* What the potential formulation does is to exploit this feature to maximum advantage, reducing a vector problem down to a scalar one, in which there is no need to fuss with components.

(iii) The reference point $\mathcal{O}$. There is an essential ambiguity in the definition of potential, since the choice of reference point $\mathcal{O}$ was arbitrary. Changing reference points amounts to adding a constant K to the potential:

$$V'(\mathbf{r}) = -\int_{\mathcal{O}'}^{\mathbf{r}} \mathbf{E}\cdot d\mathbf{l} = -\int_{\mathcal{O}'}^{\mathcal{O}} \mathbf{E}\cdot d\mathbf{l} - \int_{\mathcal{O}}^{\mathbf{r}} \mathbf{E}\cdot d\mathbf{l} = K + V(\mathbf{r}),$$

where K is the line integral of $\mathbf{E}$ from the old reference point $\mathcal{O}$ to the new one $\mathcal{O}'$. Of course, adding a constant to V will not affect the potential *difference* between two points:

$$V'(\mathbf{b}) - V'(\mathbf{a}) = V(\mathbf{b}) - V(\mathbf{a}),$$

since the K's cancel out. (Actually, it was already clear from Eq. 2.22 that the potential difference is independent of $\mathcal{O}$, because it can be written as the line integral of $\mathbf{E}$ from $\mathbf{a}$ to $\mathbf{b}$, with no reference to $\mathcal{O}$.) Nor does the ambiguity affect the gradient of V:

$$\nabla V' = \nabla V,$$

since the derivative of a constant is zero. That's why all such V's, differing only in their choice of reference point, correspond to the same field $\mathbf{E}$.

Evidently potential as such carries no real physical significance, for at any given point we can adjust its value at will by a suitable relocation of $\mathcal{O}$. In this sense it is rather like altitude: If I ask you how high Denver is, you will probably tell me its height above sea level, because that is a convenient and traditional reference point. But we could as well agree to measure altitude above Washington D.C., or Greenwich, or wherever. That would add (or, rather, subtract) a fixed amount from all our sea-level readings, but it wouldn't change anything about the real world. The only quantity of intrinsic interest is the *difference* in altitude between two points, and *that* is the same *whatever* your reference level.

Having said this, however, there *is* a "natural" spot to use for $\mathcal{O}$ in electrostatics—analogous to sea level for altitude—and that is a point infinitely far from the charge. Ordinarily, then, we "set the zero of potential at infinity." (Since $V(\mathcal{O}) = 0$, choosing a reference point is equivalent to selecting a place where V is to be zero.) But I must warn you that there is one special circumstance in which this convention fails: when the charge distribution itself extends to infinity. The symptom of trouble, in such cases, is that the potential blows up. For instance, the field of a uniformly charged plane is $(\sigma/2\epsilon_0)\hat{\mathbf{n}}$, as we found in Ex. 2.4; if we naïvely put $\mathcal{O} = \infty$, then the potential at height z above the plane becomes

$$V(z) = -\int_{\infty}^{z} \frac{1}{2\epsilon_0}\sigma\, dz = -\frac{1}{2\epsilon_0}\sigma(z - \infty).$$

The remedy is simply to choose some other reference point (in this problem you might use the origin). Notice that the difficulty occurs only in textbook problems; in "real life" there is no such thing as a charge distribution that goes on forever, and we can *always* use infinity as our reference point.

(iv) Potential obeys the superposition principle. The original superposition principle of electrodynamics pertains to the force on a test charge Q. It says that the total force on Q is the vector sum of the forces attributable to the source charges individually:

$$\mathbf{F} = \mathbf{F}_1 + \mathbf{F}_2 + \ldots$$

Dividing through by Q, we find that the electric field, too, obeys the superposition principle:

$$\mathbf{E} = \mathbf{E}_1 + \mathbf{E}_2 + \ldots$$

Integrating from the common reference point to $\mathbf{r}$, it follows that the potential also satisfies such a principle:

$$V = V_1 + V_2 + \ldots$$

That is, the potential at any given point is the sum of the potentials due to all the source charges separately. Only this time it is an *ordinary* sum, not a *vector* sum, which makes it a lot easier to work with.

(v) Units of Potential. In our units, force is measured in newtons and charge in coulombs, so electric fields are in newtons per coulomb. Accordingly, potential is measured in newton-meters per coulomb or joules per coulomb. A joule per coulomb is called a **volt**.

Example 2.6

Find the potential inside and outside a spherical shell of radius R (Fig. 2.31), which carries a uniform surface charge. Set the reference point at infinity.

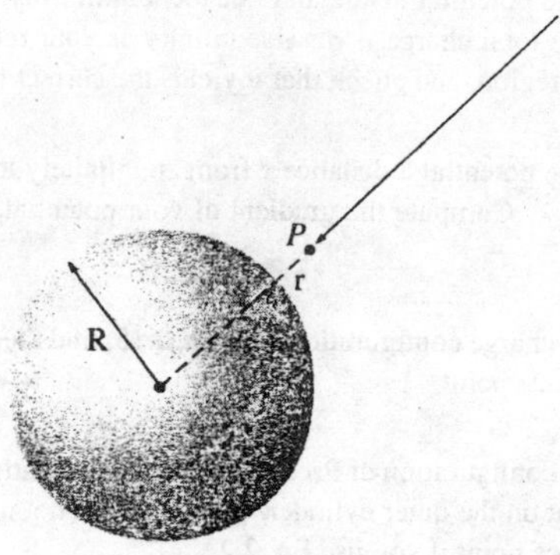

Figure 2.31

Solution: From Gauss's law, the field outside is

$$\mathbf{E} = \frac{1}{4\pi\epsilon_0}\frac{q}{r^2}\hat{\mathbf{r}},$$

where q is the total charge on the sphere. The field inside is zero. For points outside the sphere ($r > R$),

$$V(r) = -\int_{\mathcal{O}}^{\mathbf{r}} \mathbf{E}\cdot d\mathbf{l} = \frac{-1}{4\pi\epsilon_0}\int_{\infty}^{r}\frac{q}{r'^2}\,dr' = \frac{1}{4\pi\epsilon_0}\frac{q}{r'}\bigg|_{\infty}^{r} = \frac{1}{4\pi\epsilon_0}\frac{q}{r}.$$

To find the potential inside the sphere ($r < R$), we must break the integral into two sections, using in each region the field that prevails there:

$$V(r) = \frac{-1}{4\pi\epsilon_0}\int_{\infty}^{R}\frac{q}{r'^2}\,dr' - \int_{R}^{r}(0)\,dr' = \frac{1}{4\pi\epsilon_0}\frac{q}{r'}\bigg|_{\infty}^{R} + 0 = \frac{1}{4\pi\epsilon_0}\frac{q}{R}.$$

Notice that the potential is *not* zero inside the shell, even though the field *is*. V is a *constant* in this region, to be sure, so that $\nabla V = 0$—that's what matters. In problems of this type you must always *work your way in from the reference point;* that's where the potential is "nailed down." It is tempting to suppose that you could figure out the potential inside the sphere on the basis of the field there alone, but this is false: The potential inside the sphere is sensitive to what's going on outside the sphere as well. If I placed a second uniformly charged shell out at radius $R' > R$, the potential inside R would change, even though the field would still be zero. Gauss's law guarantees that charge exterior to a given point (that is, at larger r) produces no net *field* at that point, provided it is spherically or cylindrically symmetric; but there is no such rule for *potential*, when infinity is used as the reference point.

Problem 2.21 Find the potential inside and outside a uniformly charged solid sphere whose radius is R and whose total charge is q. Use infinity as your reference point. Compute the gradient of V in each region, and check that it yields the correct field. Sketch $V(r)$.

Problem 2.22 Find the potential a distance s from an infinitely long straight wire that carries a uniform line charge λ. Compute the gradient of your potential, and check that it yields the correct field.

Problem 2.23 For the charge configuration of Prob. 2.15, find the potential at the center, using infinity as your reference point.

Problem 2.24 For the configuration of Prob. 2.16, find the potential difference between a point on the axis and a point on the outer cylinder. Note that it is not necessary to commit yourself to a particular reference point if you use Eq. 2.22.

2.3.3 Poisson's Equation and Laplace's Equation

We found in Sect. 2.3.1 that the electric field can be written as the gradient of a scalar potential.

$$\mathbf{E} = -\nabla V.$$

The question arises: What do the fundamental equations for **E**,

$$\nabla \cdot \mathbf{E} = \frac{\rho}{\epsilon_0} \quad \text{and} \quad \nabla \times \mathbf{E} = 0,$$

look like, in terms of V? Well, $\nabla \cdot \mathbf{E} = \nabla \cdot (-\nabla V) = -\nabla^2 V$, so, apart from that persisting minus sign, the divergence of **E** is the Laplacian of V. Gauss's law then says that

$$\boxed{\nabla^2 V = -\frac{\rho}{\epsilon_0}.} \tag{2.24}$$

This is known as **Poisson's equation**. In regions where there is no charge, so that $\rho = 0$, Poisson's equation reduces to **Laplace's equation**,

$$\nabla^2 V = 0. \tag{2.25}$$

We'll explore these equations more fully in Chapter 3.

So much for Gauss's law. What about the curl law? This says that

$$\nabla \times \mathbf{E} = \nabla \times (-\nabla V)$$

must equal zero. But that's no condition on V—curl of gradient is *always* zero. Of course, we *used* the curl law to show that **E** could be expressed as the gradient of a scalar, so it's not really surprising that this works out: $\nabla \times \mathbf{E} = 0$ *permits* $\mathbf{E} = -\nabla V$; in return, $\mathbf{E} = -\nabla V$ *guarantees* $\nabla \times \mathbf{E} = 0$. It takes only *one* differential equation (Poisson's) to determine V, because V is a scalar; for **E** we needed *two*, the divergence and the curl.

2.3.4 The Potential of a Localized Charge Distribution

I defined V in terms of **E** (Eq. 2.21). Ordinarily, though, it's **E** that we're looking for (if we already knew **E** there wouldn't be much point in calculating V). The idea is that it might be easier to get V first, and then calculate **E** by taking the gradient. Typically, then, we know where the charge is (that is, we know ρ), and we want to find V. Now, Poisson's equation relates V and ρ, but unfortunately it's "the wrong way around": it would give us ρ, if we knew V, whereas we want V, knowing ρ. What we must do, then, is "invert" Poisson's equation. That's the program for this section, although I shall do it by roundabout means, beginning, as always, with a point charge at the origin.

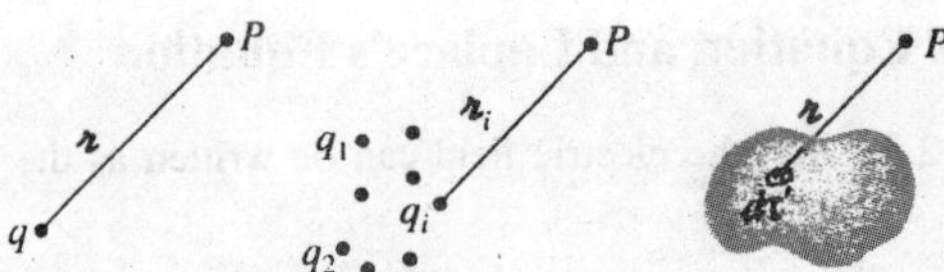

Figure 2.32

Setting the reference point at infinity, the potential of a point charge q at the origin is

$$V(r) = \frac{-1}{4\pi\epsilon_0}\int_{\infty}^{r}\frac{q}{r'^2}\,dr' = \frac{1}{4\pi\epsilon_0}\frac{q}{r'}\bigg|_{\infty}^{r} = \frac{1}{4\pi\epsilon_0}\frac{q}{r}.$$

(You see here the special virtue of using infinity for the reference point: it kills the lower limit on the integral.) Notice the sign of V; presumably the conventional minus sign in the definition of V (Eq. 2.21) was chosen precisely in order to *make* the potential of a positive charge come out positive. It is useful to remember that regions of positive charge are potential "hills," regions of negative charge are potential "valleys," and the electric field points "downhill," from plus toward minus.

In general, the potential of a point charge q is

$$V(\mathbf{r}) = \frac{1}{4\pi\epsilon_0}\frac{q}{\mathscr{r}}, \tag{2.26}$$

where $\mathscr{r}$, as always, is the distance from the charge to $\mathbf{r}$ (Fig. 2.32). Invoking the superposition principle, then, the potential of a collection of charges is

$$V(\mathbf{r}) = \frac{1}{4\pi\epsilon_0}\sum_{i=1}^{n}\frac{q_i}{\mathscr{r}_i}, \tag{2.27}$$

or, for a continuous distribution,

$$V(\mathbf{r}) = \frac{1}{4\pi\epsilon_0}\int\frac{1}{\mathscr{r}}\,dq. \tag{2.28}$$

In particular, for a volume charge, it's

$$\boxed{V(\mathbf{r}) = \frac{1}{4\pi\epsilon_0}\int\frac{\rho(\mathbf{r}')}{\mathscr{r}}\,d\tau'.} \tag{2.29}$$

This is the equation we were looking for, telling us how to compute V when we know ρ; it is, if you like, the "solution" to Poisson's equation, for a localized charge distribution.[5] I

[5] Equation 2.29 is an example of the Helmholtz theorem (Appendix B), in the context of electrostatics, where the curl of $\mathbf{E}$ is zero and its divergence is ρ/ϵ_0.

invite you to compare Eq. 2.29 with the corresponding formula for the electric *field* in terms of ρ (Eq. 2.8):

$$\mathbf{E}(\mathbf{r}) = \frac{1}{4\pi\epsilon_0}\int \frac{\rho(\mathbf{r}')}{\imath^2}\hat{\boldsymbol{\imath}}\,d\tau'.$$

The main point to notice is that the pesky unit vector $\hat{\boldsymbol{\imath}}$ is now missing, so there is no need to worry about components. Incidentally, the potentials of line and surface charges are

$$\frac{1}{4\pi\epsilon_0}\int \frac{\lambda(\mathbf{r}')}{\imath}\,dl' \quad \text{and} \quad \frac{1}{4\pi\epsilon_0}\int \frac{\sigma(\mathbf{r}')}{\imath}\,da'. \tag{2.30}$$

I should warn you that everything in this section is predicated on the assumption that the reference point is at infinity. This is hardly apparent in Eq. 2.29, but remember that we *got* that equation from the potential of a point charge at the origin, $(1/4\pi\epsilon_0)(q/r)$, which is valid only when $\mathcal{O} = \infty$. If you try to apply these formulas to one of those artificial problems in which the charge itself extends to infinity, the integral will diverge.

Example 2.7

Find the potential of a uniformly charged spherical shell of radius R (Fig. 2.33).

Solution: This is the same problem we solved in Ex. 2.6, but this time we shall do it using Eq. 2.30:

$$V(\mathbf{r}) = \frac{1}{4\pi\epsilon_0}\int \frac{\sigma}{\imath}\,da'.$$

Let's set the point $\mathbf{r}$ on the z axis and use the law of cosines to express $\imath$ in terms of the polar angle θ:

$$\imath^2 = R^2 + z^2 - 2Rz\cos\theta'.$$

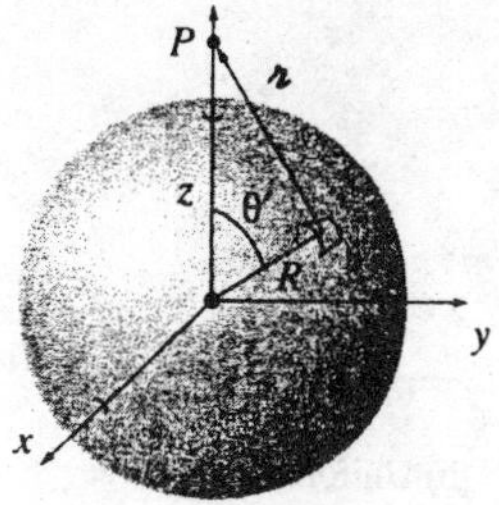

Figure 2.33

An element of surface area on this sphere is $R^2 \sin\theta'\, d\theta'\, d\phi'$, so

$$\begin{aligned}
4\pi\epsilon_0 V(z) &= \sigma \int \frac{R^2 \sin\theta'\, d\theta'\, d\phi'}{\sqrt{R^2+z^2-2Rz\cos\theta'}} \\
&= 2\pi R^2 \sigma \int_0^\pi \frac{\sin\theta'}{\sqrt{R^2+z^2-2Rz\cos\theta'}}\, d\theta' \\
&= 2\pi R^2 \sigma \left(\frac{1}{Rz}\sqrt{R^2+z^2-2Rz\cos\theta'} \right)\Bigg|_0^\pi \\
&= \frac{2\pi R\sigma}{z}\left(\sqrt{R^2+z^2+2Rz} - \sqrt{R^2+z^2-2Rz}\right) \\
&= \frac{2\pi R\sigma}{z}\left[\sqrt{(R+z)^2} - \sqrt{(R-z)^2}\right].
\end{aligned}$$

At this stage we must be very careful to take the *positive* root. For points *outside* the sphere, z is greater than R, and hence $\sqrt{(R-z)^2} = z-R$; for points *inside* the sphere, $\sqrt{(R-z)^2} = R-z$. Thus,

$$\begin{aligned}
V(z) &= \frac{R\sigma}{2\epsilon_0 z}[(R+z)-(z-R)] = \frac{R^2\sigma}{\epsilon_0 z}, \quad \text{outside;} \\
V(z) &= \frac{R\sigma}{2\epsilon_0 z}[(R+z)-(R-z)] = \frac{R\sigma}{\epsilon_0}, \quad \text{inside.}
\end{aligned}$$

In terms of the total charge on the shell, $q = 4\pi R^2\sigma$, $V(z) = (1/4\pi\epsilon_0)(q/z)$ (or, in general, $V(r) = (1/4\pi\epsilon_0)(q/r)$) for points outside the sphere, and $(1/4\pi\epsilon_0)(q/R)$ for points inside.

Of course, in this particular case, it was easier to get V by using 2.21 than 2.30, because Gauss's law gave us $\mathbf{E}$ with so little effort. But if you compare Ex. 2.7 with Prob. 2.7, you will appreciate the power of the potential formulation.

Problem 2.25 Using Eqs. 2.27 and 2.30, find the potential at a distance z above the center of the charge distributions in Fig. 2.34. In each case, compute $\mathbf{E} = -\nabla V$, and compare your answers with Prob. 2.2a, Ex. 2.1, and Prob. 2.6, respectively. Suppose that we changed the right-hand charge in Fig. 2.34a to $-q$; what then is the potential at P? What field does that suggest? Compare your answer to Prob. 2.2b, and explain carefully any discrepancy.

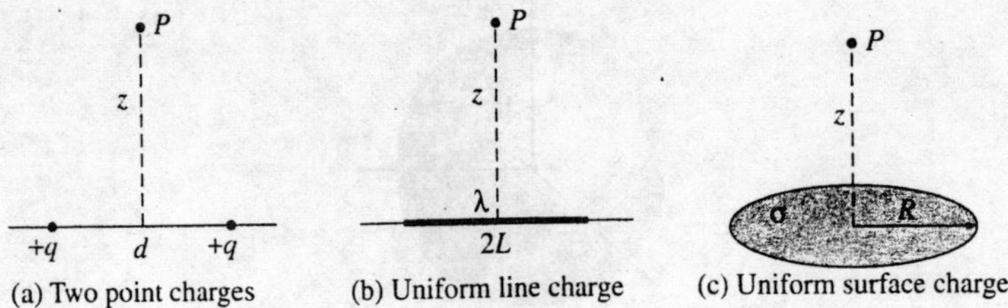

(a) Two point charges (b) Uniform line charge (c) Uniform surface charge

Figure 2.34

Problem 2.26 A conical surface (an empty ice-cream cone) carries a uniform surface charge σ. The height of the cone is h, as is the radius of the top. Find the potential difference between points **a** (the vertex) and **b** (the center of the top).

Problem 2.27 Find the potential on the axis of a uniformly charged solid cylinder, a distance z from the center. The length of the cylinder is L, its radius is R, and the charge density is ρ. Use your result to calculate the electric field at this point. (Assume that $z > L/2$.)

Problem 2.28 Use Eq. 2.29 to calculate the potential inside a uniformly charged solid sphere of radius R and total charge q. Compare your answer to Prob. 2.21.

Problem 2.29 Check that Eq. 2.29 satisfies Poisson's equation, by applying the Laplacian and using Eq. 1.102.

2.3.5 Summary; Electrostatic Boundary Conditions

In the typical electrostatic problem you are given a source charge distribution ρ, and you want to find the electric field $\mathbf{E}$ it produces. Unless the symmetry of the problem admits a solution by Gauss's law, it is generally to your advantage to calculate the potential first, as an intermediate step. These, then, are the three fundamental quantities of electrostatics: ρ, $\mathbf{E}$, and V. We have, in the course of our discussion, derived all six formulas interrelating them. These equations are neatly summarized in Fig. 2.35. We began with just two experimental observations: (1) the principle of superposition—a broad general rule applying to *all* electromagnetic forces, and (2) Coulomb's law—the fundamental law of electrostatics. From these, all else followed.

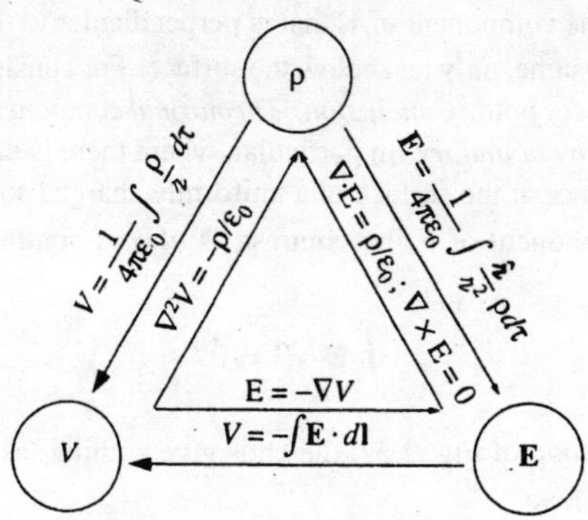

Figure 2.35

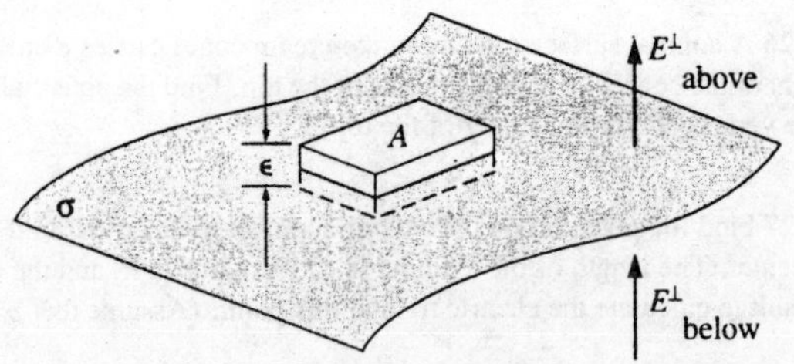

Figure 2.36

You may have noticed, in studying Exs. 2.4 and 2.5, or working problems such as 2.7, 2.11, and 2.16, that the electric field always undergoes a discontinuity when you cross a surface charge σ. In fact, it is a simple matter to find the *amount* by which **E** changes at such a boundary. Suppose we draw a wafer-thin Gaussian pillbox, extending just barely over the edge in each direction (Fig. 2.36). Gauss's law states that

$$\oint_{S} \mathbf{E} \cdot d\mathbf{a} = \frac{1}{\epsilon_0} Q_{\text{enc}} = \frac{1}{\epsilon_0} \sigma A,$$

where A is the area of the pillbox lid. (If σ varies from point to point or the surface is curved, we must pick A to be extremely small.) Now, the *sides* of the pillbox contribute nothing to the flux, in the limit as the thickness ϵ goes to zero, so we are left with

$$E^{\perp}_{\text{above}} - E^{\perp}_{\text{below}} = \frac{1}{\epsilon_0} \sigma, \tag{2.31}$$

where $E^{\perp}_{\text{above}}$ denotes the component of **E** that is perpendicular to the surface immediately above, and $E^{\perp}_{\text{below}}$ is the same, only just below the surface. For consistency, we let "upward" be the positive direction for both. *Conclusion: The normal component of* **E** *is discontinuous by an amount* σ/ϵ_0 *at any boundary.* In particular, where there is *no* surface charge, $E^{\perp}$ is continuous, as for instance at the surface of a uniformly charged solid sphere.

The *tangential* component of **E**, by contrast, is *always* continuous. For if we apply Eq. 2.19,

$$\oint \mathbf{E} \cdot d\mathbf{l} = 0,$$

to the thin rectangular loop of Fig. 2.37, the ends give nothing (as $\epsilon \to 0$), and the sides give $(E^{\parallel}_{\text{above}} l - E^{\parallel}_{\text{below}} l)$, so

$$\mathbf{E}^{\parallel}_{\text{above}} = \mathbf{E}^{\parallel}_{\text{below}}, \tag{2.32}$$

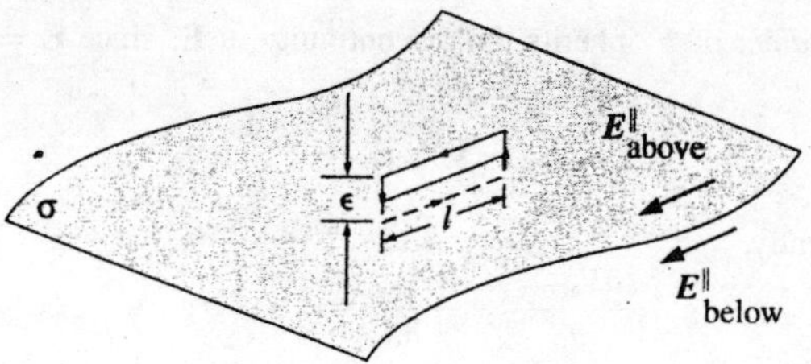

Figure 2.37

where $\mathbf{E}^{\|}$ stands for the components of **E** *parallel* to the surface. The boundary conditions on **E** (Eqs. 2.31 and 2.32) can be combined into a single formula:

$$\mathbf{E}_{\text{above}} - \mathbf{E}_{\text{below}} = \frac{\sigma}{\epsilon_0}\hat{\mathbf{n}}, \tag{2.33}$$

where $\hat{\mathbf{n}}$ is a unit vector perpendicular to the surface, pointing from "below" to "above."[6]

The potential, meanwhile, is continuous across any boundary (Fig. 2.38), since

$$V_{\text{above}} - V_{\text{below}} = -\int_{\mathbf{a}}^{\mathbf{b}} \mathbf{E} \cdot d\mathbf{l};$$

as the path length shrinks to zero, so too does the integral:

$$V_{\text{above}} = V_{\text{below}}. \tag{2.34}$$

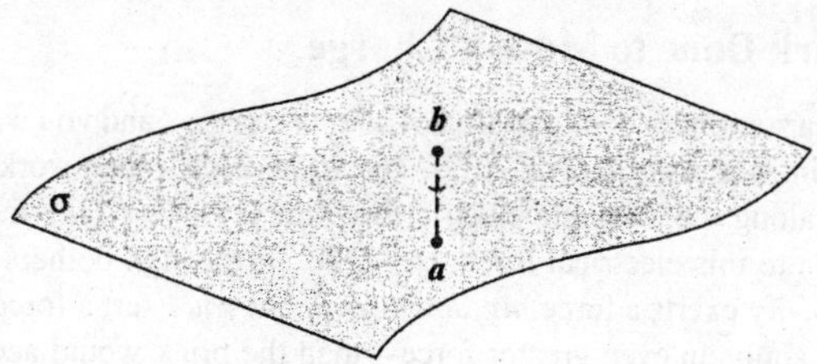

Figure 2.38

[6]Notice that it doesn't matter which side you call "above" and which "below," since reversal would switch the direction of $\hat{\mathbf{n}}$. Incidentally, if you're only interested in the field *due to the* (essentially flat) *local patch of surface charge itself,* the answer is $(\sigma/2\epsilon_0)\hat{\mathbf{n}}$ immediately above the surface, and $-(\sigma/2\epsilon_0)\hat{\mathbf{n}}$ immediately below. This follows from Ex. 2.4, for if you are close enough to the patch it "looks" like an infinite plane. Evidently the entire *discontinuity* in **E** is attributable to this local patch of charge.

However, the *gradient* of V inherits the discontinuity in $\mathbf{E}$; since $\mathbf{E} = -\nabla V$, Eq. 2.33 implies that

$$\nabla V_{\text{above}} - \nabla V_{\text{below}} = -\frac{1}{\epsilon_0}\sigma\hat{\mathbf{n}}, \tag{2.35}$$

or, more conveniently,

$$\frac{\partial V_{\text{above}}}{\partial n} - \frac{\partial V_{\text{below}}}{\partial n} = -\frac{1}{\epsilon_0}\sigma, \tag{2.36}$$

where

$$\frac{\partial V}{\partial n} = \nabla V \cdot \hat{\mathbf{n}} \tag{2.37}$$

denotes the **normal derivative** of V (that is, the rate of change in the direction perpendicular to the surface).

Please note that these boundary conditions relate the fields and potentials *just* above and *just* below the surface. For example, the derivatives in Eq. 2.36 are the *limiting* values as we approach the surface from either side.

Problem 2.30

(a) Check that the results of Exs. 2.4 and 2.5, and Prob. 2.11, are consistent with Eq. 2.33.

(b) Use Gauss's law to find the field inside and outside a long hollow cylindrical tube, which carries a uniform surface charge σ. Check that your result is consistent with Eq. 2.33.

(c) Check that the result of Ex. 2.7 is consistent with boundary conditions 2.34 and 2.36.

2.4 Work and Energy in Electrostatics

2.4.1 The Work Done to Move a Charge

Suppose you have a stationary configuration of source charges, and you want to move a test charge Q from point **a** to point **b** (Fig. 2.39). *Question:* How much work will you have to do? At any point along the path, the electric force on Q is $\mathbf{F} = Q\mathbf{E}$; the force *you* must exert, in opposition to this electrical force, is $-Q\mathbf{E}$. (If the sign bothers you, think about lifting a brick: Gravity exerts a force mg *downward,* but *you* exert a force mg *upward.* Of course, you *could* apply an even greater force—then the brick would accelerate, and part

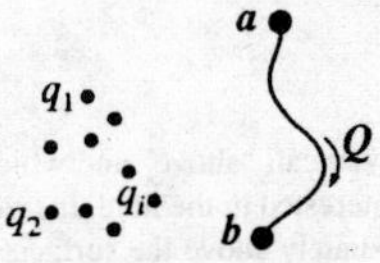

Figure 2.39

of your effort would be "wasted" generating kinetic energy. What we're interested in here is the *minimum* force you must exert to do the job.) The work is therefore

$$W = \int_{\mathbf{a}}^{\mathbf{b}} \mathbf{F} \cdot d\mathbf{l} = -Q \int_{\mathbf{a}}^{\mathbf{b}} \mathbf{E} \cdot d\mathbf{l} = Q[V(\mathbf{b}) - V(\mathbf{a})].$$

Notice that the answer is independent of the path you take from **a** to **b**; in mechanics, then, we would call the electrostatic force "conservative." Dividing through by Q, we have

$$V(\mathbf{b}) - V(\mathbf{a}) = \frac{W}{Q}. \tag{2.38}$$

In words, the *potential difference between points* **a** *and* **b** *is equal to the work per unit charge required to carry a particle from* **a** *to* **b**. In particular, if you want to bring the charge Q in from far away and stick it at point **r**, the work you must do is

$$W = Q[V(\mathbf{r}) - V(\infty)],$$

so, if you have set the reference point at infinity,

$$W = QV(\mathbf{r}). \tag{2.39}$$

In this sense *potential* is potential *energy* (the work it takes to create the system) *per unit charge* (just as the *field* is the *force* per unit charge).

2.4.2 The Energy of a Point Charge Distribution

How much work would it take to assemble an entire *collection* of point charges? Imagine bringing in the charges, one by one, from far away (Fig. 2.40). The first charge, q_1, takes *no* work, since there is no field yet to fight against. Now bring in q_2. According to Eq. 2.39, this will cost you $q_2 V_1(\mathbf{r}_2)$, where V_1 is the potential due to q_1, and $\mathbf{r}_2$ is the place we're putting q_2:

$$W_2 = \frac{1}{4\pi\epsilon_0} q_2 \left(\frac{q_1}{\mathscr{r}_{12}}\right)$$

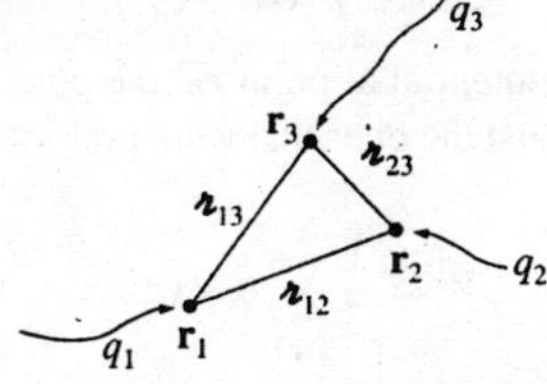

Figure 2.40

($\mathscr{r}_{12}$ is the distance between q_1 and q_2 once they are in position). Now bring in q_3; this requires work $q_3 V_{1,2}(\mathbf{r}_3)$, where $V_{1,2}$ is the potential due to charges q_1 and q_2, namely, $(1/4\pi\epsilon_0)(q_1/\mathscr{r}_{13} + q_2/\mathscr{r}_{23})$. Thus

$$W_3 = \frac{1}{4\pi\epsilon_0} q_3 \left(\frac{q_1}{\mathscr{r}_{13}} + \frac{q_2}{\mathscr{r}_{23}} \right).$$

Similarly, the extra work to bring in q_4 will be

$$W_4 = \frac{1}{4\pi\epsilon_0} q_4 \left(\frac{q_1}{\mathscr{r}_{14}} + \frac{q_2}{\mathscr{r}_{24}} + \frac{q_3}{\mathscr{r}_{34}} \right).$$

The *total* work necessary to assemble the first four charges, then, is

$$W = \frac{1}{4\pi\epsilon_0} \left(\frac{q_1 q_2}{\mathscr{r}_{12}} + \frac{q_1 q_3}{\mathscr{r}_{13}} + \frac{q_1 q_4}{\mathscr{r}_{14}} + \frac{q_2 q_3}{\mathscr{r}_{23}} + \frac{q_2 q_4}{\mathscr{r}_{24}} + \frac{q_3 q_4}{\mathscr{r}_{34}} \right).$$

You see the general rule: Take the product of each pair of charges, divide by their separation distance, and add it all up:

$$W = \frac{1}{4\pi\epsilon_0} \sum_{i=1}^{n} \sum_{\substack{j=1 \\ j>i}}^{n} \frac{q_i q_j}{\mathscr{r}_{ij}}. \tag{2.40}$$

The stipulation $j > i$ is just to remind you not to count the same pair twice. A nicer way to accomplish the same purpose is *intentionally* to count each pair twice, and then divide by 2:

$$W = \frac{1}{8\pi\epsilon_0} \sum_{i=1}^{n} \sum_{\substack{j=1 \\ j\neq i}}^{n} \frac{q_i q_j}{\mathscr{r}_{ij}} \tag{2.41}$$

(we must still avoid $i = j$, of course). Notice that in this form the answer plainly does not depend on the *order* in which you assemble the charges, since every pair occurs in the sum. Let me next pull out the factor q_i:

$$W = \frac{1}{2} \sum_{i=1}^{n} q_i \left(\sum_{\substack{j=1 \\ j\neq i}}^{n} \frac{1}{4\pi\epsilon_0} \frac{q_j}{\mathscr{r}_{ij}} \right).$$

The term in parentheses is the potential at point $\mathbf{r}_i$ (the position of q_i) due to all the other charges—all of them, now, not just the ones that were present at some stage in the building-up process. Thus,

$$W = \frac{1}{2} \sum_{i=1}^{n} q_i V(\mathbf{r}_i). \tag{2.42}$$

That's how much work it takes to assemble a configuration of point charges; it's also the amount of work you'd get back out if you dismantled the system. In the meantime, it represents energy stored in the configuration ("potential" energy, if you like, though for obvious reasons I prefer to avoid that word in this context).

Problem 2.31

(a) Three charges are situated at the corners of a square (side a), as shown in Fig. 2.41. How much work does it take to bring in another charge, $+q$, from far away and place it in the fourth corner?

(b) How much work does it take to assemble the whole configuration of four charges?

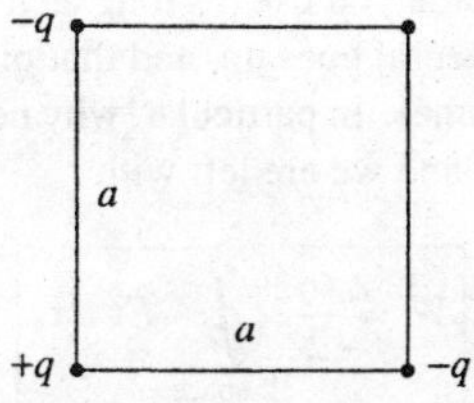

Figure 2.41

2.4.3 The Energy of a Continuous Charge Distribution

For a volume charge density ρ, Eq. 2.42 becomes

$$W = \frac{1}{2}\int \rho V \, d\tau. \tag{2.43}$$

(The corresponding integrals for line and surface charges would be $\int \lambda V \, dl$ and $\int \sigma V \, da$, respectively.) There is a lovely way to rewrite this result, in which ρ and V are eliminated in favor of $\mathbf{E}$. First use Gauss's law to express ρ in terms of $\mathbf{E}$:

$$\rho = \epsilon_0 \nabla \cdot \mathbf{E}, \quad \text{so} \quad W = \frac{\epsilon_0}{2}\int (\nabla \cdot \mathbf{E}) V \, d\tau.$$

Now use integration by parts (Eq. 1.59) to transfer the derivative from $\mathbf{E}$ to V:

$$W = \frac{\epsilon_0}{2}\left[-\int \mathbf{E} \cdot (\nabla V) \, d\tau + \oint V \mathbf{E} \cdot d\mathbf{a}\right].$$

But $\nabla V = -\mathbf{E}$, so

$$W = \frac{\epsilon_0}{2}\left(\int_{\mathcal{V}} E^2 \, d\tau + \oint_{\mathcal{S}} V \mathbf{E} \cdot d\mathbf{a}\right). \tag{2.44}$$

But what volume *is* this we're integrating over? Let's go back to the formula we started with, Eq. 2.43. From its derivation, it is clear that we should integrate over the region where the charge is located. But actually, any *larger* volume would do just as well: The "extra" territory we throw in will contribute nothing to the integral anyway, since $\rho = 0$ out there. With this in mind, let's return to Eq. 2.44. What happens *here,* as we enlarge the volume beyond the minimum necessary to trap all the charge? Well, the integral of E^2 can only increase (the integrand being positive); evidently the surface integral must decrease correspondingly to leave the sum intact. In fact, at large distances from the charge, E goes like $1/r^2$ and V like $1/r$, while the surface area grows like r^2. Roughly speaking, then, the surface integral goes down like $1/r$. Please understand that Eq. 2.44 gives you the correct energy W, *whatever* volume you use (as long as it encloses all the charge), but the contribution from the volume integral goes up, and that of the surface integral goes down, as you take larger and larger volumes. In particular, why not integrate over *all* space? Then the surface integral goes to zero, and we are left with

$$\boxed{W = \frac{\epsilon_0}{2}\int_{\text{all space}} E^2\,d\tau.} \tag{2.45}$$

Example 2.8

Find the energy of a uniformly charged spherical shell of total charge q and radius R.

Solution 1: Use Eq. 2.43, in the version appropriate to surface charges:

$$W = \frac{1}{2}\int \sigma V\,da.$$

Now, the potential at the surface of this sphere is $(1/4\pi\epsilon_0)q/R$ (a constant), so

$$W = \frac{1}{8\pi\epsilon_0}\frac{q}{R}\int \sigma\,da = \frac{1}{8\pi\epsilon_0}\frac{q^2}{R}.$$

Solution 2: Use Eq. 2.45. Inside the sphere $\mathbf{E} = 0$; outside,

$$\mathbf{E} = \frac{1}{4\pi\epsilon_0}\frac{q}{r^2}\hat{\mathbf{r}}, \quad \text{so} \quad E^2 = \frac{q^2}{(4\pi\epsilon_0)^2 r^4}.$$

Therefore,

$$\begin{aligned} W_{\text{tot}} &= \frac{\epsilon_0}{2(4\pi\epsilon_0)^2}\int_{\text{outside}}\left(\frac{q^2}{r^4}\right)(r^2\sin\theta\,dr\,d\theta\,d\phi) \\ &= \frac{1}{32\pi^2\epsilon_0}q^2 4\pi\int_R^\infty \frac{1}{r^2}\,dr = \frac{1}{8\pi\epsilon_0}\frac{q^2}{R}. \end{aligned}$$

Problem 2.32 Find the energy stored in a uniformly charged solid sphere of radius R and charge q. Do it three different ways:

(a) Use Eq. 2.43. You found the potential in Prob. 2.21.

(b) Use Eq. 2.45. Don't forget to integrate over *all space*.

(c) Use Eq. 2.44. Take a spherical volume of radius a. Notice what happens as $a \to \infty$.

Problem 2.33 Here is a fourth way of computing the energy of a uniformly charged sphere: Assemble the sphere layer by layer, each time bringing in an infinitesimal charge dq from far away and smearing it uniformly over the surface, thereby increasing the radius. How much work dW does it take to build up the radius by an amount dr? Integrate this to find the work necessary to create the entire sphere of radius R and total charge q.

2.4.4 Comments on Electrostatic Energy

(i) A perplexing "inconsistency." Equation 2.45 clearly implies that the energy of a stationary charge distribution is always *positive*. On the other hand, Eq. 2.42 (from which 2.45 was in fact derived), can be positive or negative. For instance, according to 2.42, the energy of two equal but opposite charges a distance $\imath$ apart would be $-(1/4\pi\epsilon_0)(q^2/\imath)$. What's gone wrong? Which equation is correct?

The answer is that *both* equations are correct, but they pertain to slightly different situations. Equation 2.42 does not take into account the work necessary to *make* the point charges in the first place; we *started* with point charges and simply found the work required to bring them together. This is wise policy, since Eq. 2.45 indicates that the energy of a point charge is in fact *infinite*:

$$W = \frac{\epsilon_0}{2(4\pi\epsilon_0)^2}\int\left(\frac{q^2}{r^4}\right)(r^2\sin\theta\,dr\,d\theta\,d\phi) = \frac{q^2}{8\pi\epsilon_0}\int_0^\infty \frac{1}{r^2}\,dr = \infty.$$

Equation 2.45 is more *complete*, in the sense that it tells you the *total* energy stored in a charge configuration, but Eq. 2.42 is more appropriate when you're dealing with point charges, because we prefer (for good reason!) to leave out that portion of the total energy that is attributable to the fabrication of the point charges themselves. In practice, after all, the point charges (electrons, say) are *given* to us ready-made; all *we* do is move them around. Since we did not put them together, and we cannot take them apart, it is immaterial how much work the process would involve. (Still, the infinite energy of a point charge is a recurring source of embarrassment for electromagnetic theory, afflicting the quantum version as well as the classical. We shall return to the problem in Chapter 11.)

Now, you may wonder where the inconsistency crept into an apparently water-tight derivation. The "flaw" lies between Eqs. 2.42 and 2.43: In the former, $V(\mathbf{r}_i)$ represents the potential due to all the *other* charges *but not* q_i, whereas in the latter, $V(\mathbf{r})$ is the *full* potential. For a continuous distribution there is no distinction, since the amount of charge *right at the point* $\mathbf{r}$ is vanishingly small, and its contribution to the potential is zero.

(ii) Where is the energy stored? Equations 2.43 and 2.45 offer two different ways of calculating the same thing. The first is an integral over the charge distribution; the second is an integral over the field. These can involve completely different regions. For instance, in the case of the spherical shell (Ex. 2.8) the charge is confined to the surface, whereas the electric field is present everywhere *outside* this surface. Where *is* the energy, then? Is it stored in the field, as Eq. 2.45 seems to suggest, or is it stored in the charge, as Eq. 2.43 implies? At the present level, this is simply an unanswerable question: I can tell you what the total energy is, and I can provide you with several different ways to compute it, but it is unnecessary to worry about *where* the energy is located. In the context of radiation theory (Chapter 11) it is useful (and in General Relativity it is *essential*) to regard the energy as being stored in the field, with a density

$$\frac{\epsilon_0}{2}E^2 = \text{energy per unit volume.} \tag{2.46}$$

But in electrostatics one could just as well say it is stored in the charge, with a density $\frac{1}{2}\rho V$. The difference is purely a matter of bookkeeping.

(iii) The superposition principle. Because electrostatic energy is *quadratic* in the fields, it does *not* obey a superposition principle. The energy of a compound system is *not* the sum of the energies of its parts considered separately—there are also "cross terms":

$$\begin{aligned} W_{\text{tot}} &= \frac{\epsilon_0}{2}\int E^2\,d\tau = \frac{\epsilon_0}{2}\int (\mathbf{E}_1 + \mathbf{E}_2)^2\,d\tau \\ &= \frac{\epsilon_0}{2}\int (E_1^2 + E_2^2 + 2\mathbf{E}_1\cdot\mathbf{E}_2)\,d\tau \\ &= W_1 + W_2 + \epsilon_0\int \mathbf{E}_1\cdot\mathbf{E}_2\,d\tau. \end{aligned} \tag{2.47}$$

For example, if you double the charge everywhere, you *quadruple* the total energy.

Problem 2.34 Consider two concentric spherical shells, of radii a and b. Suppose the inner one carries a charge q, and the outer one a charge $-q$ (both of them uniformly distributed over the surface). Calculate the energy of this configuration, (a) using Eq. 2.45, and (b) using Eq. 2.47 and the results of Ex. 2.8.

2.5 Conductors

2.5.1 Basic Properties

In an **insulator,** such as glass or rubber, each electron is attached to a particular atom. In a metallic **conductor,** by contrast, one or more electrons per atom are free to roam about at will through the material. (In liquid conductors such as salt water it is *ions* that do the moving.) A *perfect* conductor would be a material containing an *unlimited* supply of completely free charges. In real life there are no perfect conductors, but many substances come amazingly close. From this definition the basic electrostatic properties of ideal conductors immediately follow:

(i) $\mathbf{E} = 0$ inside a conductor. Why? Because if there *were* any field, those free charges would move, and it wouldn't be electro*statics* any more. Well... that's hardly a satisfactory explanation; maybe all it proves is that you can't have electrostatics when conductors are present. We had better examine what happens when you put a conductor into an external electric field $\mathbf{E}_0$ (Fig. 2.42). Initially, this will drive any free positive charges to the right, and negative ones to the left. (In practice it's only the negative charges—electrons—that do the moving, but when they depart the right side is left with a net positive charge—the stationary nuclei—so it doesn't really matter which charges move; the effect is the same.) When they come to the edge of the material, the charges pile up: plus on the right side, minus on the left. Now, these **induced charges** produce a field of their own, $\mathbf{E}_1$, which, as you can see from the figure, is in the *opposite direction* to $\mathbf{E}_0$. That's the crucial point, for it means that the field of the induced charges *tends to cancel off the original field.* Charge will continue to flow until this cancellation is complete, and the resultant field inside the conductor is precisely zero.[7] The whole process is practically instantaneous.

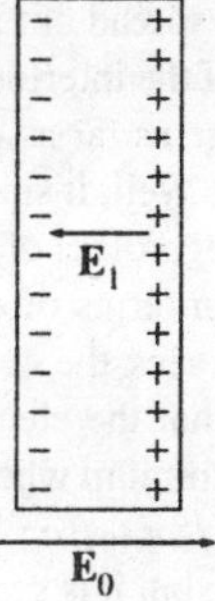

Figure 2.42

(ii) $\rho = 0$ inside a conductor. This follows from Gauss's law: $\nabla \cdot \mathbf{E} = \rho/\epsilon_0$. If $\mathbf{E} = 0$, so also is ρ. There is still charge around, but exactly as much plus charge as minus, so the *net* charge density in the interior is zero.

(iii) Any net charge resides on the surface. That's the only other place it *can* be.

(iv) A conductor is an equipotential. For if **a** and **b** are any two points within (or at the surface of) a given conductor, $V(\mathbf{b}) - V(\mathbf{a}) = -\int_{\mathbf{a}}^{\mathbf{b}} \mathbf{E} \cdot d\mathbf{l} = 0$, and hence $V(\mathbf{a}) = V(\mathbf{b})$.

---

[7] *Outside* the conductor the field is *not* zero, for here $\mathbf{E}_0$ and $\mathbf{E}_1$ do not cancel.

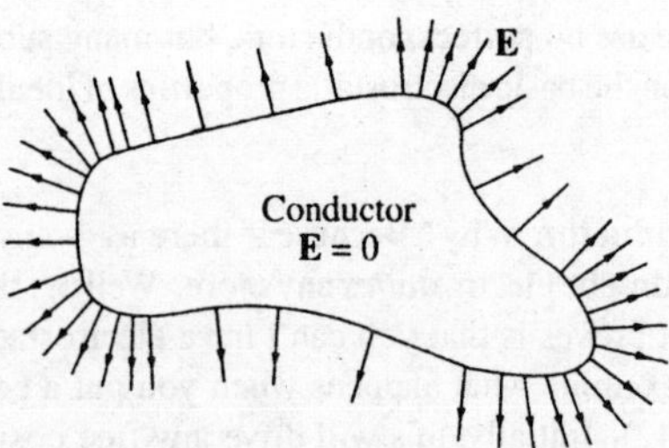

Figure 2.43

(v) E is perpendicular to the surface, just outside a conductor. Otherwise, as in **(i)**, charge will immediately flow around the surface until it kills off the tangential component (Fig. 2.43). (*Perpendicular* to the surface, charge cannot flow, of course, since it is confined to the conducting object.)

I think it is strange that the charge on a conductor flows to the surface. Because of their mutual repulsion, the charges naturally spread out as much as possible, but for *all* of them to go to the surface seems like a waste of the interior space. Surely we could do better, from the point of view of making each charge as far as possible from its neighbors, to sprinkle *some* of them throughout the volume... Well, it simply is not so. You do best to put *all* the charge on the surface, and this is true regardless of the size or shape of the conductor.[8]

The problem can also be phrased in terms of energy. Like any other free dynamical system, the charge on a conductor will seek the configuration that minimizes its potential energy. What property **(iii)** asserts is that the electrostatic energy of a solid object (with specified shape and total charge) is a minimum when that charge is spread over the surface. For instance, the energy of a sphere is $(1/8\pi\epsilon_0)(q^2/R)$ if the charge is uniformly distributed over the surface, as we found in Ex. 2.8, but it is greater, $(3/20\pi\epsilon_0)(q^2/R)$, if the charge is uniformly distributed throughout the volume (Prob. 2.32).

2.5.2 Induced Charges

If you hold a charge $+q$ near an uncharged conductor (Fig. 2.44), the two will attract one another. The reason for this is that q will pull minus charges over to the near side and repel plus charges to the far side. (Another way to think of it is that the charge moves around in such a way as to cancel off the field of q for points inside the conductor, where the total field must be zero.) Since the negative induced charge is closer to q, there is a net force of attraction. (In Chapter 3 we shall calculate this force explicitly, for the case of a spherical conductor.)

---

[8] By the way, the one- and two-dimensional analogs are quite different: The charge on a conducting *disk* does *not* all go to the perimeter (R. Friedberg, *Am. J. of Phys.* **61**, 1084 (1993)), nor does the charge on a conducting needle go to the ends (D. J. Griffiths and Y. Li, *Am. J. of Phys.* **64**, 706 (1996)). See Prob. 2.52.

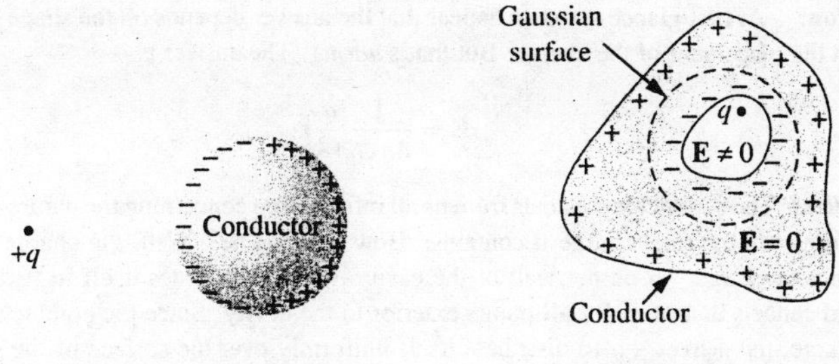

Figure 2.44 Figure 2.45

By the way, when I speak of the field, charge, or potential "inside" a conductor, I mean in the "meat" of the conductor; if there is some *cavity* in the conductor, and within that cavity there is some charge, then the field *in the cavity* will *not* be zero. But in a remarkable way the cavity and its contents are electrically isolated from the outside world by the surrounding conductor (Fig. 2.45). No external fields penetrate the conductor; they are canceled at the outer surface by the induced charge there. Similarly, the field due to charges within the cavity is killed off, for all exterior points, by the induced charge on the inner surface. (However, the compensating charge left over on the *outer* surface of the conductor effectively "communicates" the presence of q to the outside world, as we shall see in Ex. 2.9.) Incidentally, the total charge induced on the cavity wall is equal and opposite to the charge inside, for if we surround the cavity with a Gaussian surface, all points of which are in the conductor (Fig. 2.45), $\oint \mathbf{E} \cdot d\mathbf{a} = 0$, and hence (by Gauss's law) the net enclosed charge must be zero. But $Q_{\text{enc}} = q + q_{\text{induced}}$, so $q_{\text{induced}} = -q$.

Example 2.9

An uncharged spherical conductor centered at the origin has a cavity of some weird shape carved out of it (Fig. 2.46). Somewhere within the cavity is a charge q. *Question:* What is the field outside the sphere?

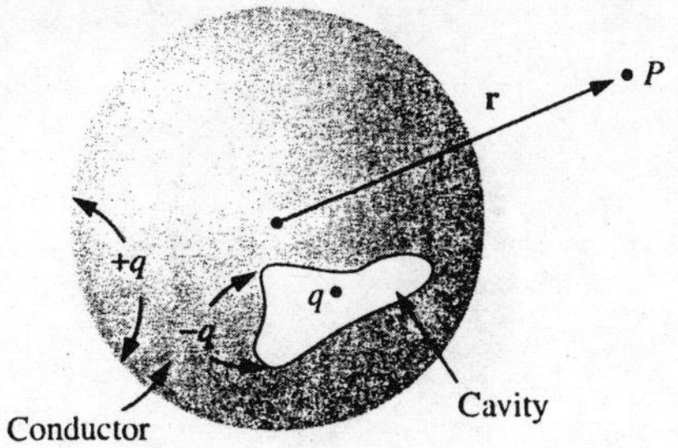

Figure 2.46

Solution: At first glance it would appear that the answer depends on the shape of the cavity and on the placement of the charge. But that's wrong: The answer is

$$\mathbf{E} = \frac{1}{4\pi\epsilon_0}\frac{q}{r^2}\hat{\mathbf{r}}$$

regardless. The conductor conceals from us all information concerning the nature of the cavity, revealing only the total charge it contains. How can this be? Well, the charge $+q$ induces an opposite charge $-q$ on the wall of the cavity, which distributes itself in such a way that its field cancels that of q, for all points exterior to the cavity. Since the conductor carries no net charge, this leaves $+q$ to distribute itself uniformly over the surface of the sphere. (It's *uniform* because the asymmetrical influence of the point charge $+q$ is negated by that of the induced charge $-q$ on the inner surface.) For points outside the sphere, then, the only thing that survives is the field of the leftover $+q$, uniformly distributed over the outer surface.

It may occur to you that in one respect this argument is open to challenge: There are actually *three* fields at work here, $\mathbf{E}_q$, $\mathbf{E}_{\text{induced}}$, and $\mathbf{E}_{\text{leftover}}$. All we know for certain is that the sum of the three is zero inside the conductor, yet I claimed that the first two *alone* cancel, while the third is separately zero there. Moreover, even if the first two cancel within the conductor, who is to say they still cancel for points outside? They do not, after all, cancel for points *inside* the cavity. I cannot give you a completely satisfactory answer at the moment, but this much at least is true: There *exists* a way of distributing $-q$ over the inner surface so as to cancel the field of q at all exterior points. For that same cavity could have been carved out of a *huge* spherical conductor with a radius of 27 miles or light years or whatever. In that case the leftover $+q$ on the outer surface is simply too far away to produce a significant field, and the other two fields would *have* to accomplish the cancellation by themselves. So we know they *can* do it ... but are we sure they *choose* to? Perhaps for small spheres nature prefers some complicated three-way cancellation. Nope: As we'll see in the uniqueness theorems of Chapter 3, electrostatics is very stingy with its options; there is always precisely one way—no more—of distributing the charge on a conductor so as to make the field inside zero. Having found a *possible* way, we are guaranteed that no alternative exists even in principle.

If a cavity surrounded by conducting material is itself empty of charge, then the field within the cavity is zero. For any field line would have to begin and end on the cavity wall, going from a plus charge to a minus charge (Fig. 2.47). Letting that field line be part of a closed loop, the rest of which is entirely inside the conductor (where $\mathbf{E} = 0$), the integral

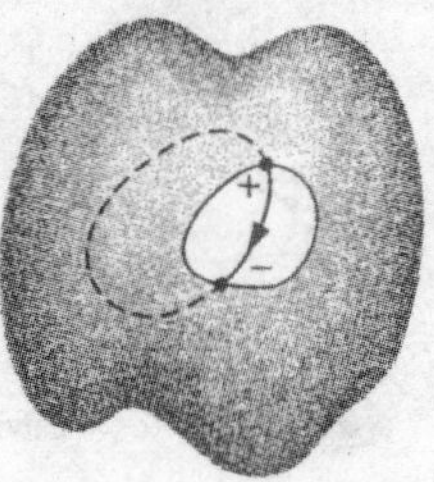

Figure 2.47

$\oint \mathbf{E} \cdot d\mathbf{l}$ is distinctly *positive,* in violation of Eq. 2.19. It follows that $\mathbf{E} = 0$ within an *empty* cavity, and there is in fact *no* charge on the surface of the cavity. (This is why you are relatively safe inside a metal car during a thunderstorm—you may get *cooked,* if lightning strikes, but you will not be *electrocuted.* The same principle applies to the placement of sensitive apparatus inside a grounded **Faraday cage,** to shield out stray electric fields. In practice, the enclosure doesn't even have to be solid conductor—chicken wire will often suffice.)

Problem 2.35 A metal sphere of radius R, carrying charge q, is surrounded by a thick concentric metal shell (inner radius a, outer radius b, as in Fig. 2.48). The shell carries no net charge.

(a) Find the surface charge density σ at R, at a, and at b.

(b) Find the potential at the center, using infinity as the reference point.

(c) Now the outer surface is touched to a grounding wire, which lowers its potential to zero (same as at infinity). How do your answers to (a) and (b) change?

Problem 2.36 Two spherical cavities, of radii a and b, are hollowed out from the interior of a (neutral) conducting sphere of radius R (Fig. 2.49). At the center of each cavity a point charge is placed—call these charges q_a and q_b.

(a) Find the surface charges σ_a, σ_b, and σ_R.

(b) What is the field outside the conductor?

(c) What is the field within each cavity?

(d) What is the force on q_a and q_b?

(e) Which of these answers would change if a third charge, q_c, were brought near the conductor?

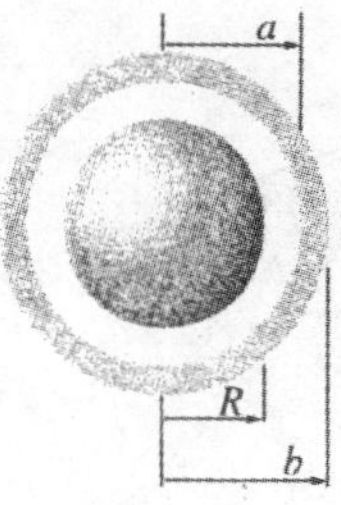

Figure 2.48

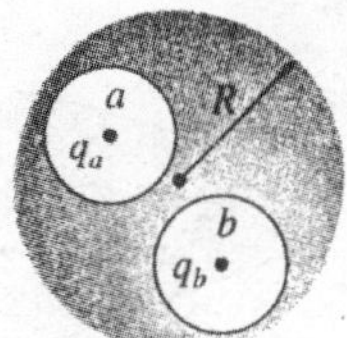

Figure 2.49

2.5.3 Surface Charge and the Force on a Conductor

Because the field inside a conductor is zero, boundary condition 2.33 requires that the field immediately *outside* is

$$\mathbf{E} = \frac{\sigma}{\epsilon_0}\hat{\mathbf{n}}, \tag{2.48}$$

consistent with our earlier conclusion that the field is normal to the surface. In terms of potential, Eq. 2.36 yields

$$\sigma = -\epsilon_0 \frac{\partial V}{\partial n}. \tag{2.49}$$

These equations enable you to calculate the surface charge on a conductor, if you can determine **E** or V; we shall use them frequently in the next chapter.

In the presence of an electric field, a surface charge will, naturally, experience a force; the force per unit area, **f**, is $\sigma\mathbf{E}$. But there's a problem here, for the electric field is *discontinuous* at a surface charge, so which value are we supposed to use: $\mathbf{E}_{\text{above}}$, $\mathbf{E}_{\text{below}}$, or something in between? The answer is that we should use the *average* of the two:

$$\mathbf{f} = \sigma\mathbf{E}_{\text{average}} = \frac{1}{2}\sigma(\mathbf{E}_{\text{above}} + \mathbf{E}_{\text{below}}). \tag{2.50}$$

Why the average? The reason is very simple, though the telling makes it sound complicated: Let's focus our attention on a small patch of surface surrounding the point in question (Fig. 2.50). Make it tiny enough so it is essentially flat and the surface charge on it is essentially constant. The *total* field consists of two parts—that attributable to the patch itself, and that due to everything else (other regions of the surface, as well as any external sources that may be present):

$$\mathbf{E} = \mathbf{E}_{\text{patch}} + \mathbf{E}_{\text{other}}.$$

Now, the patch cannot exert a force on itself, any more than you can lift yourself by standing in a basket and pulling up on the handles. The force on the patch, then, is due exclusively to $\mathbf{E}_{\text{other}}$, and *this* suffers *no* discontinuity (if we removed the patch, the field in the "hole" would be perfectly smooth). The discontinuity is due entirely to the charge on the patch,

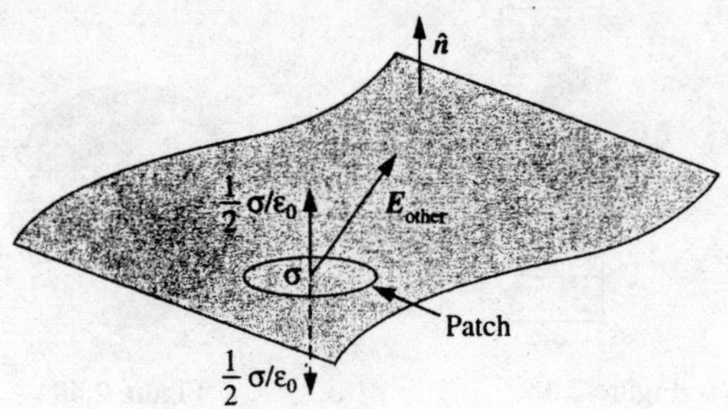

Figure 2.50

which puts out a field $(\sigma/2\epsilon_0)$ on either side, pointing away from the surface (Fig. 2.50). Thus,

$$\mathbf{E}_{\text{above}} = \mathbf{E}_{\text{other}} + \frac{\sigma}{2\epsilon_0}\hat{\mathbf{n}},$$

$$\mathbf{E}_{\text{below}} = \mathbf{E}_{\text{other}} - \frac{\sigma}{2\epsilon_0}\hat{\mathbf{n}},$$

and hence

$$\mathbf{E}_{\text{other}} = \frac{1}{2}(\mathbf{E}_{\text{above}} + \mathbf{E}_{\text{below}}) = \mathbf{E}_{\text{average}}.$$

Averaging is really just a device for removing the contribution of the patch itself.

That argument applies to *any* surface charge; in the particular case of a conductor, the field is zero inside and $(\sigma/\epsilon_0)\hat{\mathbf{n}}$ outside (Eq. 2.48), so the average is $(\sigma/2\epsilon_0)\hat{\mathbf{n}}$, and the force per unit area is

$$\mathbf{f} = \frac{1}{2\epsilon_0}\sigma^2\hat{\mathbf{n}}. \tag{2.51}$$

This amounts to an outward **electrostatic pressure** on the surface, tending to draw the conductor into the field, regardless of the sign of σ. Expressing the pressure in terms of the field just outside the surface,

$$P = \frac{\epsilon_0}{2}E^2. \tag{2.52}$$

Problem 2.37 Two large metal plates (each of area A) are held a distance d apart. Suppose we put a charge Q on each plate; what is the electrostatic pressure on the plates?

Problem 2.38 A metal sphere of radius R carries a total charge Q. What is the force of repulsion between the "northern" hemisphere and the "southern" hemisphere?

2.5.4 Capacitors

Suppose we have *two* conductors, and we put charge $+Q$ on one and $-Q$ on the other (Fig. 2.51). Since V is constant over a conductor, we can speak unambiguously of the potential difference between them:

$$V = V_+ - V_- = -\int_{(-)}^{(+)} \mathbf{E}\cdot d\mathbf{l}.$$

We don't know how the charge distributes itself over the two conductors, and calculating the field would be a mess, if their shapes are complicated, but this much we *do* know: $\mathbf{E}$ is *proportional* to Q. For $\mathbf{E}$ is given by Coulomb's law:

$$\mathbf{E} = \frac{1}{4\pi\epsilon_0}\int \frac{\rho}{\mathscr{r}^2}\hat{\mathscr{r}}\,d\tau,$$

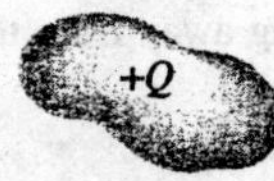

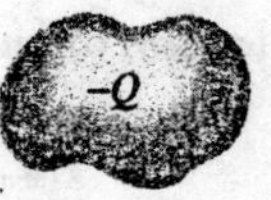

Figure 2.51

so if you double ρ, you double $\mathbf{E}$. (Wait a minute! How do we know that doubling Q (and also $-Q$) simply doubles ρ? Maybe the charge *moves around* into a completely different configuration, quadrupling ρ in some places and halving it in others, just so the *total* charge on each conductor is doubled. The *fact* is that this concern is unwarranted—doubling Q *does* double ρ everywhere; it *doesn't* shift the charge around. The proof of this will come in Chapter 3; for now you'll just have to believe me.)

Since $\mathbf{E}$ is proportional to Q, so also is V. The constant of proportionality is called the **capacitance** of the arrangement:

$$C \equiv \frac{Q}{V}. \tag{2.53}$$

Capacitance is a purely geometrical quantity, determined by the sizes, shapes, and separation of the two conductors. In SI units, C is measured in **farads** (F); a farad is a coulomb-per-volt. Actually, this turns out to be inconveniently large;[9] more practical units are the microfarad (10^{-6} F) and the picofarad (10^{-12} F).

Notice that V is, by definition, the potential of the *positive* conductor less that of the negative one; likewise, Q is the charge of the *positive* conductor. Accordingly, capacitance is an intrinsically positive quantity. (By the way, you will occasionally hear someone speak of the capacitance of a *single* conductor. In this case the "second conductor," with the negative charge, is an imaginary spherical shell of infinite radius surrounding the one conductor. It contributes nothing to the field, so the capacitance is given by Eq. 2.53, where V is the potential with infinity as the reference point.)

Example 2.10

Find the capacitance of a "parallel-plate capacitor" consisting of two metal surfaces of area A held a distance d apart (Fig. 2.52).

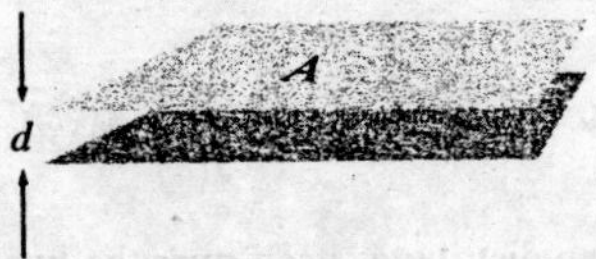

Figure 2.52

[9] In the second edition I claimed you would need a forklift to carry a 1 F capacitor. This is no longer the case—you can now buy a 1 F capacitor that fits comfortably in a soup spoon.

Solution: If we put $+Q$ on the top and $-Q$ on the bottom, they will spread out uniformly over the two surfaces, provided the area is reasonably large and the separation distance small.[10] The surface charge density, then, is $\sigma = Q/A$ on the top plate, and so the field, according to Ex. 2.5, is $(1/\epsilon_0)Q/A$. The potential difference between the plates is therefore

$$V = \frac{Q}{A\epsilon_0}d,$$

and hence

$$C = \frac{A\epsilon_0}{d}. \tag{2.54}$$

If, for instance, the plates are square with sides 1 cm long, and they are held 1 mm apart, then the capacitance is 9×10^{-13} F.

Example 2.11

Find the capacitance of two concentric spherical metal shells, with radii a and b.

Solution: Place charge $+Q$ on the inner sphere, and $-Q$ on the outer one. The field between the spheres is

$$\mathbf{E} = \frac{1}{4\pi\epsilon_0}\frac{Q}{r^2}\hat{\mathbf{r}},$$

so the potential difference between them is

$$V = -\int_b^a \mathbf{E}\cdot d\mathbf{l} = -\frac{Q}{4\pi\epsilon_0}\int_b^a \frac{1}{r^2}\,dr = \frac{Q}{4\pi\epsilon_0}\left(\frac{1}{a} - \frac{1}{b}\right).$$

As promised, V is proportional to Q; the capacitance is

$$C = \frac{Q}{V} = 4\pi\epsilon_0\frac{ab}{(b-a)}.$$

To "charge up" a capacitor, you have to remove electrons from the positive plate and carry them to the negative plate. In doing so you fight against the electric field, which is pulling them back toward the positive conductor and pushing them away from the negative one. How much work does it take, then, to charge the capacitor up to a final amount Q? Suppose that at some intermediate stage in the process the charge on the positive plate is q, so that the potential difference is q/C. According to Eq. 2.38, the work you must do to transport the next piece of charge, dq, is

$$dW = \left(\frac{q}{C}\right)dq.$$

[10] The *exact* solution is not easy—even for the simpler case of circular plates. See G. T. Carlson and B. L. Illman, *Am. J. Phys.* **62**, 1099 (1994).

The total work necessary, then, to go from $q = 0$ to $q = Q$, is

$$W = \int_0^Q \left(\frac{q}{C}\right) dq = \frac{1}{2}\frac{Q^2}{C},$$

or, since $Q = CV$,

$$W = \frac{1}{2}CV^2, \tag{2.55}$$

where V is the final potential of the capacitor.

Problem 2.39 Find the capacitance per unit length of two coaxial metal cylindrical tubes, of radii a and b (Fig. 2.53).

Figure 2.53

Problem 2.40 Suppose the plates of a parallel-plate capacitor move closer together by an infinitesimal distance ϵ, as a result of their mutual attraction.

(a) Use Eq. 2.52 to express the amount of work done by electrostatic forces, in terms of the field E, and the area of the plates, A.

(b) Use Eq. 2.46 to express the energy lost by the field in this process.

(This problem is supposed to be easy, but it contains the embryo of an alternative derivation of Eq. 2.52, using conservation of energy.)

More Problems on Chapter 2

Problem 2.41 Find the electric field at a height z above the center of a square sheet (side a) carrying a uniform surface charge σ. Check your result for the limiting cases $a \to \infty$ and $z \gg a$.

[*Answer:* $(\sigma/2\epsilon_0)\{(4/\pi)\tan^{-1}\sqrt{1+(a^2/2z^2)} - 1\}$]

Problem 2.42 If the electric field in some region is given (in spherical coordinates) by the expression

$$\mathbf{E}(\mathbf{r}) = \frac{A\,\hat{\mathbf{r}} + B\sin\theta\cos\phi\,\hat{\boldsymbol{\phi}}}{r},$$

where A and B are constants, what is the charge density? [*Answer:* $\epsilon_0(A - B\sin\phi)/r^2$]

Problem 2.43 Find the net force that the southern hemisphere of a uniformly charged sphere exerts on the northern hemisphere. Express your answer in terms of the radius R and the total charge Q. [*Answer:* $(1/4\pi\epsilon_0)(3Q^2/16R^2)$]

Problem 2.44 An inverted hemispherical bowl of radius R carries a uniform surface charge density σ. Find the potential difference between the "north pole" and the center. [*Answer:* $(R\sigma/2\epsilon_0)(\sqrt{2}-1)$]

Problem 2.45 A sphere of radius R carries a charge density $\rho(r) = kr$ (where k is a constant). Find the energy of the configuration. Check your answer by calculating it in at least two different ways. [*Answer:* $\pi k^2 R^7/7\epsilon_0$]

Problem 2.46 The electric potential of some configuration is given by the expression

$$V(\mathbf{r}) = A\frac{e^{-\lambda r}}{r},$$

where A and λ are constants. Find the electric field $\mathbf{E}(\mathbf{r})$, the charge density $\rho(r)$, and the total charge Q. [*Answer:* $\rho = \epsilon_0 A(4\pi\delta^3(\mathbf{r}) - \lambda^2 e^{-\lambda r}/r)$]

! **Problem 2.47** Two infinitely long wires running parallel to the x axis carry uniform charge densities $+\lambda$ and $-\lambda$ (Fig. 2.54).

(a) Find the potential at any point (x, y, z), using the origin as your reference.

(b) Show that the equipotential surfaces are circular cylinders, and locate the axis and radius of the cylinder corresponding to a given potential V_0.

! **Problem 2.48** In a vacuum diode, electrons are "boiled" off a hot **cathode**, at potential zero, and accelerated across a gap to the **anode**, which is held at positive potential V_0. The cloud of moving electrons within the gap (called **space charge**) quickly builds up to the point where it reduces the field at the surface of the cathode to zero. From then on a steady current I flows between the plates.

Suppose the plates are large relative to the separation ($A \gg d^2$ in Fig. 2.55), so that edge effects can be neglected. Then V, ρ, and v (the speed of the electrons) are all functions of x alone.

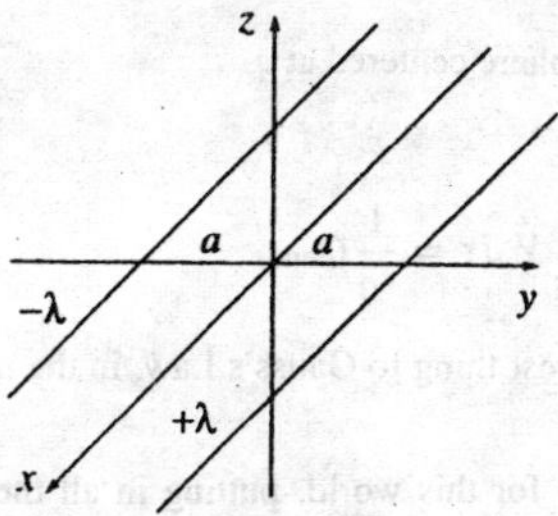

Figure 2.54

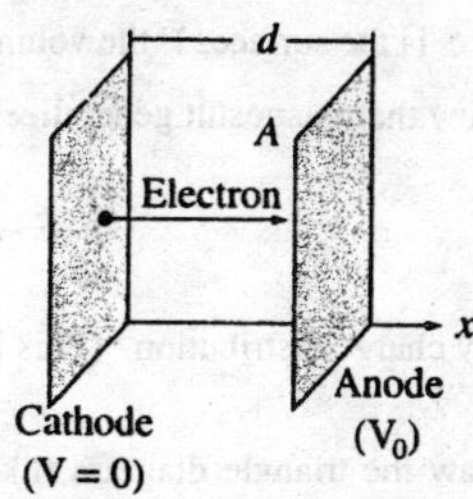

Figure 2.55

(a) Write Poisson's equation for the region between the plates.

(b) Assuming the electrons start from rest at the cathode, what is their speed at point x, where the potential is $V(x)$?

(c) In the steady state, I is independent of x. What, then, is the relation between ρ and v?

(d) Use these three results to obtain a differential equation for V, by eliminating ρ and v.

(e) Solve this equation for V as a function of x, V_0, and d. Plot $V(x)$, and compare it to the potential *without* space-charge. Also, find ρ and v as functions of x.

(f) Show that

$$I = K V_0^{3/2}, \tag{2.56}$$

and find the constant K. (Equation 2.56 is called the **Child-Langmuir law**. It holds for other geometries as well, whenever space-charge limits the current. Notice that the space-charge limited diode is *nonlinear*—it does not obey Ohm's law.)

! **Problem 2.49** Imagine that new and extraordinarily precise measurements have revealed an error in Coulomb's law. The *actual* force of interaction between two point charges is found to be

$$\mathbf{F} = \frac{1}{4\pi\epsilon_0}\frac{q_1 q_2}{\mathscr{r}^2}\left(1 + \frac{\mathscr{r}}{\lambda}\right)e^{-\mathscr{r}/\lambda}\hat{\boldsymbol{\mathscr{r}}},$$

where λ is a new constant of nature (it has dimensions of length, obviously, and is a huge number—say half the radius of the known universe—so that the correction is small, which is why no one ever noticed the discrepancy before). You are charged with the task of reformulating electrostatics to accommodate the new discovery. Assume the principle of superposition still holds.

(a) What is the electric field of a charge distribution ρ (replacing Eq. 2.8)?

(b) Does this electric field admit a scalar potential? Explain briefly how you reached your conclusion. (No formal proof necessary—just a persuasive argument.)

(c) Find the potential of a point charge q—the analog to Eq. 2.26. (If your answer to (b) was "no," better go back and change it!) Use ∞ as your reference point.

(d) For a point charge q at the origin, show that

$$\oint_{\mathcal{S}} \mathbf{E}\cdot d\mathbf{a} + \frac{1}{\lambda^2}\int_{\mathcal{V}} V\,d\tau = \frac{1}{\epsilon_0}q,$$

where $\mathcal{S}$ is the surface, $\mathcal{V}$ the volume, of any sphere centered at q.

(e) Show that this result generalizes:

$$\oint_{\mathcal{S}} \mathbf{E}\cdot d\mathbf{a} + \frac{1}{\lambda^2}\int_{\mathcal{V}} V\,d\tau = \frac{1}{\epsilon_0}Q_{\text{enc}},$$

for *any* charge distribution. (This is the next best thing to Gauss's Law, in the new "electrostatics.")

(f) Draw the triangle diagram (like Fig. 2.35) for this world, putting in all the appropriate formulas. (Think of Poisson's equation as the formula for ρ in terms of V, and Gauss's law (differential form) as an equation for ρ in terms of $\mathbf{E}$.)

Problem 2.50 Suppose an electric field $\mathbf{E}(x, y, z)$ has the form

$$E_x = ax, \qquad E_y = 0, \qquad E_z = 0$$

where a is a constant. What is the charge density? How do you account for the fact that the field points in a particular direction, when the charge density is uniform? [This is a more subtle problem than it looks, and worthy of careful thought.]

Problem 2.51 All of electrostatics follows from the $1/r^2$ character of Coulomb's law, together with the principle of superposition. An analogous theory can therefore be constructed for Newton's law of universal gravitation. What is the gravitational energy of a sphere, of mass M and radius R, assuming the density is uniform? Use your result to estimate the gravitational energy of the sun (look up the relevant numbers). The sun radiates at a rate of 3.86×10^{26} W; if all this came from stored gravitational energy, how long would the sun last? [The sun is in fact much older than that, so evidently this is *not* the source of its power.]

Problem 2.52 We know that the charge on a conductor goes to the surface, but just how it distributes itself there is not easy to determine. One famous example in which the surface charge density can be calculated explicitly is the ellipsoid:

$$\frac{x^2}{a^2} + \frac{y^2}{b^2} + \frac{z^2}{c^2} = 1.$$

In this case[11]

$$\sigma = \frac{Q}{4\pi abc}\left(\frac{x^2}{a^4} + \frac{y^2}{b^4} + \frac{z^2}{c^4}\right)^{-1/2}, \tag{2.57}$$

where Q is the total charge. By choosing appropriate values for a, b, and c, obtain (from Eq. 2.57): (a) the net (both sides) surface charge density $\sigma(r)$ on a circular disk of radius R; (b) the net surface charge density $\sigma(x)$ on an infinite conducting "ribbon" in the $x\,y$ plane, which straddles the y axis from $x = -a$ to $x = a$ (let Λ be the total charge per unit length of ribbon); (c) the net charge per unit length $\lambda(x)$ on a conducting "needle", running from $x = -a$ to $x = a$. In each case, sketch the graph of your result.

---

[11] For the derivation (which is a real *tour de force*) see W. R. Smythe, *Static and Dynamic Electricity*, 3rd ed. (New York: Hemisphere, 1989), Sect. 5.02.

Chapter 3

Special Techniques

3.1 Laplace's Equation

3.1.1 Introduction

The primary task of electrostatics is to find the electric field of a given stationary charge distribution. In principle, this purpose is accomplished by Coulomb's law, in the form of Eq. 2.8:

$$\mathbf{E}(\mathbf{r}) = \frac{1}{4\pi\epsilon_0}\int \frac{\hat{\mathscr{r}}}{\mathscr{r}^2}\rho(\mathbf{r}')\,d\tau'. \tag{3.1}$$

Unfortunately, integrals of this type can be difficult to calculate for any but the simplest charge configurations. Occasionally we can get around this by exploiting symmetry and using Gauss's law, but ordinarily the best strategy is first to calculate the *potential,* V, which is given by the somewhat more tractable Eq. 2.29:

$$V(\mathbf{r}) = \frac{1}{4\pi\epsilon_0}\int \frac{1}{\mathscr{r}}\rho(\mathbf{r}')\,d\tau'. \tag{3.2}$$

Still, even *this* integral is often too tough to handle analytically. Moreover, in problems involving conductors ρ itself may not be known in advance: since charge is free to move around, the only thing we control directly is the *total* charge (or perhaps the potential) of each conductor.

In such cases it is fruitful to recast the problem in differential form, using Poisson's equation (2.24),

$$\nabla^2 V = -\frac{1}{\epsilon_0}\rho, \tag{3.3}$$

which, together with appropriate boundary conditions, is equivalent to Eq. 3.2. Very often, in fact, we are interested in finding the potential in a region where $\rho = 0$. (If $\rho = 0$ *everywhere,* of course, then $V = 0$, and there is nothing further to say—that's not what I mean. There may be plenty of charge *elsewhere*, but we're confining our attention to places where there is no charge.) In this case Poisson's equation reduces to Laplace's equation:

$$\nabla^2 V = 0, \tag{3.4}$$

or, written out in Cartesian coordinates,

$$\frac{\partial^2 V}{\partial x^2} + \frac{\partial^2 V}{\partial y^2} + \frac{\partial^2 V}{\partial z^2} = 0. \tag{3.5}$$

This formula is so fundamental to the subject that one might almost say electrostatics *is* the study of Laplace's equation. At the same time, it is a ubiquitous equation, appearing in such diverse branches of physics as gravitation and magnetism, the theory of heat, and the study of soap bubbles. In mathematics it plays a major role in analytic function theory. To get a feel for Laplace's equation and its solutions (which are called **harmonic functions**), we shall begin with the one- and two-dimensional versions, which are easier to picture and illustrate all the essential properties of the three-dimensional case (though the one-dimensional example lacks the richness of the other two).

3.1.2 Laplace's Equation in One Dimension

Suppose V depends on only one variable, x. Then Laplace's equation becomes

$$\frac{d^2 V}{dx^2} = 0.$$

The general solution is

$$V(x) = mx + b, \tag{3.6}$$

the equation for a straight line. It contains two undetermined constants (m and b), as is appropriate for a second-order (ordinary) differential equation. They are fixed, in any particular case, by the boundary conditions of that problem. For instance, it might be specified that $V = 4$ at $x = 1$, and $V = 0$ at $x = 5$. In that case $m = -1$ and $b = 5$, so $V = -x + 5$ (see Fig. 3.1).

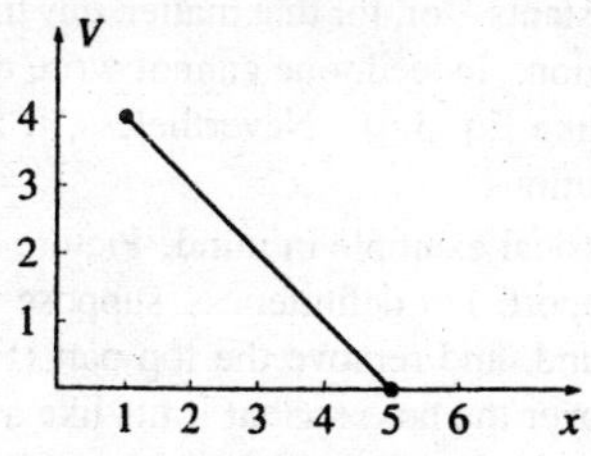

Figure 3.1

I want to call your attention to two features of this result; they may seem silly and obvious in one dimension, where I can write down the general solution explicitly, but the analogs in two and three dimensions are powerful and by no means obvious:

1. $V(x)$ is the *average* of $V(x+a)$ and $V(x-a)$, for any a:

$$V(x) = \frac{1}{2}[V(x+a) + V(x-a)].$$

   Laplace's equation is a kind of averaging instruction; it tells you to assign to the point x the average of the values to the left and to the right of x. Solutions to Laplace's equation are, in this sense, *as boring as they could possibly be,* and yet fit the end points properly.

2. Laplace's equation tolerates *no local maxima or minima;* extreme values of V must occur at the end points. Actually, this is a consequence of (1), for if there *were* a local maximum, V at that point would be greater than on either side, and therefore could not be the average. (Ordinarily, you expect the second derivative to be negative at a maximum and positive at a minimum. Since Laplace's equation requires, on the contrary, that the second derivative be zero, it seems reasonable that solutions should exhibit no extrema. However, this is not a *proof,* since there exist functions that have maxima and minima at points where the second derivative vanishes: x^4, for example, has such a minimum at the point $x = 0$.)

3.1.3 Laplace's Equation in Two Dimensions

If V depends on two variables, Laplace's equation becomes

$$\frac{\partial^2 V}{\partial x^2} + \frac{\partial^2 V}{\partial y^2} = 0.$$

This is no longer an *ordinary* differential equation (that is, one involving ordinary derivatives only); it is a *partial* differential equation. As a consequence, some of the simple rules you may be familiar with do not apply. For instance, the general solution to this equation doesn't contain just two arbitrary constants—or, for that matter, *any* finite number—despite the fact that it's a second-order equation. Indeed, one cannot write down a "general solution" (at least, not in a closed form like Eq. 3.6). Nevertheless, it is possible to deduce certain properties common to all solutions.

It may help to have a physical example in mind. Picture a thin rubber sheet (or a soap film) stretched over some support. For definiteness, suppose you take a cardboard box, cut a wavy line all the way around, and remove the top part (Fig. 3.2). Now glue a tightly stretched rubber membrane over the box, so that it fits like a drum head (it won't be a *flat* drumhead, of course, unless you chose to cut the edges off straight). Now, if you lay out coordinates (x, y) on the bottom of the box, the height $V(x, y)$ of the sheet above the point

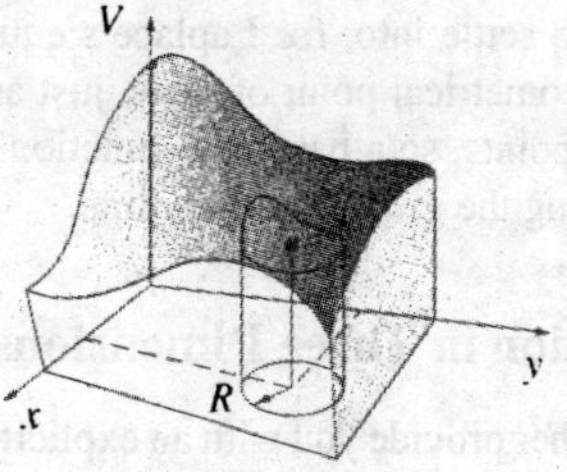

Figure 3.2

(x, y) will satisfy Laplace's equation.[1] (The one-dimensional analog would be a rubber band stretched between two points. Of course, it would form a straight line.)

Harmonic functions in two dimensions have the same properties we noted in one dimension:

1. The value of V at a point (x, y) is the average of those *around* the point. More precisely, if you draw a circle of any radius R about the point (x, y), the average value of V on the circle is equal to the value at the center:

$$V(x, y) = \frac{1}{2\pi R} \oint_{\text{circle}} V\, dl.$$

(This, incidentally, suggests the **method of relaxation** on which computer solutions to Laplace's equation are based: Starting with specified values for V at the boundary, and reasonable guesses for V on a grid of interior points, the first pass reassigns to each point the average of its nearest neighbors. The second pass repeats the process, using the corrected values, and so on. After a few iterations, the numbers begin to settle down, so that subsequent passes produce negligible changes, and a numerical solution to Laplace's equation, with the given boundary values, has been achieved.)[2]

2. V has no local maxima or minima; all extrema occur at the boundaries. (As before, this follows from (1).) Again, Laplace's equation picks the most featureless function possible, consistent with the boundary conditions: no hills, no valleys, just the smoothest surface available. For instance, if you put a ping-pong ball on the stretched rubber sheet of Fig. 3.2, it will roll over to one side and fall off—it will not find a

[1] Actually, the equation satisfied by a rubber sheet is

$$\frac{\partial}{\partial x}\left(g\frac{\partial V}{\partial x}\right) + \frac{\partial}{\partial y}\left(g\frac{\partial V}{\partial y}\right) = 0, \quad \text{where} \quad g = \left[1 + \left(\frac{\partial V}{\partial x}\right)^2 + \left(\frac{\partial V}{\partial y}\right)^2\right]^{-1/2};$$

it reduces (approximately) to Laplace's equation as long as the surface does not deviate too radically from a plane.

[2] See, for example, E. M. Purcell, *Electricity and Magnetism*, 2nd ed., problem 3.30 (p. 119) (New York: McGraw-Hill, 1985).

"pocket" somewhere to settle into, for Laplace's equation allows no such dents in the surface. From a geometrical point of view, just as a straight line is the shortest distance between two points, so a harmonic function in two dimensions minimizes the surface area spanning the given boundary line.

3.1.4 Laplace's Equation in Three Dimensions

In three dimensions I can neither provide you with an explicit solution (as in one dimension) nor offer a suggestive physical example to guide your intuition (as I did in two dimensions). Nevertheless, the same two properties remain true, and this time I will sketch a proof.

1. The value of V at point $\mathbf{r}$ is the average value of V over a spherical surface of radius R centered at $\mathbf{r}$:
$$V(\mathbf{r}) = \frac{1}{4\pi R^2} \oint_{\text{sphere}} V\, da.$$

2. As a consequence, V can have no local maxima or minima; the extreme values of V must occur at the boundaries. (For if V had a local maximum at $\mathbf{r}$, then by the very nature of maximum I could draw a sphere around $\mathbf{r}$ over which all values of V—and *a fortiori* the average—would be less than at $\mathbf{r}$.)

Proof: Let's begin by calculating the average potential over a spherical surface of radius R due to a *single* point charge q located outside the sphere. We may as well center the sphere at the origin and choose coordinates so that q lies on the z-axis (Fig. 3.3). The potential at a point on the surface is

$$V = \frac{1}{4\pi\epsilon_0}\frac{q}{\mathscr{r}},$$

where

$$\mathscr{r}^2 = z^2 + R^2 - 2zR\cos\theta,$$

so

$$\begin{aligned} V_{\text{ave}} &= \frac{1}{4\pi R^2}\frac{q}{4\pi\epsilon_0}\int [z^2 + R^2 - 2zR\cos\theta]^{-1/2} R^2 \sin\theta\, d\theta\, d\phi \\ &= \frac{q}{4\pi\epsilon_0}\frac{1}{2zR}\sqrt{z^2 + R^2 - 2zR\cos\theta}\,\Big|_0^{\pi} \\ &= \frac{q}{4\pi\epsilon_0}\frac{1}{2zR}[(z+R) - (z-R)] = \frac{1}{4\pi\epsilon_0}\frac{q}{z}. \end{aligned}$$

But this is precisely the potential due to q at the *center* of the sphere! By the superposition principle, the same goes for any *collection* of charges outside the sphere: their average potential over the sphere is equal to the net potential they produce at the center. qed

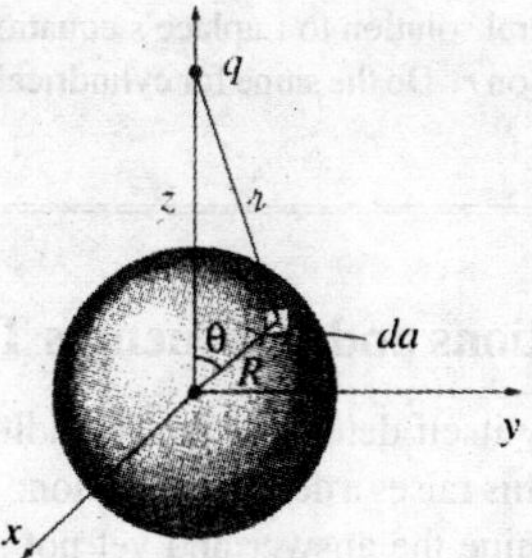

Figure 3.3

Problem 3.1 Find the average potential over a spherical surface of radius R due to a point charge q located *inside* (same as above, in other words, only with $z < R$). (In this case, of course, Laplace's equation does not hold within the sphere.) Show that, in general,

$$V_{\text{ave}} = V_{\text{center}} + \frac{Q_{\text{enc}}}{4\pi\epsilon_0 R},$$

where V_{center} is the potential at the center due to all the *external* charges, and Q_{enc} is the total enclosed charge.

Problem 3.2 In one sentence, justify **Earnshaw's Theorem:** *A charged particle cannot be held in a stable equilibrium by electrostatic forces alone.* As an example, consider the cubical arrangement of fixed charges in Fig. 3.4. It *looks,* off hand, as though a positive charge at the center would be suspended in midair, since it is repelled away from each corner. Where is the leak in this "electrostatic bottle"? [To harness nuclear fusion as a practical energy source it is necessary to heat a plasma (soup of charged particles) to fantastic temperatures—so hot that contact would vaporize any ordinary pot. Earnshaw's theorem says that electrostatic containment is also out of the question. Fortunately, it *is* possible to confine a hot plasma *magnetically.*]

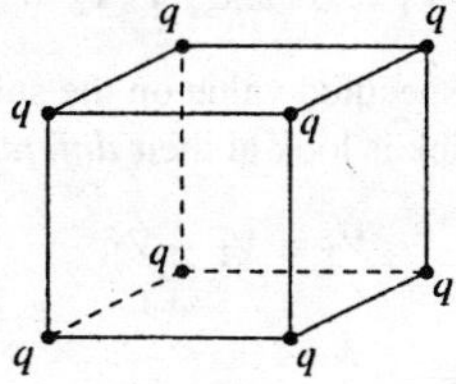

Figure 3.4

Problem 3.3 Find the general solution to Laplace's equation in spherical coordinates, for the case where V depends only on r. Do the same for cylindrical coordinates, assuming V depends only on s.

3.1.5 Boundary Conditions and Uniqueness Theorems

Laplace's equation does not by itself determine V; in addition, a suitable set of boundary conditions must be supplied. This raises a delicate question: What are appropriate boundary conditions, sufficient to determine the answer and yet not so strong as to generate inconsistencies? The one-dimensional case is easy, for here the general solution $V = mx + b$ contains two arbitrary constants, and we therefore require two boundary conditions. We might, for instance, specify the value of the function at the two ends, or we might give the value of the function and its derivative at one end, or the value at one end and the derivative at the other, and so on. But we cannot get away with *just* the value or *just* the derivative at *one* end—this is insufficient information. Nor would it do to specify the derivatives at both ends—this would either be redundant (if the two are equal) or inconsistent (if they are not).

In two or three dimensions we are confronted by a partial differential equation, and it is not so easy to see what would constitute acceptable boundary conditions. Is the shape of a taut rubber membrane, for instance, uniquely determined by the frame over which it is stretched, or, like a canning jar lid, can it snap from one stable configuration to another? The answer, as I think your intuition would suggest, is that V *is* uniquely determined by its value at the boundary (canning jars evidently don't obey Laplace's equation). However, other boundary conditions can also be used (see Prob. 3.4). The *proof* that a proposed set of boundary conditions will suffice is usually presented in the form of a **uniqueness theorem.** There are many such theorems for electrostatics, all sharing the same basic format—I'll show you the two most useful ones.[3]

First uniqueness theorem: The solution to Laplace's equation in some volume $\mathcal{V}$ is uniquely determined if V is specified on the boundary surface $\mathcal{S}$.

Proof: In Fig. 3.5 I have drawn such a region and its boundary. (There could also be "islands" inside, so long as V is given on all their surfaces; also, the outer boundary could be at infinity, where V is ordinarily taken to be zero.) Suppose there were *two* solutions to Laplace's equation:

$$\nabla^2 V_1 = 0 \quad \text{and} \quad \nabla^2 V_2 = 0,$$

both of which assume the specified value on the surface. I want to prove that they must be equal. The trick is look at their *difference:*

$$V_3 \equiv V_1 - V_2.$$

[3] I do not intend to prove the *existence* of solutions here—that's a much more difficult job. In context, the existence is generally clear on physical grounds.

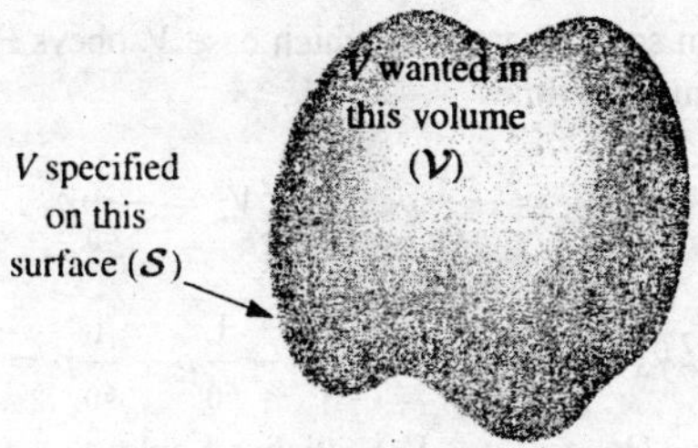

Figure 3.5

This obeys Laplace's equation,

$$\nabla^2 V_3 = \nabla^2 V_1 - \nabla^2 V_2 = 0,$$

and it takes the value *zero* on all boundaries (since V_1 and V_2 are equal there). But Laplace's equation allows no local maxima or minima—all extrema occur on the boundaries. So the maximum and minimum of V_3 are both zero. Therefore V_3 must be zero everywhere, and hence

$$V_1 = V_2. \quad \text{qed}$$

Example 3.1

Show that the potential is *constant* inside an enclosure completely surrounded by conducting material, provided there is no charge within the enclosure.

Solution: The potential on the cavity wall is some constant, V_0 (that's item (iv), in Sect. 2.5.1), so the potential inside is a function that satisfies Laplace's equation and has the constant value V_0 at the boundary. It doesn't take a genius to think of *one* solution to this problem: $V = V_0$ everywhere. The uniqueness theorem guarantees that this is the *only* solution. (It follows that the *field* inside an empty cavity is zero—the same result we found in Sect. 2.5.2 on rather different grounds.)

The uniqueness theorem is a license to your imagination. It doesn't matter *how* you come by your solution; if (a) it satisfies Laplace's equation and (b) it has the correct value on the boundaries, then it's *right.* You'll see the power of this argument when we come to the method of images.

Incidentally, it is easy to improve on the first uniqueness theorem: I assumed there was no charge inside the region in question, so the potential obeyed Laplace's equation, but we may as well throw in some charge (in which case V obeys Poisson's equation). The argument is the same, only this time

$$\nabla^2 V_1 = -\frac{1}{\epsilon_0}\rho, \qquad \nabla^2 V_2 = -\frac{1}{\epsilon_0}\rho,$$

so

$$\nabla^2 V_3 = \nabla^2 V_1 - \nabla^2 V_2 = -\frac{1}{\epsilon_0}\rho + \frac{1}{\epsilon_0}\rho = 0.$$

Once again the *difference* ($V_3 \equiv V_1 - V_2$) satisfies Laplace's equation and has the value zero on all boundaries, so $V_3 = 0$ and hence $V_1 = V_2$.

Corollary: The potential in a volume $\mathcal{V}$ is uniquely determined if (a) the charge density throughout the region, and (b) the value of V on all boundaries, are specified.

3.1.6 Conductors and the Second Uniqueness Theorem

The *simplest* way to set the boundary conditions for an electrostatic problem is to specify the value of V on all surfaces surrounding the region of interest. And this situation often occurs in practice: In the laboratory, we have conductors connected to batteries, which maintain a given potential, or to **ground**, which is the experimentalist's word for $V = 0$. However, there are other circumstances in which we do not know the *potential* at the boundary, but rather the *charges* on various conducting surfaces. Suppose I put charge Q_1 on the first conductor, Q_2 on the second, and so on—I'm not telling you how the charge distributes itself over each conducting surface, because as soon as I put it on, it moves around in a way I do not control. And for good measure, let's say there is some specified charge density ρ in the region between the conductors. Is the electric field now uniquely determined? Or are there perhaps a number of different ways the charges could arrange themselves on their respective conductors, each leading to a different field?

Second uniqueness theorem: In a volume $\mathcal{V}$ surrounded by conductors and containing a specified charge density ρ, the electric field is uniquely determined if the *total charge* on each conductor is given (Fig. 3.6). (The region as a whole can be bounded by another conductor, or else unbounded.)

Proof: Suppose there are *two* fields satisfying the conditions of the problem. Both obey Gauss's law in differential form in the space between the conductors:

$$\nabla \cdot \mathbf{E}_1 = \frac{1}{\epsilon_0}\rho, \qquad \nabla \cdot \mathbf{E}_2 = \frac{1}{\epsilon_0}\rho.$$

And both obey Gauss's law in integral form for a Gaussian surface enclosing each conductor:

$$\oint_{\substack{i\text{th conducting}\\ \text{surface}}} \mathbf{E}_1 \cdot d\mathbf{a} = \frac{1}{\epsilon_0}Q_i, \qquad \oint_{\substack{i\text{th conducting}\\ \text{surface}}} \mathbf{E}_2 \cdot d\mathbf{a} = \frac{1}{\epsilon_0}Q_i.$$

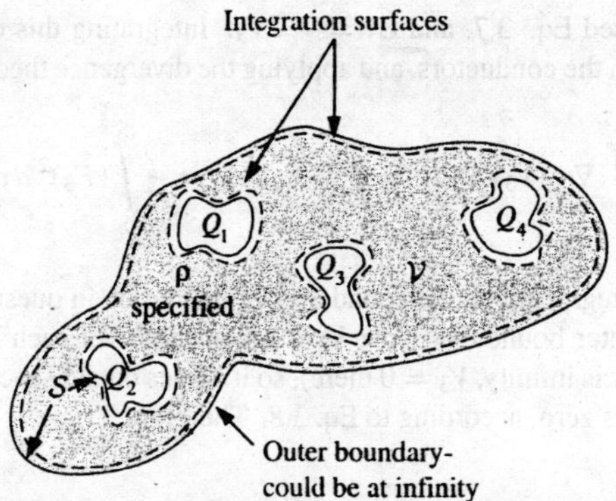

Figure 3.6

Likewise, for the outer boundary (whether this is just inside an enclosing conductor or at infinity),

$$\oint_{\substack{\text{outer}\\\text{boundary}}} \mathbf{E}_1 \cdot d\mathbf{a} = \frac{1}{\epsilon_0} Q_{\text{tot}}, \qquad \oint_{\substack{\text{outer}\\\text{boundary}}} \mathbf{E}_2 \cdot d\mathbf{a} = \frac{1}{\epsilon_0} Q_{\text{tot}}.$$

As before, we examine the difference

$$\mathbf{E}_3 \equiv \mathbf{E}_1 - \mathbf{E}_2,$$

which obeys

$$\nabla \cdot \mathbf{E}_3 = 0 \tag{3.7}$$

in the region between the conductors, and

$$\oint \mathbf{E}_3 \cdot d\mathbf{a} = 0 \tag{3.8}$$

over each boundary surface.

Now there is one final piece of information we must exploit: Although we do not know how the charge Q_i distributes itself over the ith conducting surface, we *do* know that each conductor is an equipotential, and hence V_3 is a *constant* (not necessarily the *same* constant) over each conducting surface. (It need not be *zero,* for the potentials V_1 and V_2 may not be equal—all we know for sure is that *both* are *constant* over any given conductor.) Next comes a trick. Invoking product rule number (5), we find that

$$\nabla \cdot (V_3 \mathbf{E}_3) = V_3(\nabla \cdot \mathbf{E}_3) + \mathbf{E}_3 \cdot (\nabla V_3) = -(E_3)^2.$$

Here I have used Eq. 3.7, and $\mathbf{E}_3 = -\nabla V_3$. Integrating this over the entire region between the conductors, and applying the divergence theorem to the left side:

$$\int_{\mathcal{V}} \nabla \cdot (V_3 \mathbf{E}_3)\, d\tau = \oint_{\mathcal{S}} V_3 \mathbf{E}_3 \cdot d\mathbf{a} = -\int_{\mathcal{V}} (E_3)^2\, d\tau.$$

The surface integral covers all boundaries of the region in question—the conductors and outer boundary. Now V_3 is a constant over each surface (if the outer boundary is infinity, $V_3 = 0$ there), so it comes outside each integral, and what remains is zero, according to Eq. 3.8. Therefore,

$$\int_{\mathcal{V}} (E_3)^2\, d\tau = 0.$$

But this integrand is never negative; the only way the integral can vanish is if $E_3 = 0$ everywhere. Consequently, $\mathbf{E}_1 = \mathbf{E}_2$, and the theorem is proved.

This proof was not easy, and there is a real danger that the theorem itself will seem more plausible to you than the proof. In case you think the second uniqueness theorem is "obvious," consider this example of Purcell's: Figure 3.7 shows a comfortable electrostatic configuration, consisting of four conductors with charges $\pm Q$, situated so that the plusses are near the minuses. It looks very stable. Now, what happens if we join them in pairs, by tiny wires, as indicated in Fig. 3.8? Since the positive charges are very near negative charges (which is where they *like* to be) you might well guess that *nothing* will happen—the configuration still looks stable.

Well, that sounds reasonable, but it's wrong. The configuration in Fig. 3.8 is *impossible.* For there are now effectively *two* conductors, and the total charge on each is *zero. One* possible way to distribute zero charge over these conductors is to have no accumulation of charge anywhere, and hence zero field everywhere (Fig. 3.9). By the second uniqueness theorem, this must be *the* solution: The charge will flow down the tiny wires, canceling itself off.

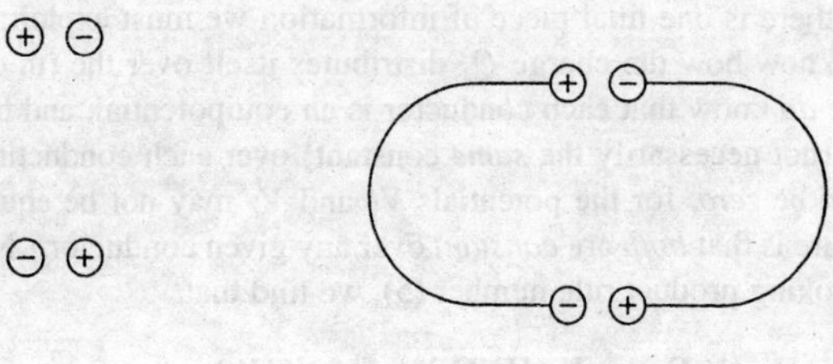

Figure 3.7 Figure 3.8

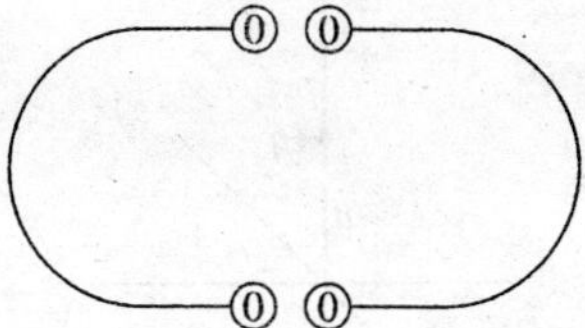

Figure 3.9

Problem 3.4 Prove that the field is uniquely determined when the charge density ρ is given and *either* V *or* the normal derivative $\partial V/\partial n$ is specified on each boundary surface. Do not assume the boundaries are conductors, or that V is constant over any given surface.

Problem 3.5 A more elegant proof of the second uniqueness theorem uses Green's identity (Prob. 1.60c), with $T = U = V_3$. Supply the details.

3.2 The Method of Images

3.2.1 The Classic Image Problem

Suppose a point charge q is held a distance d above an infinite grounded conducting plane (Fig. 3.10). *Question:* What is the potential in the region above the plane? It's not just $(1/4\pi\epsilon_0)q/\mathscr{r}$, for q will induce a certain amount of negative charge on the nearby surface of the conductor; the total potential is due in part to q directly, and in part to this induced charge. But how can we possibly calculate the potential, when we don't know how much charge is induced or how it is distributed?

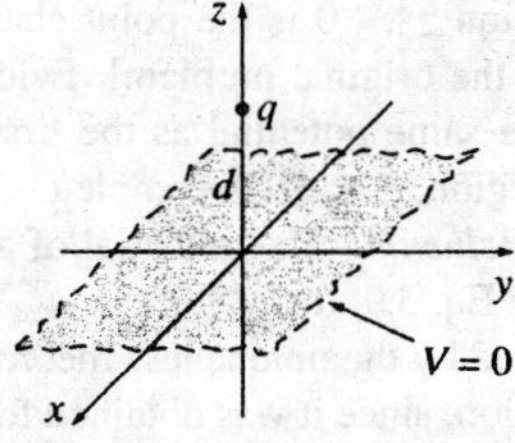

Figure 3.10

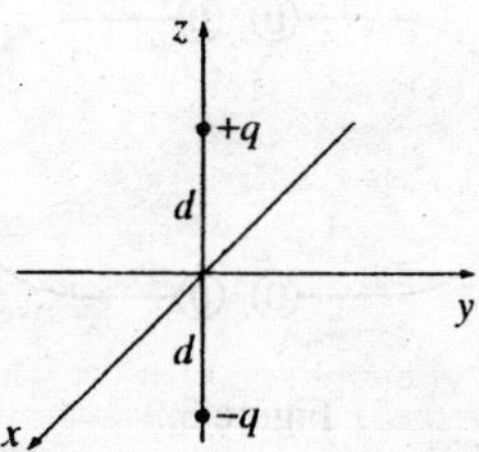

Figure 3.11

From a mathematical point of view our problem is to solve Poisson's equation in the region $z > 0$, with a single point charge q at $(0, 0, d)$, subject to the boundary conditions:

1. $V = 0$ when $z = 0$ (since the conducting plane is grounded), and
2. $V \to 0$ far from the charge (that is, for $x^2 + y^2 + z^2 \gg d^2$).

The first uniqueness theorem (actually, its corollary) guarantees that there is only one function that meets these requirements. If by trick or clever guess we can discover such a function, it's got to be the right answer.

Trick: Forget about the actual problem; we're going to study a *completely different* situation. This new problem consists of *two* point charges, $+q$ at $(0, 0, d)$ and $-q$ at $(0, 0, -d)$, and *no* conducting plane (Fig. 3.11). For this configuration I can easily write down the potential:

$$V(x, y, z) = \frac{1}{4\pi\epsilon_0}\left[\frac{q}{\sqrt{x^2 + y^2 + (z - d)^2}} - \frac{q}{\sqrt{x^2 + y^2 + (z + d)^2}}\right]. \quad (3.9)$$

(The denominators represent the distances from (x, y, z) to the charges $+q$ and $-q$, respectively.) It follows that

1. $V = 0$ when $z = 0$, and
2. $V \to 0$ for $x^2 + y^2 + z^2 \gg d^2$,

and the only charge in the region $z > 0$ is the point charge $+q$ at $(0, 0, d)$. But these are precisely the conditions of the original problem! Evidently the second configuration happens to produce exactly the same potential as the first configuration, in the "upper" region $z \geq 0$. (The "lower" region, $z < 0$, is completely different, but who cares? The upper part is all we need.) *Conclusion:* The potential of a point charge above an infinite grounded conductor is given by Eq. 3.9, for $z \geq 0$.

Notice the crucial role played by the uniqueness theorem in this argument: without it, no one would believe this solution, since it was obtained for a completely different charge distribution. But the uniqueness theorem certifies it: If it satisfies Poisson's equation in the region of interest, and assumes the correct value at the boundaries, then it must be right.

3.2.2 Induced Surface Charge

Now that we know the potential, it is a straightforward matter to compute the surface charge σ induced on the conductor. According to Eq. 2.49,

$$\sigma = -\epsilon_0 \frac{\partial V}{\partial n},$$

where $\partial V/\partial n$ is the normal derivative of V at the surface. In this case the normal direction is the z-direction, so

$$\sigma = -\epsilon_0 \left.\frac{\partial V}{\partial z}\right|_{z=0}.$$

From Eq. 3.9,

$$\frac{\partial V}{\partial z} = \frac{1}{4\pi\epsilon_0}\left\{\frac{-q(z-d)}{[x^2+y^2+(z-d)^2]^{3/2}} + \frac{q(z+d)}{[x^2+y^2+(z+d)^2]^{3/2}}\right\},$$

so

$$\sigma(x, y) = \frac{-qd}{2\pi(x^2+y^2+d^2)^{3/2}}. \tag{3.10}$$

As expected, the induced charge is negative (assuming q is positive) and greatest at $x = y = 0$.

While we're at it, let's compute the *total* induced charge

$$Q = \int \sigma\, da.$$

This integral, over the xy plane, could be done in Cartesian coordinates, with $da = dx\,dy$, but it's a little easier to use polar coordinates (r, ϕ), with $r^2 = x^2 + y^2$ and $da = r\,dr\,d\phi$. Then

$$\sigma(r) = \frac{-qd}{2\pi(r^2+d^2)^{3/2}},$$

and

$$Q = \int_0^{2\pi}\int_0^{\infty} \frac{-qd}{2\pi(r^2+d^2)^{3/2}} r\,dr\,d\phi = \left.\frac{qd}{\sqrt{r^2+d^2}}\right|_0^{\infty} = -q. \tag{3.11}$$

Evidently the total charge induced on the plane is $-q$, as (with benefit of hindsight) you can perhaps convince yourself it *had* to be.

3.2.3 Force and Energy

The charge q is attracted toward the plane, because of the negative induced charge. Let's calculate the force of attraction. Since the potential in the vicinity of q is the same as in the analog problem (the one with $+q$ and $-q$ but no conductor), so also is the field and, therefore, the force:

$$\mathbf{F} = -\frac{1}{4\pi\epsilon_0}\frac{q^2}{(2d)^2}\hat{\mathbf{z}}. \tag{3.12}$$

Beware: It is easy to get carried away, and assume that *everything* is the same in the two problems. Energy, however, is *not* the same. With the two point charges and no conductor, Eq. 2.42 gives

$$W = -\frac{1}{4\pi\epsilon_0}\frac{q^2}{2d}. \tag{3.13}$$

But for a single charge and conducting plane the energy is *half* of this:

$$W = -\frac{1}{4\pi\epsilon_0}\frac{q^2}{4d}. \tag{3.14}$$

Why half? Think of the energy stored in the fields (Eq. 2.45):

$$W = \frac{\epsilon_0}{2}\int E^2\,d\tau.$$

In the first case both the upper region ($z > 0$) and the lower region ($z < 0$) contribute—and by symmetry they contribute equally. But in the second case only the upper region contains a nonzero field, and hence the energy is half as great.

Of course, one could also determine the energy by calculating the work required to bring q in from infinity. The force required (to oppose the electrical force in Eq. 3.12) is $(1/4\pi\epsilon_0)(q^2/4z^2)\hat{\mathbf{z}}$, so

$$\begin{aligned} W &= \int_\infty^d \mathbf{F}\cdot d\mathbf{l} = \frac{1}{4\pi\epsilon_0}\int_\infty^d \frac{q^2}{4z^2}\,dz \\ &= \frac{1}{4\pi\epsilon_0}\left(-\frac{q^2}{4z}\right)\Bigg|_\infty^d = -\frac{1}{4\pi\epsilon_0}\frac{q^2}{4d}. \end{aligned}$$

As I move q toward the conductor, I do work *only on* q. It is true that induced charge is moving in over the conductor, but this costs me nothing, since the whole conductor is at potential zero. By contrast, if I simultaneously bring in *two* point charges (with no conductor), I do work on *both* of them, and the total is twice as great.

3.2.4 Other Image Problems

The method just described is not limited to a single point charge; *any* stationary charge distribution near a grounded conducting plane can be treated in the same way, by introducing its mirror image—hence the name **method of images.** (Remember that the image charges have the *opposite sign;* this is what guarantees that the xy plane will be at potential zero.) There are also some exotic problems that can be handled in similar fashion; the nicest of these is the following.

Example 3.2

A point charge q is situated a distance a from the center of a grounded conducting sphere of radius R (Fig. 3.12). Find the potential outside the sphere.

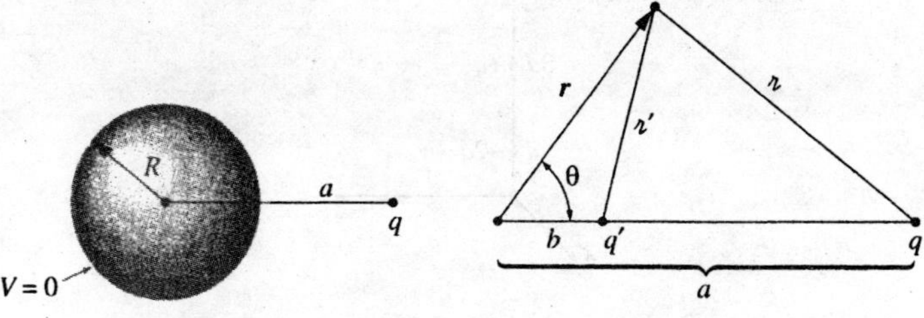

Figure 3.12 Figure 3.13

Solution: Examine the *completely different* configuration, consisting of the point charge q together with another point charge

$$q' = -\frac{R}{a}q, \tag{3.15}$$

placed a distance

$$b = \frac{R^2}{a} \tag{3.16}$$

to the right of the center of the sphere (Fig. 3.13). No conductor, now—just the two point charges. The potential of this configuration is

$$V(\mathbf{r}) = \frac{1}{4\pi\epsilon_0}\left(\frac{q}{\mathscr{r}} + \frac{q'}{\mathscr{r}'}\right), \tag{3.17}$$

where $\mathscr{r}$ and $\mathscr{r}'$ are the distances from q and q', respectively. Now, it happens (see Prob. 3.7) that this potential vanishes at all points on the sphere, and therefore fits the boundary conditions for our original problem, in the exterior region.

Conclusion: Eq. 3.17 is the potential of a point charge near a grounded conducting sphere. (Notice that b is less than R, so the "image" charge q' is safely inside the sphere—*you cannot put image charges in the region where you are calculating* V; that would change ρ, and you'd be solving Poisson's equation with the wrong source.) In particular, the force of attraction between the charge and the sphere is

$$F = \frac{1}{4\pi\epsilon_0}\frac{qq'}{(a-b)^2} = -\frac{1}{4\pi\epsilon_0}\frac{q^2Ra}{(a^2-R^2)^2}. \tag{3.18}$$

This solution is delightfully simple, but extraordinarily lucky. There's as much art as science in the method of images, for you must somehow think up the right "auxiliary problem" to look at. The first person who solved the problem this way cannot have known in advance what image charge q' to use or where to put it. Presumably, he (she?) started with an *arbitrary* charge at an *arbitrary* point inside the sphere, calculated the potential on the sphere, and then discovered that with q' and b just right the potential on the sphere vanishes. But it is really a miracle that *any* choice does the job—with a cube instead of a sphere, for example, *no* single charge *anywhere* inside would make the potential zero on the surface.

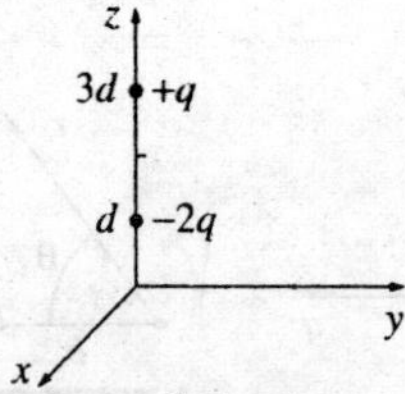

Figure 3.14

Problem 3.6 Find the force on the charge $+q$ in Fig. 3.14. (The xy plane is a grounded conductor.)

Problem 3.7

(a) Using the law of cosines, show that Eq. 3.17 can be written as follows:

$$V(r,\theta) = \frac{1}{4\pi\epsilon_0}\left[\frac{q}{\sqrt{r^2+a^2-2ra\cos\theta}} - \frac{q}{\sqrt{R^2+(ra/R)^2-2ra\cos\theta}}\right], \tag{3.19}$$

where r and θ are the usual spherical polar coordinates, with the z axis along the line through q. In this form it is obvious that $V = 0$ on the sphere, $r = R$.

(b) Find the induced surface charge on the sphere, as a function of θ. Integrate this to get the total induced charge. (What *should* it be?)

(c) Calculate the energy of this configuration.

Problem 3.8 In Ex. 3.2 we assumed that the conducting sphere was grounded ($V = 0$). But with the addition of a second image charge, the same basic model will handle the case of a sphere at *any* potential V_0 (relative, of course, to infinity). What charge should you use, and where should you put it? Find the force of attraction between a point charge q and a *neutral* conducting sphere.

Problem 3.9 A uniform line charge λ is placed on an infinite straight wire, a distance d above a grounded conducting plane. (Let's say the wire runs parallel to the x-axis and directly above it, and the conducting plane is the xy plane.)

(a) Find the potential in the region above the plane.

(b) Find the charge density σ induced on the conducting plane.

Problem 3.10 Two semi-infinite grounded conducting planes meet at right angles. In the region between them, there is a point charge q, situated as shown in Fig. 3.15. Set up the image configuration, and calculate the potential in this region. What charges do you need, and where should they be located? What is the force on q? How much work did it take to bring q in from infinity? Suppose the planes met at some angle other than 90°; would you still be able to solve the problem by the method of images? If not, for what particular angles *does* the method work?

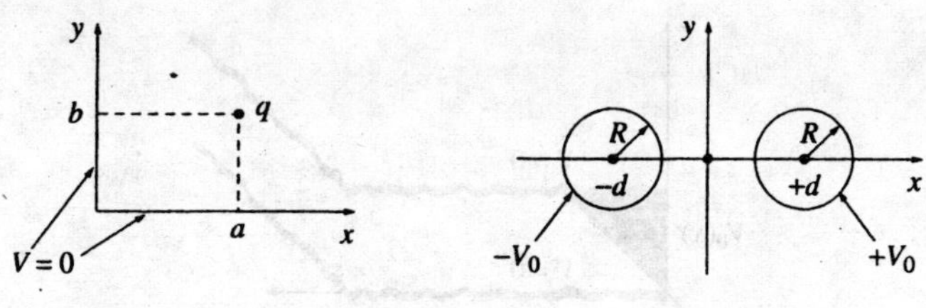

Figure 3.15 Figure 3.16

! **Problem 3.11** Two long, straight copper pipes, each of radius R, are held a distance $2d$ apart. One is at potential V_0, the other at $-V_0$ (Fig. 3.16). Find the potential everywhere. [*Suggestion:* Exploit the result of Prob. 2.47.]

3.3 Separation of Variables

In this section we shall attack Laplace's equation directly, using the method of **separation of variables,** which is the physicist's favorite tool for solving partial differential equations. The method is applicable in circumstances where the potential (V) or the charge density (σ) is specified on the boundaries of some region, and we are asked to find the potential in the interior. The basic strategy is very simple: *We look for solutions that are* products *of functions, each of which depends on only* one *of the coordinates.* The algebraic details, however, can be formidable, so I'm going to develop the method through a sequence of examples. We'll start with Cartesian coordinates and then do spherical coordinates (I'll leave the cylindrical case for you to tackle on your own, in Prob. 3.23).

3.3.1 Cartesian Coordinates

Example 3.3

Two infinite grounded metal plates lie parallel to the xz plane, one at $y = 0$, the other at $y = a$ (Fig. 3.17). The left end, at $x = 0$, is closed off with an infinite strip insulated from the two plates and maintained at a specific potential $V_0(y)$. Find the potential inside this "slot."

Solution: The configuration is independent of z, so this is really a *two*-dimensional problem. In mathematical terms, we must solve Laplace's equation,

$$\frac{\partial^2 V}{\partial x^2} + \frac{\partial^2 V}{\partial y^2} = 0, \tag{3.20}$$

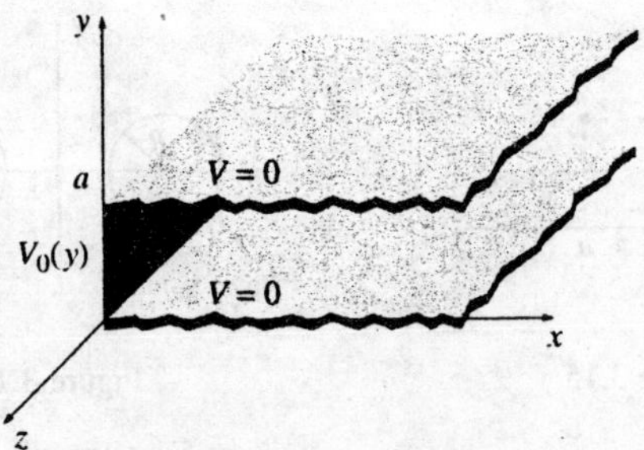

Figure 3.17

subject to the boundary conditions

$$\left.\begin{array}{lll}\text{(i)} & V=0 \text{ when } y=0, \\ \text{(ii)} & V=0 \text{ when } y=a, \\ \text{(iii)} & V=V_0(y) \text{ when } x=0, \\ \text{(iv)} & V\rightarrow 0 \text{ as } x\rightarrow\infty. \end{array}\right\} \tag{3.21}$$

(The latter, although not explicitly stated in the problem, is necessary on physical grounds: as you get farther and farther away from the "hot" strip at $x = 0$, the potential should drop to zero.) Since the potential is specified on all boundaries, the answer is uniquely determined.

The first step is to look for solutions in the form of products:

$$V(x, y) = X(x)Y(y). \tag{3.22}$$

On the face of it, this is an absurd restriction—the overwhelming majority of solutions to Laplace's equation do *not* have such a form. For example, $V(x, y) = (5x + 6y)$ satisfies Eq. 3.20, but you can't express it as the product of a function x times a function y. Obviously, we're only going to get a tiny subset of all possible solutions by this means, and it would be a *miracle* if one of them happened to fit the boundary conditions of our problem ... But hang on, because the solutions we *do* get are very special, and it turns out that by pasting them together we can construct the general solution.

Anyway, putting Eq. 3.22 into Eq. 3.20, we obtain

$$Y\frac{d^2X}{dx^2} + X\frac{d^2Y}{dy^2} = 0.$$

The next step is to "separate the variables" (that is, collect all the x-dependence into one term and all the y-dependence into another). Typically, this is accomplished by dividing through by V:

$$\frac{1}{X}\frac{d^2X}{dx^2} + \frac{1}{Y}\frac{d^2Y}{dy^2} = 0. \tag{3.23}$$

Here the first term depends only on x and the second only on y; in other words, we have an equation of the form

$$f(x) + g(y) = 0. \tag{3.24}$$

Now, there's only one way this could possibly be true: *f and g must both be constant.* For what if $f(x)$ *changed,* as you vary x—then if we held y fixed and fiddled with x, the sum $f(x) + g(x)$ would *change,* in violation of Eq. 3.24, which says it's always zero. (That's a simple but somehow rather elusive argument; don't accept it without due thought, because the whole method rides on it.) It follows from Eq. 3.23, then, that

$$\frac{1}{X}\frac{d^2X}{dx^2} = C_1 \quad \text{and} \quad \frac{1}{Y}\frac{d^2Y}{dy^2} = C_2, \qquad \text{with } C_1 + C_2 = 0. \tag{3.25}$$

One of these constants is positive, the other negative (or perhaps both are zero). In general, one must investigate all the possibilities; however, in our particular problem we need C_1 positive and C_2 negative, for reasons that will appear in a moment. Thus

$$\frac{d^2X}{dx^2} = k^2X, \qquad \frac{d^2Y}{dy^2} = -k^2Y. \tag{3.26}$$

Notice what has happened: A *partial* differential equation (3.20) has been converted into two *ordinary* differential equations (3.26). The advantage of this is obvious—ordinary differential equations are a lot easier to solve. Indeed:

$$X(x) = Ae^{kx} + Be^{-kx}, \qquad Y(y) = C\sin ky + D\cos ky,$$

so that

$$V(x, y) = (Ae^{kx} + Be^{-kx})(C\sin ky + D\cos ky). \tag{3.27}$$

This is the appropriate separable solution to Laplace's equation; it remains to impose the boundary conditions, and see what they tell us about the constants. To begin at the end, condition (iv) requires that A equal zero.[4] Absorbing B into C and D, we are left with

$$V(x, y) = e^{-kx}(C\sin ky + D\cos ky).$$

Condition (i) now demands that D equal zero, so

$$V(x, y) = Ce^{-kx}\sin ky. \tag{3.28}$$

Meanwhile (ii) yields $\sin ka = 0$, from which it follows that

$$k = \frac{n\pi}{a}, \qquad (n = 1, 2, 3, \ldots). \tag{3.29}$$

(At this point you can see why I chose C_1 positive and C_2 negative: If X were sinusoidal, we could never arrange for it to go to zero at infinity, and if Y were exponential we could not make it vanish at both 0 and a. Incidentally, $n = 0$ is no good, for in that case the potential vanishes *everywhere.* And we have already excluded negative n's.)

That's as far as we can go, using separable solutions, and unless $V_0(y)$ just happens to have the form $\sin(n\pi y/a)$ for some integer n we simply *can't fit* the final boundary condition at $x = 0$. But now comes the crucial step that redeems the method: Separation of variables has given us an *infinite set* of solutions (one for each n), and whereas none of them *by itself* satisfies

[4] I'm assuming k is positive, but this involves no loss of generality—negative k gives the same solution (3.27), only with the constants shuffled ($A \leftrightarrow B$, $C \rightarrow -C$). Occasionally (but not in this example) $k = 0$ must also be included (see Prob. 3.47).

the final boundary condition, it is possible to combine them in a way that *does.* Laplace's equation is *linear,* in the sense that if $V_1, V_2, V_3, \ldots$ satisfy it, so does any linear combination, $V = \alpha_1 V_1 + \alpha_2 V_2 + \alpha_3 V_3 + \ldots$, where $\alpha_1, \alpha_2, \ldots$ are arbitrary constants. For

$$\nabla^2 V = \alpha_1 \nabla^2 V_1 + \alpha_2 \nabla^2 V_2 + \ldots = 0\alpha_1 + 0\alpha_2 + \ldots = 0.$$

Exploiting this fact, we can patch together the separable solutions (3.28) to construct a much more general solution:

$$V(x, y) = \sum_{n=1}^{\infty} C_n e^{-n\pi x/a} \sin(n\pi y/a). \tag{3.30}$$

This still satisfies three of the boundary conditions; the question is, can we (by astute choice of the coefficients C_n) fit the final boundary condition (iii)?

$$V(0, y) = \sum_{n=1}^{\infty} C_n \sin(n\pi y/a) = V_0(y). \tag{3.31}$$

Well, you may recognize this sum—it's a Fourier sine series. And Dirichlet's theorem[5] guarantees that virtually *any* function $V_0(y)$—it can even have a finite number of discontinuities—can be expanded in such a series.

But how do we actually *determine* the coefficients C_n, buried as they are in that infinite sum? The device for accomplishing this is so lovely it deserves a name—I call it **Fourier's trick**, though it seems Euler had used essentially the same idea somewhat earlier. Here's how it goes: Multiply Eq. 3.31 by $\sin(n'\pi y/a)$ (where n' is a positive integer), and integrate from 0 to a:

$$\sum_{n=1}^{\infty} C_n \int_0^a \sin(n\pi y/a) \sin(n'\pi y/a)\, dy = \int_0^a V_0(y) \sin(n'\pi y/a)\, dy. \tag{3.32}$$

You can work out the integral on the left for yourself; the answer is

$$\int_0^a \sin(n\pi y/a) \sin(n'\pi y/a)\, dy = \begin{cases} 0, & \text{if } n' \neq n, \\ \dfrac{a}{2}, & \text{if } n' = n. \end{cases} \tag{3.33}$$

Thus all the terms in the series drop out, save only the one where $n' = n$, and the left side of Eq. 3.32, reduces to $(a/2)C_{n'}$. *Conclusion:*[6]

$$C_n = \frac{2}{a} \int_0^a V_0(y) \sin(n\pi y/a)\, dy. \tag{3.34}$$

That *does* it: Eq. 3.30 is the solution, with coefficients given by Eq. 3.34. As a concrete example, suppose the strip at $x = 0$ is a metal plate with constant potential V_0 (remember, it's insulated from the grounded plates at $y = 0$ and $y = a$). Then

$$C_n = \frac{2V_0}{a} \int_0^a \sin(n\pi y/a)\, dy = \frac{2V_0}{n\pi}(1 - \cos n\pi) = \begin{cases} 0, & \text{if } n \text{ is even,} \\ \dfrac{4V_0}{n\pi}, & \text{if } n \text{ is odd.} \end{cases} \tag{3.35}$$

---

[5] Boas, M., *Mathematical Methods in the Physical Sciences,* 2nd ed. (New York: John Wiley, 1983).

[6] For aesthetic reasons I've dropped the prime; Eq. 3.34 holds for $n = 1, 2, 3, \ldots$, and it doesn't matter (obviously) what letter you use for the "dummy" index.

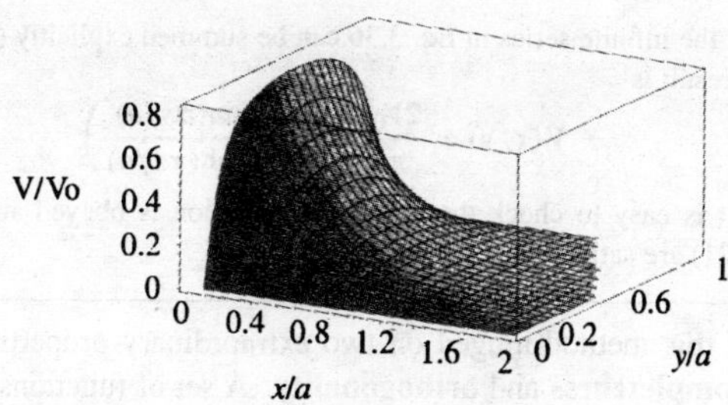

Figure 3.18

Evidently,

$$V(x, y) = \frac{4V_0}{\pi} \sum_{n=1,3,5...} \frac{1}{n} e^{-n\pi x/a} \sin(n\pi y/a). \tag{3.36}$$

Figure 3.18 is a plot of this potential; Fig. 3.19 shows how the first few terms in the Fourier series combine to make a better and better approximation to the constant V_0: (a) is $n = 1$ only, (b) includes n up to 5, (c) is the sum of the first 10 terms, and (d) is the sum of the first 100 terms.

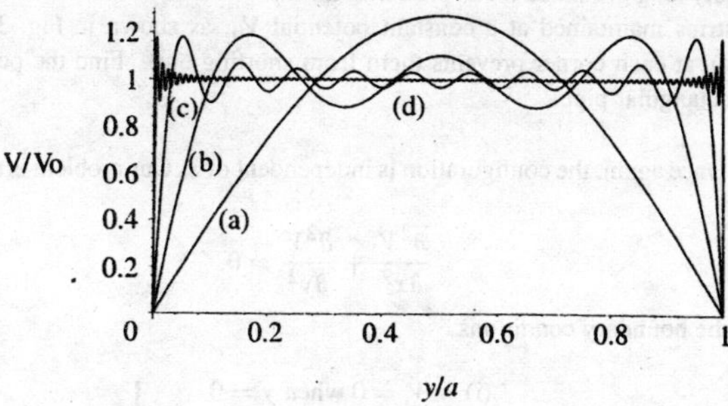

Figure 3.19

Incidentally, the infinite series in Eq. 3.36 can be summed explicitly (try your hand at it, if you like); the result is

$$V(x, y) = \frac{2V_0}{\pi} \tan^{-1}\left(\frac{\sin(\pi y/a)}{\sinh(\pi x/a)}\right). \tag{3.37}$$

In this form it is easy to check that Laplace's equation is obeyed and the four boundary conditions (3.21) are satisfied.

The success of this method hinged on two extraordinary properties of the separable solutions (3.28): **completeness** and **orthogonality**. A set of functions $f_n(y)$ is said to be **complete** if any other function $f(y)$ can be expressed as a linear combination of them:

$$f(y) = \sum_{n=1}^{\infty} C_n f_n(y). \tag{3.38}$$

The functions $\sin(n\pi y/a)$ are complete on the interval $0 \leq y \leq a$. It was this fact, guaranteed by Dirichlet's theorem, that assured us Eq. 3.31 could be satisfied, given the proper choice of the coefficients C_n. (The *proof* of completeness, for a particular set of functions, is an extremely difficult business, and I'm afraid physicists tend to *assume* it's true and leave the checking to others.) A set of functions is **orthogonal** if the integral of the product of any two different members of the set is zero:

$$\int_0^a f_n(y) f_{n'}(y)\, dy = 0 \qquad \text{for } n' \neq n. \tag{3.39}$$

The sine functions are orthogonal (Eq. 3.33); this is the property on which Fourier's trick is based, allowing us to kill off all terms but one in the infinite series and thereby solve for the coefficients C_n. (Proof of orthogonality is generally quite simple, either by direct integration or by analysis of the differential equation from which the functions came.)

Example 3.4

Two infinitely long grounded metal plates, again at $y = 0$ and $y = a$, are connected at $x = \pm b$ by metal strips maintained at a constant potential V_0, as shown in Fig. 3.20 (a thin layer of insulation at each corner prevents them from shorting out). Find the potential inside the resulting rectangular pipe.

Solution: Once again, the configuration is independent of z. Our problem is to solve Laplace's equation

$$\frac{\partial^2 V}{\partial x^2} + \frac{\partial^2 V}{\partial y^2} = 0,$$

subject to the boundary conditions

$$\left.\begin{array}{ll} \text{(i)} & V = 0 \text{ when } y = 0, \\ \text{(ii)} & V = 0 \text{ when } y = a, \\ \text{(iii)} & V = V_0 \text{ when } x = b, \\ \text{(iv)} & V = V_0 \text{ when } x = -b. \end{array}\right\} \tag{3.40}$$

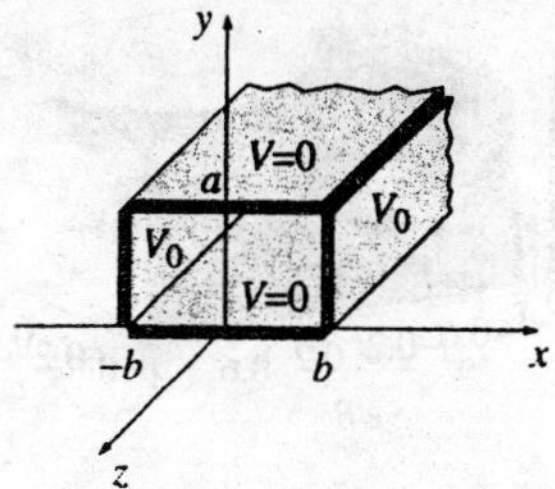

Figure 3.20

The argument runs as before, up to Eq. 3.27:

$$V(x, y) = (Ae^{kx} + Be^{-kx})(C \sin ky + D \cos ky).$$

This time, however, we cannot set $A = 0$; the region in question does not extend to $x = \infty$, so e^{kx} is perfectly acceptable. On the other hand, the situation is *symmetric* with respect to x, so $V(-x, y) = V(x, y)$, and it follows that $A = B$. Using

$$e^{kx} + e^{-kx} = 2 \cosh kx,$$

and absorbing $2A$ into C and D, we have

$$V(x, y) = \cosh kx \,(C \sin ky + D \cos ky).$$

Boundary conditions (i) and (ii) require, as before, that $D = 0$ and $k = n\pi/a$, so

$$V(x, y) = C \cosh(n\pi x/a) \sin(n\pi y/a). \tag{3.41}$$

Because $V(x, y)$ is even in x, it will automatically meet condition (iv) if it fits (iii). It remains, therefore, to construct the general linear combination,

$$V(x, y) = \sum_{n=1}^{\infty} C_n \cosh(n\pi x/a) \sin(n\pi y/a),$$

and pick the coefficients C_n in such a way as to satisfy condition (iii):

$$V(b, y) = \sum_{n=1}^{\infty} C_n \cosh(n\pi b/a) \sin(n\pi y/a) = V_0.$$

This is the same problem in Fourier analysis that we faced before; I quote the result from Eq. 3.35:

$$C_n \cosh(n\pi b/a) = \begin{cases} 0, & \text{if } n \text{ is even} \\ \dfrac{4V_0}{n\pi}, & \text{if } n \text{ is odd} \end{cases}$$

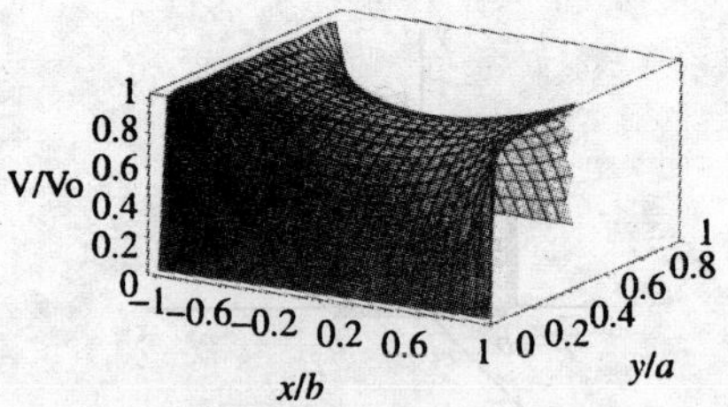

Figure 3.21

Conclusion: The potential in this case is given by

$$V(x, y) = \frac{4V_0}{\pi} \sum_{n=1,3,5...} \frac{1}{n} \frac{\cosh(n\pi x/a)}{\cosh(n\pi b/a)} \sin(n\pi y/a). \tag{3.42}$$

This function is shown in Fig. 3.21.

Example 3.5

An infinitely long rectangular metal pipe (sides a and b) is grounded, but one end, at $x = 0$, is maintained at a specified potential $V_0(y, z)$, as indicated in Fig. 3.22. Find the potential inside the pipe.

Solution: This is a genuinely three-dimensional problem,

$$\frac{\partial^2 V}{\partial x^2} + \frac{\partial^2 V}{\partial y^2} + \frac{\partial^2 V}{\partial z^2} = 0, \tag{3.43}$$

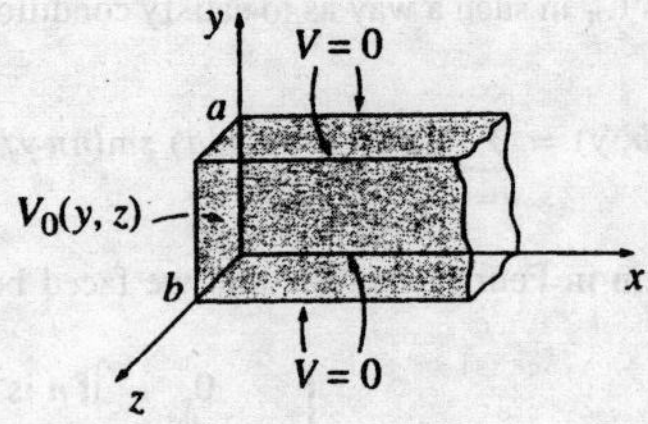

Figure 3.22

subject to the boundary conditions

$$\left.\begin{array}{ll}\text{(i)} & V=0 \text{ when } y=0,\\ \text{(ii)} & V=0 \text{ when } y=a,\\ \text{(iii)} & V=0 \text{ when } z=0,\\ \text{(iv)} & V=0 \text{ when } z=b,\\ \text{(v)} & V\to 0 \text{ as } x\to\infty,\\ \text{(vi)} & V=V_0(y,z) \text{ when } x=0.\end{array}\right\} \tag{3.44}$$

As always, we look for solutions that are products:

$$V(x,y,z)=X(x)Y(y)Z(z). \tag{3.45}$$

Putting this into Eq. 3.43, and dividing by V, we find

$$\frac{1}{X}\frac{d^2X}{dx^2}+\frac{1}{Y}\frac{d^2Y}{dy^2}+\frac{1}{Z}\frac{d^2Z}{dz^2}=0.$$

It follows that

$$\frac{1}{X}\frac{d^2X}{dx^2}=C_1,\quad \frac{1}{Y}\frac{d^2Y}{dy^2}=C_2,\quad \frac{1}{Z}\frac{d^2Z}{dz^2}=C_3,\quad \text{with } C_1+C_2+C_3=0.$$

Our previous experience (Ex. 3.3) suggests that C_1 must be positive, C_2 and C_3 negative. Setting $C_2=-k^2$ and $C_3=-l^2$, we have $C_1=k^2+l^2$, and hence

$$\frac{d^2X}{dx^2}=(k^2+l^2)X,\quad \frac{d^2Y}{dy^2}=-k^2Y,\quad \frac{d^2Z}{dz^2}=-l^2Z. \tag{3.46}$$

Once again, separation of variables has turned a *partial* differential equation into *ordinary* differential equations. The solutions are

$$\begin{aligned} X(x) &= Ae^{\sqrt{k^2+l^2}\,x}+Be^{-\sqrt{k^2+l^2}\,x},\\ Y(y) &= C\sin ky+D\cos ky,\\ Z(z) &= E\sin lz+F\cos lz.\end{aligned}$$

Boundary condition (v) implies $A=0$, (i) gives $D=0$, and (iii) yields $F=0$, whereas (ii) and (iv) require that $k=n\pi/a$ and $l=m\pi/b$, where n and m are positive integers. Combining the remaining constants, we are left with

$$V(x,y,z)=Ce^{-\pi\sqrt{(n/a)^2+(m/b)^2}\,x}\sin(n\pi y/a)\sin(m\pi z/b). \tag{3.47}$$

This solution meets all the boundary conditions except (vi). It contains *two* unspecified integers (n and m), and the most general linear combination is a *double* sum:

$$V(x,y,z)=\sum_{n=1}^{\infty}\sum_{m=1}^{\infty}C_{n,m}e^{-\pi\sqrt{(n/a)^2+(m/b)^2}\,x}\sin(n\pi y/a)\sin(m\pi z/b). \tag{3.48}$$

We hope to fit the remaining boundary condition,

$$V(0,y,z)=\sum_{n=1}^{\infty}\sum_{m=1}^{\infty}C_{n,m}\sin(n\pi y/a)\sin(m\pi z/b)=V_0(y,z), \tag{3.49}$$

by appropriate choice of the coefficients $C_{n,m}$. To determine these constants, we multiply by $\sin(n'\pi y/a)\,\sin(m'\pi z/b)$, where n' and m' are arbitrary positive integers, and integrate:

$$\sum_{n=1}^{\infty}\sum_{m=1}^{\infty} C_{n,m}\int_0^a \sin(n\pi y/a)\,\sin(n'\pi y/a)\,dy\int_0^b \sin(m\pi z/b)\,\sin(m'\pi z/b)\,dz$$

$$= \int_0^a\int_0^b V_0(y,z)\sin(n'\pi y/a)\,\sin(m'\pi z/b)\,dy\,dz.$$

Quoting Eq. 3.33, the left side is $(ab/4)C_{n',m'}$, so

$$C_{n,m} = \frac{4}{ab}\int_0^a\int_0^b V_0(y,z)\sin(n\pi y/a)\,\sin(m\pi z/b)\,dy\,dz. \tag{3.50}$$

Equation 3.48, with the coefficients given by Eq. 3.50, is the solution to our problem.

For instance, if the end of the tube is a conductor at *constant* potential V_0,

$$\begin{aligned} C_{n,m} &= \frac{4V_0}{ab}\int_0^a \sin(n\pi y/a)\,dy\int_0^b \sin(m\pi z/b)\,dz \\ &= \begin{cases} 0, & \text{if } n \text{ or } m \text{ is even,} \\ \dfrac{16V_0}{\pi^2 nm}, & \text{if } n \text{ and } m \text{ are odd.} \end{cases} \end{aligned} \tag{3.51}$$

In this case

$$V(x,y,z) = \frac{16V_0}{\pi^2}\sum_{n,m=1,3,5\ldots}^{\infty}\frac{1}{nm}e^{-\pi\sqrt{(n/a)^2+(m/b)^2}\,x}\sin(n\pi y/a)\,\sin(m\pi z/b). \tag{3.52}$$

Notice that the successive terms decrease rapidly; a reasonable approximation would be obtained by keeping only the first few.

Problem 3.12 Find the potential in the infinite slot of Ex. 3.3 if the boundary at $x = 0$ consists of two metal strips: one, from $y = 0$ to $y = a/2$, is held at a constant potential V_0, and the other, from $y = a/2$ to $y = a$, is at potential $-V_0$.

Problem 3.13 For the infinite slot (Ex. 3.3) determine the charge density $\sigma(y)$ on the strip at $x = 0$, assuming it is a conductor at constant potential V_0.

Problem 3.14 A rectangular pipe, running parallel to the z-axis (from $-\infty$ to $+\infty$), has three grounded metal sides, at $y = 0$, $y = a$, and $x = 0$. The fourth side, at $x = b$, is maintained at a specified potential $V_0(y)$.

(a) Develop a general formula for the potential within the pipe.

(b) Find the potential explicitly, for the case $V_0(y) = V_0$ (a constant).

Problem 3.15 A cubical box (sides of length a) consists of five metal plates, which are welded together and grounded (Fig. 3.23). The top is made of a separate sheet of metal, insulated from the others, and held at a constant potential V_0. Find the potential inside the box.

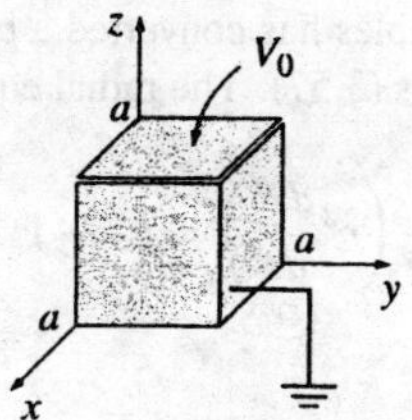

Figure 3.23

3.3.2 Spherical Coordinates

In the examples considered so far, Cartesian coordinates were clearly appropriate, since the boundaries were *planes*. For *round* objects spherical coordinates are more natural. In the spherical system, Laplace's equation reads:

$$\frac{1}{r^2}\frac{\partial}{\partial r}\left(r^2\frac{\partial V}{\partial r}\right)+\frac{1}{r^2\sin\theta}\frac{\partial}{\partial\theta}\left(\sin\theta\frac{\partial V}{\partial\theta}\right)+\frac{1}{r^2\sin^2\theta}\frac{\partial^2 V}{\partial\phi^2}=0. \tag{3.53}$$

I shall assume the problem has **azimuthal symmetry**, so that V *is independent of* ϕ;[7] in that case Eq. 3.53 reduces to

$$\frac{\partial}{\partial r}\left(r^2\frac{\partial V}{\partial r}\right)+\frac{1}{\sin\theta}\frac{\partial}{\partial\theta}\left(\sin\theta\frac{\partial V}{\partial\theta}\right)=0. \tag{3.54}$$

As before, we look for solutions that are products:

$$V(r,\theta)=R(r)\Theta(\theta). \tag{3.55}$$

Putting this into Eq. 3.54, and dividing by V,

$$\frac{1}{R}\frac{d}{dr}\left(r^2\frac{dR}{dr}\right)+\frac{1}{\Theta\sin\theta}\frac{d}{d\theta}\left(\sin\theta\frac{d\Theta}{d\theta}\right)=0. \tag{3.56}$$

Since the first term depends only on r, and the second only on θ, it follows that each must be a constant:

$$\frac{1}{R}\frac{d}{dr}\left(r^2\frac{dR}{dr}\right)=l(l+1),\qquad \frac{1}{\Theta\sin\theta}\frac{d}{d\theta}\left(\sin\theta\frac{d\Theta}{d\theta}\right)=-l(l+1). \tag{3.57}$$

Here $l(l+1)$ is just a fancy way of writing the separation constant—you'll see in a minute why this is convenient.

[7]The general case, for ϕ-dependent potentials, is treated in all the graduate texts. See, for instance, J. D. Jackson's *Classical Electrodynamics*, 3rd ed., Chapter 3 (New York: John Wiley, 1999).

As always, separation of variables has converted a *partial* differential equation (3.54) into *ordinary* differential equations (3.57). The radial equation,

$$\frac{d}{dr}\left(r^2\frac{dR}{dr}\right) = l(l+1)R, \tag{3.58}$$

has the general solution

$$R(r) = Ar^l + \frac{B}{r^{l+1}}, \tag{3.59}$$

as you can easily check; A and B are the two arbitrary constants to be expected in the solution of a second-order differential equation. But the angular equation,

$$\frac{d}{d\theta}\left(\sin\theta\frac{d\Theta}{d\theta}\right) = -l(l+1)\sin\theta\,\Theta, \tag{3.60}$$

is not so simple. The solutions are **Legendre polynomials** in the variable $\cos\theta$:

$$\Theta(\theta) = P_l(\cos\theta). \tag{3.61}$$

$P_l(x)$ is most conveniently defined by the **Rodrigues formula:**

$$P_l(x) \equiv \frac{1}{2^l l!}\left(\frac{d}{dx}\right)^l (x^2-1)^l. \tag{3.62}$$

The first few Legendre polynomials are listed in Table 3.1.

$P_0(x)$	=	1
$P_1(x)$	=	x
$P_2(x)$	=	$(3x^2-1)/2$
$P_3(x)$	=	$(5x^3-3x)/2$
$P_4(x)$	=	$(35x^4-30x^2+3)/8$
$P_5(x)$	=	$(63x^5-70x^3+15x)/8$

Table 3.1 Legendre Polynomials

Notice that $P_l(x)$ is (as the name suggests) an lth-order *polynomial* in x; it contains only *even* powers, if l is even, and *odd* powers, if l is odd. The factor in front $(1/2^l l!)$ was chosen in order that

$$P_l(1) = 1. \tag{3.63}$$

The Rodrigues formula obviously works only for nonnegative integer values of l. Moreover, it provides us with only *one* solution. But Eq. 3.60 is *second*-order, and it should possess *two* independent solutions, for *every* value of l. It turns out that these "other solutions" blow up at $\theta = 0$ and/or $\theta = \pi$, and are therefore unacceptable on physical grounds.[8] For instance, the second solution for $l = 0$ is

$$\Theta(\theta) = \ln\left(\tan\frac{\theta}{2}\right). \tag{3.64}$$

You might want to check for yourself that this satisfies Eq. 3.60.

In the case of azimuthal symmetry, then, the most general *separable* solution to Laplace's equation, consistent with minimal physical requirements, is

$$V(r,\theta) = \left(Ar^l + \frac{B}{r^{l+1}}\right) P_l(\cos\theta).$$

(There was no need to include an overall constant in Eq. 3.61 because it can be absorbed into A and B at this stage.) As before, separation of variables yields an infinite set of solutions, one for each l. The *general* solution is the linear combination of separable solutions:

$$\boxed{V(r,\theta) = \sum_{l=0}^{\infty}\left(A_l r^l + \frac{B_l}{r^{l+1}}\right) P_l(\cos\theta).} \tag{3.65}$$

The following examples illustrate the power of this important result.

Example 3.6

The potential $V_0(\theta)$ is specified on the surface of a hollow sphere, of radius R. Find the potential inside the sphere.

Solution: In this case $B_l = 0$ for all l—otherwise the potential would blow up at the origin. Thus,

$$V(r,\theta) = \sum_{l=0}^{\infty} A_l r^l P_l(\cos\theta). \tag{3.66}$$

At $r = R$ this must match the specified function $V_0(\theta)$:

$$V(R,\theta) = \sum_{l=0}^{\infty} A_l R^l P_l(\cos\theta) = V_0(\theta). \tag{3.67}$$

Can this equation be satisfied, for an appropriate choice of coefficients A_l? *Yes:* The Legendre polynomials (like the sines) constitute a complete set of functions, on the interval $-1 \le x \le 1$

---

[8] In rare cases where the z axis is for some reason inaccessible, these "other solutions" may have to be considered.

$(0 \leq \theta \leq \pi)$. How do we determine the constants? Again, by Fourier's trick, for the Legendre polynomials (like the sines) are *orthogonal* functions:[9]

$$\int_{-1}^{1} P_l(x) P_{l'}(x)\, dx = \int_0^{\pi} P_l(\cos\theta) P_{l'}(\cos\theta) \sin\theta\, d\theta = \begin{cases} 0, & \text{if } l' \neq l, \\ \dfrac{2}{2l+1}, & \text{if } l' = l. \end{cases} \tag{3.68}$$

Thus, multiplying Eq. 3.67 by $P_{l'}(\cos\theta)\sin\theta$ and integrating, we have

$$A_{l'} R^{l'} \frac{2}{2l'+1} = \int_0^{\pi} V_0(\theta) P_{l'}(\cos\theta) \sin\theta\, d\theta,$$

or

$$A_l = \frac{2l+1}{2R^l} \int_0^{\pi} V_0(\theta) P_l(\cos\theta) \sin\theta\, d\theta. \tag{3.69}$$

Equation 3.66 is the solution to our problem, with the coefficients given by Eq. 3.69.

It can be difficult to evaluate integrals of the form 3.69 analytically, and in practice it is often easier to solve Eq. 3.67 "by eyeball."[10] For instance, suppose we are told that the potential on the sphere is

$$V_0(\theta) = k \sin^2(\theta/2), \tag{3.70}$$

where k is a constant. Using the half-angle formula, we rewrite this as

$$V_0(\theta) = \frac{k}{2}(1 - \cos\theta) = \frac{k}{2}[P_0(\cos\theta) - P_1(\cos\theta)].$$

Putting this into Eq. 3.67, we read off immediately that $A_0 = k/2$, $A_1 = -k/(2R)$, and all other A_l's vanish. Evidently,

$$V(r,\theta) = \frac{k}{2}\left[r^0 P_0(\cos\theta) - \frac{r^1}{R} P_1(\cos\theta)\right] = \frac{k}{2}\left(1 - \frac{r}{R}\cos\theta\right). \tag{3.71}$$

Example 3.7

The potential $V_0(\theta)$ is again specified on the surface of a sphere of radius R, but this time we are asked to find the potential *outside*, assuming there is no charge there.

Solution: In this case it's the A_l's that must be zero (or else V would not go to zero at ∞), so

$$V(r,\theta) = \sum_{l=0}^{\infty} \frac{B_l}{r^{l+1}} P_l(\cos\theta). \tag{3.72}$$

---

[9] M. Boas, *Mathematical Methods in the Physical Sciences*, 2nd ed., Section 12.7 (New York: John Wiley, 1983).

[10] This is certainly true whenever $V_0(\theta)$ can be expressed as a polynomial in $\cos\theta$. The degree of the polynomial tells us the highest l we require, and the leading coefficient determines the corresponding A_l. Subtracting off $A_l R^l P_l(\cos\theta)$ and repeating the process, we systematically work our way down to A_0. Notice that if V_0 is an *even* function of $\cos\theta$, then only even terms will occur in the sum (and likewise for odd functions).

At the surface of the sphere we require that

$$V(R,\theta) = \sum_{l=0}^{\infty} \frac{B_l}{R^{l+1}} P_l(\cos\theta) = V_0(\theta).$$

Multiplying by $P_{l'}(\cos\theta)\sin\theta$ and integrating—exploiting, again, the orthogonality relation 3.68—we have

$$\frac{B_{l'}}{R^{l'+1}} \frac{2}{2l'+1} = \int_0^{\pi} V_0(\theta) P_{l'}(\cos\theta)\sin\theta\, d\theta,$$

or

$$B_l = \frac{2l+1}{2} R^{l+1} \int_0^{\pi} V_0(\theta) P_l(\cos\theta)\sin\theta\, d\theta. \tag{3.73}$$

Equation 3.72, with the coefficients given by Eq. 3.73, is the solution to our problem.

Example 3.8

An uncharged metal sphere of radius R is placed in an otherwise uniform electric field $\mathbf{E} = E_0\hat{\mathbf{z}}$. [The field will push positive charge to the "northern" surface of the sphere, leaving a negative charge on the "southern" surface (Fig. 3.24). This induced charge, in turn, distorts the field in the neighborhood of the sphere.] Find the potential in the region outside the sphere.

Solution: The sphere is an equipotential—we may as well set it to zero. Then by symmetry the entire xy plane is at potential zero. This time, however, V does *not* go to zero at large z. In fact, far from the sphere the field is $E_0\hat{\mathbf{z}}$, and hence

$$V \to -E_0 z + C.$$

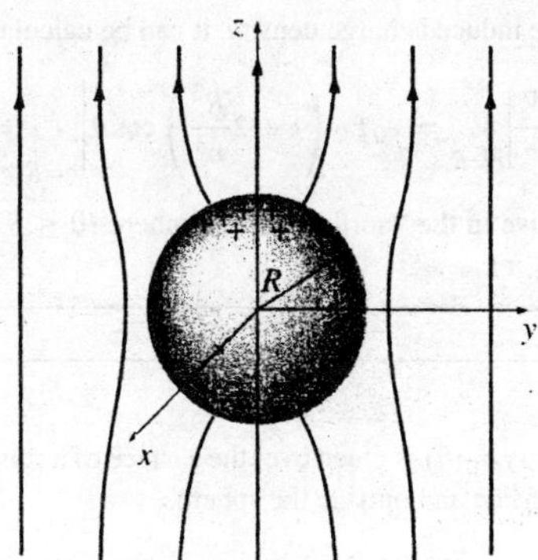

Figure 3.24

Since $V = 0$ in the equatorial plane, the constant C must be zero. Accordingly, the boundary conditions for this problem are

$$\left.\begin{array}{ll}\text{(i)} & V = 0 \quad \text{when } r = R, \\ \text{(ii)} & V \to -E_0 r\cos\theta \quad \text{for } r \gg R.\end{array}\right\} \tag{3.74}$$

We must fit these boundary conditions with a function of the form 3.65.

The first condition yields

$$A_l R^l + \frac{B_l}{R^{l+1}} = 0,$$

or

$$B_l = -A_l R^{2l+1}, \tag{3.75}$$

so

$$V(r,\theta) = \sum_{l=0}^{\infty} A_l \left(r^l - \frac{R^{2l+1}}{r^{l+1}} \right) P_l(\cos\theta).$$

For $r \gg R$, the second term in parentheses is negligible, and therefore condition (ii) requires that

$$\sum_{l=0}^{\infty} A_l r^l P_l(\cos\theta) = -E_0 r\cos\theta.$$

Evidently, only one term is present: $l = 1$. In fact, since $P_1(\cos\theta) = \cos\theta$, we can read off immediately

$$A_1 = -E_0, \quad \text{all other } A_l\text{'s zero.}$$

Conclusion:

$$V(r,\theta) = -E_0\left(r - \frac{R^3}{r^2} \right)\cos\theta. \tag{3.76}$$

The first term ($-E_0 r\cos\theta$) is due to the external field; the contribution attributable to the induced charge is evidently

$$E_0 \frac{R^3}{r^2}\cos\theta.$$

If you want to know the induced charge density, it can be calculated in the usual way:

$$\sigma(\theta) = -\epsilon_0 \frac{\partial V}{\partial r}\bigg|_{r=R} = \epsilon_0 E_0 \left(1 + 2\frac{R^3}{r^3} \right)\cos\theta \bigg|_{r=R} = 3\epsilon_0 E_0 \cos\theta. \tag{3.77}$$

As expected, it is positive in the "northern" hemisphere ($0 \le \theta \le \pi/2$) and negative in the "southern" ($\pi/2 \le \theta \le \pi$).

Example 3.9

A specified charge density $\sigma_0(\theta)$ is glued over the surface of a spherical shell of radius R. Find the resulting potential inside and outside the sphere.

Solution: You could, of course, do this by direct integration:

$$V = \frac{1}{4\pi\epsilon_0}\int \frac{\sigma_0}{\mathscr{r}}\, da,$$

but separation of variables is often easier. For the interior region we have

$$V(r,\theta)=\sum_{l=0}^{\infty}A_l r^l P_l(\cos\theta) \qquad (r\le R) \tag{3.78}$$

(no B_l terms—they blow up at the origin); in the exterior region

$$V(r,\theta)=\sum_{l=0}^{\infty}\frac{B_l}{r^{l+1}}P_l(\cos\theta) \qquad (r\ge R) \tag{3.79}$$

(no A_l terms—they don't go to zero at infinity). These two functions must be joined together by the appropriate boundary conditions at the surface itself. First, the potential is *continuous* at $r=R$ (Eq. 2.34):

$$\sum_{l=0}^{\infty}A_l R^l P_l(\cos\theta)=\sum_{l=0}^{\infty}\frac{B_l}{R^{l+1}}P_l(\cos\theta). \tag{3.80}$$

It follows that the coefficients of like Legendre polynomials are equal:

$$B_l=A_l R^{2l+1}. \tag{3.81}$$

(To prove that formally, multiply both sides of Eq. 3.80 by $P_{l'}(\cos\theta)\sin\theta$ and integrate from 0 to π, using the orthogonality relation 3.68.) Second, the radial derivative of V suffers a discontinuity at the surface (Eq. 2.36):

$$\left.\left(\frac{\partial V_{\text{out}}}{\partial r}-\frac{\partial V_{\text{in}}}{\partial r}\right)\right|_{r=R}=-\frac{1}{\epsilon_0}\sigma_0(\theta). \tag{3.82}$$

Thus

$$-\sum_{l=0}^{\infty}(l+1)\frac{B_l}{R^{l+2}}P_l(\cos\theta)-\sum_{l=0}^{\infty}lA_l R^{l-1}P_l(\cos\theta)=-\frac{1}{\epsilon_0}\sigma_0(\theta),$$

or, using Eq. 3.81:

$$\sum_{l=0}^{\infty}(2l+1)A_l R^{l-1}P_l(\cos\theta)=\frac{1}{\epsilon_0}\sigma_0(\theta). \tag{3.83}$$

From here, the coefficients can be determined using Fourier's trick:

$$A_l=\frac{1}{2\epsilon_0 R^{l-1}}\int_0^{\pi}\sigma_0(\theta)P_l(\cos\theta)\sin\theta\,d\theta. \tag{3.84}$$

Equations 3.78 and 3.79 constitute the solution to our problem, with the coefficients given by Eqs. 3.81 and 3.84.

For instance, if

$$\sigma_0(\theta)=k\cos\theta=kP_1(\cos\theta), \tag{3.85}$$

for some constant k, then all the A_l's are zero except for $l=1$, and

$$A_1=\frac{k}{2\epsilon_0}\int_0^{\pi}[P_1(\cos\theta)]^2\sin\theta\,d\theta=\frac{k}{3\epsilon_0}.$$

The potential inside the sphere is therefore

$$V(r,\theta) = \frac{k}{3\epsilon_0} r\cos\theta \qquad (r \le R), \tag{3.86}$$

whereas outside the sphere

$$V(r,\theta) = \frac{kR^3}{3\epsilon_0}\frac{1}{r^2}\cos\theta \qquad (r \ge R). \tag{3.87}$$

In particular, if $\sigma_0(\theta)$ is the induced charge on a metal sphere in an external field $E_0\hat{\mathbf{z}}$, so that $k = 3\epsilon_0 E_0$ (Eq. 3.77), then the potential inside is $E_0 r\cos\theta = E_0 z$, and the *field* is $-E_0\hat{\mathbf{z}}$—exactly right to cancel off the external field, as of course it *should* be. Outside the sphere the potential due to this surface charge is

$$E_0\frac{R^3}{r^2}\cos\theta,$$

consistent with our conclusion in Ex. 3.8.

Problem 3.16 Derive $P_3(x)$ from the Rodrigues formula, and check that $P_3(\cos\theta)$ satisfies the angular equation (3.60) for $l = 3$. Check that P_3 and P_1 are orthogonal by explicit integration.

Problem 3.17

(a) Suppose the potential is a *constant* V_0 over the surface of the sphere. Use the results of Ex. 3.6 and Ex. 3.7 to find the potential inside and outside the sphere. (Of course, you know the answers in advance—this is just a consistency check on the method.)

(b) Find the potential inside and outside a spherical shell that carries a *uniform* surface charge σ_0, using the results of Ex. 3.9.

Problem 3.18 The potential at the surface of a sphere (radius R) is given by

$$V_0 = k\cos 3\theta,$$

where k is a constant. Find the potential inside and outside the sphere, as well as the surface charge density $\sigma(\theta)$ on the sphere. (Assume there's no charge inside or outside the sphere.)

Problem 3.19 Suppose the potential $V_0(\theta)$ at the surface of a sphere is specified, and there is no charge inside or outside the sphere. Show that the charge density on the sphere is given by

$$\sigma(\theta) = \frac{\epsilon_0}{2R}\sum_{l=0}^{\infty}(2l+1)^2 C_l P_l(\cos\theta), \tag{3.88}$$

where

$$C_l = \int_0^{\pi} V_0(\theta) P_l(\cos\theta)\sin\theta\, d\theta. \tag{3.89}$$

Problem 3.20 Find the potential outside a *charged* metal sphere (charge Q, radius R) placed in an otherwise uniform electric field $\mathbf{E}_0$. Explain clearly where you are setting the zero of potential.

Problem 3.21 In Prob. 2.25 you found the potential on the axis of a uniformly charged disk:

$$V(r, 0) = \frac{\sigma}{2\epsilon_0}\left(\sqrt{r^2 + R^2} - r\right).$$

(a) Use this, together with the fact that $P_l(1) = 1$, to evaluate the first three terms in the expansion (3.72) for the potential of the disk at points *off* the axis, assuming $r > R$.

(b) Find the potential for $r < R$ by the same method, using (3.66). [*Note:* You must break the interior region up into two hemispheres, above and below the disk. Do *not* assume the coefficients A_l are the same in both hemispheres.]

Problem 3.22 A spherical shell of radius R carries a uniform surface charge σ_0 on the "northern" hemisphere and a uniform surface charge $-\sigma_0$ on the "southern" hemisphere. Find the potential inside and outside the sphere, calculating the coefficients explicitly up to A_6 and B_6.

Problem 3.23 Solve Laplace's equation by separation of variables in *cylindrical* coordinates, assuming there is no dependence on z (cylindrical symmetry). [Make sure you find *all* solutions to the radial equation; in particular, your result must accommodate the case of an infinite line charge, for which (of course) we already know the answer.]

Problem 3.24 Find the potential outside an infinitely long metal pipe, of radius R, placed at right angles to an otherwise uniform electric field $\mathbf{E}_0$. Find the surface charge induced on the pipe. [Use your result from Prob. 3.23.]

Problem 3.25 Charge density

$$\sigma(\phi) = a \sin 5\phi$$

(where a is a constant) is glued over the surface of an infinite cylinder of radius R (Fig. 3.25). Find the potential inside and outside the cylinder. [Use your result from Prob. 3.23.]

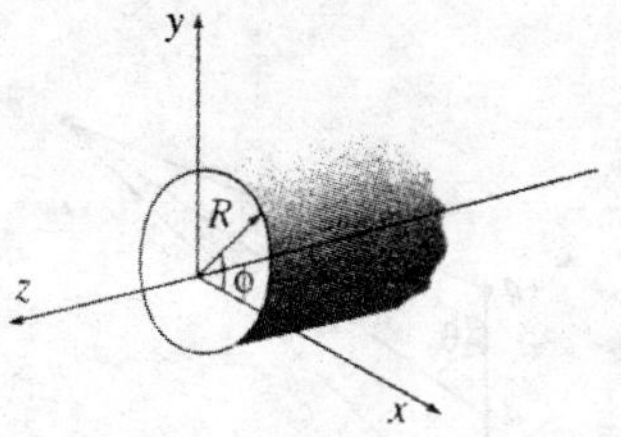

Figure 3.25

3.4 Multipole Expansion

3.4.1 Approximate Potentials at Large Distances

If you are very far away from a localized charge distribution, it "looks" like a point charge, and the potential is—to good approximation—$(1/4\pi\epsilon_0)Q/r$, where Q is the total charge. We have often used this as a check on formulas for V. But what if Q is *zero?* You might reply that the potential is then approximately zero, and of course, you're *right,* in a sense (indeed, the potential at large r is *pretty small* even if Q is *not* zero). But we're looking for something a bit more informative than that.

Example 3.10

A (physical) **electric dipole** consists of two equal and opposite charges ($\pm q$) separated by a distance d. Find the approximate potential at points far from the dipole.

Solution: Let $\imath_-$ be the distance from $-q$ and $\imath_+$ the distance from $+q$ (Fig. 3.26). Then

$$V(\mathbf{r}) = \frac{1}{4\pi\epsilon_0}\left(\frac{q}{\imath_+} - \frac{q}{\imath_-}\right),$$

and (from the law of cosines)

$$\imath_\pm^2 = r^2 + (d/2)^2 \mp rd\cos\theta = r^2\left(1 \mp \frac{d}{r}\cos\theta + \frac{d^2}{4r^2}\right).$$

We're interested in the régime $r \gg d$, so the third term is negligible, and the binomial expansion yields

$$\frac{1}{\imath_\pm} \cong \frac{1}{r}\left(1 \mp \frac{d}{r}\cos\theta\right)^{-1/2} \cong \frac{1}{r}\left(1 \pm \frac{d}{2r}\cos\theta\right).$$

Thus

$$\frac{1}{\imath_+} - \frac{1}{\imath_-} \cong \frac{d}{r^2}\cos\theta,$$

and hence

$$V(\mathbf{r}) \cong \frac{1}{4\pi\epsilon_0}\frac{qd\cos\theta}{r^2}. \tag{3.90}$$

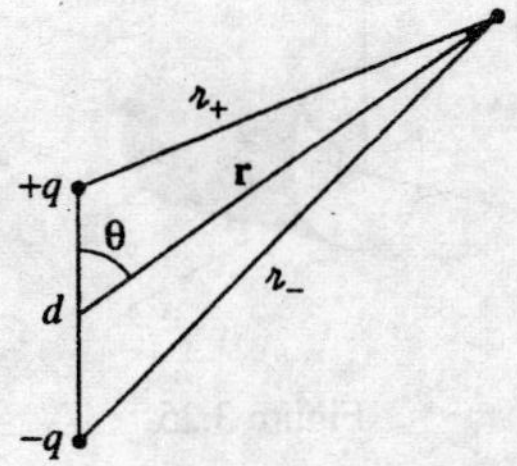

Figure 3.26

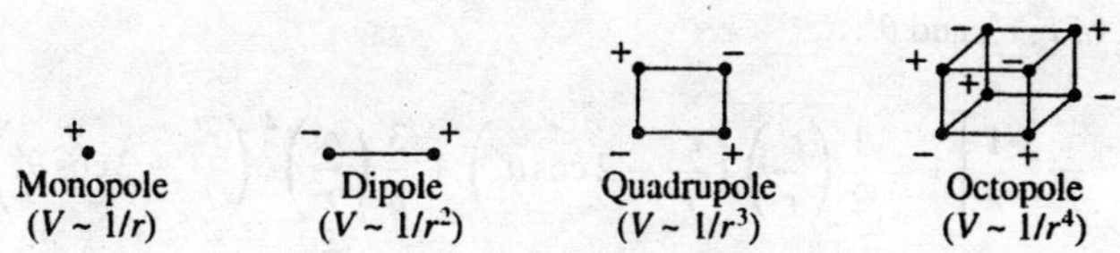

Monopole ($V \sim 1/r$) Dipole ($V \sim 1/r^2$) Quadrupole ($V \sim 1/r^3$) Octopole ($V \sim 1/r^4$)

Figure 3.27

Evidently the potential of a dipole goes like $1/r^2$ at large r; as we might have anticipated, it falls off more rapidly than the potential of a point charge. Incidentally, if we put together a pair of equal and opposite *dipoles* to make a **quadrupole**, the potential goes like $1/r^3$; for back-to-back *quadrupoles* (an **octopole**) it goes like $1/r^4$; and so on. Figure 3.27 summarizes this hierarchy; for completeness I have included the electric **monopole** (point charge), whose potential, of course, goes like $1/r$.

Example 3.10 pertained to a very special charge configuration. I propose now to develop a *systematic expansion for the potential of an arbitrary localized charge distribution, in powers of* $1/r$. Figure 3.28 defines the appropriate variables; the potential at $\mathbf{r}$ is given by

$$V(\mathbf{r}) = \frac{1}{4\pi\epsilon_0} \int \frac{1}{\mathscr{r}} \rho(\mathbf{r}')\, d\tau'. \tag{3.91}$$

Using the law of cosines,

$$\mathscr{r}^2 = r^2 + (r')^2 - 2rr'\cos\theta' = r^2 \left[1 + \left(\frac{r'}{r}\right)^2 - 2\left(\frac{r'}{r}\right)\cos\theta'\right],$$

or

$$\mathscr{r} = r\sqrt{1+\epsilon} \tag{3.92}$$

where

$$\epsilon \equiv \left(\frac{r'}{r}\right)\left(\frac{r'}{r} - 2\cos\theta'\right).$$

For points well outside the charge distribution, ϵ is much less than 1, and this invites a binomial expansion:

$$\frac{1}{\mathscr{r}} = \frac{1}{r}(1+\epsilon)^{-1/2} = \frac{1}{r}\left(1 - \frac{1}{2}\epsilon + \frac{3}{8}\epsilon^2 - \frac{5}{16}\epsilon^3 + \dots\right), \tag{3.93}$$

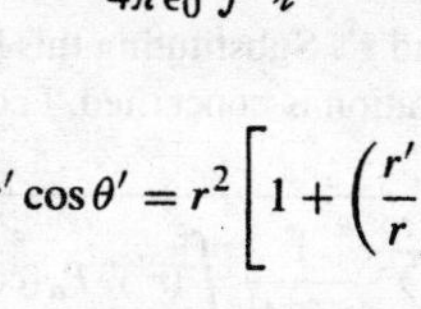

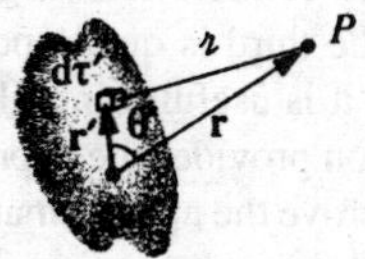

Figure 3.28

or, in terms of r, r', and θ':

$$\begin{aligned}\frac{1}{\imath} &= \frac{1}{r}\left[1-\frac{1}{2}\left(\frac{r'}{r}\right)\left(\frac{r'}{r}-2\cos\theta'\right)+\frac{3}{8}\left(\frac{r'}{r}\right)^2\left(\frac{r'}{r}-2\cos\theta'\right)^2\right.\\ &\quad\left.-\frac{5}{16}\left(\frac{r'}{r}\right)^3\left(\frac{r'}{r}-2\cos\theta'\right)^3+\ldots\right]\\ &= \frac{1}{r}\left[1+\left(\frac{r'}{r}\right)(\cos\theta')+\left(\frac{r'}{r}\right)^2(3\cos^2\theta'-1)/2\right.\\ &\quad\left.+\left(\frac{r'}{r}\right)^3(5\cos^3\theta'-3\cos\theta')/2+\ldots\right].\end{aligned}$$

In the last step I have collected together like powers of (r'/r); surprisingly, their coefficients (the terms in parentheses) are Legendre polynomials! The remarkable result[11] is that

$$\frac{1}{\imath}=\frac{1}{r}\sum_{n=0}^{\infty}\left(\frac{r'}{r}\right)^n P_n(\cos\theta'), \tag{3.94}$$

where θ' is the angle between $\mathbf{r}$ and $\mathbf{r}'$. Substituting this back into Eq. 3.91, and noting that r is a constant, as far as the integration is concerned, I conclude that

$$\boxed{V(\mathbf{r})=\frac{1}{4\pi\epsilon_0}\sum_{n=0}^{\infty}\frac{1}{r^{(n+1)}}\int(r')^n P_n(\cos\theta')\rho(\mathbf{r}')\,d\tau',} \tag{3.95}$$

or, more explicitly,

$$\begin{aligned}V(\mathbf{r}) &= \frac{1}{4\pi\epsilon_0}\left[\frac{1}{r}\int\rho(\mathbf{r}')\,d\tau'+\frac{1}{r^2}\int r'\cos\theta'\rho(\mathbf{r}')\,d\tau'\right.\\ &\quad\left.+\frac{1}{r^3}\int(r')^2\left(\frac{3}{2}\cos^2\theta'-\frac{1}{2}\right)\rho(\mathbf{r}')\,d\tau'+\ldots\right].\end{aligned} \tag{3.96}$$

This is the desired result—the **multipole expansion** of V in powers of $1/r$. The first term ($n=0$) is the monopole contribution (it goes like $1/r$); the second ($n=1$) is the dipole (it goes like $1/r^2$); the third is quadrupole; the fourth octopole; and so on. As it stands, Eq. 3.95 is *exact*, but it is useful primarily as an *approximation* scheme: the lowest nonzero term in the expansion provides the approximate potential at large r, and the successive terms tell us how to improve the approximation if greater precision is required.

---

[11] Incidentally, this affords a second way of obtaining the Legendre polynomials (the first being Rodrigues' formula); $1/\imath$ is called the **generating function** for Legendre polynomials.

Problem 3.26 A sphere of radius R, centered at the origin, carries charge density

$$\rho(r, \theta) = k\frac{R}{r^2}(R - 2r)\sin\theta,$$

where k is a constant, and r, θ are the usual spherical coordinates. Find the approximate potential for points on the z axis, far from the sphere.

3.4.2 The Monopole and Dipole Terms

Ordinarily, the multipole expansion is dominated (at large r) by the monopole term:

$$V_{\text{mon}}(\mathbf{r}) = \frac{1}{4\pi\epsilon_0}\frac{Q}{r}, \tag{3.97}$$

where $Q = \int \rho\, d\tau$ is the total charge of the configuration. This is just what we expected for the approximate potential at large distances from the charge. Incidentally, for a *point* charge *at the origin,* V_{mon} represents the *exact* potential everywhere, not merely a first approximation at large r; in this case all the higher multipoles vanish.

If the total charge is zero, the dominant term in the potential will be the dipole (unless, of course, it *also* vanishes):

$$V_{\text{dip}}(\mathbf{r}) = \frac{1}{4\pi\epsilon_0}\frac{1}{r^2}\int r'\cos\theta' \rho(\mathbf{r}')\, d\tau'.$$

Since θ' is the angle between $\mathbf{r}'$ and $\mathbf{r}$ (Fig. 3.28),

$$r'\cos\theta' = \hat{\mathbf{r}}\cdot\mathbf{r}',$$

and the dipole potential can be written more succinctly:

$$V_{\text{dip}}(\mathbf{r}) = \frac{1}{4\pi\epsilon_0}\frac{1}{r^2}\hat{\mathbf{r}}\cdot\int \mathbf{r}'\rho(\mathbf{r}')\, d\tau'.$$

This integral, which does not depend on $\mathbf{r}$ at all, is called **dipole moment** of the distribution:

$$\boxed{\mathbf{p} \equiv \int \mathbf{r}'\rho(\mathbf{r}')\, d\tau',} \tag{3.98}$$

and the dipole contribution to the potential simplifies to

$$\boxed{V_{\text{dip}}(\mathbf{r}) = \frac{1}{4\pi\epsilon_0}\frac{\mathbf{p}\cdot\hat{\mathbf{r}}}{r^2}.} \tag{3.99}$$

The dipole moment is determined by the geometry (size, shape, and density) of the charge distribution. Equation 3.98 translates in the usual way (Sect. 2.1.4) for point, line, and surface charges. Thus, the dipole moment of a collection of *point* charges is

$$\mathbf{p} = \sum_{i=1}^{n} q_i \mathbf{r}'_i. \tag{3.100}$$

For the "physical" dipole (equal and opposite charges, $\pm q$)

$$\mathbf{p} = q\mathbf{r}'_+ - q\mathbf{r}'_- = q(\mathbf{r}'_+ - \mathbf{r}'_-) = q\mathbf{d}, \tag{3.101}$$

where $\mathbf{d}$ is the vector from the negative charge to the positive one (Fig. 3.29).

Is this consistent with what we got for a *physical* dipole, in Ex. 3.10? Yes: If you put Eq. 3.100 into Eq. 3.99, you recover Eq. 3.90. Notice, however, that this is only the *approximate* potential of the physical dipole—evidently there are higher multipole contributions. Of course, as you go farther and farther away, V_{dip} becomes a better and better approximation, since the higher terms die off more rapidly with increasing r. By the same token, at a fixed r the dipole approximation improves as you shrink the separation d. To construct a "pure" dipole whose potential is given *exactly* by Eq. 3.99, you'd have to let d approach zero. Unfortunately, you then lose the dipole term *too,* unless you simultaneously arrange for q to go to infinity! A *physical* dipole becomes a *pure* dipole, then, in the rather artificial limit $d \to 0$, $q \to \infty$, with the product $qd = p$ held fixed. (When someone uses the word "dipole," you can't always tell whether they mean a *physical* dipole (with finite separation between the charges) or a *pure* (point) dipole. If in doubt, assume that d is small enough (compared to r) that you can safely apply Eq. 3.99.)

Dipole moments are *vectors,* and they add accordingly: if you have two dipoles, $\mathbf{p}_1$ and $\mathbf{p}_2$, the total dipole moment is $\mathbf{p}_1 + \mathbf{p}_2$. For instance, with four charges at the corners of a square, as shown in Fig. 3.30, the net dipole moment is zero. You can see this by combining the charges in pairs (vertically, $\downarrow + \uparrow = 0$, or horizontally, $\rightarrow + \leftarrow = 0$) or by adding up the four contributions individually, using Eq. 3.100. This is a *quadrupole,* as I indicated earlier, and its potential is dominated by the quadrupole term in the multipole expansion.)

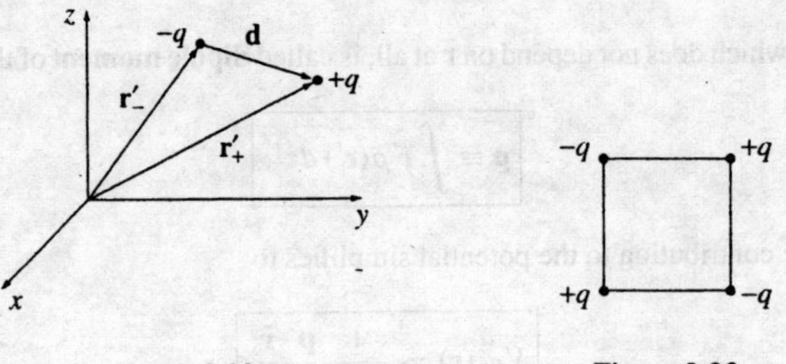

Figure 3.29 Figure 3.30

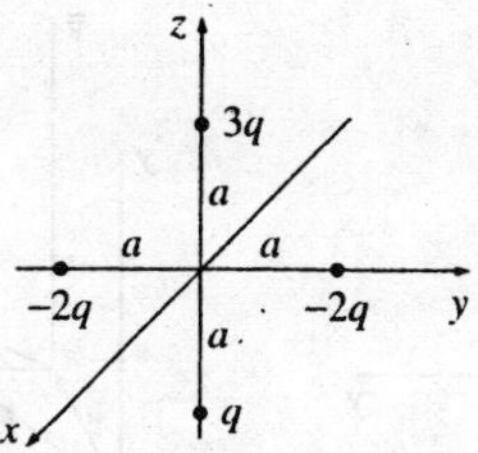

Figure 3.31

Problem 3.27 Four particles (one of charge q, one of charge $3q$, and two of charge $-2q$) are placed as shown in Fig. 3.31, each a distance a from the origin. Find a simple approximate formula for the potential, valid at points far from the origin. (Express your answer in spherical coordinates.)

Problem 3.28 In Ex. 3.9 we derived the exact potential for a spherical shell of radius R, which carries a surface charge $\sigma = k\cos\theta$.

(a) Calculate the dipole moment of this charge distribution.

(b) Find the approximate potential, at points far from the sphere, and compare the exact answer (3.87). What can you conclude about the higher multipoles?

Problem 3.29 For the dipole in Ex. 3.10, expand $1/\mathscr{r}_\pm$ to order $(d/r)^3$, and use this to determine the quadrupole and octopole terms in the potential.

3.4.3 Origin of Coordinates in Multipole Expansions

I mentioned earlier that a point charge at the origin constitutes a "pure" monopole. If it is *not* at the origin, it's no longer a pure monopole. For instance, the charge in Fig. 3.32 has a dipole moment $\mathbf{p} = qd\hat{\mathbf{y}}$, and a corresponding dipole term in its potential. The monopole potential $(1/4\pi\epsilon_0)q/r$ is not quite correct for this configuration; rather, the exact potential is $(1/4\pi\epsilon_0)q/\mathscr{r}$. The multipole expansion is, remember, a series in inverse powers of r (the distance to the *origin*), and when we expand $1/\mathscr{r}$ we get *all* powers, not just the first.

So moving the origin (or, what amounts to the same thing, moving the *charge*) can radically alter a multipole expansion. The **monopole moment** Q does not change, since the total charge is obviously independent of the coordinate system. (In Fig. 3.32 the monopole term was unaffected when we moved q away from the origin—it's just that it was no longer the whole story: a dipole term—and for that matter all higher poles—appeared as well.) Ordinarily, the dipole moment *does* change when you shift the origin, but there is an important exception: *If the total charge is zero, then the dipole moment is independent of*

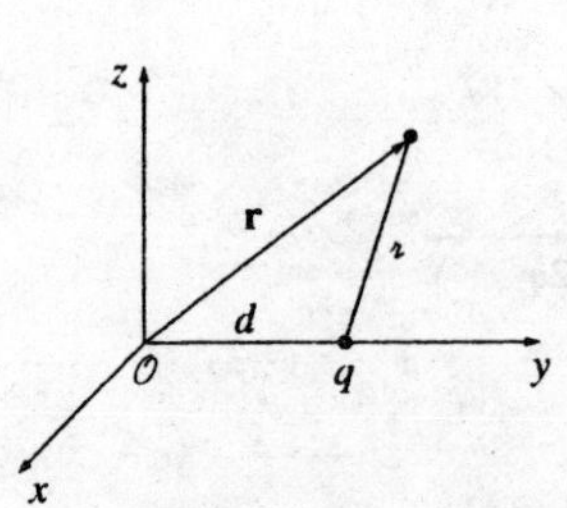

Figure 3.32

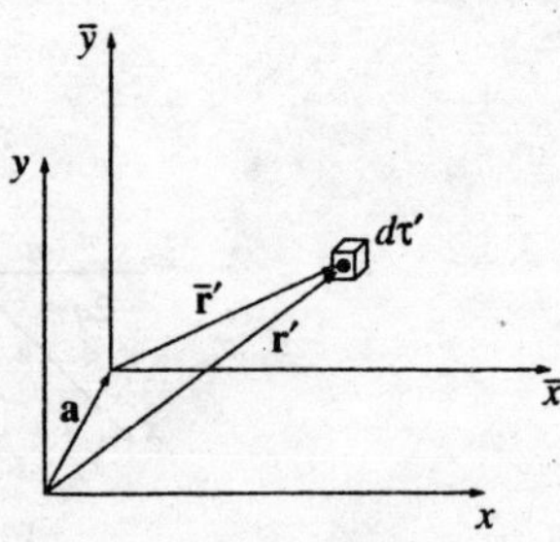

Figure 3.33

the choice of origin. For suppose we displace the origin by an amount $\mathbf{a}$ (Fig. 3.33). The new dipole moment is then

$$\begin{aligned}\bar{\mathbf{p}} &= \int \bar{\mathbf{r}}' \rho(\mathbf{r}')\, d\tau' = \int (\mathbf{r}' - \mathbf{a}) \rho(\mathbf{r}')\, d\tau' \\ &= \int \mathbf{r}' \rho(\mathbf{r}')\, d\tau' - \mathbf{a} \int \rho(\mathbf{r}')\, d\tau' = \mathbf{p} - Q\mathbf{a}.\end{aligned}$$

In particular, if $Q = 0$, then $\bar{\mathbf{p}} = \mathbf{p}$. So if someone asks for the dipole moment in Fig. 3.34(a), you can answer with confidence "$q\mathbf{d}$," but if you're asked for the dipole moment in Fig. 3.34(b) the appropriate response would be: "With respect to *what origin*?"

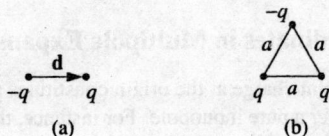

Figure 3.34

Problem 3.30 Two point charges, $3q$ and $-q$, are separated by a distance a. For each of the arrangements in Fig. 3.35, find (i) the monopole moment, (ii) the dipole moment, and (iii) the approximate potential (in spherical coordinates) at large r (include both the monopole and dipole contributions).

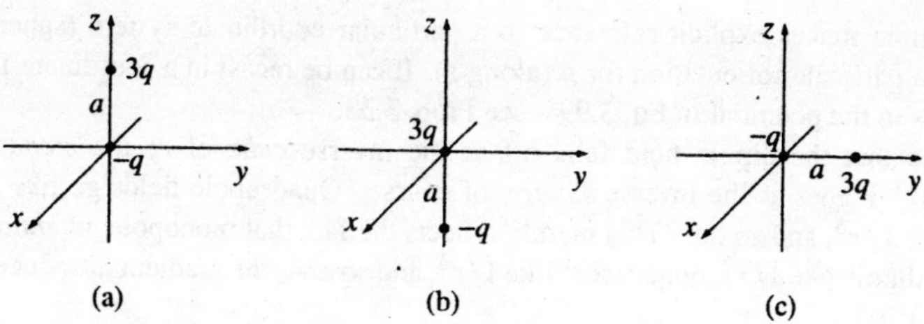

Figure 3.35

3.4.4 The Electric Field of a Dipole

So far we have worked only with *potentials*. Now I would like to calculate the electric *field* of a (pure) dipole. If we choose coordinates so that $\mathbf{p}$ lies at the origin and points in the z direction (Fig. 3.36), then the potential at r, θ is (Eq. 3.99):

$$V_{\text{dip}}(r, \theta) = \frac{\hat{\mathbf{r}} \cdot \mathbf{p}}{4\pi\epsilon_0 r^2} = \frac{p\cos\theta}{4\pi\epsilon_0 r^2}. \tag{3.102}$$

To get the field, we take the negative gradient of V:

$$\begin{aligned} E_r &= -\frac{\partial V}{\partial r} = \frac{2p\cos\theta}{4\pi\epsilon_0 r^3}, \\ E_\theta &= -\frac{1}{r}\frac{\partial V}{\partial \theta} = \frac{p\sin\theta}{4\pi\epsilon_0 r^3}, \\ E_\phi &= -\frac{1}{r\sin\theta}\frac{\partial V}{\partial \phi} = 0. \end{aligned}$$

Thus

$$\boxed{\mathbf{E}_{\text{dip}}(r, \theta) = \frac{p}{4\pi\epsilon_0 r^3}(2\cos\theta\,\hat{\mathbf{r}} + \sin\theta\,\hat{\boldsymbol{\theta}}).} \tag{3.103}$$

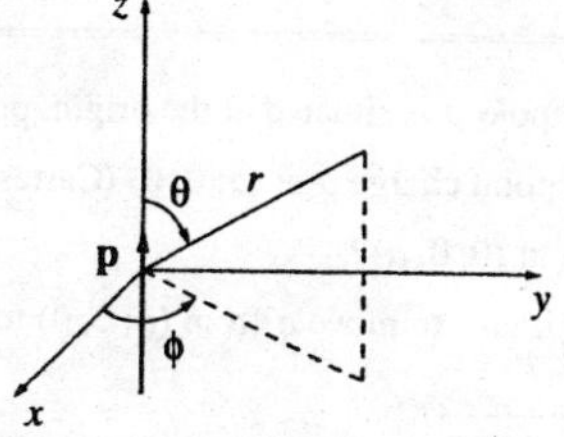

Figure 3.36

This formula makes explicit reference to a particular coordinate system (spherical) and assumes a particular orientation for **p** (along z). It can be recast in a coordinate-free form, analogous to the potential in Eq. 3.99—see Prob. 3.33.

Notice that the dipole field falls off as the inverse *cube* of r; the *monopole* field $(Q/4\pi\epsilon_0 r^2)\hat{\mathbf{r}}$ goes as the inverse *square,* of course. Quadrupole fields go like $1/r^4$, octopole like $1/r^5$, and so on. (This merely reflects the fact that monopole *potentials* fall off like $1/r$, dipole like $1/r^2$, quadrupole like $1/r^3$, and so on—the gradient introduces another factor of $1/r$.)

Figure 3.37(a) shows the field lines of a "pure" dipole (Eq. 3.103). For comparison, I have also sketched the field lines for a "physical" dipole, in Fig. 3.37(b). Notice how similar the two pictures become if you blot out the central region; up close, however, they are entirely different. Only for points $r \gg d$ does Eq. 3.103 represent a valid approximation to the field of a physical dipole. As I mentioned earlier, this régime can be reached either by going to large r or by squeezing the charges very close together.[12]

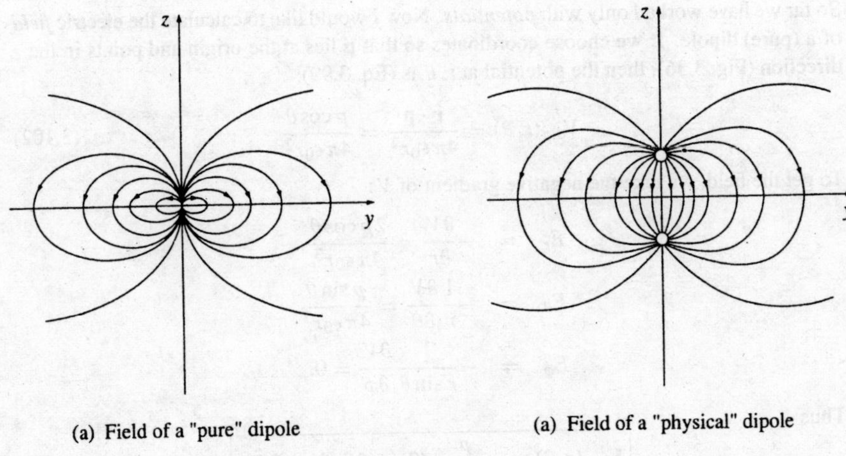

(a) Field of a "pure" dipole

(a) Field of a "physical" dipole

Figure 3.37

Problem 3.31 A "pure" dipole p is situated at the origin, pointing in the z direction.

(a) What is the force on a point charge q at $(a, 0, 0)$ (Cartesian coordinates)?

(b) What is the force on q at $(0, 0, a)$?

(c) How much work does it take to move q from $(a, 0, 0)$ to $(0, 0, a)$?

[12]Even in the limit, there remains an infinitesimal region at the origin where the field of a physical dipole points in the "wrong" direction, as you can see by "walking" down the z axis in Fig. 3.35(b). If you want to explore this subtle and important point, work Prob. 3.42.

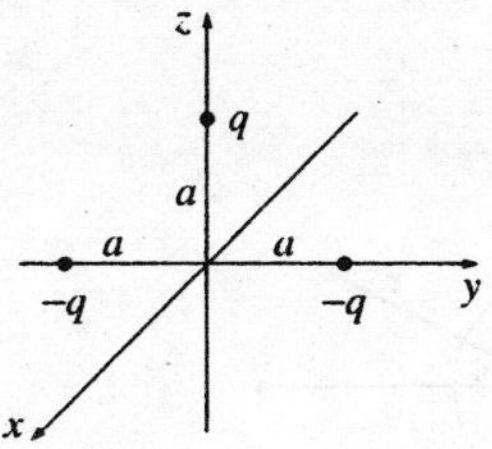

Figure 3.38

Problem 3.32 Three point charges are located as shown in Fig. 3.38, each a distance a from the origin. Find the approximate electric field at points far from the origin. Express your answer in spherical coordinates, and include the two lowest orders in the multipole expansion.

Problem 3.33 Show that the electric field of a ("pure") dipole (Eq. 3.103) can be written in the coordinate-free form

$$\boxed{\mathbf{E}_{\text{dip}}(\mathbf{r}) = \frac{1}{4\pi\epsilon_0}\frac{1}{r^3}[3(\mathbf{p}\cdot\hat{\mathbf{r}})\hat{\mathbf{r}} - \mathbf{p}].} \tag{3.104}$$

More Problems on Chapter 3

Problem 3.34 A point charge q of mass m is released from rest at a distance d from an infinite grounded conducting plane. How long will it take for the charge to hit the plane? [*Answer:* $(\pi d/q)\sqrt{2\pi\epsilon_0 md}$.]

Problem 3.35 Two infinite parallel grounded conducting planes are held a distance a apart. A point charge q is placed in the region between them, a distance x from one plate. Find the force on q. Check that your answer is correct for the special cases $a \to \infty$ and $x = a/2$. (Obtaining the induced surface charge is not so easy. See B. G. Dick, *Am. J. Phys.* **41,** 1289 (1973), M. Zahn, *Am. J. Phys.* **44,** 1132 (1976), J. Pleines and S. Mahajan, *Am. J. Phys.* **45,** 868 (1977), and Prob. 3.44 below.)

Problem 3.36 Two long straight wires, carrying opposite uniform line charges $\pm\lambda$, are situated on either side of a long conducting cylinder (Fig. 3.39). The cylinder (which carries no net charge) has radius R, and the wires are a distance a from the axis. Find the potential at point $\mathbf{r}$.

$$\left[\textit{Answer: } V(s,\phi) = \frac{\lambda}{4\pi\epsilon_0}\ln\left\{\frac{(s^2+a^2+2sa\cos\phi)[(sa/R)^2+R^2-2sa\cos\phi]}{(s^2+a^2-2sa\cos\phi)[(sa/R)^2+R^2+2sa\cos\phi]}\right\}\right]$$

Problem 3.37 A conducting sphere of radius a, at potential V_0, is surrounded by a thin concentric spherical shell of radius b, over which someone has glued a surface charge

$$\sigma(\theta) = k\cos\theta,$$

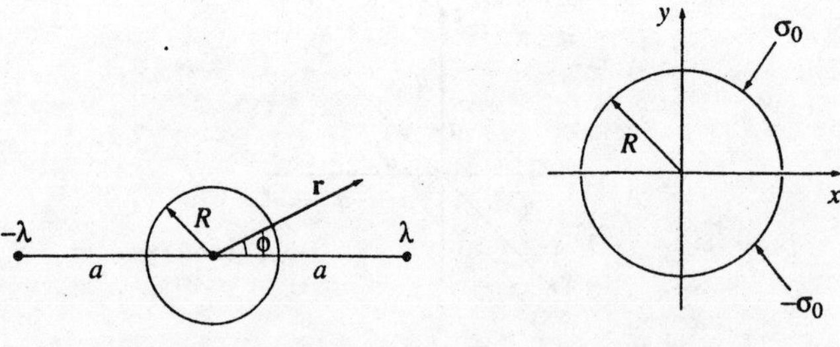

Figure 3.39 Figure 3.40

where k is a constant, and θ is the usual spherical coordinate.

(a) Find the potential in each region: (i) $r > b$, and (ii) $a < r < b$.

(b) Find the induced surface charge $\sigma_i(\theta)$ on the conductor.

(c) What is the total charge of this system? Check that your answer is consistent with the behavior of V at large r.

$$\left[\textit{Answer: } V(r,\theta) = \begin{cases} aV_0/r + (b^3 - a^3)k\cos\theta/3r^2\epsilon_0, & r \geq b \\ aV_0/r + (r^3 - a^3)k\cos\theta/3r^2\epsilon_0, & r \leq b \end{cases} \right]$$

Problem 3.38 A charge $+Q$ is distributed uniformly along the z axis from $z = -a$ to $z = +a$. Show that the electric potential at a point $\mathbf{r}$ is given by

$$V(r,\theta) = \frac{q}{4\pi\epsilon_0}\frac{1}{r}\left[1 + \frac{1}{3}\left(\frac{a}{r}\right)^2 P_2(\cos\theta) + \frac{1}{5}\left(\frac{a}{r}\right)^4 P_4(\cos\theta) + \ldots\right],$$

for $r > a$.

Problem 3.39 A long cylindrical shell of radius R carries a uniform surface charge σ_0 on the upper half and an opposite charge $-\sigma_0$ on the lower half (Fig. 3.40). Find the electric potential inside and outside the cylinder.

Problem 3.40 A thin insulating rod, running from $z = -a$ to $z = +a$, carries the indicated line charges. In each case, find the leading term in the multipole expansion of the potential: (a) $\lambda = k\cos(\pi z/2a)$, (b) $\lambda = k\sin(\pi z/a)$, (c) $\lambda = k\cos(\pi z/a)$, where k is a constant.

Problem 3.41 Show that the *average* field inside a sphere of radius R, due to all the charge within the sphere, is

$$\mathbf{E}_{\text{ave}} = -\frac{1}{4\pi\epsilon_0}\frac{\mathbf{p}}{R^3}, \tag{3.105}$$

where $\mathbf{p}$ is the total dipole moment. There are several ways to prove this delightfully simple result. Here's one method:

(a) Show that the average field due to a single charge q at point $\mathbf{r}$ inside the sphere is the same as the field at $\mathbf{r}$ due to a uniformly charged sphere with $\rho = -q/(\frac{4}{3}\pi R^3)$, namely

$$\frac{1}{4\pi\epsilon_0}\frac{1}{(\frac{4}{3}\pi R^3)}\int \frac{q}{\imath^2}\hat{\imath}\, d\tau',$$

where $\imath$ is the vector from $\mathbf{r}$ to $d\tau'$.

(b) The latter can be found from Gauss's law (see Prob. 2.12). Express the answer in terms of the dipole moment of q.

(c) Use the superposition principle to generalize to an arbitrary charge distribution.

(d) While you're at it, show that the average field over the sphere due to all the charges *outside* is the same as the field they produce at the center.

Problem 3.42 Using Eq. 3.103, calculate the average electric field of a dipole, over a spherical volume of radius R, centered at the origin. Do the angular intervals first. [*Note:* You *must* express $\hat{\mathbf{r}}$ and $\hat{\boldsymbol{\theta}}$ in terms of $\hat{\mathbf{x}}$, $\hat{\mathbf{y}}$, and $\hat{\mathbf{z}}$ (see back cover) before integrating. If you don't understand why, reread the discussion in Sect. 1.4.1.] Compare your answer with the general theorem Eq. 3.105. The discrepancy here is related to the fact that the field of a dipole blows up at $r = 0$. The angular integral is zero, but the radial integral is infinite, so we really don't know *what* to make of the answer. To resolve this dilemma, let's say that Eq. 3.103 applies *outside a tiny sphere of radius* ϵ—its contribution to E_{ave} is then *unambiguously* zero, and the whole answer has to come from the field *inside* the ϵ-sphere.

(b) What must the field *inside* the ϵ-sphere be, in order for the general theorem (3.105) to hold? [*Hint:* since ϵ is arbitrarily small, we're talking about something that is infinite at $r = 0$ and whose integral over an infinitesimal volume is finite.] [*Answer:* $-(\mathbf{p}/3\epsilon_0)\delta^3(\mathbf{r})$]

[Evidently, the *true* field of a dipole is

$$\mathbf{E}_{\text{dip}}(\mathbf{r}) = \frac{1}{4\pi\epsilon_0}\frac{1}{r^3}[3(\mathbf{p}\cdot\hat{\mathbf{r}})\hat{\mathbf{r}} - \mathbf{p}] - \frac{1}{3\epsilon_0}\mathbf{p}\,\delta^3(\mathbf{r}). \tag{3.106}$$

You may well wonder how we missed the delta-function term when we calculated the field back in Sect. 3.4.4. The answer is that the differentiation leading to Eq. 3.103 is perfectly valid *except* at $r = 0$, but we should have known (from our experience in Sect. 1.5.1) that the point $r = 0$ is problematic. See C. P. Frahm, *Am. J. Phys.* **51**, 826 (1983), or more recently R. Estrada and R. P. Kanwal, *Am. J. Phys.* **63**, 278 (1995). For further details and applications, see D. J. Griffiths, *Am. J. Phys.* **50**, 698 (1982).]

Problem 3.43

(a) Suppose a charge distribution $\rho_1(\mathbf{r})$ produces a potential $V_1(\mathbf{r})$, and some other charge distribution $\rho_2(\mathbf{r})$ produces a potential $V_2(\mathbf{r})$. [The two situations may have nothing in common, for all I care—perhaps number 1 is a uniformly charged sphere and number 2 is a parallel-plate capacitor. Please understand that ρ_1 and ρ_2 are not present *at the same time;* we are talking about two *different problems,* one in which only ρ_1 is present, and another in which only ρ_2 is present.] Prove **Green's reciprocity theorem:**

$$\int_{\text{all space}} \rho_1 V_2\, d\tau = \int_{\text{all space}} \rho_2 V_1\, d\tau.$$

Figure 3.41

[*Hint:* Evaluate $\int \mathbf{E}_1 \cdot \mathbf{E}_2 \, d\tau$ two ways, first writing $\mathbf{E}_1 = -\nabla V_1$ and using integration-by-parts to transfer the derivative to $\mathbf{E}_2$, then writing $\mathbf{E}_2 = -\nabla V_2$ and transferring the derivative to $\mathbf{E}_1$.]

(b) Suppose now that you have two separated conductors (Fig. 3.41). If you charge up conductor a by amount Q (leaving b uncharged) the resulting potential of b is, say, V_{ab}. On the other hand, if you put that same charge Q on conductor b (leaving a uncharged) the potential of a would be V_{ba}. Use Green's reciprocity theorem to show that $V_{ab} = V_{ba}$ (an astonishing result, since we assumed nothing about the shapes or placement of the conductors).

Problem 3.44 Use Green's reciprocity theorem (Prob. 3.43) to solve the following two problems. [*Hint:* for distribution 1, use the actual situation; for distribution 2, remove q, and set one of the conductors at potential V_0.]

(a) Both plates of a parallel-plate capacitor are grounded, and a point charge q is placed between them at a distance x from plate 1. The plate separation is d. Find the induced charge on each plate. [*Answer:* $Q_1 = q(x/d - 1)$; $Q_2 = -qx/d$]

(b) Two concentric spherical conducting shells (radii a and b) are grounded, and a point charge q is placed between them (at radius r). Find the induced charge on each sphere.

Problem 3.45

(a) Show that the quadrupole term in the multipole expansion can be written

$$V_{\text{quad}}(\mathbf{r}) = \frac{1}{4\pi\epsilon_0}\frac{1}{2r^3}\sum_{i,j=1}^{3}\hat{r}_i\hat{r}_j Q_{ij},$$

where

$$Q_{ij} \equiv \int [3r'_i r'_j - (r')^2\delta_{ij}]\rho(\mathbf{r}')\,d\tau'.$$

Here

$$\delta_{ij} = \begin{cases} 1 & \text{if } i = j \\ 0 & \text{if } i \neq j \end{cases}$$

is the **Kronecker delta**, and Q_{ij} is the **quadrupole moment** of the charge distribution. Notice the hierarchy:

$$V_{\text{mon}} = \frac{1}{4\pi\epsilon_0}\frac{Q}{r}; \quad V_{\text{dip}} = \frac{1}{4\pi\epsilon_0}\frac{\sum \hat{r}_i p_i}{r^2}; \quad V_{\text{quad}} = \frac{1}{4\pi\epsilon_0}\frac{\frac{1}{2}\sum \hat{r}_i\hat{r}_j Q_{ij}}{r^3}; \ldots$$

The monopole moment (Q) is a scalar, the dipole moment ($\mathbf{p}$) is a vector, the quadrupole moment (Q_{ij}) is a second-rank tensor, and so on.

(b) Find all nine components of Q_{ij} for the configuration in Fig. 3.30 (assume the square has side a and lies in the xy plane, centered at the origin).

(c) Show that the quadrupole moment is independent of origin if the monopole and dipole moments both vanish. (This works all the way up the hierarchy—the lowest nonzero multipole moment is always independent of origin.)

(d) How would you define the **octopole moment**? Express the octopole term in the multipole expansion in terms of the octopole moment.

Problem 3.46 In Ex. 3.8 we determined the electric field outside a spherical conductor (radius R) placed in a uniform external field $\mathbf{E}_0$. Solve the problem now using the method of images, and check that your answer agrees with Eq. 3.76. [*Hint:* Use Ex. 3.2, but put another charge, $-q$, diametrically opposite q. Let $a \to \infty$, with $(1/4\pi\epsilon_0)(2q/a^2) = -E_0$ held constant.]

! **Problem 3.47**For the infinite rectangular pipe in Ex. 3.4, suppose the potential on the bottom ($y = 0$) and the two sides ($x = \pm b$) is zero, but the potential on the top ($y = a$) is a nonzero constant V_0. Find the potential inside the pipe. [*Note:* This is a rotated version of Prob. 3.14(b), but set it up as in Ex. 3.4 using sinusoidal functions in y and hyperbolics in x. It is an unusual case in which $k = 0$ must be included. Begin by finding the general solution to Eq. 3.26 when $k = 0$. For further discussion see S. Hassani, *Am. J. Phys.* **59**, 470 (1991).]
$\left[\textit{Answer}: V_0\left(\frac{y}{a} + \frac{2}{\pi}\sum_{n=1}^{\infty}\frac{(-1)^n}{n}\frac{\cosh(n\pi x/a)}{\cosh(n\pi b/a)}\sin(n\pi y/a)\right)\right]$

! **Problem 3.48**

(a) A long metal pipe of square cross-section (side a) is grounded on three sides, while the fourth (which is insulated from the rest) is maintained at constant potential V_0. Find the net charge per unit length on the side *opposite* to V_0. [*Hint:* Use your answer to Prob. 3.14 or Prob. 3.47.]

(b) A long metal pipe of circular cross-section (radius R) is divided (lengthwise) into four equal sections, three of them grounded and the fourth maintained at constant potential V_0. Find the net charge per unit length on the section opposite to V_0. [*Answer to both (a) and (b):* $\lambda = -\epsilon_0 V_0 \ln 2$][13]

Problem 3.49 An ideal electric dipole is situated at the origin, and points in the z direction, as in Fig. 3.36. An electric charge is released from rest at a point in the xy plane. Show that it swings back and forth in a semi-circular arc, as though it were a pendulum supported at the origin. [This charming result is due to R. S. Jones, *Am. J. Phys.* **63**, 1042 (1995).]

[13] These are special cases of the **Thompson-Lampard theorem**; see J. D. Jackson, *Am. J. Phys.* **67**, 107 (1999).

Chapter 4

Electric Fields in Matter

4.1 Polarization

4.1.1 Dielectrics

In this chapter we shall study electric fields in matter. Matter, of course, comes in many varieties—solids, liquids, gases, metals, woods, glasses—and these substances do not all respond in the same way to electrostatic fields. Nevertheless, *most* everyday objects belong (at least, in good approximation) to one of two large classes: **conductors** and **insulators** (or **dielectrics**). We have already talked about conductors; these are substances that contain an "unlimited" supply of charges that are free to move about through the material. In practice what this ordinarily means is that many of the electrons (one or two per atom in a typical metal) are not associated with any particular nucleus, but roam around at will. In dielectrics, by contrast, *all charges are attached to specific atoms or molecules*—they're on a tight leash, and all they can do is move a bit *within* the atom or molecule. Such microscopic displacements are not as dramatic as the wholesale rearrangement of charge in a conductor, but their cumulative effects account for the characteristic behavior of dielectric materials. There are actually two principal mechanisms by which electric fields can distort the charge distribution of a dielectric atom or molecule: *stretching* and *rotating*. In the next two sections I'll discuss these processes.

4.1.2 Induced Dipoles

What happens to a neutral atom when it is placed in an electric field **E**? Your first guess might well be: "Absolutely nothing—since the atom is not charged, the field has no effect on it." But that is incorrect. Although the atom as a whole is electrically neutral, there *is* a positively charged core (the nucleus) and a negatively charged electron cloud surrounding it. These two regions of charge within the atom are influenced by the field: the nucleus is pushed in the direction of the field, and the electrons the opposite way. In principle, if the field is large enough, it can pull the atom apart completely, "ionizing" it (the substance then becomes a conductor). With less extreme fields, however, an equilibrium is soon established, for if the center of the electron cloud does not coincide with the nucleus, these positive and negative charges attract one another, and this holds the atoms together. The two opposing forces—**E** pulling the electrons and nucleus apart, their mutual attraction drawing them together—reach a balance, leaving the atom **polarized,** with plus charge shifted slightly one way, and minus the other. The atom now has a tiny dipole moment **p**, which points in the *same direction as* **E**. Typically, this induced dipole moment is approximately proportional to the field (as long as the latter is not too strong):

$$\mathbf{p} = \alpha \mathbf{E}. \tag{4.1}$$

The constant of proportionality α is called **atomic polarizability.** Its value depends on the detailed structure of the atom in question. Table 4.1 lists some experimentally determined atomic polarizabilities.

H	He	Li	Be	C	Ne	Na	Ar	K	Cs
0.667	0.205	24.3	5.60	1.76	0.396	24.1	1.64	43.4	59.6

Table 4.1 Atomic Polarizabilities ($\alpha/4\pi\epsilon_0$, in units of 10^{-30} m^3).
Source: Handbook of Chemistry and Physics, 78th ed.
(Boca Raton: CRC Press, Inc., 1997).

Example 4.1

A primitive model for an atom consists of a point nucleus ($+q$) surrounded by a uniformly charged spherical cloud ($-q$) of radius a (Fig. 4.1). Calculate the atomic polarizability of such an atom.

Solution: In the presence of an external field **E**, the nucleus will be shifted slightly to the right and the electron cloud to the left, as shown in Fig. 4.2. (Because the actual displacements

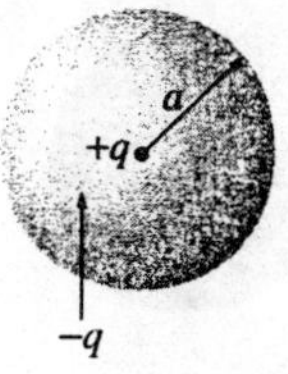

Figure 4.1

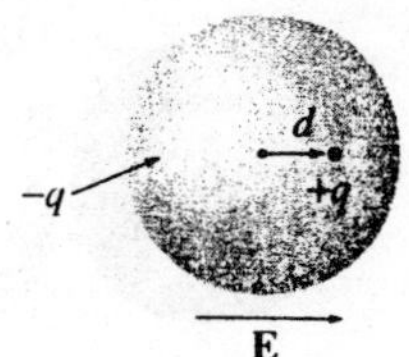

Figure 4.2

involved are extremely small, as you'll see in Prob. 4.1, it is reasonable to assume that the electron cloud retains its spherical shape.) Say that equilibrium occurs when the nucleus is displaced a distance d from the center of the sphere. At that point the external field pushing the nucleus to the right exactly balances the internal field pulling it to the left: $E = E_e$, where E_e is the field produced by the electron cloud. Now the field at a distance d from the center of a uniformly charged sphere is

$$E_e = \frac{1}{4\pi\epsilon_0}\frac{qd}{a^3}$$

(Prob. 2.12). At equilibrium, then,

$$E = \frac{1}{4\pi\epsilon_0}\frac{qd}{a^3}, \quad \text{or } p = qd = (4\pi\epsilon_0 a^3)E.$$

The atomic polarizability is therefore

$$\alpha = 4\pi\epsilon_0 a^3 = 3\epsilon_0 v, \tag{4.2}$$

where v is the volume of the atom. Although this atomic model is extremely crude, the result (4.2) is not too bad—it's accurate to within a factor of four or so for many simple atoms.

For molecules the situation is not quite so simple, because frequently they polarize more readily in some directions than others. Carbon dioxide (Fig. 4.3), for instance, has a polarizability of 4.5×10^{-40} $C^2 \cdot m/N$ when you apply the field along the axis of the molecule, but only 2×10^{-40} for fields perpendicular to this direction. When the field is at some *angle* to the axis, you must resolve it into parallel and perpendicular components, and multiply each by the pertinent polarizability:

$$\mathbf{p} = \alpha_\perp \mathbf{E}_\perp + \alpha_\parallel \mathbf{E}_\parallel.$$

In this case the induced dipole moment may not even be in the same *direction* as **E**. And CO_2 is relatively simple, as molecules go, since the atoms at least arrange themselves in a straight line; for a completely asymmetrical molecule Eq. 4.1 is replaced by the most general linear relation between **E** and **p**:

$$\left.\begin{aligned} p_x &= \alpha_{xx}E_x + \alpha_{xy}E_y + \alpha_{xz}E_z \\ p_y &= \alpha_{yx}E_x + \alpha_{yy}E_y + \alpha_{yz}E_z \\ p_z &= \alpha_{zx}E_x + \alpha_{zy}E_y + \alpha_{zz}E_z \end{aligned}\right\} \tag{4.3}$$

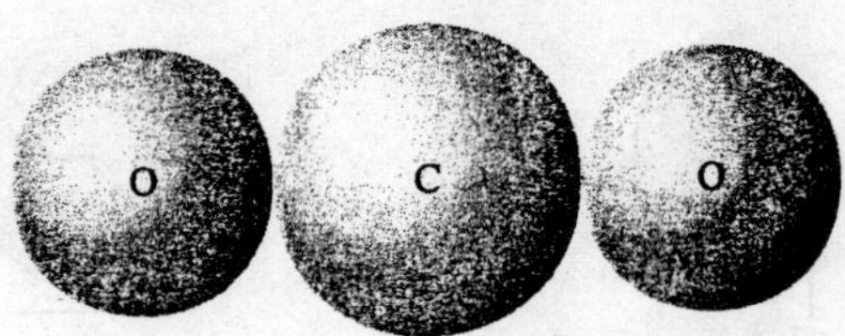

Figure 4.3

The set of nine constants α_{ij} constitute the **polarizability tensor** for the molecule. Their actual values depend on the orientation of the axes you chose, though it is always possible to choose "principal" axes such that all the off-diagonal terms (α_{xy}, α_{zx}, etc.) vanish, leaving just three nonzero polarizabilities: α_{xx}, α_{yy}, and α_{zz}.

Problem 4.1 A hydrogen atom (with the Bohr radius of half an angstrom) is situated between two metal plates 1 mm apart, which are connected to opposite terminals of a 500 V battery. What fraction of the atomic radius does the separation distance d amount to, roughly? Estimate the voltage you would need with this apparatus to ionize the atom. [Use the value of α in Table 4.1. *Moral:* The displacements we're talking about are *minute*, even on an atomic scale.]

Problem 4.2 According to quantum mechanics, the electron cloud for a hydrogen atom in the ground state has a charge density.

$$\rho(r) = \frac{q}{\pi a^3} e^{-2r/a},$$

where q is the charge of the electron and a is the Bohr radius. Find the atomic polarizability of such an atom. [*Hint:* First calculate the electric field of the electron cloud, $E_e(r)$; then expand the exponential, assuming $r \ll a$. For a more sophisticated approach, see W. A. Bowers, *Am. J. Phys.* **54**, 347 (1986).]

Problem 4.3 According to Eq. 4.1, the induced dipole moment of an atom is proportional to the external field. This is a "rule of thumb," not a fundamental law, and it is easy to concoct exceptions—in theory. Suppose, for example, the charge density of the electron cloud were proportional to the distance from the center, out to a radius R. To what power of E would p be proportional in that case? Find the condition on $\rho(r)$ such that Eq. 4.1 will hold in the weak-field limit.

Problem 4.4 A point charge q is situated a large distance r from a neutral atom of polarizability α. Find the force of attraction between them.

4.1.3 Alignment of Polar Molecules

The neutral atom discussed in Sect. 4.1.2 had no dipole moment to start with—$\mathbf{p}$ was *induced* by the applied field. Some molecules have built-in, permanent dipole moments. In the water molecule, for example, the electrons tend to cluster around the oxygen atom (Fig. 4.4), and since the molecule is bent at 105°, this leaves a negative charge at the vertex and a net positive charge at the opposite end. (The dipole moment of water is unusually large: 6.1×10^{-30} C·m; in fact, this is what accounts for its effectiveness as a solvent.) What happens when such molecules (called **polar molecules**) are placed in an electric field?

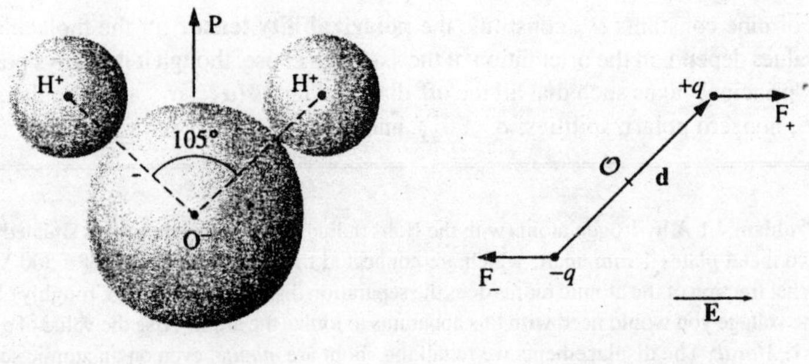

Figure 4.4

Figure 4.5

If the field is uniform, the *force* on the positive end, $\mathbf{F}_+ = q\mathbf{E}$, exactly cancels the force on the negative end, $\mathbf{F}_- = -q\mathbf{E}$ (Fig. 4.5). However, there will be a *torque:*

$$\begin{aligned}\mathbf{N} &= (\mathbf{r}_+ \times \mathbf{F}_+) + (\mathbf{r}_- \times \mathbf{F}_-)\\ &= [(\mathbf{d}/2) \times (q\mathbf{E})] + [(-\mathbf{d}/2) \times (-q\mathbf{E})] = q\mathbf{d} \times \mathbf{E}.\end{aligned}$$

Thus a dipole $\mathbf{p} = q\mathbf{d}$ in a uniform field $\mathbf{E}$ experiences a torque

$$\boxed{\mathbf{N} = \mathbf{p} \times \mathbf{E}.} \tag{4.4}$$

Notice that $\mathbf{N}$ is in such a direction as to line $\mathbf{p}$ up *parallel* to $\mathbf{E}$; a polar molecule that is free to rotate will swing around until it points in the direction of the applied field.

If the field is *non*uniform, so that $\mathbf{F}_+$ does not exactly balance $\mathbf{F}_-$, there will be a net *force* on the dipole, in addition to the torque. Of course, $\mathbf{E}$ must change rather abruptly for there to be significant variation in the space of one molecule, so this is not ordinarily a major consideration in discussing the behavior of dielectrics. Nevertheless, the formula for the force on a dipole in a nonuniform field is of some interest:

$$\mathbf{F} = \mathbf{F}_+ + \mathbf{F}_- = q(\mathbf{E}_+ - \mathbf{E}_-) = q(\Delta\mathbf{E}),$$

where $\Delta\mathbf{E}$ represents the difference between the field at the plus end and the field at the minus end. Assuming the dipole is very short, we may use Eq. 1.35 to approximate the small change in E_x:

$$\Delta E_x \equiv (\nabla E_x) \cdot \mathbf{d},$$

with corresponding formulas for E_y and E_z. More compactly,

$$\Delta\mathbf{E} = (\mathbf{d} \cdot \nabla)\mathbf{E},$$

and therefore[1]

$$\boxed{\mathbf{F} = (\mathbf{p} \cdot \nabla)\mathbf{E}.} \tag{4.5}$$

For a "perfect" dipole of infinitesimal length, Eq. 4.4 gives the torque *about the center of the dipole* even in a *non*uniform field; about any *other* point $\mathbf{N} = (\mathbf{p} \times \mathbf{E}) + (\mathbf{r} \times \mathbf{F})$.

Problem 4.5 In Fig. 4.6, $\mathbf{p}_1$ and $\mathbf{p}_2$ are (perfect) dipoles a distance r apart. What is the torque on $\mathbf{p}_1$ due to $\mathbf{p}_2$? What is the torque on $\mathbf{p}_2$ due to $\mathbf{p}_1$? [In each case I want the torque on the dipole *about its own center*. If it bothers you that the answers are not equal and opposite, see Prob. 4.29.]

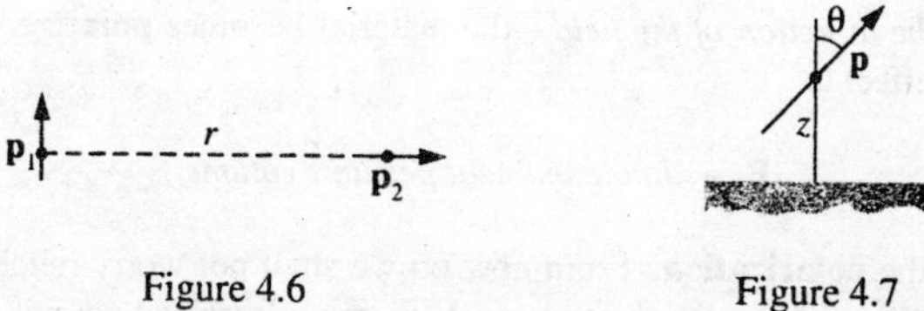

Figure 4.6 Figure 4.7

Problem 4.6 A (perfect) dipole $\mathbf{p}$ is situated a distance z above an infinite grounded conducting plane (Fig. 4.7). The dipole makes an angle θ with the perpendicular to the plane. Find the torque on $\mathbf{p}$. If the dipole is free to rotate, in what orientation will it come to rest?

Problem 4.7 Show that the energy of an ideal dipole $\mathbf{p}$ in an electric field $\mathbf{E}$ is given by

$$\boxed{U = -\mathbf{p} \cdot \mathbf{E}.} \tag{4.6}$$

Problem 4.8 Show that the interaction energy of two dipoles separated by a displacement $\mathbf{r}$ is

$$U = \frac{1}{4\pi\epsilon_0}\frac{1}{r^3}[\mathbf{p}_1 \cdot \mathbf{p}_2 - 3(\mathbf{p}_1 \cdot \hat{\mathbf{r}})(\mathbf{p}_2 \cdot \hat{\mathbf{r}})]. \tag{4.7}$$

[*Hint:* use Prob. 4.7 and Eq. 3.104.]

Problem 4.9 A dipole $\mathbf{p}$ is a distance r from a point charge q, and oriented so that $\mathbf{p}$ makes an angle θ with the vector $\mathbf{r}$ from q to $\mathbf{p}$.

(a) What is the force on $\mathbf{p}$?

(b) What is the force on q?

---

[1] In the present context Eq. 4.5 could be written more conveniently as $\mathbf{F} = \nabla(\mathbf{p} \cdot \mathbf{E})$. However, it is safer to stick with $(\mathbf{p} \cdot \nabla)\mathbf{E}$, because we will be applying the formula to materials in which the dipole moment (per unit volume) is itself a function of position and this second expression would imply (incorrectly) that $\mathbf{p}$ *too* is to be differentiated.

4.1.4 Polarization

In the previous two sections we have considered the effect of an external electric field on an individual atom or molecule. We are now in a position to answer (qualitatively) the original question: What happens to a piece of dielectric material when it is placed in an electric field? If the substance consists of neutral atoms (or nonpolar molecules), the field will induce in each a tiny dipole moment, pointing in the same direction as the field.[2] If the material is made up of polar molecules, each permanent dipole will experience a torque, tending to line it up along the field direction. (Random thermal motions compete with this process, so the alignment is never complete, especially at higher temperatures, and disappears almost at once when the field is removed.)

Notice that these two mechanisms produce the same basic result: *a lot of little dipoles pointing along the direction of the field*—the material becomes **polarized**. A convenient measure of this effect is

$$\mathbf{P} \equiv \textit{dipole moment per unit volume},$$

which is called the **polarization**. From now on we shall not worry much about how the polarization *got* there. Actually, the two mechanisms I described are not as clear-cut as I tried to pretend. Even in polar molecules there will be some polarization by displacement (though generally it is a lot easier to rotate a molecule than to stretch it, so the second mechanism dominates). It's even possible in some materials to "freeze in" polarization, so that it persists after the field is removed. But let's forget for a moment about the *cause* of the polarization and study the field that a chunk of polarized material *itself* produces. Then, in Sect. 4.3, we'll put it all together: the original field, which was *responsible* for **P**, plus the new field, which is *due* to **P**.

4.2 The Field of a Polarized Object

4.2.1 Bound Charges

Suppose we have a piece of polarized material—that is, an object containing a lot of microscopic dipoles lined up. The dipole moment per unit volume **P** is given. *Question:* What is the field produced by this object (not the field that may have *caused* the polarization, but the field the polarization *itself* causes)? Well, we know what the field of an individual dipole looks like, so why not chop the material up into infinitesimal dipoles and integrate to get the total? As usual it's easier to work with the potential. For a single dipole **p** we have equation (Eq. 3.99),

$$V(\mathbf{r}) = \frac{1}{4\pi\epsilon_0}\frac{\hat{\boldsymbol{\imath}}\cdot\mathbf{p}}{\imath^2}, \tag{4.8}$$

[2] In asymmetric molecules the induced dipole moment may not be parallel to the field, but if the molecules are randomly oriented, the perpendicular contributions will *average* to zero. Within a single crystal, the orientations are certainly *not* random, and we would have to treat this case separately.

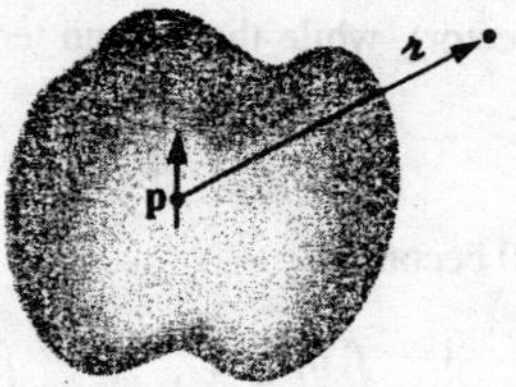

Figure 4.8

where $\boldsymbol{\imath}$ is the vector from the dipole to the point at which we are evaluating the potential (Fig. 4.8). In the present context we have a dipole moment $\mathbf{p} = \mathbf{P}\,d\tau'$ in each volume element $d\tau'$, so the total potential is

$$V(\mathbf{r}) = \frac{1}{4\pi\epsilon_0}\int_{\mathcal{V}} \frac{\hat{\boldsymbol{\imath}}\cdot\mathbf{P}(\mathbf{r}')}{\imath^2}\,d\tau'. \tag{4.9}$$

That *does* it, in principle. But a little sleight-of-hand casts this integral into a much more illuminating form. Observing that

$$\nabla'\left(\frac{1}{\imath}\right) = \frac{\hat{\boldsymbol{\imath}}}{\imath^2},$$

where (unlike Prob. 1.13) the differentiation is with respect to the *source* coordinates ($\mathbf{r}'$), we have

$$V = \frac{1}{4\pi\epsilon_0}\int_{\mathcal{V}} \mathbf{P}\cdot\nabla'\left(\frac{1}{\imath}\right)\,d\tau'.$$

Integrating by parts, using product rule number 5, gives

$$V = \frac{1}{4\pi\epsilon_0}\left[\int_{\mathcal{V}} \nabla'\cdot\left(\frac{\mathbf{P}}{\imath}\right)d\tau' - \int_{\mathcal{V}} \frac{1}{\imath}(\nabla'\cdot\mathbf{P})\,d\tau'\right],$$

or, using the divergence theorem,

$$V = \frac{1}{4\pi\epsilon_0}\oint_{\mathcal{S}} \frac{1}{\imath}\mathbf{P}\cdot d\mathbf{a}' - \frac{1}{4\pi\epsilon_0}\int_{\mathcal{V}} \frac{1}{\imath}(\nabla'\cdot\mathbf{P})\,d\tau'. \tag{4.10}$$

The first term looks like the potential of a surface charge

$$\boxed{\sigma_b \equiv \mathbf{P}\cdot\hat{\mathbf{n}}} \tag{4.11}$$

(where $\hat{\mathbf{n}}$ is the normal unit vector), while the second term looks like the potential of a volume charge

$$\boxed{\rho_b \equiv -\nabla \cdot \mathbf{P}.} \tag{4.12}$$

With these definitions, Eq. 4.10 becomes

$$V(\mathbf{r}) = \frac{1}{4\pi\epsilon_0} \oint_{\mathcal{S}} \frac{\sigma_b}{\imath} \, da' + \frac{1}{4\pi\epsilon_0} \int_{\mathcal{V}} \frac{\rho_b}{\imath} \, d\tau'. \tag{4.13}$$

What this means is that the potential (and hence also the field) of a polarized object is the same as that produced by a volume charge density $\rho_b = -\nabla \cdot \mathbf{P}$ plus a surface charge density $\sigma_b = \mathbf{P} \cdot \hat{\mathbf{n}}$. Instead of integrating the contributions of all the infinitesimal dipoles, as in Eq. 4.9, we just find those **bound charges**, and then calculate the fields *they* produce, in the same way we calculate the field of any other volume and surface charges (for example, using Gauss's law).

Example 4.2

Find the electric field produced by a uniformly polarized sphere of radius R.

Solution: We may as well choose the z axis to coincide with the direction of polarization (Fig. 4.9). The volume bound charge density ρ_b is zero, since $\mathbf{P}$ is uniform, but

$$\sigma_b = \mathbf{P} \cdot \hat{\mathbf{n}} = P \cos\theta,$$

where θ is the usual spherical coordinate. What we want, then, is the field produced by a charge density $P\cos\theta$ plastered over the surface of a sphere. But we have already computed the potential of such a configuration in Ex. 3.9:

$$V(r,\theta) = \begin{cases} \dfrac{P}{3\epsilon_0} r \cos\theta, & \text{for } r \leq R, \\[2ex] \dfrac{P}{3\epsilon_0} \dfrac{R^3}{r^2} \cos\theta, & \text{for } r \geq R. \end{cases}$$

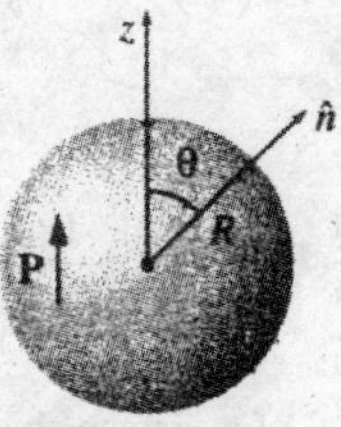

Figure 4.9

Since $r\cos\theta = z$, the *field* inside the sphere is uniform,

$$\mathbf{E} = -\nabla V = -\frac{P}{3\epsilon_0}\hat{\mathbf{z}} = -\frac{1}{3\epsilon_0}\mathbf{P}, \quad \text{for } r < R. \tag{4.14}$$

This remarkable result will be very useful in what follows. Outside the sphere the potential is identical to that of a perfect dipole at the origin,

$$V = \frac{1}{4\pi\epsilon_0}\frac{\mathbf{p}\cdot\hat{\mathbf{r}}}{r^2}, \quad \text{for } r \geq R, \tag{4.15}$$

whose dipole moment is, not surprisingly, equal to the total dipole moment of the sphere:

$$\mathbf{p} = \tfrac{4}{3}\pi R^3 \mathbf{P}. \tag{4.16}$$

The field of the uniformly polarized sphere is shown in Fig. 4.10.

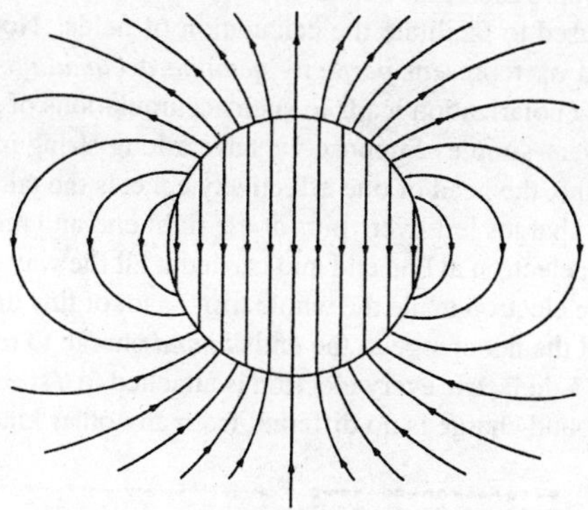

Figure 4.10

Problem 4.10 A sphere of radius R carries a polarization

$$\mathbf{P}(\mathbf{r}) = k\mathbf{r},$$

where k is a constant and $\mathbf{r}$ is the vector from the center.

(a) Calculate the bound charges σ_b and ρ_b.

(b) Find the field inside and outside the sphere.

Problem 4.11 A short cylinder, of radius a and length L, carries a "frozen-in" uniform polarization $\mathbf{P}$, parallel to its axis. Find the bound charge, and sketch the electric field (i) for $L \gg a$, (ii) for $L \ll a$, and (iii) for $L \approx a$. [This device is known as a **bar electret**; it is the electrical analog to a bar magnet. In practice, only very special materials—barium titanate is the most "familiar" example—will hold a permanent electric polarization. That's why you can't buy electrets at the toy store.]

Problem 4.12 Calculate the potential of a uniformly polarized sphere (Ex. 4.2) directly from Eq. 4.9.

4.2.2 Physical Interpretation of Bound Charges

In the last section we found that the field of a polarized object is identical to the field that would be produced by a certain distribution of "bound charges," σ_b and ρ_b. But this conclusion emerged in the course of abstract manipulations on the integral in Eq. 4.9, and left us with no clue as to the physical meaning of these bound charges. Indeed, some authors give you the impression that bound charges are in some sense "fictitious"—mere bookkeeping devices used to facilitate the calculation of fields. Nothing could be farther from the truth; ρ_b and σ_b represent *perfectly genuine accumulations of charge.* In this section I'll explain how polarization leads to such accumulations of charge.

The basic idea is very simple: Suppose we have a long string of dipoles, as shown in Fig. 4.11. Along the line, the head of one effectively cancels the tail of its neighbor, but at the ends there are two charges left over: plus at the right end and minus at the left. It is as if we had peeled off an electron at one end and carried it all the way down to the other end, though in fact no single electron made the whole trip—a lot of tiny displacements add up to one large one. We call the net charge at the ends *bound* charge to remind ourselves that it cannot be removed; in a dielectric every electron is attached to a specific atom or molecule. But apart from that, bound charge is no different from any other kind.

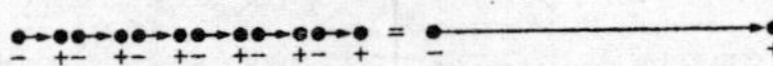

Figure 4.11

To calculate the actual *amount* of bound charge resulting from a given polarization, examine a "tube" of dielectric parallel to $\mathbf{P}$. The dipole moment of the tiny chunk shown in Fig. 4.12 is $P(Ad)$, where A is the cross-sectional area of the tube and d is the length of the chunk. In terms of the charge (q) at the end, this same dipole moment can be written qd. The bound charge that piles up at the right end of the tube is therefore

$$q = PA.$$

If the ends have been sliced off perpendicularly, the surface charge density is

$$\sigma_b = \frac{q}{A} = P.$$

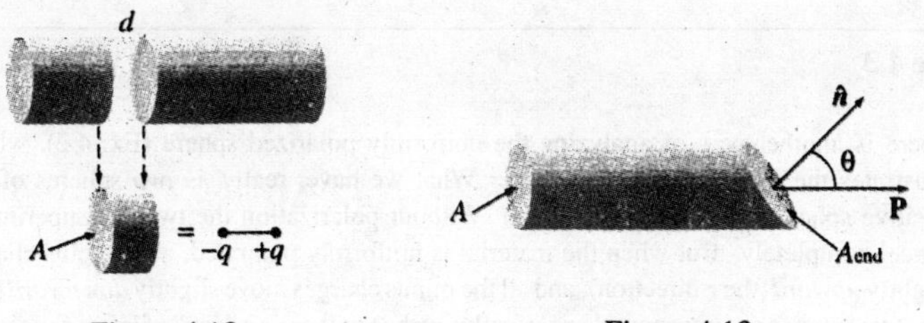

Figure 4.12 Figure 4.13

For an oblique cut (Fig. 4.13), the *charge* is still the same, but $A \doteq A_{\text{end}} \cos\theta$, so

$$\sigma_b = \frac{q}{A_{\text{end}}} = P\cos\theta = \mathbf{P}\cdot\hat{\mathbf{n}}.$$

The effect of the polarization, then, is to paint a bound charge $\sigma_b = \mathbf{P}\cdot\hat{\mathbf{n}}$ over the surface of the material. This is exactly what we found by more rigorous means in Sect. 4.2.1. But now we know where the bound charge *comes* from.

If the polarization is nonuniform we get accumulations of bound charge *within* the material as well as on the surface. A glance at Fig. 4.14 suggests that a diverging **P** results in a pileup of negative charge. Indeed, the net bound charge $\int \rho_b\, d\tau$ in a given volume is equal and opposite to the amount that has been pushed out through the surface. The latter (by the same reasoning we used before) is $\mathbf{P}\cdot\hat{\mathbf{n}}$ per unit area, so

$$\int_{\mathcal{V}} \rho_b\, d\tau = -\oint_{\mathcal{S}} \mathbf{P}\cdot d\mathbf{a} = -\int_{\mathcal{V}} (\nabla\cdot\mathbf{P})\, d\tau.$$

Since this is true for *any* volume, we have

$$\rho_b = -\nabla\cdot\mathbf{P},$$

confirming, again, the more rigorous conclusion of Sect. 4.2.1.

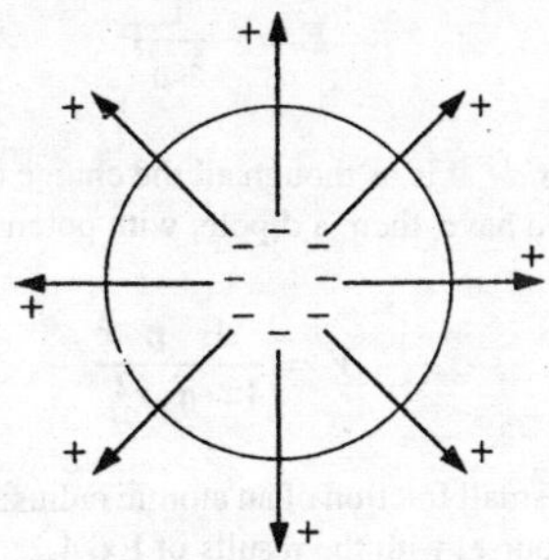

Figure 4.14

Example 4.3

There is another way of analyzing the uniformly polarized sphere (Ex. 4.2), which nicely illustrates the idea of a bound charge. What we have, really, is *two* spheres of charge: a positive sphere and a negative sphere. Without polarization the two are superimposed and cancel completely. But when the material is uniformly polarized, all the plus charges move slightly *upward* (the z direction), and all the minus charges move slightly *downward* (Fig. 4.15). The two spheres no longer overlap perfectly: at the top there's a "cap" of leftover positive charge and at the bottom a cap of negative charge. This "leftover" charge is precisely the bound surface charge σ_b.

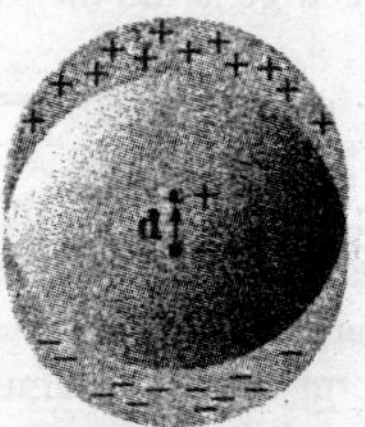

Figure 4.15

In Prob. 2.18 you calculated the field in the region of overlap between two uniformly charged spheres; the answer was

$$\mathbf{E} = -\frac{1}{4\pi\epsilon_0}\frac{q\mathbf{d}}{R^3},$$

where q is the total charge of the positive sphere, $\mathbf{d}$ is the vector from the negative center to the positive center, and R is the radius of the sphere. We can express this in terms of the polarization of the sphere, $\mathbf{p} = q\mathbf{d} = (\frac{4}{3}\pi R^3)\mathbf{P}$, as

$$\mathbf{E} = -\frac{1}{3\epsilon_0}\mathbf{P}.$$

Meanwhile, for points *outside*, it is as though all the charge on each sphere were concentrated at the respective center. We have, then, a dipole, with potential

$$V = \frac{1}{4\pi\epsilon_0}\frac{\mathbf{p}\cdot\hat{\mathbf{r}}}{r^2}.$$

(Remember that $\mathbf{d}$ is some small fraction of an atomic radius; Fig. 4.15 is grossly exaggerated.) These answers agree, of course, with the results of Ex. 4.2.

Problem 4.13 A very long cylinder, of radius a, carries a uniform polarization $\mathbf{P}$ perpendicular to its axis. Find the electric field inside the cylinder. Show that the field *outside* the cylinder can be expressed in the form

$$\mathbf{E}(\mathbf{r}) = \frac{a^2}{2\epsilon_0 s^2}[2(\mathbf{P}\cdot\hat{\mathbf{s}})\hat{\mathbf{s}} - \mathbf{P}].$$

[*Careful:* I said "uniform," not "radial"!]

Problem 4.14 When you polarize a neutral dielectric, charge moves a bit, but the *total* remains zero. This fact should be reflected in the bound charges σ_b and ρ_b. Prove from Eqs. 4.11 and 4.12 that the total bound charge vanishes.

4.2.3 The Field Inside a Dielectric

I have been sloppy about the distinction between "pure" dipoles and "physical" dipoles. In developing the theory of bound charges, I assumed we were working with the pure kind—indeed, I started with Eq. 4.8, the formula for the potential of a pure dipole. And yet, an actual polarized dielectric consists of *physical* dipoles, albeit extremely tiny ones. What is more, I presumed to represent discrete molecular dipoles by a continuous density function $\mathbf{P}$. How can I justify this method? *Outside* the dielectric there is no real problem: here we are far away from the molecules ($\mathscr{r}$ is many times greater than the separation distance between plus and minus charges), so the dipole potential dominates overwhelmingly and the detailed "graininess" of the source is blurred by distance. *Inside* the dielectric, however, we can hardly pretend to be far from all the dipoles, and the procedure I used in Sect. 4.2.1 is open to serious challenge.

In fact, when you stop to think about it, the electric field inside matter must be fantastically complicated, on the microscopic level. If you happen to be very near an electron, the field is gigantic, whereas a short distance away it may be small or point in a totally different direction. Moreover, an instant later, as the atoms move about, the field will have altered entirely. This true **microscopic** field would be utterly impossible to calculate, nor would it be of much interest if you could. Just as, for macroscopic purposes, we regard water as a continuous fluid, ignoring its molecular structure, so also we can ignore the microscopic bumps and wrinkles in the electric field inside matter, and concentrate on the **macroscopic** field. This is defined as the *average* field over regions large enough to contain many thousands of atoms (so that the uninteresting microscopic fluctuations are smoothed over), and yet small enough to ensure that we do not wash out any significant large-scale variations in the field. (In practice, this means we must average over regions much smaller than the dimensions of the object itself.) Ordinarily, the macroscopic field is what people *mean* when they speak of "the" field inside matter.[3]

[3]In case the introduction of the macroscopic field sounds suspicious to you, let me point out that you do *exactly* the same averaging whenever you speak of the *density* of a material.

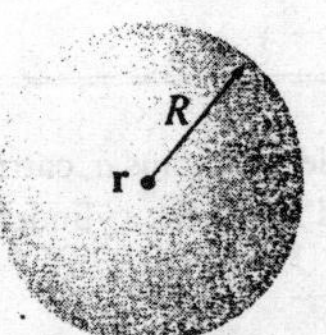

Figure 4.16

It remains to show that the macroscopic field is what we actually obtain when we use the methods of Sect. 4.2.1. The argument is subtle, so hang on. Suppose I want to calculate the macroscopic field at some point **r** within a dielectric (Fig. 4.16). I know I must average the true (microscopic) field over an appropriate volume, so let me draw a small sphere about **r**, of radius, say, a thousand times the size of a molecule. The macroscopic field at **r**, then, consists of two parts: the average field over the sphere due to all charges *outside*, plus the average due to all charges *inside:*

$$\mathbf{E} = \mathbf{E}_{\text{out}} + \mathbf{E}_{\text{in}}.$$

Now you proved in Prob. 3.41(d) that the average field (over a sphere), produced by charges *outside*, is equal to the field they produce at the center, so $\mathbf{E}_{\text{out}}$ is the field at **r** due to the dipoles exterior to the sphere. These are far enough away that we can safely use Eq. 4.9:

$$V_{\text{out}} = \frac{1}{4\pi\epsilon_0} \int_{\text{outside}} \frac{\hat{\boldsymbol{\imath}} \cdot \mathbf{P}(\mathbf{r}')}{\imath^2} \, d\tau'. \tag{4.17}$$

The dipoles *inside* the sphere are too close to treat in this fashion. But fortunately all we need is their *average* field, and that, according to Eq. 3.105, is

$$\mathbf{E}_{\text{in}} = -\frac{1}{4\pi\epsilon_0} \frac{\mathbf{p}}{R^3},$$

regardless of the details of the charge distribution within the sphere. The only relevant quantity is the total dipole moment, $\mathbf{p} = (\frac{4}{3}\pi R^3)\,\mathbf{P}$:

$$\mathbf{E}_{\text{in}} = -\frac{1}{3\epsilon_0} \mathbf{P}. \tag{4.18}$$

Now, by assumption the sphere is small enough that **P** does not vary significantly over its volume, so the term *left out* of the integral in Eq. 4.17 corresponds to the field at the center of a *uniformly* polarized sphere, to wit: $-(1/3\epsilon_0)\mathbf{P}$ (Eq. 4.14). But this is precisely what $\mathbf{E}_{\text{in}}$ (Eq. 4.18) puts back in! The macroscopic field, then, is given by the potential

$$V(\mathbf{r}) = \frac{1}{4\pi\epsilon_0} \int \frac{\hat{\boldsymbol{\imath}} \cdot \mathbf{P}(\mathbf{r}')}{\imath^2} \, d\tau', \tag{4.19}$$

where the integral runs over the *entire* volume of the dielectric. This is, of course, what we used in Sect. 4.2.1; without realizing it, we were correctly calculating the averaged, macroscopic field, for points inside the dielectric.

You may have to reread the last couple of paragraphs for the argument to sink in. Notice that it all revolves around the curious fact that the average field over *any* sphere (due to the charge inside) is the same as the field at the center of a *uniformly polarized* sphere with the same total dipole moment. This means that no matter how crazy the actual microscopic charge configuration, we can replace it by a nice smooth distribution of perfect dipoles, if all we want is the macroscopic (average) field. Incidentally, while the argument ostensibly relies on the spherical shape I chose to average over, the macroscopic field is certainly independent of the geometry of the averaging region, and this is reflected in the final answer, Eq. 4.19. Presumably, one could reproduce the same argument for a cube or an ellipsoid or whatever—the calculation might be more difficult, but the conclusion would be the same.

4.3 The Electric Displacement

4.3.1 Gauss's Law in the Presence of Dielectrics

In Sect. 4.2 we found that the effect of polarization is to produce accumulations of bound charge, $\rho_b = -\nabla \cdot \mathbf{P}$ within the dielectric and $\sigma_b = \mathbf{P} \cdot \hat{\mathbf{n}}$ on the surface. The field due to polarization of the medium is just the field of this bound charge. We are now ready to put it all together: the field attributable to bound charge plus the field due to everything *else* (which, for want of a better term, we call **free charge**). The free charge might consist of electrons on a conductor or ions embedded in the dielectric material or whatever; any charge, in other words, that is *not* a result of polarization. Within the dielectric, then, the total charge density can be written:

$$\rho = \rho_b + \rho_f, \tag{4.20}$$

and Gauss's law reads

$$\epsilon_0 \nabla \cdot \mathbf{E} = \rho = \rho_b + \rho_f = -\nabla \cdot \mathbf{P} + \rho_f,$$

where $\mathbf{E}$ is now the *total* field, not just that portion generated by polarization.

It is convenient to combine the two divergence terms:

$$\nabla \cdot (\epsilon_0 \mathbf{E} + \mathbf{P}) = \rho_f.$$

The expression in parentheses, designated by the letter $\mathbf{D}$,

$$\boxed{\mathbf{D} \equiv \epsilon_0 \mathbf{E} + \mathbf{P},} \tag{4.21}$$

is known as the **electric displacement**. In terms of $\mathbf{D}$, Gauss's law reads

$$\boxed{\nabla \cdot \mathbf{D} = \rho_f,} \tag{4.22}$$

or, in integral form,

$$\oint \mathbf{D} \cdot d\mathbf{a} = Q_{f_{\text{enc}}}, \tag{4.23}$$

where $Q_{f_{\text{enc}}}$ denotes the total free charge enclosed in the volume. This is a particularly useful way to express Gauss's law, in the context of dielectrics, because *it makes reference only to free charges,* and free charge is the stuff we control. Bound charge comes along for the ride: when we put the free charge in place, a certain polarization automatically ensues, by the mechanisms of Sect. 4.1, and this polarization produces the bound charge. In a typical problem, therefore, we know ρ_f, but we do not (initially) know ρ_b; Eq. 4.23 lets us go right to work with the information at hand. In particular, whenever the requisite symmetry is present, we can immediately calculate **D** by the standard Gauss's law methods.

Example 4.4

A long straight wire, carrying uniform line charge λ, is surrounded by rubber insulation out to a radius a (Fig. 4.17). Find the electric displacement.

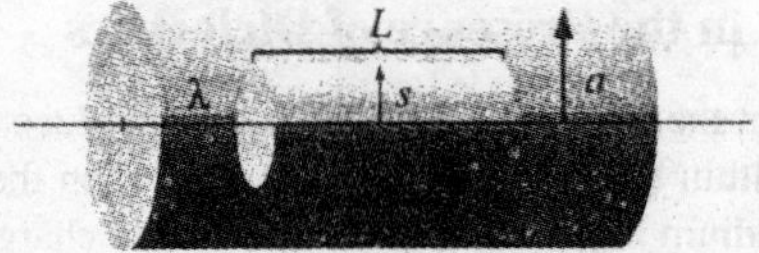

Figure 4.17

Solution: Drawing a cylindrical Gaussian surface, of radius s and length L, and applying Eq. 4.23, we find

$$D(2\pi s L) = \lambda L.$$

Therefore,

$$\mathbf{D} = \frac{\lambda}{2\pi s}\hat{\mathbf{s}}. \tag{4.24}$$

Notice that this formula holds both within the insulation and outside it. In the latter region, $\mathbf{P} = 0$, so

$$\mathbf{E} = \frac{1}{\epsilon_0}\mathbf{D} = \frac{\lambda}{2\pi\epsilon_0 s}\hat{\mathbf{s}}, \quad \text{for } s > a.$$

Inside the rubber the electric field cannot be determined, since we do not know **P**.

It may have appeared to you that I left out the surface bound charge σ_b in deriving Eq. 4.22, and in a sense that is true. We cannot apply Gauss's law precisely *at* the surface of a dielectric, for here ρ_b blows up, taking the divergence of **E** with it. But everywhere *else* the logic is sound, and in fact if we picture the edge of the dielectric as having some finite thickness within which the polarization tapers off to zero (probably a more realistic model than an abrupt cut-off anyway), then there *is* no surface bound charge; ρ_b varies rapidly but smoothly within this "skin," and Gauss's law can be safely applied *everywhere*. At any rate, the integral form (Eq. 4.23) is free from this "defect."

Problem 4.15 A thick spherical shell (inner radius a, outer radius b) is made of dielectric material with a "frozen-in" polarization

$$\mathbf{P}(\mathbf{r}) = \frac{k}{r}\,\hat{\mathbf{r}},$$

where k is a constant and r is the distance from the center (Fig. 4.18). (There is no *free* charge in the problem.) Find the electric field in all three regions by two different methods:

(a) Locate all the bound charge, and use Gauss's law (Eq. 2.13) to calculate the field it produces.

(b) Use Eq. 4.23 to find $\mathbf{D}$, and then get $\mathbf{E}$ from Eq. 4.21. [Notice that the second method is much faster, and avoids any explicit reference to the bound charges.]

Problem 4.16 Suppose the field inside a large piece of dielectric is $\mathbf{E}_0$, so that the electric displacement is $\mathbf{D}_0 = \epsilon_0\mathbf{E}_0 + \mathbf{P}$.

(a) Now a small spherical cavity (Fig. 4.19a) is hollowed out of the material. Find the field at the center of the cavity in terms of $\mathbf{E}_0$ and $\mathbf{P}$. Also find the displacement at the center of the cavity in terms of $\mathbf{D}_0$ and $\mathbf{P}$.

(b) Do the same for a long needle-shaped cavity running parallel to $\mathbf{P}$ (Fig. 4.19b).

(c) Do the same for a thin wafer-shaped cavity perpendicular to $\mathbf{P}$ (Fig. 4.19c).

[Assume the cavities are small enough that $\mathbf{P}$, $\mathbf{E}_0$, and $\mathbf{D}_0$ are essentially uniform. *Hint:* Carving out a cavity is the same as superimposing an object of the same shape but opposite polarization.]

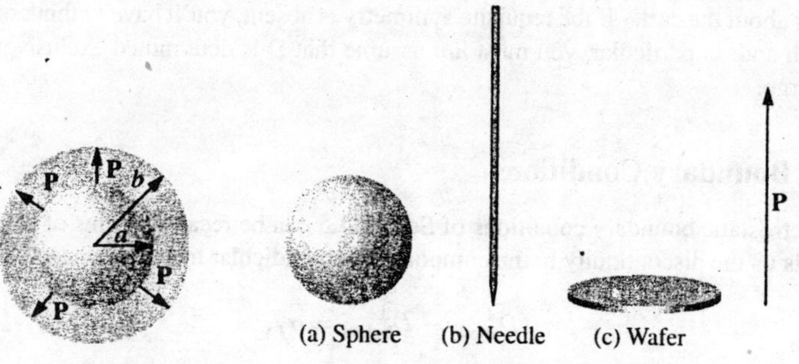

Figure 4.18

Figure 4.19

4.3.2 A Deceptive Parallel

Equation 4.22 looks just like Gauss's law, only the *total* charge density ρ is replaced by the *free* charge density ρ_f, and $\mathbf{D}$ is substituted for $\epsilon_0\mathbf{E}$. For this reason, you may be tempted to conclude that $\mathbf{D}$ is "just like" $\mathbf{E}$ (apart from the factor ϵ_0), except that its source is ρ_f instead of ρ: "To solve problems involving dielectrics, you just forget all about the bound charge—calculate the field as you ordinarily would, only call the answer $\mathbf{D}$ instead of $\mathbf{E}$." This reasoning is seductive, but the conclusion is false; in particular, there is no "Coulomb's law" for $\mathbf{D}$:

$$\mathbf{D}(\mathbf{r}) \neq \frac{1}{4\pi}\int \frac{\hat{\boldsymbol{\mathscr{r}}}}{\mathscr{r}^2}\rho_f(\mathbf{r}')\,d\tau'.$$

The parallel between $\mathbf{E}$ and $\mathbf{D}$ is more subtle than that.

For the divergence alone is insufficient to determine a vector field; you need to know the curl as well. One tends to forget this in the case of electrostatic fields because the curl of $\mathbf{E}$ is always zero. But the curl of $\mathbf{D}$ is *not* always zero.

$$\nabla \times \mathbf{D} = \epsilon_0(\nabla \times \mathbf{E}) + (\nabla \times \mathbf{P}) = \nabla \times \mathbf{P}, \tag{4.25}$$

and there is no reason, in general, to suppose that the curl of $\mathbf{P}$ vanishes. Sometimes it does, as in Ex. 4.4 and Prob. 4.15, but more often it does not. The bar electret of Prob. 4.11 is a case in point: here there is no free charge anywhere, so if you really believe that the only source of $\mathbf{D}$ is ρ_f, you will be forced to conclude that $\mathbf{D} = 0$ everywhere, and hence that $\mathbf{E} = (-1/\epsilon_0)\mathbf{P}$ inside and $\mathbf{E} = 0$ outside the electret, which is obviously wrong. (I leave it for you to find the place where $\nabla \times \mathbf{P} \neq 0$ in this problem.) Because $\nabla \times \mathbf{D} \neq 0$, moreover, $\mathbf{D}$ cannot be expressed as the gradient of a scalar—there is no "potential" for $\mathbf{D}$.

Advice: When you are asked to compute the electric displacement, first look for symmetry. If the problem exhibits spherical, cylindrical, or plane symmetry, then you can get $\mathbf{D}$ directly from Eq. 4.23 by the usual Gauss's law methods. (Evidently in such cases $\nabla \times \mathbf{P}$ is automatically zero, but since symmetry alone dictates the answer you're not really obliged to worry about the curl.) If the requisite symmetry is absent, you'll have to think of another approach and, in particular, you must *not* assume that $\mathbf{D}$ is determined exclusively by the free charge.

4.3.3 Boundary Conditions

The electrostatic boundary conditions of Sect. 2.3.5 can be recast in terms of $\mathbf{D}$. Equation 4.23 tells us the discontinuity in the component perpendicular to an interface:

$$D^{\perp}_{\text{above}} - D^{\perp}_{\text{below}} = \sigma_f, \tag{4.26}$$

while Eq. 4.25 gives the discontinuity in parallel components:

$$\mathbf{D}^{\parallel}_{\text{above}} - \mathbf{D}^{\parallel}_{\text{below}} = \mathbf{P}^{\parallel}_{\text{above}} - \mathbf{P}^{\parallel}_{\text{below}}. \tag{4.27}$$

In the presence of dielectrics these are sometimes more useful than the corresponding boundary conditions on **E** (Eqs. 2.31 and 2.23):

$$E^{\perp}_{\text{above}} - E^{\perp}_{\text{below}} = \frac{1}{\epsilon_0}\sigma, \tag{4.28}$$

and

$$\mathbf{E}^{\parallel}_{\text{above}} - \mathbf{E}^{\parallel}_{\text{below}} = 0. \tag{4.29}$$

You might try applying them, for example, to Probs. 4.16 and 4.17.

Problem 4.17 For the bar electret of Prob. 4.11, make three careful sketches: one of **P**, one of **E**, and one of **D**. Assume L is about $2a$. [*Hint:* **E** lines terminate on charges; **D** lines terminate on *free* charges.]

4.4 Linear Dielectrics

4.4.1 Susceptibility, Permittivity, Dielectric Constant

In Sects. 4.2 and 4.3 we did not commit ourselves as to the *cause* of **P**; we dealt only with the *effects* of polarization. From the qualitative discussion of Sect. 4.1, though, we know that the polarization of a dielectric ordinarily results from an electric field, which lines up the atomic or molecular dipoles. For many substances, in fact, the polarization is *proportional* to the field, provided **E** is not too strong:

$$\mathbf{P} = \epsilon_0 \chi_e \mathbf{E}. \tag{4.30}$$

The constant of proportionality, χ_e, is called the **electric susceptibility** of the medium (a factor of ϵ_0 has been extracted to make χ_e dimensionless). The value of χ_e depends on the microscopic structure of the substance in question (and also on external conditions such as temperature). I shall call materials that obey Eq. 4.30 **linear dielectrics.**[4]

Note that **E** in Eq. 4.30 is the *total* field; it may be due in part to free charges and in part to the polarization itself. If, for instance, we put a piece of dielectric into an external field $\mathbf{E}_0$, we cannot compute **P** directly from Eq. 4.30; the external field will polarize the material, and this polarization will produce its own field, which then contributes to the total field, and this in turn modifies the polarization, which ... Breaking out of this infinite regress is not always easy. You'll see some examples in a moment. The simplest approach is to begin with the *displacement,* at least in those cases where **D** can be deduced directly from the free charge distribution.

---

[4]In modern optical applications, especially, *non*linear materials have become increasingly important. For these there is a second term in the formula for **P** as a function of **E**—typically a *cubic* one. In general, Eq. 4.30 can be regarded as the first (nonzero) term in the Taylor expansion of **P** in powers of **E**.

In linear media we have

$$\mathbf{D} = \epsilon_0\mathbf{E} + \mathbf{P} = \epsilon_0\mathbf{E} + \epsilon_0\chi_e\mathbf{E} = \epsilon_0(1+\chi_e)\mathbf{E}, \tag{4.31}$$

so **D** is *also* proportional to **E**:

$$\mathbf{D} = \epsilon\mathbf{E}, \tag{4.32}$$

where

$$\epsilon \equiv \epsilon_0(1+\chi_e). \tag{4.33}$$

This new constant ϵ is called the **permittivity** of the material. (In vacuum, where there is no matter to polarize, the susceptibility is zero, and the permittivity is ϵ_0. That's why ϵ_0 is called the **permittivity of free space.** I dislike the term, for it suggests that the vacuum is just a special kind of linear dielectric, in which the permittivity happens to have the value $8.85 \times 10^{-12}\ \mathrm{C^2/N{\cdot}m^2}$.) If you remove a factor of ϵ_0, the remaining dimensionless quantity

$$\epsilon_r \equiv 1 + \chi_e = \frac{\epsilon}{\epsilon_0} \tag{4.34}$$

is called the **relative permittivity**, or **dielectric constant**, of the material. Dielectric constants for some common substances are listed in Table 4.2. Of course, the permittivity and the dielectric constant do not convey any information that was not already available in the susceptibility, nor is there anything essentially new in Eq. 4.32; the *physics* of linear dielectrics is all contained in Eq. 4.30.[5]

Material	Dielectric Constant	Material	Dielectric Constant
Vacuum	1	Benzene	2.28
Helium	1.000065	Diamond	5.7
Neon	1.00013	Salt	5.9
Hydrogen	1.00025	Silicon	11.8
Argon	1.00052	Methanol	33.0
Air (dry)	1.00054	Water	80.1
Nitrogen	1.00055	Ice (-30° C)	99
Water vapor (100° C)	1.00587	$KTaNbO_3$ (0° C)	34,000

Table 4.2 Dielectric Constants (unless otherwise specified, values given are for 1 atm, 20° C). *Source: Handbook of Chemistry and Physics*, 78th ed. (Boca Raton: CRC Press, Inc., 1997).

[5]As long as we are engaged in this orgy of unnecessary terminology and notation, I might as well mention that formulas for **D** in terms of **E** (Eq. 4.32, in the case of linear dielectrics) are called **constitutive relations.**

Figure 4.20

Example 4.5

A metal sphere of radius a carries a charge Q (Fig. 4.20). It is surrounded, out to radius b, by linear dielectric material of permittivity ϵ. Find the potential at the center (relative to infinity).

Solution: To compute V, we need to know $\mathbf{E}$; to find $\mathbf{E}$, we might first try to locate the bound charge; we could get the bound charge from $\mathbf{P}$, but we can't calculate $\mathbf{P}$ unless we already know $\mathbf{E}$ (Eq. 4.30). We seem to be in a bind. What we *do* know is the *free* charge Q, and fortunately the arrangement is spherically symmetric, so let's begin by calculating $\mathbf{D}$, using Eq. 4.23:

$$\mathbf{D} = \frac{Q}{4\pi r^2}\hat{\mathbf{r}}, \qquad \text{for all points } r > a.$$

(Inside the metal sphere, of course, $\mathbf{E} = \mathbf{P} = \mathbf{D} = 0$.) Once we know $\mathbf{D}$, it is a trivial matter to obtain $\mathbf{E}$, using Eq. 4.32:

$$\mathbf{E} = \begin{cases} \dfrac{Q}{4\pi\epsilon r^2}\hat{\mathbf{r}}, & \text{for } a < r < b, \\[2ex] \dfrac{Q}{4\pi\epsilon_0 r^2}\hat{\mathbf{r}}, & \text{for } r > b. \end{cases}$$

The potential at the center is therefore

$$\begin{aligned} V &= -\int_{\infty}^{0} \mathbf{E}\cdot d\mathbf{l} = -\int_{\infty}^{b}\left(\frac{Q}{4\pi\epsilon_0 r^2}\right)dr - \int_{b}^{a}\left(\frac{Q}{4\pi\epsilon r^2}\right)dr - \int_{a}^{0}(0)\,dr \\ &= \frac{Q}{4\pi}\left(\frac{1}{\epsilon_0 b} + \frac{1}{\epsilon a} - \frac{1}{\epsilon b}\right). \end{aligned}$$

As it turns out, it was not necessary for us to compute the polarization or the bound charge explicitly, though this can easily be done:

$$\mathbf{P} = \epsilon_0\chi_e\mathbf{E} = \frac{\epsilon_0\chi_e Q}{4\pi\epsilon r^2}\hat{\mathbf{r}},$$

in the dielectric, and hence

$$\rho_b = -\nabla \cdot \mathbf{P} = 0,$$

while

$$\sigma_b = \mathbf{P} \cdot \hat{\mathbf{n}} = \begin{cases} \dfrac{\epsilon_0 \chi_e Q}{4\pi\epsilon b^2}, & \text{at the outer surface,} \\ \dfrac{-\epsilon_0 \chi_e Q}{4\pi\epsilon a^2}, & \text{at the inner surface.} \end{cases}$$

Notice that the surface bound charge at a is *negative* ($\hat{\mathbf{n}}$ points outward *with respect to the dielectric,* which is $+\hat{\mathbf{r}}$ at b but $-\hat{\mathbf{r}}$ at a). This is natural, since the charge on the metal sphere attracts its opposite in all the dielectric molecules. It is this layer of negative charge that reduces the field, within the dielectric, from $1/4\pi\epsilon_0(Q/r^2)\hat{\mathbf{r}}$ to $1/4\pi\epsilon(Q/r^2)\hat{\mathbf{r}}$. In this respect a dielectric is rather like an imperfect conductor: on a *conducting* shell the induced surface charge would be such as to cancel the field of Q *completely* in the region $a < r < b$; the dielectric does the best it can, but the cancellation is only partial.

You might suppose that linear dielectrics would escape the defect in the parallel between **E** and **D**. Since **P** and **D** are now proportional to **E**, does it not follow that their curls, like **E**'s, must vanish? Unfortunately, it does *not,* for the line integral of **P** around a closed path that *straddles the boundary between one type of material and another* need not be zero, even though the integral of **E** around the same loop *must* be. The reason is that the proportionality factor $\epsilon_0\chi_e$ is different on the two sides. For instance, at the interface between a polarized dielectric and the vacuum (Fig. 4.21), **P** is zero on one side but not on the other. Around this loop $\oint \mathbf{P} \cdot d\mathbf{l} \neq 0$, and hence, by Stokes' theorem, the curl of **P** cannot vanish everywhere within the loop (in fact, it is *infinite* at the boundary).

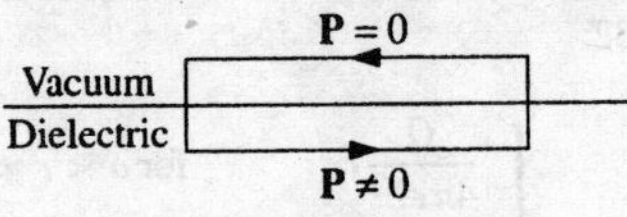

Figure 4.21

Of course, if the space is *entirely* filled with a homogeneous[6] linear dielectric, then this objection is void; in this rather special circumstance

$$\nabla \cdot \mathbf{D} = \rho_f \quad \text{and} \quad \nabla \times \mathbf{D} = 0,$$

so **D** can be found from the free charge just as though the dielectric were not there:

$$\mathbf{D} = \epsilon_0 \mathbf{E}_{\text{vac}},$$

where $\mathbf{E}_{\text{vac}}$ is the field the same free charge distribution would produce in the absence of any dielectric. According to Eqs. 4.32 and 4.34, therefore,

$$\mathbf{E} = \frac{1}{\epsilon}\mathbf{D} = \frac{1}{\epsilon_r}\mathbf{E}_{\text{vac}}. \tag{4.35}$$

[6]A **homogeneous** medium is one whose properties (in this case the susceptibility) do not vary with position.

Conclusion: When all space is filled with a homogeneous linear dielectric, the field everywhere is simply reduced by a factor of one over the dielectric constant. (Actually, it is not necessary for the dielectric to fill *all* space: in regions where the field is zero anyway, it can hardly matter whether the dielectric is present or not, since there's no polarization in any event.)

For example, if a free charge q is embedded in a large dielectric, the field it produces is

$$\mathbf{E} = \frac{1}{4\pi\epsilon}\frac{q}{r^2}\hat{\mathbf{r}} \tag{4.36}$$

(that's ϵ, not ϵ_0), and the force it exerts on nearby charges is reduced accordingly. But it's not that there is anything wrong with Coulomb's law; rather, the polarization of the medium partially "shields" the charge, by surrounding it with bound charge of the opposite sign (Fig. 4.22).[7]

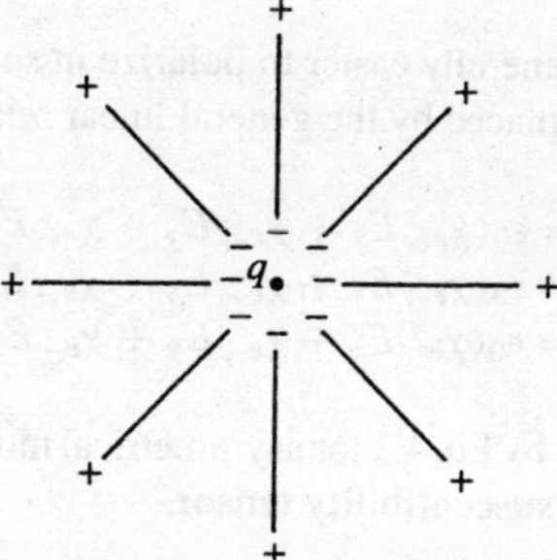

Figure 4.22

Example 4.6

A parallel-plate capacitor (Fig. 4.23) is filled with insulating material of dielectric constant ϵ_r. What effect does this have on its capacitance?

Solution: Since the field is confined to the space between the plates, the dielectric will reduce **E**, and hence also the potential difference V, by a factor $1/\epsilon_r$. Accordingly, the capacitance $C = Q/V$ is *increased by a factor of the dielectric constant,*

$$C = \epsilon_r C_{\text{vac}}. \tag{4.37}$$

This is, in fact, a common way to beef up a capacitor.

---

[7]In *quantum* electrodynamics the vacuum itself can be polarized, and this means that the effective (or "renormalized") charge of the electron, as you might measure it in the laboratory, is not its true ("bare") value, and in fact depends slightly on how far away you are!

Figure 4.23

By the way, a *crystal* is generally easier to polarize in some directions than in others,[8] and in this case Eq. 4.30 is replaced by the general linear relation

$$\left.\begin{array}{l} P_x = \epsilon_0(\chi_{e_{xx}} E_x + \chi_{e_{xy}} E_y + \chi_{e_{xz}} E_z) \\ P_y = \epsilon_0(\chi_{e_{yx}} E_x + \chi_{e_{yy}} E_y + \chi_{e_{yz}} E_z) \\ P_z = \epsilon_0(\chi_{e_{zx}} E_x + \chi_{e_{zy}} E_y + \chi_{e_{zz}} E_z) \end{array}\right\}, \tag{4.38}$$

just as Eq. 4.1 was superseded by Eq. 4.3 for asymmetrical molecules. The nine coefficients, $\chi_{e_{xx}}, \chi_{e_{xy}}, \ldots$, constitute the **susceptibility tensor.**

Problem 4.18 The space between the plates of a parallel-plate capacitor (Fig. 4.24) is filled with two slabs of linear dielectric material. Each slab has thickness a, so the total distance between the plates is $2a$. Slab 1 has a dielectric constant of 2, and slab 2 has a dielectric constant of 1.5. The free charge density on the top plate is σ and on the bottom plate $-\sigma$.

(a) Find the electric displacement **D** in each slab.

(b) Find the electric field **E** in each slab.

(c) Find the polarization **P** in each slab.

(d) Find the potential difference between the plates.

(e) Find the location and amount of all bound charge.

(f) Now that you know all the charge (free and bound), recalculate the field in each slab, and confirm your answer to (b).

[8] A medium is said to be **isotropic** if its properties (such as susceptibility) are the same in all directions. Thus Eq. 4.30 is the special case of Eq. 4.38 that holds for isotropic media. Physicists tend to be sloppy with their language, and unless otherwise indicated the term "linear dielectric" certainly means "isotropic linear dielectric," and probably means "homogeneous isotropic linear dielectric."

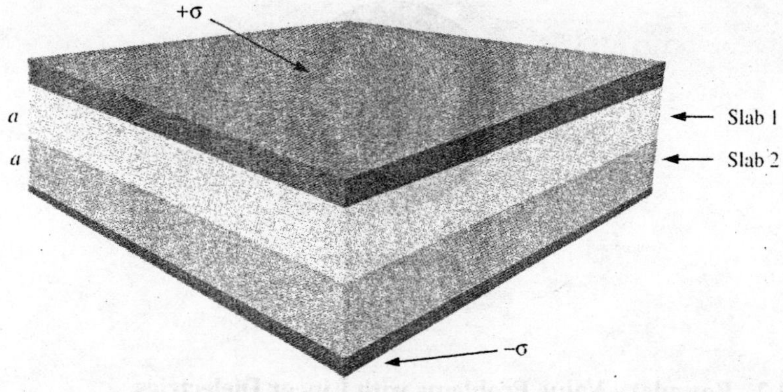

Figure 4.24

Problem 4.19 Suppose you have enough linear dielectric material, of dielectric constant ϵ_r, to *half*-fill a parallel-plate capacitor (Fig. 4.25). By what fraction is the capacitance increased when you distribute the material as in Fig. 4.25(a)? How about Fig. 4.25(b)? For a given potential difference V between the plates, find $\mathbf{E}$, $\mathbf{D}$, and $\mathbf{P}$, in each region, and the free and bound charge on all surfaces, for both cases.

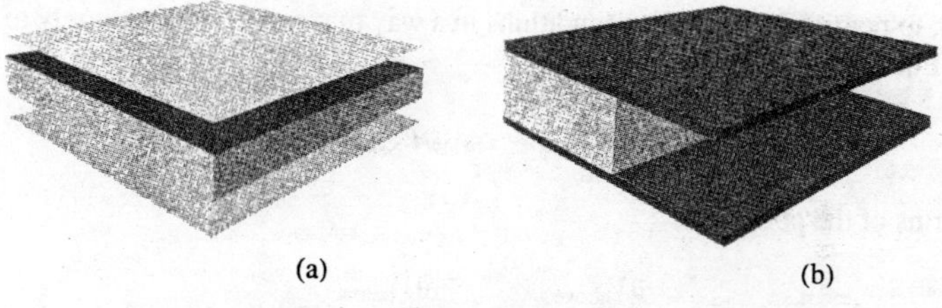

Figure 4.25

Problem 4.20 A sphere of linear dielectric material has embedded in it a uniform free charge density ρ. Find the potential at the center of the sphere (relative to infinity), if its radius is R and its dielectric constant is ϵ_r.

Problem 4.21 A certain coaxial cable consists of a copper wire, radius a, surrounded by a concentric copper tube of inner radius c (Fig. 4.26). The space between is partially filled (from b out to c) with material of dielectric constant ϵ_r, as shown. Find the capacitance per unit length of this cable.

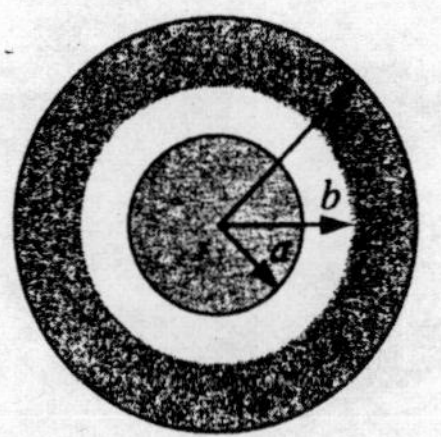

Figure 4.26

4.4.2 Boundary Value Problems with Linear Dielectrics

In a homogeneous linear dielectric the bound charge density (ρ_b) is proportional to the free charge density (ρ_f):[9]

$$\rho_b = -\nabla \cdot \mathbf{P} = -\nabla \cdot \left(\epsilon_0 \frac{\chi_e}{\epsilon} \mathbf{D}\right) = -\left(\frac{\chi_e}{1+\chi_e}\right)\rho_f. \tag{4.39}$$

In particular, unless free charge is actually embedded in the material, $\rho = 0$, and any net charge must reside at the surface. Within such a dielectric, then, the potential obeys Laplace's equation, and all the machinery of Chapter 3 carries over. It is convenient, however, to rewrite the boundary conditions in a way that makes reference only to the free charge. Equation 4.26 says

$$\epsilon_{\text{above}} E^{\perp}_{\text{above}} - \epsilon_{\text{below}} E^{\perp}_{\text{below}} = \sigma_f, \tag{4.40}$$

or (in terms of the potential),

$$\epsilon_{\text{above}} \frac{\partial V_{\text{above}}}{\partial n} - \epsilon_{\text{below}} \frac{\partial V_{\text{below}}}{\partial n} = -\sigma_f, \tag{4.41}$$

whereas the potential itself is, of course, continuous (Eq. 2.34):

$$V_{\text{above}} = V_{\text{below}}. \tag{4.42}$$

Example 4.7

A sphere of homogeneous linear dielectric material is placed in an otherwise uniform electric field $\mathbf{E}_0$ (Fig. 4.27). Find the electric field inside the sphere.

---

[9]This does not apply to the surface charge (σ_b), because χ_e is not independent of position (obviously) at the boundary.

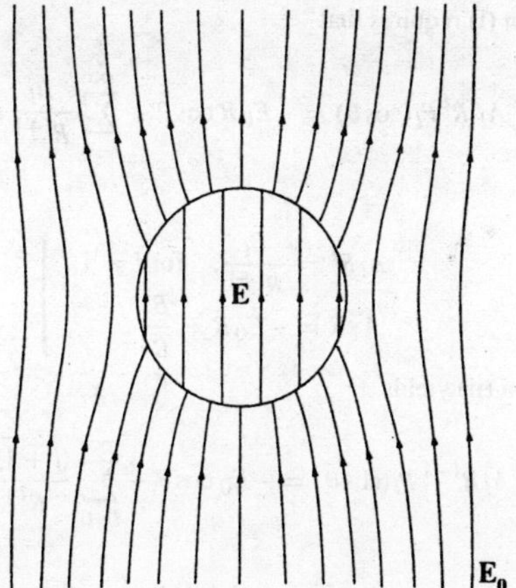

Figure 4.27

Solution: This is reminiscent of Ex. 3.8, in which an uncharged *conducting* sphere was introduced into a uniform field. In that case the field of the induced charge completely canceled $\mathbf{E}_0$ within the sphere; in a *dielectric*, the cancellation (from the bound charge) is only partial.

Our problem is to solve Laplace's equation, for $V_{\text{in}}(r, \theta)$ when $r \leq R$, and $V_{\text{out}}(r, \theta)$ when $r \geq R$, subject to the boundary conditions

$$\left.\begin{array}{lrcll} \text{(i)} & V_{\text{in}} & = & V_{\text{out}}, & \text{at } r = R, \\ \text{(ii)} & \epsilon \dfrac{\partial V_{\text{in}}}{\partial r} & = & \epsilon_0 \dfrac{\partial V_{\text{out}}}{\partial r}, & \text{at } r = R, \\ \text{(iii)} & V_{\text{out}} & \rightarrow & -E_0 r \cos\theta, & \text{for } r \gg R. \end{array}\right\} \tag{4.43}$$

(The second of these follows from Eq. 4.41, since there is no free charge at the surface.) Inside the sphere Eq. 3.65 says

$$V_{\text{in}}(r, \theta) = \sum_{l=0}^{\infty} A_l r^l P_l(\cos\theta); \tag{4.44}$$

outside the sphere, in view of (iii), we have

$$V_{\text{out}}(r, \theta) = -E_0 r \cos\theta + \sum_{l=0}^{\infty} \frac{B_l}{r^{l+1}} P_l(\cos\theta). \tag{4.45}$$

Boundary condition (i) requires that

$$\sum_{l=0}^{\infty} A_l R^l P_l(\cos\theta) = -E_0 R \cos\theta + \sum_{l=0}^{\infty} \frac{B_l}{R^{l+1}} P_l(\cos\theta),$$

so[10]

$$\left.\begin{aligned} A_l R^l &= \frac{B_l}{R^{l+1}}, \quad \text{for } l \neq 1, \\ A_1 R &= -E_0 R + \frac{B_1}{R^2}. \end{aligned}\right\} \tag{4.46}$$

Meanwhile, condition (ii) yields

$$\epsilon_r \sum_{l=0}^{\infty} l A_l R^{l-1} P_l(\cos\theta) = -E_0 \cos\theta - \sum_{l=0}^{\infty} \frac{(l+1)B_l}{R^{l+2}} P_l(\cos\theta),$$

so

$$\left.\begin{aligned} \epsilon_r l A_l R^{l-1} &= -\frac{(l+1)B_l}{R^{l+2}}, \quad \text{for } l \neq 1, \\ \epsilon_r A_1 &= -E_0 - \frac{2B_1}{R^3}. \end{aligned}\right\} \tag{4.47}$$

It follows that

$$\left.\begin{aligned} &A_l = B_l = 0, \quad \text{for } l \neq 1, \\ &A_1 = -\frac{3}{\epsilon_r+2} E_0 \quad B_1 = \frac{\epsilon_r-1}{\epsilon_r+2} R^3 E_0. \end{aligned}\right\} \tag{4.48}$$

Evidently

$$V_{\text{in}}(r,\theta) = -\frac{3E_0}{\epsilon_r+2} r\cos\theta = -\frac{3E_0}{\epsilon_r+2} z,$$

and hence the field inside the sphere is (surprisingly) *uniform*:

$$\mathbf{E} = \frac{3}{\epsilon_r+2}\mathbf{E}_0. \tag{4.49}$$

Example 4.8

Suppose the entire region below the plane $z = 0$ in Fig. 4.28 is filled with uniform linear dielectric material of susceptibility χ_e. Calculate the force on a point charge q situated a distance d above the origin.

Solution: The surface bound charge on the xy plane is of opposite sign to q, so the force will be attractive. (In view of Eq. 4.39, there is no volume bound charge.) Let us first calculate σ_b, using Eqs. 4.11 and 4.30.

$$\sigma_b = \mathbf{P}\cdot\hat{\mathbf{n}} = P_z = \epsilon_0 \chi_e E_z,$$

[10]Remember, $P_1(\cos\theta) = \cos\theta$, and the coefficients must be equal for each l, as you could prove by multiplying by $P_{l'}(\cos\theta)\sin\theta$, integrating from 0 to π, and invoking the orthogonality of the Legendre polynomials (Eq. 3.68).

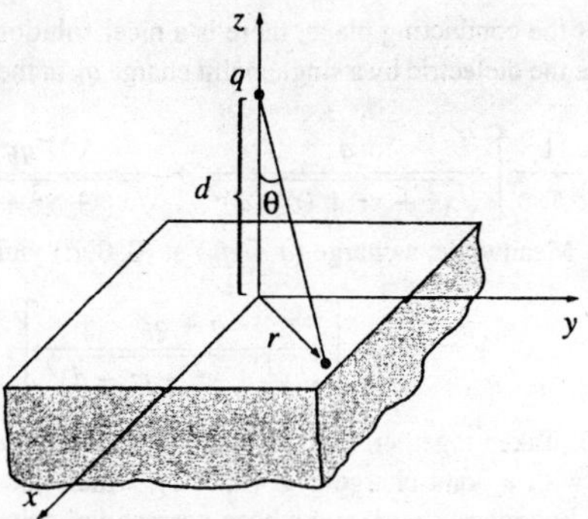

Figure 4.28

where E_z is the z-component of the total field just inside the dielectric, at $z = 0$. This field is due in part to q and in part to the bound charge itself. From Coulomb's law, the former contribution is

$$-\frac{1}{4\pi\epsilon_0}\frac{q}{(r^2+d^2)}\cos\theta = -\frac{1}{4\pi\epsilon_0}\frac{qd}{(r^2+d^2)^{3/2}},$$

where $r = \sqrt{x^2+y^2}$ is the distance from the origin. The z component of the field of the bound charge, meanwhile, is $-\sigma_b/2\epsilon_0$ (see footnote 6, p. 89). Thus

$$\sigma_b = \epsilon_0\chi_e\left[-\frac{1}{4\pi\epsilon_0}\frac{qd}{(r^2+d^2)^{3/2}} - \frac{\sigma_b}{2\epsilon_0}\right],$$

which we can solve for σ_b:

$$\sigma_b = -\frac{1}{2\pi}\left(\frac{\chi_e}{\chi_e+2}\right)\frac{qd}{(r^2+d^2)^{3/2}}. \tag{4.50}$$

Apart from the factor $\chi_e/(\chi_e+2)$, this is exactly the same as the induced charge on an infinite *conducting* plane under similar circumstances (Eq. 3.10).[11] Evidently the *total* bound charge is

$$q_b = -\left(\frac{\chi_e}{\chi_e+2}\right)q. \tag{4.51}$$

We could, of course, obtain the field of σ_b by direct integration

$$\mathbf{E} = \frac{1}{4\pi\epsilon_0}\int\left(\frac{\hat{\mathbf{\imath}}}{\imath^2}\right)\sigma_b\,da.$$

[11] For some purposes a conductor can be regarded as the limiting case of a linear dielectric, with $\chi_e \to \infty$. This is often a useful check—try applying it to Exs. 4.5, 4.6, and 4.7.

But as in the case of the conducting plane, there is a nicer solution by the method of images. Indeed, if we replace the dielectric by a single point charge q_b at the image position $(0, 0, -d)$, we have

$$V = \frac{1}{4\pi\epsilon_0}\left[\frac{q}{\sqrt{x^2+y^2+(z-d)^2}} + \frac{q_b}{\sqrt{x^2+y^2+(z+d)^2}}\right], \tag{4.52}$$

in the region $z > 0$. Meanwhile, a charge $(q + q_b)$ at $(0, 0, d)$ yields the potential

$$V = \frac{1}{4\pi\epsilon_0}\left[\frac{q+q_b}{\sqrt{x^2+y^2+(z-d)^2}}\right], \tag{4.53}$$

for the region $z < 0$. Taken together, Eqs. 4.52 and 4.53 constitute a function which satisfies Poisson's equation with a point charge q at $(0, 0, d)$, which goes to zero at infinity, which is continuous at the boundary $z = 0$, and whose normal derivative exhibits the discontinuity appropriate to a surface charge σ_b at $z = 0$:

$$-\epsilon_0\left(\left.\frac{\partial V}{\partial z}\right|_{z=0^+} - \left.\frac{\partial V}{\partial z}\right|_{z=0^-}\right) = -\frac{1}{2\pi}\left(\frac{\chi_e}{\chi_e+2}\right)\frac{qd}{(x^2+y^2+d^2)^{3/2}}.$$

Accordingly, this is the correct potential for our problem. In particular, the force on q is:

$$\mathbf{F} = \frac{1}{4\pi\epsilon_0}\frac{qq_b}{(2d)^2}\hat{\mathbf{z}} = -\frac{1}{4\pi\epsilon_0}\left(\frac{\chi_e}{\chi_e+2}\right)\frac{q^2}{4d^2}\hat{\mathbf{z}}. \tag{4.54}$$

I do not claim to have provided a compelling *motivation* for Eqs. 4.52 and 4.53—like all image solutions, this one owes its justification to the fact that it *works:* it solves Poisson's equation, and it meets the boundary conditions. Still, discovering an image solution is not entirely a matter of guesswork. There are at least two "rules of the game": (1) You must never put an image charge into the region where you're computing the potential. (Thus Eq. 4.52 gives the potential for $z > 0$, but this image charge q_b is at $z = -d$; when we turn to the region $z < 0$ (Eq. 4.53), the image charge $(q + q_b)$ is at $z = +d$.) (2) The image charges must add up to the correct total in each region. (That's how I knew to use q_b to account for the charge in the region $z \le 0$, and $(q + q_b)$ to cover the region $z \ge 0$.)

Problem 4.22 A very long cylinder of linear dielectric material is placed in an otherwise uniform electric field $\mathbf{E}_0$. Find the resulting field within the cylinder. (The radius is a, the susceptibility χ_e, and the axis is perpendicular to $\mathbf{E}_0$.)

Problem 4.23 Find the field inside a sphere of linear dielectric material in an otherwise uniform electric field $\mathbf{E}_0$ (Ex. 4.7) by the following method of successive approximations: First pretend the field inside is just $\mathbf{E}_0$, and use Eq. 4.30 to write down the resulting polarization $\mathbf{P}_0$. This polarization generates a field of its own, $\mathbf{E}_1$ (Ex. 4.2), which in turn modifies the polarization by an amount $\mathbf{P}_1$, which further changes the field by an amount $\mathbf{E}_2$, and so on. The resulting field is $\mathbf{E}_0 + \mathbf{E}_1 + \mathbf{E}_2 + \cdots$. Sum the series, and compare your answer with Eq. 4.49.

Problem 4.24 An uncharged conducting sphere of radius a is coated with a thick insulating shell (dielectric constant ϵ_r) out to radius b. This object is now placed in an otherwise uniform electric field $\mathbf{E}_0$. Find the electric field in the insulator.

! **Problem 4.25** Suppose the region *above* the xy plane in Ex. 4.8 is *also* filled with linear dielectric but of a different susceptibility χ'_e. Find the potential everywhere.

4.4.3 Energy in Dielectric Systems

It takes work to charge up a capacitor (Eq. 2.55):

$$W = \frac{1}{2}CV^2.$$

If the capacitor is filled with linear dielectric, its capacitance exceeds the vacuum value by a factor of the dielectric constant,

$$C = \epsilon_r C_{\text{vac}},$$

as we found in Ex. 4.6. Evidently the work necessary to charge a dielectric-filled capacitor is increased by the same factor. The reason is pretty clear: you have to pump on more (free) charge to achieve a given potential, because part of the field is canceled off by the bound charges.

In Chapter 2, I derived a general formula for the energy stored in any electrostatic system (Eq. 2.45):

$$W = \frac{\epsilon_0}{2}\int E^2\,d\tau. \tag{4.55}$$

The case of the dielectric-filled capacitor suggests that this should be changed to

$$W = \frac{\epsilon_0}{2}\int \epsilon_r E^2\,d\tau = \frac{1}{2}\int \mathbf{D}\cdot\mathbf{E}\,d\tau,$$

in the presence of linear dielectrics. To *prove* it, suppose the dielectric material is fixed in position, and we bring in the free charge, a bit at a time. As ρ_f is increased by an amount $\Delta\rho_f$, the polarization will change and with it the bound charge distribution; but we're interested only in the work done on the incremental *free* charge:

$$\Delta W = \int (\Delta\rho_f)V\,d\tau. \tag{4.56}$$

Since $\nabla\cdot\mathbf{D} = \rho_f$, $\Delta\rho_f = \nabla\cdot(\Delta\mathbf{D})$, where $\Delta\mathbf{D}$ is the resulting change in $\mathbf{D}$, so

$$\Delta W = \int [\nabla\cdot(\Delta\mathbf{D})]V\,d\tau.$$

Now

$$\nabla\cdot[(\Delta\mathbf{D})V] = [\nabla\cdot(\Delta\mathbf{D})]V + \Delta\mathbf{D}\cdot(\nabla V),$$

and hence (integrating by parts):

$$\Delta W = \int \nabla\cdot[(\Delta\mathbf{D})V]\,d\tau + \int (\Delta\mathbf{D})\cdot\mathbf{E}\,d\tau.$$

The divergence theorem turns the first term into a surface integral, which vanishes if we integrate over all of space. Therefore, the work done is equal to

$$\Delta W = \int (\Delta \mathbf{D}) \cdot \mathbf{E}\, d\tau. \tag{4.57}$$

So far, this applies to *any* material. Now, if the medium is a linear dielectric, then $\mathbf{D} = \epsilon \mathbf{E}$, so

$$\tfrac{1}{2}\Delta(\mathbf{D}\cdot\mathbf{E}) = \tfrac{1}{2}\Delta(\epsilon E^2) = \epsilon(\Delta\mathbf{E})\cdot\mathbf{E} = (\Delta\mathbf{D})\cdot\mathbf{E}$$

(for infinitesimal increments). Thus

$$\Delta W = \Delta\left(\frac{1}{2}\int \mathbf{D}\cdot\mathbf{E}\, d\tau\right).$$

The total work done, then, as we build the free charge up from zero to the final configuration, is

$$W = \frac{1}{2}\int \mathbf{D}\cdot\mathbf{E}\, d\tau, \tag{4.58}$$

as anticipated.[12]

It may puzzle you that Eq. 4.55, which we derived quite generally in Chapter 2, does not seem to apply in the presence of dielectrics, where it is replaced by Eq. 4.58. The point is not that one or the other of these equations is *wrong,* but rather that they speak to somewhat different questions. The distinction is subtle, so let's go right back to the beginning: What do we *mean* by "the energy of a system"? *Answer:* It is the work required to assemble the system. Very well—but when dielectrics are involved there are two quite different ways one might construe this process: (1) We bring in all the charges (free *and* bound), one by one, with tweezers, and glue each one down in its proper final location. If *this* is what you mean by "assemble the system," then Eq. 4.55 is your formula for the energy stored. Notice, however, that this will *not* include the work involved in stretching and twisting the dielectric molecules (if we picture the positive and negative charges as held together by tiny springs, it does not include the spring energy, $\frac{1}{2}kx^2$, associated with polarizing each molecule).[13] (2) With the unpolarized dielectric in place, we bring in the *free* charges, one by one, allowing the dielectric to respond as it sees fit. If *this* is what you mean by "assemble the system" (and ordinarily it *is,* since free charge is what we actually push around), then Eq. 4.58 is the formula you want. In this case the "spring" energy *is* included, albeit indirectly, because the force you must apply to the *free* charge depends on the disposition of the *bound* charge; as you move the free charge you are automatically stretching those "springs." To put it another

[12]In case you are wondering why I did not do this more simply by the method of Sect. 2.4.3, starting with $W = \frac{1}{2}\int \rho_f V\, d\tau$, the reason is that *this* formula is untrue, in general. Study the derivation of Eq. 2.42 and you will see that it applies only to the *total* charge. For *linear* dielectrics it happens to hold for the free charge alone, but this is scarcely obvious a priori and, in fact, is most easily confirmed by working backward from Eq. 4.58.

[13]The "spring" itself may be electrical in nature, but it is still not included in Eq. 4.55, if $\mathbf{E}$ is taken to be the *macroscopic* field.

way, in method (2) the total energy of the system consists of three parts: the electrostatic energy of the free charge, the electrostatic energy of the bound charge, and the "spring" energy:

$$W_{\text{tot}} = W_{\text{free}} + W_{\text{bound}} + W_{\text{spring}}.$$

The last two are equal and opposite (in procedure (2) the bound charges are always in equilibrium, and hence the *net* work done on them is zero); thus method (2), in calculating W_{free}, actually delivers W_{tot}, whereas method (1), by calculating $W_{\text{free}}+W_{\text{bound}}$, leaves out W_{spring}.

Incidentally, it is sometimes alleged that Eq. 4.58 represents the energy even for *non*linear dielectrics, but this is false: To proceed beyond Eq. 4.57 one must assume linearity. In fact, for dissipative systems the whole notion of "stored energy" loses its meaning, because the work done depends not only on the final configuration but on *how it got there.* If the molecular "springs" are allowed to have some *friction,* for instance, then W_{spring} can be made as large as you like, by assembling the charges in such a way that the spring is obliged to expand and contract many times before reaching its final state. In particular, you get nonsensical results if you try to apply Eq. 4.58 to electrets, with frozen-in polarization (see Prob. 4.27).

Problem 4.26 A spherical conductor, of radius a, carries a charge Q (Fig. 4.29). It is surrounded by linear dielectric material of susceptibility χ_e, out to radius b. Find the energy of this configuration (Eq. 4.58).

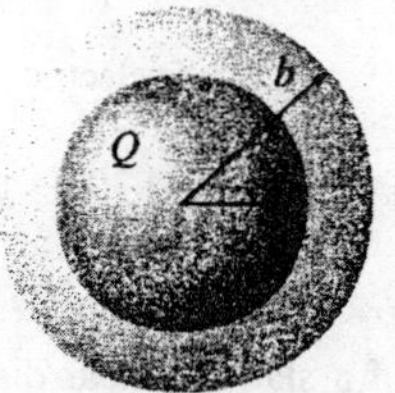

Figure 4.29

Problem 4.27 Calculate W, using both Eq. 4.55 and Eq. 4.58, for a sphere of radius R with frozen-in uniform polarization $\mathbf{P}$ (Ex. 4.2). Comment on the discrepancy. Which (if either) is the "true" energy of the system?

4.4.4 Forces on Dielectrics

Just as a conductor is attracted into an electric field (Eq. 2.51), so too is a dielectric—and for essentially the same reason: the bound charge tends to accumulate near the free charge of the opposite sign. But the calculation of forces on dielectrics can be surprisingly tricky.

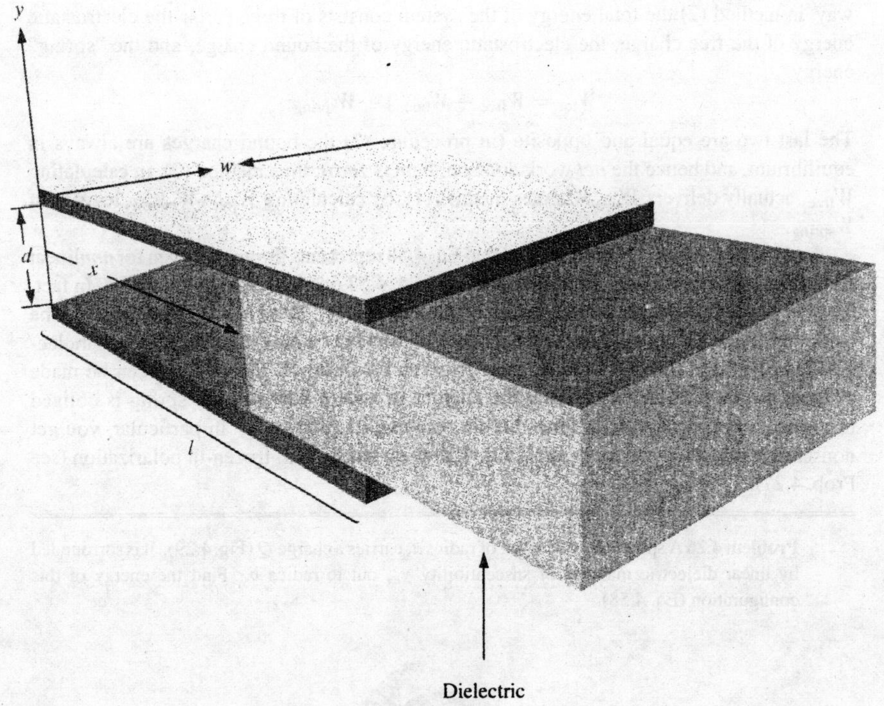

Figure 4.30

Consider, for example, the case of a slab of linear dielectric material, partially inserted between the plates of a parallel-plate capacitor (Fig. 4.30). We have always pretended that the field is uniform inside a parallel-plate capacitor, and zero outside. If this were literally true, there would be no net force on the dielectric at all, since the field everywhere would be perpendicular to the plates. However, there is in reality a **fringing field** around the edges, which for most purposes can be ignored but in this case is responsible for the whole effect. (Indeed, the field *could* not terminate abruptly at the edge of the capacitor, for if it did the line integral of **E** around the closed loop shown in Fig. 4.31 would not be zero.) It is this nonuniform fringing field that pulls the dielectric into the capacitor.

Fringing fields are notoriously difficult to calculate; luckily, we can avoid this altogether, by the following ingenious method. Let W be the energy of the system—it depends, of course, on the amount of overlap. If I pull the dielectric out an infinitesimal distance dx, the energy is changed by an amount equal to the work done:

$$dW = F_{\text{me}}\, dx, \tag{4.59}$$

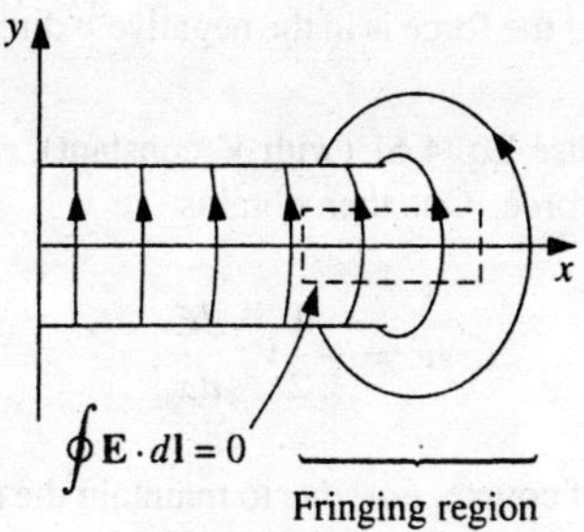

Figure 4.31

where F_{me} is the force I must exert, to counteract the electrical force F on the dielectric: $F_{me} = -F$. Thus the electrical force on the slab is

$$F = -\frac{dW}{dx}. \tag{4.60}$$

Now, the energy stored in the capacitor is

$$W = \tfrac{1}{2}CV^2, \tag{4.61}$$

and the capacitance in this case is

$$C = \frac{\epsilon_0 w}{d}(\epsilon_r l - \chi_e x), \tag{4.62}$$

where l is the length of the plates (Fig. 4.30). Let's assume that the total charge on the plates ($Q = CV$) is held constant, as the dielectric moves. In terms of Q,

$$W = \frac{1}{2}\frac{Q^2}{C}, \tag{4.63}$$

so

$$F = -\frac{dW}{dx} = \frac{1}{2}\frac{Q^2}{C^2}\frac{dC}{dx} = \frac{1}{2}V^2\frac{dC}{dx}. \tag{4.64}$$

But

$$\frac{dC}{dx} = -\frac{\epsilon_0 \chi_e w}{d},$$

and hence

$$F = -\frac{\epsilon_0 \chi_e w}{2d}V^2. \tag{4.65}$$

(The minus sign indicates that the force is in the negative x direction; the dielectric is pulled *into* the capacitor.)

It is a common error to use Eq. 4.61 (with V constant), rather than Eq. 4.63 (with Q constant), in computing the force. One then obtains

$$F = -\frac{1}{2}V^2\frac{dC}{dx},$$

which is off by a sign. It is, of course, *possible* to maintain the capacitor at a fixed potential, by connecting it up to a battery. But in that case the *battery also does work* as the dielectric moves; instead of Eq. 4.59, we now have

$$dW = F_{\text{me}}\,dx + V\,dQ, \tag{4.66}$$

where $V\,dQ$ is the work done by the battery. It follows that

$$F = -\frac{dW}{dx} + V\frac{dQ}{dx} = -\frac{1}{2}V^2\frac{dC}{dx} + V^2\frac{dC}{dx} = \frac{1}{2}V^2\frac{dC}{dx}, \tag{4.67}$$

the same as before (Eq. 4.64), with the *correct* sign. (Please understand, the force on the dielectric cannot possibly depend on whether you plan to hold Q constant or V constant—it is determined entirely by the distribution of charge, free and bound. It's simpler to *calculate* the force assuming constant Q, because then you don't have to worry about work done by the battery; but if you insist, it can be done correctly either way.)

Notice that we were able to determine the force *without knowing anything about the fringing fields that are ultimately responsible for it!* Of course, it's built into the whole structure of electrostatics that $\nabla \times \mathbf{E} = 0$, and hence that the fringing fields must be present; we're not really getting something for nothing here—just cleverly exploiting the internal consistency of the theory. The energy stored in the fringing fields themselves (which was not accounted for in this derivation) stays constant, as the slab moves; what *does* change is the energy well *inside* the capacitor, where the field is nice and uniform.

Problem 4.28 Two long coaxial cylindrical metal tubes (inner radius a, outer radius b) stand vertically in a tank of dielectric oil (susceptibility χ_e, mass density ρ). The inner one is maintained at potential V, and the outer one is grounded (Fig. 4.32). To what height (h) does the oil rise in the space between the tubes?

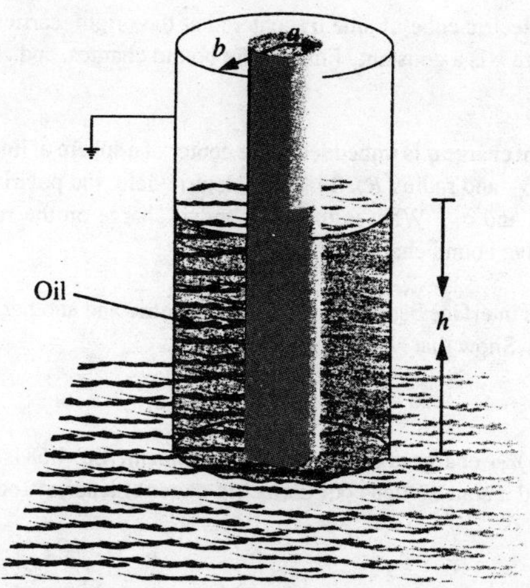

Figure 4.32

More Problems on Chapter 4

Problem 4.29

(a) For the configuration in Prob. 4.5, calculate the *force* on $\mathbf{p}_2$ due to $\mathbf{p}_1$, and the force on $\mathbf{p}_1$ due to $\mathbf{p}_2$. Are the answers consistent with Newton's third law?

(b) Find the total torque on $\mathbf{p}_2$ *with respect to the center of* $\mathbf{p}_1$, and compare it with the torque on $\mathbf{p}_1$ *about that same point.* [*Hint:* combine your answer to (a) with the result of Prob. 4.5.]

Problem 4.30 An electric dipole $\mathbf{p}$, pointing in the y direction, is placed midway between two large conducting plates, as shown in Fig. 4.33. Each plate makes a small angle θ with respect to the x axis, and they are maintained at potentials $\pm V$. What is the *direction* of the net force on $\mathbf{p}$? (There's nothing to *calculate*, here, but do explain your answer qualitatively.)

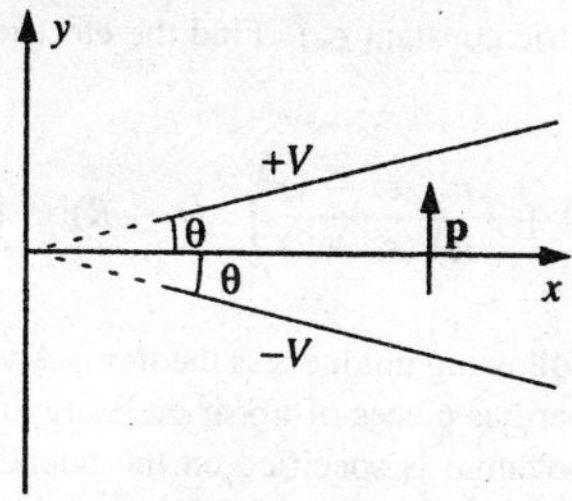

Figure 4.33

Problem 4.31 A dielectric cube of side a, centered at the origin, carries a "frozen-in" polarization $\mathbf{P} = k\mathbf{r}$, where k is a constant. Find all the bound charges, and check that they add up to zero.

Problem 4.32 A point charge q is imbedded at the center of a sphere of linear dielectric material (with susceptibility χ_e and radius R). Find the electric field, the polarization, and the bound charge densities, ρ_b and σ_b. What is the total bound charge on the surface? Where is the compensating negative bound charge located?

Problem 4.33 At the interface between one linear dielectric and another the electric field lines bend (see Fig. 4.34). Show that

$$\tan\theta_2 / \tan\theta_1 = \epsilon_2/\epsilon_1, \tag{4.68}$$

assuming there is no *free* charge at the boundary. [*Comment:* Eq. 4.68 is reminiscent of Snell's law in optics. Would a convex "lens" of dielectric material tend to "focus," or "defocus," the electric field?]

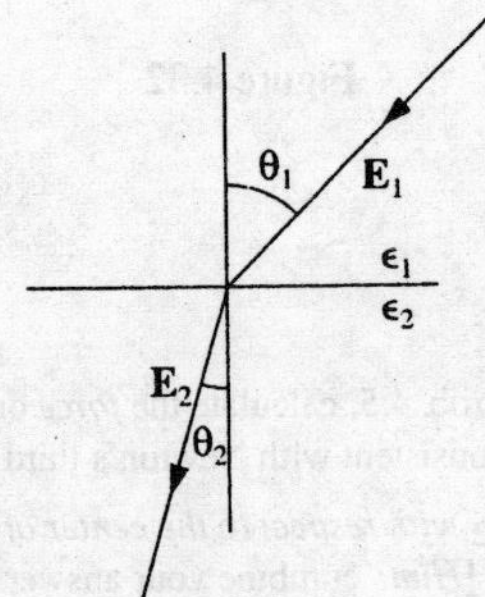

Figure 4.34

! **Problem 4.34** A point dipole $\mathbf{p}$ is imbedded at the center of a sphere of linear dielectric material (with radius R and dielectric constant ϵ_r). Find the electric potential inside and outside the sphere.

$$\left[\textit{Answer}: \quad \frac{p\cos\theta}{4\pi\epsilon r^2}\left(1+2\frac{r^3}{R^3}\frac{(\epsilon_r-1)}{(\epsilon_r+2)}\right), \; (r \le R); \quad \frac{p\cos\theta}{4\pi\epsilon_0 r^2}\left(\frac{3}{\epsilon_r+2}\right), \; (r \ge R)\right]$$

Problem 4.35 Prove the following uniqueness theorem: A volume $\mathcal{V}$ contains a specified free charge distribution, and various pieces of linear dielectric material, with the susceptibility of each one given. If the potential is specified on the boundaries $\mathcal{S}$ of $\mathcal{V}$ ($V = 0$ at infinity would be suitable) then the potential throughout $\mathcal{V}$ is uniquely determined. [*Hint:* integrate $\nabla \cdot (V_3 \mathbf{D}_3)$ over $\mathcal{V}$.]

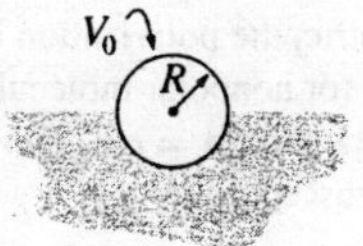

Figure 4.35

Problem 4.36 A conducting sphere at potential V_0 is half embedded in linear dielectric material of susceptibility χ_e, which occupies the region $z < 0$ (Fig. 4.35). *Claim:* the potential everywhere is exactly the same as it would have been in the absence of the dielectric! Check this claim, as follows:

(a) Write down the formula for the proposed potential $V(r)$, in terms of V_0, R, and r. Use it to determine the field, the polarization, the bound charge, and the free charge distribution on the sphere.

(b) Show that the total charge configuration would indeed produce the potential $V(r)$.

(c) Appeal to the uniqueness theorem in Prob. 4.35 to complete the argument.

(d) Could you solve the configurations in Fig. 4.36 with the same potential? If not, explain *why*.

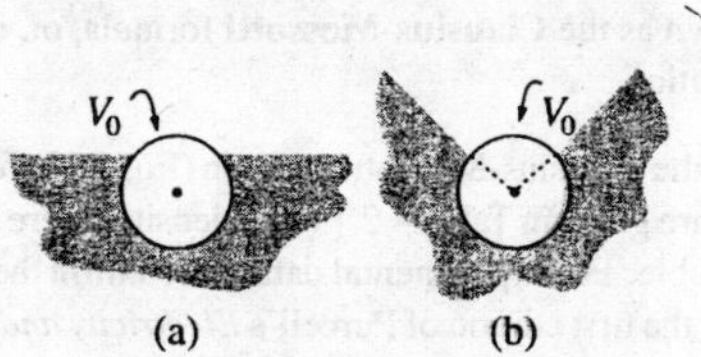

Figure 4.36

Problem 4.37 According to Eq. 4.5, the force on a single dipole is $(\mathbf{p} \cdot \nabla)\mathbf{E}$, so the *net* force on a dielectric object is

$$\mathbf{F} = \int (\mathbf{P} \cdot \nabla)\mathbf{E}_{\text{ext}}\, d\tau. \tag{4.69}$$

[Here $\mathbf{E}_{\text{ext}}$ is the field of everything *except* the dielectric. You might assume that it wouldn't matter if you used the *total* field; after all, the dielectric can't exert a force on *itself.* However, because the field of the dielectric is discontinuous at the location of any bound surface charge, the derivative introduces a spurious delta function, and you must either add a compensating surface term, or (better) stick with $\mathbf{E}_{\text{ext}}$, which suffers no such discontinuity.] Use Eq. 4.69 to determine the force on a tiny sphere of radius R, composed of linear dielectric material of susceptibility χ_e, which is situated a distance s from a fine wire carrying a uniform line charge λ.

! **Problem 4.38** In a linear dielectric, the polarization is proportional to the field: $\mathbf{P} = \epsilon_0 \chi_e \mathbf{E}$. If the material consists of atoms (or nonpolar molecules), the induced dipole moment of each one is likewise proportional to the field $\mathbf{p} = \alpha \mathbf{E}$. *Question:* What is the relation between the atomic polarizability α and the susceptibility χ_e?

Since $\mathbf{P}$ (the dipole moment per unit volume) is $\mathbf{p}$ (the dipole moment per atom) times N (the number of atoms per unit volume), $\mathbf{P} = N\mathbf{p} = N\alpha\mathbf{E}$, one's first inclination is to say that

$$\chi_e = \frac{N\alpha}{\epsilon_0}. \tag{4.70}$$

And in fact this is not far off, if the density is low. But closer inspection reveals a subtle problem, for the field $\mathbf{E}$ in Eq. 4.30 is the *total macroscopic* field in the medium, whereas the field in Eq. 4.1 is due to everything *except* the particular atom under consideration (polarizability was defined for an isolated atom subject to a specified external field); call this field $\mathbf{E}_{\text{else}}$. Imagine that the space allotted to each atom is a sphere of radius R, and show that

$$\mathbf{E} = \left(1 - \frac{N\alpha}{3\epsilon_0}\right)\mathbf{E}_{\text{else}}. \tag{4.71}$$

Use this to conclude that

$$\chi_e = \frac{N\alpha/\epsilon_0}{1 - N\alpha/3\epsilon_0},$$

or

$$\alpha = \frac{3\epsilon_0}{N}\left(\frac{\epsilon_r - 1}{\epsilon_r + 2}\right). \tag{4.72}$$

Equation 4.72 is known as the **Clausius-Mossotti** formula, or, in its application to optics, the **Lorentz-Lorenz** equation.

Problem 4.39 Check the Clausius-Mossotti relation (Eq. 4.72) for the gases listed in Table 4.1. (Dielectric constants are given in Table 4.2.) (The densities here are so small that Eqs. 4.70 and 4.72 are indistinguishable. For experimental data that confirm the Clausius-Mossotti correction term see, for instance, the first edition of Purcell's *Electricity and Magnetism,* Problem 9.28.)[14]

! **Problem 4.40** The Clausius-Mossotti equation (Prob. 4.38) tells you how to calculate the susceptibility of a *nonpolar* substance, in terms of the atomic polarizability α. The Langevin equation tells you how to calculate the susceptibility of a *polar* substance, in terms of the permanent molecular dipole moment p. Here's how it goes:

(a) The energy of a dipole in an external field $\mathbf{E}$ is $u = -\mathbf{p}\cdot\mathbf{E}$ (Eq. 4.6); it ranges from $-pE$ to $+pE$, depending on the orientation. Statistical mechanics says that for a material in equilibrium at absolute temperature T, the probability of a given molecule having energy u is proportional to the Boltzmann factor,

$$\exp(-u/kT).$$

The average energy of the dipoles is therefore

$$<u> = \frac{\int u e^{-(u/kT)}\, du}{\int e^{-(u/kT)}\, du},$$

[14] E. M. Purcell, *Electricity and Magnetism* (Berkeley Physics Course, Vol. 2), (New York: McGraw-Hill, 1963).

where the integrals run from $-pE$ to $+pE$. Use this to show that the polarization of a substance containing N molecules per unit volume is

$$P = Np[\coth(pE/kT) - (kT/pE)]. \tag{4.73}$$

That's the **Langevin formula**. Sketch P/Np as a function of pE/kT.

(b) Notice that for large fields/low temperatures, virtually *all* the molecules are lined up, and the material is *non*linear. Ordinarily, however, kT is much greater than pE. Show that in this régime the material *is* linear, and calculate its susceptibility, in terms of N, p, T, and k. Compute the susceptibility of water at 20°C, and compare the experimental value in Table 4.2. (The dipole moment of water is 6.1×10^{-30} C·m.) This is rather far off, because we have again neglected the distinction between **E** and $\mathbf{E}_{\text{else}}$. The agreement is better in low-density gases, for which the difference between **E** and $\mathbf{E}_{\text{else}}$ is negligible. Try it for water vapor at 100°C and 1 atm.

Chapter 5

Magnetostatics

5.1 The Lorentz Force Law

5.1.1 Magnetic Fields

Remember the basic problem of classical electrodynamics: We have a collection of charges $q_1, q_2, q_3, \ldots$ (the "source" charges), and we want to calculate the force they exert on some other charge Q (the "test" charge). (See Fig. 5.1.) According to the principle of superposition, it is sufficient to find the force of a *single* source charge—the total is then the vector sum of all the individual forces. Up to now we have confined our attention to the simplest case, *electrostatics*, in which the source charge is *at rest* (though the test charge need not be). The time has come to consider the forces between charges *in motion*.

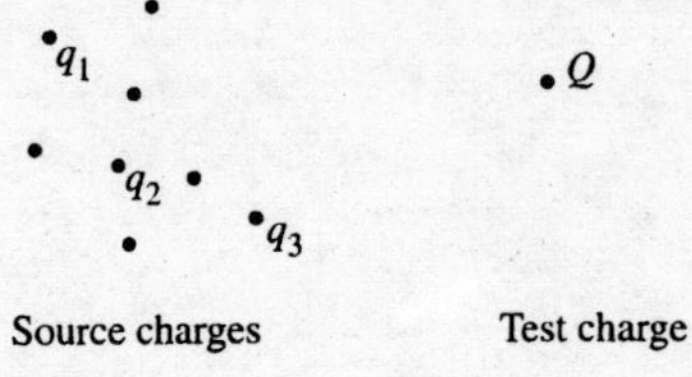

Figure 5.1

To give you some sense of what is in store, imagine that I set up the following demonstration: Two wires hang from the ceiling, a few centimeters apart; when I turn on a current, so that it passes up one wire and back down the other, the wires jump apart—they evidently repel one another (Fig. 5.2(a)). How do you explain this? Well, you might suppose that the battery (or whatever drives the current) is actually charging up the wire, and that the force is simply due to the electrical repulsion of like charges. But this explanation is incorrect. I could hold up a test charge near these wires and there would be *no* force on it,

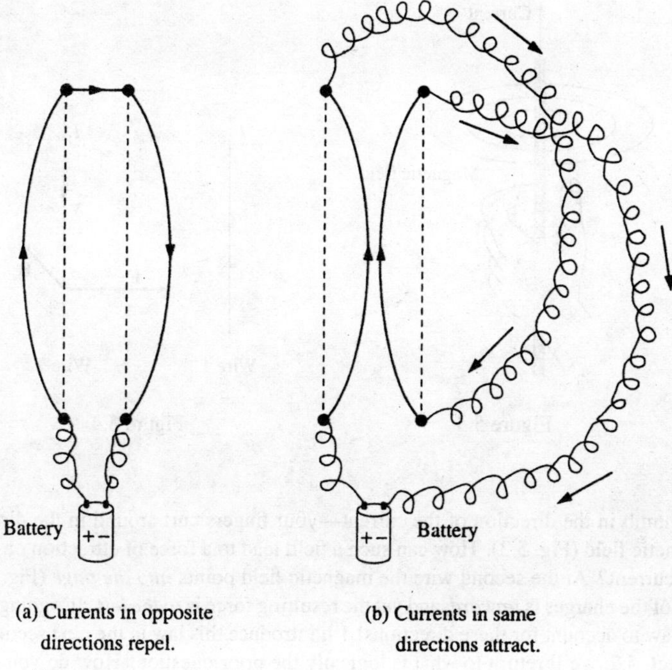

(a) Currents in opposite directions repel.

(b) Currents in same directions attract.

Figure 5.2

for the wires are in fact electrically neutral. (It's true that electrons are flowing down the line—that's what a current *is*—but there are just as many stationary plus charges as moving minus charges on any given segment.) Moreover, I could hook up my demonstration so as to make the current flow up *both* wires (Fig. 5.2(b)); in this case they are found to *attract!*

Whatever force accounts for the attraction of parallel currents and the repulsion of antiparallel ones is *not* electrostatic in nature. It is our first encounter with a *magnetic* force. Whereas a *stationary* charge produces only an electric field **E** in the space around it, a *moving* charge generates, in addition, a magnetic field **B**. In fact, magnetic fields are a lot easier to detect, in practice—all you need is a Boy Scout compass. How these devices work is irrelevant at the moment; it is enough to know that the needle points in the direction of the local magnetic field. Ordinarily, this means *north*, in response to the *earth's* magnetic field, but in the laboratory, where typical fields may be hundreds of times stronger than that, the compass indicates the direction of whatever magnetic field is present.

Now, if you hold up a tiny compass in the vicinity of a current-carrying wire, you quickly discover a very peculiar thing: The field does not point *toward* the wire, nor *away* from it, but rather it *circles around the wire*. In fact, if you grab the wire with your right

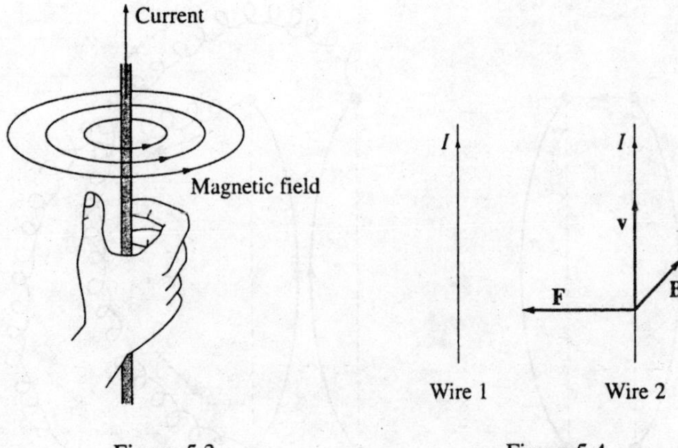

Figure 5.3

Figure 5.4

hand—thumb in the direction of the current—your fingers curl around in the direction of the magnetic field (Fig. 5.3). How can such a field lead to a force of attraction on a nearby parallel current? At the second wire the magnetic field points *into the page* (Fig. 5.4), the velocity of the charges is *upward*, and yet the resulting force is *to the left*. It's going to take a strange law to account for these directions! I'll introduce this law in the next section. Later on, in Sect. 5.2, we'll return to what is logically the prior question: How do you calculate the magnetic field of the first wire?

5.1.2 Magnetic Forces

It may have occurred to you that the combination of directions in Fig. 5.4 is just right for a cross product. In fact, the magnetic force in a charge Q, moving with velocity $\mathbf{v}$ in a magnetic field $\mathbf{B}$, is[1]

$$\boxed{\mathbf{F}_{\text{mag}} = Q(\mathbf{v} \times \mathbf{B}).} \tag{5.1}$$

This is known as the **Lorentz force law**. In the presence of both electric *and* magnetic fields, the net force on Q would be

$$\mathbf{F} = Q[\mathbf{E} + (\mathbf{v} \times \mathbf{B})]. \tag{5.2}$$

I do not pretend to have *derived* Eq. 5.1, of course; it is a fundamental axiom of the theory, whose justification is to be found in experiments such as the one I described in Sect. 5.1.1. Our main job from now on is to calculate the magnetic field $\mathbf{B}$ (and for that matter the electric field $\mathbf{E}$ as well, for the rules are more complicated when the source charges are in motion). But before we proceed, it is worthwhile to take a closer look at the Lorentz force law itself; it is a peculiar law, and it leads to some truly bizarre particle trajectories.

[1] Since $\mathbf{F}$ and $\mathbf{v}$ are vectors, $\mathbf{B}$ is actually a *pseudo*vector.

Example 5.1

Cyclotron motion

The archetypical motion of a charged particle in a magnetic field is circular, with the magnetic force providing the centripetal acceleration. In Fig. 5.5, a uniform magnetic field points *into* the page; if the charge Q moves counterclockwise, with speed v, around a circle of radius R, the magnetic force (5.1) points *inward*, and has a fixed magnitude QvB—just right to sustain uniform circular motion:

$$QvB = m\frac{v^2}{R}, \quad \text{or} \quad p = QBR, \tag{5.3}$$

where m is the particle's mass and $p = mv$ is its momentum. Equation 5.3 is known as the **cyclotron formula** because it describes the motion of a particle in a cyclotron—the first of the modern particle accelerators. It also suggests a simple experimental technique for finding the momentum of a particle: send it through a region of known magnetic field, and measure the radius of its circular trajectory. This is in fact the standard means for determining the momenta of elementary particles.

Incidentally, I assumed that the charge moves in a plane perpendicular to **B**. If it starts out with some additional speed $v_\parallel$ *parallel* to **B**, this component of the motion is unaffected by the magnetic field, and the particle moves in a *helix* (Fig. 5.6). The radius is still given by Eq. 5.3, but the velocity in question is now the component perpendicular to **B**, $v_\perp$.

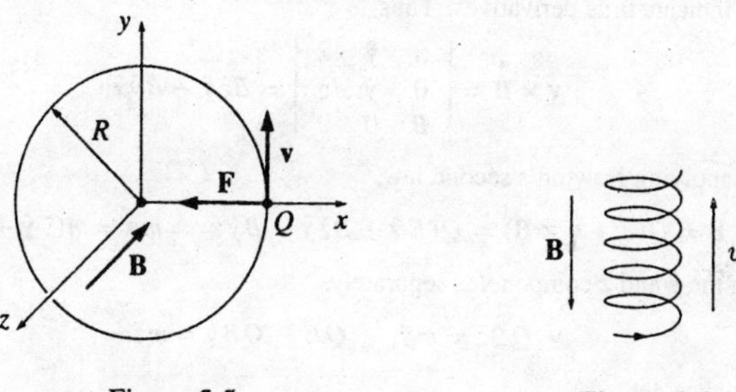

Figure 5.5 Figure 5.6

Example 5.2

Cycloid Motion

A more exotic trajectory occurs if we include a uniform electric field, at right angles to the magnetic one. Suppose, for instance, that **B** points in the x-direction, and **E** in the z-direction, as shown in Fig. 5.7. A particle at rest is released from the origin; what path will it follow?

Solution: Let's think it through qualitatively, first. Initially, the particle is at rest, so the magnetic force is zero, and the electric field accelerates the charge in the z-direction. As it picks up speed, a magnetic force develops which, according to Eq. 5.1, pulls the charge around

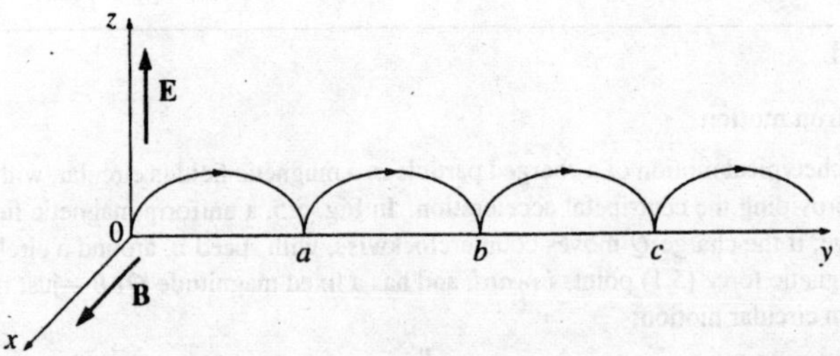

Figure 5.7

to the right. The faster it goes, the stronger F_{mag} becomes; eventually, it curves the particle back around towards the y axis. At this point the charge is moving *against* the electrical force, so it begins to slow down—the magnetic force then decreases, and the electrical force takes over, bringing the charge to rest at point a, in Fig. 5.7. There the entire process commences anew, carrying the particle over to point b, and so on.

Now let's do it quantitatively. There being no force in the x-direction, the position of the particle at any time t can be described by the vector $(0, y(t), z(t))$; the velocity is therefore

$$\mathbf{v} = (0, \dot{y}, \dot{z}),$$

where dots indicate time derivatives. Thus

$$\mathbf{v} \times \mathbf{B} = \begin{vmatrix} \hat{\mathbf{x}} & \hat{\mathbf{y}} & \hat{\mathbf{z}} \\ 0 & \dot{y} & \dot{z} \\ B & 0 & 0 \end{vmatrix} = B\dot{z}\,\hat{\mathbf{y}} - B\dot{y}\,\hat{\mathbf{z}},$$

and hence, applying Newton's second law,

$$\mathbf{F} = Q(\mathbf{E} + \mathbf{v} \times \mathbf{B}) = Q(E\,\hat{\mathbf{z}} + B\dot{z}\,\hat{\mathbf{y}} - B\dot{y}\,\hat{\mathbf{z}}) = m\mathbf{a} = m(\ddot{y}\,\hat{\mathbf{y}} + \ddot{z}\,\hat{\mathbf{z}}).$$

Or, treating the $\hat{\mathbf{y}}$ and $\hat{\mathbf{z}}$ components separately,

$$QB\dot{z} = m\ddot{y}, \quad QE - QB\dot{y} = m\ddot{z}.$$

For convenience, let

$$\omega \equiv \frac{QB}{m}. \tag{5.4}$$

(This is the **cyclotron frequency**, at which the particle would revolve in the absence of any electric field.) Then the equations of motion take the form

$$\ddot{y} = \omega\dot{z}, \quad \ddot{z} = \omega\left(\frac{E}{B} - \dot{y}\right). \tag{5.5}$$

Their general solution[2] is

$$\left.\begin{aligned} y(t) &= C_1 \cos\omega t + C_2 \sin\omega t + (E/B)t + C_3, \\ z(t) &= C_2 \cos\omega t - C_1 \sin\omega t + C_4. \end{aligned}\right\} \tag{5.6}$$

[2] As coupled differential equations, they are easily solved by differentiating the first and using the second to eliminate $\ddot{z}$.

But the particle started from rest ($\dot{y}(0) = \dot{z}(0) = 0$), at the origin ($y(0) = z(0) = 0$); these four conditions determine the constants $C_1, C_2, C_3,$ and C_4:

$$y(t) = \frac{E}{\omega B}(\omega t - \sin \omega t), \quad z(t) = \frac{E}{\omega B}(1 - \cos \omega t). \tag{5.7}$$

In this form the answer is not terribly enlightening, but if we let

$$R \equiv \frac{E}{\omega B}, \tag{5.8}$$

and eliminate the sines and cosines by exploiting the trigonometric identity $\sin^2 \omega t + \cos^2 \omega t = 1$, we find that

$$(y - R\omega t)^2 + (z - R)^2 = R^2. \tag{5.9}$$

This is the formula for a *circle*, of radius R, whose center $(0, R\omega t, R)$ travels in the y-direction at a constant speed,

$$v = \omega R = \frac{E}{B}. \tag{5.10}$$

The particle moves as though it were a spot on the rim of a wheel, rolling down the y axis at speed v. The curve generated in this way is called a **cycloid**. Notice that the overall motion is *not* in the direction of $\mathbf{E}$, as you might suppose, but perpendicular to it.

One feature of the magnetic force law (Eq. 5.1) warrants special attention:

Magnetic forces do no work.

For if Q moves an amount $d\mathbf{l} = \mathbf{v}\,dt$, the work done is

$$dW_{\text{mag}} = \mathbf{F}_{\text{mag}} \cdot d\mathbf{l} = Q(\mathbf{v} \times \mathbf{B}) \cdot \mathbf{v}\,dt = 0. \tag{5.11}$$

This follows because $(\mathbf{v} \times \mathbf{B})$ is perpendicular to $\mathbf{v}$, so $(\mathbf{v} \times \mathbf{B}) \cdot \mathbf{v} = 0$. Magnetic forces may alter the *direction* in which a particle moves, but they cannot speed it up or slow it down. The fact that magnetic forces do no work is an elementary and direct consequence of the Lorentz force law, but there are many situations in which it *appears* so manifestly false that one's confidence is bound to waver. When a magnetic crane lifts the carcass of a junked car, for instance, *something* is obviously doing work, and it seems perverse to deny that the magnetic force is responsible. Well, perverse or not, deny it we must, and it can be a very subtle matter to figure out what agency *does* deserve the credit in such circumstances. I'll show you several examples as we go along.

Problem 5.1 A particle of charge q enters a region of uniform magnetic field $\mathbf{B}$ (pointing *into* the page). The field deflects the particle a distance d above the original line of flight, as shown in Fig. 5.8. Is the charge positive or negative? In terms of a, d, B and q, find the momentum of the particle.

Problem 5.2 Find and sketch the trajectory of the particle in Ex. 5.2, if it starts at the origin with velocity

(*a*) $\mathbf{v}(0) = (E/B)\hat{\mathbf{y}}$,

(*b*) $\mathbf{v}(0) = (E/2B)\hat{\mathbf{y}}$,

(*c*) $\mathbf{v}(0) = (E/B)(\hat{\mathbf{y}} + \hat{\mathbf{z}})$.

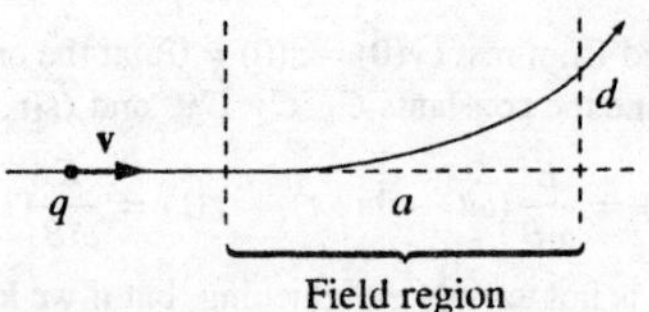

Figure 5.8

Problem 5.3 In 1897 J. J. Thomson "discovered" the electron by measuring the charge-to-mass ratio of "cathode rays" (actually, streams of electrons, with charge q and mass m) as follows:

(a) First he passed the beam through uniform crossed electric and magnetic fields $\mathbf{E}$ and $\mathbf{B}$ (mutually perpendicular, and both of them perpendicular to the beam), and adjusted the electric field until he got zero deflection. What, then, was the speed of the particles (in terms of E and B)?

(b) Then he turned off the electric field, and measured the radius of curvature, R, of the beam, as deflected by the magnetic field alone. In terms of E, B, and R, what is the charge-to-mass ratio (q/m) of the particles?

5.1.3 Currents

The **current** in a wire is the *charge per unit time* passing a given point. By definition, negative charges moving to the left count the same as positive ones to the right. This conveniently reflects the *physical* fact that almost all phenomena involving moving charges depend on the *product* of charge and velocity—if you change the sign of q *and* $\mathbf{v}$, you get the same answer, so it doesn't really matter which you have. (The Lorentz force law is a case in point; the Hall effect (Prob. 5.39) is a notorious exception.) In practice, it is ordinarily the negatively charged electrons that do the moving—in the direction *opposite* the electric current. To avoid the petty complications this entails, I shall often pretend it's the positive charges that move, as in fact everyone assumed they did for a century or so after Benjamin Franklin established his unfortunate convention.[3] Current is measured in coulombs-per-second, or **amperes** (A):

$$1\ \text{A} = 1\ \text{C/s}. \tag{5.12}$$

A line charge λ traveling down a wire at speed v (Fig. 5.9) constitutes a current

$$I = \lambda v, \tag{5.13}$$

because a segment of length $v\Delta t$, carrying charge $\lambda v \Delta t$, passes point P in a time interval Δt. Current is actually a *vector:*

$$\mathbf{I} = \lambda \mathbf{v}; \tag{5.14}$$

[3] If we called the electron plus and the proton minus, the problem would never arise. In the context of Franklin's experiments with cat's fur and glass rods, the choice was completely arbitrary.

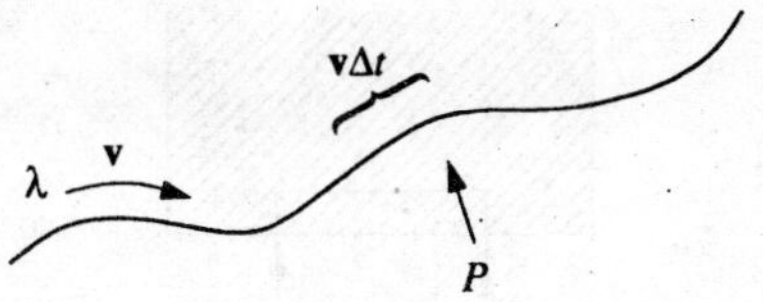

Figure 5.9

since the path of the flow is dictated by the shape of the wire, most people don't bother to display the vectorial character of **I** explicitly, but when it comes to surface and volume currents we cannot afford to be so casual, and for the sake of notational consistency it is a good idea to acknowledge this right from the start. A neutral wire, of course, contains as many stationary positive charges as mobile negative ones. The former do not contribute to the current—the charge density λ in Eq. 5.13 refers only to the *moving* charges. In the unusual situation where *both* types move, $\mathbf{I} = \lambda_+\mathbf{v}_+ + \lambda_-\mathbf{v}_-$.

The magnetic force on a segment of current-carrying wire is evidently

$$\mathbf{F}_{\text{mag}} = \int (\mathbf{v} \times \mathbf{B})\, dq = \int (\mathbf{v} \times \mathbf{B})\lambda\, dl = \int (\mathbf{I} \times \mathbf{B})\, dl. \tag{5.15}$$

Inasmuch as **I** and $d\mathbf{l}$ both point in the same direction, we can just as well write this as

$$\boxed{\mathbf{F}_{\text{mag}} = \int I\,(d\mathbf{l} \times \mathbf{B}).} \tag{5.16}$$

Typically, the current is constant (in magnitude) along the wire, and in that case I comes outside the integral:

$$\mathbf{F}_{\text{mag}} = I \int (d\mathbf{l} \times \mathbf{B}). \tag{5.17}$$

Example 5.3

A rectangular loop of wire, supporting a mass m, hangs vertically with one end in a uniform magnetic field **B**, which points into the page in the shaded region of Fig. 5.10. For what current I, in the loop, would the magnetic force upward exactly balance the gravitational force downward?

Solution: First of all, the current must circulate clockwise, in order for $(\mathbf{I} \times \mathbf{B})$ in the horizontal segment to point upward. The force is

$$F_{\text{mag}} = IBa,$$

where a is the width of the loop. (The magnetic forces on the two vertical segments cancel.) For F_{mag} to balance the weight (mg), we must therefore have

$$I = \frac{mg}{Ba}. \tag{5.18}$$

The weight just *hangs* there, suspended in mid-air!

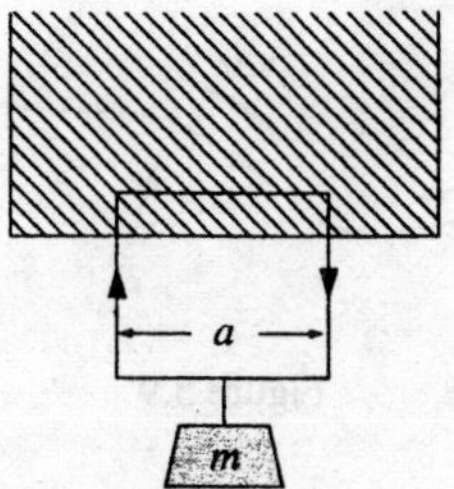

Figure 5.10

What happens if we now *increase* the current? Then the upward magnetic force *exceeds* the downward force of gravity, and the loop rises, lifting the weight. *Somebody's* doing work, and it sure looks as though the magnetic force is responsible. Indeed, one is tempted to write

$$W_{\text{mag}} = F_{\text{mag}} h = IBah, \tag{5.19}$$

where h is the distance the loop rises. But we know that magnetic forces *never* do work. What's going on here?

Well, when the loop starts to rise, the charges in the wire are no longer moving horizontally—their velocity now acquires an upward component u, the speed of the loop (Fig. 5.11), in addition to the horizontal component w associated with the current ($I = \lambda w$). The magnetic force, which is always perpendicular to the velocity, no longer points straight up, but tilts back. It is perpendicular to the *net* displacement of the charge (which is in the direction of $\mathbf{v}$), and therefore *it does no work on* q. It does have a vertical *component* (qwB); indeed, the net vertical force on all the charge (λa) in the upper segment of the loop is

$$F_{\text{vert}} = \lambda a w B = IBa \tag{5.20}$$

(as before); but now it also has a *horizontal* component (quB), which opposes the flow of current. Whoever is in charge of maintaining that current, therefore, must now *push* those charges along, against the backward component of the magnetic force.

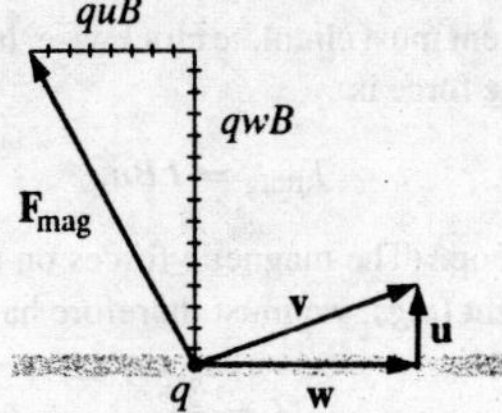

Figure 5.11

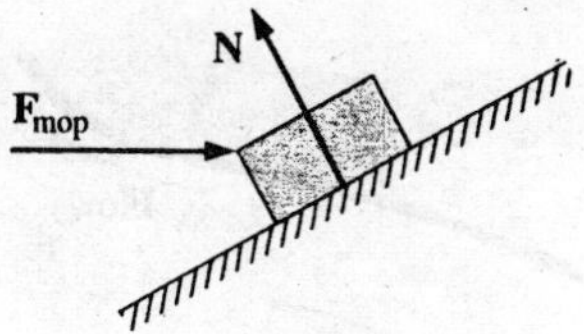

Figure 5.12

The total horizontal force on the top segment is evidently

$$F_{\text{horiz}} = \lambda a u B. \tag{5.21}$$

In a time dt the charges move a (horizontal) distance $w\,dt$, so the work done by this agency (presumably a battery or a generator) is

$$W_{\text{battery}} = \lambda a B \int u w\, dt = I B a h,$$

which is precisely what we naïvely attributed to the *magnetic* force in Eq. 5.19. Was work done in this process? Absolutely! Who *did* it? The battery! What, then, was the role of the magnetic force? Well, it *redirected* the horizontal force of the battery into the *vertical* motion of the loop and the weight.

It may help to consider a mechanical analogy. Imagine you're pushing a trunk up a frictionless ramp, by pushing on it horizontally with a mop (Fig. 5.12). The normal force (**N**) does no work, because it is perpendicular to the displacement. But it *does* have a vertical component (which in fact is what lifts the trunk), and a (backward) horizontal component (which you have to overcome by pushing on the mop). Who is doing the work here? *You* are, obviously—and yet your *force* (which is purely horizontal) is not (at least, not directly) what lifts the box. The normal force plays the same passive (but crucial) role as the magnetic force in Ex. 5.3: while doing no work itself, it *redirects* the efforts of the active agent (you, or the battery, as the case may be), from horizontal to vertical.

When charge flows over a *surface*, we describe it by the **surface current density, K**, defined as follows: Consider a "ribbon" of infinitesimal width $dl_\perp$, running parallel to the flow (Fig. 5.13). If the current in this ribbon is $d\mathbf{I}$, the surface current density is

$$\mathbf{K} \equiv \frac{d\mathbf{I}}{dl_\perp}. \tag{5.22}$$

In words, K is the *current per unit width-perpendicular-to-flow*. In particular, if the (mobile) surface charge density is σ and its velocity is $\mathbf{v}$, then

$$\mathbf{K} = \sigma \mathbf{v}. \tag{5.23}$$

In general, **K** will vary from point to point over the surface, reflecting variations in σ and/or **v**. The magnetic force on the surface current is

$$\mathbf{F}_{\text{mag}} = \int (\mathbf{v} \times \mathbf{B}) \sigma\, da = \int (\mathbf{K} \times \mathbf{B})\, da. \tag{5.24}$$

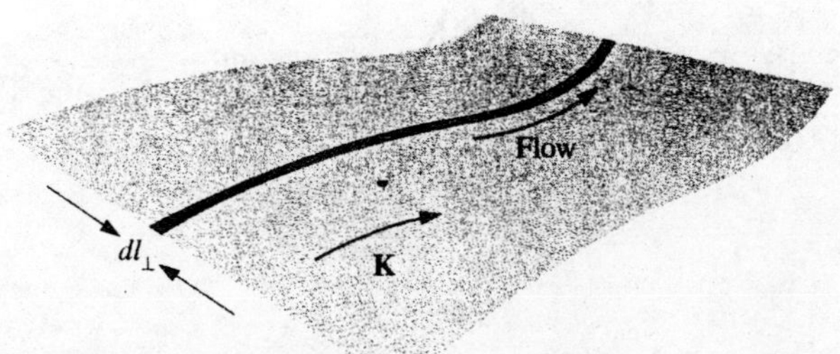

Figure 5.13

Caveat: Just as **E** suffers a discontinuity at a surface *charge*, so **B** is discontinuous at a surface *current*. In Eq. 5.24, you must be careful to use the *average* field, just as we did in Sect. 2.5.3.

When the flow of charge is distributed throughout a three-dimensional region, we describe it by the **volume current density**, **J**, defined as follows: Consider a "tube" of infinitesimal cross section $da_\perp$, running parallel to the flow (Fig. 5.14). If the current in this tube is $d\mathbf{I}$, the volume current density is

$$\mathbf{J} \equiv \frac{d\mathbf{I}}{da_\perp}. \tag{5.25}$$

In words, J is the ***current per unit area-perpendicular-to-flow***. If the (mobile) volume charge density is ρ and the velocity is **v**, then

$$\mathbf{J} = \rho\mathbf{v}. \tag{5.26}$$

The magnetic force on a volume current is therefore

$$\mathbf{F}_{\text{mag}} = \int (\mathbf{v} \times \mathbf{B})\rho\, d\tau = \int (\mathbf{J} \times \mathbf{B})\, d\tau. \tag{5.27}$$

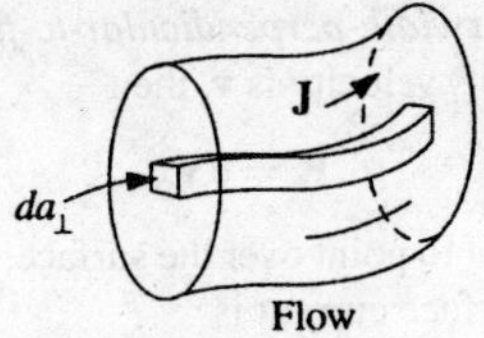

Figure 5.14

Figure 5.15

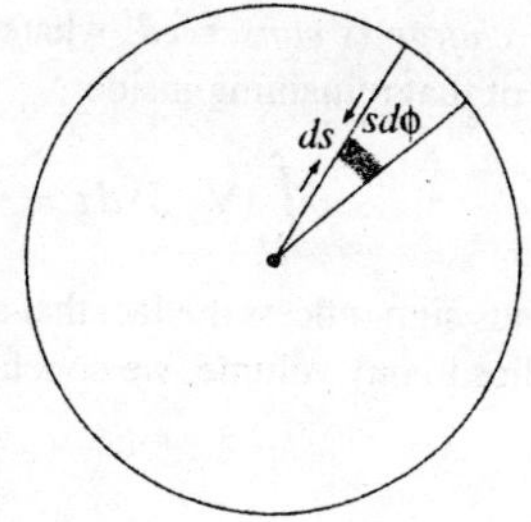

Figure 5.16

Example 5.4

(a) A current I is uniformly distributed over a wire of circular cross section, with radius a (Fig. 5.15). Find the volume current density J.

Solution: The area-perpendicular-to-flow is πa^2, so

$$J = \frac{I}{\pi a^2}.$$

This was trivial because the current density was uniform.

(b) Suppose the current density in the wire is proportional to the distance from the axis,

$$J = ks$$

(for some constant k). Find the total current in the wire.

Solution: Because J varies with s, we must *integrate* Eq. 5.25. The current in the shaded patch (Fig. 5.16) is $J\,da_\perp$, and $da_\perp = s\,ds\,d\phi$. So,

$$I = \int (ks)(s\,ds\,d\phi) = 2\pi k \int_0^a s^2\,ds = \frac{2\pi k a^3}{3}.$$

According to Eq. 5.25, the current crossing a surface S can be written as

$$I = \int_S J\,da_\perp = \int_S \mathbf{J} \cdot d\mathbf{a}. \tag{5.28}$$

(The dot product serves neatly to pick out the appropriate component of $d\mathbf{a}$.) In particular, the total charge per unit time leaving a volume $\mathcal{V}$ is

$$\oint_S \mathbf{J} \cdot d\mathbf{a} = \int_{\mathcal{V}} (\nabla \cdot \mathbf{J})\,d\tau.$$

Because charge is conserved, whatever flows out through the surface must come at the expense of that remaining inside:

$$\int_{\mathcal{V}} (\nabla \cdot \mathbf{J})\, d\tau = -\frac{d}{dt}\int_{\mathcal{V}} \rho\, d\tau = -\int_{\mathcal{V}} \left(\frac{\partial \rho}{\partial t}\right) d\tau.$$

(The minus sign reflects the fact that an *outward* flow *decreases* the charge left in $\mathcal{V}$.) Since this applies to *any* volume, we conclude that

$$\boxed{\nabla \cdot \mathbf{J} = -\frac{\partial \rho}{\partial t}.} \qquad (5.29)$$

This is the precise mathematical statement of local charge conservation; it is called the **continuity equation**.

For future reference, let me summarize the "dictionary" we have implicitly developed for translating equations into the forms appropriate to point, line, surface, and volume currents:

$$\sum_{i=1}^{n} (\;\;) q_i \mathbf{v}_i \sim \int_{\text{line}} (\;\;) \mathbf{I}\, dl \sim \int_{\text{surface}} (\;\;) \mathbf{K}\, da \sim \int_{\text{volume}} (\;\;) \mathbf{J}\, d\tau. \qquad (5.30)$$

This correspondence, which is analogous to $q \sim \lambda\, dl \sim \sigma\, da \sim \rho\, d\tau$ for the various charge distributions, generates Eqs. 5.15, 5.24, and 5.27 from the original Lorentz force law (5.1).

Problem 5.4 Suppose that the magnetic field in some region has the form

$$\mathbf{B} = kz\,\hat{\mathbf{x}}$$

(where k is a constant). Find the force on a square loop (side a), lying in the yz plane and centered at the origin, if it carries a current I, flowing counterclockwise, when you look down the x axis.

Problem 5.5 A current I flows down a wire of radius a.

(a) If it is uniformly distributed over the surface, what is the surface current density K?

(b) If it is distributed in such a way that the volume current density is inversely proportional to the distance from the axis, what is J?

Problem 5.6

(a) A phonograph record carries a uniform density of "static electricity" σ. If it rotates at angular velocity ω, what is the surface current density K at a distance r from the center?

(b) A uniformly charged solid sphere, of radius R and total charge Q, is centered at the origin and spinning at a constant angular velocity ω about the z axis. Find the current density $\mathbf{J}$ at any point (r, θ, ϕ) within the sphere.

Problem 5.7 For a configuration of charges and currents confined within a volume $\mathcal{V}$, show that

$$\int_{\mathcal{V}} \mathbf{J}\, d\tau = d\mathbf{p}/dt,$$

where $\mathbf{p}$ is the total dipole moment. [*Hint:* evaluate $\int_{\mathcal{V}} \nabla \cdot (x\mathbf{J})\, d\tau$.]

5.2 The Biot-Savart Law

5.2.1 Steady Currents

Stationary charges produce electric fields that are constant in time; hence the term **electrostatics**.[4] *Steady currents* produce magnetic fields that are constant in time; the theory of steady currents is called **magnetostatics**.

Stationary charges	⇒	**constant electric fields: electrostatics.**
Steady currents	⇒	**constant magnetic fields: magnetostatics.**

By **steady current** I mean a continuous flow that has been going on forever, without change and without charge piling up anywhere. (Some people call them "stationary currents"; to my ear, that's a contradiction in terms.) Of course, there's no such thing in practice as a *truly* steady current, any more than there is a *truly* stationary charge. In this sense both electrostatics and magnetostatics describe artificial worlds that exist only in textbooks. However, they represent suitable *approximations* as long as the actual fluctuations are reasonably slow; in fact, for most purposes magnetostatics applies very well to household currents, which alternate 60 times a second!

Notice that a moving *point* charge *cannot possibly constitute a steady current.* If it's here one instant, it's gone the next. This may seem like a minor thing to *you*, but it's a major headache for *me*. I developed each topic in electrostatics by starting out with the simple case of a point charge at rest; then I generalized to an arbitrary charge distribution by invoking the superposition principle. This approach is not open to us in magnetostatics because a moving point charge does not produce a static field in the first place. We are *forced* to deal with extended current distributions, right from the start, and as a result the arguments are bound to be more cumbersome.

When a steady current flows in a wire, its magnitude I must be the same all along the line; otherwise, charge would be piling up somewhere, and it wouldn't be a steady current. By the same token, $\partial\rho/\partial t = 0$ in magnetostatics, and hence the continuity equation (5.29) becomes

$$\nabla \cdot \mathbf{J} = 0. \tag{5.31}$$

5.2.2 The Magnetic Field of a Steady Current

The magnetic field of a steady line current is given by the **Biot-Savart law**:

$$\boxed{\mathbf{B}(\mathbf{r}) = \frac{\mu_0}{4\pi}\int \frac{\mathbf{I} \times \hat{\boldsymbol{\imath}}}{\imath^2}\, dl' = \frac{\mu_0}{4\pi} I \int \frac{d\mathbf{l}' \times \hat{\boldsymbol{\imath}}}{\imath^2}.} \tag{5.32}$$

[4]Actually, it is not necessary that the charges be stationary, but only that the charge *density* at each point be constant. For example, the sphere in Prob. 5.6b produces an electrostatic field $1/4\pi\epsilon_0(Q/r^2)\hat{\mathbf{r}}$, even though it is rotating, because ρ does not depend on t.

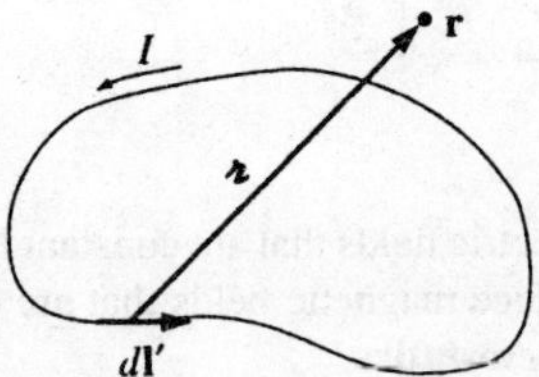

Figure 5.17

The integration is along the current path, in the direction of the flow; $d\mathbf{l}'$ is an element of length along the wire, and $\boldsymbol{\mathscr{r}}$, as always, is the vector from the source to the point $\mathbf{r}$ (Fig. 5.17). The constant μ_0 is called the **permeability of free space:**[5]

$$\mu_0 = 4\pi \times 10^{-7}\ \mathrm{N/A^2}. \tag{5.33}$$

These units are such that $\mathbf{B}$ itself comes out in newtons per ampere-meter (as required by the Lorentz force law), or **teslas** (T):[6]

$$1\ \mathrm{T} = 1\ \mathrm{N/(A\cdot m)}. \tag{5.34}$$

As the starting point for magnetostatics, the Biot-Savart law plays a role analogous to Coulomb's law in electrostatics. Indeed, the $1/\mathscr{r}^2$ dependence is common to both laws.

Example 5.5

Find the magnetic field a distance s from a long straight wire carrying a steady current I (Fig. 5.18).

Solution: In the diagram, $(d\mathbf{l}' \times \hat{\boldsymbol{\mathscr{r}}})$ points *out* of the page, and has the magnitude

$$dl' \sin\alpha = dl' \cos\theta.$$

Also, $l' = s\tan\theta$, so

$$dl' = \frac{s}{\cos^2\theta}\, d\theta,$$

and $s = \mathscr{r}\cos\theta$, so

$$\frac{1}{\mathscr{r}^2} = \frac{\cos^2\theta}{s^2}.$$

---

[5]This is an exact number, not an empirical constant. It serves (via Eq. 5.37) to define the ampere, and the ampere in turn defines the coulomb.

[6]For some reason, in this one case the cgs unit (the **gauss**) is more commonly used than the SI unit: 1 tesla = 10^4 gauss. The earth's magnetic field is about half a gauss; a fairly strong laboratory magnetic field is, say, 10,000 gauss.

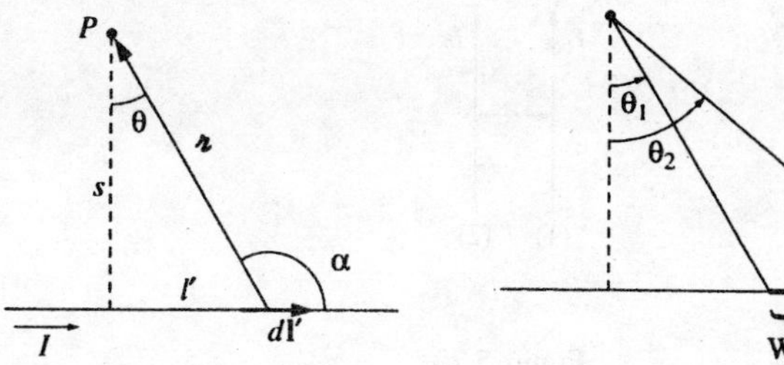

Figure 5.18 Figure 5.19

Thus

$$B = \frac{\mu_0 I}{4\pi}\int_{\theta_1}^{\theta_2}\left(\frac{\cos^2\theta}{s^2}\right)\left(\frac{s}{\cos^2\theta}\right)\cos\theta\,d\theta$$

$$= \frac{\mu_0 I}{4\pi s}\int_{\theta_1}^{\theta_2}\cos\theta\,d\theta = \frac{\mu_0 I}{4\pi s}(\sin\theta_2 - \sin\theta_1). \tag{5.35}$$

Equation 5.35 gives the field of any straight segment of wire, in terms of the initial and final angles θ_1 and θ_2 (Fig. 5.19). Of course, a finite segment by itself could never support a steady current (where would the charge go when it got to the end?), but it might be a *piece* of some closed circuit, and Eq. 5.35 would then represent its contribution to the total field. In the case of an infinite wire, $\theta_1 = -\pi/2$ and $\theta_2 = \pi/2$, so we obtain

$$B = \frac{\mu_0 I}{2\pi s}. \tag{5.36}$$

Notice that the field is inversely proportional to the distance from the wire—just like the electric field of an infinite line charge. In the region *below* the wire, **B** points *into* the page, and in general, it "circles around" the wire, in accordance with the right-hand rule stated earlier (Fig. 5.3).

As an application, let's find the force of attraction between two long, parallel wires a distance d apart, carrying currents I_1 and I_2 (Fig. 5.20). The field at (2) due to (1) is

$$B = \frac{\mu_0 I_1}{2\pi d},$$

and it points into the page. The Lorentz force law (in the form appropriate to line currents, Eq. 5.17) predicts a force directed towards (1), of magnitude

$$F = I_2\left(\frac{\mu_0 I_1}{2\pi d}\right)\int dl.$$

The *total* force, not surprisingly, is infinite, but the force per unit length is

$$f = \frac{\mu_0}{2\pi}\frac{I_1 I_2}{d}. \tag{5.37}$$

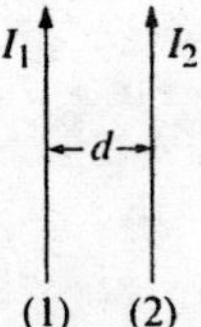

Figure 5.20

If the currents are antiparallel (one up, one down), the force is repulsive—consistent again with the qualitative observations in Sect. 5.1.1.

Example 5.6

Find the magnetic field a distance z above the center of a circular loop of radius R, which carries a steady current I (Fig. 5.21).

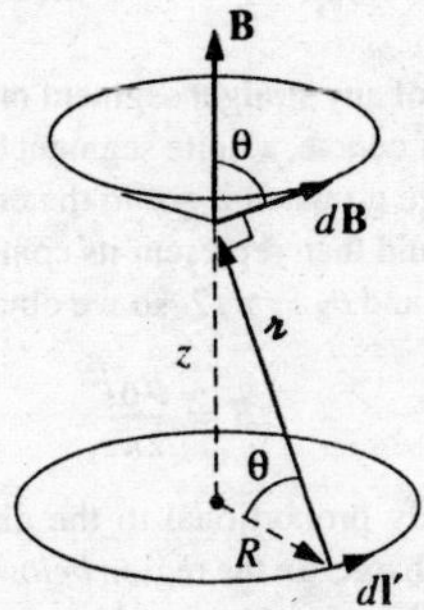

Figure 5.21

Solution: The field $d\mathbf{B}$ attributable to the segment $d\mathbf{l}'$ points as shown. As we integrate $d\mathbf{l}'$ around the loop, $d\mathbf{B}$ sweeps out a cone. The horizontal components cancel, and the vertical components combine to give

$$B(z) = \frac{\mu_0}{4\pi} I \int \frac{dl'}{\imath^2} \cos\theta.$$

(Notice that $d\mathbf{l}'$ and $\hat{\boldsymbol{\imath}}$ are perpendicular, in this case; the factor of $\cos\theta$ projects out the vertical component.) Now, $\cos\theta$ and $\imath^2$ are constants, and $\int dl'$ is simply the circumference, $2\pi R$, so

$$B(z) = \frac{\mu_0 I}{4\pi}\left(\frac{\cos\theta}{\imath^2}\right) 2\pi R = \frac{\mu_0 I}{2}\frac{R^2}{(R^2+z^2)^{3/2}}. \tag{5.38}$$

For surface and volume currents the Biot-Savart law becomes

$$\mathbf{B}(\mathbf{r}) = \frac{\mu_0}{4\pi}\int \frac{\mathbf{K}(\mathbf{r}') \times \hat{\boldsymbol{\imath}}}{\imath^2}\, da' \quad \text{and} \quad \mathbf{B}(\mathbf{r}) = \frac{\mu_0}{4\pi}\int \frac{\mathbf{J}(\mathbf{r}') \times \hat{\boldsymbol{\imath}}}{\imath^2}\, d\tau', \tag{5.39}$$

respectively. You might be tempted to write down the corresponding formula for a moving point charge, using the "dictionary" 5.30:

$$\mathbf{B}(\mathbf{r}) = \frac{\mu_0}{4\pi}\frac{q\mathbf{v} \times \hat{\boldsymbol{\imath}}}{\imath^2}, \tag{5.40}$$

but this is simply *wrong*.[7] As I mentioned earlier, a point charge does not constitute a steady current, and the Biot-Savart law, which only holds for steady currents, does *not* correctly determine its field.

Incidentally, the superposition principle applies to magnetic fields just as it does to electric fields: If you have a *collection* of source currents, the net field is the (vector) sum of the fields due to each of them taken separately.

Problem 5.8

(a) Find the magnetic field at the center of a square loop, which carries a steady current I. Let R be the distance from center to side (Fig. 5.22).

(b) Find the field at the center of a regular n-sided polygon, carrying a steady current I. Again, let R be the distance from the center to any side.

(c) Check that your formula reduces to the field at the center of a circular loop, in the limit $n \to \infty$.

Problem 5.9 Find the magnetic field at point P for each of the steady current configurations shown in Fig. 5.23.

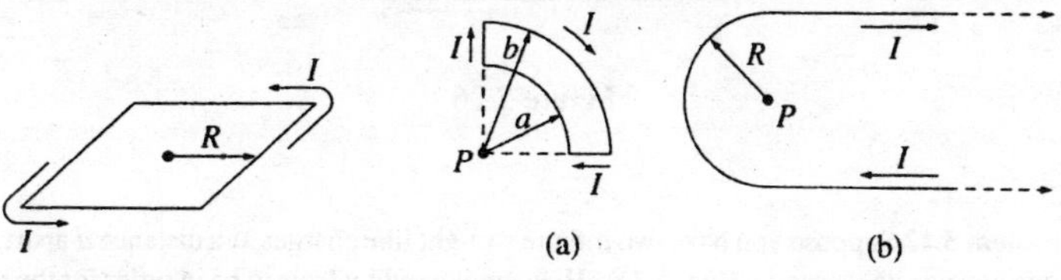

Figure 5.22

Figure 5.23

---

[7] I say this loud and clear to emphasize the point of principle; actually, Eq. 5.40 is *approximately* right for nonrelativistic charges ($v \ll c$), under conditions where retardation can be neglected (see Ex. 10.4).

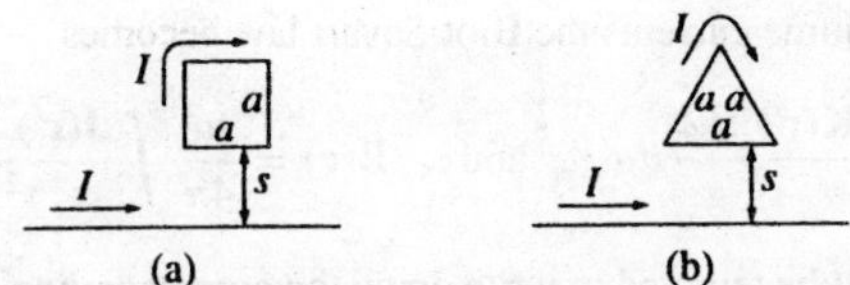

Figure 5.24

Problem 5.10

(a) Find the force on a square loop placed as shown in Fig. 5.24(a), near an infinite straight wire. Both the loop and the wire carry a steady current I.

(b) Find the force on the triangular loop in Fig. 5.24(b).

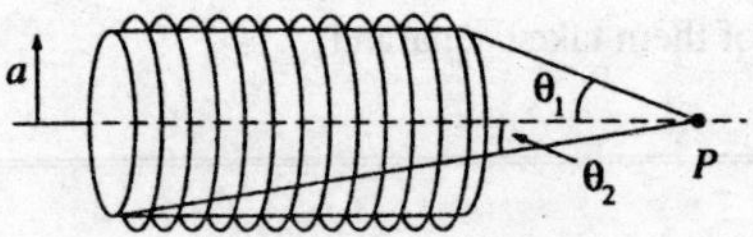

Figure 5.25

Problem 5.11 Find the magnetic field at point P on the axis of a tightly wound solenoid (helical coil) consisting of n turns per unit length wrapped around a cylindrical tube of radius a and carrying current I (Fig. 5.25). Express your answer in terms of θ_1 and θ_2 (it's easiest that way). Consider the turns to be essentially circular, and use the result of Ex. 5.6. What is the field on the axis of an *infinite* solenoid (infinite in both directions)?

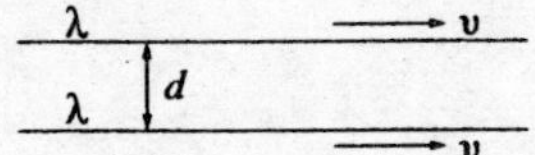

Figure 5.26

Problem 5.12 Suppose you have two infinite straight line charges λ, a distance d apart, moving along at a constant speed v (Fig. 5.26). How great would v have to be in order for the magnetic attraction to balance the electrical repulsion? Work out the actual number... Is this a reasonable sort of speed?[8]

[8] If you've studied special relativity, you may be tempted to look for complexities in this problem that are not really there—λ and v are both measured *in the laboratory frame*, and this is *ordinary electrostatics* (see footnote 4).

5.3 The Divergence and Curl of B

5.3.1 Straight-Line Currents

The magnetic field of an infinite straight wire is shown in Fig. 5.27 (the current is coming *out* of the page). At a glance, it is clear that this field has a nonzero curl (something you'll *never* see in an *electro*static field); let's *calculate* it.

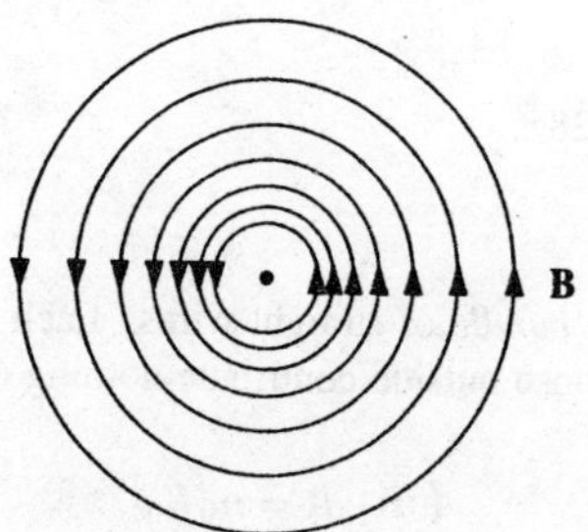

Figure 5.27

According to Eq. 5.36, the integral of **B** around a circular path of radius s, centered at the wire, is

$$\oint \mathbf{B} \cdot d\mathbf{l} = \oint \frac{\mu_0 I}{2\pi s} \, dl = \frac{\mu_0 I}{2\pi s} \oint dl = \mu_0 I.$$

Notice that the answer is independent of s; that's because B *de*creases at the same rate as the circumference *in*creases. In fact, it doesn't have to be a circle; *any* old loop that encloses the wire would give the same answer. For if we use cylindrical coordinates (s, ϕ, z), with the current flowing along the z axis,

$$\mathbf{B} = \frac{\mu_0 I}{2\pi s} \hat{\boldsymbol{\phi}}, \tag{5.41}$$

and $d\mathbf{l} = ds\,\hat{\mathbf{s}} + s\,d\phi\,\hat{\boldsymbol{\phi}} + dz\,\hat{\mathbf{z}}$, so

$$\oint \mathbf{B} \cdot d\mathbf{l} = \frac{\mu_0 I}{2\pi} \oint \frac{1}{s} s\,d\phi = \frac{\mu_0 I}{2\pi} \int_0^{2\pi} d\phi = \mu_0 I.$$

This assumes the loop encircles the wire exactly once; if it went around twice, the ϕ would run from 0 to 4π, and if it didn't enclose the wire at all, then ϕ would go from ϕ_1 to ϕ_2 and back again, with $\int d\phi = 0$ (Fig. 5.28).

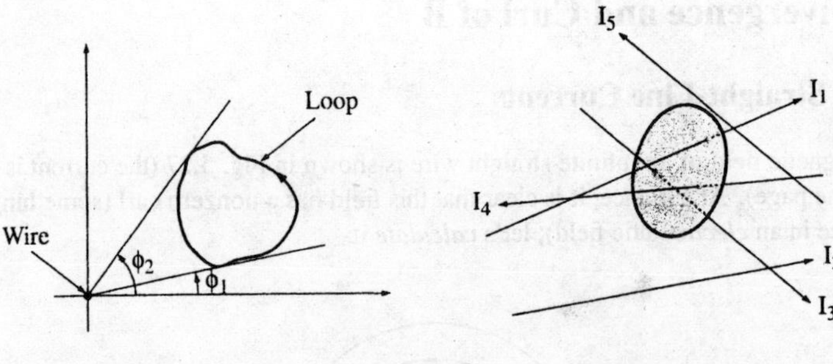

Figure 5.28

Figure 5.29

Now suppose we have a *bundle* of straight wires. Each wire that passes through our loop contributes $\mu_0 I$, and those outside contribute nothing (Fig. 5.29). The line integral will then be

$$\oint \mathbf{B} \cdot d\mathbf{l} = \mu_0 I_{\text{enc}}, \tag{5.42}$$

where I_{enc} stands for the total current enclosed by the integration path. If the flow of charge is represented by a volume current density $\mathbf{J}$, the enclosed current is

$$I_{\text{enc}} = \int \mathbf{J} \cdot d\mathbf{a}, \tag{5.43}$$

with the integral taken over the surface bounded by the loop. Applying Stokes' theorem to Eq. 5.42, then,

$$\int (\nabla \times \mathbf{B}) \cdot d\mathbf{a} = \mu_0 \int \mathbf{J} \cdot d\mathbf{a},$$

and hence

$$\nabla \times \mathbf{B} = \mu_0 \mathbf{J}. \tag{5.44}$$

With minimal labor we have actually obtained the general formula for the curl of **B**. But our derivation is seriously flawed by the restriction to infinite straight line currents (and combinations thereof). Most current configurations *cannot* be constructed out of infinite straight wires, and we have no right to assume that Eq. 5.44 applies to them. So the next section is devoted to the formal derivation of the divergence and curl of **B**, starting from the Biot-Savart law itself.

5.3.2 The Divergence and Curl of B

The Biot-Savart law for the general case of a volume current reads

$$\mathbf{B}(\mathbf{r}) = \frac{\mu_0}{4\pi} \int \frac{\mathbf{J}(\mathbf{r}') \times \hat{\boldsymbol{\mathscr{r}}}}{\mathscr{r}^2} d\tau'. \tag{5.45}$$

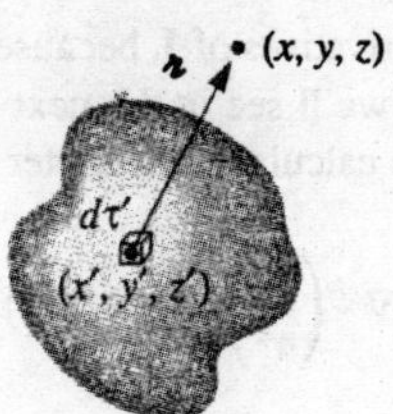

Figure 5.30

This formula gives the magnetic field at a point $\mathbf{r} = (x, y, z)$ in terms of an integral over the current distribution $\mathbf{J}(x', y', z')$ (Fig. 5.30). It is best to be absolutely explicit at this stage:

$$\mathbf{B} \text{ is a function of } (x, y, z),$$

$$\mathbf{J} \text{ is a function of } (x', y', z'),$$

$$\boldsymbol{\mathscr{r}} = (x - x')\,\hat{\mathbf{x}} + (y - y')\,\hat{\mathbf{y}} + (z - z')\,\hat{\mathbf{z}},$$

$$d\tau' = dx'\,dy'\,dz'.$$

The integration is over the *primed* coordinates; the divergence and the curl are to be taken with respect to the *unprimed* coordinates.

Applying the divergence to Eq. 5.45, we obtain:

$$\nabla \cdot \mathbf{B} = \frac{\mu_0}{4\pi} \int \nabla \cdot \left(\mathbf{J} \times \frac{\hat{\boldsymbol{\mathscr{r}}}}{\mathscr{r}^2} \right) d\tau'. \tag{5.46}$$

Invoking product rule number (6),

$$\nabla \cdot \left(\mathbf{J} \times \frac{\hat{\boldsymbol{\mathscr{r}}}}{\mathscr{r}^2} \right) = \frac{\hat{\boldsymbol{\mathscr{r}}}}{\mathscr{r}^2} \cdot (\nabla \times \mathbf{J}) - \mathbf{J} \cdot \left(\nabla \times \frac{\hat{\boldsymbol{\mathscr{r}}}}{\mathscr{r}^2} \right). \tag{5.47}$$

But $\nabla \times \mathbf{J} = 0$, because $\mathbf{J}$ doesn't depend on the unprimed variables (x, y, z), whereas $\nabla \times (\hat{\boldsymbol{\mathscr{r}}}/\mathscr{r}^2) = 0$ (Prob. 1.62), so

$$\boxed{\nabla \cdot \mathbf{B} = 0.} \tag{5.48}$$

Evidently, the *divergence* of the magnetic field is *zero*.

Applying the curl to Eq. 5.45, we obtain:

$$\nabla \times \mathbf{B} = \frac{\mu_0}{4\pi} \int \nabla \times \left(\mathbf{J} \times \frac{\hat{\boldsymbol{\mathscr{r}}}}{\mathscr{r}^2} \right) d\tau'. \tag{5.49}$$

Again, our strategy is to expand the integrand, using the appropriate product rule—in this case number 8:

$$\nabla \times \left(\mathbf{J} \times \frac{\hat{\boldsymbol{\mathscr{r}}}}{\mathscr{r}^2} \right) = \mathbf{J} \left(\nabla \cdot \frac{\hat{\boldsymbol{\mathscr{r}}}}{\mathscr{r}^2} \right) - (\mathbf{J} \cdot \nabla) \frac{\hat{\boldsymbol{\mathscr{r}}}}{\mathscr{r}^2}. \tag{5.50}$$

(I have dropped terms involving derivatives of **J**, because **J** does not depend on x, y, z.) The second term integrates to zero, as we'll see in the next paragraph. The first term involves the divergence we were at pains to calculate in Chapter 1 (Eq. 1.100):

$$\nabla \cdot \left(\frac{\hat{\boldsymbol{\imath}}}{\imath^2}\right) = 4\pi\delta^3(\boldsymbol{\imath}). \tag{5.51}$$

Thus

$$\nabla \times \mathbf{B} = \frac{\mu_0}{4\pi}\int \mathbf{J}(\mathbf{r}')4\pi\delta^3(\mathbf{r}-\mathbf{r}')\,d\tau' = \mu_0\mathbf{J}(\mathbf{r}),$$

which confirms that Eq. 5.44 is not restricted to straight-line currents, but holds quite generally in magnetostatics.

To complete the argument, however, we must check that the second term in Eq. 5.50 integrates to zero. Because the derivative acts only on $\hat{\boldsymbol{\imath}}/\imath^2$, we can switch from ∇ to ∇' at the cost of a minus sign:[9]

$$-(\mathbf{J}\cdot\nabla)\frac{\hat{\boldsymbol{\imath}}}{\imath^2} = (\mathbf{J}\cdot\nabla')\frac{\hat{\boldsymbol{\imath}}}{\imath^2}. \tag{5.52}$$

The x component, in particular, is

$$(\mathbf{J}\cdot\nabla')\left(\frac{x-x'}{\imath^3}\right) = \nabla'\cdot\left[\frac{(x-x')}{\imath^3}\mathbf{J}\right] - \left(\frac{x-x'}{\imath^3}\right)(\nabla'\cdot\mathbf{J})$$

(using product rule 5). Now, for *steady* currents the divergence of **J** is zero (Eq. 5.31), so

$$\left[-(\mathbf{J}\cdot\nabla)\frac{\hat{\boldsymbol{\imath}}}{\imath^2}\right]_x = \nabla'\cdot\left[\frac{(x-x')}{\imath^3}\mathbf{J}\right],$$

and therefore this contribution to the integral (5.49) can be written

$$\int_{\mathcal{V}} \nabla'\cdot\left[\frac{(x-x')}{\imath^3}\mathbf{J}\right]d\tau' = \oint_{\mathcal{S}} \frac{(x-x')}{\imath^3}\mathbf{J}\cdot d\mathbf{a}'. \tag{5.53}$$

(The reason for switching from ∇ to ∇' was precisely to permit this integration by parts.) But what region are we integrating over? Well, it's the volume that appears in the Biot-Savart law (5.45)—large enough, that is, to include all the current. You can make it *bigger* than that, if you like; $\mathbf{J} = 0$ out there anyway, so it will add nothing to the integral. The essential point is that *on the boundary* the current is *zero* (all current is safely *inside*) and hence the surface integral (5.53) vanishes.[10]

---

[9]The point here is that $\boldsymbol{\imath}$ depends only on the *difference* between the coordinates, and $(\partial/\partial x)f(x-x') = -(\partial/\partial x')f(x-x')$.

[10]If **J** itself extends to infinity (as in the case of an infinite straight wire), the surface integral is still typically zero, though the analysis calls for greater care.

5.3.3 Applications of Ampère's Law

The equation for the curl of **B**

$$\boxed{\nabla \times \mathbf{B} = \mu_0 \mathbf{J},} \tag{5.54}$$

is called **Ampère's law** (in differential form). It can be converted to integral form by the usual device of applying one of the fundamental theorems—in this case Stokes' theorem:

$$\int (\nabla \times \mathbf{B}) \cdot d\mathbf{a} = \oint \mathbf{B} \cdot d\mathbf{l} = \mu_0 \int \mathbf{J} \cdot d\mathbf{a}.$$

Now, $\int \mathbf{J} \cdot d\mathbf{a}$ is the total current passing through the surface (Fig. 5.31), which we call I_{enc} (the **current enclosed** by the **amperian loop**). Thus

$$\boxed{\oint \mathbf{B} \cdot d\mathbf{l} = \mu_0 I_{\text{enc}}.} \tag{5.55}$$

This is the integral version of Ampère's law; it generalizes Eq. 5.42 to *arbitrary* steady currents. Notice that Eq. 5.55 inherits the sign ambiguity of Stokes' theorem (Sect. 1.3.5): Which *way* around the loop am I supposed to go? And which *direction* through the surface corresponds to a "positive" current? The resolution, as always, is the right-hand rule: If the fingers of your right hand indicate the direction of integration around the boundary, then your thumb defines the direction of a positive current.

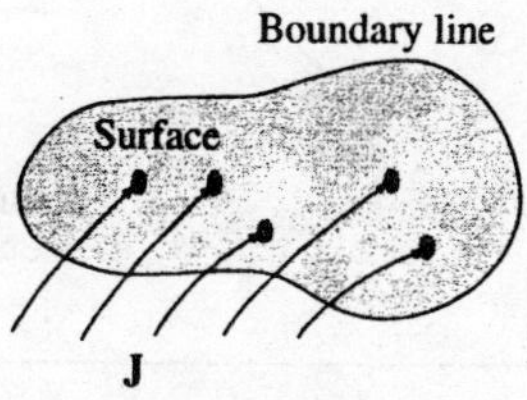

Figure 5.31

Just as the Biot-Savart law plays a role in magnetostatics that Coulomb's law assumed in electrostatics, so Ampère's plays the role of Gauss's:

$$\begin{cases} \text{Electrostatics}: & \text{Coulomb} & \rightarrow & \text{Gauss}, \\ \text{Magnetostatics}: & \text{Biot-Savart} & \rightarrow & \text{Ampère}. \end{cases}$$

In particular, for currents with appropriate symmetry, Ampère's law in integral form offers a lovely and extraordinarily efficient means for calculating the magnetic field.

Example 5.7

Find the magnetic field a distance s from a long straight wire (Fig. 5.32), carrying a steady current I (the same problem we solved in Ex. 5.5, using the Biot-Savart law).

Solution: We know the direction of **B** is "circumferential," circling around the wire as indicated by the right hand rule. By symmetry, the magnitude of **B** is constant around an amperian loop of radius s, centered on the wire. So Ampère's law gives

$$\oint \mathbf{B} \cdot d\mathbf{l} = B \oint dl = B 2\pi s = \mu_0 I_{\text{enc}} = \mu_0 I,$$

or

$$B = \frac{\mu_0 I}{2\pi s}.$$

This is the same answer we got before (Eq. 5.36), but it was obtained this time with far less effort.

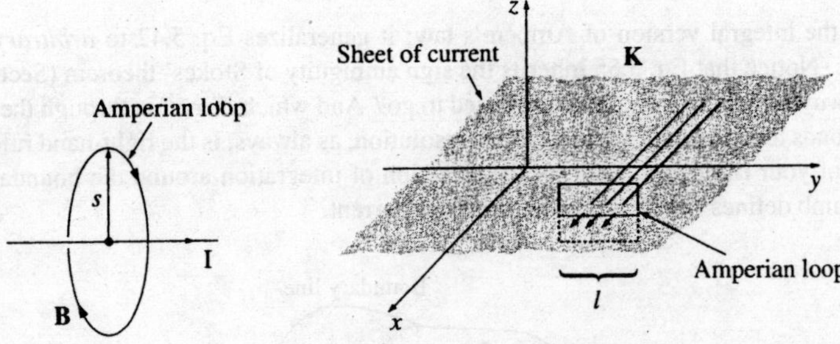

Figure 5.32 Figure 5.33

Example 5.8

Find the magnetic field of an infinite uniform surface current $\mathbf{K} = K\,\hat{\mathbf{x}}$, flowing over the xy plane (Fig. 5.33).

Solution: First of all, what is the *direction* of **B**? Could it have any x-component? *No*: A glance at the Biot-Savart law (5.39) reveals that **B** is *perpendicular* to **K**. Could it have a z-component? *No again*. You could confirm this by noting that any vertical contribution from a filament at $+y$ is canceled by the corresponding filament at $-y$. But there is a nicer argument: Suppose the field pointed *away* from the plane. By reversing the direction of the current, I could make it point *toward* the plane (in the Biot-Savart law, changing the sign of the current switches the sign of the field). But the z-component of **B** cannot possibly depend on the *direction* of the current in the xy plane. (Think about it!) So **B** can only have a y-component, and a quick check with your right hand should convince you that it points to the *left* above the plane and to the *right* below it.

With this in mind we draw a rectangular amperian loop as shown in Fig. 5.33, parallel to the yz plane and extending an equal distance above and below the surface. Applying Ampère's law, we find

$$\oint \mathbf{B} \cdot d\mathbf{l} = 2Bl = \mu_0 I_{\text{enc}} = \mu_0 K l,$$

(one Bl comes from the top segment, and the other from the bottom), so $B = (\mu_0/2)K$, or, more precisely,

$$\mathbf{B} = \begin{cases} +(\mu_0/2)K\,\hat{\mathbf{y}} & \text{for } z < 0, \\ -(\mu_0/2)K\,\hat{\mathbf{y}} & \text{for } z > 0. \end{cases} \tag{5.56}$$

Notice that the field is independent of the distance from the plane, just like the *electric* field of a uniform surface *charge* (Ex. 2.4).

Example 5.9

Find the magnetic field of a very long solenoid, consisting of n closely wound turns per unit length on a cylinder of radius R and carrying a steady current I (Fig. 5.34). [The point of making the windings so close is that one can then pretend each turn is circular. If this troubles you (after all, there is a net current I in the direction of the solenoid's axis, no matter *how* tight the winding), picture instead a sheet of aluminum foil wrapped around the cylinder, carrying the equivalent uniform surface current $K = nI$ (Fig. 5.35). Or make a double winding, going up to one end and then—always in the same sense—going back down again, thereby eliminating the net longitudinal current. But, in truth, this is all unnecessary fastidiousness, for the field inside a solenoid is huge (relatively speaking), and the field of the longitudinal current is at most a tiny refinement.]

Solution: First of all, what is the *direction* of **B**? Could it have a radial component? *No.* For suppose B_s were *positive*; if we reversed the direction of the current, B_s would then be *negative*. But switching I is physically equivalent to turning the solenoid upside down, and

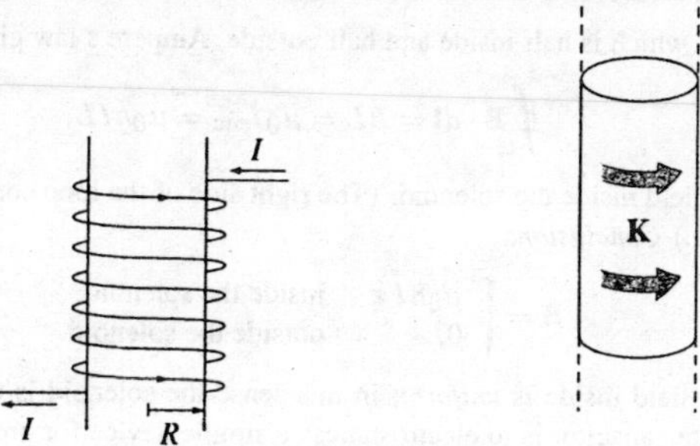

Figure 5.34 Figure 5.35

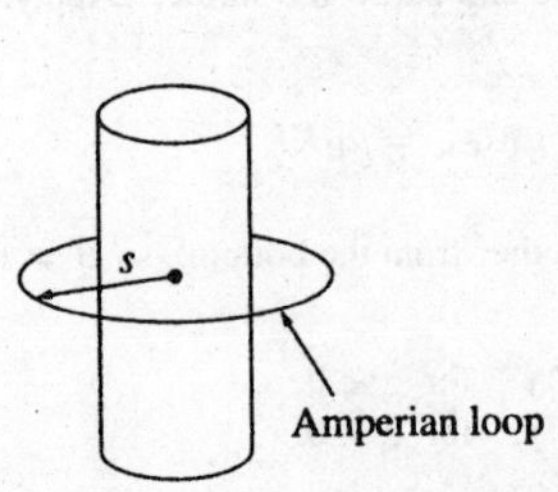

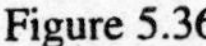

Figure 5.36

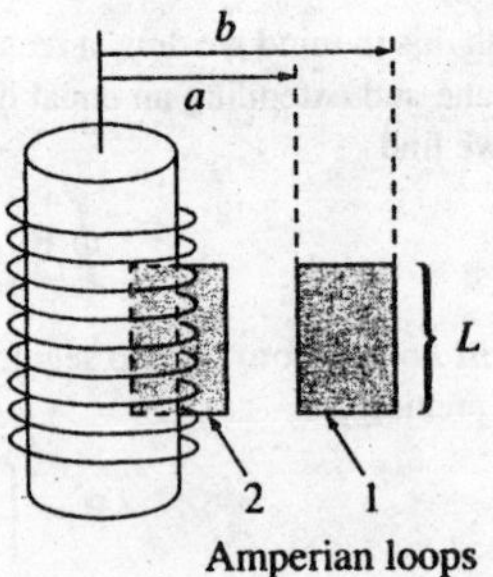

Figure 5.37

that certainly should not alter the radial field. How about a "circumferential" component? *No.* For B_ϕ would be constant around an amperian loop concentric with the solenoid (Fig. 5.36), and hence

$$\oint \mathbf{B} \cdot d\mathbf{l} = B_\phi(2\pi s) = \mu_0 I_{\text{enc}} = 0,$$

since the loop encloses no current.

So the magnetic field of an infinite, closely wound solenoid runs *parallel to the axis*. From the right hand rule, we expect that it points upward inside the solenoid and downward outside. Moreover, it certainly approaches zero as you go very far away. With this in mind, let's apply Ampère's law to the two rectangular loops in Fig. 5.37. Loop 1 lies entirely outside the solenoid, with its sides at distances a and b from the axis:

$$\oint \mathbf{B} \cdot d\mathbf{l} = [B(a) - B(b)]L = \mu_0 I_{\text{enc}} = 0,$$

so

$$B(a) = B(b).$$

Evidently the *field outside does not depend on the distance from the axis*. But we know that it goes to *zero* for large s. It must therefore be zero *everywhere*! (This astonishing result can also be derived from the Biot-Savart law, of course, but it's much more difficult. See Prob. 5.44.)

As for loop 2, which is half inside and half outside, Ampère's law gives

$$\oint \mathbf{B} \cdot d\mathbf{l} = BL = \mu_0 I_{\text{enc}} = \mu_0 n I L,$$

where B is the field inside the solenoid. (The right side of the loop contributes nothing, since $B = 0$ out there.) *Conclusion:*

$$\mathbf{B} = \begin{cases} \mu_0 n I\, \hat{\mathbf{z}}, & \text{inside the solenoid,} \\ 0, & \text{outside the solenoid.} \end{cases} \tag{5.57}$$

Notice that the field inside is *uniform*; in this sense the solenoid is to magnetostatics what the parallel-plate capacitor is to electrostatics: a simple device for producing strong uniform fields.

Like Gauss's law, Ampère's law is always *true* (for steady currents), but it is not always *useful*. Only when the symmetry of the problem enables you to pull B outside the integral $\oint \mathbf{B} \cdot d\mathbf{l}$ can you calculate the magnetic field from Ampère's law. When it *does* work, it's by far the fastest method; when it doesn't, you have to fall back on the Biot-Savart law. The current configurations that can be handled by Ampère's law are

1. Infinite straight lines (prototype: Ex. 5.7).
2. Infinite planes (prototype: Ex. 5.8).
3. Infinite solenoids (prototype: Ex. 5.9).
4. Toroids (prototype: Ex. 5.10).

The last of these is a surprising and elegant application of Ampère's law; it is treated in the following example. As in Exs. 5.8 and 5.9, the hard part is figuring out the *direction* of the field (which we will now have done, once and for all, for each of the four geometries); the actual application of Ampère's law takes only one line.

Example 5.10

A toroidal coil consists of a circular ring, or "donut," around which a long wire is wrapped (Fig. 5.38). The winding is uniform and tight enough so that each turn can be considered a closed loop. The cross-sectional shape of the coil is immaterial. I made it rectangular in Fig. 5.38 for the sake of simplicity, but it could just as well be circular or even some weird asymmetrical form, as in Fig. 5.39, just as long as the shape remains the same all the way around the ring. In that case it follows that the *magnetic field of the toroid is circumferential at all points, both inside and outside the coil.*

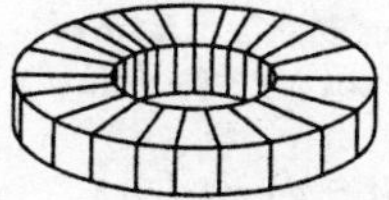

Figure 5.38

Proof: According to the Biot-Savart law, the field at $\mathbf{r}$ due to the current element at $\mathbf{r}'$ is

$$d\mathbf{B} = \frac{\mu_0}{4\pi}\frac{\mathbf{I} \times \boldsymbol{\mathscr{r}}}{\mathscr{r}^3}\, dl'.$$

We may as well put $\mathbf{r}$ in the xz plane (Fig. 5.39), so its Cartesian components are $(x, 0, z)$, while the source coordinates are

$$\mathbf{r}' = (s'\cos\phi', s'\sin\phi', z').$$

Then

$$\boldsymbol{\mathscr{r}} = (x - s'\cos\phi', -s'\sin\phi', z - z').$$

Since the current has no ϕ component, $\mathbf{I} = I_s\,\hat{\mathbf{s}} + I_z\,\hat{\mathbf{z}}$, or (in Cartesian coordinates)

$$\mathbf{I} = (I_s\cos\phi', I_s\sin\phi', I_z).$$

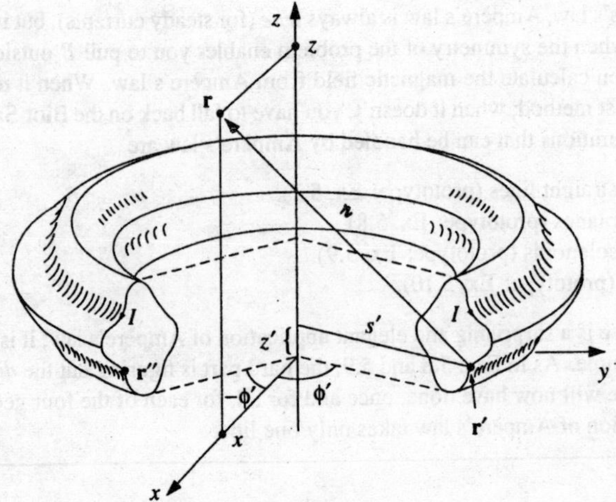

Figure 5.39

Accordingly,

$$\mathbf{I} \times \boldsymbol{\mathscr{r}} = \begin{bmatrix} \hat{\mathbf{x}} & \hat{\mathbf{y}} & \hat{\mathbf{z}} \\ I_s \cos\phi' & I_s \sin\phi' & I_z \\ (x - s'\cos\phi') & (-s'\sin\phi') & (z - z') \end{bmatrix}$$

$$= [\sin\phi'(I_s(z - z') + s' I_z)]\,\hat{\mathbf{x}}$$

$$+ [I_z(x - s'\cos\phi') - I_s\cos\phi'(z - z')]\,\hat{\mathbf{y}} + [-I_s x \sin\phi']\,\hat{\mathbf{z}}.$$

But there is a symmetrically situated current element at $\mathbf{r}''$, with the same s', the same $\mathscr{r}$, the same dl', the same I_s, and the same I_z, *but negative* ϕ' (Fig. 5.39). Because $\sin\phi'$ changes sign, the $\hat{\mathbf{x}}$ and $\hat{\mathbf{z}}$ contributions from $\mathbf{r}'$ and $\mathbf{r}''$ cancel, leaving only a $\hat{\mathbf{y}}$ term. Thus the field at $\mathbf{r}$ is in the $\hat{\mathbf{y}}$ direction, and in general the field points in the $\hat{\boldsymbol{\phi}}$ direction. qed

Now that we know the field is circumferential, determining its magnitude is ridiculously easy. Just apply Ampère's law to a circle of radius s about the axis of the toroid:

$$B 2\pi s = \mu_0 I_{\text{enc}},$$

and hence

$$\mathbf{B}(\mathbf{r}) = \begin{cases} \dfrac{\mu_0 N I}{2\pi s}\hat{\boldsymbol{\phi}}, & \text{for points inside the coil,} \\ 0, & \text{for points outside the coil,} \end{cases} \tag{5.58}$$

where N is the total number of turns.

Problem 5.13 A steady current I flows down a long cylindrical wire of radius a (Fig. 5.40). Find the magnetic field, both inside and outside the wire, if

(a) The current is uniformly distributed over the outside surface of the wire.

(b) The current is distributed in such a way that J is proportional to s, the distance from the axis.

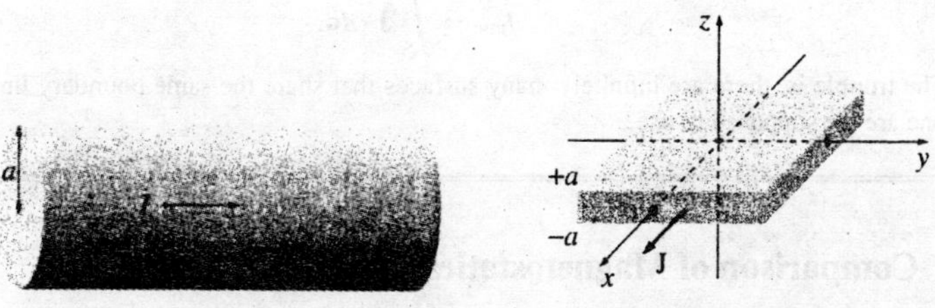

Figure 5.40 Figure 5.41

Problem 5.14 A thick slab extending from $z = -a$ to $z = +a$ carries a uniform volume current $\mathbf{J} = J\,\hat{\mathbf{x}}$ (Fig. 5.41). Find the magnetic field, as a function of z, both inside and outside the slab.

Problem 5.15 Two long coaxial solenoids each carry current I, but in opposite directions, as shown in Fig. 5.42. The inner solenoid (radius a) has n_1 turns per unit length, and the outer one (radius b) has n_2. Find $\mathbf{B}$ in each of the three regions: (i) inside the inner solenoid, (ii) between them, and (iii) outside both.

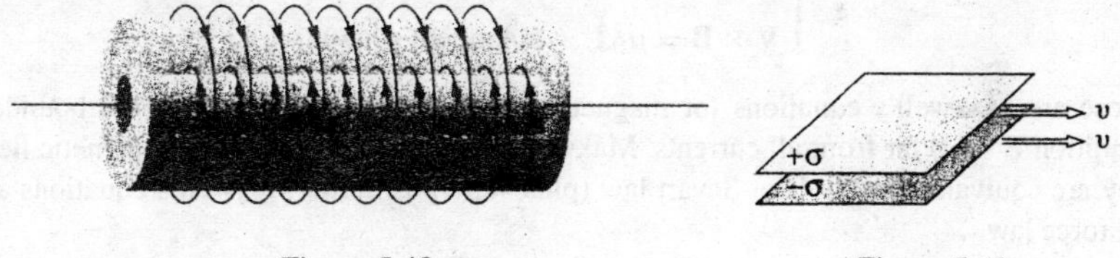

Figure 5.42 Figure 5.43

Problem 5.16 A large parallel-plate capacitor with uniform surface charge σ on the upper plate and $-\sigma$ on the lower is moving with a constant speed v, as shown in Fig. 5.43.

(a) Find the magnetic field between the plates and also above and below them.

(b) Find the magnetic force per unit area on the upper plate, including its direction.

(c) At what speed v would the magnetic force balance the electrical force?[11]

[11] See footnote 8.

! **Problem 5.17** Show that the magnetic field of an infinite solenoid runs parallel to the axis, *regardless of the cross-sectional shape of the coil*, as long as that shape is constant along the length of the solenoid. What is the magnitude of the field, inside and outside of such a coil? Show that the toroid field (5.58) reduces to the solenoid field, when the radius of the donut is so large that a segment can be considered essentially straight.

Problem 5.18 In calculating the current enclosed by an amperian loop, one must, in general, evaluate an integral of the form

$$I_{\text{enc}} = \int_{\mathcal{S}} \mathbf{J} \cdot d\mathbf{a}.$$

The trouble is, there are infinitely many surfaces that share the same boundary line. Which one are we supposed to use?

5.3.4 Comparison of Magnetostatics and Electrostatics

The divergence and curl of the *electrostatic* field are

$$\begin{cases} \nabla \cdot \mathbf{E} = \dfrac{1}{\epsilon_0}\rho, & \text{(Gauss's law);} \\ \nabla \times \mathbf{E} = 0, & \text{(no name).} \end{cases}$$

These are **Maxwell's equations** for electrostatics. Together with the boundary condition $\mathbf{E} \to 0$ far from all charges, Maxwell's equations determine the field, if the source charge density ρ is given; they contain essentially the same information as Coulomb's law plus the principle of superposition. The divergence and curl of the *magnetostatic* field are

$$\begin{cases} \nabla \cdot \mathbf{B} = 0, & \text{(no name);} \\ \nabla \times \mathbf{B} = \mu_0 \mathbf{J}, & \text{(Ampère's law).} \end{cases}$$

These are Maxwell's equations for magnetostatics. Again, together with the boundary condition $\mathbf{B} \to 0$ far from all currents, Maxwell's equations determine the magnetic field; they are equivalent to the Biot-Savart law (plus superposition). Maxwell's equations and the force law

$$\mathbf{F} = Q(\mathbf{E} + \mathbf{v} \times \mathbf{B})$$

constitute the most elegant formulation of electrostatics and magnetostatics.

The electric field *diverges away from* a (positive) charge; the magnetic field line *curls around* a current (Fig. 5.44). Electric field lines originate on positive charges and terminate on negative ones; magnetic field lines do not begin or end anywhere—to do so would require a nonzero divergence. They either form closed loops or extend out to infinity. To put it another way, *there are no point sources for* **B**, as there are for **E**; there exists no magnetic analog to electric charge. This is the physical content of the statement $\nabla \cdot \mathbf{B} = 0$. Coulomb and others believed that magnetism was produced by **magnetic charges** (**magnetic monopoles**, as we would now call them), and in some older books you will still

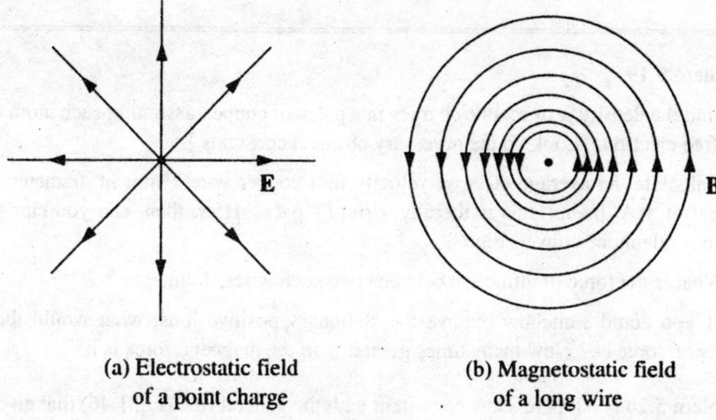

(a) Electrostatic field of a point charge

(b) Magnetostatic field of a long wire

Figure 5.44

find references to a magnetic version of Coulomb's law, giving the force of attraction or repulsion between them. It was Ampère who first speculated that all magnetic effects are attributable to *electric* charges *in motion* (currents). As far as we know, Ampère was right; nevertheless, it remains an open experimental question whether magnetic monopoles exist in nature (they are obviously pretty *rare*, or somebody would have found one[12]), and in fact some recent elementary particle theories *require* them. For our purposes, though, **B** is divergenceless and there are no magnetic monopoles. It takes a *moving* electric charge to *produce* a magnetic field, and it takes another moving electric charge to "feel" a magnetic field.

Typically, electric forces are enormously larger than magnetic ones. That's not something you can tell from the theory as such; it has to do with the sizes of the fundamental constants ϵ_0 and μ_0. In general, it is only when both the source charges and the test charge are moving at velocities comparable to the speed of light that the magnetic force approaches the electric force in strength. (Problems 5.12 and 5.16 illustrate this rule.) How is it, then, that we ever notice magnetic effects at all? The answer is that both in the production of a magnetic field (Biot-Savart) and in its detection (Lorentz) it is the *current* (charge times velocity) that enters, and we can compensate for a smallish velocity by pouring huge quantities of charge down the wire. Ordinarily, this charge would simultaneously generate so large an *electric* force as to swamp the magnetic one. But if we arrange to keep the wire *neutral*, by embedding in it an equal amount of opposite charge at rest, the electric field cancels out, leaving the magnetic field to stand alone. It sounds very elaborate, but of course this is precisely what happens in an ordinary current carrying wire.

[12]An apparent detection (B. Cabrera, *Phys. Rev. Lett.* **48**, 1378 (1982)) has never been reproduced—and not for want of trying. For a delightful brief history of ideas about magnetism, see Chapter 1 in D. C. Mattis, *The Theory of Magnetism* (New York: Harper and Row, 1965).

Problem 5.19

(a) Find the density ρ of mobile charges in a piece of copper, assuming each atom contributes one free electron. [Look up the necessary physical constants.]

(b) Calculate the average electron velocity in a copper wire 1 mm in diameter, carrying a current of 1 A. [*Note*: this is literally a *snail's* pace. How, then, can you carry on a long distance telephone conversation?]

(c) What is the force of attraction between two such wires, 1 cm apart?

(d) If you could somehow remove the stationary positive ions, what would the electrical repulsion force be? How many times greater than the magnetic force is it?

Problem 5.20 Is Ampère's law consistent with the general rule (Eq. 1.46) that divergence-of-curl is always zero? Show that Ampère's law *cannot* be valid, in general, outside magnetostatics. Is there any such "defect" in the other three Maxwell equations?

Problem 5.21 Suppose there *did* exist magnetic monopoles. How would you modify Maxwell's equations and the force law, to accommodate them? If you think there are several plausible options, list them, and suggest how you might decide experimentally which one is right.

5.4 Magnetic Vector Potential

5.4.1 The Vector Potential

Just as $\nabla \times \mathbf{E} = 0$ permitted us to introduce a scalar potential (V) in electrostatics,

$$\mathbf{E} = -\nabla V,$$

so $\nabla \cdot \mathbf{B} = 0$ invites the introduction of a *vector* potential $\mathbf{A}$ in magnetostatics:

$$\boxed{\mathbf{B} = \nabla \times \mathbf{A}.} \tag{5.59}$$

The former is authorized by Theorem 1 (of Sect. 1.6.2), the latter by Theorem 2 (the proof of Theorem 2 is developed in Prob. 5.30). The potential formulation automatically takes care of $\nabla \cdot \mathbf{B} = 0$ (since the divergence of a curl is *always* zero); there remains Ampère's law:

$$\nabla \times \mathbf{B} = \nabla \times (\nabla \times \mathbf{A}) = \nabla(\nabla \cdot \mathbf{A}) - \nabla^2 \mathbf{A} = \mu_0 \mathbf{J}. \tag{5.60}$$

Now, the electric potential had a built-in ambiguity: you can add to V any function whose gradient is zero (which is to say, any *constant*), without altering the *physical* quantity $\mathbf{E}$. Likewise, you can add to the magnetic potential any function whose *curl* vanishes (which is to say, the *gradient of any scalar*), with no effect on $\mathbf{B}$. We can exploit this freedom to eliminate the divergence of $\mathbf{A}$:

$$\boxed{\nabla \cdot \mathbf{A} = 0.} \tag{5.61}$$

To prove that this is always possible, suppose that our original potential, $\mathbf{A}_o$, is *not* divergenceless. If we add to it the gradient of λ ($\mathbf{A} = \mathbf{A}_o + \nabla\lambda$), the new divergence is

$$\nabla \cdot \mathbf{A} = \nabla \cdot \mathbf{A}_o + \nabla^2 \lambda.$$

We can accommodate Eq. 5.61, then, if a function λ can be found that satisfies

$$\nabla^2 \lambda = -\nabla \cdot \mathbf{A}_o.$$

But this is *mathematically* identical to Poisson's equation (2.24),

$$\nabla^2 V = -\frac{\rho}{\epsilon_0},$$

with $\nabla \cdot \mathbf{A}_o$ in place of ρ/ϵ_0 as the "source." And we *know* how to solve Poisson's equation—that's what electrostatics is all about ("given the charge distribution, find the potential"). In particular, if ρ goes to zero at infinity, the solution is Eq. 2.29:

$$V = \frac{1}{4\pi\epsilon_0} \int \frac{\rho}{\mathscr{r}}\, d\tau',$$

and by the same token, if $\nabla \cdot \mathbf{A}_o$ goes to zero at infinity, then

$$\lambda = \frac{1}{4\pi} \int \frac{\nabla \cdot \mathbf{A}_o}{\mathscr{r}}\, d\tau'.$$

If $\nabla \cdot \mathbf{A}_o$ does *not* go to zero at infinity, we'll have to use other means to discover the appropriate λ, just as we get the electric potential by other means when the charge distribution extends to infinity. But the *essential* point remains: *It is always possible to make the vector potential divergenceless.* To put it the other way around: The definition $\mathbf{B} = \nabla \times \mathbf{A}$ specifies the *curl* of $\mathbf{A}$, but it doesn't say anything about the *divergence*—we are at liberty to pick that as we see fit, and zero is ordinarily the simplest choice.

With this condition on $\mathbf{A}$, Ampère's law (5.60) becomes

$$\boxed{\nabla^2 \mathbf{A} = -\mu_0 \mathbf{J}.} \tag{5.62}$$

This *again* is nothing but Poisson's equation—or rather, it is *three* Poisson's equations, one for each Cartesian[13] component. Assuming $\mathbf{J}$ goes to zero at infinity, we can read off the solution:

$$\boxed{\mathbf{A}(\mathbf{r}) = \frac{\mu_0}{4\pi} \int \frac{\mathbf{J}(\mathbf{r}')}{\mathscr{r}}\, d\tau'.} \tag{5.63}$$

[13]In Cartesian coordinates, $\nabla^2 \mathbf{A} = (\nabla^2 A_x)\hat{\mathbf{x}} + (\nabla^2 A_y)\hat{\mathbf{y}} + (\nabla^2 A_z)\hat{\mathbf{z}}$, so Eq. 5.62 reduces to $\nabla^2 A_x = -\mu_0 J_x$, $\nabla^2 A_y = -\mu_0 J_y$, and $\nabla^2 A_z = -\mu_0 J_z$. In curvilinear coordinates the unit vectors *themselves* are functions of position, and must be differentiated, so it is *not* the case, for example, that $\nabla^2 A_r = -\mu_0 J_r$. The safest way to calculate the Laplacian of a *vector*, in terms of its curvilinear components, is to use $\nabla^2 \mathbf{A} = \nabla(\nabla \cdot \mathbf{A}) - \nabla \times (\nabla \times \mathbf{A})$. Remember also that even if you *calculate* integrals such as 5.63 using curvilinear coordinates, you must first express $\mathbf{J}$ in terms of its *Cartesian* components (see Sect. 1.4.1).

For line and surface currents,

$$\mathbf{A} = \frac{\mu_0}{4\pi}\int \frac{\mathbf{I}}{\imath}\,dl' = \frac{\mu_0 I}{4\pi}\int \frac{1}{\imath}\,d\mathbf{l}'; \qquad \mathbf{A} = \frac{\mu_0}{4\pi}\int \frac{\mathbf{K}}{\imath}\,da'. \tag{5.64}$$

(If the current does *not* go to zero at infinity, we have to find other ways to get **A**; some of these are explored in Ex. 5.12 and in the problems at the end of the section.)

It must be said that A is not as *useful* as V. For one thing, it's still a *vector*, and although Eqs. 5.63 and 5.64 are somewhat easier to work with than the Biot-Savart law, you still have to fuss with components. It would be nice if we could get away with a *scalar* potential,

$$\mathbf{B} = -\nabla U, \tag{5.65}$$

but this is incompatible with Ampère's law, since the curl of a gradient is always zero. (A **magnetostatic scalar potential** *can* be used, if you stick scrupulously to simply-connected, current-free regions, but as a theoretical tool it is of limited interest. See Prob. 5.28.) Moreover, since magnetic forces do no work, **A** does not admit a simple physical interpretation in terms of potential energy per unit charge. (In some contexts it can be interpreted as *momentum* per unit charge.[14]) Nevertheless, the vector potential has substantial theoretical importance, as we shall see in Chapter 10.

Example 5.11

A spherical shell, of radius R, carrying a uniform surface charge σ, is set spinning at angular velocity $\boldsymbol{\omega}$. Find the vector potential it produces at point **r** (Fig. 5.45).

Solution: It might seem natural to align the polar axis along $\boldsymbol{\omega}$, but in fact the integration is easier if we let **r** lie on the z axis, so that $\boldsymbol{\omega}$ is tilted at an angle ψ. We may as well orient the x axis so that $\boldsymbol{\omega}$ lies in the xz plane, as shown in Fig. 5.46. According to Eq. 5.64,

$$\mathbf{A}(\mathbf{r}) = \frac{\mu_0}{4\pi}\int \frac{\mathbf{K}(\mathbf{r}')}{\imath}\,da',$$

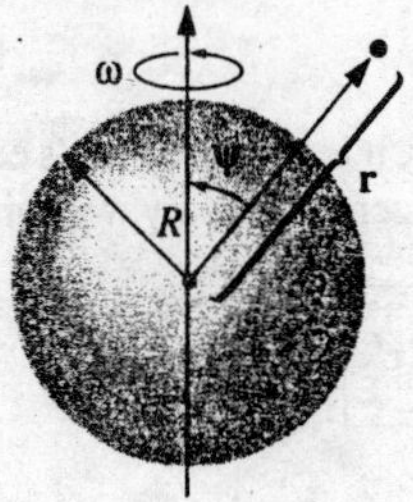

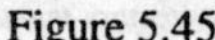

Figure 5.45

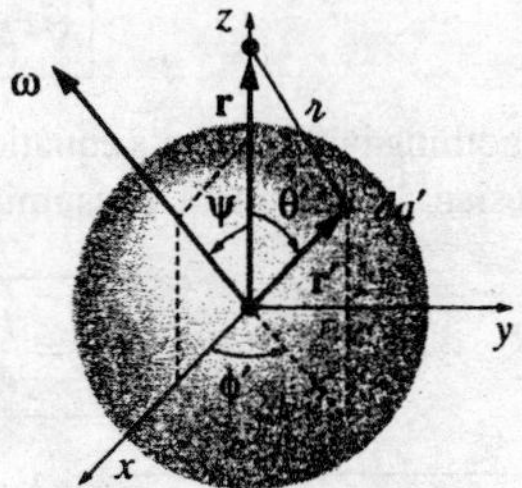

Figure 5.46

[14] M. D. Semon and J. R. Taylor, *Am. J. Phys.* **64**, 1361 (1996).

where $\mathbf{K} = \sigma\mathbf{v}$, $\imath = \sqrt{R^2 + r^2 - 2Rr\cos\theta'}$, and $da' = R^2 \sin\theta'\, d\theta'\, d\phi'$. Now the velocity of a point $\mathbf{r}'$ in a rotating rigid body is given by $\boldsymbol{\omega} \times \mathbf{r}'$; in this case,

$$\mathbf{v} = \boldsymbol{\omega} \times \mathbf{r}' = \begin{vmatrix} \hat{\mathbf{x}} & \hat{\mathbf{y}} & \hat{\mathbf{z}} \\ \omega\sin\psi & 0 & \omega\cos\psi \\ R\sin\theta'\cos\phi' & R\sin\theta'\sin\phi' & R\cos\theta' \end{vmatrix}$$

$$= R\omega[-(\cos\psi\sin\theta'\sin\phi')\,\hat{\mathbf{x}} + (\cos\psi\sin\theta'\cos\phi' - \sin\psi\cos\theta')\,\hat{\mathbf{y}} + (\sin\psi\sin\theta'\sin\phi')\,\hat{\mathbf{z}}].$$

Notice that each of these terms, save one, involves either $\sin\phi'$ or $\cos\phi'$. Since

$$\int_0^{2\pi} \sin\phi'\, d\phi' = \int_0^{2\pi} \cos\phi'\, d\phi' = 0,$$

such terms contribute nothing. There remains

$$\mathbf{A}(\mathbf{r}) = -\frac{\mu_0 R^3 \sigma\omega\sin\psi}{2}\left(\int_0^{\pi} \frac{\cos\theta'\sin\theta'}{\sqrt{R^2 + r^2 - 2Rr\cos\theta'}}\, d\theta'\right)\hat{\mathbf{y}}.$$

Letting $u \equiv \cos\theta'$, the integral becomes

$$\int_{-1}^{+1} \frac{u}{\sqrt{R^2 + r^2 - 2Rru}}\, du = -\frac{(R^2 + r^2 + Rru)}{3R^2r^2}\sqrt{R^2 + r^2 - 2Rru}\,\Bigg|_{-1}^{+1}$$

$$= -\frac{1}{3R^2r^2}\left[(R^2 + r^2 + Rr)|R - r| - (R^2 + r^2 - Rr)(R + r)\right].$$

If the point $\mathbf{r}$ lies *inside* the sphere, then $R > r$, and this expression reduces to $(2r/3R^2)$; if $\mathbf{r}$ lies *outside* the sphere, so that $R < r$, it reduces to $(2R/3r^2)$. Noting that $(\boldsymbol{\omega} \times \mathbf{r}) = -\omega r \sin\psi\, \hat{\mathbf{y}}$, we have, finally,

$$\mathbf{A}(\mathbf{r}) = \begin{cases} \dfrac{\mu_0 R\sigma}{3}(\boldsymbol{\omega} \times \mathbf{r}), & \text{for points } \textit{inside} \text{ the sphere,} \\[2ex] \dfrac{\mu_0 R^4\sigma}{3r^3}(\boldsymbol{\omega} \times \mathbf{r}), & \text{for points } \textit{outside} \text{ the sphere.} \end{cases} \tag{5.66}$$

Having evaluated the integral, I revert to the "natural" coordinates of Fig. 5.45, in which $\boldsymbol{\omega}$ coincides with the z axis and the point $\mathbf{r}$ is at (r, θ, ϕ):

$$\mathbf{A}(r, \theta, \phi) = \begin{cases} \dfrac{\mu_0 R\omega\sigma}{3} r\sin\theta\,\hat{\boldsymbol{\phi}}, & (r \le R), \\[2ex] \dfrac{\mu_0 R^4\omega\sigma}{3}\dfrac{\sin\theta}{r^2}\hat{\boldsymbol{\phi}}, & (r \ge R). \end{cases} \tag{5.67}$$

Curiously, the field inside this spherical shell is *uniform*:

$$\mathbf{B} = \nabla \times \mathbf{A} = \frac{2\mu_0 R\omega\sigma}{3}(\cos\theta\,\hat{\mathbf{r}} - \sin\theta\,\hat{\boldsymbol{\theta}}) = \frac{2}{3}\mu_0\sigma R\omega\hat{\mathbf{z}} = \frac{2}{3}\mu_0\sigma R\boldsymbol{\omega}. \tag{5.68}$$

Example 5.12

Find the vector potential of an infinite solenoid with n turns per unit length, radius R, and current I.

Solution: This time we cannot use Eq. 5.64, since the current itself extends to infinity. But here's a cute method that does the job. Notice that

$$\oint \mathbf{A} \cdot d\mathbf{l} = \int (\nabla \times \mathbf{A}) \cdot d\mathbf{a} = \int \mathbf{B} \cdot d\mathbf{a} = \Phi, \tag{5.69}$$

where Φ is the flux of $\mathbf{B}$ through the loop in question. This is reminiscent of Ampère's law in the integral form (5.55),

$$\oint \mathbf{B} \cdot d\mathbf{l} = \mu_0 I_{\text{enc}}.$$

In fact, it's the same equation, with $\mathbf{B} \to \mathbf{A}$ and $\mu_0 I_{\text{enc}} \to \Phi$. If symmetry permits, we can determine $\mathbf{A}$ from Φ in the same way we got $\mathbf{B}$ from I_{enc}, in Sect. 5.3.3. The present problem (with a uniform longitudinal magnetic field $\mu_0 n I$ inside the solenoid and no field outside) is analogous to the Ampère's law problem of a fat wire carrying a uniformly distributed current. The vector potential is "circumferential" (it mimics the magnetic field of the wire); using a circular "amperian loop" at radius s *inside* the solenoid, we have

$$\oint \mathbf{A} \cdot d\mathbf{l} = A(2\pi s) = \int \mathbf{B} \cdot d\mathbf{a} = \mu_0 n I(\pi s^2),$$

so

$$\mathbf{A} = \frac{\mu_0 n I}{2} s\,\hat{\boldsymbol{\phi}}, \quad \text{for } s < R. \tag{5.70}$$

For an amperian loop *outside* the solenoid, the flux is

$$\int \mathbf{B} \cdot d\mathbf{a} = \mu_0 n I(\pi R^2),$$

since the field only extends out to R. Thus

$$\mathbf{A} = \frac{\mu_0 n I}{2} \frac{R^2}{s}\,\hat{\boldsymbol{\phi}}, \quad \text{for } s > R. \tag{5.71}$$

If you have any doubts about this answer, *check* it: Does $\nabla \times \mathbf{A} = \mathbf{B}$? Does $\nabla \cdot \mathbf{A} = 0$? If so, we're done.

Typically, the direction of $\mathbf{A}$ will mimic the direction of the current. For instance, both were azimuthal in Exs. 5.11 and 5.12. Indeed, if all the current flows in *one* direction, then Eq. 5.63 suggests that $\mathbf{A}$ *must* point that way too. Thus the potential of a finite segment of straight wire (Prob. 5.22) is in the direction of the current. Of course, if the current extends to infinity you can't use Eq. 5.63 in the first place (see Probs. 5.25 and 5.26). Moreover, you can always add an arbitrary constant vector to $\mathbf{A}$—this is analogous to changing the reference point for V, and it won't affect the divergence or curl of $\mathbf{A}$, which is all that matters (in Eq. 5.63 we have chosen the constant so that $\mathbf{A}$ goes to zero at infinity). In principle you could even use a vector potential that is not divergenceless, in which case all bets are off. Despite all these caveats, the essential point remains: *Ordinarily* the direction of $\mathbf{A}$ will match the direction of the current.

Problem 5.22 Find the magnetic vector potential of a finite segment of straight wire, carrying a current I. [Put the wire on the z axis, from z_1 to z_2, and use Eq. 5.64.] Check that your answer is consistent with Eq. 5.35.

Problem 5.23 What current density would produce the vector potential, $\mathbf{A} = k\,\hat{\boldsymbol{\phi}}$ (where k is a constant), in cylindrical coordinates?

Problem 5.24 If $\mathbf{B}$ is *uniform*, show that $\mathbf{A}(\mathbf{r}) = -\frac{1}{2}(\mathbf{r} \times \mathbf{B})$ works. That is, check that $\nabla \cdot \mathbf{A} = 0$ and $\nabla \times \mathbf{A} = \mathbf{B}$. Is this result unique, or are there other functions with the same divergence and curl?

Problem 5.25

(a) By whatever means you can think of (short of looking it up), find the vector potential a distance s from an infinite straight wire carrying a current I. Check that $\nabla \cdot \mathbf{A} = 0$ and $\nabla \times \mathbf{A} = \mathbf{B}$.

(b) Find the magnetic potential *inside* the wire, if it has radius R and the current is uniformly distributed.

Problem 5.26 Find the vector potential above and below the plane surface current in Ex. 5.8.

Problem 5.27

(a) Check that Eq. 5.63 is consistent with Eq. 5.61, by applying the *divergence*.

(b) Check that Eq. 5.63 is consistent with Eq. 5.45, by applying the *curl*.

(c) Check that Eq. 5.63 is consistent with Eq. 5.62, by applying the *Laplacian*.

Problem 5.28 Suppose you want to define a magnetic scalar potential U (Eq. 5.65), in the vicinity of a current-carrying wire. First of all, you must stay away from the wire itself (there $\nabla \times \mathbf{B} \neq 0$); but that's not enough. Show, by applying Ampère's law to a path that starts at **a** and circles the wire, returning to **b** (Fig. 5.47), that the scalar potential cannot be single-valued (that is, $U(\mathbf{a}) \neq U(\mathbf{b})$, even if they represent the same physical point). As an example, find the scalar potential for an infinite straight wire. (To avoid a multivalued potential, you must restrict yourself to simply-connected regions that remain on one side or the other of every wire, never allowing you to go all the way around.)

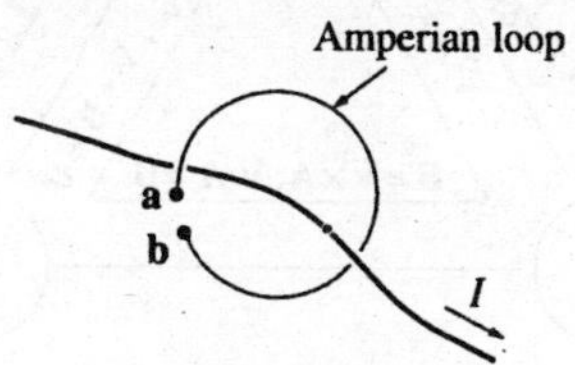

Figure 5.47

Problem 5.29 Use the results of Ex. 5.11 to find the field inside a uniformly charged sphere, of total charge Q and radius R, which is rotating at a constant angular velocity $\boldsymbol{\omega}$.

Problem 5.30

(a) Complete the proof of Theorem 2, Sect. 1.6.2. That is, show that any divergenceless vector field $\mathbf{F}$ can be written as the curl of a vector potential $\mathbf{A}$. What you have to do is find A_x, A_y, and A_z such that: (i) $\partial A_z/\partial y - \partial A_y/\partial z = F_x$; (ii) $\partial A_x/\partial z - \partial A_z/\partial x = F_y$; and (iii) $\partial A_y/\partial x - \partial A_x/\partial y = F_z$. Here's one way to do it: Pick $A_x = 0$, and solve (ii) and (iii) for A_y and A_z. Note that the "constants of integration" here are themselves functions of y and z—they're constant only with respect to x. Now plug these expressions into (i), and use the fact that $\nabla \cdot \mathbf{F} = 0$ to obtain

$$A_y = \int_0^x F_z(x', y, z)\, dx'; \quad A_z = \int_0^y F_x(0, y', z)\, dy' - \int_0^x F_y(x', y, z)\, dx'.$$

(b) By direct differentiation, check that the $\mathbf{A}$ you obtained in part (a) satisfies $\nabla \times \mathbf{A} = \mathbf{F}$. Is $\mathbf{A}$ divergenceless? [This was a very asymmetrical construction, and it would be surprising if it *were*—although we know that there *exists* a vector whose curl is $\mathbf{F}$ *and* whose divergence is zero.]

(c) As an example, let $\mathbf{F} = y\,\hat{\mathbf{x}} + z\,\hat{\mathbf{y}} + x\,\hat{\mathbf{z}}$. Calculate $\mathbf{A}$, and confirm that $\nabla \times \mathbf{A} = \mathbf{F}$. (For further discussion see Prob. 5.51.)

5.4.2 Summary; Magnetostatic Boundary Conditions

In Chapter 2, I drew a triangular diagram to summarize the relations among the three fundamental quantities of electrostatics: the charge density ρ, the electric field $\mathbf{E}$, and the potential V. A similar diagram can be constructed for magnetostatics (Fig. 5.48), relating

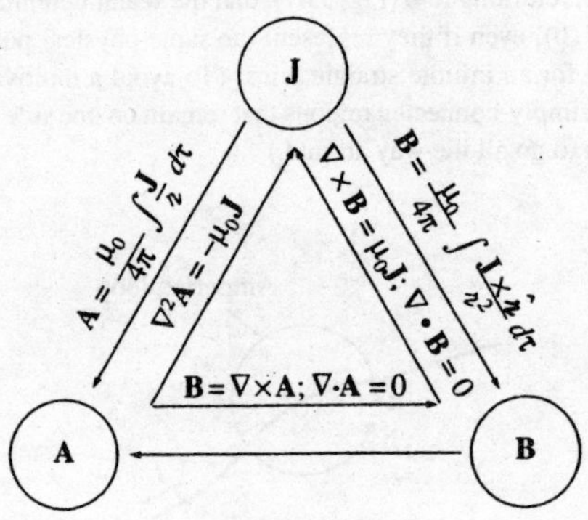

Figure 5.48

the current density **J**, the field **B**, and the potential **A**. There is one "missing link" in the diagram: the equation for **A** in terms of **B**. It's unlikely you would ever need such a formula, but in case you are interested, see Probs. 5.50 and 5.51.

Just as the electric field suffers a discontinuity at a surface *charge*, so the magnetic field is discontinuous at a surface *current*. Only this time it is the *tangential* component that changes. For if we apply Eq. 5.48, in the integral form

$$\oint \mathbf{B} \cdot d\mathbf{a} = 0,$$

to a wafer-thin pillbox straddling the surface (Fig. 5.49), we get

$$B^{\perp}_{\text{above}} = B^{\perp}_{\text{below}}. \tag{5.72}$$

As for the tangential components, an amperian loop running perpendicular to the current (Fig. 5.50) yields

$$\oint \mathbf{B} \cdot d\mathbf{l} = (B^{\parallel}_{\text{above}} - B^{\parallel}_{\text{below}})l = \mu_0 I_{\text{enc}} = \mu_0 K l,$$

or

$$B^{\parallel}_{\text{above}} - B^{\parallel}_{\text{below}} = \mu_0 K. \tag{5.73}$$

Thus the component of **B** that is parallel to the surface but perpendicular to the current is discontinuous in the amount $\mu_0 K$. A similar amperian loop running *parallel* to the current reveals that the *parallel* component is *continuous*. These results can be summarized in a single formula:

$$\mathbf{B}_{\text{above}} - \mathbf{B}_{\text{below}} = \mu_0(\mathbf{K} \times \hat{\mathbf{n}}), \tag{5.74}$$

where $\hat{\mathbf{n}}$ is a unit vector perpendicular to the surface, pointing "upward."

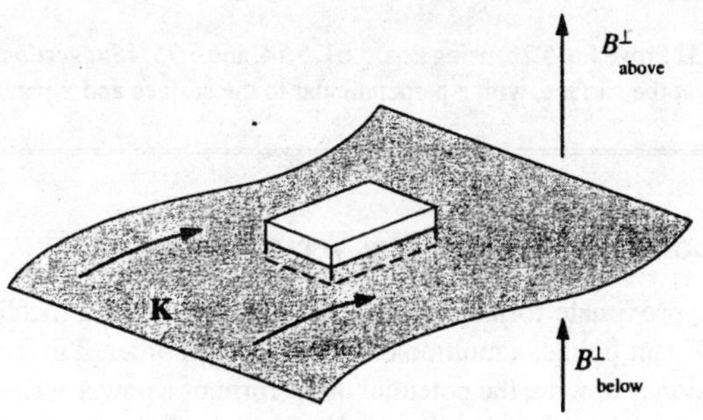

Figure 5.49

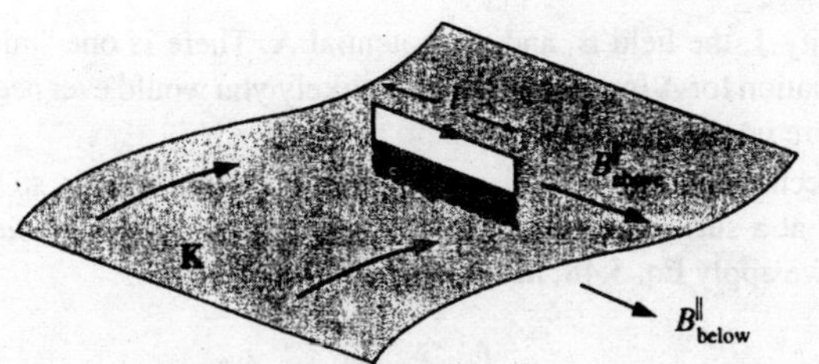

Figure 5.50

Like the scalar potential in electrostatics, the vector potential is continuous across any boundary:

$$\mathbf{A}_{\text{above}} = \mathbf{A}_{\text{below}}, \tag{5.75}$$

for $\nabla \cdot \mathbf{A} = 0$ guarantees[15] that the *normal* component is continuous, and $\nabla \times \mathbf{A} = \mathbf{B}$, in the form

$$\oint \mathbf{A} \cdot d\mathbf{l} = \int \mathbf{B} \cdot d\mathbf{a} = \Phi,$$

means that the tangential components are continuous (the flux through an amperian loop of vanishing thickness is zero). But the *derivative* of **A** inherits the discontinuity of **B**:

$$\frac{\partial \mathbf{A}_{\text{above}}}{\partial n} - \frac{\partial \mathbf{A}_{\text{below}}}{\partial n} = -\mu_0 \mathbf{K}. \tag{5.76}$$

Problem 5.31

(a) Check Eq. 5.74 for the configuration in Ex. 5.9.

(b) Check Eqs. 5.75 and 5.76 for the configuration in Ex. 5.11.

Problem 5.32 Prove Eq. 5.76, using Eqs. 5.61, 5.74, and 5.75. [*Suggestion:* I'd set up Cartesian coordinates at the surface, with z perpendicular to the surface and x parallel to the current.]

5.4.3 Multipole Expansion of the Vector Potential

If you want an approximate formula for the vector potential of a localized current distribution, valid at distant points, a multipole expansion is in order. Remember: the idea of a multipole expansion is to write the potential in the form of a power series in $1/r$, where r is the distance to the point in question (Fig. 5.51); if r is sufficiently large, the series will be

[15]Note that Eqs. 5.75 and 5.76 presuppose that **A** is divergenceless.

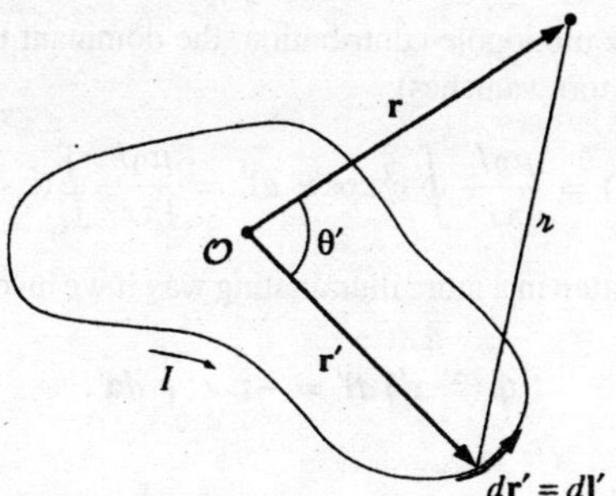

Figure 5.51

dominated by the lowest nonvanishing contribution, and the higher terms can be ignored. As we found in Sect. 3.4.1 (Eq. 3.94),

$$\frac{1}{\imath} = \frac{1}{\sqrt{r^2 + (r')^2 - 2rr'\cos\theta'}} = \frac{1}{r}\sum_{n=0}^{\infty}\left(\frac{r'}{r}\right)^n P_n(\cos\theta'). \tag{5.77}$$

Accordingly, the vector potential of a current loop can be written

$$\mathbf{A}(\mathbf{r}) = \frac{\mu_0 I}{4\pi}\oint \frac{1}{\imath}\, d\mathbf{l}' = \frac{\mu_0 I}{4\pi}\sum_{n=0}^{\infty}\frac{1}{r^{n+1}}\oint (r')^n P_n(\cos\theta')\, d\mathbf{l}', \tag{5.78}$$

or, more explicitly:

$$\begin{aligned}\mathbf{A}(\mathbf{r}) = \; & \frac{\mu_0 I}{4\pi}\left[\frac{1}{r}\oint d\mathbf{l}' + \frac{1}{r^2}\oint r'\cos\theta'\, d\mathbf{l}'\right.\\ & \left. + \frac{1}{r^3}\oint (r')^2\left(\frac{3}{2}\cos^2\theta' - \frac{1}{2}\right) d\mathbf{l}' + \cdots\right].\end{aligned} \tag{5.79}$$

As in the multipole expansion of V, we call the first term (which goes like $1/r$) the **monopole** term, the second (which goes like $1/r^2$) **dipole**, the third **quadrupole**, and so on.

Now, it happens that the *magnetic monopole term is always zero*, for the integral is just the total vector displacement around a closed loop:

$$\oint d\mathbf{l}' = 0. \tag{5.80}$$

This reflects the fact that there are (apparently) no magnetic monopoles in nature (an assumption contained in Maxwell's equation $\nabla \cdot \mathbf{B} = 0$, on which the entire theory of vector potential is predicated).

In the absence of any monopole contribution, the dominant term is the dipole (except in the rare case where it, too, vanishes):

$$\mathbf{A}_{\text{dip}}(\mathbf{r}) = \frac{\mu_0 I}{4\pi r^2}\oint r' \cos\theta' \, d\mathbf{l}' = \frac{\mu_0 I}{4\pi r^2}\oint (\hat{\mathbf{r}} \cdot \mathbf{r}')\, d\mathbf{l}'. \tag{5.81}$$

This integral can be rewritten in a more illuminating way if we invoke Eq. 1.108, with $\mathbf{c} = \hat{\mathbf{r}}$:

$$\oint (\hat{\mathbf{r}} \cdot \mathbf{r}')\, d\mathbf{l}' = -\hat{\mathbf{r}} \times \int d\mathbf{a}'. \tag{5.82}$$

Then

$$\boxed{\mathbf{A}_{\text{dip}}(\mathbf{r}) = \frac{\mu_0}{4\pi}\frac{\mathbf{m} \times \hat{\mathbf{r}}}{r^2},} \tag{5.83}$$

where $\mathbf{m}$ is the **magnetic dipole moment**:

$$\boxed{\mathbf{m} \equiv I \int d\mathbf{a} = I\mathbf{a}.} \tag{5.84}$$

Here $\mathbf{a}$ is the "vector area" of the loop (Prob. 1.61); if the loop is *flat*, $\mathbf{a}$ is the ordinary area enclosed, with the direction assigned by the usual right hand rule (fingers in the direction of the current).

Example 5.13

Find the magnetic dipole moment of the "bookend-shaped" loop shown in Fig. 5.52. All sides have length w, and it carries a current I.

Solution: This wire could be considered the superposition of two plane square loops (Fig. 5.53). The "extra" sides (AB) cancel when the two are put together, since the currents flow in opposite directions. The net magnetic dipole moment is

$$\mathbf{m} = Iw^2\,\hat{\mathbf{y}} + Iw^2\,\hat{\mathbf{z}};$$

its magnitude is $\sqrt{2}Iw^2$, and it points along the 45° line $z = y$.

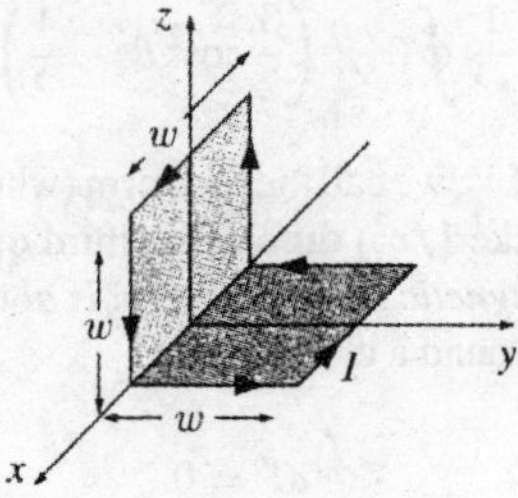

Figure 5.52

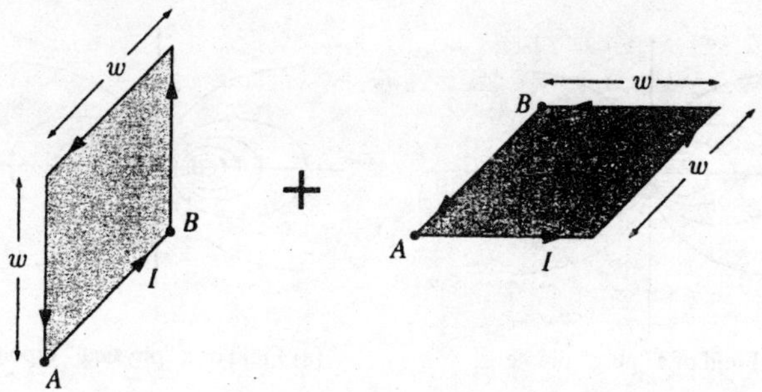

Figure 5.53

It is clear from Eq. 5.84 that the magnetic dipole moment is independent of the choice of origin. You may remember that the *electric* dipole moment is independent of the origin only when the total charge vanishes (Sect. 3.4.3). Since the *magnetic* monopole moment is *always* zero, it is not really surprising that the magnetic dipole moment is always independent of origin.

Although the dipole term *dominates* the multipole expansion (unless $\mathbf{m} = 0$), and thus offers a good approximation to the true potential, it is not ordinarily the *exact* potential; there will be quadrupole, octopole, and higher contributions. You might ask, is it possible to devise a current distribution whose potential is "pure" dipole—for which Eq. 5.83 is *exact*? Well, yes and no: like the electrical analog, it can be done, but the model is a bit contrived. To begin with, you must take an *infinitesimally small* loop at the origin, but then, in order to keep the dipole moment finite, you have to crank the current up to infinity, with the product $m = Ia$ held fixed. In practice, the dipole potential is a suitable approximation whenever the distance r greatly exceeds the size of the loop.

The magnetic *field* of a (pure) dipole is easiest to calculate if we put $\mathbf{m}$ at the origin and let it point in the z-direction (Fig. 5.54). According to Eq. 5.83, the potential at point

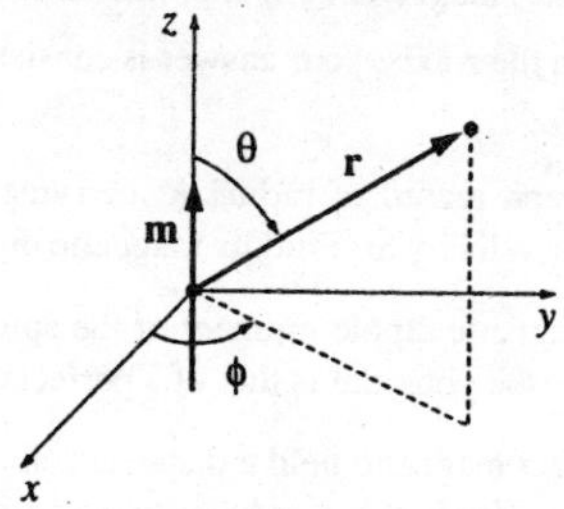

Figure 5.54

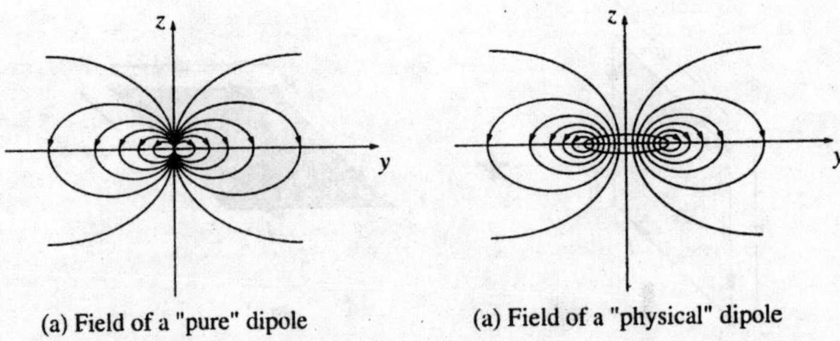

(a) Field of a "pure" dipole

(a) Field of a "physical" dipole

Figure 5.55

(r, θ, ϕ) is

$$\mathbf{A}_{\text{dip}}(\mathbf{r}) = \frac{\mu_0}{4\pi} \frac{m \sin\theta}{r^2} \hat{\boldsymbol{\phi}}, \tag{5.85}$$

and hence

$$\mathbf{B}_{\text{dip}}(\mathbf{r}) = \nabla \times \mathbf{A} = \frac{\mu_0 m}{4\pi r^3} (2\cos\theta\, \hat{\mathbf{r}} + \sin\theta\, \hat{\boldsymbol{\theta}}). \tag{5.86}$$

Surprisingly, this is *identical* in structure to the field of an *electric* dipole (Eq. 3.103)! (Up close, however, the field of a *physical* magnetic dipole—a small current loop—looks quite different from the field of a physical electric dipole—plus and minus charges a short distance apart. Compare Fig. 5.55 with Fig. 3.37.)

• **Problem 5.33** Show that the magnetic field of a dipole can be written in coordinate-free form:

$$\mathbf{B}_{\text{dip}}(\mathbf{r}) = \frac{\mu_0}{4\pi} \frac{1}{r^3} [3(\mathbf{m} \cdot \hat{\mathbf{r}})\hat{\mathbf{r}} - \mathbf{m}]. \tag{5.87}$$

Problem 5.34 A circular loop of wire, with radius R, lies in the xy plane, centered at the origin, and carries a current I running counterclockwise as viewed from the positive z axis.

(a) What is its magnetic dipole moment?

(b) What is the (approximate) magnetic field at points far from the origin?

(c) Show that, for points on the z axis, your answer is consistent with the *exact* field (Ex. 5.6), when $z \gg R$.

Problem 5.35 A phonograph record of radius R, carrying a uniform surface charge σ, is rotating at constant angular velocity ω. Find its magnetic dipole moment.

Problem 5.36 Find the magnetic dipole moment of the spinning spherical shell in Ex. 5.11. Show that for points $r > R$ the potential is that of a perfect dipole.

Problem 5.37 Find the exact magnetic field a distance z above the center of a square loop of side w, carrying a current I. Verify that it reduces to the field of a dipole, with the appropriate dipole moment, when $z \gg w$.

More Problems on Chapter 5

Problem 5.38 It may have occurred to you that since parallel currents attract, the current within a single wire should contract into a tiny concentrated stream along the axis. Yet in practice the current typically distributes itself quite uniformly over the wire. How do you account for this? If the positive charges (density ρ_+) are at rest, and the negative charges (density ρ_-) move at speed v (and none of these depends on the distance from the axis), show that $\rho_- = -\rho_+\gamma^2$, where $\gamma \equiv 1/\sqrt{1-(v/c)^2}$ and $c^2 = 1/\mu_0\epsilon_0$. If the wire as a whole is neutral, where is the compensating charge located?[16] [Notice that for typical velocities (see Prob. 5.19) the two charge densities are essentially unchanged by the current (since $\gamma \approx 1$). In **plasmas**, however, where the positive charges are *also* free to move, this so-called **pinch effect** can be very significant.]

Problem 5.39 A current I flows to the right through a rectangular bar of conducting material, in the presence of a uniform magnetic field **B** pointing out of the page (Fig. 5.56).

(a) If the moving charges are *positive*, in which direction are they deflected by the magnetic field? This deflection results in an accumulation of charge on the upper and lower surfaces of the bar, which in turn produces an electric force to counteract the magnetic one. Equilibrium occurs when the two exactly cancel. (This phenomenon is known as the **Hall effect**.)

(b) Find the resulting potential difference (the **Hall voltage**) between the top and bottom of the bar, in terms of B, v (the speed of the charges), and the relevant dimensions of the bar.[17]

(c) How would your analysis change if the moving charges were *negative*? [The Hall effect is the classic way of determining the sign of the mobile charge carriers in a material.]

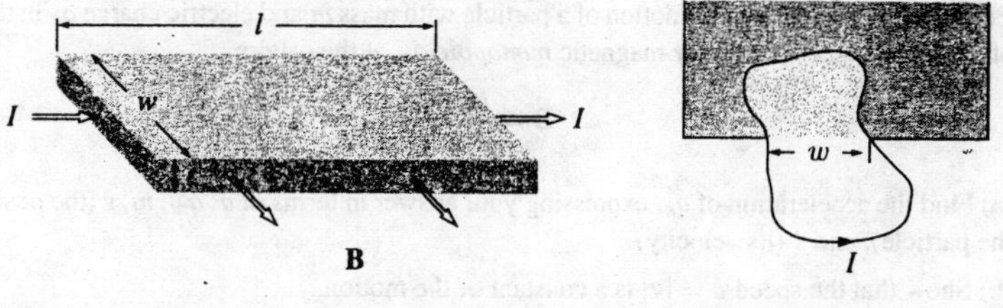

Figure 5.56

Figure 5.57

Problem 5.40 A plane wire loop of irregular shape is situated so that part of it is in a uniform magnetic field **B** (in Fig. 5.57 the field occupies the shaded region, and points perpendicular to the plane of the loop). The loop carries a current I. Show that the net magnetic force on the loop is $F = IBw$, where w is the chord subtended. Generalize this result to the case where the magnetic field region itself has an irregular shape. What is the direction of the force?

[16] For further discussion, see D. C. Gabuzda, *Am. J. Phys.* **61**, 360 (1993).

[17] The potential *within* the bar makes an interesting boundary-value problem. See M. J. Moelter, J. Evans, and G. Elliot, *Am. J. Phys.* **66**, 668(1998).

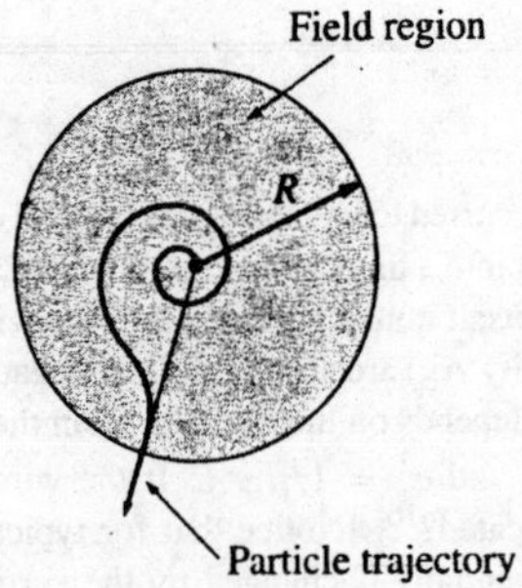

Figure 5.58

Problem 5.41 A circularly symmetrical magnetic field (**B** depends only on the distance from the axis), pointing perpendicular to the page, occupies the shaded region in Fig. 5.58. If the total flux ($\int \mathbf{B} \cdot d\mathbf{a}$) is zero, show that a charged particle that starts out at the center will emerge from the field region on a *radial* path (provided it escapes at all—if the initial velocity is too great, it may simply circle around forever). On the reverse trajectory, a particle fired at the center from outside will hit its target, though it may follow a weird route getting there. [*Hint:* Calculate the total angular momentum acquired by the particle, using the Lorentz force law.]

Problem 5.42 Calculate the magnetic force of attraction between the northern and southern hemispheres of a spinning charged spherical shell (Ex. 5.11). [*Answer:* $(\pi/4)\mu_0\sigma^2\omega^2 R^4$.]

! **Problem 5.43** Consider the motion of a particle with mass m and electric charge q_e in the field of a (hypothetical) stationary magnetic *monopole* q_m at the origin:

$$\mathbf{B} = \frac{\mu_0}{4\pi}\frac{q_m}{r^2}\,\hat{\mathbf{r}}.$$

(a) Find the acceleration of q_e, expressing your answer in terms of q, q_m, m, $\mathbf{r}$ (the position of the particle), and $\mathbf{v}$ (its velocity).

(b) Show that the speed $v = |\mathbf{v}|$ is a constant of the motion.

(c) Show that the vector quantity

$$\mathbf{Q} \equiv m(\mathbf{r} \times \mathbf{v}) - \frac{\mu_0 q_e q_m}{4\pi}\,\hat{\mathbf{r}}$$

is a constant of the motion. [*Hint:* differentiate it with respect to time, and prove—using the equation of motion from (a)—that the derivative is zero.]

(d) Choosing spherical coordinates (r, θ, ϕ), with the polar (z) axis along $\mathbf{Q}$,

(i) calculate $\mathbf{Q} \cdot \hat{\boldsymbol{\phi}}$, and show that θ is a constant of the motion (so q_e moves on the surface of a cone—something Poincaré first discovered in 1896)[18];

[18] In point of fact the charge follows a *geodesic* on the cone. The original paper is H. Poincaré, ***Comptes rendus de l'Academie des Sciences*** **123**, 530 (1896); for a more modern treatment see B. Rossi and S. Olbert, ***Introduction to the Physics of Space*** (New York: McGraw-Hill, 1970).

(ii) calculate $\mathbf{Q} \cdot \hat{\mathbf{r}}$, and show that the magnitude of $\mathbf{Q}$ is

$$Q = \frac{\mu_0}{4\pi} \left| \frac{q_e q_m}{\cos\theta} \right|;$$

(iii) calculate $\mathbf{Q} \cdot \hat{\boldsymbol{\theta}}$, show that

$$\frac{d\phi}{dt} = \frac{k}{r^2},$$

and determine the constant k.

(e) By expressing v^2 in spherical coordinates, obtain the equation for the trajectory, in the form

$$\frac{dr}{d\phi} = f(r)$$

(that is: determine the function $f(r)$).

(f) Solve this equation for $r(\phi)$.

! **Problem 5.44** Use the Biot-Savart law (most conveniently in the form 5.39 appropriate to surface currents) to find the field inside and outside an infinitely long solenoid of radius R, with n turns per unit length, carrying a steady current I.

Problem 5.45 A semicircular wire carries a steady current I (it must be hooked up to some *other* wires to complete the circuit, but we're not concerned with them here). Find the magnetic field at a point P on the other semicircle (Fig. 5.59). [*Answer:* $(\mu_0 I/8\pi R)\ln\{\tan\left(\frac{\theta+\pi}{4}\right)/\tan\left(\frac{\theta}{4}\right)\}$]

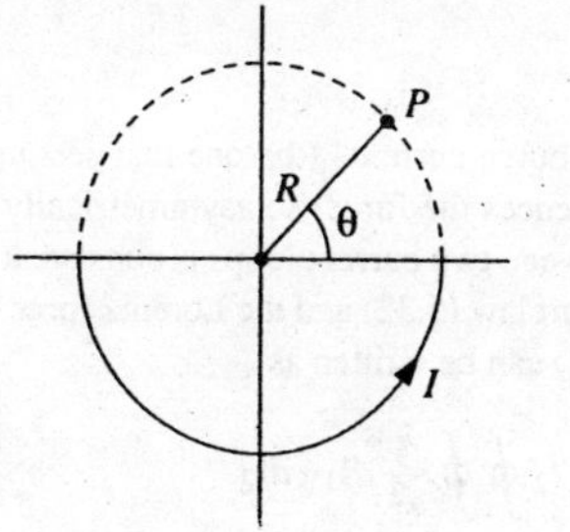

Figure 5.59

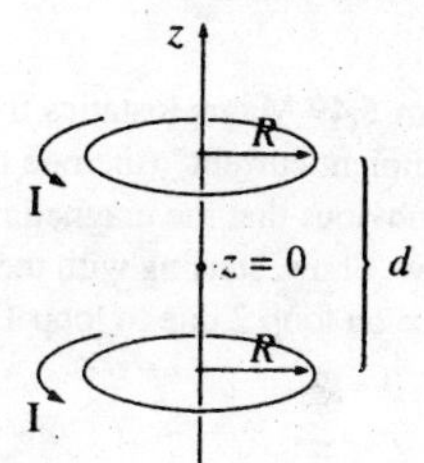

Figure 5.60

Problem 5.46 The magnetic field on the axis of a circular current loop (Eq. 5.38) is far from uniform (it falls off sharply with increasing z). You can produce a more nearly uniform field by using *two* such loops a distance d apart (Fig. 5.60).

(a) Find the field (B) as a function of z, and show that $\partial B/\partial z$ is zero at the point midway between them ($z = 0$). Now, if you pick d just right the *second* derivative of B will *also* vanish at the midpoint. This arrangement is known as a **Helmholtz coil**; it's a convenient way of producing relatively uniform fields in the laboratory.

(b) Determine d such that $\partial^2 B/\partial z^2 = 0$ at the midpoint, and find the resulting magnetic field at the center. [*Answer:* $8\mu_0 I/5\sqrt{5}R$]

Problem 5.47 Find the magnetic field at a point $z > R$ on the axis of (a) the rotating disk and (b) the rotating sphere, in Prob. 5.6.

Problem 5.48 Suppose you wanted to find the field of a circular loop (Ex. 5.6) at a point **r** that is *not* directly above the center (Fig. 5.61). You might as well choose your axes so that **r** lies in the yz plane at $(0, y, z)$. The source point is $(R\cos\phi', R\sin\phi', 0)$, and ϕ' runs from 0 to 2π. Set up the integrals from which you could calculate B_x, B_y, and B_z, and evaluate B_x explicitly.

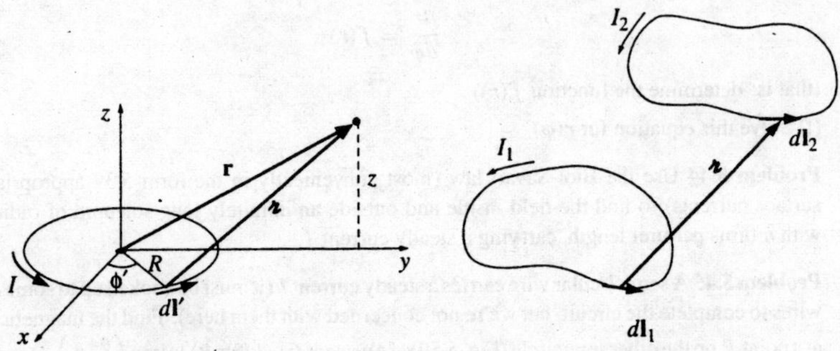

Figure 5.61

Figure 5.62

Problem 5.49 Magnetostatics treats the "source current" (the one that sets up the field) and the "recipient current" (the one that experiences the force) so asymmetrically that it is by no means obvious that the magnetic force between two current loops is consistent with Newton's third law. Show, starting with the Biot-Savart law (5.32) and the Lorentz force law (5.16), that the force on loop 2 due to loop 1 (Fig. 5.62) can be written as

$$\mathbf{F}_2 = -\frac{\mu_0}{4\pi} I_1 I_2 \oint\oint \frac{\hat{\boldsymbol{\mathscr{r}}}}{\mathscr{r}^2}\, d\mathbf{l}_1 \cdot d\mathbf{l}_2. \tag{5.88}$$

In this form it is clear that $\mathbf{F}_2 = -\mathbf{F}_1$, since $\hat{\boldsymbol{\mathscr{r}}}$ changes direction when the roles of 1 and 2 are interchanged. (If you seem to be getting an "extra" term, it will help to note that $d\mathbf{l}_2 \cdot \hat{\boldsymbol{\mathscr{r}}} = d\mathscr{r}$.)

Problem 5.50

(a) One way to fill in the "missing link" in Fig. 5.48 is to exploit the analogy between the defining equations for **A** ($\nabla \cdot \mathbf{A} = 0$, $\nabla \times \mathbf{A} = \mathbf{B}$) and Maxwell's equations for **B** ($\nabla \cdot \mathbf{B} = 0$, $\nabla \times \mathbf{B} = \mu_0 \mathbf{J}$). Evidently **A** depends on **B** in exactly the same way that **B** depends on $\mu_0\mathbf{J}$ (to wit: the Biot-Savart law). Use this observation to write down the formula for **A** in terms of **B**.

(b) The electrical analog to your result in (a) is

$$V(\mathbf{r}) = -\frac{1}{4\pi}\int \frac{\mathbf{E}(\mathbf{r}') \cdot \hat{\boldsymbol{\mathscr{r}}}}{\mathscr{r}^2}\, d\tau'.$$

Derive it, by exploiting the appropriate analogy.

Problem 5.51 Another way to fill in the "missing link" in Fig. 5.48 is to look for a magnetostatic analog to Eq. 2.21. The obvious candidate would be

$$\mathbf{A}(\mathbf{r}) = \int_{\mathcal{O}}^{\mathbf{r}} (\mathbf{B} \times d\mathbf{l}).$$

(a) Test this formula for the simplest possible case—uniform **B** (use the origin as your reference point). Is the result consistent with Prob. 5.24? You could cure this problem by throwing in a factor of $\frac{1}{2}$, but the flaw in this equation runs deeper.

(b) Show that $\int (\mathbf{B} \times d\mathbf{l})$ is *not* independent of path, by calculating $\oint (\mathbf{B} \times d\mathbf{l})$ around the rectangular loop shown in Fig. 5.63.

As far as I know[19] the best one can do along these lines is the pair of equations

(i) $V(\mathbf{r}) = -\mathbf{r} \cdot \int_0^1 \mathbf{E}(\lambda \mathbf{r})\, d\lambda$,

(ii) $\mathbf{A}(\mathbf{r}) = -\mathbf{r} \times \int_0^1 \lambda \mathbf{B}(\lambda \mathbf{r})\, d\lambda$.

[Equation (i) amounts to selecting a *radial* path for the integral in Eq. 2.21; equation (ii) constitutes a more "symmetrical" solution to Prob. 5.30.]

(c) Use (ii) to find the vector potential for *uniform* **B**.

(d) Use (ii) to find the vector potential of an infinite straight wire carrying a steady current I. Does (ii) automatically satisfy $\nabla \cdot \mathbf{A} = 0$? [*Answer:* $(\mu_0 I / 2\pi s)(z\,\hat{\mathbf{s}} - s\,\hat{\mathbf{z}})$]

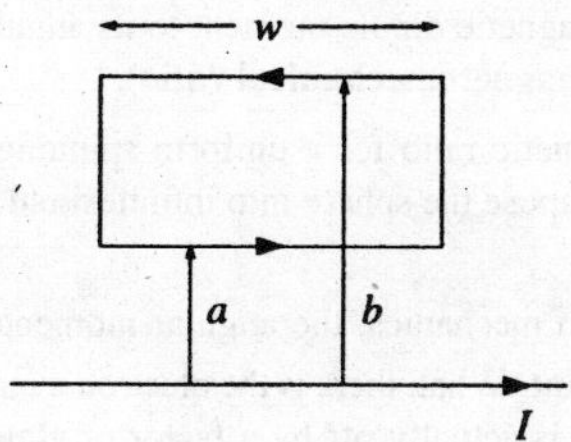

Figure 5.63

Problem 5.52

(a) Construct the scalar potential $U(\mathbf{r})$ for a "pure" magnetic dipole **m**.

(b) Construct a scalar potential for the spinning spherical shell (Ex. 5.11). [*Hint:* for $r > R$ this is a pure dipole field, as you can see by comparing Eqs. 5.67 and 5.85.]

(c) Try doing the same for the interior of a *solid* spinning sphere. [*Hint:* if you solved Prob. 5.29, you already know the *field*; set it equal to $-\nabla U$, and solve for U. What's the trouble?]

[19] R. L. Bishop and S. I. Goldberg, *Tensor Analysis on Manifolds*, Section 4.5 (New York: Macmillan, 1968).

Problem 5.53 Just as $\nabla \cdot \mathbf{B} = 0$ allows us to express $\mathbf{B}$ as the curl of a vector potential ($\mathbf{B} = \nabla \times \mathbf{A}$), so $\nabla \cdot \mathbf{A} = 0$ permits us to write $\mathbf{A}$ itself as the curl of a "higher" potential: $\mathbf{A} = \nabla \times \mathbf{W}$. (And this hierarchy can be extended ad infinitum.)

(a) Find the general formula for $\mathbf{W}$ (as an integral over $\mathbf{B}$), which holds when $\mathbf{B} \to 0$ at ∞.

(b) Determine $\mathbf{W}$ for the case of a *uniform* magnetic field $\mathbf{B}$. [*Hint:* see Prob. 5.24.]

(c) Find $\mathbf{W}$ inside and outside an infinite solenoid. [*Hint:* see Ex. 5.12.]

Problem 5.54 Prove the following uniqueness theorem: If the current density $\mathbf{J}$ is specified throughout a volume $\mathcal{V}$, and *either* the potential $\mathbf{A}$ *or* the magnetic field $\mathbf{B}$ is specified on the surface $\mathcal{S}$ bounding $\mathcal{V}$, then the magnetic field itself is uniquely determined throughout $\mathcal{V}$. [*Hint:* First use the divergence theorem to show that

$$\int \{(\nabla \times \mathbf{U}) \cdot (\nabla \times \mathbf{V}) - \mathbf{U} \cdot [\nabla \times (\nabla \times \mathbf{V})]\}\, d\tau = \oint [\mathbf{U} \times (\nabla \times \mathbf{V})] \cdot d\mathbf{a},$$

for arbitrary vector functions $\mathbf{U}$ and $\mathbf{V}$.]

Problem 5.55 A magnetic dipole $\mathbf{m} = -m_0\,\hat{\mathbf{z}}$ is situated at the origin, in an otherwise uniform magnetic field $\mathbf{B} = B_0\,\hat{\mathbf{z}}$. Show that there exists a spherical surface, centered at the origin, through which no magnetic field lines pass. Find the radius of this sphere, and sketch the field lines, inside and out.

Problem 5.56 A thin uniform donut, carrying charge Q and mass M, rotates about its axis as shown in Fig. 5.64.

(a) Find the ratio of its magnetic dipole moment to its angular momentum. This is called the **gyromagnetic ratio** (or **magnetomechanical ratio**).

(b) What is the gyromagnetic ratio for a uniform spinning sphere? [This requires no new calculation; simply decompose the sphere into infinitesimal rings, and apply the result of part (a).]

(c) According to quantum mechanics, the angular momentum of a spinning electron is $\frac{1}{2}\hbar$, where $\hbar$ is Planck's constant. What, then, is the electron's magnetic dipole moment, in $\mathrm{A} \cdot m^2$? [This semiclassical value is actually off by a factor of almost exactly 2. Dirac's relativistic electron theory got the 2 right, and Feynman, Schwinger, and Tomonaga later calculated tiny further corrections. The determination of the electron's magnetic dipole moment remains the finest achievement of quantum electrodynamics, and exhibits perhaps the most stunningly precise agreement between theory and experiment in all of physics. Incidentally, the quantity $(e\hbar/2m)$, where e is the charge of the electron and m is its mass, is called the **Bohr magneton**.]

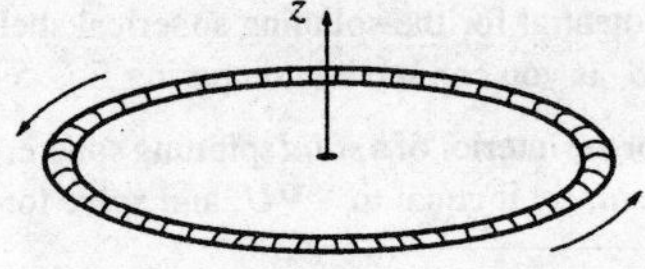

Figure 5.64

• **Problem 5.57**

(a) Prove that the average magnetic field, over a sphere of radius R, due to steady currents within the sphere, is

$$\mathbf{B}_{\text{ave}} = \frac{\mu_0}{4\pi}\frac{2\mathbf{m}}{R^3}, \tag{5.89}$$

where $\mathbf{m}$ is the total dipole moment of the sphere. Contrast the electrostatic result, Eq. 3.105. [This is tough, so I'll give you a start:

$$\mathbf{B}_{\text{ave}} = \frac{1}{\frac{4}{3}\pi R^3}\int \mathbf{B}\, d\tau.$$

Write $\mathbf{B}$ as $(\nabla \times \mathbf{A})$, and apply Prob. 1.60b. Now put in Eq. 5.63, and do the surface integral first, showing that

$$\int \frac{1}{\mathscr{r}}\, d\mathbf{a} = \frac{4}{3}\pi \mathbf{r}'$$

(see Fig. 5.65). Use Eq. 5.91, if you like.]

(b) Show that the average magnetic field due to steady currents *outside* the sphere is the same as the field they produce at the center.

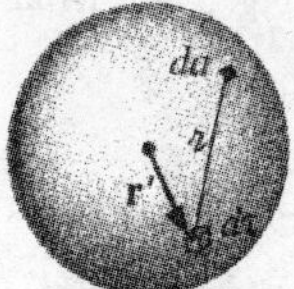

Figure 5.65

Problem 5.58 A uniformly charged solid sphere of radius R carries a total charge Q, and is set spinning with angular velocity ω about the z axis.

(a) What is the magnetic dipole moment of the sphere?

(b) Find the average magnetic field within the sphere (see Prob. 5.57).

(c) Find the approximate vector potential at a point (r, θ) where $r \gg R$.

(d) Find the *exact* potential at a point (r, θ) outside the sphere, and check that it is consistent with (c). [*Hint:* refer to Ex. 5.11.]

(e) Find the magnetic field at a point (r, θ) *inside* the sphere, and check that it is consistent with (b).

Problem 5.59 Using Eq. 5.86, calculate the average magnetic field of a dipole over a sphere of radius R centered at the origin. Do the angular integrals first. Compare your answer with the general theorem in Prob. 5.57. Explain the discrepancy, and indicate how Eq. 5.87 can be corrected to resolve the ambiguity at $r = 0$. (If you get stuck, refer to Prob. 3.42.)

Evidently the *true* field of a magnetic dipole is

$$\mathbf{B}_{\text{dip}}(\mathbf{r}) = \frac{\mu_0}{4\pi}\frac{1}{r^3}[3(\mathbf{m}\cdot\hat{\mathbf{r}})\hat{\mathbf{r}} - \mathbf{m}] + \frac{2\mu_0}{3}\mathbf{m}\delta^3(\mathbf{r}). \tag{5.90}$$

Compare the electrostatic analog, Eq. 3.106. [Incidentally, the delta-function term is responsible for the **hyperfine splitting** in atomic spectra—see, for example, D. J. Griffiths, *Am. J. Phys.* **50**, 698 (1982).]

Problem 5.60 I worked out the multipole expansion for the vector potential of a *line* current because that's the most common type, and in some respects the easiest to handle. For a *volume* current **J**:

(a) Write down the multipole expansion, analogous to Eq. 5.78.

(b) Write down the monopole potential, and prove that it vanishes.

(c) Using Eqs. 1.107 and 5.84, show that the dipole moment can be written

$$\mathbf{m} = \tfrac{1}{2}\int (\mathbf{r}\times\mathbf{J})\,d\tau. \tag{5.91}$$

Problem 5.61 A thin glass rod of radius R and length L carries a uniform surface charge σ. It is set spinning about its axis, at an angular velocity ω. Find the magnetic field at a distance $s \gg R$ from the center of the rod (Fig. 5.66). [*Hint:* treat it as a stack of magnetic dipoles.] [*Answer:* $\mu_0\omega\sigma LR^3/4[s^2 + (L/2)^2]^{3/2}$]

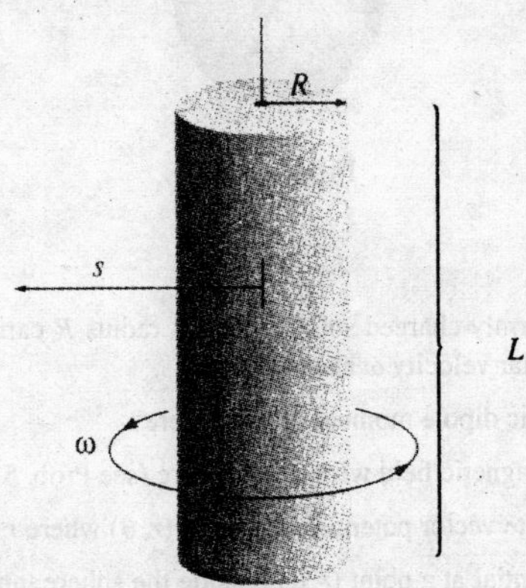

Figure 5.66

Chapter 6

Magnetic Fields in Matter

6.1 Magnetization

6.1.1 Diamagnets, Paramagnets, Ferromagnets

If you ask the average person what "magnetism" is, you will probably be told about horseshoe magnets, compass needles, and the North Pole—none of which has any obvious connection with moving charges or current-carrying wires. Yet all magnetic phenomena are due to electric charges in motion, and in fact, if you could examine a piece of magnetic material on an atomic scale you *would* find tiny currents: electrons orbiting around nuclei and electrons spinning about their axes. For macroscopic purposes, these current loops are so small that we may treat them as magnetic dipoles. Ordinarily, they cancel each other out because of the random orientation of the atoms. But when a magnetic field is applied, a net alignment of these magnetic dipoles occurs, and the medium becomes magnetically polarized, or **magnetized**.

Unlike electric polarization, which is almost always in the same direction as **E**, some materials acquire a magnetization *parallel* to **B** (*para*magnets) and some *opposite* to **B** (*dia*magnets). A few substances (called *ferro*magnets, in deference to the most common example, iron) retain their magnetization even after the external field has been removed—for these the magnetization is not determined by the *present* field but by the whole magnetic "history" of the object. Permanent magnets made of iron are the most familiar examples of magnetism, though from a theoretical point of view they are the most complicated; I'll save ferromagnetism for the end of the chapter, and begin with qualitative models of paramagnetism and diamagnetism.

6.1.2 Torques and Forces on Magnetic Dipoles

A magnetic dipole experiences a torque in a magnetic field, just as an electric dipole does in an electric field. Let's calculate the torque on a rectangular current loop in a uniform field **B**. (Since any current loop could be built up from infinitesimal rectangles, with all

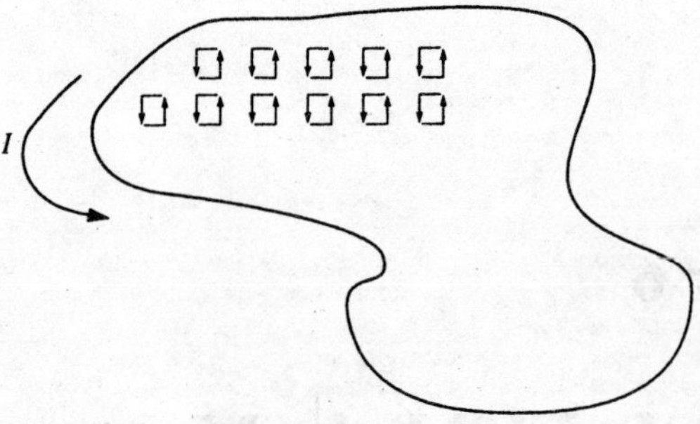

Figure 6.1

the "internal" sides canceling, as indicated in Fig. 6.1, there is no actual loss of generality in using this shape; but if you prefer to start from scratch with an arbitrary shape, see Prob. 6.2.) Center the loop at the origin, and tilt it an angle θ from the z axis towards the y axis (Fig. 6.2). Let $\mathbf{B}$ point in the z direction. The forces on the two sloping sides cancel (they tend to *stretch* the loop, but they don't *rotate* it). The forces on the "horizontal" sides are likewise equal and opposite (so the net *force* on the loop is zero), but they do generate a torque:

$$\mathbf{N} = aF \sin\theta\, \hat{\mathbf{x}}.$$

The magnitude of the force on each of these segments is

$$F = IbB,$$

and therefore

$$\mathbf{N} = IabB \sin\theta\, \hat{\mathbf{x}} = mB \sin\theta\, \hat{\mathbf{x}},$$

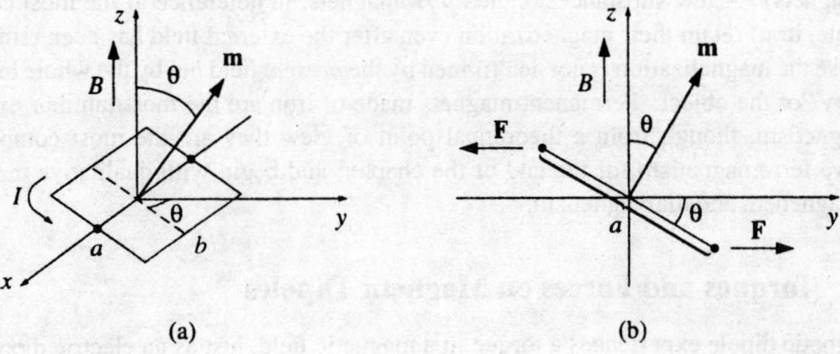

Figure 6.2

or

$$\boxed{\mathbf{N} = \mathbf{m} \times \mathbf{B},} \tag{6.1}$$

where $m = Iab$ is the magnetic dipole moment of the loop. Equation 6.1 gives the exact torque on *any* localized current distribution, in the presence of a *uniform* field; in a *nonuniform* field it is the exact torque (about the center) for a *perfect* dipole of infinitesimal size.

Notice that Eq. 6.1 is identical in form to the electrical analog, Eq. 4.4: $\mathbf{N} = \mathbf{p} \times \mathbf{E}$. In particular, the torque is again in such a direction as to line the dipole up *parallel* to the field. It is this torque that accounts for **paramagnetism**. Since every electron constitutes a magnetic dipole (picture it, if you wish, as a tiny spinning sphere of charge), you might expect paramagnetism to be a universal phenomenon. Actually, the laws of quantum mechanics (specifically, the Pauli exclusion principle) dictate that the electrons within a given atom lock together in pairs with opposing spins, and this effectively neutralizes the torque on the combination. As a result, paramagnetism normally occurs in atoms or molecules with an odd number of electrons, where the "extra" unpaired member is subject to the magnetic torque. Even here the alignment is far from complete, since random thermal collisions tend to destroy the order.

In a *uniform* field, the net *force* on a current loop is zero:

$$\mathbf{F} = I \oint (d\mathbf{l} \times \mathbf{B}) = I \left(\oint d\mathbf{l} \right) \times \mathbf{B} = 0;$$

the constant $\mathbf{B}$ comes outside the integral, and the net displacement $\oint d\mathbf{l}$ around a closed loop vanishes. In a *nonuniform* field this is no longer the case. For example, suppose a circular wire of radius R, carrying a current I, is suspended above a short solenoid in the "fringing" region (Fig. 6.3). Here $\mathbf{B}$ has a radial component, and there is a net downward force on the loop (Fig. 6.4):

$$F = 2\pi I R B \cos\theta. \tag{6.2}$$

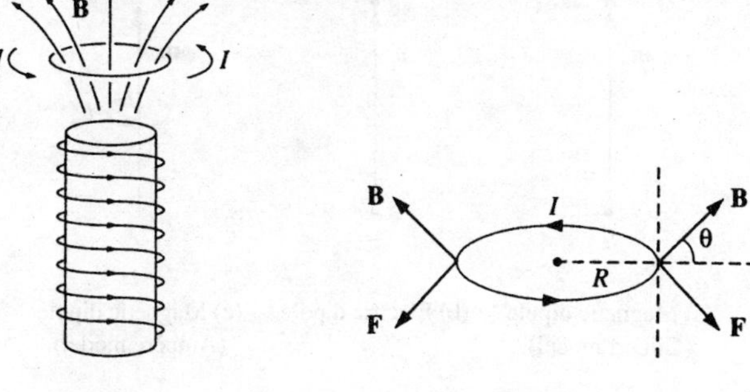

Figure 6.3 Figure 6.4

For an *infinitesimal* loop, with dipole moment **m**, in a field **B**, the force is

$$\boxed{\mathbf{F} = \nabla(\mathbf{m} \cdot \mathbf{B})} \tag{6.3}$$

(see Prob. 6.4). Once again the magnetic formula is identical to its electrical "twin," provided we agree to write the latter in the form $\mathbf{F} = \nabla(\mathbf{p} \cdot \mathbf{E})$.

If you're starting to get a sense of *déjà vu*, perhaps you will have more respect for those early physicists who thought magnetic dipoles consisted of positive and negative magnetic "charges" (north and south "poles," they called them), separated by a small distance, just like electric dipoles (Fig. 6.5(a)). They wrote down a "Coulomb's law" for the attraction and repulsion of these poles, and developed the whole of magnetostatics in exact analogy to electrostatics. It's not a bad model, for many purposes—it gives the correct field of a dipole (at least, away from the origin), the right torque on a dipole (at least, on a *stationary* dipole), and the proper force on a dipole (at least, in the absence of external currents). But it's bad physics, because *there's no such thing* as a single magnetic north pole or south pole. If you break a bar magnet in half, you don't get a north pole in one hand and a south pole in the other; you get two complete magnets. Magnetism is *not* due to magnetic monopoles, but rather to *moving electric charges*; magnetic dipoles are tiny current loops (Fig. 6.5(c)), and it's an extraordinary thing, really, that the formulas involving **m** bear any resemblance at all to the corresponding formulas for **p**. Sometimes it is easier to think in terms of the "Gilbert" model of a magnetic dipole (separated monopoles) instead of the physically correct "Ampère" model (current loop). Indeed, this picture occasionally offers a quick and clever solution to an otherwise cumbersome problem (you just copy the corresponding result from electrostatics, changing **p** to **m**, $1/\epsilon_0$ to μ_0, and **E** to **B**). But whenever the *close-up* features of the dipole come into play, the two models can yield strikingly different answers. My advice is to use the Gilbert model, if you like, to get an intuitive "feel" for a problem, but never rely on it for quantitative results.

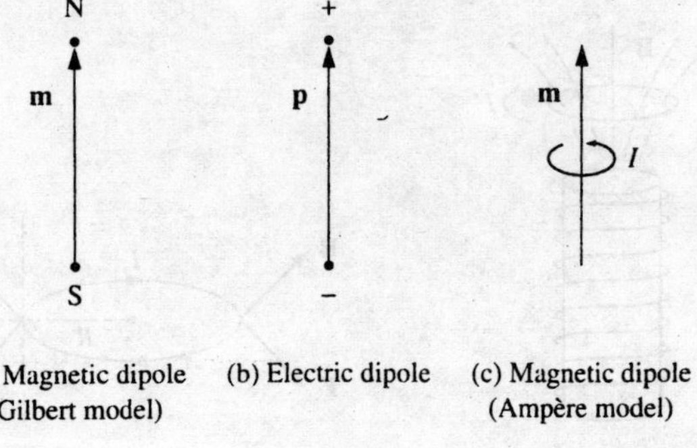

Figure 6.5

Problem 6.1 Calculate the torque exerted on the square loop shown in Fig. 6.6, due to the circular loop (assume r is much larger than a or b). If the square loop is free to rotate, what will its equilibrium orientation be?

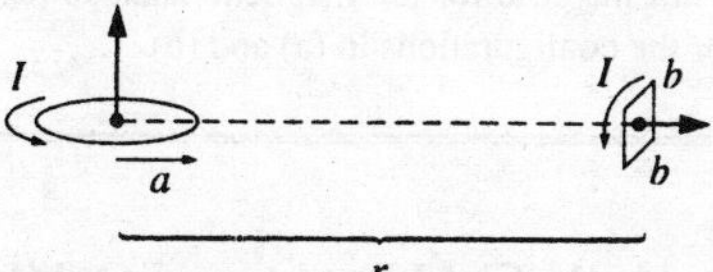

Figure 6.6

Problem 6.2 Starting from the Lorentz force law, in the form of Eq. 5.16, show that the torque on *any* steady current distribution (not just a square loop) in a uniform field $\mathbf{B}$ is $\mathbf{m} \times \mathbf{B}$.

Problem 6.3 Find the force of attraction between two magnetic dipoles, $\mathbf{m}_1$ and $\mathbf{m}_2$, oriented as shown in Fig. 6.7, a distance r apart, (a) using Eq. 6.2, and (b) using Eq. 6.3.

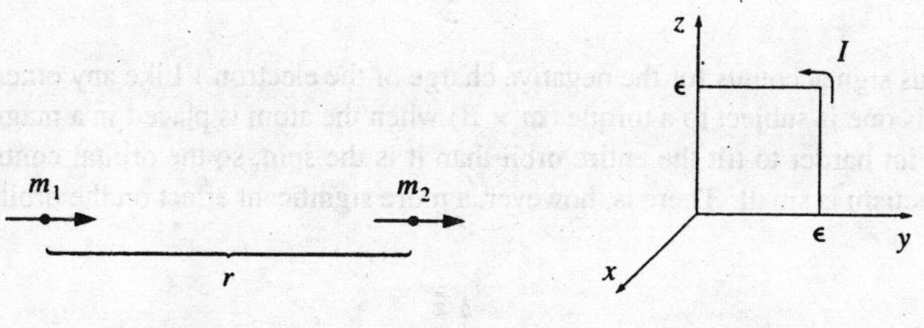

Figure 6.7

Figure 6.8

Problem 6.4 Derive Eq. 6.3. [Here's one way to do it: Assume the dipole is an infinitesimal square, of side ϵ (if it's not, chop it up into squares, and apply the argument to each one). Choose axes as shown in Fig. 6.8, and calculate $\mathbf{F} = I \int (d\mathbf{l} \times \mathbf{B})$ along each of the four sides. Expand $\mathbf{B}$ in a Taylor series—on the right side, for instance,

$$\mathbf{B} = \mathbf{B}(0, \epsilon, z) \cong \mathbf{B}(0, 0, z) + \epsilon \left.\frac{\partial \mathbf{B}}{\partial y}\right|_{(0,0,z)}$$

For a more sophisticated method, see Prob. 6.22.]

Problem 6.5 A uniform current density $\mathbf{J} = J_0\,\hat{\mathbf{z}}$ fills a slab straddling the yz plane, from $x = -a$ to $x = +a$. A magnetic dipole $\mathbf{m} = m_0\,\hat{\mathbf{x}}$ is situated at the origin.

(a) Find the force on the dipole, using Eq. 6.3.

(b) Do the same for a dipole pointing in the y direction: $\mathbf{m} = m_0\hat{\mathbf{y}}$.

(c) In the *electrostatic* case the expressions $\mathbf{F} = \nabla(\mathbf{p} \cdot \mathbf{E})$ and $\mathbf{F} = (\mathbf{p} \cdot \nabla)\mathbf{E}$ are equivalent (prove it), but this is *not* the case for the magnetic analogs (explain why). As an example, calculate $(\mathbf{m} \cdot \nabla)\mathbf{B}$ for the configurations in (a) and (b).

6.1.3 Effect of a Magnetic Field on Atomic Orbits

Electrons not only *spin*; they also *revolve* around the nucleus—for simplicity, let's assume the orbit is a circle of radius R (Fig. 6.9). Although technically this orbital motion does not constitute a steady current, in practice the period $T = 2\pi R/v$ is so short that unless you blink awfully fast, it's going to *look* like a steady current:

$$I = \frac{e}{T} = \frac{ev}{2\pi R}.$$

Accordingly, the orbital dipole moment $(I\pi R^2)$ is

$$\mathbf{m} = -\frac{1}{2}evR\,\hat{\mathbf{z}}. \tag{6.4}$$

(The minus sign accounts for the negative charge of the electron.) Like any other magnetic dipole, this one is subject to a torque $(\mathbf{m} \times \mathbf{B})$ when the atom is placed in a magnetic field. But it's a lot harder to tilt the entire orbit than it is the spin, so the orbital contribution to paramagnetism is small. There is, however, a more significant effect on the orbital motion:

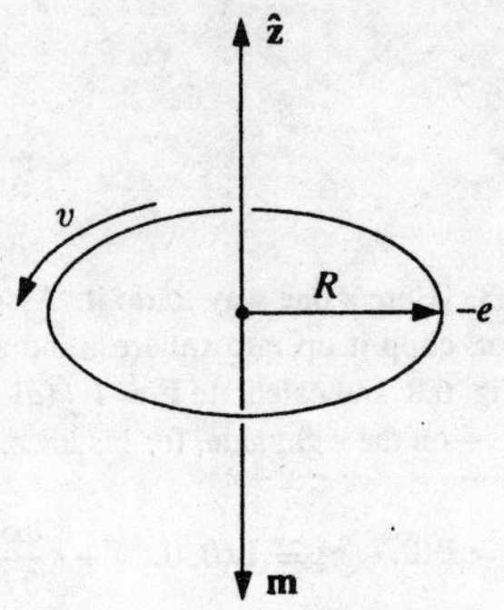

Figure 6.9

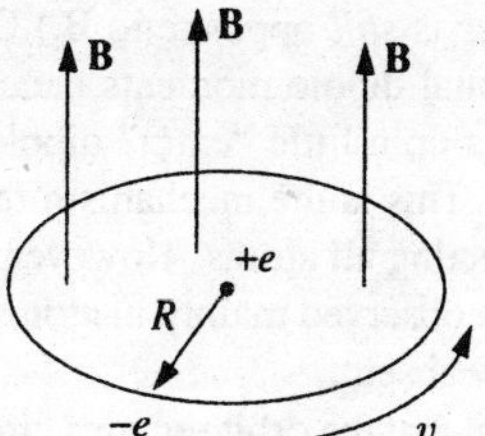

Figure 6.10

The electron *speeds up* or *slows down*, depending on the orientation of **B**. For whereas the centripetal acceleration v^2/R is ordinarily sustained by electrical forces alone,[1]

$$\frac{1}{4\pi\epsilon_0}\frac{e^2}{R^2} = m_e\frac{v^2}{R}, \tag{6.5}$$

in the presence of a magnetic field there is an additional force, $-e(\mathbf{v}\times\mathbf{B})$. For the sake of argument, let's say that **B** is perpendicular to the plane of the orbit, as shown in Fig. 6.10; then

$$\frac{1}{4\pi\epsilon_0}\frac{e^2}{R^2} + e\bar{v}B = m_e\frac{\bar{v}^2}{R}. \tag{6.6}$$

Under these conditions, the new speed $\bar{v}$ is *greater* than v:

$$e\bar{v}B = \frac{m_e}{R}(\bar{v}^2 - v^2) = \frac{m_e}{R}(\bar{v}+v)(\bar{v}-v),$$

or, assuming the change $\Delta v = \bar{v} - v$ is small,

$$\Delta v = \frac{eRB}{2m_e}. \tag{6.7}$$

When **B** is turned on, then, the electron speeds up.[2]

A change in orbital speed means a change in the dipole moment (6.4):

$$\Delta\mathbf{m} = -\frac{1}{2}e(\Delta v)R\,\hat{\mathbf{z}} = -\frac{e^2R^2}{4m_e}\mathbf{B}. \tag{6.8}$$

Notice that *the change in* **m** *is opposite to the direction of* **B**. (An electron circling the other way would have a dipole moment pointing *upward*, but such an orbit would be *slowed*

[1] To avoid confusion with the magnetic dipole moment m, I'll write the electron mass with subscript: m_e.

[2] I said earlier (Eq. 5.11) that magnetic fields do no work, and are incapable of speeding a particle up. I stand by that. However, as we shall see in Chapter 7, a *changing* magnetic field induces an *electric* field, and it is the latter that accelerates the electrons in this instance.

down by the field, so the *change* is *still* opposite to **B**.) Ordinarily, the electron orbits are randomly oriented, and the orbital dipole moments cancel out. But in the presence of a magnetic field, each atom picks up a little "extra" dipole moment, and these increments are all *antiparallel* to the field. This is the mechanism responsible for **diamagnetism**. It is a universal phenomenon, affecting all atoms. However, it is typically much weaker than paramagnetism, and is therefore observed mainly in atoms with *even* numbers of electrons, where paramagnetism is usually absent.

In deriving Eq. 6.8 I assumed that the orbit remains circular, with its original radius R. I cannot offer a justification for this at the present stage. If the atom is stationary while the field is turned on, then my assumption can be proved—this is not magneto*statics*, however, and the details will have to await Chapter 7 (see Prob. 7.49). If the atom is moved into the field, the situation is enormously more complicated. But never mind—I'm only trying to give you a qualitative account of diamagnetism. Assume, if you prefer, that the velocity remains the same while the *radius* changes—the formula (6.8) is altered (by a factor of 2), but the *conclusion* is unaffected. The truth is that this classical model is fundamentally flawed (diamagnetism is really a *quantum* phenomenon), so there's not much point in refining the details.[3] What *is* important is the *empirical* fact that in diamagnetic materials the induced dipole moments point *opposite* to the magnetic field.

6.1.4 Magnetization

In the presence of a magnetic field, matter becomes *magnetized*; that is, upon microscopic examination it will be found to contain many tiny dipoles, with a net alignment along some direction. We have discussed two mechanisms that account for this magnetic polarization: (1) paramagnetism (the dipoles associated with the spins of unpaired electrons experience a torque tending to line them up parallel to the field) and (2) diamagnetism (the orbital speed of the electrons is altered in such a way as to change the orbital dipole moment in a direction opposite to the field). Whatever the *cause*, we describe the state of magnetic polarization by the vector quantity

$$\mathbf{M} \equiv \textit{magnetic dipole moment per unit volume.} \tag{6.9}$$

M is called the **magnetization**; it plays a role analogous to the polarization **P** in electrostatics. In the following section, we will not worry about how the magnetization *got* there—it could be paramagnetism, diamagnetism, or even ferromagnetism—we shall take **M** as *given*, and calculate the field this magnetization itself produces.

Incidentally, it may have surprised you to learn that materials other than the famous ferromagnetic trio (iron, nickel, and cobalt) are affected by a magnetic field *at all*. You cannot, of course, pick up a piece of wood or aluminum with a magnet. The reason is that diamagnetism and paramagnetism are extremely weak: It takes a delicate experiment and a powerful magnet to detect them at all. If you were to suspend a piece of paramagnetic

[3] S. L. O'Dell and R. K. P. Zia, *Am. J. Phys.* **54**, 32, (1986); R. Peierls, *Surprises in Theoretical Physics*, Section 4.3 (Princeton, N.J.: Princeton University Press, 1979); R. P. Feynman, R. B. Leighton, and M. Sands, *The Feynman Lectures on Physics*, Vol. 2, Sec. 34–36 (New York: Addison-Wesley, 1966).

material above a solenoid, as in Fig. 6.3, the induced magnetization would be upward, and hence the force downward. By contrast, the magnetization of a diamagnetic object would be downward and the force upward. In general, when a sample is placed in a region of nonuniform field, the *paramagnet is attracted into the field*, whereas the *diamagnet is repelled away*. But the actual forces are pitifully weak—in a typical experimental arrangement the force on a comparable sample of iron would be 10^4 or 10^5 times as great. That's why it was reasonable for us to calculate the field inside a piece of copper wire, say, in Chapter 5, without worrying about the effects of magnetization.

Problem 6.6 Of the following materials, which would you expect to be paramagnetic and which diamagnetic? Aluminum, copper, copper chloride ($CuCl_2$), carbon, lead, nitrogen (N_2), salt (NaCl), sodium, sulfur, water. (Actually, copper is slightly *dia*magnetic; otherwise they're all what you'd expect.)

6.2 The Field of a Magnetized Object

6.2.1 Bound Currents

Suppose we have a piece of magnetized material; the magnetic dipole moment per unit volume, **M**, is given. What field does this object produce? Well, the vector potential of a single dipole **m** is given by Eq. 5.83:

$$\mathbf{A}(\mathbf{r}) = \frac{\mu_0}{4\pi}\frac{\mathbf{m}\times\hat{\boldsymbol{\imath}}}{\imath^2}. \tag{6.10}$$

In the magnetized object, each volume element $d\tau'$ carries a dipole moment $\mathbf{M}\,d\tau'$, so the total vector potential is (Fig. 6.11)

$$\mathbf{A}(\mathbf{r}) = \frac{\mu_0}{4\pi}\int\frac{\mathbf{M}(\mathbf{r}')\times\hat{\boldsymbol{\imath}}}{\imath^2}\,d\tau'. \tag{6.11}$$

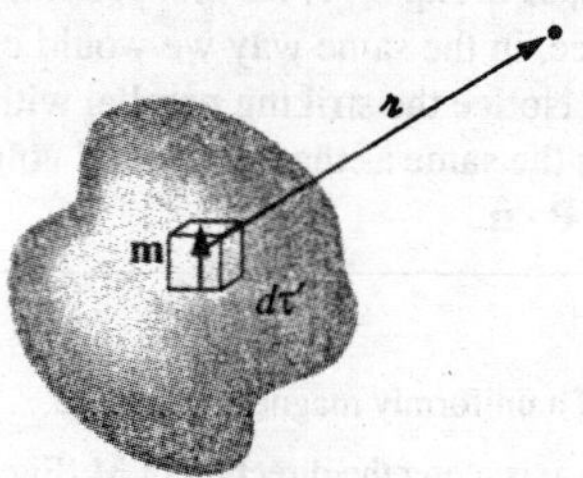

Figure 6.11

That *does* it, in principle. But as in the electrical case (Sect. 4.2.1), the integral can be cast in a more illuminating form by exploiting the identity

$$\nabla' \frac{1}{\mathscr{r}} = \frac{\hat{\mathscr{r}}}{\mathscr{r}^2}.$$

With this,

$$\mathbf{A}(\mathbf{r}) = \frac{\mu_0}{4\pi} \int \left[\mathbf{M}(\mathbf{r}') \times \left(\nabla' \frac{1}{\mathscr{r}} \right) \right] d\tau'.$$

Integrating by parts, using product rule 7, gives

$$\mathbf{A}(\mathbf{r}) = \frac{\mu_0}{4\pi} \left\{ \int \frac{1}{\mathscr{r}} [\nabla' \times \mathbf{M}(\mathbf{r}')]\, d\tau' - \int \nabla' \times \left[\frac{\mathbf{M}(\mathbf{r}')}{\mathscr{r}} \right] d\tau' \right\}.$$

Problem 1.60(b) invites us to express the latter as a surface integral,

$$\mathbf{A}(\mathbf{r}) = \frac{\mu_0}{4\pi} \int \frac{1}{\mathscr{r}} [\nabla' \times \mathbf{M}(\mathbf{r}')]\, d\tau' + \frac{\mu_0}{4\pi} \oint \frac{1}{\mathscr{r}} [\mathbf{M}(\mathbf{r}') \times d\mathbf{a}']. \tag{6.12}$$

The first term looks just like the potential of a *volume* current,

$$\boxed{\mathbf{J}_b = \nabla \times \mathbf{M},} \tag{6.13}$$

while the second looks like the potential of a surface current,

$$\boxed{\mathbf{K}_b = \mathbf{M} \times \hat{\mathbf{n}},} \tag{6.14}$$

where $\hat{\mathbf{n}}$ is the normal unit vector. With these definitions,

$$\mathbf{A}(\mathbf{r}) = \frac{\mu_0}{4\pi} \int_{\mathcal{V}} \frac{\mathbf{J}_b(\mathbf{r}')}{\mathscr{r}}\, d\tau' + \frac{\mu_0}{4\pi} \oint_{\mathcal{S}} \frac{\mathbf{K}_b(\mathbf{r}')}{\mathscr{r}}\, da'. \tag{6.15}$$

What this means is that the potential (and hence also the field) of a magnetized object is the same as would be produced by a volume current $\mathbf{J}_b = \nabla \times \mathbf{M}$ throughout the material, plus a surface current $\mathbf{K}_b = \mathbf{M} \times \hat{\mathbf{n}}$, on the boundary. Instead of integrating the contributions of all the infinitesimal dipoles, as in Eq. 6.11, we first determine these **bound currents**, and then find the field *they* produce, in the same way we would calculate the field of any other volume and surface currents. Notice the striking parallel with the electrical case: there the field of a polarized object was the same as that of a bound volume charge $\rho_b = -\nabla \cdot \mathbf{P}$ plus a bound surface charge $\sigma_b = \mathbf{P} \cdot \hat{\mathbf{n}}$.

Example 6.1

Find the magnetic field of a uniformly magnetized sphere.

Solution: Choosing the z axis along the direction of $\mathbf{M}$ (Fig. 6.12), we have

$$\mathbf{J}_b = \nabla \times \mathbf{M} = 0, \quad \mathbf{K}_b = \mathbf{M} \times \hat{\mathbf{n}} = M \sin\theta\, \hat{\boldsymbol{\phi}}.$$

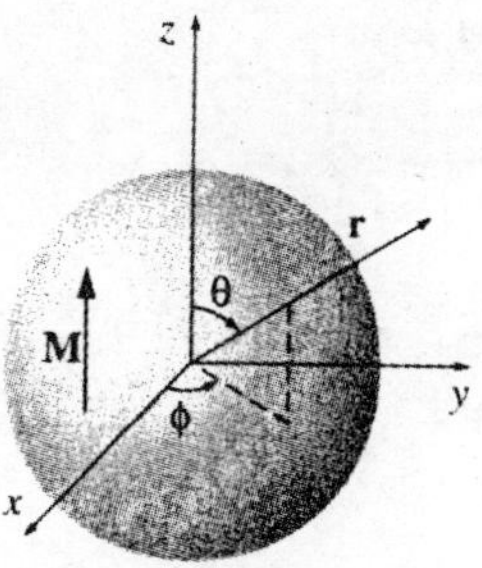

Figure 6.12

Now, a rotating spherical shell, of uniform surface charge σ, corresponds to a surface current density

$$\mathbf{K} = \sigma \mathbf{v} = \sigma \omega R \sin\theta \,\hat{\boldsymbol{\phi}}.$$

It follows, therefore, that the field of a uniformly magnetized sphere is identical to the field of a spinning spherical shell, with the identification $\sigma R\omega \rightarrow \mathbf{M}$. Referring back to Ex. 5.11, I conclude that

$$\mathbf{B} = \frac{2}{3}\mu_0 \mathbf{M}, \tag{6.16}$$

inside the sphere, whereas the field outside is the same as that of a pure dipole,

$$\mathbf{m} = \frac{4}{3}\pi R^3 \mathbf{M}.$$

Notice that the internal field is *uniform*, like the *electric* field inside a uniformly *polarized* sphere (Eq. 4.14), although the actual *formulas* for the two cases are curiously different ($\frac{2}{3}$ in place of $-\frac{1}{3}$). The external fields are also analogous: pure dipole in both instances.

Problem 6.7 An infinitely long circular cylinder carries a uniform magnetization **M** parallel to its axis. Find the magnetic field (due to **M**) inside and outside the cylinder.

Problem 6.8 A long circular cylinder of radius R carries a magnetization $\mathbf{M} = ks^2\,\hat{\boldsymbol{\phi}}$, where k is a constant, s is the distance from the axis, and $\hat{\boldsymbol{\phi}}$ is the usual azimuthal unit vector (Fig. 6.13). Find the magnetic field due to **M**, for points inside and outside the cylinder.

Problem 6.9 A short circular cylinder of radius a and length L carries a "frozen-in" uniform magnetization **M** parallel to its axis. Find the bound current, and sketch the magnetic field of the cylinder. (Make three sketches: one for $L \gg a$, one for $L \ll a$, and one for $L \approx a$.) Compare this **bar magnet** with the bar electret of Prob. 4.11.

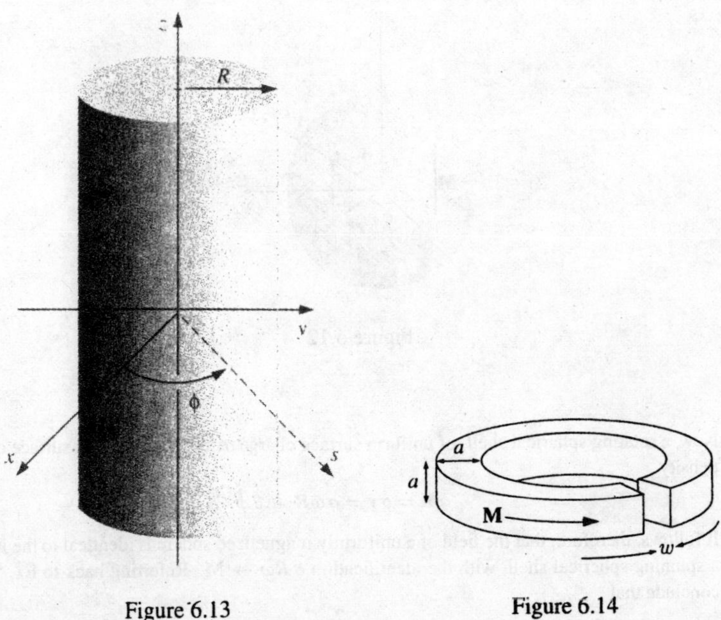

Figure 6.13

Figure 6.14

Problem 6.10 An iron rod of length L and square cross section (side a), is given a uniform longitudinal magnetization $\mathbf{M}$, and then bent around into a circle with a narrow gap (width w), as shown in Fig. 6.14. Find the magnetic field at the center of the gap, assuming $w \ll a \ll L$. [*Hint:* treat it as the superposition of a *complete* torus plus a square loop with reversed current.]

6.2.2 Physical Interpretation of Bound Currents

In the last section we found that the field of a magnetized object is identical to the field that would be produced by a certain distribution of "bound" currents, $\mathbf{J}_b$ and $\mathbf{K}_b$. I want to show you how these bound currents arise physically. This will be a *heuristic* argument—the *rigorous* derivation has already been given. Figure 6.15 depicts a thin slab of uniformly magnetized material, with the dipoles represented by tiny current loops. Notice that all the "internal" currents cancel: every time there is one going to the right, a contiguous one is going to the left. However, at the edge there is *no adjacent loop to do the canceling*. The whole thing, then, is equivalent to a single ribbon of current I flowing around the boundary (Fig. 6.16).

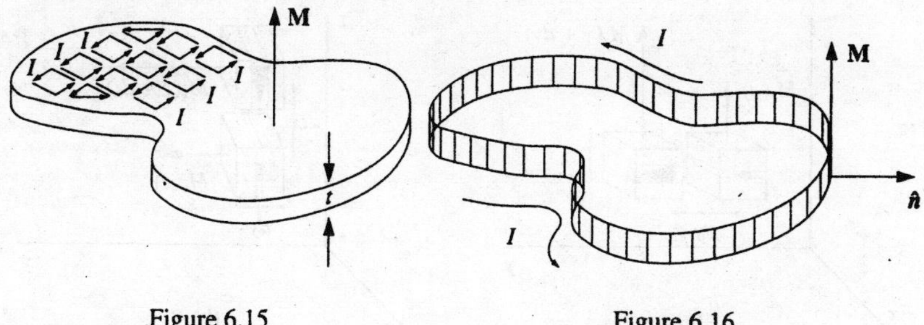

Figure 6.15 Figure 6.16

What *is* this current, in terms of **M**? Say that each of the tiny loops has area a and thickness t (Fig. 6.17). In terms of the magnetization M, its dipole moment is $m = Mat$. In terms of the circulating current I, however, $m = Ia$. Therefore $I = Mt$, so the surface current is $K_b = I/t = M$. Using the outward-drawn unit vector $\hat{\mathbf{n}}$ (Fig. 6.16), the direction of $\mathbf{K}_b$ is conveniently indicated by the cross product:

$$\mathbf{K}_b = \mathbf{M} \times \hat{\mathbf{n}}.$$

(This expression also records the fact that there is *no* current on the top or bottom surface of the slab; here **M** is parallel to $\hat{\mathbf{n}}$, so the cross product vanishes.)

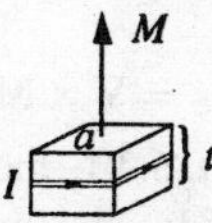

Figure 6.17

This bound surface current is exactly what we obtained in Sect. 6.2.1. It is a peculiar *kind* of current, in the sense that no single charge makes the whole trip—on the contrary, each charge moves only in a tiny little loop within a single atom. Nevertheless, the net effect is a macroscopic current flowing over the surface of the magnetized object. We call it a "bound" current to remind ourselves that every charge is attached to a particular atom, but it's a perfectly genuine current, and it produces a magnetic field in the same way any other current does.

When the magnetization is *non*uniform, the internal currents no longer cancel. Figure 6.18a shows two adjacent chunks of magnetized material, with a larger arrow on the one to the right suggesting greater magnetization at that point. On the surface where they join there is a net current in the x-direction, given by

$$I_x = [M_z(y + dy) - M_z(y)]\, dz = \frac{\partial M_z}{\partial y}\, dy\, dz.$$

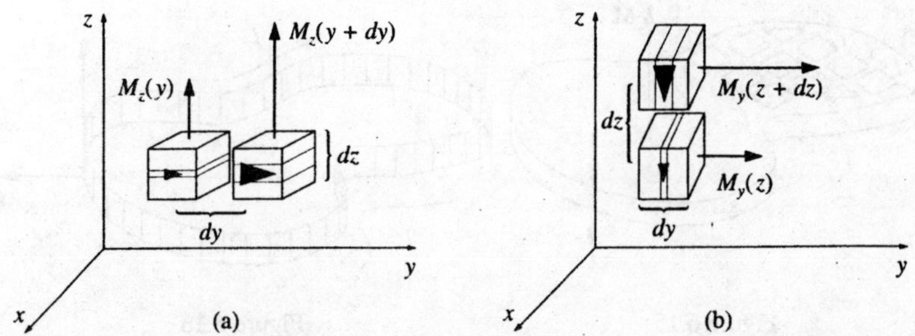

Figure 6.18

The corresponding volume current density is therefore

$$(J_b)_x = \frac{\partial M_z}{\partial y}.$$

By the same token, a nonuniform magnetization in the y-direction would contribute an amount $-\partial M_y/\partial z$ (Fig. 6.18b), so

$$(J_b)_x = \frac{\partial M_z}{\partial y} - \frac{\partial M_y}{\partial z}.$$

In general, then,

$$\mathbf{J}_b = \nabla \times \mathbf{M},$$

consistent, again, with the result of Sect. 6.2.1. Incidentally, like any other steady current, $\mathbf{J}_b$ should obey the conservation law 5.31:

$$\nabla \cdot \mathbf{J}_b = 0.$$

Does it? *Yes*, for the divergence of a curl is *always* zero.

6.2.3 The Magnetic Field Inside Matter

Like the electric field, the actual *microscopic* magnetic field inside matter fluctuates wildly from point to point and instant to instant. When we speak of "the" magnetic field in matter, we mean the *macroscopic* field: the average over regions large enough to contain many atoms. (The magnetization **M** is "smoothed out" in the same sense.) It is this macroscopic field one obtains when the methods of Sect. 6.2.1 are applied to points inside magnetized material, as you can prove for yourself in the following problem.

Problem 6.11 In Sect, 6.2.1, we began with the potential of a *perfect* dipole (Eq. 6.10), whereas in *fact* we are dealing with *physical* dipoles. Show, by the method of Sect. 4.2.3, that we nonetheless get the correct macroscopic field.

6.3 The Auxiliary Field H

6.3.1 Ampère's law in Magnetized Materials

In Sect. 6.2 we found that the effect of magnetization is to establish bound currents $\mathbf{J}_b = \nabla \times \mathbf{M}$ within the material and $\mathbf{K}_b = \mathbf{M} \times \hat{\mathbf{n}}$ on the surface. The field due to magnetization of the medium is just the field produced by these bound currents. We are now ready to put everything together: the field attributable to bound currents, plus the field due to everything else—which I shall call the **free current**. The free current might flow through wires imbedded in the magnetized substance or, if the latter is a conductor, through the material itself. In any event, the total current can be written as

$$\mathbf{J} = \mathbf{J}_b + \mathbf{J}_f. \tag{6.17}$$

There is no new physics in Eq. 6.17; it is simply a *convenience* to separate the current into these two parts because they *got* there by quite different means: the free current is there because somebody hooked up a wire to a battery—it involves actual transport of charge; the bound current is there because of magnetization—it results from the conspiracy of many aligned atomic dipoles.

In view of Eqs. 6.13 and 6.17, Ampère's law can be written

$$\frac{1}{\mu_0}(\nabla \times \mathbf{B}) = \mathbf{J} = \mathbf{J}_f + \mathbf{J}_b = \mathbf{J}_f + (\nabla \times \mathbf{M}),$$

or, collecting together the two curls:

$$\nabla \times \left(\frac{1}{\mu_0}\mathbf{B} - \mathbf{M}\right) = \mathbf{J}_f.$$

The quantity in parentheses is designated by the letter **H**:

$$\boxed{\mathbf{H} \equiv \frac{1}{\mu_0}\mathbf{B} - \mathbf{M}.} \tag{6.18}$$

In terms of **H**, then, Ampère's law reads

$$\boxed{\nabla \times \mathbf{H} = \mathbf{J}_f,} \tag{6.19}$$

or, in integral form,

$$\oint \mathbf{H} \cdot d\mathbf{l} = I_{f_{\text{enc}}}, \tag{6.20}$$

where $I_{f_{\text{enc}}}$ is the total *free* current passing through the Amperian loop.

H plays a role in magnetostatics analogous to **D** in electrostatics: Just as **D** allowed us to write *Gauss's* law in terms of the free *charge* alone, **H** permits us to express *Ampère's* law in terms of the free *current* alone—and free current is what we control directly. Bound current, like bound charge, comes along for the ride—the material gets magnetized, and this results in bound currents; we cannot turn them on or off independently, as we can free currents. In applying Eq. 6.20 all we need to worry about is the *free* current, which we know about because we *put* it there. In particular, when symmetry permits, we can calculate **H** immediately from Eq. 6.20 by the usual Ampère's law methods. (For example, Probs. 6.7 and 6.8 can be done in one line by noting that $\mathbf{H} = 0$.)

Example 6.2

A long copper rod of radius R carries a uniformly distributed (free) current I (Fig. 6.19). Find H inside and outside the rod.

Solution: Copper is weakly diamagnetic, so the dipoles will line up *opposite* to the field. This results in a bound current running *antiparallel* to I within the wire and *parallel* to I along the surface (see Fig. 6.20). Just how *great* these bound currents will be we are not yet in a position

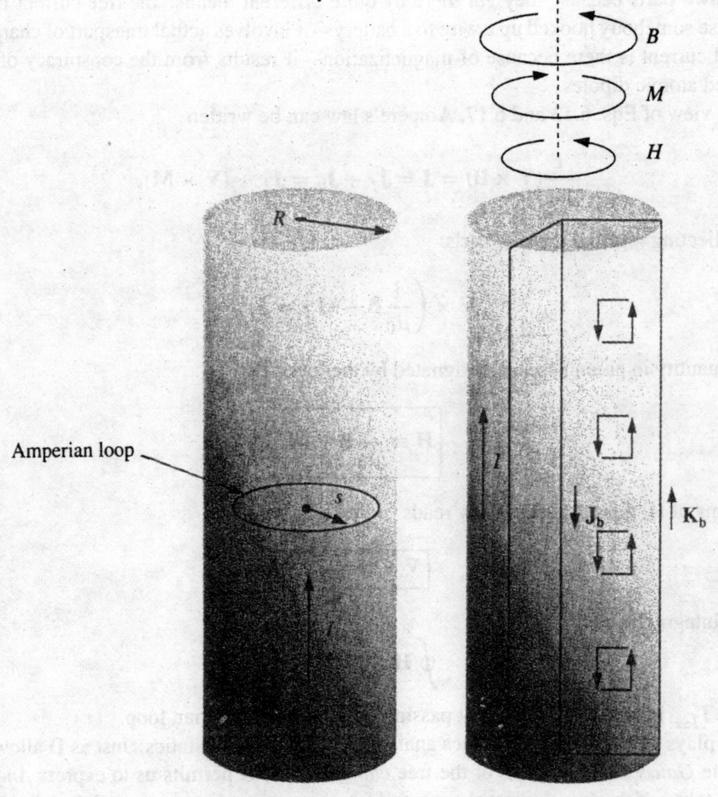

Figure 6.19 Figure 6.20

to say—but in order to calculate **H** it is sufficient to realize that all the currents are longitudinal, so **B**, **M**, and therefore also **H**, are circumferential. Applying Eq. 6.20 to an Amperian loop of radius $s < R$,

$$H(2\pi s) = I_{f_{\text{enc}}} = I\frac{\pi s^2}{\pi R^2},$$

so

$$\mathbf{H} = \frac{I}{2\pi R^2} s\,\hat{\boldsymbol{\phi}} \quad (s \leq R), \tag{6.21}$$

within the wire. Meanwhile, outside the wire

$$\mathbf{H} = \frac{I}{2\pi s}\hat{\boldsymbol{\phi}} \quad (s \geq R). \tag{6.22}$$

In the latter region (as always, in empty space) $\mathbf{M} = 0$, so

$$\mathbf{B} = \mu_0 \mathbf{H} = \frac{\mu_0 I}{2\pi s}\hat{\boldsymbol{\phi}} \quad (s \geq R),$$

the same as for a *non*magnetized wire (Ex. 5.7). ***Inside*** the wire **B** cannot be determined at this stage, since we have no way of knowing **M** (though in practice the magnetization in copper is so slight that for most purposes we can ignore it altogether).

As it turns out, **H** is a more useful quantity than **D**. In the laboratory you will frequently hear people talking about **H** (more often even than **B**), but you will never hear anyone speak of **D** (only **E**). The reason is this: To build an electromagnet you run a certain (free) current through a coil. The *current* is the thing you read on the dial, and this determines **H** (or at any rate, the line integral of **H**); **B** depends on the specific materials you used and even, if iron is present, on the history of your magnet. On the other hand, if you want to set up an *electric* field, you do *not* plaster a known free charge on the plates of a parallel plate capacitor; rather, you connect them to a battery of known *voltage*. It's the *potential difference* you read on your dial, and that determines **E** (or at any rate, the line integral of **E**); **D** depends on the details of the dielectric you're using. If it were easy to measure charge, and hard to measure potential, then you'd find experimentalists talking about **D** instead of **E**. So the relative familiarity of **H**, as contrasted with **D**, derives from purely practical considerations; theoretically, they're all on equal footing.

Many authors call **H**, not **B**, the "magnetic field." Then they have to invent a new word for **B**: the "flux density," or magnetic "induction" (an absurd choice, since that term already has at least two other meanings in electrodynamics). Anyway, **B** is indisputably the fundamental quantity, so I shall continue to call it the "magnetic field," as everyone does in the spoken language. **H** has no sensible name: just call it "**H**".[4]

---

[4]For those who disagree, I quote A. Sommerfeld's *Electrodynamics* (New York: Academic Press, 1952), p. 45: "The unhappy term 'magnetic field' for **H** should be avoided as far as possible. It seems to us that this term has led into error none less than Maxwell himself . . . "

Problem 6.12 An infinitely long cylinder, of radius R, carries a "frozen-in" magnetization, parallel to the axis,

$$\mathbf{M} = ks\,\hat{\mathbf{z}},$$

where k is a constant and s is the distance from the axis; there is no free current anywhere. Find the magnetic field inside and outside the cylinder by two different methods:

(a) As in Sect. 6.2, locate all the bound currents, and calculate the field they produce.

(b) Use Ampère's law (in the form of Eq. 6.20) to find $\mathbf{H}$, and then get $\mathbf{B}$ from Eq. 6.18. (Notice that the second method is much faster, and avoids any explicit reference to the bound currents.)

Problem 6.13 Suppose the field inside a large piece of magnetic material is $\mathbf{B}_0$, so that $\mathbf{H}_0 = (1/\mu_0)\mathbf{B}_0 - \mathbf{M}$.

(a) Now a small spherical cavity is hollowed out of the material (Fig. 6.21). Find the field at the center of the cavity, in terms of $\mathbf{B}_0$ and $\mathbf{M}$. Also find $\mathbf{H}$ at the center of the cavity, in terms of $\mathbf{H}_0$ and $\mathbf{M}$.

(b) Do the same for a long needle-shaped cavity running parallel to $\mathbf{M}$.

(c) Do the same for a thin wafer-shaped cavity perpendicular to $\mathbf{M}$.

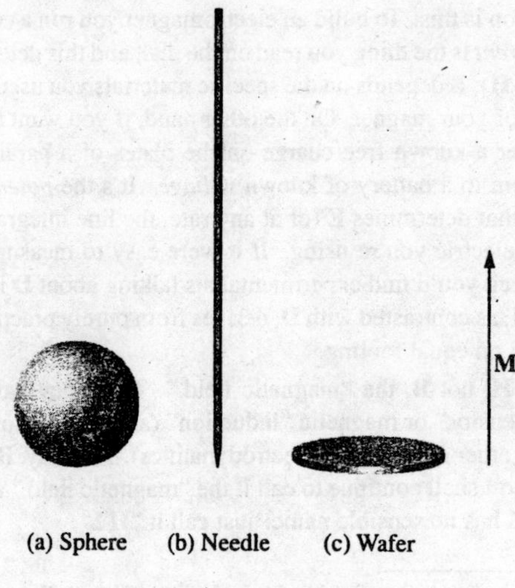

(a) Sphere (b) Needle (c) Wafer

Figure 6.21

Assume the cavities are small enough so that $\mathbf{M}$, $\mathbf{B}_0$, and $\mathbf{H}_0$ are essentially constant. Compare Prob. 4.16. [*Hint:* Carving out a cavity is the same as superimposing an object of the same shape but opposite magnetization.]

6.3.2 A Deceptive Parallel

Equation 6.19 looks just like Ampère's original law (5.54), only the *total* current is replaced by the *free* current, and $\mathbf{B}$ is replaced by $\mu_0\mathbf{H}$. As in the case of $\mathbf{D}$, however, I must warn you against reading too much into this correspondence. It does *not* say that $\mu_0\mathbf{H}$ is "just like $\mathbf{B}$, only its source is $\mathbf{J}_f$ instead of $\mathbf{J}$." For the curl alone does not determine a vector field—you must know the divergence as well. And whereas $\nabla \cdot \mathbf{B} = 0$, the divergence of $\mathbf{H}$ is *not*, in general, zero. In fact, from Eq. 6.18

$$\nabla \cdot \mathbf{H} = -\nabla \cdot \mathbf{M}. \tag{6.23}$$

Only when the divergence of $\mathbf{M}$ vanishes is the parallel between $\mathbf{B}$ and $\mu_0\mathbf{H}$ faithful.

If you think I'm being pedantic, consider the example of the bar magnet—a short cylinder of iron that carries a permanent uniform magnetization $\mathbf{M}$ parallel to its axis. (See Probs. 6.9 and 6.14.) In this case there is no free current anywhere, and a naïve application of Eq. 6.20 might lead you to suppose that $\mathbf{H} = 0$, and hence that $\mathbf{B} = \mu_0\mathbf{M}$ inside the magnet and $\mathbf{B} = 0$ outside, which is nonsense. It is quite true that the *curl* of $\mathbf{H}$ vanishes everywhere, but the divergence does not. (Can you see where $\nabla \cdot \mathbf{M} = 0$?) *Advice*: When you are asked to find $\mathbf{B}$ or $\mathbf{H}$ in a problem involving magnetic materials, first look for symmetry. If the problem exhibits cylindrical, plane, solenoidal, or toroidal symmetry, then you can get $\mathbf{H}$ directly from Eq. 6.20 by the usual Ampère's law methods. (Evidently, in such cases $\nabla \cdot \mathbf{M}$ is automatically zero, since the free current alone determines the answer.) If the requisite symmetry is absent, you'll have to think of another approach, and in particular you must *not* assume that $\mathbf{H}$ is zero just because you see no free current.

6.3.3 Boundary Conditions

The magnetostatic boundary conditions of Sect. 5.4.2 can be rewritten in terms of $\mathbf{H}$ and the *free* current. From Eq. 6.23 it follows that

$$H^{\perp}_{\text{above}} - H^{\perp}_{\text{below}} = -(M^{\perp}_{\text{above}} - M^{\perp}_{\text{below}}), \tag{6.24}$$

while Eq. 6.19 says

$$\mathbf{H}^{\parallel}_{\text{above}} - \mathbf{H}^{\parallel}_{\text{below}} = \mathbf{K}_f \times \hat{\mathbf{n}}. \tag{6.25}$$

In the presence of materials these are sometimes more useful than the corresponding boundary conditions on **B** (Eqs. 5.72 and 5.73):

$$B^{\perp}_{\text{above}} - B^{\perp}_{\text{below}} = 0, \tag{6.26}$$

and

$$\mathbf{B}^{\parallel}_{\text{above}} - \mathbf{B}^{\parallel}_{\text{below}} = \mu_0(\mathbf{K} \times \hat{\mathbf{n}}). \tag{6.27}$$

You might want to check them, for Ex. 6.2 or Prob. 6.14.

Problem 6.14 For the bar magnet of Prob. 6.9, make careful sketches of **M**, **B**, and **H**, assuming L is about $2a$. Compare Prob. 4.17.

Problem 6.15 If $\mathbf{J}_f = 0$ everywhere, the curl of **H** vanishes (Eq. 6.19), and we can express **H** as the gradient of a scalar potential W:

$$\mathbf{H} = -\nabla W.$$

According to Eq. 6.23, then,

$$\nabla^2 W = (\nabla \cdot \mathbf{M}),$$

so W obeys Poisson's equation, with $\nabla \cdot \mathbf{M}$ as the "source." This opens up all the machinery of Chapter 3. As an example, find the field inside a uniformly magnetized sphere (Ex. 6.1) by separation of variables. [*Hint:* $\nabla \cdot \mathbf{M} = 0$ everywhere except at the surface ($r = R$), so W satisfies Laplace's equation in the regions $r < R$ and $r > R$; use Eq. 3.65, and from Eq. 6.24 figure out the appropriate boundary condition on W.]

6.4 Linear and Nonlinear Media

6.4.1 Magnetic Susceptibility and Permeability

In paramagnetic and diamagnetic materials, the magnetization is sustained by the field; when **B** is removed, **M** disappears. In fact, for most substances the magnetization is *proportional* to the field, provided the field is not too strong. For notational consistency with the electrical case (Eq. 4.30), I *should* express the proportionality thus:

$$\mathbf{M} = \frac{1}{\mu_0}\chi_m \mathbf{B} \quad \text{(incorrect!)}. \tag{6.28}$$

But custom dictates that it be written in terms of **H**, instead of **B**:

$$\boxed{\mathbf{M} = \chi_m \mathbf{H}.} \tag{6.29}$$

The constant of proportionality χ_m is called the **magnetic susceptibility**; it is a dimensionless quantity that varies from one substance to another—positive for paramagnets and negative for diamagnets. Typical values are around 10^{-5} (see Table 6.1).

Material	Susceptibility	Material	Susceptibility
Diamagnetic:		*Paramagnetic:*	
Bismuth	-1.6×10^{-4}	Oxygen	1.9×10^{-6}
Gold	-3.4×10^{-5}	Sodium	8.5×10^{-6}
Silver	-2.4×10^{-5}	Aluminum	2.1×10^{-5}
Copper	-9.7×10^{-6}	Tungsten	7.8×10^{-5}
Water	-9.0×10^{-6}	Platinum	2.8×10^{-4}
Carbon Dioxide	-1.2×10^{-8}	Liquid Oxygen (−200° C)	3.9×10^{-3}
Hydrogen	-2.2×10^{-9}	Gadolinium	4.8×10^{-1}

Table 6.1 Magnetic Susceptibilities (unless otherwise specified, values are for 1 atm, 20° C). *Source: Handbook of Chemistry and Physics,* 67th ed. (Boca Raton: CRC Press, Inc., 1986).

Materials that obey Eq. 6.29 are called **linear media**. In view of Eq. 6.18,

$$\mathbf{B} = \mu_0(\mathbf{H} + \mathbf{M}) = \mu_0(1 + \chi_m)\mathbf{H}, \tag{6.30}$$

for linear media. Thus **B** is *also* proportional to **H**:[5]

$$\mathbf{B} = \mu\mathbf{H}, \tag{6.31}$$

where

$$\mu \equiv \mu_0(1 + \chi_m). \tag{6.32}$$

μ is called the **permeability** of the material.[6] In a vacuum, where there is no matter to magnetize, the susceptibility χ_m vanishes, and the permeability is μ_0. That's why μ_0 is called the **permeability of free space.**

Example 6.3

An infinite solenoid (n turns per unit length, current I) is filled with linear material of susceptibility χ_m. Find the magnetic field inside the solenoid.

Solution: Since **B** is due in part to bound currents (which we don't yet know), we cannot compute it directly. However, this is one of those symmetrical cases in which we can get **H** from the free current alone, using Ampère's law in the form of Eq. 6.20:

$$\mathbf{H} = nI\,\hat{\mathbf{z}}$$

[5]Physically, therefore, Eq. 6.28 would say exactly the same as Eq. 6.29, only the constant χ_m would have a different value. Equation 6.29 is a little more convenient, because experimentalists find it handier to work with **H** than **B**.

[6]If you factor out μ_0, what's left is called the **relative permeability**: $\mu_r \equiv 1 + \chi_m = \mu/\mu_0$. By the way, formulas for **H** in terms of **B** (Eq. 6.31, in the case of linear media) are called **constitutive relations**, just like those for **D** in terms of **E**.

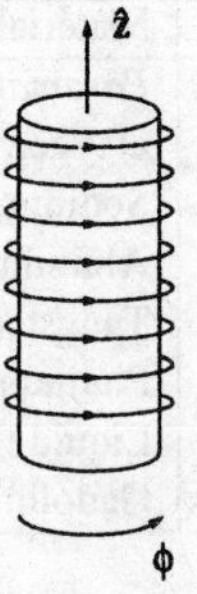

Figure 6.22

(Fig. 6.22). According to Eq. 6.31, then,

$$\mathbf{B} = \mu_0(1 + \chi_m)nI\,\hat{\mathbf{z}}.$$

If the medium is paramagnetic, the field is slightly enhanced; if it's diamagnetic, the field is somewhat reduced. This reflects the fact that the bound surface current

$$\mathbf{K}_b = \mathbf{M} \times \hat{\mathbf{n}} = \chi_m(\mathbf{H} \times \hat{\mathbf{n}}) = \chi_m nI\,\hat{\boldsymbol{\phi}}$$

is in the same direction as I, in the former case ($\chi_m > 0$), and opposite in the latter ($\chi_m < 0$).

You might suppose that linear media avoid the defect in the parallel between **B** and **H**: since **M** and **H** are now proportional to **B**, does it not follow that their divergence, like **B**'s, must always vanish? Unfortunately, it does *not*; at the *boundary* between two materials of different permeability the divergence of **M** can actually be infinite. For instance, at the end of a cylinder of linear paramagnetic material, **M** is zero on one side but not on the other. For the "Gaussian pillbox" shown in Fig. 6.23, $\oint \mathbf{M} \cdot d\mathbf{a} \neq 0$, and hence, by the divergence theorem, $\nabla \cdot \mathbf{M}$ cannot vanish everywhere within.

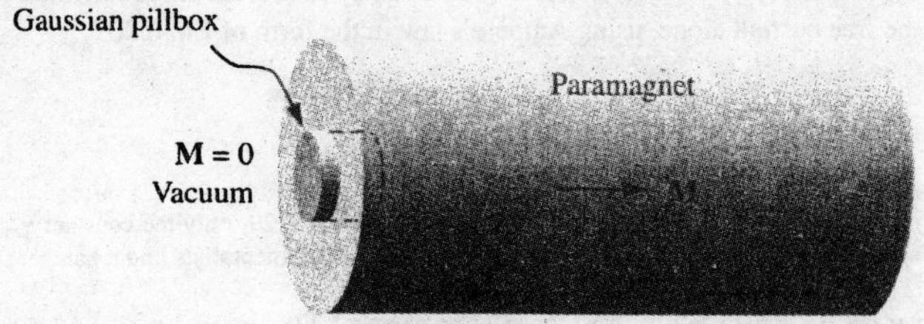

Figure 6.23

Incidentally, the volume bound current density in a homogeneous linear material is proportional to the *free* current density:

$$\mathbf{J}_b = \nabla \times \mathbf{M} = \nabla \times (\chi_m \mathbf{H}) = \chi_m \mathbf{J}_f. \tag{6.33}$$

In particular, unless free current actually flows *through* the material, all bound current will be at the surface.

Problem 6.16 A coaxial cable consists of two very long cylindrical tubes, separated by linear insulating material of magnetic susceptibility χ_m. A current I flows down the inner conductor and returns along the outer one; in each case the current distributes itself uniformly over the surface (Fig. 6.24). Find the magnetic field in the region between the tubes. As a check, calculate the magnetization and the bound currents, and confirm that (together, of course, with the free currents) they generate the correct field.

Figure 6.24

Problem 6.17 A current I flows down a long straight wire of radius a. If the wire is made of linear material (copper, say, or aluminum) with susceptibility χ_m, and the current is distributed uniformly, what is the magnetic field a distance s from the axis? Find all the bound currents. What is the *net* bound current flowing down the wire?

! **Problem 6.18** A sphere of linear magnetic material is placed in an otherwise uniform magnetic field $\mathbf{B}_0$. Find the new field inside the sphere. [*Hint:* See Prob. 6.15 or Prob. 4.23.]

Problem 6.19 On the basis of the naïve model presented in Sect. 6.1.3, estimate the magnetic susceptibility of a diamagnetic metal such as copper. Compare your answer with the empirical value in Table 6.1, and comment on any discrepancy.

6.4.2 Ferromagnetism

In a linear medium the alignment of atomic dipoles is maintained by a magnetic field imposed from the outside. Ferromagnets—which are emphatically *not* linear[7]—require no external fields to sustain the magnetization; the alignment is "frozen in." Like paramagnetism, ferromagnetism involves the magnetic dipoles associated with the spins of unpaired electrons. The new feature, which makes ferromagnetism so different from paramagnetism, is the interaction between nearby dipoles: In a ferromagnet, *each dipole "likes" to point in the same direction as its neighbors*. The *reason* for this preference is essentially quantum mechanical, and I shall not endeavor to explain it here; it is enough to know that the correlation is so strong as to align virtually 100% of the unpaired electron spins. If you could somehow magnify a piece of iron and "see" the individual dipoles as tiny arrows, it would look something like Fig. 6.25, with all the spins pointing the same way.

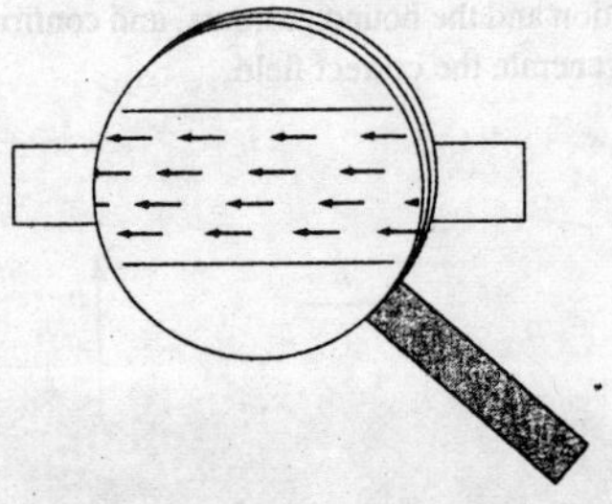

Figure 6.25

But if that is true, why isn't every wrench and nail a powerful magnet? The answer is that the alignment occurs in relatively small patches, called **domains**. Each domain contains billions of dipoles, all lined up (these domains are actually *visible* under a microscope, using suitable etching techniques—see Fig. 6.26), but the domains *themselves* are randomly oriented. The household wrench contains an enormous number of domains, and their magnetic fields cancel, so the wrench as a whole is not magnetized. (Actually, the orientation of domains is not *completely* random; within a given crystal there may be some preferential alignment along the crystal axes. But there will be just as many domains pointing one way as the other, so there is still no large-scale magnetization. Moreover, the crystals themselves are randomly oriented within any sizable chunk of metal.)

How, then, would you produce a **permanent magnet**, such as they sell in toy stores? If you put a piece of iron into a strong magnetic field, the torque $\mathbf{N} = \mathbf{m} \times \mathbf{B}$ tends to align the dipoles parallel to the field. Since they like to stay parallel to their neighbors, most of the dipoles will resist this torque. However, at the *boundary* between two domains, there

[7]In this sense it is misleading to speak of the susceptibility or permeability of a ferromagnet. The terms *are* used for such materials, but they refer to the proportionality factor between a *differential* increase in $\mathbf{H}$ and the resulting *differential* change in $\mathbf{M}$ (or $\mathbf{B}$); moreover, they are not *constants*, but functions of $\mathbf{H}$.

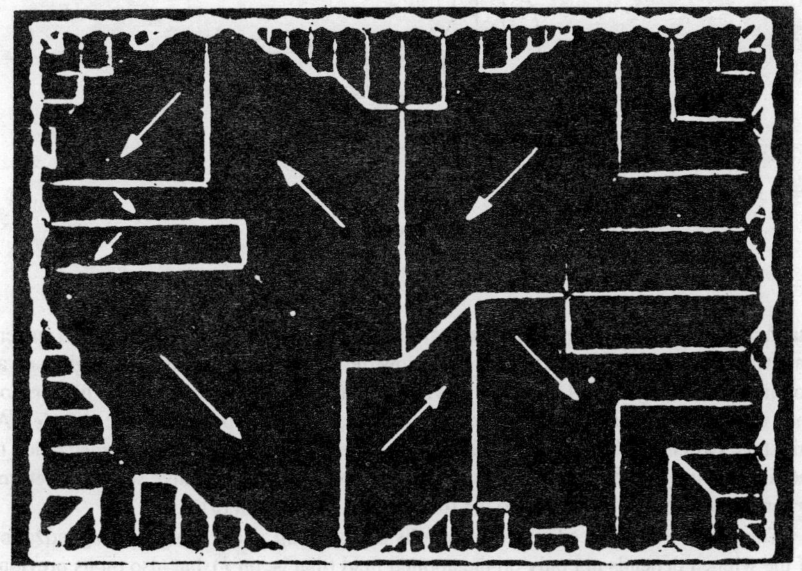

Ferromagnetic domains. (Photo courtesy of R. W. DeBlois)

Figure 6.26

are *competing* neighbors, and the torque will throw its weight on the side of the domain most nearly parallel to the field; this domain will win over some converts, at the expense of the less favorably oriented one. The net effect of the magnetic field, then, is to *move the domain boundaries*. Domains parallel to the field grow, and the others shrink. If the field is strong enough, one domain takes over entirely, and the iron is said to be "saturated."

It turns out that this process (the shifting of domain boundaries in response to an external field) is not entirely reversible: When the field is switched off, there will be *some* return to randomly oriented domains, but it is far from complete—there remains a preponderance of domains in the original direction. The object is now a permanent magnet.

A simple way to accomplish this, in practice, is to wrap a coil of wire around the object to be magnetized (Fig. 6.27). Run a current I through the coil; this provides the external magnetic field (pointing to the left in the diagram). As you increase the current, the field increases, the domain boundaries move, and the magnetization grows. Eventually, you reach the saturation point, with all the dipoles aligned, and a further increase in current has no effect on $\mathbf{M}$ (Fig. 6.28, point b).

Now suppose you *reduce* the current. Instead of retracing the path back to $M = 0$, there is only a *partial* return to randomly oriented domains. M decreases, but even with the current off there is some residual magnetization (point c). The wrench is now a permanent

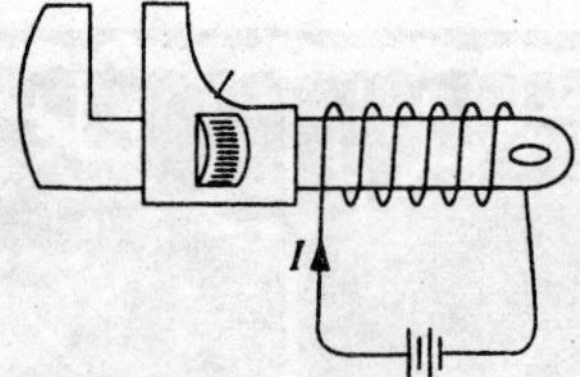

Figure 6.27

magnet. If you want to eliminate the remaining magnetization, you'll have to run a current backwards through the coil (a negative I). Now the external field points to the right, and as you increase I (negatively), M drops down to zero (point d). If you turn I still higher, you soon reach saturation in the other direction—all the dipoles now pointing to the *right* (e). At this stage switching off the current will leave the wrench with a permanent magnetization to the right (point f). To complete the story, turn I on again in the positive sense: M returns to zero (point g), and eventually to the forward saturation point (b).

The path we have traced out is called a **hysteresis loop**. Notice that the magnetization of the wrench depends not only on the applied field (that is, on I), but also on its previous magnetic "history."[8] For instance, at three different times in our experiment the current was zero (a, c, and f), yet the magnetization was different for each of them. Actually, it is customary to draw hysteresis loops as plots of B against H, rather than M against I. (If our coil is approximated by a long solenoid, with n turns per unit length, then $H = nI$, so H and I are proportional. Meanwhile, $\mathbf{B} = \mu_0(\mathbf{H} + \mathbf{M})$, but in practice M is huge compared to H, so to all intents and purposes $\mathbf{B}$ is proportional to $\mathbf{M}$.)

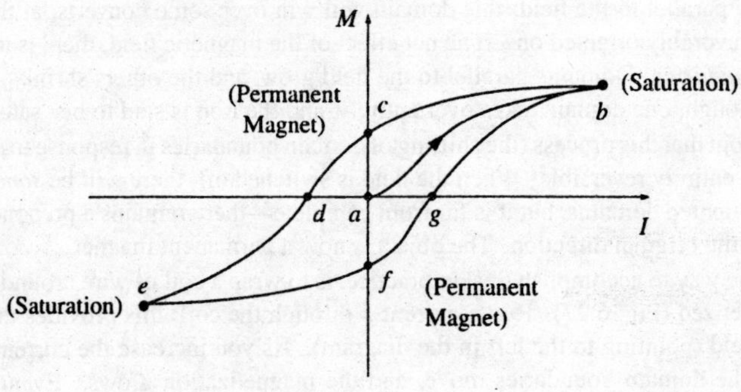

Figure 6.28

[8] Etymologically, the word *hysteresis* has nothing to do with the word *history*—nor with the word *hysteria*. It derives from a Greek verb meaning "to lag behind."

To make the units consistent (teslas), I have plotted ($\mu_0 H$) horizontally (Fig. 6.29); notice, however, that the vertical scale is 10^4 times greater than the horizontal one. Roughly speaking, $\mu_0\mathbf{H}$ is the field our coil *would* have produced in the absence of any iron; $\mathbf{B}$ is what we *actually* got, and compared to $\mu_0\mathbf{H}$ it is gigantic. A little current goes a long way when you have ferromagnetic materials around. That's why anyone who wants to make a powerful electromagnet will wrap the coil around an iron core. It doesn't take much of an external field to move the domain boundaries, and as soon as you've done that, you have all the dipoles in the iron working with you.

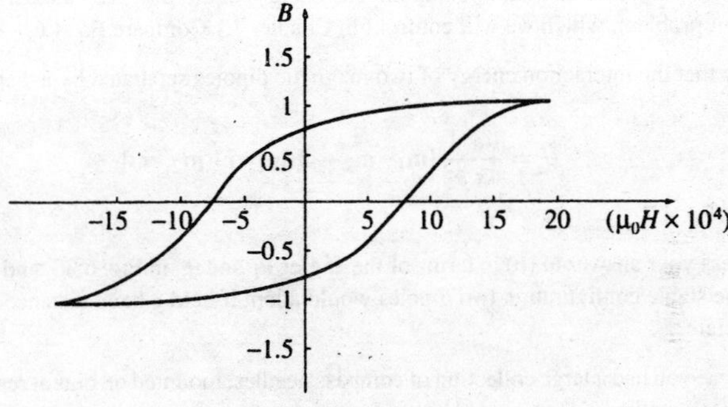

Figure 6.29

One final point concerning ferromagnetism: It all follows, remember, from the fact that the dipoles within a given domain line up parallel to one another. Random thermal motions compete with this ordering, but as long as the temperature doesn't get too high, they cannot budge the dipoles out of line. It's not surprising, though, that *very* high temperatures do destroy the alignment. What *is* surprising is that this occurs at a precise temperature (770° C, for iron). Below this temperature (called the **Curie point**), iron is ferromagnetic; above, it is paramagnetic. The Curie point is rather like the boiling point or the freezing point in that there is no *gradual* transition from ferro- to para-magnetic behavior, any more than there is between water and ice. These abrupt changes in the properties of a substance, occurring at sharply defined temperatures, are known in statistical mechanics as **phase transitions**.

Problem 6.20 How would you go about *de*magnetizing a permanent magnet (such as the wrench we have been discussing, at point c in the hysteresis loop)? That is, how could you restore it to its original state, with $M = 0$ at $I = 0$?

Problem 6.21

(a) Show that the energy of a magnetic dipole in a magnetic field $\mathbf{B}$ is given by

$$\boxed{U = -\mathbf{m}\cdot\mathbf{B}.} \tag{6.34}$$

Figure 6.30

[Assume that the *magnitude* of the dipole moment is *fixed*, and all you have to do is move it into place and rotate it into its final orientation. The energy required to keep the current flowing is a different problem, which we will confront in Chapter 7.] Compare Eq. 4.6.

(b) Show that the interaction energy of two magnetic dipoles separated by a displacement **r** is given by

$$U = \frac{\mu_0}{4\pi}\frac{1}{r^3}[\mathbf{m}_1 \cdot \mathbf{m}_2 - 3(\mathbf{m}_1 \cdot \hat{\mathbf{r}})(\mathbf{m}_2 \cdot \hat{\mathbf{r}})]. \tag{6.35}$$

Compare Eq. 4.7.

(c) Express your answer to (b) in terms of the angles θ_1 and θ_2 in Fig. 6.30, and use the result to find the stable configuration two dipoles would adopt if held a fixed distance apart, but left free to rotate.

(d) Suppose you had a large collection of compass needles, mounted on pins at regular intervals along a straight line. How would they point (assuming the earth's magnetic field can be neglected)? [A rectangular array of compass needles also aligns itself spontaneously, and this is sometimes used as a demonstration of "ferromagnetic" behavior on a large scale. It's a bit of a fraud, however, since the mechanism here is purely classical, and much weaker than the quantum mechanical **exchange forces** that are actually responsible for ferromagnetism.]

More Problems on Chapter 6

! **Problem 6.22** In Prob. 6.4 you calculated the force on a dipole by "brute force." Here's a more elegant approach. First write $\mathbf{B}(\mathbf{r})$ as a Taylor expansion about the center of the loop:

$$\mathbf{B}(\mathbf{r}) \cong \mathbf{B}(\mathbf{r}_0) + [(\mathbf{r} - \mathbf{r}_0) \cdot \nabla_0]\mathbf{B}(\mathbf{r}_0),$$

where $\mathbf{r}_0$ is the position of the dipole and ∇_0 denotes differentiation with respect to $\mathbf{r}_0$. Put this into the Lorentz force law (Eq. 5.16) to obtain

$$\mathbf{F} = I \oint d\mathbf{l} \times [(\mathbf{r} \cdot \nabla_0)\mathbf{B}(\mathbf{r}_0)].$$

Or, numbering the Cartesian coordinates from 1 to 3:

$$F_i = I \sum_{j,k,l=1}^{3} \epsilon_{ijk} \left\{ \oint r_l \, dl_j \right\} [\nabla_{0_l} B_k(\mathbf{r}_0)],$$

where ϵ_{ijk} is the **Levi-Civita symbol** (+1 if ijk = 123, 231, or 312; −1 if ijk = 132, 213, or 321; 0 otherwise), in terms of which the cross-product can be written $(\mathbf{A} \times \mathbf{B})_i = \sum_{j,k=1}^{3} \epsilon_{ijk} A_j B_k$. Use Eq. 1.108 to evaluate the integral. Note that

$$\sum_{j=1}^{3} \epsilon_{ijk}\epsilon_{ljm} = \delta_{il}\delta_{km} - \delta_{im}\delta_{kl},$$

where δ_{ij} is the Kronecker delta (Prob. 3.45).

Problem 6.23 Notice the following parallel:

$$\begin{cases} \nabla \cdot \mathbf{D} = 0, & \nabla \times \mathbf{E} = 0, & \epsilon_0 \mathbf{E} = \mathbf{D} - \mathbf{P}, & \text{(no free charge);} \\ \nabla \cdot \mathbf{B} = 0, & \nabla \times \mathbf{H} = 0, & \mu_0 \mathbf{H} = \mathbf{B} - \mu_0 \mathbf{M}, & \text{(no free current).} \end{cases}$$

Thus, the transcription $\mathbf{D} \to \mathbf{B}$, $\mathbf{E} \to \mathbf{H}$, $\mathbf{P} \to \mu_0 \mathbf{M}$, $\epsilon_0 \to \mu_0$ turns an electrostatic problem into an analogous magnetostatic one. Use this observation, together with your knowledge of the electrostatic results, to rederive

(a) the magnetic field inside a uniformly magnetized sphere (Eq. 6.16);

(b) the magnetic field inside a sphere of linear magnetic material in an otherwise uniform magnetic field (Prob. 6.18);

(c) the average magnetic field over a sphere, due to steady currents within the sphere (Eq. 5.89).

Problem 6.24 Compare Eqs. 2.15, 4.9, and 6.11. Notice that if ρ, $\mathbf{P}$, and $\mathbf{M}$ are *uniform*, the *same integral* is involved in all three:

$$\int \frac{\hat{\mathbf{\scriptstyle r}}}{\scriptstyle r^2}\, d\tau'.$$

Therefore, if you happen to know the electric field of a uniformly *charged* object, you can immediately write down the scalar potential of a uniformly *polarized* object, and the vector potential of a uniformly *magnetized* object, of the same shape. Use this observation to obtain V inside and outside a uniformly polarized sphere (Ex. 4.2), and $\mathbf{A}$ inside and outside a uniformly magnetized sphere (Ex. 6.1).

Problem 6.25 A familiar toy consists of donut-shaped permanent magnets (magnetization parallel to the axis), which slide frictionlessly on a vertical rod (Fig. 6.31). Treat the magnets as dipoles, with mass m_d and dipole moment $\mathbf{m}$.

(a) If you put two back-to-back magnets on the rod, the upper one will "float"—the magnetic force upward balancing the gravitational force downward. At what height (z) does it float?

(b) If you now add a *third* magnet (parallel to the bottom one), what is the *ratio* of the two heights? (Determine the actual number, to three significant digits.)
[*Answer:* (a) $[3\mu_0 m^2/2\pi m_d g]^{1/4}$; (b) 0.8501]

Problem 6.26 At the interface between one linear magnetic material and another the magnetic field lines bend (see Fig. 6.32). Show that $\tan\theta_2/\tan\theta_1 = \mu_2/\mu_1$, assuming there is no free current at the boundary. Compare Eq. 4.68.

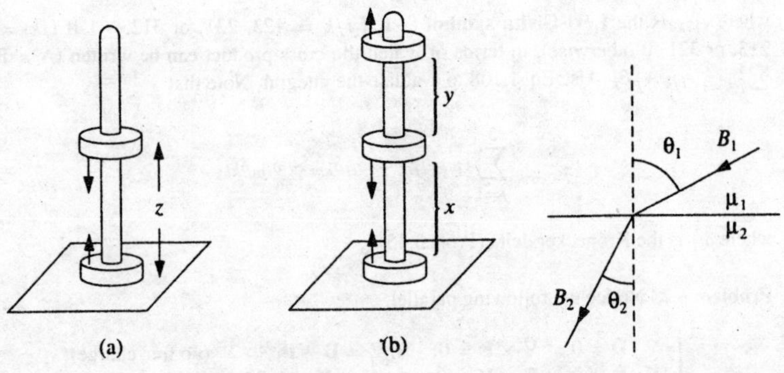

Figure 6.31

Figure 6.32

! **Problem 6.27** A magnetic dipole **m** is imbedded at the center of a sphere (radius R) of linear magnetic material (permeability μ). Show that the magnetic field inside the sphere ($0 < r \leq R$) is

$$\frac{\mu}{4\pi}\left\{\frac{1}{r^3}[3(\mathbf{m}\cdot\hat{\mathbf{r}})\hat{\mathbf{r}} - \mathbf{m}] + \frac{2(\mu_0 - \mu)\mathbf{m}}{(2\mu_0 + \mu)R^3}\right\}.$$

What is the field *outside* the sphere?

Problem 6.28 You are asked to referee a grant application, which proposes to determine whether the magnetization of iron is due to "Ampère" dipoles (current loops) or "Gilbert" dipoles (separated magnetic monopoles). The experiment will involve a cylinder of iron (radius R and length $L = 10R$), uniformly magnetized along the direction of the axis. If the dipoles are Ampère-type, the magnetization is equivalent to a surface bound current $\mathbf{K}_b = M\,\hat{\boldsymbol{\phi}}$; if they are Gilbert-type, the magnetization is equivalent to surface monopole densities $\sigma_b = \pm M$ at the two ends. Unfortunately, these two configurations produce identical magnetic fields, at exterior points. However, the *interior* fields are radically different—in the first case **B** is in the *same* general direction as **M**, whereas in the second it is roughly *opposite* to **M**. The applicant proposes to measure this internal field by carving out a small cavity and finding the torque on a tiny compass needle placed inside.

Assuming that the obvious technical difficulties can be overcome, and that the question itself is worthy of study, would you advise funding this experiment? If so, what shape cavity would you recommend? If not, what is wrong with the proposal? [*Hint:* refer to Probs. 4.11, 4.16, 6.9, and 6.13.]

Chapter 7

Electrodynamics

7.1 Electromotive Force

7.1.1 Ohm's Law

To make a current flow, you have to *push* on the charges. How *fast* they move, in response to a given push, depends on the nature of the material. For most substances, the current density $\mathbf{J}$ is proportional to the *force per unit charge*, $\mathbf{f}$:

$$\mathbf{J} = \sigma \mathbf{f}. \tag{7.1}$$

The proportionality factor σ (not to be confused with surface charge) is an empirical constant that varies from one material to another; it's called the **conductivity** of the medium. Actually, the handbooks usually list the *reciprocal* of σ, called the **resistivity**: $\rho = 1/\sigma$ (not to be confused with charge density—I'm sorry, but we're running out of Greek letters, and this is the standard notation). Some typical values are listed in Table 7.1. Notice that even *insulators* conduct slightly, though the conductivity of a metal is astronomically greater—by a factor of 10^{22} or so. In fact, for most purposes metals can be regarded as **perfect conductors**, with $\sigma = \infty$.

In principle, the force that drives the charges to produce the current could be anything—chemical, gravitational, or trained ants with tiny harnesses. For *our* purposes, though, it's usually an electromagnetic force that does the job. In this case Eq. 7.1 becomes

$$\mathbf{J} = \sigma(\mathbf{E} + \mathbf{v} \times \mathbf{B}). \tag{7.2}$$

Ordinarily, the velocity of the charges is sufficiently small that the second term can be ignored:

$$\boxed{\mathbf{J} = \sigma \mathbf{E}.} \tag{7.3}$$

(However, in plasmas, for instance, the magnetic contribution to $\mathbf{f}$ can be significant.) Equation 7.3 is called **Ohm's law**, though the physics behind it is really contained in Eq. 7.1, of which 7.3 is just a special case.

Material	Resistivity	Material	Resistivity
Conductors:		*Semiconductors:*	
Silver	1.59×10^{-8}	Salt water (saturated)	4.4×10^{-2}
Copper	1.68×10^{-8}	Germanium	4.6×10^{-1}
Gold	2.21×10^{-8}	Diamond	2.7
Aluminum	2.65×10^{-8}	Silicon	2.5×10^{3}
Iron	9.61×10^{-8}	*Insulators:*	
Mercury	9.58×10^{-7}	Water (pure)	2.5×10^{5}
Nichrome	1.00×10^{-6}	Wood	$10^{8} - 10^{11}$
Manganese	1.44×10^{-6}	Glass	$10^{10} - 10^{14}$
Graphite	1.4×10^{-5}	Quartz (fused)	$\sim 10^{16}$

Table 7.1 Resistivities, in ohm-meters (all values are for 1 atm, 20° C).
Source: Handbook of Chemistry and Physics, 78th ed.
(Boca Raton: CRC Press, Inc., 1997).

I know: you're confused because I said $\mathbf{E} = 0$ inside a conductor (Sect. 2.5.1). But that's for *stationary* charges ($\mathbf{J} = 0$). Moreover, for *perfect* conductors $\mathbf{E} = \mathbf{J}/\sigma = 0$ even if current *is* flowing. In practice, metals are such good conductors that the electric field required to drive current in them is negligible. Thus we routinely treat the connecting wires in electric circuits (for example) as equipotentials. **Resistors**, by contrast, are made from *poorly* conducting materials.

Example 7.1

A cylindrical resistor of cross-sectional area A and length L is made from material with conductivity σ. (See Fig. 7.1; as indicated, the cross section need not be circular, but I *do* assume it is the same all the way down.) If the potential is constant over each end, and the potential difference between the ends is V, what current flows?

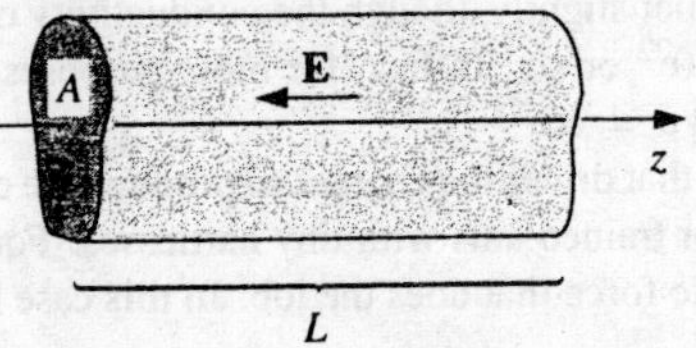

Figure 7.1

Solution: As it turns out, the electric field is *uniform* within the wire (I'll *prove* this in a moment). It follows from Eq. 7.3 that the current density is also uniform, so

$$I = JA = \sigma EA = \frac{\sigma A}{L} V.$$

Example 7.2

Two long cylinders (radii a and b) are separated by material of conductivity σ (Fig. 7.2). If they are maintained at a potential difference V, what current flows from one to the other, in a length L?

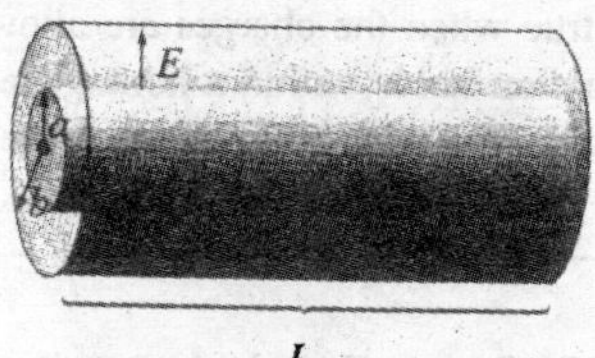

Figure 7.2

Solution: The field between the cylinders is

$$\mathbf{E} = \frac{\lambda}{2\pi\epsilon_0 s}\hat{\mathbf{s}},$$

where λ is the charge per unit length on the inner cylinder. The current is therefore

$$I = \int \mathbf{J}\cdot d\mathbf{a} = \sigma\int \mathbf{E}\cdot d\mathbf{a} = \frac{\sigma}{\epsilon_0}\lambda L.$$

(The integral is over any surface enclosing the inner cylinder.) Meanwhile, the potential difference between the cylinders is

$$V = -\int_b^a \mathbf{E}\cdot d\mathbf{l} = \frac{\lambda}{2\pi\epsilon_0}\ln\left(\frac{b}{a}\right),$$

so

$$I = \frac{2\pi\sigma L}{\ln(b/a)}V.$$

As these examples illustrate, the total current flowing from one **electrode** to the other is proportional to the potential difference between them:

$$\boxed{V = IR.} \tag{7.4}$$

This, of course, is the more familiar version of Ohm's law. The constant of proportionality R is called the **resistance**; it's a function of the geometry of the arrangement and the conductivity of the medium between the electrodes. (In Ex. 7.1, $R = (L/\sigma A)$; in Ex. 7.2, $R = \ln(b/a)/2\pi\sigma L$.) Resistance is measured in **ohms** (Ω): an ohm is a volt per ampere. Notice that the proportionality between V and I is a direct consequence of Eq. 7.3: if you want to double V, you simply double the charge everywhere—but that doubles $\mathbf{E}$, which doubles $\mathbf{J}$, which doubles I.

For *steady* currents and *uniform* conductivity,

$$\nabla \cdot \mathbf{E} = \frac{1}{\sigma} \nabla \cdot \mathbf{J} = 0, \tag{7.5}$$

(Eq. 5.31), and therefore the charge density is zero; any unbalanced charge resides on the *surface*. (We proved this long ago, for the case of *stationary* charges, using the fact that $\mathbf{E} = 0$; evidently, it is still true when the charges are allowed to move.) It follows, in particular, that Laplace's equation holds within a homogeneous ohmic material carrying a steady current, so all the tools and tricks of Chapter 3 are available for computing the potential.

Example 7.3

I asserted that the field in Ex. 7.1 is *uniform*. Let's *prove* it.

Solution: Within the cylinder V obeys Laplace's equation. What are the boundary conditions? At the left end the potential is constant—we may as well set it equal to zero. At the right end the potential is likewise constant—call it V_0. On the cylindrical surface, $\mathbf{J} \cdot \hat{\mathbf{n}} = 0$, or else charge would be leaking out into the surrounding space (which we take to be nonconducting). Therefore $\mathbf{E} \cdot \hat{\mathbf{n}} = 0$, and hence $\partial V / \partial n = 0$. With V or its normal derivative specified on all surfaces, the potential is uniquely determined (Prob. 3.4). But it's *easy* to guess *one* potential that obeys Laplace's equation and fits these boundary conditions:

$$V(z) = \frac{V_0 z}{L},$$

where z is measured along the axis. The uniqueness theorem guarantees that this is *the* solution. The corresponding field is

$$\mathbf{E} = -\nabla V = -\frac{V_0}{L} \hat{\mathbf{z}},$$

which is indeed uniform. qed

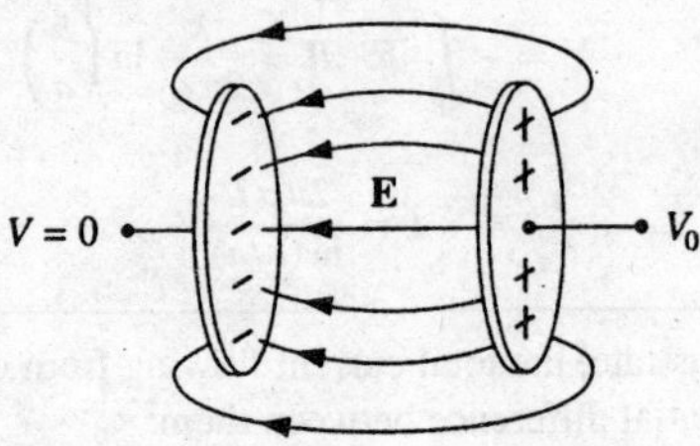

Figure 7.3

Contrast the *enormously* more difficult problem that arises if the conducting material is removed, leaving only a metal plate at either end (Fig. 7.3). Evidently in the present case charge arranges itself over the surface of the wire in just such a way as to produce a nice uniform field within.[1]

---

[1] *Calculating* this surface charge is not easy. See, for example, J. D. Jackson, *Am. J. Phys.* **64**, 855 (1996). Nor is it a simple matter to determine the field *outside* the wire—see Prob. 7.57.

I don't suppose there is any formula in physics more widely known than Ohm's law, and yet it's not really a true law, in the sense of Gauss's law or Ampère's law; rather, it is a "rule of thumb" that applies pretty well to many substances. You're not going to win a Nobel prize for finding an exception. In fact, when you stop to think about it, it's a little surprising that Ohm's law *ever* holds. After all, a given field $\mathbf{E}$ produces a force $q\mathbf{E}$ (on a charge q), and according to Newton's second law the charge will accelerate. But if the charges are *accelerating*, why doesn't the current *increase* with time, growing larger and larger the longer you leave the field on? Ohm's law implies, on the contrary, that a constant field produces a constant *current*, which suggests a constant *velocity*. Isn't that a contradiction of Newton's law?

No, for we are forgetting the frequent collisions electrons make as they pass down the wire. It's a little like this: Suppose you're driving down a street with a stop sign at every intersection, so that, although you accelerate constantly in between, you are obliged to start all over again with each new block. Your *average* speed is then a constant, in spite of the fact that (save for the periodic abrupt stops) you are always accelerating. If the length of a block is λ and your acceleration is a, the time it takes to go a block is

$$t = \sqrt{\frac{2\lambda}{a}},$$

and hence the average velocity is

$$v_{\text{ave}} = \frac{1}{2}at = \sqrt{\frac{\lambda a}{2}}.$$

But wait! That's no good *either!* It says that the velocity is proportional to the *square root* of the acceleration, and therefore that the current should be proportional to the *square root* of the field! There's another twist to the story: The charges in practice are already moving quite fast because of their thermal energy. But the thermal velocities have random directions, and average to zero. The net **drift velocity** we're concerned with is a tiny extra bit (Prob. 5.19). So the time between collisions is actually much shorter than we supposed; in fact,

$$t = \frac{\lambda}{v_{\text{thermal}}},$$

and therefore

$$v_{\text{ave}} = \frac{1}{2}at = \frac{a\lambda}{2v_{\text{thermal}}}.$$

If there are n molecules per unit volume and f free electrons per molecule, each with charge q and mass m, the current density is

$$\mathbf{J} = nfq\mathbf{v}_{\text{ave}} = \frac{nfq\lambda}{2v_{\text{thermal}}}\frac{\mathbf{F}}{m} = \left(\frac{nf\lambda q^2}{2mv_{\text{thermal}}}\right)\mathbf{E}. \tag{7.6}$$

I don't claim that the term in parentheses is an accurate formula for the conductivity,[2] but it

[2]This classical model (due to Drude) bears little resemblance to the modern quantum theory of conductivity. See, for instance, D. Park's *Introduction to the Quantum Theory*, 3rd ed., Chap. 15 (New York: McGraw-Hill, 1992).

does indicate the basic ingredients, and it correctly predicts that conductivity is proportional to the density of the moving charges and (ordinarily) decreases with increasing temperature.

As a result of all the collisions, the work done by the electrical force is converted into heat in the resistor. Since the work done per unit charge is V and the charge flowing per unit time is I, the power delivered is

$$\boxed{P = VI = I^2 R.} \tag{7.7}$$

This is the **Joule heating law**. With I in amperes and R in ohms, P comes out in watts (joules per second).

Problem 7.1 Two concentric metal spherical shells, of radius a and b, respectively, are separated by weakly conducting material of conductivity σ (Fig. 7.4a).

(a) If they are maintained at a potential difference V, what current flows from one to the other?

(b) What is the resistance between the shells?

(c) Notice that if $b \gg a$ the outer radius (b) is irrelevant. How do you account for that? Exploit this observation to determine the current flowing between two metal spheres, each of radius a, immersed deep in the sea and held quite far apart (Fig. 7.4b), if the potential difference between them is V. (This arrangement can be used to measure the conductivity of sea water.)

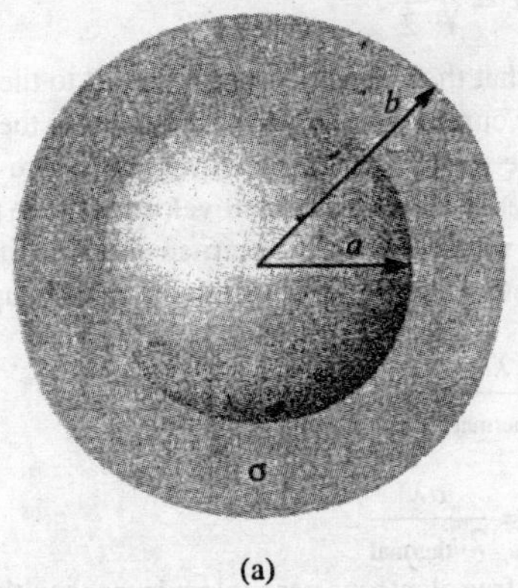

(a)

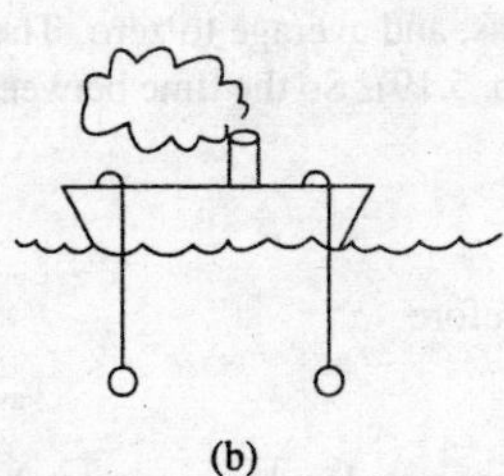

(b)

Figure 7.4

Problem 7.2 A capacitor C has been charged up to potential V_0; at time $t = 0$ it is connected to a resistor R, and begins to discharge (Fig. 7.5a).

(a) Determine the charge on the capacitor as a function of time, $Q(t)$. What is the current through the resistor, $I(t)$?

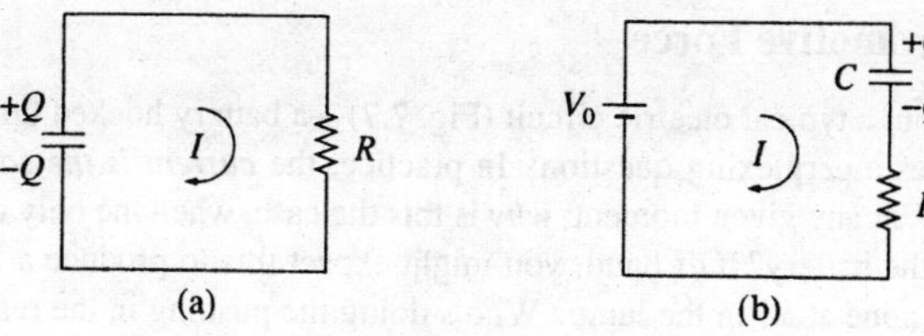

Figure 7.5

(b) What was the original energy stored in the capacitor (Eq. 2.55)? By integrating Eq. 7.7, confirm that the heat delivered to the resistor is equal to the energy lost by the capacitor.

Now imagine *charging up* the capacitor, by connecting it (and the resistor) to a battery of fixed voltage V_0, at time $t = 0$ (Fig. 7.5b).

(c) Again, determine $Q(t)$ and $I(t)$.

(d) Find the total energy output of the battery ($\int V_0 I\, dt$). Determine the heat delivered to the resistor. What is the final energy stored in the capacitor? What fraction of the work done by the battery shows up as energy in the capacitor? [Notice that the answer is independent of R!]

Problem 7.3

(a) Two metal objects are embedded in weakly conducting material of conductivity σ (Fig. 7.6). Show that the resistance between them is related to the capacitance of the arrangement by

$$R = \frac{\epsilon_0}{\sigma C}.$$

(b) Suppose you connected a battery between 1 and 2 and charged them up to a potential difference V_0. If you then disconnect the battery, the charge will gradually leak off. Show that $V(t) = V_0 e^{-t/\tau}$, and find the **time constant**, τ, in terms of ϵ_0 and σ.

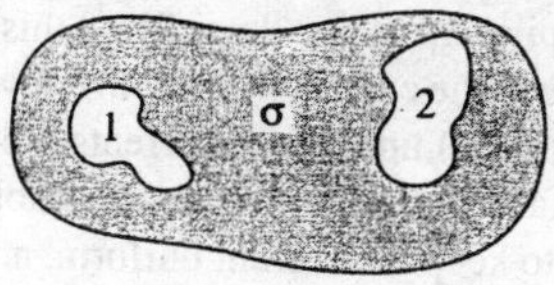

Figure 7.6

Problem 7.4 Suppose the conductivity of the material separating the cylinders in Ex. 7.2 is not uniform; specifically, $\sigma(s) = k/s$, for some constant k. Find the resistance between the cylinders. [*Hint:* Because σ is a function of position, Eq. 7.5 does not hold, the charge density is not zero in the resistive medium, and **E** does not go like $1/s$. But we *do* know that for steady currents I is the same across each cylindrical surface. Take it from there.]

7.1.2 Electromotive Force

If you think about a typical electric circuit (Fig. 7.7)—a battery hooked up to a light bulb, say—there arises a perplexing question: In practice, the *current is the same all the way around the loop*, at any given moment; *why* is this the case, when the only obvious driving force is inside the battery? Off hand, you might expect this to produce a large current in the battery and none at all in the lamp. Who's doing the pushing in the rest of the circuit, and how does it happen that this push is exactly right to produce the same current in each segment? What's more, given that the charges in a typical wire move (literally) at a *snail's* pace (see Prob. 5.19), why doesn't it take half an hour for the news to reach the light bulb? How do all the charges know to start moving at the same instant?

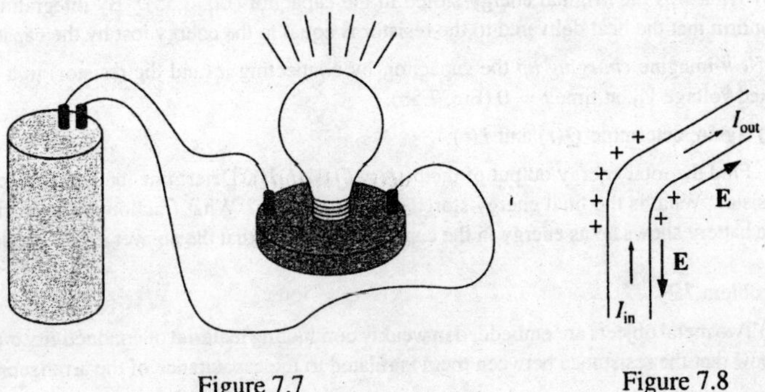

Figure 7.7 Figure 7.8

Answer: If the current is *not* the same all the way around (for instance, during the first split second after the switch is closed), then charge is piling up somewhere, and—here's the crucial point—the electric field of this accumulating charge is in such a direction as to even out the flow. Suppose, for instance, that the current *into* the bend in Fig. 7.8 is greater than the current *out*. Then charge piles up at the "knee," and this produces a field aiming *away* from the kink. This field *opposes* the current flowing in (slowing it down) and *promotes* the current flowing out (speeding it up) until these currents are equal, at which point there is no further accumulation of charge, and equilibrium is established. It's a beautiful system, automatically self-correcting to keep the current uniform, and it does it all so quickly that, in practice, you can safely assume the current is the same all around the circuit even in systems that oscillate at radio frequencies.

The upshot of all this is that there are really *two* forces involved in driving current around a circuit: the *source*, $\mathbf{f}_s$, which is ordinarily confined to one portion of the loop (a battery, say), and the *electrostatic* force, which serves to smooth out the flow and communicate the influence of the source to distant parts of the circuit:

$$\mathbf{f} = \mathbf{f}_s + \mathbf{E}. \tag{7.8}$$

The physical agency responsible for $\mathbf{f}_s$ can be any one of many different things: in a battery it's a chemical force; in a piezoelectric crystal mechanical pressure is converted into an electrical impulse; in a thermocouple it's a temperature gradient that does the job; in a photoelectric cell it's light; and in a Van de Graaff generator the electrons are literally loaded onto a conveyer belt and swept along. Whatever the *mechanism*, its net effect is determined by the line integral of $\mathbf{f}$ around the circuit:

$$\boxed{\mathcal{E} \equiv \oint \mathbf{f} \cdot d\mathbf{l} = \oint \mathbf{f}_s \cdot d\mathbf{l}.} \tag{7.9}$$

(Because $\oint \mathbf{E} \cdot d\mathbf{l} = 0$ for electrostatic fields, it doesn't matter whether you use $\mathbf{f}$ or $\mathbf{f}_s$.) $\mathcal{E}$ is called the **electromotive force**, or **emf**, of the circuit. It's a lousy term, since this is not a *force* at all—it's the *integral* of a *force per unit charge*. Some people prefer the word **electromotance**, but emf is so ingrained that I think we'd better stick with it.

Within an ideal source of emf (a resistanceless battery,[3] for instance), the *net* force on the charges is *zero* (Eq. 7.1 with $\sigma = \infty$), so $\mathbf{E} = -\mathbf{f}_s$. The potential difference between the terminals (a and b) is therefore

$$V = -\int_a^b \mathbf{E} \cdot d\mathbf{l} = \int_a^b \mathbf{f}_s \cdot d\mathbf{l} = \oint \mathbf{f}_s \cdot d\mathbf{l} = \mathcal{E} \tag{7.10}$$

(we can extend the integral to the entire loop because $\mathbf{f}_s = 0$ outside the source). The function of a battery, then, is to establish and maintain a voltage difference equal to the electromotive force (a 6 V battery, for example, holds the positive terminal 6 V above the negative terminal). The resulting electrostatic field drives current around the rest of the circuit (notice, however, that *inside* the battery $\mathbf{f}_s$ drives current in the direction *opposite* to $\mathbf{E}$).

Because it's the line integral of $\mathbf{f}_s$, $\mathcal{E}$ can be interpreted as the *work done, per unit charge*, by the source—indeed, in some books electromotive force is *defined* this way. However, as you'll see in the next section, there is some subtlety involved in this interpretation, so I prefer Eq. 7.9.

Problem 7.5 A battery of emf $\mathcal{E}$ and internal resistance r is hooked up to a variable "load" resistance, R. If you want to deliver the maximum possible power to the load, what resistance R should you choose? (You can't change $\mathcal{E}$ and r, of course.)

Problem 7.6 A rectangular loop of wire is situated so that one end (height h) is between the plates of a parallel-plate capacitor (Fig. 7.9), oriented parallel to the field $\mathbf{E}$. The other end is way outside, where the field is essentially zero. What is the emf in this loop? If the total resistance is R, what current flows? Explain. [*Warning:* this is a trick question, so be careful; if you have invented a perpetual motion machine, there's probably something wrong with it.]

---

[3] *Real* batteries have a certain **internal resistance**, r, and the potential difference between their terminals is $\mathcal{E} - Ir$, when a current I is flowing. For an illuminating discussion of how batteries work, see D. Roberts, *Am. J. Phys.* **51**, 829 (1983).

Figure 7.9

7.1.3 Motional emf

In the last section I listed several possible sources of electromotive force in a circuit, batteries being the most familiar. But I did not mention the most common one of all: the **generator**. Generators exploit **motional emf**'s, which arise when you *move a wire through a magnetic field*. Figure 7.10 shows a primitive model for a generator. In the shaded region there is a uniform magnetic field **B**, pointing into the page, and the resistor R represents whatever it is (maybe a light bulb or a toaster) we're trying to drive current through. If the entire loop is pulled to the right with speed v, the charges in segment ab experience a magnetic force whose vertical component qvB drives current around the loop, in the clockwise direction. The emf is

$$\mathcal{E} = \oint \mathbf{f}_{\mathrm{mag}} \cdot d\mathbf{l} = vBh, \tag{7.11}$$

where h is the width of the loop. (The horizontal segments bc and ad contribute nothing, since the force here is perpendicular to the wire.)

Notice that the integral you perform to calculate $\mathcal{E}$ (Eq. 7.9 or 7.11) is carried out at *one instant of time*—take a "snapshot" of the loop, if you like, and work from that. Thus $d\mathbf{l}$, for the segment ab in Fig. 7.10, points straight up, even though the loop is moving to the right. You can't quarrel with this—it's simply the way emf is *defined*—but it *is* important to be *clear* about it.

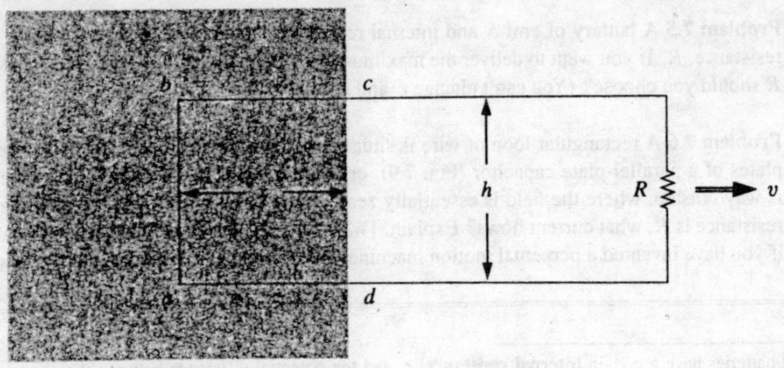

Figure 7.10

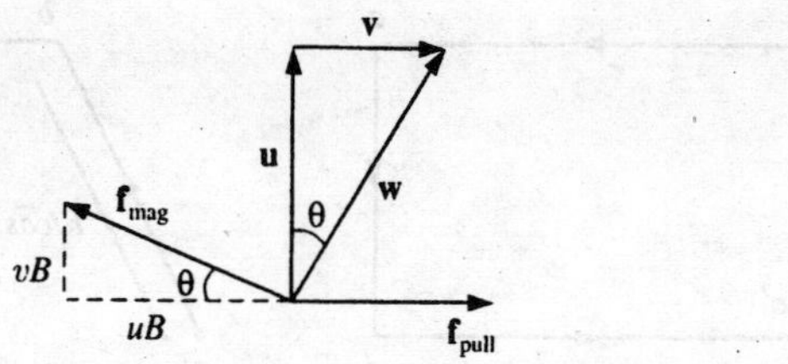

Figure 7.11

In particular, although the magnetic force is responsible for establishing the emf, it is certainly *not* doing any work—magnetic forces *never* do work. Who, then, *is* supplying the energy that heats the resistor? *Answer:* The person who's pulling on the loop! With the current flowing, charges in segment ab have a vertical velocity (call it $\mathbf{u}$) in addition to the horizontal velocity $\mathbf{v}$ they inherit from the motion of the loop. Accordingly, the magnetic force has a component quB to the left. To counteract this, the person pulling on the wire must exert a force per unit charge

$$f_{\text{pull}} = uB$$

to the *right* (Fig. 7.11). This force is transmitted to the charge by the structure of the wire. Meanwhile, the particle is actually *moving* in the direction of the resultant velocity $\mathbf{w}$, and the distance it goes is $(h/\cos\theta)$. The work done per unit charge is therefore

$$\int \mathbf{f}_{\text{pull}} \cdot d\mathbf{l} = (uB)\left(\frac{h}{\cos\theta}\right)\sin\theta = vBh = \mathcal{E}$$

($\sin\theta$ coming from the dot product). As it turns out, then, the *work done per unit charge is exactly equal to the emf,* though the integrals are taken along entirely different paths (Fig. 7.12) and completely different forces are involved. To calculate the emf you integrate around the loop at *one instant*, but to calculate the work done you follow a charge in its motion around the loop; $\mathbf{f}_{\text{pull}}$ contributes nothing to the emf, because it is perpendicular to the wire, whereas $\mathbf{f}_{\text{mag}}$ contributes nothing to work because it is perpendicular to the motion of the charge.[4]

There is a particularly nice way of expressing the emf generated in a moving loop. Let Φ be the flux of $\mathbf{B}$ through the loop:

$$\Phi \equiv \int \mathbf{B} \cdot d\mathbf{a}. \tag{7.12}$$

For the rectangular loop in Fig. 7.10,

$$\Phi = Bhx.$$

[4]For further discussion, see E. P. Mosca, *Am. J. Phys.* **42**, 295 (1974).

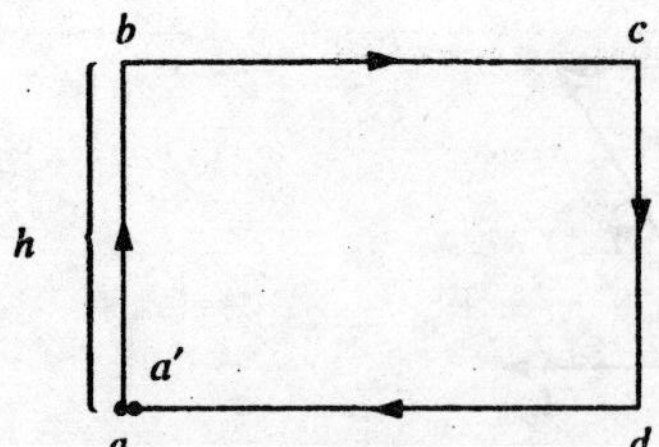

(a) Integration for computing $\mathcal{E}$ (follow the wire at one instant of time).

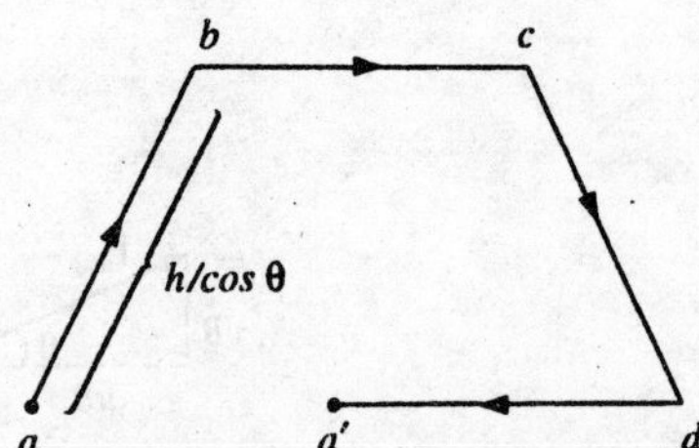

(b) Integration path for calculating work done (follow the charge around the loop).

Figure 7.12

As the loop moves, the flux decreases:

$$\frac{d\Phi}{dt} = Bh\frac{dx}{dt} = -Bhv.$$

(The minus sign accounts for the fact that dx/dt is negative.) But this is precisely the emf (Eq. 7.11); evidently the emf generated in the loop is minus the rate of change of flux through the loop:

$$\boxed{\mathcal{E} = -\frac{d\Phi}{dt}.} \tag{7.13}$$

This is the **flux rule** for motional emf. Apart from its delightful simplicity, it has the virtue of applying to *non*rectangular loops moving in *arbitrary* directions through *non*uniform magnetic fields; in fact, the loop need not even maintain a fixed shape.

Proof: Figure 7.13 shows a loop of wire at time t and also a short time dt later. Suppose we compute the flux at time t, using surface $\mathcal{S}$, and the flux at time $t + dt$, using the surface consisting of $\mathcal{S}$ plus the "ribbon" that connects the new position of the loop to the old. The *change* in flux, then, is

$$d\Phi = \Phi(t + dt) - \Phi(t) = \Phi_{\text{ribbon}} = \int_{\text{ribbon}} \mathbf{B} \cdot d\mathbf{a}.$$

Focus your attention on point P: in time dt it moves to P'. Let $\mathbf{v}$ be the velocity of the *wire*, and $\mathbf{u}$ the velocity of a charge *down* the wire; $\mathbf{w} = \mathbf{v} + \mathbf{u}$ is the resultant velocity of a charge at P. The infinitesimal element of area on the ribbon can be written as

$$d\mathbf{a} = (\mathbf{v} \times d\mathbf{l})\, dt$$

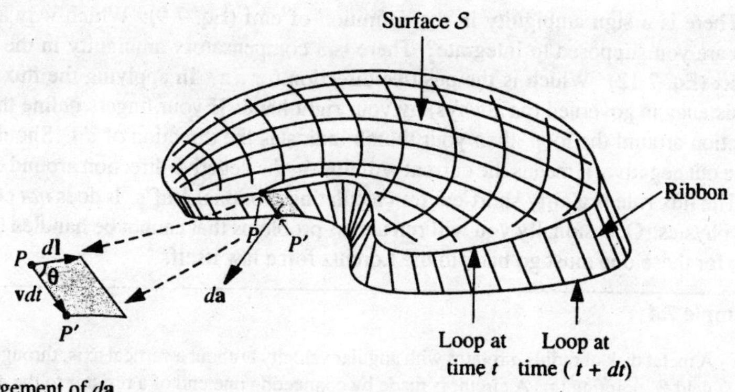

Figure 7.13

(see inset in Fig. 7.13). Therefore

$$\frac{d\Phi}{dt} = \oint \mathbf{B} \cdot (\mathbf{v} \times d\mathbf{l}).$$

Since $\mathbf{w} = (\mathbf{v} + \mathbf{u})$ and $\mathbf{u}$ is parallel to $d\mathbf{l}$, we can also write this as

$$\frac{d\Phi}{dt} = \oint \mathbf{B} \cdot (\mathbf{w} \times d\mathbf{l}).$$

Now, the scalar triple-product can be rewritten:

$$\mathbf{B} \cdot (\mathbf{w} \times d\mathbf{l}) = -(\mathbf{w} \times \mathbf{B}) \cdot d\mathbf{l},$$

so

$$\frac{d\Phi}{dt} = -\oint (\mathbf{w} \times \mathbf{B}) \cdot d\mathbf{l}.$$

But $(\mathbf{w} \times \mathbf{B})$ is the magnetic force per unit charge, $\mathbf{f}_{\text{mag}}$, so

$$\frac{d\Phi}{dt} = -\oint \mathbf{f}_{\text{mag}} \cdot d\mathbf{l},$$

and the integral of $\mathbf{f}_{\text{mag}}$ is the emf

$$\mathcal{E} = -\frac{d\Phi}{dt}. \quad \text{qed}$$

There is a sign ambiguity in the definition of emf (Eq. 7.9): Which *way* around the loop are you supposed to integrate? There is a compensatory ambiguity in the definition of *flux* (Eq. 7.12): Which is the positive direction for $d\mathbf{a}$? In applying the flux rule, sign consistency is governed (as always) by your right hand: If your fingers define the positive direction around the loop, then your thumb indicates the direction of $d\mathbf{a}$. Should the emf come out negative, it means the current will flow in the negative direction around the circuit.

The flux rule is a nifty short-cut for calculating motional emf's. It does *not* contain any new physics. Occasionally you will run across problems that cannot be handled by the flux rule; for these one must go back to the Lorentz force law itself.

Example 7.4

A metal disk of radius a rotates with angular velocity ω about a vertical axis, through a uniform field $\mathbf{B}$, pointing up. A circuit is made by connecting one end of a resistor to the axle and the other end to a sliding contact, which touches the outer edge of the disk (Fig. 7.14). Find the current in the resistor.

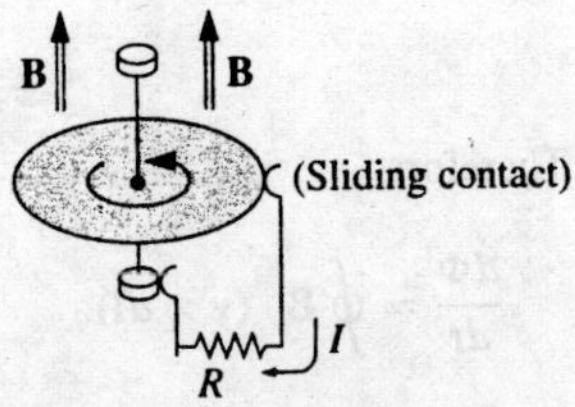

Figure 7.14

Solution: The speed of a point on the disk at a distance s from the axis is $v = \omega s$, so the force per unit charge is $\mathbf{f}_{\text{mag}} = \mathbf{v} \times \mathbf{B} = \omega s B\,\hat{\mathbf{s}}$. The emf is therefore

$$\mathcal{E} = \int_0^a f_{\text{mag}}\, ds = \omega B \int_0^a s\, ds = \frac{\omega B a^2}{2},$$

and the current is

$$I = \frac{\mathcal{E}}{R} = \frac{\omega B a^2}{2R}.$$

The trouble with the flux rule is that it assumes the current flows along a well-defined path, whereas in this example the current spreads out over the whole disk. It's not even clear what the "flux through the circuit" would *mean* in this context. Even more tricky is the case of **eddy currents**. Take a chunk of aluminum (say), and shake it around in a nonuniform magnetic field. Currents will be generated in the material, and you will feel a kind of "viscous drag"—as though you were pulling the block through molasses (this is the force I called $\mathbf{f}_{\text{pull}}$ in the discussion of motional emf). Eddy currents are notoriously difficult to calculate,[5] but easy and dramatic to demonstrate. You may have witnessed the classic experiment in which an

[5]See, for example, W. M. Saslow, *Am. J. Phys.*, **60**, 693 (1992).

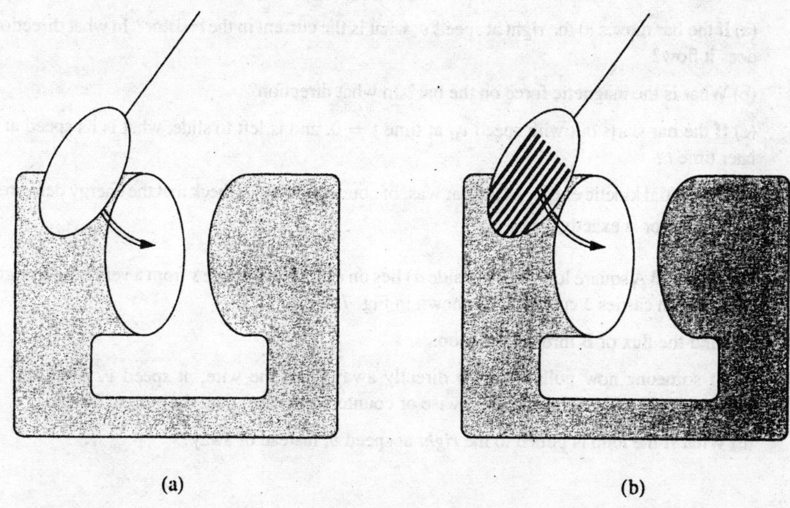

Figure 7.15

aluminum disk mounted as a pendulum on a horizontal axis swings down and passes between the poles of a magnet (Fig. 7.15a). When it enters the field region it suddenly slows way down. To confirm that eddy currents are responsible, one repeats the process using a disk that has many slots cut in it, to prevent the flow of large-scale currents (Fig. 7.15b). This time the disk swings freely, unimpeded by the field.

Problem 7.7 A metal bar of mass m slides frictionlessly on two parallel conducting rails a distance l apart (Fig. 7.16). A resistor R is connected across the rails and a uniform magnetic field **B**, pointing into the page, fills the entire region.

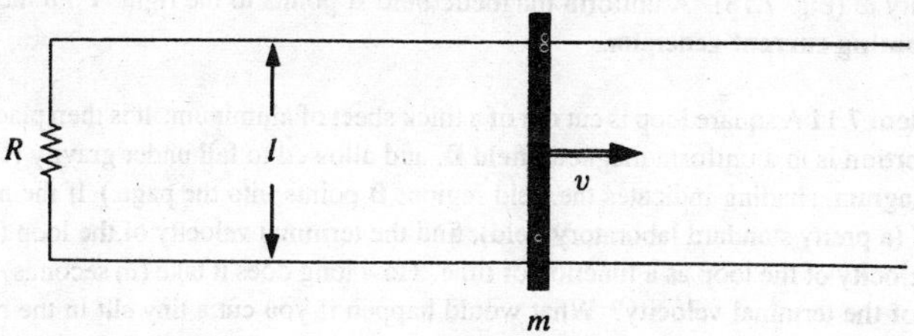

Figure 7.16

(a) If the bar moves to the right at speed v, what is the current in the resistor? In what direction does it flow?

(b) What is the magnetic force on the bar? In what direction?

(c) If the bar starts out with speed v_0 at time $t = 0$, and is left to slide, what is its speed at a later time t?

(d) The initial kinetic energy of the bar was, of course, $\frac{1}{2}mv_0{}^2$. Check that the energy delivered to the resistor is exactly $\frac{1}{2}mv_0{}^2$.

Problem 7.8 A square loop of wire (side a) lies on a table, a distance s from a very long straight wire, which carries a current I, as shown in Fig. 7.17.

(a) Find the flux of **B** through the loop.

(b) If someone now pulls the loop directly away from the wire, at speed v, what emf is generated? In what direction (clockwise or counterclockwise) does the current flow?

(c) What if the loop is pulled to the *right* at speed v, instead of away?

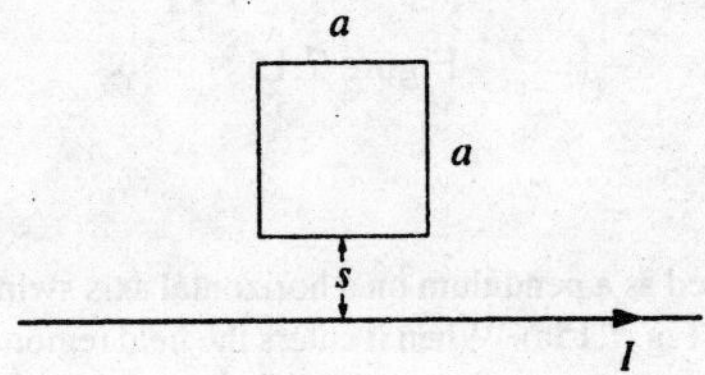

Figure 7.17

Problem 7.9 An infinite number of different surfaces can be fit to a given boundary line, and yet, in defining the magnetic flux through a loop, $\Phi = \int \mathbf{B} \cdot d\mathbf{a}$, I never specified the particular surface to be used. Justify this apparent oversight.

Problem 7.10 A square loop (side a) is mounted on a vertical shaft and rotated at angular velocity ω (Fig. 7.18). A uniform magnetic field **B** points to the right. Find the $\mathcal{E}(t)$ for this **alternating current** generator.

Problem 7.11 A square loop is cut out of a thick sheet of aluminum. It is then placed so that the top portion is in a uniform magnetic field **B**, and allowed to fall under gravity (Fig. 7.19). (In the diagram, shading indicates the field region; **B** points into the page.) If the magnetic field is 1 T (a pretty standard laboratory field), find the terminal velocity of the loop (in m/s). Find the velocity of the loop as a function of time. How long does it take (in seconds) to reach, say, 90% of the terminal velocity? What would happen if you cut a tiny slit in the ring, breaking the circuit? [*Note:* The dimensions of the loop cancel out; determine the actual *numbers*, in the units indicated.]

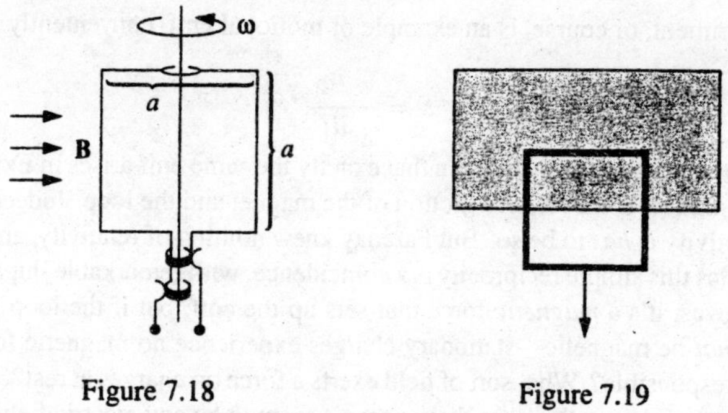

Figure 7.18 Figure 7.19

7.2 Electromagnetic Induction

7.2.1 Faraday's Law

In 1831 Michael Faraday reported on a series of experiments, including three that (with some violence to history) can be characterized as follows:

Experiment 1. He pulled a loop of wire to the right through a magnetic field (Fig. 7.20a). A current flowed in the loop.

Experiment 2. He moved the *magnet* to the *left*, holding the loop still (Fig. 7.20b). Again, a current flowed in the loop.

Experiment 3. With both the loop and the magnet at rest (Fig. 7.20c), he changed the *strength* of the field (he used an electromagnet, and varied the current in the coil). Once again, current flowed in the loop.

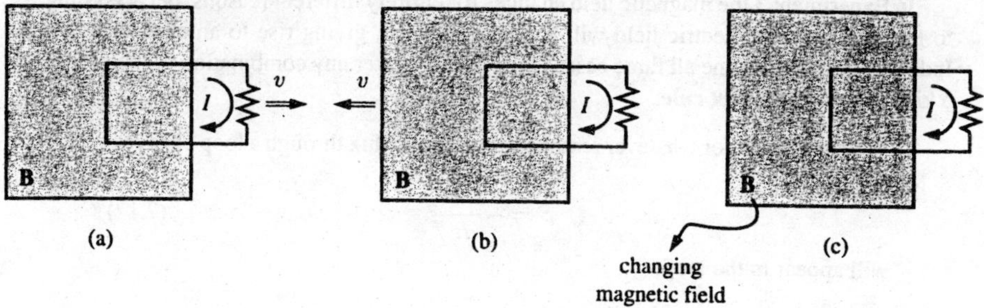

Figure 7.20

The first experiment, of course, is an example of motional emf, conveniently expressed by the flux rule:

$$\mathcal{E} = -\frac{d\Phi}{dt}.$$

I don't think it will surprise you to learn that exactly the same emf arises in Experiment 2—all that really matters is the *relative* motion of the magnet and the loop. Indeed, in the light of special relativity is *has* to be so. But Faraday knew nothing of relativity, and in classical electrodynamics this simple reciprocity is a coincidence, with remarkable implications. For if the *loop* moves, it's a *magnetic* force that sets up the emf, but if the loop is *stationary*, the force *cannot* be magnetic—stationary charges experience no magnetic forces. In that case, what *is* responsible? What sort of field exerts a force on charges at rest? Well, *electric* fields do, of course, but in this case there doesn't seem to be any electric field in sight.

Faraday had an ingenious inspiration:

A changing magnetic field induces an electric field.

It is this "induced" electric field that accounts for the emf in Experiment 2.[6] Indeed, if (as Faraday found empirically) the emf is again equal to the rate of change of the flux,

$$\mathcal{E} = \oint \mathbf{E} \cdot d\mathbf{l} = -\frac{d\Phi}{dt}, \tag{7.14}$$

then **E** is related to the change in **B** by the equation

$$\oint \mathbf{E} \cdot d\mathbf{l} = -\int \frac{\partial \mathbf{B}}{\partial t} \cdot d\mathbf{a}. \tag{7.15}$$

This is **Faraday's law**, in integral form. We can convert it to differential form by applying Stokes' theorem:

$$\boxed{\nabla \times \mathbf{E} = -\frac{\partial \mathbf{B}}{\partial t}.} \tag{7.16}$$

Note that Faraday's law reduces to the old rule $\oint \mathbf{E} \cdot d\mathbf{l} = 0$ (or, in differential form, $\nabla \times \mathbf{E} = 0$) in the static case (constant **B**) as, of course, it should.

In Experiment 3 the magnetic field changes for entirely different reasons, but according to Faraday's law an electric field will again be induced, giving rise to an emf $-d\Phi/dt$. Indeed, one can subsume all three cases (and for that matter any combination of them) into a kind of **universal flux rule**:

> Whenever (and for whatever reason) the magnetic flux through a loop changes, an emf
>
> $$\mathcal{E} = -\frac{d\Phi}{dt} \tag{7.17}$$
>
> will appear in the loop.

[6]You might argue that the magnetic field in Experiment 2 is not really *changing*—just *moving*. What I mean is that if you sit at a *fixed location*, the field *does* change, as the magnet passes by.

Many people call *this* "Faraday's law." Maybe I'm overly fastidious, but I find this confusing. There are really *two* totally different mechanisms underlying Eq. 7.17, and to identify them both as "Faraday's law" is a little like saying that because identical twins look alike we ought to call them by the same name. In Faraday's first experiment it's the Lorentz force law at work; the emf is *magnetic*. But in the other two it's an *electric* field (induced by the changing magnetic field) that does the job. Viewed in this light, it is quite astonishing that all three processes yield the same formula for the emf. In fact, it was precisely this "coincidence" that led Einstein to the special theory of relativity—he sought a deeper understanding of what is, in classical electrodynamics, a peculiar accident. But that's a story for Chapter 12. In the meantime I shall reserve the term "Faraday's law" for electric fields induced by changing magnetic fields, and I do *not* regard Experiment 1 as an instance of Faraday's law.

Example 7.5

A long cylindrical magnet of length L and radius a carries a uniform magnetization $\mathbf{M}$ parallel to its axis. It passes at constant velocity v through a circular wire ring of slightly larger diameter (Fig. 7.21). Graph the emf induced in the ring, as a function of time.

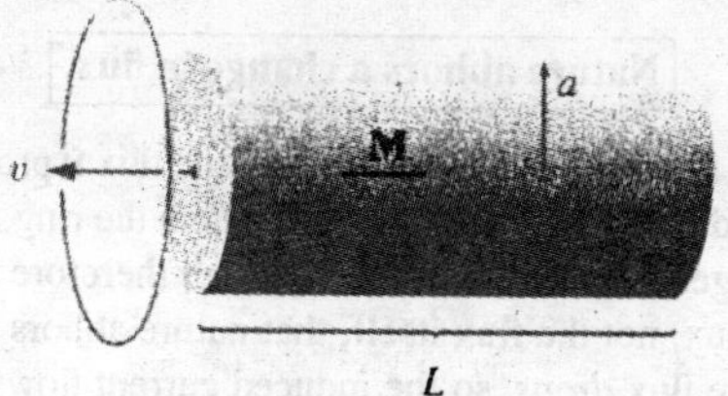

Figure 7.21

Solution: The magnetic field is the same as that of a long solenoid with surface current $\mathbf{K}_b = M\,\hat{\boldsymbol{\phi}}$. So the field inside is $\mathbf{B} = \mu_0 \mathbf{M}$, except near the ends, where it starts to spread out. The flux through the ring is zero when the magnet is far away; it builds up to a maximum of $\mu_0 M \pi a^2$ as the leading end passes through; and it drops back to zero as the trailing end emerges (Fig. 7.22a). The emf is (minus) the derivative of Φ with respect to time, so it consists of two spikes, as shown in Fig. 7.22b.

Keeping track of the *signs* in Faraday's law can be a real headache. For instance, in Ex. 7.5 we would like to know which *way* around the ring the induced current flows. In principle, the right-hand rule does the job (we called Φ positive to the left, in Fig. 7.22a, so the positive direction for current in the ring is counterclockwise, as viewed from the left; since the first spike in Fig. 7.22b is *negative*, the first current pulse flows *clockwise*, and the second counterclockwise). But there's a handy rule, called **Lenz's law**, whose sole purpose is to help you get the directions right:[7]

[7]Lenz's law applies to *motional* emf's, too, but for them it is usually easier to get the direction of the current from the Lorentz force law.

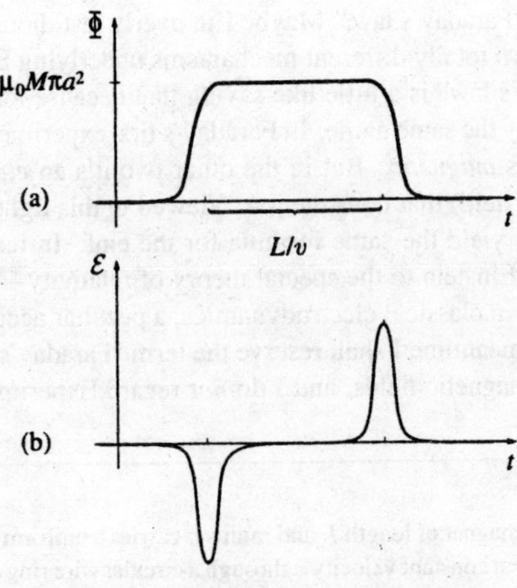

Figure 7.22

Nature abhors a change in flux.

The induced current will flow in such a direction that the flux *it* produces tends to cancel the change. (As the front end of the magnet in Ex. 7.5 enters the ring, the flux increases, so the current in the ring must generate a field to the *right*—it therefore flows *clockwise*.) Notice that it is the *change* in flux, not the flux itself, that nature abhors (when the tail end of the magnet exits the ring, the flux *drops*, so the induced current flows *counterclockwise*, in an effort to restore it). Faraday induction is a kind of "inertial" phenomenon: A conducting loop "likes" to maintain a constant flux through it; if you try to *change* the flux, the loop responds by sending a current around in such a direction as to frustrate your efforts. (It doesn't *succeed* completely; the flux produced by the induced current is typically only a tiny fraction of the original. All Lenz's law tells you is the *direction* of the flow.)

Example 7.6

The "jumping ring" demonstration. If you wind a solenoidal coil around an iron core (the iron is there to beef up the magnetic field), place a metal ring on top, and plug it in, the ring will jump several feet in the air (Fig. 7.23). Why?

Solution: *Before* you turned on the current, the flux through the ring was *zero*. *Afterward* a flux appeared (upward, in the diagram), and the emf generated in the ring led to a current (in the ring) which, according to Lenz's law, was in such a direction that *its* field tended to cancel this new flux. This means that the current in the loop is *opposite* to the current in the solenoid. And opposite currents repel, so the ring flies off.[8]

---

[8]For further discussion of the jumping ring (and the related "floating ring"), see C. S. Schneider and J. P. Ertel, *Am. J. Phys.* **66**, 686 (1998).

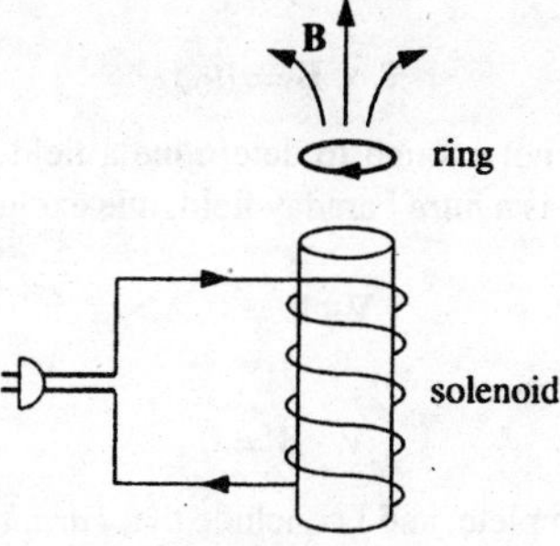

Figure 7.23

Problem 7.12 A long solenoid, of radius a, is driven by an alternating current, so that the field inside is sinusoidal: $\mathbf{B}(t) = B_0 \cos(\omega t)\,\hat{\mathbf{z}}$. A circular loop of wire, of radius $a/2$ and resistance R, is placed inside the solenoid, and coaxial with it. Find the current induced in the loop, as a function of time.

Problem 7.13 A square loop of wire, with sides of length a, lies in the first quadrant of the xy plane, with one corner at the origin. In this region there is a nonuniform time-dependent magnetic field $\mathbf{B}(y,t) = ky^3t^2\,\hat{\mathbf{z}}$ (where k is a constant). Find the emf induced in the loop.

Problem 7.14 As a lecture demonstration a short cylindrical bar magnet is dropped down a vertical aluminum pipe of slightly larger diameter, about 2 meters long. It takes several seconds to emerge at the bottom, whereas an otherwise identical piece of *unmagnetized* iron makes the trip in a fraction of a second. Explain why the magnet falls more slowly.

7.2.2 The Induced Electric Field

What Faraday's discovery tells us is that there are really two distinct kinds of electric fields: those attributable directly to electric charges, and those associated with changing magnetic fields.[9] The former can be calculated (in the static case) using Coulomb's law; the latter can be found by exploiting the analogy between Faraday's law,

$$\nabla \times \mathbf{E} = -\frac{\partial \mathbf{B}}{\partial t},$$

[9]You could, I suppose, introduce an entirely new word to denote the field generated by a changing $\mathbf{B}$. Electrodynamics would then involve *three* fields: E-fields, produced by electric charges [$\nabla \cdot \mathbf{E} = (1/\epsilon_0)\rho,\ \nabla \times \mathbf{E} = 0$]; B-fields, produced by electric currents [$\nabla \cdot \mathbf{B} = 0,\ \nabla \times \mathbf{B} = \mu_0 \mathbf{J}$]; and G-fields, produced by changing magnetic fields [$\nabla \cdot \mathbf{G} = 0,\ \nabla \times \mathbf{G} = -\partial \mathbf{B}/\partial t$]. Because E and G exert *forces* in the same way [$\mathbf{F} = q(\mathbf{E}+\mathbf{G})$], it is tidier to regard their sum as a *single* entity and call the whole thing "the electric field."

and Ampère's law,

$$\nabla \times \mathbf{B} = \mu_0 \mathbf{J}.$$

Of course, the curl alone is not enough to determine a field—you must also specify the divergence. But as long as **E** is a *pure* Faraday field, due exclusively to a changing **B** (with $\rho = 0$), Gauss's law says

$$\nabla \cdot \mathbf{E} = 0,$$

while for magnetic fields, of course,

$$\nabla \cdot \mathbf{B} = 0$$

always. So the parallel is complete, and I conclude that *Faraday-induced electric fields are determined by* $-(\partial \mathbf{B}/\partial t)$ *in exactly the same way as magnetostatic fields are determined by* $\mu_0 \mathbf{J}$.

In particular, if symmetry permits, we can use all the tricks associated with Ampère's law in integral form,

$$\oint \mathbf{B} \cdot d\mathbf{l} = \mu_0 I_{\text{enc}},$$

only this time it's *Faraday's* law in integral form:

$$\oint \mathbf{E} \cdot d\mathbf{l} = -\frac{d\Phi}{dt}. \tag{7.18}$$

The rate of change of (magnetic) flux through the Amperian loop plays the role formerly assigned to $\mu_0 I_{\text{enc}}$.

Example 7.7

A uniform magnetic field $\mathbf{B}(t)$, pointing straight up, fills the shaded circular region of Fig. 7.24. If **B** is changing with time, what is the induced electric field?

Solution: **E** points in the circumferential direction, just like the *magnetic* field inside a long straight wire carrying a uniform *current* density. Draw an Amperian loop of radius s, and apply Faraday's law:

$$\oint \mathbf{E} \cdot d\mathbf{l} = E(2\pi s) = -\frac{d\Phi}{dt} = -\frac{d}{dt}\left(\pi s^2 B(t)\right) = -\pi s^2 \frac{dB}{dt}.$$

Therefore

$$\mathbf{E} = -\frac{s}{2}\frac{dB}{dt}\,\hat{\boldsymbol{\phi}}.$$

If **B** is *increasing*, **E** runs *clockwise*, as viewed from above.

Example 7.8

A line charge λ is glued onto the rim of a wheel of radius b, which is then suspended horizontally, as shown in Fig. 7.25, so that it is free to rotate (the spokes are made of some nonconducting material—wood, maybe). In the central region, out to radius a, there is a uniform magnetic field $\mathbf{B}_0$, pointing up. Now someone turns the field off. What happens?

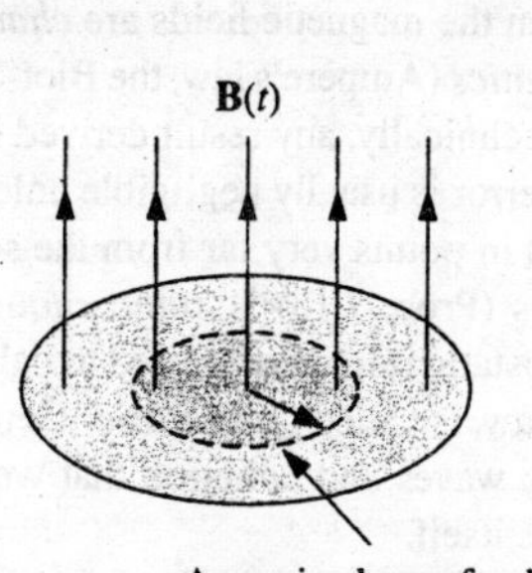

Figure 7.24

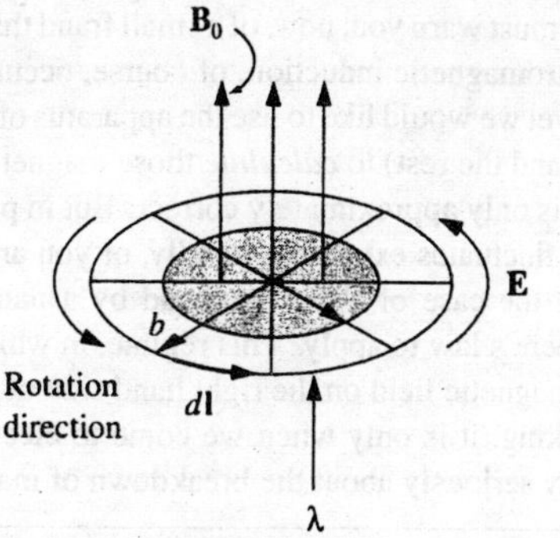

$\mathbf{B}_0$

$\mathbf{E}$

b

Rotation direction

$d\mathbf{l}$

λ

Figure 7.25

Solution: The changing magnetic field will induce an electric field, curling around the axis of the wheel. This electric field exerts a force on the charges at the rim, and the wheel starts to turn. According to Lenz's law, it will rotate in such a direction that *its* field tends to restore the upward flux. The motion, then, is counterclockwise, as viewed from above.

Quantitatively, Faraday's law says

$$\oint \mathbf{E} \cdot d\mathbf{l} = -\frac{d\Phi}{dt} = -\pi a^2 \frac{dB}{dt}.$$

Now, the torque on a segment of length $d\mathbf{l}$ is $(\mathbf{r} \times \mathbf{F})$, or $b\lambda E\, dl$. The total torque on the wheel is therefore

$$N = b\lambda \oint E\, dl = -b\lambda\pi a^2 \frac{dB}{dt},$$

and the angular momentum imparted to the wheel is

$$\int N dt = -\lambda\pi a^2 b \int_{B_0}^{0} dB = \lambda\pi a^2 b B_0.$$

It doesn't matter how fast or slow you turn off the field; the ultimate angular velocity of the wheel is the same regardless. (If you find yourself wondering where this angular momentum *came* from, you're getting ahead of the story! Wait for the next chapter.)

A final word on this example: It's the *electric* field that did the rotating. To convince you of this I deliberately set things up so that the *magnetic* field is always *zero* at the location of the charge (on the rim). The experimenter may tell you she never put in any electric fields—all she did was switch off the magnetic field. But when she did that, an electric field automatically appeared, and it's this electric field that turned the wheel.

I must warn you, now, of a small fraud that tarnishes many applications of Faraday's law: Electromagnetic induction, of course, occurs only when the magnetic fields are *changing*, and yet we would like to use the apparatus of magneto*statics* (Ampère's law, the Biot-Savart law, and the rest) to *calculate* those magnetic fields. Technically, any result derived in this way is only approximately correct. But in practice the error is usually negligible unless the field fluctuates extremely rapidly, or you are interested in points very far from the source. Even the case of a wire snipped by a pair of scissors (Prob. 7.18) is *static enough* for Ampère's law to apply. This régime, in which magnetostatic rules can be used to calculate the magnetic field on the right hand side of Faraday's law, is called **quasistatic**. Generally speaking, it is only when we come to electromagnetic waves and radiation that we must worry seriously about the breakdown of magnetostatics itself.

Example 7.9

An infinitely long straight wire carries a slowly varying current $I(t)$. Determine the induced electric field, as a function of the distance s from the wire.[10]

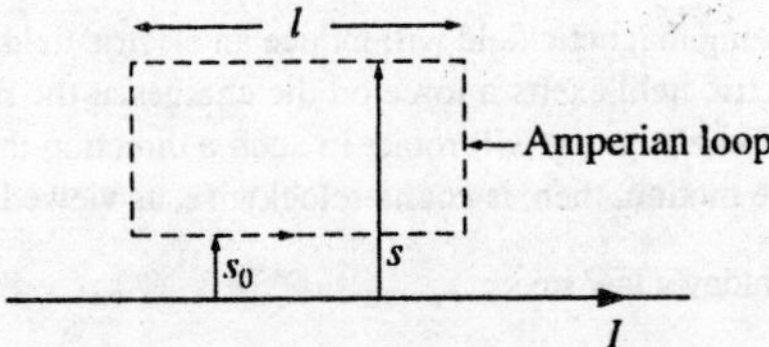

Figure 7.26

Solution: In the quasistatic approximation, the magnetic field is $(\mu_0 I/2\pi s)$, and it circles around the wire. Like the **B**-field of a solenoid, **E** here runs parallel to the axis. For the rectangular "Amperian loop" in Fig. 7.26, Faraday's law gives:

$$\begin{aligned}\oint \mathbf{E}\cdot d\mathbf{l} &= E(s_0)l - E(s)l = -\frac{d}{dt}\int \mathbf{B}\cdot d\mathbf{a} \\ &= -\frac{\mu_0 l}{2\pi}\frac{dI}{dt}\int_{s_0}^{s}\frac{1}{s'}\,ds' = -\frac{\mu_0 l}{2\pi}\frac{dI}{dt}(\ln s - \ln s_0).\end{aligned}$$

Thus

$$\mathbf{E}(s) = \left[\frac{\mu_0}{2\pi}\frac{dI}{dt}\ln s + K\right]\hat{\mathbf{z}}, \tag{7.19}$$

where K is a constant (that is to say, it is independent of s—it might still be a function of t). The actual *value* of K depends on the whole history of the function $I(t)$—we'll see some examples in Chapter 10.

---

[10] This example is artificial, and not just in the usual sense of involving infinite wires, but in a more subtle respect. It assumes that the current is the same (at any given instant) all the way down the line. This is a safe assumption for the *short* wires in typical electric circuits, but not (in practice) for *long* wires (transmission lines), unless you supply a distributed and synchronized driving mechanism. But never mind—the problem doesn't inquire how you would *produce* such a current; it only asks what *fields* would result if you *did*. (Variations on this problem are discussed in M. A. Heald, *Am. J. Phys.* **54**, 1142 (1986), and references cited therein.)

Equation 7.19 has the peculiar implication that E blows up as s goes to infinity. *That* can't be true ... What's gone wrong? *Answer:* We have overstepped the limits of the quasistatic approximation. As we shall see in Chapter 9, electromagnetic "news" travels at the speed of light, and at large distances **B** depends not on the current *now*, but on the current *as it was* at some earlier time (indeed, a whole *range* of earlier times, since different points on the wire are different distances away). If τ is the time it takes I to change substantially, then the quasistatic approximation should hold only for

$$s \ll c\tau, \tag{7.20}$$

and hence Eq. 7.19 simply does not apply, at extremely large s.

Problem 7.15 A long solenoid with radius a and n turns per unit length carries a time-dependent current $I(t)$ in the $\hat{\boldsymbol{\phi}}$ direction. Find the electric field (magnitude and direction) at a distance s from the axis (both inside and outside the solenoid), in the quasistatic approximation.

Problem 7.16 An alternating current $I = I_0 \cos(\omega t)$ flows down a long straight wire, and returns along a coaxial conducting tube of radius a.

(a) In what *direction* does the induced electric field point (radial, circumferential, or longitudinal)?

(b) Assuming that the field goes to zero as $s \to \infty$, find $\mathbf{E}(s, t)$. [Incidentally, this is not at all the way electric fields *actually* behave in coaxial cables, for reasons suggested in footnote 10. See Sect. 9.5.3, or J. G. Cherveniak, *Am. J. Phys.*, **54**, 946 (1986), for a more realistic treatment.]

Problem 7.17 A long solenoid of radius a, carrying n turns per unit length, is looped by a wire with resistance R, as shown in Fig. 7.27.

(a) If the current in the solenoid is increasing at a constant rate ($dI/dt = k$), what current flows in the loop, and which way (left or right) does it pass through the resistor?

(b) If the current I in the solenoid is constant but the solenoid is pulled out of the loop, turned around, and reinserted, what total charge passes through the resistor?

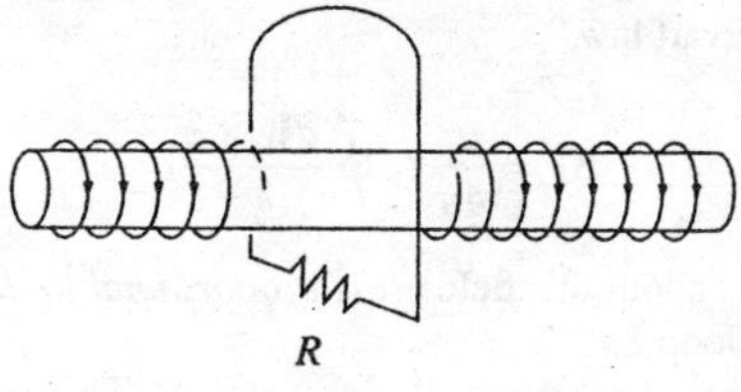

Figure 7.27

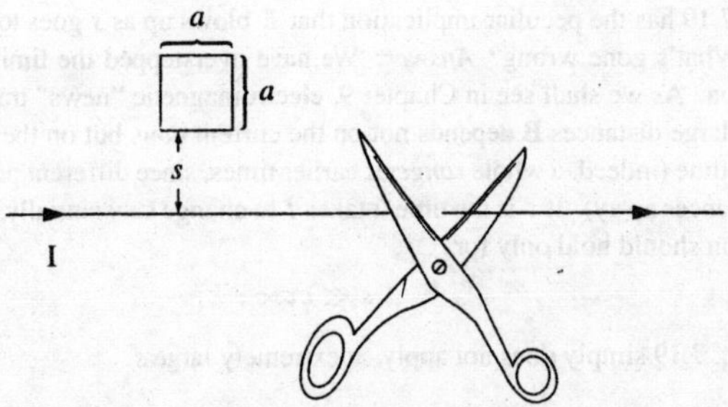

Figure 7.28

Problem 7.18 A square loop, side a, resistance R, lies a distance s from an infinite straight wire that carries current I (Fig. 7.28). Now someone cuts the wire, so that I drops to zero. In what direction does the induced current in the square loop flow, and what total charge passes a given point in the loop during the time this current flows? If you don't like the scissors model, turn the current down *gradually*:

$$I(t) = \begin{cases} (1-\alpha t)I, & \text{for } 0 \le t \le 1/\alpha, \\ 0, & \text{for } t > 1/\alpha. \end{cases}$$

Problem 7.19 A toroidal coil has a rectangular cross section, with inner radius a, outer radius $a+w$, and height h. It carries a total of N tightly wound turns, and the current is increasing at a constant rate ($dI/dt = k$). If w and h are both much less than a, find the electric field at a point z above the center of the toroid. [*Hint:* exploit the analogy between Faraday fields and magnetostatic fields, and refer to Ex. 5.6.]

7.2.3 Inductance

Suppose you have two loops of wire, at rest (Fig. 7.29). If you run a steady current I_1 around loop 1, it produces a magnetic field $\mathbf{B}_1$. Some of the field lines pass through loop 2; let Φ_2 be the flux of $\mathbf{B}_1$ through 2. You might have a tough time actually *calculating* $\mathbf{B}_1$, but a glance at the Biot-Savart law,

$$\mathbf{B}_1 = \frac{\mu_0}{4\pi} I_1 \oint \frac{d\mathbf{l}_1 \times \hat{\boldsymbol{\mathscr{r}}}}{\mathscr{r}^2},$$

reveals one significant fact about this field: *It is proportional to the current I_1*. Therefore, so too is the flux through loop 2:

$$\Phi_2 = \int \mathbf{B}_1 \cdot d\mathbf{a}_2.$$

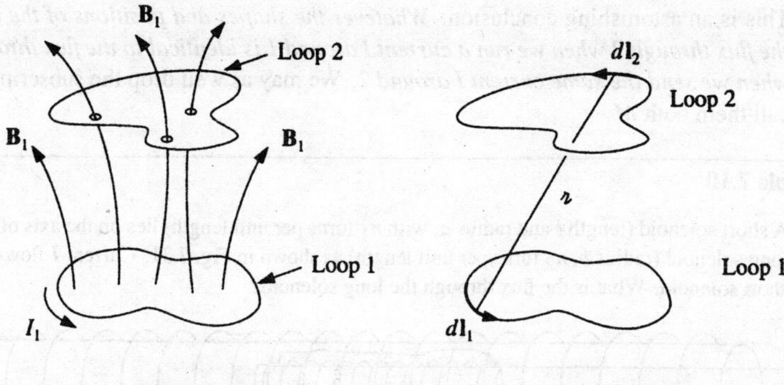

Figure 7.29 Figure 7.30

Thus

$$\Phi_2 = M_{21} I_1, \tag{7.21}$$

where M_{21} is the constant of proportionality; it is known as the **mutual inductance** of the two loops.

There is a cute formula for the mutual inductance, which you can derive by expressing the flux in terms of the vector potential and invoking Stokes' theorem:

$$\Phi_2 = \int \mathbf{B}_1 \cdot d\mathbf{a}_2 = \int (\nabla \times \mathbf{A}_1) \cdot d\mathbf{a}_2 = \oint \mathbf{A}_1 \cdot d\mathbf{l}_2.$$

Now, according to Eq. 5.63,

$$\mathbf{A}_1 = \frac{\mu_0 I_1}{4\pi} \oint \frac{d\mathbf{l}_1}{\mathscr{r}},$$

and hence

$$\Phi_2 = \frac{\mu_0 I_1}{4\pi} \oint \left(\oint \frac{d\mathbf{l}_1}{\mathscr{r}} \right) \cdot d\mathbf{l}_2.$$

Evidently

$$M_{21} = \frac{\mu_0}{4\pi} \oint \oint \frac{d\mathbf{l}_1 \cdot d\mathbf{l}_2}{\mathscr{r}}. \tag{7.22}$$

This is the **Neumann formula**; it involves a double line integral—one integration around loop 1, the other around loop 2 (Fig. 7.30). It's not very useful for practical calculations, but it does reveal two important things about mutual inductance:

1. M_{21} is a purely geometrical quantity, having to do with the sizes, shapes, and relative positions of the two loops.

2. The integral in Eq. 7.22 is unchanged if we switch the roles of loops 1 and 2; it follows that

$$M_{21} = M_{12}. \tag{7.23}$$

This is an astonishing conclusion: *Whatever the shapes and positions of the loops, the flux through 2 when we run a current I around 1 is identical to the flux through 1 when we send the same current I around 2.* We may as well drop the subscripts and call them both M.

Example 7.10

A short solenoid (length l and radius a, with n_1 turns per unit length) lies on the axis of a very long solenoid (radius b, n_2 turns per unit length) as shown in Fig. 7.31. Current I flows in the short solenoid. What is the flux through the long solenoid?

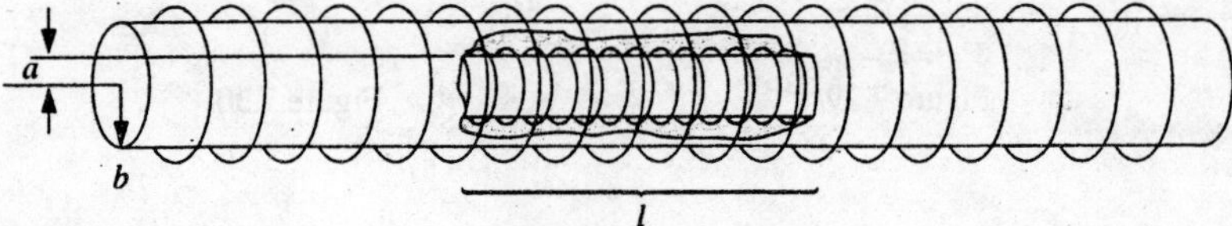

Figure 7.31

Solution: Since the inner solenoid is short, it has a very complicated field; moreover, it puts a different amount of flux through each turn of the outer solenoid. It would be a *miserable* task to compute the total flux this way. However, if we exploit the equality of the mutual inductances, the problem becomes very easy. Just look at the reverse situation: run the current I through the *outer* solenoid, and calculate the flux through the *inner* one. The field inside the long solenoid is constant:

$$B = \mu_0 n_2 I$$

(Eq. 5.57), so the flux through a single loop of the short solenoid is

$$B\pi a^2 = \mu_0 n_2 I \pi a^2.$$

There are $n_1 l$ turns in all, so the total flux through the inner solenoid is

$$\Phi = \mu_0 \pi a^2 n_1 n_2 l I.$$

This is also the flux a current I in the *short* solenoid would put through the *long* one, which is what we set out to find. Incidentally, the mutual inductance, in this case, is

$$M = \mu_0 \pi a^2 n_1 n_2 l.$$

Suppose now that you *vary* the current in loop 1. The flux through loop 2 will vary accordingly, and Faraday's law says this changing flux will induce an emf in loop 2:

$$\mathcal{E}_2 = -\frac{d\Phi_2}{dt} = -M\frac{dI_1}{dt}. \tag{7.24}$$

(In quoting Eq. 7.21—which was based on the Biot-Savart law—I am tacitly assuming that the currents change slowly enough for the configuration to be considered quasistatic.) What

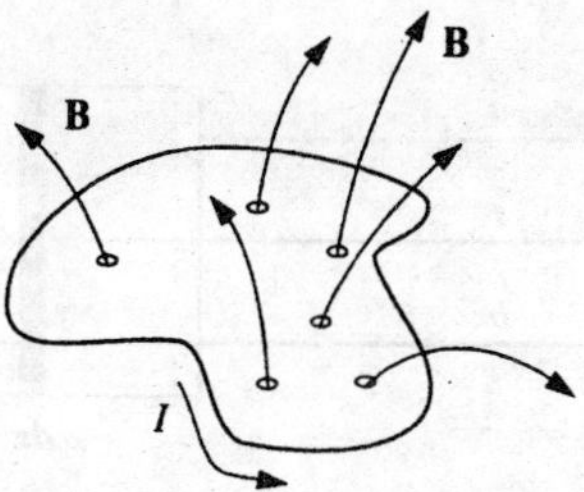

Figure 7.32

a remarkable thing: Every time you change the current in loop 1, an induced current flows in loop 2—even though there are no wires connecting them!

Come to think of it, a changing current not only induces an emf in any nearby loops, it also induces an emf in the source loop *itself* (Fig 7.32). Once again, the field (and therefore also the flux) is proportional to the current:

$$\Phi = LI. \tag{7.25}$$

The constant of proportionality L is called the **self-inductance** (or simply the **inductance**) of the loop. As with M, it depends on the geometry (size and shape) of the loop. If the current changes, the emf induced in the loop is

$$\mathcal{E} = -L\frac{dI}{dt}. \tag{7.26}$$

Inductance is measured in **henries** (H); a henry is a volt-second per ampere.

Example 7.11

Find the self-inductance of a toroidal coil with rectangular cross section (inner radius a, outer radius b, height h), which carries a total of N turns.

Solution: The magnetic field inside the toroid is (Eq. 5.58)

$$B = \frac{\mu_0 N I}{2\pi s}.$$

The flux through a single turn (Fig. 7.33) is

$$\int \mathbf{B} \cdot d\mathbf{a} = \frac{\mu_0 N I}{2\pi} h \int_a^b \frac{1}{s}\, ds = \frac{\mu_0 N I h}{2\pi} \ln\left(\frac{b}{a}\right).$$

The *total* flux is N times this, so the self-inductance (Eq. 7.25) is

$$L = \frac{\mu_0 N^2 h}{2\pi} \ln\left(\frac{b}{a}\right). \tag{7.27}$$

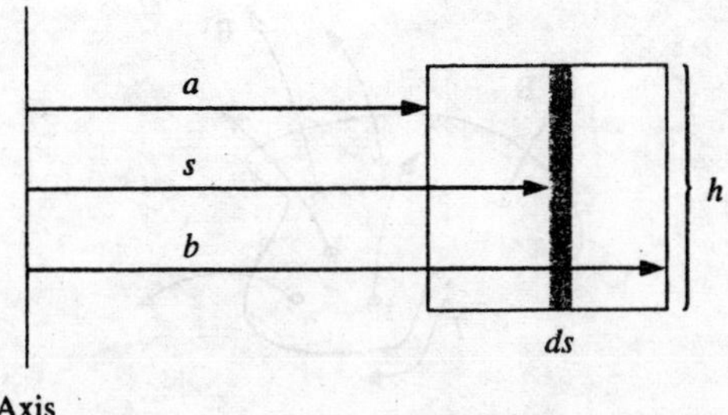

Figure 7.33

Inductance (like capacitance) is an intrinsically *positive* quantity. Lenz's law, which is enforced by the minus sign in Eq. 7.26, dictates that the emf is in such a direction as to *oppose* any *change in current*. For this reason, it is called a **back emf**. Whenever you try to alter the current in a wire, you must fight against this back emf. Thus inductance plays somewhat the same role in electric circuits that *mass* plays in mechanical systems: The greater L is, the harder it is to change the current, just as the larger the mass, the harder it is to change an object's velocity.

Example 7.12

Suppose a current I is flowing around a loop when someone suddenly cuts the wire. The current drops "instantaneously" to zero. This generates a whopping back emf, for although I may be small, dI/dt is enormous. That's why you often draw a spark when you unplug an iron or toaster—electromagnetic induction is desperately trying to keep the current going, even if it has to jump the gap in the circuit.

Nothing so dramatic occurs when you plug *in* a toaster or iron. In this case induction opposes the sudden *increase* in current, prescribing instead a smooth and continuous buildup. Suppose, for instance, that a battery (which supplies a constant emf $\mathcal{E}_0$) is connected to a circuit of resistance R and inductance L (Fig. 7.34). What current flows?

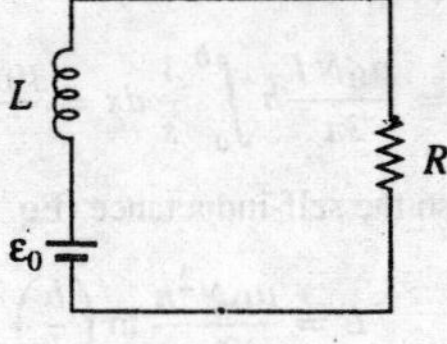

Figure 7.34

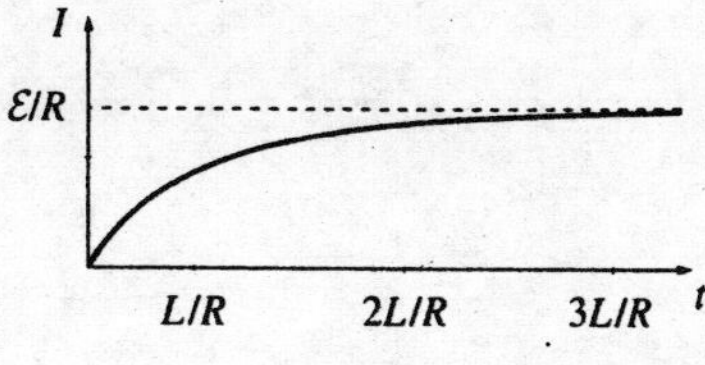

Figure 7.35

Solution: The total emf in this circuit is that provided by the battery plus that resulting from the self-inductance. Ohm's law, then, says[11]

$$\mathcal{E}_0 - L\frac{dI}{dt} = IR.$$

This is a first-order differential equation for I as a function of time. The general solution, as you can easily derive for yourself, is

$$I(t) = \frac{\mathcal{E}_0}{R} + ke^{-(R/L)t},$$

where k is a constant to be determined by the initial conditions. In particular, if the circuit is "plugged in" at time $t = 0$ (so $I(0) = 0$), then k has the value $-\mathcal{E}_0/R$, and

$$I(t) = \frac{\mathcal{E}_0}{R}\left[1 - e^{-(R/L)t}\right]. \tag{7.28}$$

This function is plotted in Fig. 7.35. Had there been no inductance in the circuit, the current would have jumped immediately to $\mathcal{E}_0/R$. In practice, *every* circuit has *some* self-inductance, and the current approaches $\mathcal{E}_0/R$ asymptotically. The quantity $\tau \equiv L/R$ is called the **time constant**; it tells you how long the current takes to reach a substantial fraction (roughly two-thirds) of its final value.

Problem 7.20 A small loop of wire (radius a) lies a distance z above the center of a large loop (radius b), as shown in Fig. 7.36. The planes of the two loops are parallel, and perpendicular to the common axis.

(a) Suppose current I flows in the big loop. Find the flux through the little loop. (The little loop is so small that you may consider the field of the big loop to be essentially constant.)

(b) Suppose current I flows in the little loop. Find the flux through the big loop. (The little loop is so small that you may treat it as a magnetic dipole.)

(c) Find the mutual inductances, and confirm that $M_{12} = M_{21}$.

---

[11] Notice that $-L(dI/dt)$ goes on the *left* side of the equation—it is part of the emf that (together with $\mathcal{E}_0$) establishes the voltage across the resistor (Eq. 7.10).

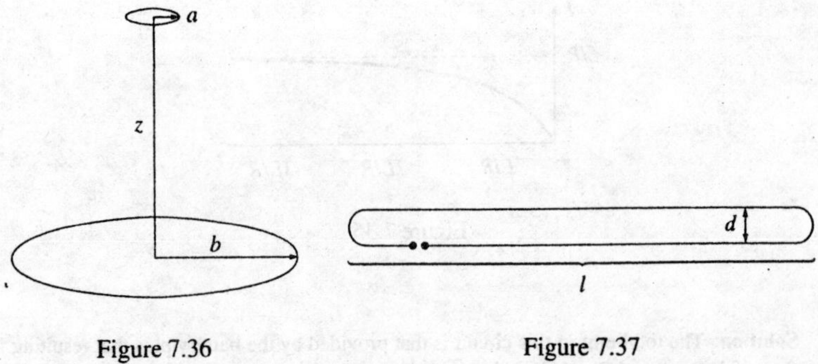

Figure 7.36 Figure 7.37

Problem 7.21 A square loop of wire, of side a, lies midway between two long wires, $3a$ apart, and in the same plane. (Actually, the long wires are sides of a large rectangular loop, but the short ends are so far away that they can be neglected.) A clockwise current I in the square loop is gradually increasing: $dI/dt = k$ (a constant). Find the emf induced in the big loop. Which way will the induced current flow?

Problem 7.22 Find the self-inductance per unit length of a long solenoid, of radius R, carrying n turns per unit length.

Problem 7.23 Try to compute the self-inductance of the "hairpin" loop shown in Fig. 7.37. (Neglect the contribution from the ends; most of the flux comes from the long straight section.) You'll run into a snag that is characteristic of many self-inductance calculations. To get a definite answer, assume the wire has a tiny radius ϵ, and ignore any flux through the wire itself.

Problem 7.24 An alternating current $I_0 \cos(\omega t)$ (amplitude 0.5 A, frequency 60 Hz) flows down a straight wire, which runs along the axis of a toroidal coil with rectangular cross section (inner radius 1 cm, outer radius 2 cm, height 1 cm, 1000 turns). The coil is connected to a 500 Ω resistor.

(a) In the quasistatic approximation, what emf is induced in the toroid? Find the current, $I_r(t)$, in the resistor.

(b) Calculate the back emf in the coil, due to the current $I_r(t)$. What is the ratio of the amplitudes of this back emf and the "direct" emf in (a)?

Problem 7.25 A capacitor C is charged up to a potential V and connected to an inductor L, as shown schematically in Fig. 7.38. At time $t = 0$ the switch S is closed. Find the current in the circuit as a function of time. How does your answer change if a resistor R is included in series with C and L?

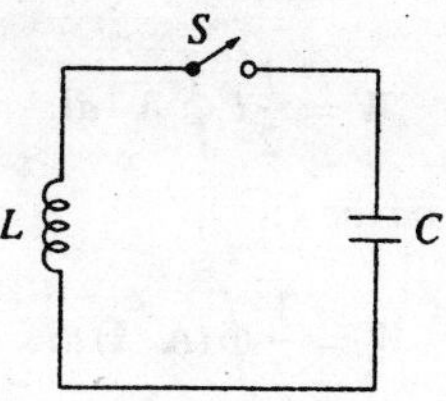

Figure 7.38

7.2.4 Energy in Magnetic Fields

It takes a certain amount of *energy* to start a current flowing in a circuit. I'm not talking about the energy delivered to the resistors and converted into heat—that is irretrievably lost as far as the circuit is concerned and can be large or small, depending on how long you let the current run. What I am concerned with, rather, is the work you must do *against the back emf* to get the current going. This is a *fixed* amount, and it is *recoverable*: you get it back when the current is turned off. In the meantime it represents energy latent in the circuit; as we'll see in a moment, it can be regarded as energy stored in the magnetic field.

The work done on a unit charge, against the back emf, in one trip around the circuit is $-\mathcal{E}$ (the minus sign records the fact that this is the work done *by you against* the emf, not the work done by the emf). The amount of charge per unit time passing down the wire is I. So the total work done per unit time is

$$\frac{dW}{dt} = -\mathcal{E}I = LI\frac{dI}{dt}.$$

If we start with zero current and build it up to a final value I, the work done (integrating the last equation over time) is

$$\boxed{W = \frac{1}{2}LI^2.} \tag{7.29}$$

It does not depend on how *long* we take to crank up the current, only on the geometry of the loop (in the form of L) and the final current I.

There is a nicer way to write W, which has the advantage that it is readily generalized to surface and volume currents. Remember that the flux Φ through the loop is equal to LI (Eq. 7.25). On the other hand,

$$\Phi = \int_S \mathbf{B} \cdot d\mathbf{a} = \int_S (\nabla \times \mathbf{A}) \cdot d\mathbf{a} = \oint_{\mathcal{P}} \mathbf{A} \cdot d\mathbf{l},$$

where $\mathcal{P}$ is the perimeter of the loop and S is any surface bounded by $\mathcal{P}$. Thus,

$$LI = \oint \mathbf{A} \cdot d\mathbf{l},$$

and therefore

$$W = \frac{1}{2} I \oint \mathbf{A} \cdot d\mathbf{l}.$$

The vector sign might as well go on the I:

$$W = \frac{1}{2} \oint (\mathbf{A} \cdot \mathbf{I})\, dl. \tag{7.30}$$

In this form, the generalization to volume currents is obvious:

$$W = \frac{1}{2} \int_{\mathcal{V}} (\mathbf{A} \cdot \mathbf{J})\, d\tau. \tag{7.31}$$

But we can do even better, and express W entirely in terms of the magnetic field: Ampère's law, $\nabla \times \mathbf{B} = \mu_0 \mathbf{J}$, lets us eliminate $\mathbf{J}$:

$$W = \frac{1}{2\mu_0} \int \mathbf{A} \cdot (\nabla \times \mathbf{B})\, d\tau. \tag{7.32}$$

Integration by parts enables us to move the derivative from $\mathbf{B}$ to $\mathbf{A}$; specifically, product rule 6 states that

$$\nabla \cdot (\mathbf{A} \times \mathbf{B}) = \mathbf{B} \cdot (\nabla \times \mathbf{A}) - \mathbf{A} \cdot (\nabla \times \mathbf{B}),$$

so

$$\mathbf{A} \cdot (\nabla \times \mathbf{B}) = \mathbf{B} \cdot \mathbf{B} - \nabla \cdot (\mathbf{A} \times \mathbf{B}).$$

Consequently,

$$\begin{aligned} W &= \frac{1}{2\mu_0} \left[\int B^2\, d\tau - \int \nabla \cdot (\mathbf{A} \times \mathbf{B})\, d\tau \right] \\ &= \frac{1}{2\mu_0} \left[\int_{\mathcal{V}} B^2\, d\tau - \oint_{\mathcal{S}} (\mathbf{A} \times \mathbf{B}) \cdot d\mathbf{a} \right], \end{aligned} \tag{7.33}$$

where $\mathcal{S}$ is the surface bounding the volume $\mathcal{V}$.

Now, the integration in Eq. 7.31 is to be taken over the *entire volume occupied by the current*. But any region *larger* than this will do just as well, for $\mathbf{J}$ is zero out there anyway. In Eq. 7.33 the larger the region we pick the greater is the contribution from the volume integral, and therefore the smaller is that of the surface integral (this makes sense: as the surface gets farther from the current, both $\mathbf{A}$ and $\mathbf{B}$ decrease). In particular, if we agree to integrate over *all* space, then the surface integral goes to zero, and we are left with

$$\boxed{W = \frac{1}{2\mu_0} \int_{\text{all space}} B^2\, d\tau.} \tag{7.34}$$

In view of this result, we say the energy is "stored in the magnetic field," in the amount $(B^2/2\mu_0)$ per unit volume. This is a nice way to think of it, though someone looking at Eq. 7.31 might prefer to say that the energy is stored in the *current distribution*, in the amount $\frac{1}{2}(\mathbf{A}\cdot\mathbf{J})$ per unit volume. The distinction is one of bookkeeping; the important quantity is the total energy W, and we shall not worry about where (if anywhere) the energy is "located."

You might find it strange that it takes energy to set up a magnetic field—after all, magnetic fields *themselves* do no work. The point is that producing a magnetic field, where previously there was none, requires *changing* the field, and a changing **B**-field, according to Faraday, induces an *electric* field. The latter, of course, *can* do work. In the beginning there is no **E**, and at the end there is no **E**; but in between, while **B** is building up, there *is* an **E**, and it is against *this* that the work is done. (You see why I could not calculate the energy stored in a magnetostatic field back in Chapter 5.) In the light of this, it is extraordinary how similar the magnetic energy formulas are to their electrostatic counterparts:

$$W_{\text{elec}} = \frac{1}{2}\int (V\rho)\,d\tau = \frac{\epsilon_0}{2}\int E^2\,d\tau, \qquad \text{(2.43 and 2.45)}$$

$$W_{\text{mag}} = \frac{1}{2}\int (\mathbf{A}\cdot\mathbf{J})\,d\tau = \frac{1}{2\mu_0}\int B^2\,d\tau. \qquad \text{(7.31 and 7.34)}$$

Example 7.13

A long coaxial cable carries current I (the current flows down the surface of the inner cylinder, radius a, and back along the outer cylinder, radius b) as shown in Fig. 7.39. Find the magnetic energy stored in a section of length l.

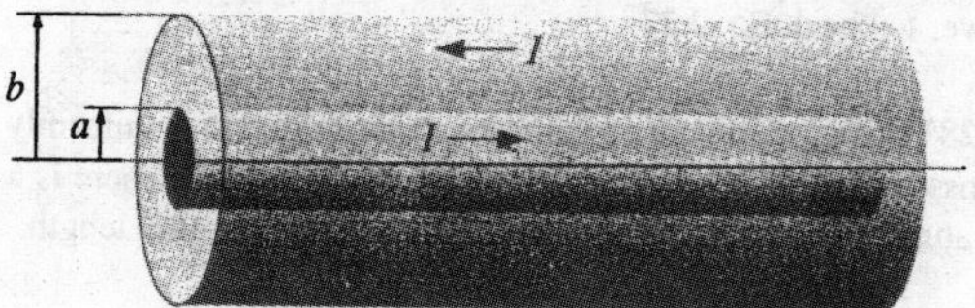

Figure 7.39

Solution: According to Ampère's law, the field between the cylinders is

$$\mathbf{B} = \frac{\mu_0 I}{2\pi s}\hat{\boldsymbol{\phi}}.$$

Elsewhere, the field is zero. Thus, the energy per unit volume is

$$\frac{1}{2\mu_0}\left(\frac{\mu_0 I}{2\pi s}\right)^2 = \frac{\mu_0 I^2}{8\pi^2 s^2}.$$

The energy in a cylindrical shell of length l, radius s, and thickness ds, then, is

$$\left(\frac{\mu_0 I^2}{8\pi^2 s^2}\right) 2\pi l s\,ds = \frac{\mu_0 I^2 l}{4\pi}\left(\frac{ds}{s}\right).$$

Integrating from a to b, we have:

$$W = \frac{\mu_0 I^2 l}{4\pi} \ln\left(\frac{b}{a}\right).$$

By the way, this suggests a very simple way to calculate the self-inductance of the cable. According to Eq. 7.29, the energy can also be written as $\frac{1}{2}LI^2$. Comparing the two expressions,[12]

$$L = \frac{\mu_0 l}{2\pi} \ln\left(\frac{b}{a}\right).$$

This method of calculating self-inductance is especially useful when the current is not confined to a single path, but spreads over some surface or volume. In such cases different parts of the current may circle different amounts of flux, and it can be very tricky to get L directly from Eq. 7.25.

Problem 7.26 Find the energy stored in a section of length l of a long solenoid (radius R, current I, n turns per unit length), (a) using Eq. 7.29 (you found L in Prob. 7.22); (b) using Eq. 7.30 (we worked out **A** in Ex. 5.12); (c) using Eq. 7.34; (d) using Eq. 7.33 (take as your volume the cylindrical tube from radius $a < R$ out to radius $b > R$).

Problem 7.27 Calculate the energy stored in the toroidal coil of Ex. 7.11, by applying Eq. 7.34. Use the answer to check Eq. 7.27.

Problem 7.28 A long cable carries current in one direction uniformly distributed over its (circular) cross section. The current returns along the surface (there is a very thin insulating sheath separating the currents). Find the self-inductance per unit length.

Problem 7.29 Suppose the circuit in Fig. 7.40 has been connected for a long time when suddenly, at time $t = 0$, switch S is thrown, bypassing the battery.

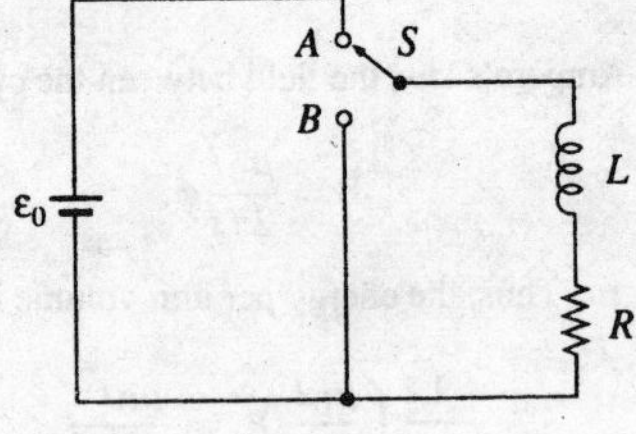

Figure 7.40

[12]Notice the similarity to Eq. 7.27—in a sense, the rectangular toroid *is* a short coaxial cable, turned on its side.

Figure 7.41

(a) What is the current at any subsequent time t?

(b) What is the total energy delivered to the resistor?

(c) Show that this is equal to the energy originally stored in the inductor.

Problem 7.30 Two tiny wire loops, with areas $\mathbf{a}_1$ and $\mathbf{a}_2$, are situated a displacement $\mathbf{\mathscr{r}}$ apart (Fig. 7.41).

(a) Find their mutual inductance. [*Hint:* Treat them as magnetic dipoles, and use Eq. 5.87.] Is your formula consistent with Eq. 7.23?

(b) Suppose a current I_1 is flowing in loop 1, and we propose to turn on a current I_2 in loop 2. How much work must be done, against the mutually induced emf, to keep the current I_1 flowing in loop 1? In light of this result, comment on Eq. 6.35.

7.3 Maxwell's Equations

7.3.1 Electrodynamics Before Maxwell

So far, we have encountered the following laws, specifying the divergence and curl of electric and magnetic fields:

$$\text{(i)} \quad \nabla \cdot \mathbf{E} = \frac{1}{\epsilon_0}\rho \quad \text{(Gauss's law)},$$

$$\text{(ii)} \quad \nabla \cdot \mathbf{B} = 0 \quad \text{(no name)},$$

$$\text{(iii)} \quad \nabla \times \mathbf{E} = -\frac{\partial \mathbf{B}}{\partial t} \quad \text{(Faraday's law)},$$

$$\text{(iv)} \quad \nabla \times \mathbf{B} = \mu_0 \mathbf{J} \quad \text{(Ampère's law)}.$$

These equations represent the state of electromagnetic theory over a century ago, when Maxwell began his work. They were not written in so compact a form in those days, but their physical content was familiar. Now, it happens there is a fatal inconsistency in these formulas. It has to do with the old rule that divergence of curl is always zero. If you apply the divergence to number (iii), everything works out:

$$\nabla \cdot (\nabla \times \mathbf{E}) = \nabla \cdot \left(-\frac{\partial \mathbf{B}}{\partial t}\right) = -\frac{\partial}{\partial t}(\nabla \cdot \mathbf{B}).$$

The left side is zero because divergence of curl is zero; the right side is zero by virtue of equation (ii). But when you do the same thing to number (iv), you get into trouble:

$$\nabla \cdot (\nabla \times \mathbf{B}) = \mu_0(\nabla \cdot \mathbf{J}); \tag{7.35}$$

the left side must be zero, but the right side, in general, is *not*. For *steady* currents, the divergence of $\mathbf{J}$ is zero, but evidently when we go beyond magnetostatics Ampère's law cannot be right.

There's another way to see that Ampère's law is bound to fail for nonsteady currents. Suppose we're in the process of charging up a capacitor (Fig. 7.42). In integral form, Ampère's law reads

$$\oint \mathbf{B} \cdot d\mathbf{l} = \mu_0 I_{enc}.$$

I want to apply it to the Amperian loop shown in the diagram. How do I determine I_{enc}? Well, it's the total current passing through the loop, or, more precisely, the current piercing a surface that has the loop for its boundary. In this case, the *simplest* surface lies in the plane of the loop—the wire punctures this surface, so $I_{enc} = I$. Fine—but what if I draw instead the balloon-shaped surface in Fig. 7.42? *No* current passes through *this* surface, and I conclude that $I_{enc} = 0$! We never had this problem in magnetostatics because the conflict arises only when charge is piling up somewhere (in this case, on the capacitor plates). But *for nonsteady currents* (such as this one) *"the current enclosed by a loop" is an ill-defined notion*, since it depends entirely on what surface you use. (If this seems pedantic to you—"obviously one should use the planar surface"—remember that the Amperian loop could be some contorted shape that doesn't even lie in a plane.)

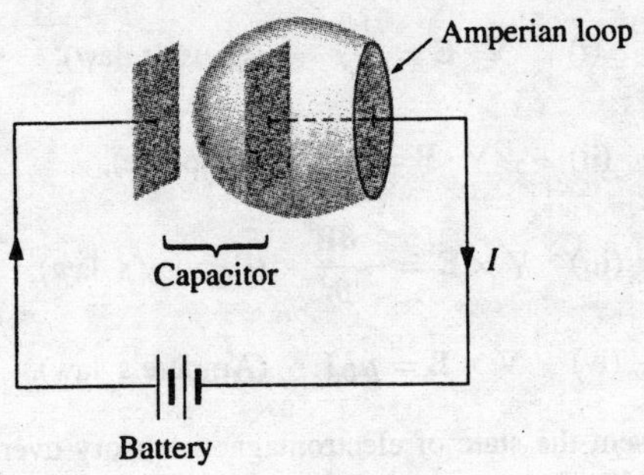

Figure 7.42

Of course, we had no right to *expect* Ampère's law to hold outside of magnetostatics; after all, we derived it from the Biot-Savart law. However, in Maxwell's time there was no *experimental* reason to doubt that Ampère's law was of wider validity. The flaw was a purely theoretical one, and Maxwell fixed it by purely theoretical arguments.

7.3.2 How Maxwell Fixed Ampère's Law

The problem is on the right side of Eq. 7.35, which *should be* zero, but *isn't*. Applying the continuity equation (5.29) and Gauss's law, the offending term can be rewritten:

$$\nabla \cdot \mathbf{J} = -\frac{\partial \rho}{\partial t} = -\frac{\partial}{\partial t}(\epsilon_0 \nabla \cdot \mathbf{E}) = -\nabla \cdot \left(\epsilon_0 \frac{\partial \mathbf{E}}{\partial t}\right).$$

It might occur to you that if we were to combine $\epsilon_0(\partial \mathbf{E}/\partial t)$ with $\mathbf{J}$, in Ampère's law, it would be just right to kill off the extra divergence:

$$\boxed{\nabla \times \mathbf{B} = \mu_0 \mathbf{J} + \mu_0 \epsilon_0 \frac{\partial \mathbf{E}}{\partial t}.} \tag{7.36}$$

(Maxwell himself had other reasons for wanting to add this quantity to Ampère's law. To him the rescue of the continuity equation was a happy dividend rather than a primary motive. But today we recognize this argument as a far more compelling one than Maxwell's, which was based on a now-discredited model of the ether.)[13]

Such a modification changes nothing, as far as magneto*statics* is concerned: when $\mathbf{E}$ is constant, we still have $\nabla \times \mathbf{B} = \mu_0 \mathbf{J}$. In fact, Maxwell's term is hard to detect in ordinary electromagnetic experiments, where it must compete for recognition with $\mathbf{J}$; that's why Faraday and the others never discovered it in the laboratory. However, it plays a crucial role in the propagation of electromagnetic waves, as we'll see in Chapter 9.

Apart from curing the defect in Ampère's law, Maxwell's term has a certain aesthetic appeal: Just as a changing *magnetic* field induces an *electric* field (Faraday's law), so

A changing electric field induces a magnetic field.

Of course, theoretical convenience and aesthetic consistency are only *suggestive*—there might, after all, be other ways to doctor up Ampère's law. The real confirmation of Maxwell's theory came in 1888 with Hertz's experiments on electromagnetic waves.

Maxwell called his extra term the **displacement current**:

$$\mathbf{J}_d \equiv \epsilon_0 \frac{\partial \mathbf{E}}{\partial t}. \tag{7.37}$$

It's a misleading name, since $\epsilon_0(\partial \mathbf{E}/\partial t)$ has nothing to do with current, except that it adds to $\mathbf{J}$ in Ampère's law. Let's see now how the displacement current resolves the paradox of the charging capacitor (Fig. 7.42). If the capacitor plates are very close together (I didn't

---

[13] For the history of this subject, see A. M. Bork, *Am. J. Phys.* **31**, 854 (1963).

draw them that way, but the calculation is simpler if you assume this), then the electric field between them is

$$E = \frac{1}{\epsilon_0}\sigma = \frac{1}{\epsilon_0}\frac{Q}{A},$$

where Q is the charge on the plate and A is its area. Thus, between the plates

$$\frac{\partial E}{\partial t} = \frac{1}{\epsilon_0 A}\frac{dQ}{dt} = \frac{1}{\epsilon_0 A}I.$$

Now, Eq. 7.36 reads, in integral form,

$$\oint \mathbf{B}\cdot d\mathbf{l} = \mu_0 I_{enc} + \mu_0\epsilon_0 \int \left(\frac{\partial \mathbf{E}}{\partial t}\right)\cdot d\mathbf{a}. \tag{7.38}$$

If we choose the *flat* surface, then $E = 0$ and $I_{enc} = I$. If, on the other hand, we use the balloon-shaped surface, then $I_{enc} = 0$, but $\int(\partial \mathbf{E}/\partial t)\cdot d\mathbf{a} = I/\epsilon_0$. So we get the same answer for either surface, though in the first case it comes from the genuine current and in the second from the displacement current.

Problem 7.31 A fat wire, radius a, carries a constant current I, uniformly distributed over its cross section. A narrow gap in the wire, of width $w \ll a$, forms a parallel-plate capacitor, as shown in Fig. 7.43. Find the magnetic field in the gap, at a distance $s < a$ from the axis.

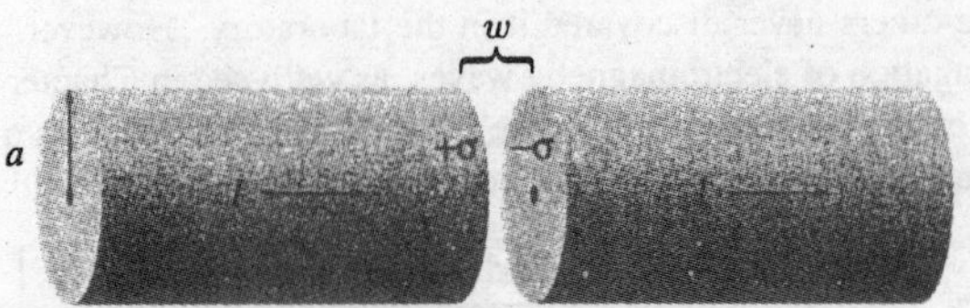

Figure 7.43

Problem 7.32 The preceding problem was an artificial model for the charging capacitor, designed to avoid complications associated with the current spreading out over the surface of the plates. For a more realistic model, imagine *thin* wires that connect to the centers of the plates (Fig. 7.44a). Again, the current I is constant, the radius of the capacitor is a, and the separation of the plates is $w \ll a$. Assume that the current flows out over the plates in such a way that the surface charge is uniform, at any given time, and is zero at $t = 0$.

(a) Find the electric field between the plates, as a function of t.

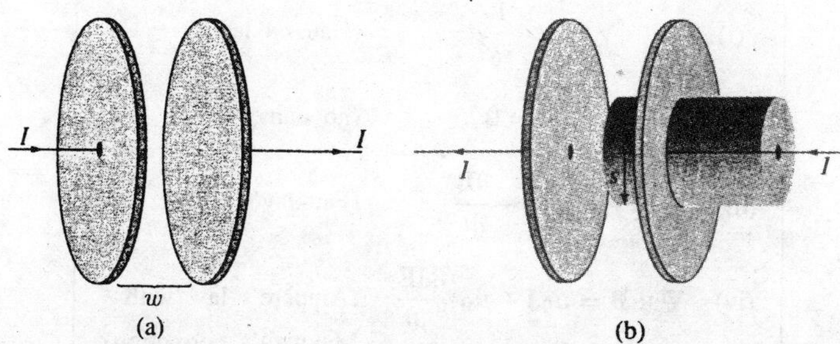

Figure 7.44

(b) Find the displacement current through a circle of radius s in the plane midway between the plates. Using this circle as your "Amperian loop," and the flat surface that spans it, find the magnetic field at a distance s from the axis.

(c) Repeat part (b), but this time use the cylindrical surface in Fig. 7.44b, which is open at the right end and extends to the left through the plate and terminates outside the capacitor. Notice that the displacement current through this surface is zero, and there are two contributions to I_{enc}.[14]

Problem 7.33 Refer to Prob. 7.16, to which the correct answer was

$$\mathbf{E}(s, t) = \frac{\mu_0 I_0 \omega}{2\pi} \sin(\omega t) \ln\left(\frac{a}{s}\right) \hat{\mathbf{z}}.$$

(a) Find the displacement current density $\mathbf{J}_d$.

(b) Integrate it to get the total displacement current,

$$I_d = \int \mathbf{J}_d \cdot d\mathbf{a}.$$

(c) Compare I_d and I. (What's their ratio?) If the outer cylinder were, say, 2 mm in diameter, how high would the frequency have to be, for I_d to be 1% of I? [This problem is designed to indicate why Faraday never discovered displacement currents, and why it is ordinarily safe to ignore them unless the frequency is extremely high.]

[14]This problem raises an interesting quasi-philosophical question: If you measure **B** in the laboratory, have you detected the effects of displacement current (as (b) would suggest), or merely confirmed the effects of ordinary currents (as (c) implies)? See D. F. Bartlett, *Am. J. Phys.* **58**, 1168 (1990).

7.3.3 Maxwell's Equations

In the last section we put the finishing touches on Maxwell's equations:

$$\begin{array}{llll} \text{(i)} & \nabla \cdot \mathbf{E} = \dfrac{1}{\epsilon_0}\rho & \text{(Gauss's law),} \\ \text{(ii)} & \nabla \cdot \mathbf{B} = 0 & \text{(no name),} \\ \text{(iii)} & \nabla \times \mathbf{E} = -\dfrac{\partial \mathbf{B}}{\partial t} & \text{(Faraday's law),} \\ \text{(iv)} & \nabla \times \mathbf{B} = \mu_0 \mathbf{J} + \mu_0\epsilon_0 \dfrac{\partial \mathbf{E}}{\partial t} & \text{(Ampère's law with Maxwell's correction).} \end{array} \tag{7.39}$$

Together with the force law,

$$\mathbf{F} = q(\mathbf{E} + \mathbf{v} \times \mathbf{B}), \tag{7.40}$$

they summarize the entire theoretical content of classical electrodynamics[15] (save for some special properties of matter, which we encountered in Chapters 4 and 6). Even the continuity equation,

$$\nabla \cdot \mathbf{J} = -\frac{\partial \rho}{\partial t}, \tag{7.41}$$

which is the mathematical expression of conservation of charge, can be derived from Maxwell's equations by applying the divergence to number (iv).

I have written Maxwell's equations in the traditional way, which emphasizes that they specify the divergence and curl of **E** and **B**. In this form they reinforce the notion that electric fields can be produced *either* by charges (ρ) *or* by changing magnetic fields ($\partial \mathbf{B}/\partial t$), and magnetic fields can be produced *either* by currents (**J**) *or* by changing electric fields ($\partial \mathbf{E}/\partial t$). Actually, this is somewhat misleading, because when you come right down to it $\partial \mathbf{B}/\partial t$ and $\partial \mathbf{E}/\partial t$ are *themselves* due to charges and currents. I think it is logically preferable to write

$$\left.\begin{array}{ll} \text{(i)}\ \nabla \cdot \mathbf{E} = \dfrac{1}{\epsilon_0}\rho, & \text{(iii)}\ \nabla \times \mathbf{E} + \dfrac{\partial \mathbf{B}}{\partial t} = 0, \\ \text{(ii)}\ \nabla \cdot \mathbf{B} = 0, & \text{(iv)}\ \nabla \times \mathbf{B} - \mu_0\epsilon_0 \dfrac{\partial \mathbf{E}}{\partial t} = \mu_0 \mathbf{J}, \end{array}\right\} \tag{7.42}$$

with the fields (**E** and **B**) on the left and the sources (ρ and **J**) on the right. This notation emphasizes that all electromagnetic fields are ultimately attributable to charges and currents. Maxwell's equations tell you how *charges* produce *fields*; reciprocally, the force law tells you how *fields* affect *charges*.

---

[15] Like any differential equations, Maxwell's must be supplemented by suitable *boundary conditions*. Because these are typically "obvious" from the context (e.g. **E** and **B** go to zero at large distances from a localized charge distribution), it is easy to forget that they play an essential role.

Problem 7.34 Suppose

$$\mathbf{E}(\mathbf{r}, t) = -\frac{1}{4\pi\epsilon_0}\frac{q}{r^2}\theta(vt - r)\hat{\mathbf{r}}; \quad \mathbf{B}(\mathbf{r}, t) = 0$$

(the theta function is defined in Prob. 1.45b). Show that these fields satisfy all of Maxwell's equations, and determine ρ and $\mathbf{J}$. Describe the physical situation that gives rise to these fields.

7.3.4 Magnetic Charge

There is a pleasing symmetry about Maxwell's equations; it is particularly striking in free space, where ρ and $\mathbf{J}$ vanish:

$$\left.\begin{array}{ll} \nabla \cdot \mathbf{E} = 0, & \nabla \times \mathbf{E} = -\dfrac{\partial \mathbf{B}}{\partial t}, \\ \nabla \cdot \mathbf{B} = 0, & \nabla \times \mathbf{B} = \mu_0\epsilon_0\dfrac{\partial \mathbf{E}}{\partial t}. \end{array}\right\}$$

If you replace $\mathbf{E}$ by $\mathbf{B}$ and $\mathbf{B}$ by $-\mu_0\epsilon_0\mathbf{E}$, the first pair of equations turns into the second, and vice versa. This symmetry[16] between $\mathbf{E}$ and $\mathbf{B}$ is spoiled, though, by the charge term in Gauss's law and the current term in Ampère's law. You can't help wondering why the corresponding quantities are "missing" from $\nabla \cdot \mathbf{B} = 0$ and $\nabla \times \mathbf{E} = -\partial\mathbf{B}/\partial t$. What if we had

$$\left.\begin{array}{ll} \text{(i) } \nabla \cdot \mathbf{E} = \dfrac{1}{\epsilon_0}\rho_e, & \text{(iii) } \nabla \times \mathbf{E} = -\mu_0\mathbf{J}_m - \dfrac{\partial \mathbf{B}}{\partial t}, \\ \text{(ii) } \nabla \cdot \mathbf{B} = \mu_0\rho_m, & \text{(iv) } \nabla \times \mathbf{B} = \mu_0\mathbf{J}_e + \mu_0\epsilon_0\dfrac{\partial \mathbf{E}}{\partial t}. \end{array}\right\} \tag{7.43}$$

Then ρ_m would represent the density of magnetic "charge," and ρ_e the density of electric charge; $\mathbf{J}_m$ would be the current of magnetic charge, and $\mathbf{J}_e$ the current of electric charge. Both charges would be conserved:

$$\nabla \cdot \mathbf{J}_m = -\frac{\partial \rho_m}{\partial t}, \quad \text{and} \quad \nabla \cdot \mathbf{J}_e = -\frac{\partial \rho_e}{\partial t}. \tag{7.44}$$

The former follows by application of the divergence to (iii), the latter by taking the divergence of (iv).

In a sense, Maxwell's equations *beg* for magnetic charge to exist—it would fit in so nicely. And yet, in spite of a diligent search, no one has ever found any.[17] As far as we know, ρ_m is zero everywhere, and so is $\mathbf{J}_m$; $\mathbf{B}$ is *not* on equal footing with $\mathbf{E}$: there exist

[16] Don't be distracted by the pesky constants μ_0 and ϵ_0; these are present only because the SI system measures **E** and **B** in different units, and would not occur, for instance, in the Gaussian system.

[17] For an extensive bibliography, see A. S. Goldhaber and W. P. Trower, *Am. J. Phys.* **58**, 429 (1990).

stationary sources for **E** (electric charges) but none for **B**. (This is reflected in the fact that magnetic multipole expansions have no monopole term, and magnetic dipoles consist of current loops, not separated north and south "poles.") Apparently God just didn't *make* any magnetic charge. (In the quantum theory of electrodynamics, by the way, it's a more than merely aesthetic shame that magnetic charge does not seem to exist: Dirac showed that the *existence* of *magnetic* charge would explain why *electric* charge is *quantized*. See Prob. 8.12.)

Problem 7.35 Assuming that "Coulomb's law" for magnetic charges (q_m) reads

$$\mathbf{F} = \frac{\mu_0}{4\pi}\frac{q_{m_1}q_{m_2}}{\mathscr{r}^2}\hat{\mathscr{r}}, \tag{7.45}$$

work out the force law for a monopole q_m moving with velocity **v** through electric and magnetic fields **E** and **B**. [For interesting commentary, see W. Rindler, *Am. J. Phys.* **57**, 993 (1989).]

Problem 7.36 Suppose a magnetic monopole q_m passes through a resistanceless loop of wire with self-inductance L. What current is induced in the loop? [This is one of the methods used to search for monopoles in the laboratory; see B. Cabrera, *Phys. Rev. Lett.* **48**, 1378 (1982).]

7.3.5 Maxwell's Equations in Matter

Maxwell's equations in the form 7.39 are complete and correct as they stand. However, when you are working with materials that are subject to electric and magnetic polarization there is a more convenient way to *write* them. For inside polarized matter there will be accumulations of "bound" charge and current over which you exert no direct control. It would be nice to reformulate Maxwell's equations in such a way as to make explicit reference only to those sources we control directly: the "free" charges and currents.

We have already learned, from the static case, that an electric polarization **P** produces a bound charge density

$$\rho_b = -\nabla\cdot\mathbf{P} \tag{7.46}$$

(Eq. 4.12). Likewise, a magnetic polarization (or "magnetization") **M** results in a bound current

$$\mathbf{J}_b = \nabla\times\mathbf{M} \tag{7.47}$$

(Eq. 6.13). There's just one new feature to consider in the *non*static case: Any *change* in the electric polarization involves a flow of (bound) charge (call it $\mathbf{J}_p$), which must be included in the total current. For suppose we examine a tiny chunk of polarized material (Fig. 7.45.) The polarization introduces a charge density $\sigma_b = P$ at one end and $-\sigma_b$ at the other (Eq. 4.11). If P now *increases* a bit, the charge on each end increases accordingly, giving a net current

$$dI = \frac{\partial\sigma_b}{\partial t}da_\perp = \frac{\partial P}{\partial t}da_\perp.$$

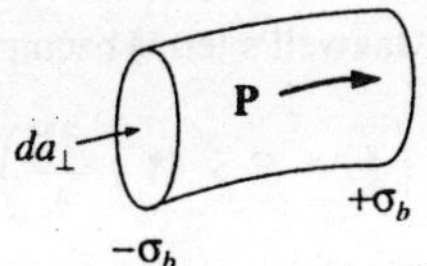

Figure 7.45

The current density, therefore, is

$$\mathbf{J}_p = \frac{\partial \mathbf{P}}{\partial t}. \tag{7.48}$$

This **polarization current** has nothing whatever to do with the *bound* current $\mathbf{J}_b$. The latter is associated with *magnetization* of the material and involves the spin and orbital motion of electrons; $\mathbf{J}_p$, by contrast, is the result of the linear motion of charge when the electric polarization changes. If **P** points to the right and is increasing, then each plus charge moves a bit to the right and each minus charge to the left; the cumulative effect is the polarization current $\mathbf{J}_p$. In this connection, we ought to check that Eq. 7.48 is consistent with the continuity equation:

$$\nabla \cdot \mathbf{J}_p = \nabla \cdot \frac{\partial \mathbf{P}}{\partial t} = \frac{\partial}{\partial t}(\nabla \cdot \mathbf{P}) = -\frac{\partial \rho_b}{\partial t}.$$

Yes: The continuity equation *is* satisfied; in fact, $\mathbf{J}_p$ is essential to account for the conservation of bound charge. (Incidentally, a changing *magnetization* does *not* lead to any analogous accumulation of charge or current. The bound current $\mathbf{J}_b = \nabla \times \mathbf{M}$ varies in response to changes in **M**, to be sure, but that's about it.)

In view of all this, the total charge density can be separated into two parts:

$$\rho = \rho_f + \rho_b = \rho_f - \nabla \cdot \mathbf{P}, \tag{7.49}$$

and the current density into *three* parts:

$$\mathbf{J} = \mathbf{J}_f + \mathbf{J}_b + \mathbf{J}_p = \mathbf{J}_f + \nabla \times \mathbf{M} + \frac{\partial \mathbf{P}}{\partial t}. \tag{7.50}$$

Gauss's law can now be written as

$$\nabla \cdot \mathbf{E} = \frac{1}{\epsilon_0}(\rho_f - \nabla \cdot \mathbf{P}),$$

or

$$\nabla \cdot \mathbf{D} = \rho_f, \tag{7.51}$$

where **D**, as in the static case, is given by

$$\mathbf{D} \equiv \epsilon_0 \mathbf{E} + \mathbf{P}. \tag{7.52}$$

Meanwhile, Ampère's law (with Maxwell's term) becomes

$$\nabla \times \mathbf{B} = \mu_0 \left(\mathbf{J}_f + \nabla \times \mathbf{M} + \frac{\partial \mathbf{P}}{\partial t} \right) + \mu_0 \epsilon_0 \frac{\partial \mathbf{E}}{\partial t},$$

or

$$\nabla \times \mathbf{H} = \mathbf{J}_f + \frac{\partial \mathbf{D}}{\partial t}, \tag{7.53}$$

where, as before,

$$\mathbf{H} \equiv \frac{1}{\mu_0}\mathbf{B} - \mathbf{M}. \tag{7.54}$$

Faraday's law and $\nabla \cdot \mathbf{B} = 0$ are not affected by our separation of charge and current into free and bound parts, since they do not involve ρ or $\mathbf{J}$.

In terms of *free* charges and currents, then, Maxwell's equations read

$$\begin{array}{ll} \text{(i)}\ \nabla \cdot \mathbf{D} = \rho_f, & \text{(iii)}\ \nabla \times \mathbf{E} = -\dfrac{\partial \mathbf{B}}{\partial t}, \\ \text{(ii)}\ \nabla \cdot \mathbf{B} = 0, & \text{(iv)}\ \nabla \times \mathbf{H} = \mathbf{J}_f + \dfrac{\partial \mathbf{D}}{\partial t}. \end{array} \tag{7.55}$$

Some people regard these as the "true" Maxwell's equations, but please understand that they are in *no way* more "general" than 7.39; they simply reflect a convenient division of charge and current into free and nonfree parts. And they have the disadvantage of hybrid notation, since they contain both **E** and **D**, both **B** and **H**. They must be supplemented, therefore, by appropriate **constitutive relations**, giving **D** and **H** in terms of **E** and **B**. These depend on the nature of the material; for linear media

$$\mathbf{P} = \epsilon_0 \chi_e \mathbf{E}, \quad \text{and} \quad \mathbf{M} = \chi_m \mathbf{H}, \tag{7.56}$$

so

$$\mathbf{D} = \epsilon \mathbf{E}, \quad \text{and} \quad \mathbf{H} = \frac{1}{\mu}\mathbf{B}, \tag{7.57}$$

where $\epsilon \equiv \epsilon_0(1 + \chi_e)$ and $\mu \equiv \mu_0(1 + \chi_m)$. Incidentally, you'll remember that **D** is called the electric "displacement"; that's why the second term in the Ampère/Maxwell equation (iv) is called the **displacement current**, generalizing Eq. 7.37:

$$\mathbf{J}_d = \frac{\partial \mathbf{D}}{\partial t}. \tag{7.58}$$

Problem 7.37 Sea water at frequency $\nu = 4 \times 10^8$ Hz has permittivity $\epsilon = 81\epsilon_0$, permeability $\mu = \mu_0$, and resistivity $\rho = 0.23\ \Omega \cdot \text{m}$. What is the ratio of conduction current to displacement current? [*Hint:* consider a parallel-plate capacitor immersed in sea water and driven by a voltage $V_0 \cos(2\pi\nu t)$.]

7.3.6 Boundary Conditions

In general, the fields **E, B, D,** and **H** will be discontinuous at a boundary between two different media, or at a surface that carries charge density σ or current density **K**. The explicit form of these discontinuities can be deduced from Maxwell's equations (7.55), in their integral form

$$\left.\begin{array}{ll}\text{(i)} & \oint_{\mathcal{S}} \mathbf{D}\cdot d\mathbf{a} = Q_{f_{\text{enc}}} \\ \text{(ii)} & \oint_{\mathcal{S}} \mathbf{B}\cdot d\mathbf{a} = 0\end{array}\right\}\quad \text{over any closed surface } \mathcal{S}.$$

$$\left.\begin{array}{ll}\text{(iii)} & \oint_{\mathcal{P}} \mathbf{E}\cdot d\mathbf{l} = -\dfrac{d}{dt}\int_{\mathcal{S}} \mathbf{B}\cdot d\mathbf{a} \\ \text{(iv)} & \oint_{\mathcal{P}} \mathbf{H}\cdot d\mathbf{l} = I_{f_{\text{enc}}} + \dfrac{d}{dt}\int_{\mathcal{S}} \mathbf{D}\cdot d\mathbf{a}\end{array}\right\}\quad \begin{array}{l}\text{for any surface } \mathcal{S} \\ \text{bounded by the} \\ \text{closed loop } \mathcal{P}.\end{array}$$

Applying (i) to a tiny, wafer-thin Gaussian pillbox extending just slightly into the material on either side of the boundary, we obtain (Fig. 7.46):

$$\mathbf{D}_1\cdot\mathbf{a} - \mathbf{D}_2\cdot\mathbf{a} = \sigma_f\, a.$$

(The positive direction for **a** is *from* 2 *toward* 1. The edge of the wafer contributes nothing in the limit as the thickness goes to zero, nor does any *volume* change density.) Thus, the component of **D** that is perpendicular to the interface is discontinuous in the amount

$$\boxed{D_1^{\perp} - D_2^{\perp} = \sigma_f.} \tag{7.59}$$

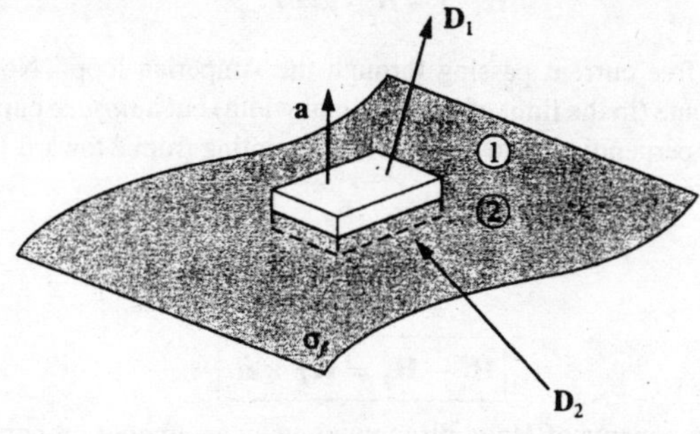

Figure 7.46

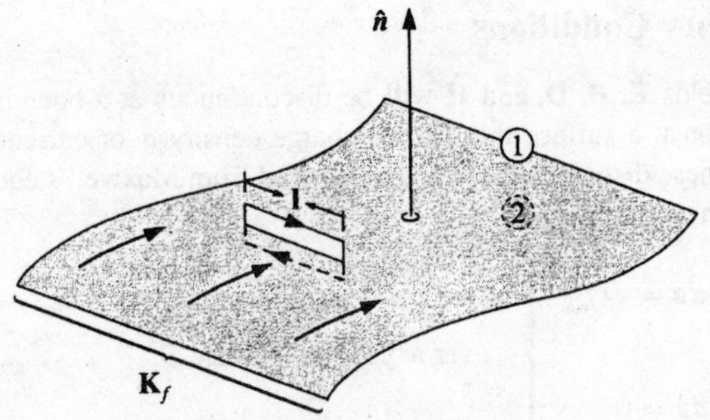

Figure 7.47

Identical reasoning, applied to equation (ii), yields

$$\boxed{B_1^\perp - B_2^\perp = 0.} \tag{7.60}$$

Turning to (iii), a very thin Amperian loop straddling the surface (Fig. 7.47) gives

$$\mathbf{E}_1 \cdot \mathbf{l} - \mathbf{E}_2 \cdot \mathbf{l} = -\frac{d}{dt}\int_S \mathbf{B} \cdot d\mathbf{a}.$$

But in the limit as the width of the loop goes to zero, the flux vanishes. (I have already dropped the contribution of the two ends to $\oint \mathbf{E} \cdot d\mathbf{l}$, on the same grounds.) Therefore,

$$\boxed{\mathbf{E}_1^\parallel - \mathbf{E}_2^\parallel = 0.} \tag{7.61}$$

That is, the components of $\mathbf{E}$ *parallel* to the interface are continuous across the boundary. By the same token, (iv) implies

$$\mathbf{H}_1 \cdot \mathbf{l} - \mathbf{H}_2 \cdot \mathbf{l} = I_{f_{\text{enc}}},$$

where $I_{f_{\text{enc}}}$ is the free current passing through the Amperian loop. No *volume* current density will contribute (in the limit of infinitesimal width) but a *surface* current can. In fact, if $\hat{\mathbf{n}}$ is a unit vector perpendicular to the interface (pointing from 2 toward 1), so that $(\hat{\mathbf{n}} \times \mathbf{l})$ is normal to the Amperian loop, then

$$I_{f_{\text{enc}}} = \mathbf{K}_f \cdot (\hat{\mathbf{n}} \times \mathbf{l}) = (\mathbf{K}_f \times \hat{\mathbf{n}}) \cdot \mathbf{l},$$

and hence

$$\boxed{\mathbf{H}_1^\parallel - \mathbf{H}_2^\parallel = \mathbf{K}_f \times \hat{\mathbf{n}}.} \tag{7.62}$$

So the *parallel* components of $\mathbf{H}$ are discontinuous by an amount proportional to the free surface current density.

Equations 7.59-62 are the general boundary conditions for electrodynamics. In the case of *linear* media, they can be expressed in terms of **E** and **B** alone:

$$\left.\begin{array}{ll}\text{(i)}\ \epsilon_1 E_1^{\perp} - \epsilon_2 E_2^{\perp} = \sigma_f, & \text{(iii)}\ \mathbf{E}_1^{\parallel} - \mathbf{E}_2^{\parallel} = 0, \\ \text{(ii)}\ B_1^{\perp} - B_2^{\perp} = 0, & \text{(iv)}\ \frac{1}{\mu_1}\mathbf{B}_1^{\parallel} - \frac{1}{\mu_2}\mathbf{B}_2^{\parallel} = \mathbf{K}_f \times \hat{\mathbf{n}}.\end{array}\right\} \tag{7.63}$$

In particular, if there is no free charge or free current at the interface, then

$$\boxed{\begin{array}{ll}\text{(i)}\ \epsilon_1 E_1^{\perp} - \epsilon_2 E_2^{\perp} = 0, & \text{(iii)}\ \mathbf{E}_1^{\parallel} - \mathbf{E}_2^{\parallel} = 0, \\ \text{(ii)}\ B_1^{\perp} - B_2^{\perp} = 0, & \text{(iv)}\ \frac{1}{\mu_1}\mathbf{B}_1^{\parallel} - \frac{1}{\mu_2}\mathbf{B}_2^{\parallel} = 0.\end{array}} \tag{7.64}$$

As we shall see in Chapter 9, these equations are the basis for the theory of reflection and refraction.

More Problems on Chapter 7

Problem 7.38 Two very large metal plates are held a distance d apart, one at potential zero, the other at potential V_0 (Fig. 7.48). A metal sphere of radius a ($a \ll d$) is sliced in two, and one hemisphere placed on the grounded plate, so that its potential is likewise zero. If the region between the plates is filled with weakly conducting material of uniform conductivity σ, what current flows to the hemisphere? [*Answer:* $(3\pi a^2\sigma/d)V_0$. *Hint:* study Ex. 3.8.]

! **Problem 7.39** Two long, straight copper pipes, each of radius a, are held a distance $2d$ apart (see Fig. 7.49). One is at potential V_0, the other at $-V_0$. The space surrounding the pipes is filled with weakly conducting material of conductivity σ. Find the current, per unit length, which flows from one pipe to the other. [*Hint:* refer to Prob. 3.11.]

Problem 7.40 A common textbook problem asks you to calculate the resistance of a cone-shaped object, of resistivity ρ, with length L, radius a at one end, and radius b at the other (Fig. 7.50). The two ends are flat, and are taken to be equipotentials. The suggested method is to slice it into circular disks of width dz, find the resistance of each disk, and integrate to get the total.

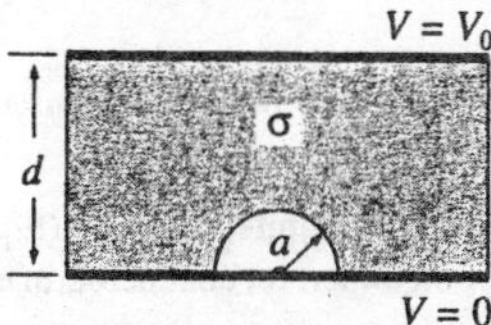

Figure 7.48

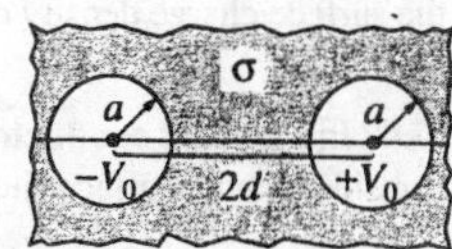

Figure 7.49

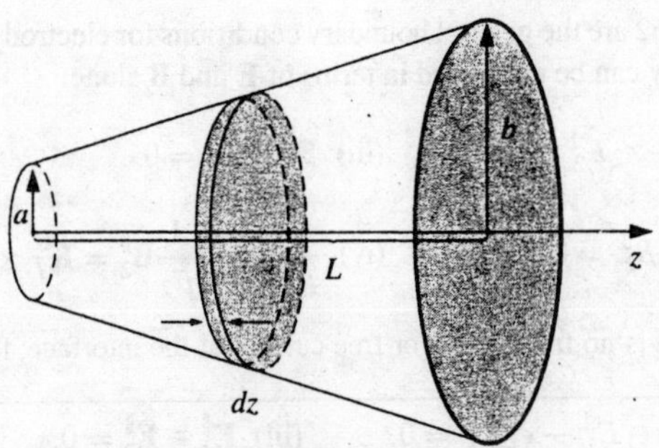

Figure 7.50

(a) Calculate R this way.

(b) Explain why this method is fundamentally flawed. [See J. D. Romano and R. H. Price, *Am. J. Phys.* **64**, 1150 (1996).]

(c) Suppose the ends are, instead, spherical surfaces, centered at the apex of the cone. Calculate the resistance in that case. (Let L be the distance between the centers of the circular perimeters of the end caps.) [*Answer:* $(\rho/2\pi ab)(b-a)^2/(\sqrt{L^2+(b-a)^2}-L)$]

Problem 7.41 A rare case in which the electrostatic field **E** for a circuit can actually be *calculated* is the following [M. A. Heald, *Am. J. Phys.* **52**, 522 (1984)]: Imagine an infinitely long cylindrical sheet, of uniform resistivity and radius a. A slot (corresponding to the battery) is maintained at $\pm V_0/2$, at $\phi = \pm\pi$, and a steady current flows over the surface, as indicated in Fig. 7.51. According to Ohm's law, then,

$$V(a, \phi) = \frac{V_0\phi}{2\pi}, \quad (-\pi < \phi < +\pi).$$

(a) Use separation of variables in cylindrical coordinates to determine $V(s, \phi)$ inside and outside the cylinder. [*Answer:* $(V_0/\pi)\tan^{-1}[(s\sin\phi)/(a+s\cos\phi)]$, $(s < a)$; $(V_0/\pi)\tan^{-1}[(a\sin\phi)/(s+a\cos\phi)]$, $(s > a)$]

(b) Find the surface charge density on the cylinder. [*Answer:* $(\epsilon_0 V_0/\pi a)\tan(\phi/2)$]

Problem 7.42 In a **perfect conductor**, the conductivity is infinite, so $\mathbf{E} = 0$ (Eq. 7.3), and any net charge resides on the surface (just as it does for an *im*perfect conductor, in electro*statics*).

(a) Show that the magnetic field is constant ($\partial\mathbf{B}/\partial t = 0$), inside a perfect conductor.

(b) Show that the magnetic flux through a perfectly conducting loop is constant.

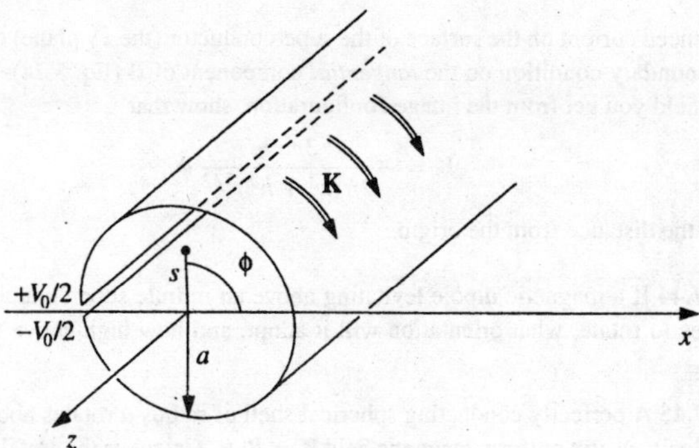

Figure 7.51

A **superconductor** is a perfect conductor with the *additional* property that the (constant) **B** inside is in fact *zero*. (This "flux exclusion" is known as the **Meissner effect.**[18])

(c) Show that the current in a superconductor is confined to the surface.

(d) Superconductivity is lost above a certain critical temperature (T_c), which varies from one material to another. Suppose you had a sphere (radius a) above its critical temperature, and you held it in a uniform magnetic field $B_0\hat{\mathbf{z}}$ while cooling it below T_c. Find the induced surface current density **K**, as a function of the polar angle θ.

Problem 7.43 A familiar demonstration of superconductivity (Prob. 7.42) is the levitation of a magnet over a piece of superconducting material. This phenomenon can be analyzed using the method of images.[19] Treat the magnet as a perfect dipole **m**, a height z above the origin (and constrained to point in the z direction), and pretend that the superconductor occupies the entire half-space below the xy plane. Because of the Meissner effect, $\mathbf{B} = 0$ for $z \leq 0$, and since **B** is divergenceless, the normal (z) component is continuous, so $B_z = 0$ just *above* the surface. This boundary condition is met by the image configuration in which an identical dipole is placed at $-z$, as a stand-in for the superconductor; the two arrangements therefore produce the same magnetic field in the region $z > 0$.

(a) Which way should the image dipole point ($+z$ or $-z$)?

(b) Find the force on the magnet due to the induced currents in the superconductor (which is to say, the force due to the image dipole). Set it equal to Mg (where M is the mass of the magnet) to determine the height h at which the magnet will "float." [*Hint:* refer to Prob. 6.3.]

---

[18] The Meissner effect is sometimes referred to as "perfect diamagnetism," in the sense that the field inside is not merely *reduced*, but canceled entirely. However, the surface currents responsible for this are *free*, not bound, so the actual *mechanism* is quite different.

[19] W. M. Saslow, *Am. J. Phys.* **59**, 16 (1991).

(c) The induced current on the surface of the superconductor (the xy plane) can be determined from the boundary condition on the *tangential* component of $\mathbf{B}$ (Eq. 5.74): $\mathbf{B} = \mu_0(\mathbf{K} \times \hat{\mathbf{z}})$. Using the field you get from the image configuration, show that

$$\mathbf{K} = -\frac{3mrh}{2\pi(r^2+h^2)^{5/2}}\,\hat{\boldsymbol{\phi}},$$

where r is the distance from the origin.

! **Problem 7.44** If a magnetic dipole levitating above an infinite superconducting plane (Prob. 7.43) is free to rotate, what orientation will it adopt, and how high above the surface will it float?

Problem 7.45 A perfectly conducting spherical shell of radius a rotates about the z axis with angular velocity ω, in a uniform magnetic field $\mathbf{B} = B_0\,\hat{\mathbf{z}}$. Calculate the emf developed between the "north pole" and the equator. [*Answer:* $\frac{1}{2}B_0\omega a^2$]

! **Problem 7.46** Refer to Prob. 7.11 (and use the result of Prob. 5.40, if it helps):

(a) Does the square ring fall faster in the orientation shown (Fig. 7.19), or when rotated 45° about an axis coming out of the page? Find the ratio of the two terminal velocities. If you dropped the loop, which orientation would it assume in falling? [*Answer:* $(\sqrt{2} - 2y/l)^2$, where l is the length of a side, and y is the height of the center above the edge of the magnetic field, in the rotated configuration.]

(b) How long does is take a *circular* ring to cross the bottom of the magnetic field, at its (changing) terminal velocity?

Problem 7.47

(a) Use the analogy between Faraday's law and Ampère's law, together with the Biot-Savart law, to show that

$$\mathbf{E}(\mathbf{r}, t) = -\frac{1}{4\pi}\frac{\partial}{\partial t}\int \frac{\mathbf{B}(\mathbf{r}', t) \times \hat{\boldsymbol{\mathscr{r}}}}{\mathscr{r}^2}\,d\tau', \tag{7.65}$$

for Faraday-induced electric fields.

(b) Referring to Prob. 5.50a, show that

$$\mathbf{E} = -\frac{\partial \mathbf{A}}{\partial t}, \tag{7.66}$$

where $\mathbf{A}$ is the vector potential. Check this result by taking the curl of both sides.

(c) A spherical shell of radius R carries a uniform surface charge σ. It spins about a fixed axis at an angular velocity $\omega(t)$ that changes slowly with time. Find the electric field inside and outside the sphere. [*Hint:* There are *two* contributions here: the Coulomb field due to the charge, and the Faraday field due to the changing $\mathbf{B}$. Refer to Ex. 5.11, and use Eq. 7.66.]

Problem 7.48 Electrons undergoing cyclotron motion can be speeded up by increasing the magnetic field; the accompanying electric field will impart tangential acceleration. This is the principle of the **betatron**. One would like to keep the radius of the orbit constant during the process. Show that this can be achieved by designing a magnet such that the average field over the area of the orbit is twice the field at the circumference (Fig. 7.52). Assume the electrons start from rest in zero field, and that the apparatus is symmetric about the center of the orbit. (Assume also that the electron velocity remains well below the speed of light, so that nonrelativistic mechanics applies.) [*Hint:* differentiate Eq. 5.3 with respect to time, and use $F = ma = qE$.]

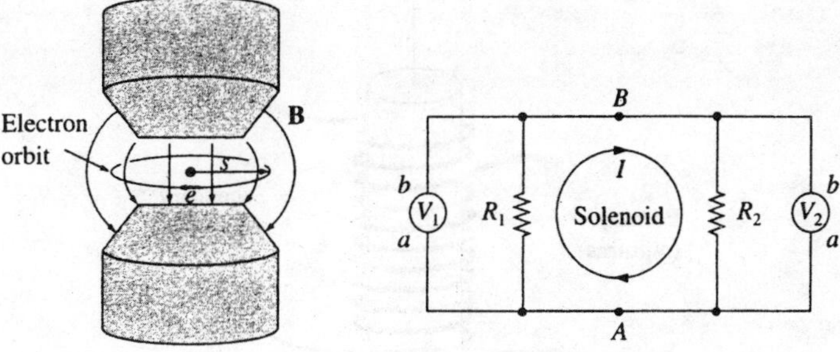

Figure 7.52 Figure 7.53

Problem 7.49 An atomic electron (charge q) circles about the nucleus (charge Q) in an orbit of radius r; the centripetal acceleration is provided, of course, by the Coulomb attraction of opposite charges. Now a small magnetic field dB is slowly turned on, perpendicular to the plane of the orbit. Show that the increase in kinetic energy, dT, imparted by the induced electric field, is just right to sustain circular motion *at the same radius* r. (That's why, in my discussion of diamagnetism, I assumed the radius is fixed. See Sect. 6.1.3 and the references cited there.)

Problem 7.50 The current in a long solenoid is increasing linearly with time, so that the flux is proportional to t: $\Phi = \alpha t$. Two voltmeters are connected to diametrically opposite points (A and B), together with resistors (R_1 and R_2), as shown in Fig. 7.53. What is the reading on each voltmeter? Assume that these are *ideal* voltmeters that draw negligible current (they have huge internal resistance), and that a voltmeter registers $\int_a^b \mathbf{E} \cdot d\mathbf{l}$ between the terminals and through the meter. [*Answer:* $V_1 = \alpha R_1/(R_1 + R_2)$; $V_2 = -\alpha R_2/(R_1 + R_2)$. Notice that $V_1 \neq V_2$, even though they are connected to the same points! See R. H. Romer, *Am. J. Phys.* **50**, 1089 (1982).]

Problem 7.51 In the discussion of motional emf (Sect. 7.1.3) I assumed that the wire loop (Fig. 7.10) has a resistance R; the current generated is then $I = vBh/R$. But what if the wire is made out of perfectly conducting material, so that R is *zero*? In that case the current is limited only by the back emf associated with the self-inductance L of the loop (which would ordinarily be negligible in comparison with IR). Show that in this régime the loop (mass m) executes simple harmonic motion, and find its frequency.[20] [*Answer:* $\omega = Bh/\sqrt{mL}$]

Problem 7.52

(a) Use the Neumann formula (Eq. 7.22) to calculate the mutual inductance of the configuration in Fig. 7.36, assuming a is very small ($a \ll b$, $a \ll z$). Compare your answer to Prob. 7.20.
(b) For the general case (*not* assuming a is small) show that

$$M = \frac{\mu_0 \pi \beta}{2} \sqrt{ab\beta} \left(1 + \frac{15}{8}\beta^2 + \dots\right),$$

[20] For a collection of related problems, see W. M. Saslow, *Am. J. Phys.* **55**, 986 (1987), and R. H. Romer, *Eur. J. Phys.* **11**, 103 (1990).

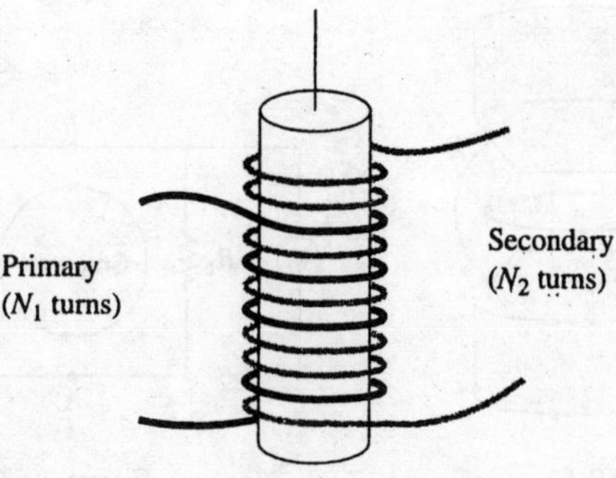

Figure 7.54

where

$$\beta \equiv \frac{ab}{z^2 + a^2 + b^2}.$$

Problem 7.53 Two coils are wrapped around a cylindrical form in such a way that the *same flux passes through every turn of both coils.* (In practice this is achieved by inserting an iron core through the cylinder; this has the effect of concentrating the flux.) The "primary" coil has N_1 turns and the secondary has N_2 (Fig. 7.54). If the current I in the primary is changing, show that the emf in the secondary is given by

$$\frac{\mathcal{E}_2}{\mathcal{E}_1} = \frac{N_2}{N_1}, \tag{7.67}$$

where $\mathcal{E}_1$ is the (back) emf of the primary. [This is a primitive **transformer**—a device for raising or lowering the emf of an alternating current source. By choosing the appropriate number of turns, any desired secondary emf can be obtained. If you think this violates the conservation of energy, check out Prob. 7.54.]

Problem 7.54 A transformer (Prob. 7.53) takes an input AC voltage of amplitude V_1, and delivers an output voltage of amplitude V_2, which is determined by the turns ratio ($V_2/V_1 = N_2/N_1$). If $N_2 > N_1$ the output voltage is greater than the input voltage. Why doesn't this violate conservation of energy? *Answer:* Power is the product of voltage and current; evidently if the voltage goes *up*, the current must come *down*. The purpose of this problem is to see exactly how this works out, in a simplified model.

(a) In an ideal transformer the same flux passes through all turns of the primary and of the secondary. Show that in this case $M^2 = L_1 L_2$, where M is the mutual inductance of the coils, and L_1, L_2 are their individual self-inductances.

(b) Suppose the primary is driven with AC voltage $V_{in} = V_1 \cos(\omega t)$, and the secondary is connected to a resistor, R. Show that the two currents satisfy the relations

$$L_1 \frac{dI_1}{dt} + M\frac{dI_2}{dt} = V_1 \cos(\omega t); \quad L_2 \frac{dI_2}{dt} + M\frac{dI_1}{dt} = -I_2 R.$$

(c) Using the result in (a), solve these equations for $I_1(t)$ and $I_2(t)$. (Assume I_1 has no DC component.)

(d) Show that the output voltage ($V_{\text{out}} = I_2 R$) divided by the input voltage (V_{in}) is equal to the turns ratio: $V_{\text{out}}/V_{\text{in}} = N_2/N_1$.

(e) Calculate the input power ($P_{\text{in}} = V_{\text{in}} I_1$) and the output power ($P_{\text{out}} = V_{\text{out}} I_2$), and show that their averages over a full cycle are equal.

Problem 7.55 Suppose $\mathbf{J}(\mathbf{r})$ is constant in time but $\rho(\mathbf{r}, t)$ is *not*—conditions that might prevail, for instance, during the charging of a capacitor.

(a) Show that the charge density at any particular point is a linear function of time:

$$\rho(\mathbf{r}, t) = \rho(\mathbf{r}, 0) + \dot{\rho}(\mathbf{r}, 0)t,$$

where $\dot{\rho}(\mathbf{r}, 0)$ is the time derivative of ρ at $t = 0$.

This is *not* an electrostatic or magnetostatic configuration;[21] nevertheless—rather surprisingly—both Coulomb's law (in the form of Eq. 2.8) and the Biot-Savart law (Eq. 5.39) hold, as you can confirm by showing that they satisfy Maxwell's equations. In particular:

(b) Show that

$$\mathbf{B}(\mathbf{r}) = \frac{\mu_0}{4\pi} \int \frac{\mathbf{J}(\mathbf{r}') \times \hat{\boldsymbol{\imath}}}{\imath^2} d\tau'$$

obeys Ampère's law *with Maxwell's displacement current term.*

Problem 7.56 The magnetic field of an infinite straight wire carrying a steady current I can be obtained from the *displacement* current term in the Ampère/Maxwell law, as follows: Picture the current as consisting of a uniform line charge λ moving along the z axis at speed v (so that $I = \lambda v$), with a tiny gap of length ϵ, which reaches the origin at time $t = 0$. In the next instant (up to $t = \epsilon/v$) there is no *real* current passing through a circular Amperian loop in the xy plane, but there *is* a *displacement* current, due to the "missing" charge in the gap.

(a) Use Coulomb's law to calculate the z component of the electric field, for points in the xy plane a distance s from the origin, due to a segment of wire with uniform density $-\lambda$ extending from $z_1 = vt - \epsilon$ to $z_2 = vt$.

(b) Determine the flux of this electric field through a circle of radius a in the xy plane.

(c) Find the displacement current through this circle. Show that I_d is equal to I, in the limit as the gap width (ϵ) goes to zero. [For a slightly different approach to the same problem, see W. K. Terry, *Am. J. Phys.* **50**, 742 (1982).]

Problem 7.57 The *magnetic* field outside a long straight wire carrying a steady current I is (of course)

$$\mathbf{B} = \frac{\mu_0}{2\pi} \frac{I}{s} \hat{\boldsymbol{\phi}}.$$

[21] Some authors *would* regard this as magnetostatic, since $\mathbf{B}$ is independent of t. For them, the Biot-Savart law is a general rule of magnetostatics, but $\nabla \cdot \mathbf{J} = 0$ and $\nabla \times \mathbf{B} = \mu_0 \mathbf{J}$ apply only under the *additional* assumption that ρ is constant. In such a formulation Maxwell's displacement term can (in this very special case) be *derived* from the Biot-Savart law, by the method of part (b). See D. F. Bartlett, *Am. J. Phys.* **58**, 1168 (1990); D. J. Griffiths and M. A. Heald, *Am. J. Phys.* **59**, 111 (1991).

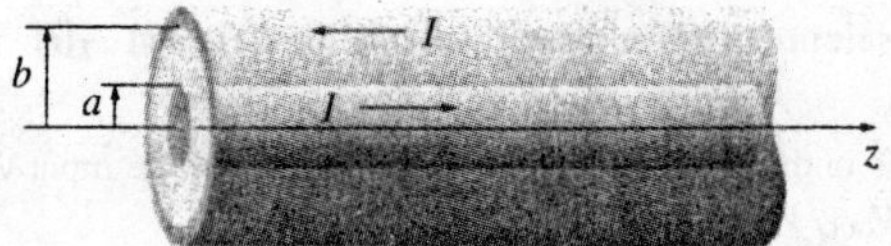

Figure 7.55

The *electric field inside* the wire is uniform:

$$\mathbf{E} = \frac{I\rho}{\pi a^2}\,\hat{\mathbf{z}},$$

where ρ is the resistivity and a is the radius (see Exs. 7.1 and 7.3). *Question:* What is the electric field *outside* the wire? This is a famous problem, first analyzed by Sommerfeld, and known in its most recent incarnation as "Merzbacher's puzzle."[22] The answer depends on how you complete the circuit. Suppose the current returns along a perfectly conducting grounded coaxial cylinder of radius b (Fig. 7.55). In the region $a < s < b$, the potential $V(s, z)$ satisfies Laplace's equation, with the boundary conditions

$$\text{(i)}\;\; V(a, z) = -\frac{I\rho z}{\pi a^2}; \quad \text{(ii)}\;\; V(b, z) = 0.$$

Unfortunately, this does not suffice to determine the answer—we still need to specify boundary conditions at the two ends. In the literature it is customary to sweep this ambiguity under the rug by simply *asserting* (in so many words) that $V(s, z)$ is proportional to z: $V(s, z) = zf(s)$. On this assumption:

(a) Determine $V(s, z)$.

(b) Find $\mathbf{E}(s, z)$.

(c) Calculate the surface charge density $\sigma(z)$ on the wire.

[*Answer:* $V = (-Iz\rho/\pi a^2)[\ln(s/b)/\ln(a/b)]$ This is a *peculiar* result, since E_s and $\sigma(z)$ are *not* independent of z—as one would certainly expect for a truly *infinite* wire.]

Problem 7.58 A certain transmission line is constructed from two thin metal "ribbons," of width w, a very small distance $h \ll w$ apart. The current travels down one strip and back along the other. In each case it spreads out uniformly over the surface of the ribbon.

(a) Find the capacitance per unit length, $\mathcal{C}$.

(b) Find the inductance per unit length, $\mathcal{L}$.

(c) What is the product $\mathcal{LC}$, numerically? [$\mathcal{L}$ and $\mathcal{C}$ will, of course, vary from one kind of transmission line to another, but their *product* is a universal constant—check, for example, the cable in Ex. 7.13—provided the space between the conductors is a vacuum. In the theory of transmission lines, this product is related to the speed with which a pulse propagates down the line: $v = 1/\sqrt{\mathcal{LC}}$.]

[22] A. Sommerfeld, *Electrodynamics*, p. 125 (New York: Academic Press, 1952); E. Merzbacher, *Am. J. Phys.* **48**, 104 (1980); further references in M. A. Heald, *Am. J. Phys.* **52**, 522 (1984).

(d) If the strips are insulated from one another by a nonconducting material of permittivity ϵ and permeability μ, what then is the product $\mathcal{LC}$? What is the propagation speed? [*Hint:* see Ex. 4.6; by what factor does L change when an inductor is immersed in linear material of permeability μ?]

Problem 7.59 Prove **Alfven's theorem**: In a perfectly conducting fluid (say, a gas of free electrons), the magnetic flux through any closed loop moving with the fluid is constant in time. (The magnetic field lines are, as it were, "frozen" into the fluid.)

(a) Use Ohm's law, in the form of Eq. 7.2, together with Faraday's law, to prove that if $\sigma = \infty$ and $\mathbf{J}$ is finite, then

$$\frac{\partial \mathbf{B}}{\partial t} = \nabla \times (\mathbf{v} \times \mathbf{B}).$$

(b) Let $\mathcal{S}$ be the surface bounded by the loop ($\mathcal{P}$) at time t, and $\mathcal{S}'$ a surface bounded by the loop in its new position ($\mathcal{P}'$) at time $t + dt$ (see Fig. 7.56). The change in flux is

$$d\Phi = \int_{\mathcal{S}'} \mathbf{B}(t + dt) \cdot d\mathbf{a} - \int_{\mathcal{S}} \mathbf{B}(t) \cdot d\mathbf{a}.$$

Show that

$$\int_{\mathcal{S}'} \mathbf{B}(t + dt) \cdot d\mathbf{a} + \int_{\mathcal{R}} \mathbf{B}(t + dt) \cdot d\mathbf{a} = \int_{\mathcal{S}} \mathbf{B}(t + dt) \cdot d\mathbf{a}$$

(where $\mathcal{R}$ is the "ribbon" joining $\mathcal{P}$ and $\mathcal{P}'$), and hence that

$$d\Phi = dt \int_{\mathcal{S}} \frac{\partial \mathbf{B}}{\partial t} \cdot d\mathbf{a} - \int_{\mathcal{R}} \mathbf{B}(t + dt) \cdot d\mathbf{a}$$

(for infinitesimal dt). Use the method of Sect. 7.1.3 to rewrite the second integral as

$$dt \oint_{\mathcal{P}} (\mathbf{B} \times \mathbf{v}) \cdot d\mathbf{l},$$

and invoke Stokes' theorem to conclude that

$$\frac{d\Phi}{dt} = \int_{\mathcal{S}} \left(\frac{\partial \mathbf{B}}{\partial t} - \nabla \times (\mathbf{v} \times \mathbf{B}) \right) \cdot d\mathbf{a}.$$

Together with the result in (a), this proves the theorem.

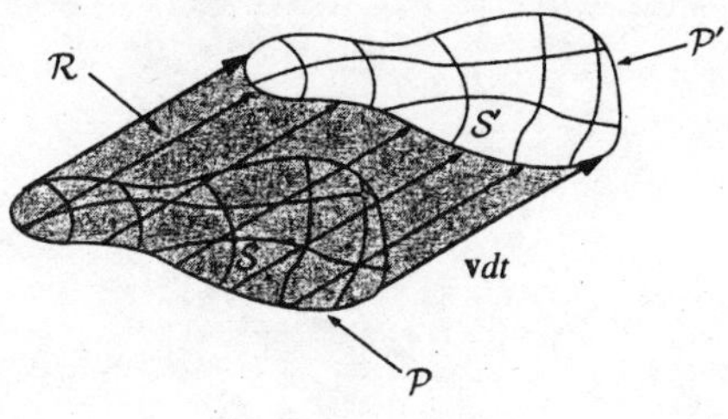

Figure 7.56

Problem 7.60

(a) Show that Maxwell's equations with magnetic charge (Eq. 7.43) are invariant under the **duality transformation**

$$\left.\begin{aligned} \mathbf{E}' &= \mathbf{E}\cos\alpha + c\mathbf{B}\sin\alpha, \\ c\mathbf{B}' &= c\mathbf{B}\cos\alpha - \mathbf{E}\sin\alpha, \\ cq_e' &= cq_e\cos\alpha + q_m\sin\alpha, \\ q_m' &= q_m\cos\alpha - cq_e\sin\alpha, \end{aligned}\right\} \tag{7.68}$$

where $c \equiv 1/\sqrt{\epsilon_0\mu_0}$ and α is an arbitrary rotation angle in "E/B-space." Charge and current densities transform in the same way as q_e and q_m. [This means, in particular, that if you know the fields produced by a configuration of *electric* charge, you can immediately (using $\alpha = 90°$) write down the fields produced by the corresponding arrangement of *magnetic* charge.]

(b) Show that the force law (Prob. 7.35)

$$\mathbf{F} = q_e(\mathbf{E} + \mathbf{v}\times\mathbf{B}) + q_m(\mathbf{B} - \frac{1}{c^2}\mathbf{v}\times\mathbf{E}) \tag{7.69}$$

is also invariant under the duality transformation.

Intermission

All of our cards are now on the table, and in a sense my job is done. In the first seven chapters we assembled electrodynamics piece by piece, and now, with Maxwell's equations in their final form, the theory is complete. There are no more laws to be learned, no further generalizations to be considered, and (with perhaps one exception) no lurking inconsistencies to be resolved. If yours is a one-semester course, this would be a reasonable place to stop.

But in another sense we have just arrived at the starting point. We are at last in possession of a full deck, and we know the rules of the game—it's time to deal. This is the fun part, in which one comes to appreciate the extraordinary power and richness of electrodynamics. In a full-year course there should be plenty of time to cover the remaining chapters, and perhaps to supplement them with a unit on plasma physics, say, or AC circuit theory, or even a little General Relativity. But if you have room only for one topic, I'd recommend Chapter 9, on Electromagnetic Waves (you'll probably want to skim Chapter 8 as preparation). This is the segue to Optics, and is historically the most important application of Maxwell's theory.

Chapter 8

Conservation Laws

8.1 Charge and Energy

8.1.1 The Continuity Equation

In this chapter we study conservation of energy, momentum, and angular momentum, in electrodynamics. But I want to begin by reviewing the conservation of *charge*, because it is the paradigm for all conservation laws. What precisely does conservation of charge tell us? That the total charge in the universe is constant? Well, sure—that's **global** conservation of charge; but **local** conservation of charge is a much stronger statement: If the total charge in some volume changes, then exactly that amount of charge must have passed in or out through the surface. The tiger can't simply rematerialize outside the cage; if it got from inside to outside it must have found a hole in the fence.

Formally, the charge in a volume $\mathcal{V}$ is

$$Q(t) = \int_{\mathcal{V}} \rho(\mathbf{r}, t)\, d\tau, \tag{8.1}$$

and the current flowing out through the boundary $\mathcal{S}$ is $\oint_{\mathcal{S}} \mathbf{J} \cdot d\mathbf{a}$, so local conservation of charge says

$$\frac{dQ}{dt} = -\oint_{\mathcal{S}} \mathbf{J} \cdot d\mathbf{a}. \tag{8.2}$$

Using Eq. 8.1 to rewrite the left side, and invoking the divergence theorem on the right, we have

$$\int_{\mathcal{V}} \frac{\partial \rho}{\partial t}\, d\tau = -\int_{\mathcal{V}} \nabla \cdot \mathbf{J}\, d\tau, \tag{8.3}$$

and since this is true for *any* volume, it follows that

$$\boxed{\frac{\partial \rho}{\partial t} = -\nabla \cdot \mathbf{J}.} \tag{8.4}$$

This is, of course, the continuity equation—the precise mathematical statement of local conservation of charge. As I indicated earlier, it can be derived from Maxwell's equations—conservation of charge is not an *independent* assumption, but a *consequence* of the laws of electrodynamics.

The purpose of this chapter is to construct the corresponding equations for conservation of energy and conservation of momentum. In the process (and perhaps more important) we will learn how to express the energy density and the momentum density (the analogs to ρ), as well as the energy "current" and the momentum "current" (analogous to **J**).

8.1.2 Poynting's Theorem

In Chapter 2, we found that the work necessary to assemble a static charge distribution (against the Coulomb repulsion of like charges) is (Eq. 2.45)

$$W_e = \frac{\epsilon_0}{2}\int E^2\,d\tau,$$

where **E** is the resulting electric field. Likewise, the work required to get currents going (against the back emf) is (Eq. 7.34)

$$W_m = \frac{1}{2\mu_0}\int B^2\,d\tau,$$

where **B** is the resulting magnetic field. This suggests that the total energy stored in electromagnetic fields is

$$U_{em} = \frac{1}{2}\int\left(\epsilon_0 E^2 + \frac{1}{\mu_0}B^2\right)d\tau. \tag{8.5}$$

I propose to derive Eq. 8.5 more generally, now, in the context of the energy conservation law for electrodynamics.

Suppose we have some charge and current configuration which, at time t, produces fields **E** and **B**. In the next instant, dt, the charges move around a bit. *Question*: How much work, dW, is done by the electromagnetic forces acting on these charges in the interval dt? According to the Lorentz force law, the work done on a charge q is

$$\mathbf{F}\cdot d\mathbf{l} = q(\mathbf{E} + \mathbf{v}\times\mathbf{B})\cdot\mathbf{v}\,dt = q\mathbf{E}\cdot\mathbf{v}\,dt.$$

Here, $q = \rho\,d\tau$ and $\rho\mathbf{v} = \mathbf{J}$, so the rate at which work is done on all the charges in a volume $\mathcal{V}$ is

$$\frac{dW}{dt} = \int_{\mathcal{V}}(\mathbf{E}\cdot\mathbf{J})\,d\tau. \tag{8.6}$$

Evidently $\mathbf{E}\cdot\mathbf{J}$ is the work done per unit time, per unit volume—which is to say, the *power* delivered per unit volume. We can express this quantity in terms of the fields alone, using the Ampère-Maxwell law to eliminate **J**:

$$\mathbf{E}\cdot\mathbf{J} = \frac{1}{\mu_0}\mathbf{E}\cdot(\nabla\times\mathbf{B}) - \epsilon_0\mathbf{E}\cdot\frac{\partial\mathbf{E}}{\partial t}.$$

From product rule 6,

$$\nabla \cdot (\mathbf{E} \times \mathbf{B}) = \mathbf{B} \cdot (\nabla \times \mathbf{E}) - \mathbf{E} \cdot (\nabla \times \mathbf{B}).$$

Invoking Faraday's law ($\nabla \times \mathbf{E} = -\partial \mathbf{B}/\partial t$), it follows that

$$\mathbf{E} \cdot (\nabla \times \mathbf{B}) = -\mathbf{B} \cdot \frac{\partial \mathbf{B}}{\partial t} - \nabla \cdot (\mathbf{E} \times \mathbf{B}).$$

Meanwhile,

$$\mathbf{B} \cdot \frac{\partial \mathbf{B}}{\partial t} = \frac{1}{2}\frac{\partial}{\partial t}(B^2), \quad \text{and} \quad \mathbf{E} \cdot \frac{\partial \mathbf{E}}{\partial t} = \frac{1}{2}\frac{\partial}{\partial t}(E^2), \tag{8.7}$$

so

$$\mathbf{E} \cdot \mathbf{J} = -\frac{1}{2}\frac{\partial}{\partial t}\left(\epsilon_0 E^2 + \frac{1}{\mu_0}B^2\right) - \frac{1}{\mu_0}\nabla \cdot (\mathbf{E} \times \mathbf{B}). \tag{8.8}$$

Putting this into Eq. 8.6, and applying the divergence theorem to the second term, we have

$$\frac{dW}{dt} = -\frac{d}{dt}\int_{\mathcal{V}} \frac{1}{2}\left(\epsilon_0 E^2 + \frac{1}{\mu_0}B^2\right) d\tau - \frac{1}{\mu_0}\oint_{\mathcal{S}} (\mathbf{E} \times \mathbf{B}) \cdot d\mathbf{a}, \tag{8.9}$$

where $\mathcal{S}$ is the surface bounding $\mathcal{V}$. This is **Poynting's theorem**; it is the "work-energy theorem" of electrodynamics. The first integral on the right is the total energy stored in the fields, U_{em} (Eq. 8.5). The second term evidently represents the rate at which energy is carried out of $\mathcal{V}$, across its boundary surface, by the electromagnetic fields. Poynting's theorem says, then, that *the work done on the charges by the electromagnetic force is equal to the decrease in energy stored in the field, less the energy that flowed out through the surface.*

The *energy per unit time, per unit area*, transported by the fields is called the **Poynting vector**:

$$\boxed{\mathbf{S} \equiv \frac{1}{\mu_0}(\mathbf{E} \times \mathbf{B}).} \tag{8.10}$$

Specifically, $\mathbf{S} \cdot d\mathbf{a}$ is the energy per unit time crossing the infinitesimal surface $d\mathbf{a}$—the energy *flux*, if you like (so $\mathbf{S}$ is the **energy flux density**).[1] We will see many applications of the Poynting vector in Chapters 9 and 11, but for the moment I am mainly interested in using it to express Poynting's theorem more compactly:

$$\frac{dW}{dt} = -\frac{dU_{em}}{dt} - \oint_{\mathcal{S}} \mathbf{S} \cdot d\mathbf{a}. \tag{8.11}$$

[1] If you're very fastidious, you'll notice a small gap in the logic here: We know from Eq. 8.9 that $\oint \mathbf{S} \cdot d\mathbf{a}$ is the total power passing through a *closed* surface, but this does not prove that $\int \mathbf{S} \cdot d\mathbf{a}$ is the power passing through any *open* surface (there could be an extra term that integrates to zero over all closed surfaces). This is, however, the obvious and natural interpretation; as always, the precise *location* of energy is not really determined in electrodynamics (see Sect. 2.4.4).

Of course, the work W done on the charges will increase their mechanical energy (kinetic, potential, or whatever). If we let u_{mech} denote the mechanical energy density, so that

$$\frac{dW}{dt} = \frac{d}{dt}\int_{\mathcal{V}} u_{\text{mech}}\, d\tau, \tag{8.12}$$

and use u_{em} for the energy density of the fields,

$$\boxed{u_{\text{em}} = \frac{1}{2}\left(\epsilon_0 E^2 + \frac{1}{\mu_0}B^2\right),} \tag{8.13}$$

then

$$\frac{d}{dt}\int_{\mathcal{V}} (u_{\text{mech}} + u_{\text{em}})\, d\tau = -\oint_{\mathcal{S}} \mathbf{S}\cdot d\mathbf{a} = -\int_{\mathcal{V}} (\nabla\cdot\mathbf{S})\, d\tau,$$

and hence

$$\boxed{\frac{\partial}{\partial t}(u_{\text{mech}} + u_{\text{em}}) = -\nabla\cdot\mathbf{S}.} \tag{8.14}$$

This is the differential version of Poynting's theorem. Compare it with the continuity equation, expressing conservation of charge (Eq. 8.4):

$$\frac{\partial\rho}{\partial t} = -\nabla\cdot\mathbf{J};$$

the charge density is replaced by the energy density (mechanical plus electromagnetic), and the current density is replaced by the Poynting vector. The latter represents the flow of *energy* in exactly the same way that $\mathbf{J}$ describes the flow of *charge*.[2]

Example 8.1

When current flows down a wire, work is done, which shows up as Joule heating of the wire (Eq. 7.7). Though there are certainly *easier* ways to do it, the energy per unit time delivered to the wire can be calculated using the Poynting vector. Assuming it's uniform, the electric field parallel to the wire is

$$E = \frac{V}{L},$$

where V is the potential difference between the ends and L is the length of the wire (Fig. 8.1). The magnetic field is "circumferential"; at the surface (radius a) it has the value

$$B = \frac{\mu_0 I}{2\pi a}.$$

Accordingly, the magnitude of the Poynting vector is

$$S = \frac{1}{\mu_0}\frac{V}{L}\frac{\mu_0 I}{2\pi a} = \frac{VI}{2\pi a L},$$

---

[2] In the presence of linear media, one is typically interested only in the work done on *free* charges and currents (see Sect. 4.4.3). In that case the appropriate energy density is $\frac{1}{2}(\mathbf{E}\cdot\mathbf{D} + \mathbf{B}\cdot\mathbf{H})$, and the Poynting vector becomes $(\mathbf{E}\times\mathbf{H})$. See J. D. Jackson, *Classical Electrodynamics*, 3rd. ed., Sect. 6.7 (New York: John Wiley, 1999).

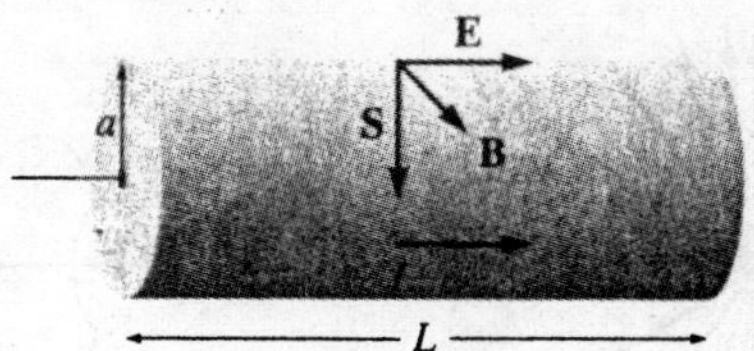

Figure 8.1

and it points radially inward. The energy per unit time passing in through the surface of the wire is therefore

$$\int \mathbf{S} \cdot d\mathbf{a} = S(2\pi a L) = VI,$$

which is exactly what we concluded, on much more direct grounds, in Sect. 7.1.1.

Problem 8.1 Calculate the power (energy per unit time) transported down the cables of Ex. 7.13 and Prob. 7.58, assuming the two conductors are held at potential difference V, and carry current I (down one and back up the other).

Problem 8.2 Consider the charging capacitor in Prob. 7.31.

(a) Find the electric and magnetic fields in the gap, as functions of the distance s from the axis and the time t. (Assume the charge is zero at $t = 0$.)

(b) Find the energy density u_{em} and the Poynting vector $\mathbf{S}$ in the gap. Note especially the *direction* of $\mathbf{S}$. Check that Eq. 8.14 is satisfied.

(c) Determine the total energy in the gap, as a function of time. Calculate the total power flowing into the gap, by integrating the Poynting vector over the appropriate surface. Check that the power input is equal to the rate of increase of energy in the gap (Eq. 8.9—in this case $W = 0$, because there is no charge in the gap). [If you're worried about the fringing fields, do it for a volume of radius $b < a$ well inside the gap.]

8.2 Momentum

8.2.1 Newton's Third Law in Electrodynamics

Imagine a point charge q traveling in along the x axis at a constant speed v. Because it is moving, its electric field is *not* given by Coulomb's law; nevertheless, $\mathbf{E}$ still points radially outward from the instantaneous position of the charge (Fig. 8.2a), as we'll see in Chapter 10. Since, moreover, a moving point charge does not constitute a steady current, its magnetic

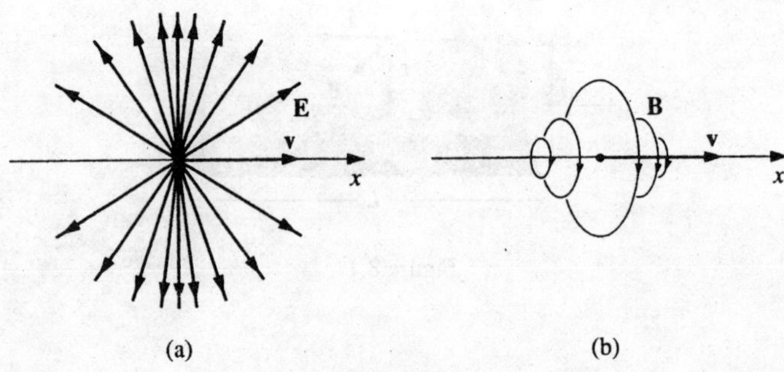

Figure 8.2

field is *not* given by the Biot-Savart law. Nevertheless, it's a fact that **B** still circles around the axis in a manner suggested by the right-hand rule (Fig. 8.2b); again, the proof will come in Chapter 10.

Now suppose this charge encounters an identical one, proceeding in at the same speed along the y axis. Of course, the electromagnetic force between them would tend to drive them off the axes, but let's assume that they're mounted on tracks, or something, so they're forced to maintain the same direction and the same speed (Fig. 8.3). The electric force between them is repulsive, but how about the magnetic force? Well, the magnetic field of q_1 points into the page (at the position of q_2), so the magnetic force on q_2 is toward the *right*, whereas the magnetic field of q_2 is *out* of the page (at the position of q_1), and the magnetic force on q_1 is *upward*. *The electromagnetic force of q_1 on q_2 is equal but not opposite to the force of q_2 on q_1, in violation of Newton's third law.* In electro*statics* and magneto*statics* the third law holds, but in electro*dynamics* it does *not*.

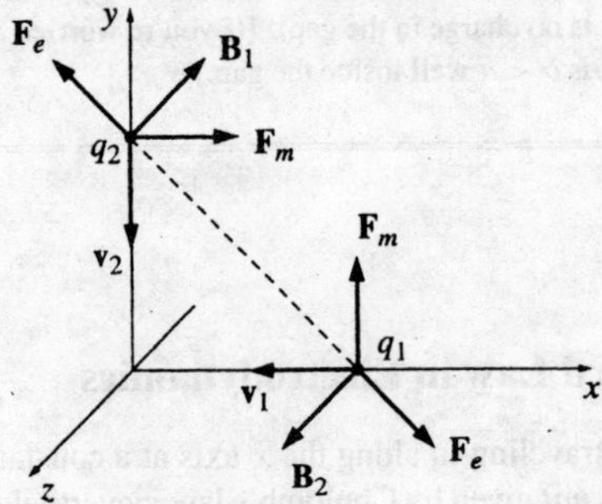

Figure 8.3

Well, that's an interesting curiosity, but then, how often does one actually use the third law, in practice? *Answer:* All the time! For the proof of conservation of momentum rests on the cancellation of internal forces, which follows from the third law. When you tamper with the third law, you are placing conservation of momentum in jeopardy, and there is no principle in physics more sacred than *that.*

Momentum conservation is rescued in electrodynamics by the realization that *the fields themselves carry momentum.* This is not so surprising when you consider that we have already attributed *energy* to the fields. In the case of the two point charges in Fig. 8.3, whatever momentum is lost to the particles is gained by the fields. Only when the field momentum is added to the mechanical momentum of the charges is momentum conservation restored. You'll see how this works out quantitatively in the following sections.

8.2.2 Maxwell's Stress Tensor

Let's calculate the total electromagnetic force on the charges in volume $\mathcal{V}$:

$$\mathbf{F} = \int_{\mathcal{V}} (\mathbf{E} + \mathbf{v} \times \mathbf{B})\rho \, d\tau = \int_{\mathcal{V}} (\rho\mathbf{E} + \mathbf{J} \times \mathbf{B}) \, d\tau. \tag{8.15}$$

The *force per unit volume* is evidently

$$\mathbf{f} = \rho\mathbf{E} + \mathbf{J} \times \mathbf{B}. \tag{8.16}$$

As before, I propose to write this in terms of fields alone, eliminating ρ and $\mathbf{J}$ by using Maxwell's equations (i) and (iv):

$$\mathbf{f} = \epsilon_0(\nabla \cdot \mathbf{E})\mathbf{E} + \left(\frac{1}{\mu_0}\nabla \times \mathbf{B} - \epsilon_0\frac{\partial \mathbf{E}}{\partial t}\right) \times \mathbf{B}.$$

Now

$$\frac{\partial}{\partial t}(\mathbf{E} \times \mathbf{B}) = \left(\frac{\partial \mathbf{E}}{\partial t} \times \mathbf{B}\right) + \left(\mathbf{E} \times \frac{\partial \mathbf{B}}{\partial t}\right),$$

and Faraday's law says

$$\frac{\partial \mathbf{B}}{\partial t} = -\nabla \times \mathbf{E},$$

so

$$\frac{\partial \mathbf{E}}{\partial t} \times \mathbf{B} = \frac{\partial}{\partial t}(\mathbf{E} \times \mathbf{B}) + \mathbf{E} \times (\nabla \times \mathbf{E}).$$

Thus

$$\mathbf{f} = \epsilon_0[(\nabla \cdot \mathbf{E})\mathbf{E} - \mathbf{E} \times (\nabla \times \mathbf{E})] - \frac{1}{\mu_0}[\mathbf{B} \times (\nabla \times \mathbf{B})] - \epsilon_0\frac{\partial}{\partial t}(\mathbf{E} \times \mathbf{B}). \tag{8.17}$$

Just to make things look more symmetrical, let's throw in a term $(\nabla \cdot \mathbf{B})\mathbf{B}$; since $\nabla \cdot \mathbf{B} = 0$, this costs us nothing. Meanwhile, product rule 4 says

$$\nabla(E^2) = 2(\mathbf{E} \cdot \nabla)\mathbf{E} + 2\mathbf{E} \times (\nabla \times \mathbf{E}),$$

so

$$\mathbf{E} \times (\nabla \times \mathbf{E}) = \frac{1}{2}\nabla(E^2) - (\mathbf{E} \cdot \nabla)\mathbf{E},$$

and the same goes for $\mathbf{B}$. Therefore,

$$\begin{aligned}\mathbf{f} = \;& \epsilon_0[(\nabla \cdot \mathbf{E})\mathbf{E} + (\mathbf{E} \cdot \nabla)\mathbf{E}] + \frac{1}{\mu_0}[(\nabla \cdot \mathbf{B})\mathbf{B} + (\mathbf{B} \cdot \nabla)\mathbf{B}] \\ & -\frac{1}{2}\nabla\left(\epsilon_0 E^2 + \frac{1}{\mu_0}B^2\right) - \epsilon_0\frac{\partial}{\partial t}(\mathbf{E} \times \mathbf{B}).\end{aligned} \tag{8.18}$$

Ugly! But it can be simplified by introducing the **Maxwell stress tensor,**

$$T_{ij} \equiv \epsilon_0\left(E_i E_j - \frac{1}{2}\delta_{ij}E^2\right) + \frac{1}{\mu_0}\left(B_i B_j - \frac{1}{2}\delta_{ij}B^2\right). \tag{8.19}$$

The indices i and j refer to the coordinates x, y, and z, so the stress tensor has a total of nine components (T_{xx}, T_{yy}, T_{xz}, T_{yx}, and so on). The **Kronecker delta,** δ_{ij}, is 1 if the indices are the same ($\delta_{xx} = \delta_{yy} = \delta_{zz} = 1$) and zero otherwise ($\delta_{xy} = \delta_{xz} = \delta_{yz} = 0$). Thus

$$\begin{aligned}T_{xx} = \;& \frac{1}{2}\epsilon_0(E_x^2 - E_y^2 - E_z^2) + \frac{1}{2\mu_0}(B_x^2 - B_y^2 - B_z^2), \\ T_{xy} = \;& \epsilon_0(E_x E_y) + \frac{1}{\mu_0}(B_x B_y),\end{aligned}$$

and so on. Because it carries *two* indices, where a vector has only one, T_{ij} is sometimes written with a double arrow: $\overleftrightarrow{\mathbf{T}}$. One can form the dot product of $\overleftrightarrow{\mathbf{T}}$ with a vector $\mathbf{a}$:

$$(\mathbf{a} \cdot \overleftrightarrow{\mathbf{T}})_j = \sum_{i=x,y,z} a_i T_{ij}; \tag{8.20}$$

the resulting object, which has one remaining index, is itself a vector. In particular, the divergence of $\overleftrightarrow{\mathbf{T}}$ has as its jth component

$$\begin{aligned}(\nabla \cdot \overleftrightarrow{\mathbf{T}})_j \;=\; & \epsilon_0\left[(\nabla \cdot \mathbf{E})E_j + (\mathbf{E} \cdot \nabla)E_j - \frac{1}{2}\nabla_j E^2\right] \\ + \; & \frac{1}{\mu_0}\left[(\nabla \cdot \mathbf{B})B_j + (\mathbf{B} \cdot \nabla)B_j - \frac{1}{2}\nabla_j B^2\right].\end{aligned}$$

Thus the force per unit volume (Eq. 8.18) can be written in the much simpler form

$$\mathbf{f} = \nabla \cdot \overleftrightarrow{\mathbf{T}} - \epsilon_0\mu_0\frac{\partial \mathbf{S}}{\partial t}, \tag{8.21}$$

where $\mathbf{S}$ is the Poynting vector (Eq. 8.10).

The *total* force on the charges in $\mathcal{V}$ (Eq. 8.15) is evidently

$$\mathbf{F} = \oint_{\mathcal{S}} \overleftrightarrow{\mathbf{T}} \cdot d\mathbf{a} - \epsilon_0\mu_0 \frac{d}{dt}\int_{\mathcal{V}} \mathbf{S}\, d\tau. \tag{8.22}$$

(I used the divergence theorem to convert the first term to a surface integral.) In the *static* case (or, more generally, whenever $\int \mathbf{S}\,d\tau$ is independent of time), the second term drops out, and the electromagnetic force on the charge configuration can be expressed entirely in terms of the stress tensor at the boundary. Physically, $\overleftrightarrow{\mathbf{T}}$ is the force per unit area (or **stress**) acting on the surface. More precisely, T_{ij} is the force (per unit area) in the ith direction acting on an element of surface oriented in the jth direction—"diagonal" elements (T_{xx}, T_{yy}, T_{zz}) represent *pressures*, and "off-diagonal" elements (T_{xy}, T_{xz}, etc.) are *shears*.

Example 8.2

Determine the net force on the "northern" hemisphere of a uniformly charged solid sphere of radius R and charge Q (the same as Prob. 2.43, only this time we'll use the Maxwell stress tensor and Eq. 8.22).

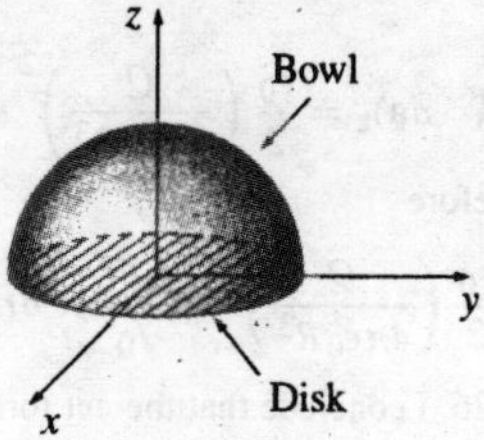

Figure 8.4

Solution: The boundary surface consists of two parts—a hemispherical "bowl" at radius R, and a circular disk at $\theta = \pi/2$ (Fig. 8.4). For the bowl,

$$d\mathbf{a} = R^2 \sin\theta\, d\theta\, d\phi\, \hat{\mathbf{r}}$$

and

$$\mathbf{E} = \frac{1}{4\pi\epsilon_0}\frac{Q}{R^2}\hat{\mathbf{r}}.$$

In Cartesian components,

$$\hat{\mathbf{r}} = \sin\theta\cos\phi\,\hat{\mathbf{x}} + \sin\theta\sin\phi\,\hat{\mathbf{y}} + \cos\theta\,\hat{\mathbf{z}},$$

so

$$T_{zx} = \epsilon_0 E_z E_x = \epsilon_0\left(\frac{Q}{4\pi\epsilon_0 R^2}\right)^2 \sin\theta\cos\theta\cos\phi,$$

$$T_{zy} = \epsilon_0 E_z E_y = \epsilon_0\left(\frac{Q}{4\pi\epsilon_0 R^2}\right)^2 \sin\theta\cos\theta\sin\phi,$$

$$T_{zz} = \frac{\epsilon_0}{2}(E_z^2 - E_x^2 - E_y^2) = \frac{\epsilon_0}{2}\left(\frac{Q}{4\pi\epsilon_0 R^2}\right)^2 (\cos^2\theta - \sin^2\theta). \tag{8.23}$$

The net force is obviously in the z-direction, so it suffices to calculate

$$(\overleftrightarrow{\mathbf{T}} \cdot d\mathbf{a})_z = T_{zx}\, da_x + T_{zy}\, da_y + T_{zz}\, da_z = \frac{\epsilon_0}{2}\left(\frac{Q}{4\pi\epsilon_0 R}\right)^2 \sin\theta\cos\theta\, d\theta\, d\phi.$$

The force on the "bowl" is therefore

$$F_{\text{bowl}} = \frac{\epsilon_0}{2}\left(\frac{Q}{4\pi\epsilon_0 R}\right)^2 2\pi \int_0^{\pi/2} \sin\theta\cos\theta\, d\theta = \frac{1}{4\pi\epsilon_0}\frac{Q^2}{8R^2}. \tag{8.24}$$

Meanwhile, for the equatorial disk,

$$d\mathbf{a} = -r\, dr\, d\phi\, \hat{\mathbf{z}}, \tag{8.25}$$

and (since we are now *inside* the sphere)

$$\mathbf{E} = \frac{1}{4\pi\epsilon_0}\frac{Q}{R^3}\mathbf{r} = \frac{1}{4\pi\epsilon_0}\frac{Q}{R^3} r(\cos\phi\,\hat{\mathbf{x}} + \sin\phi\,\hat{\mathbf{y}}).$$

Thus

$$T_{zz} = \frac{\epsilon_0}{2}(E_z^2 - E_x^2 - E_y^2) = -\frac{\epsilon_0}{2}\left(\frac{Q}{4\pi\epsilon_0 R^3}\right)^2 r^2,$$

and hence

$$(\overleftrightarrow{\mathbf{T}} \cdot d\mathbf{a})_z = \frac{\epsilon_0}{2}\left(\frac{Q}{4\pi\epsilon_0 R^3}\right)^2 r^3\, dr\, d\phi.$$

The force on the disk is therefore

$$F_{\text{disk}} = \frac{\epsilon_0}{2}\left(\frac{Q}{4\pi\epsilon_0 R^3}\right)^2 2\pi \int_0^R r^3\, dr = \frac{1}{4\pi\epsilon_0}\frac{Q^2}{16R^2}. \tag{8.26}$$

Combining Eqs. 8.24 and 8.26, I conclude that the net force on the northern hemisphere is

$$F = \frac{1}{4\pi\epsilon_0}\frac{3Q^2}{16R^2}. \tag{8.27}$$

Incidentally, in applying Eq. 8.22, *any* volume that encloses all of the charge in question (and no *other* charge) will do the job. For example, in the present case we could use the whole region $z > 0$. In that case the boundary surface consists of the entire xy plane (plus a hemisphere at $r = \infty$, but $E = 0$ out there anyway, so it contributes nothing). In place of the "bowl," we now have the outer portion of the plane ($r > R$). Here

$$T_{zz} = -\frac{\epsilon_0}{2}\left(\frac{Q}{4\pi\epsilon_0}\right)^2 \frac{1}{r^4}$$

(Eq. 8.23 with $\theta = \pi/2$ and $R \to r$), and $d\mathbf{a}$ is given by Eq. 8.25, so

$$(\overleftrightarrow{\mathbf{T}} \cdot d\mathbf{a})_z = \frac{\epsilon_0}{2}\left(\frac{Q}{4\pi\epsilon_0}\right)^2 \frac{1}{r^3}\, dr\, d\phi,$$

and the contribution from the plane for $r > R$ is

$$\frac{\epsilon_0}{2}\left(\frac{Q}{2\pi\epsilon_0}\right)^2 2\pi \int_R^\infty \frac{1}{r^3}\, dr = \frac{1}{4\pi\epsilon_0}\frac{Q^2}{8R^2},$$

the same as for the bowl (Eq. 8.24).

I hope you didn't get too bogged down in the details of Ex. 8.2. If so, take a moment to appreciate what happened. We were calculating the force on a solid object, but instead of doing a *volume* integral, as you might expect, Eq. 8.22 allowed us to set it up as a *surface* integral; somehow the stress tensor sniffs out what is going on inside.

! **Problem 8.3** Calculate the force of magnetic attraction between the northern and southern hemispheres of a uniformly charged spinning spherical shell, with radius R, angular velocity ω, and surface charge density σ. [This is the same as Prob. 5.42, but this time use the Maxwell stress tensor and Eq. 8.22.]

Problem 8.4

(a) Consider two equal point charges q, separated by a distance $2a$. Construct the plane equidistant from the two charges. By integrating Maxwell's stress tensor over this plane, determine the force of one charge on the other.

(b) Do the same for charges that are opposite in sign.

8.2.3 Conservation of Momentum

According to Newton's second law, the force on an object is equal to the rate of change of its momentum:

$$\mathbf{F} = \frac{d\mathbf{p}_{\text{mech}}}{dt}.$$

Equation 8.22 can therefore be written in the form

$$\frac{d\mathbf{p}_{\text{mech}}}{dt} = -\epsilon_0\mu_0 \frac{d}{dt}\int_{\mathcal{V}} \mathbf{S}\, d\tau + \oint_{\mathcal{S}} \overleftrightarrow{\mathbf{T}} \cdot d\mathbf{a}, \tag{8.28}$$

where $\mathbf{p}_{\text{mech}}$ is the total (mechanical) momentum of the particles contained in the volume $\mathcal{V}$. This expression is similar in structure to Poynting's theorem (Eq. 8.9), and it invites an analogous interpretation: The first integral represents *momentum stored in the electromagnetic fields themselves*:

$$\mathbf{p}_{\text{em}} = \mu_0\epsilon_0 \int_{\mathcal{V}} \mathbf{S}\, d\tau, \tag{8.29}$$

while the second integral is the *momentum per unit time flowing in through the surface*. Equation 8.28 is the general statement of *conservation of momentum* in electrodynamics: Any increase in the *total* momentum (mechanical plus electromagnetic) is equal to the momentum brought in by the fields. (If $\mathcal{V}$ is *all* of space, then *no* momentum flows in or out, and $\mathbf{p}_{\text{mech}} + \mathbf{p}_{\text{em}}$ is constant.)

As in the case of conservation of charge and conservation of energy, conservation of momentum can be given a differential formulation. Let $\boldsymbol{\wp}_{\text{mech}}$ be the density of *mechanical* momentum, and $\boldsymbol{\wp}_{\text{em}}$ the density of momentum in the fields:

$$\boxed{\boldsymbol{\wp}_{\text{em}} = \mu_0\epsilon_0 \mathbf{S}.} \tag{8.30}$$

Then Eq. 8.28, in differential form, says

$$\frac{\partial}{\partial t}(\wp_{\text{mech}} + \wp_{\text{em}}) = \nabla \cdot \overleftrightarrow{\mathbf{T}}. \tag{8.31}$$

Evidently $-\overleftrightarrow{\mathbf{T}}$ is the **momentum flux density**, playing the role of **J** (current density) in the continuity equation, or **S** (energy flux density) in Poynting's theorem. Specifically, $-T_{ij}$ is the momentum in the i direction crossing a surface oriented in the j direction, per unit area, per unit time. Notice that the Poynting vector has appeared in two quite different roles: **S** itself is the energy per unit area, per unit time, transported by the electromagnetic fields, while $\mu_0\epsilon_0\mathbf{S}$ is the momentum per unit volume stored in those fields. Similarly, $\overleftrightarrow{\mathbf{T}}$ plays a dual role: $\overleftrightarrow{\mathbf{T}}$ itself is the electromagnetic stress (force per unit area) acting on a surface, and $-\overleftrightarrow{\mathbf{T}}$ describes the flow of momentum (the momentum current density) transported by the fields.

Example 8.3

A long coaxial cable, of length l, consists of an inner conductor (radius a) and an outer conductor (radius b). It is connected to a battery at one end and a resistor at the other (Fig. 8.5). The inner conductor carries a uniform charge per unit length λ, and a steady current I to the right; the outer conductor has the opposite charge and current. What is the electromagnetic momentum stored in the fields?

Solution: The fields are

$$\mathbf{E} = \frac{1}{2\pi\epsilon_0}\frac{\lambda}{s}\,\hat{\mathbf{s}}, \qquad \mathbf{B} = \frac{\mu_0}{2\pi}\frac{I}{s}\,\hat{\boldsymbol{\phi}}.$$

The Poynting vector is therefore

$$\mathbf{S} = \frac{\lambda I}{4\pi^2\epsilon_0 s^2}\,\hat{\mathbf{z}}.$$

Evidently energy is flowing down the line, from the battery to the resistor. In fact, the power transported is

$$P = \int \mathbf{S}\cdot d\mathbf{a} = \frac{\lambda I}{4\pi^2\epsilon_0}\int_a^b \frac{1}{s^2}2\pi s\,ds = \frac{\lambda I}{2\pi\epsilon_0}\ln(b/a) = IV,$$

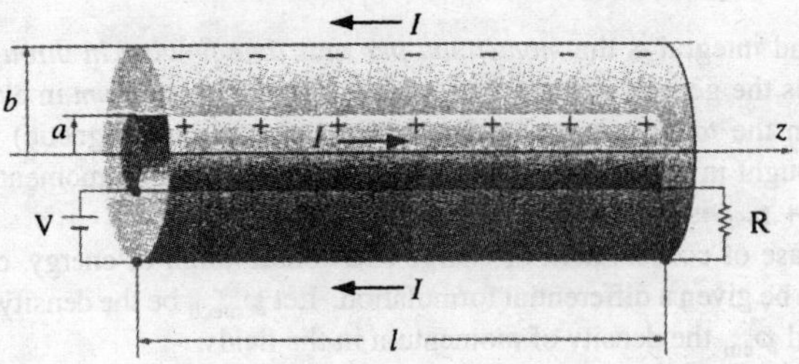

Figure 8.5

as it should be. But that's not what we're interested in right now. The *momentum* in the fields is

$$\mathbf{p}_{em} = \mu_0\epsilon_0 \int \mathbf{S}\,d\tau = \frac{\mu_0\lambda I}{4\pi^2}\,\hat{\mathbf{z}} \int_a^b \frac{1}{s^2} l 2\pi s\,ds = \frac{\mu_0\lambda I l}{2\pi}\ln(b/a)\,\hat{\mathbf{z}}.$$

This is an astonishing result. The cable is not moving, and the fields are static, and yet we are asked to believe that there is momentum in the system. If something tells you this cannot be the whole story, you have sound intuitions. In fact, if the center of mass of a localized system is at rest, its total momentum *must* be zero. In this case it turns out that there is "hidden" mechanical momentum associated with the flow of current, and this exactly cancels the momentum in the fields. But *locating* the **hidden momentum** is not easy, and it is actually a relativistic effect, so I shall save it for Chapter 12 (Ex. 12.12).

Suppose now that we turn up the resistance, so the current decreases. The changing magnetic field will induce an electric field (Eq. 7.19):

$$\mathbf{E} = \left[\frac{\mu_0}{2\pi}\frac{dI}{dt}\ln s + K\right]\hat{\mathbf{z}}.$$

This field exerts a force on $\pm\lambda$:

$$\mathbf{F} = \lambda l\left[\frac{\mu_0}{2\pi}\frac{dI}{dt}\ln a + K\right]\hat{\mathbf{z}} - \lambda l\left[\frac{\mu_0}{2\pi}\frac{dI}{dt}\ln b + K\right]\hat{\mathbf{z}} = -\frac{\mu_0\lambda l}{2\pi}\frac{dI}{dt}\ln(b/a)\,\hat{\mathbf{z}}.$$

The total momentum imparted to the cable, as the current drops from I to 0, is therefore

$$\mathbf{p}_{mech} = \int \mathbf{F}\,dt = \frac{\mu_0\lambda I l}{2\pi}\ln(b/a)\,\hat{\mathbf{z}},$$

which is precisely the momentum originally stored in the fields. (The cable will not recoil, however, because an equal and opposite impulse is delivered by the simultaneous disappearance of the hidden momentum.)

Problem 8.5 Consider an infinite parallel-plate capacitor, with the lower plate (at $z = -d/2$) carrying the charge density $-\sigma$, and the upper plate (at $z = +d/2$) carrying the charge density $+\sigma$.

(a) Determine all nine elements of the stress tensor, in the region between the plates. Display your answer as a 3×3 matrix:

$$\begin{pmatrix} T_{xx} & T_{xy} & T_{xz} \\ T_{yx} & T_{yy} & T_{yz} \\ T_{zx} & T_{zy} & T_{zz} \end{pmatrix}$$

(b) Use Eq. 8.22 to determine the force per unit area on the top plate. Compare Eq. 2.51.

(c) What is the momentum per unit area, per unit time, crossing the xy plane (or any other plane parallel to that one, between the plates)?

(d) At the plates this momentum is absorbed, and the plates recoil (unless there is some nonelectrical force holding them in position). Find the recoil force per unit area on the top

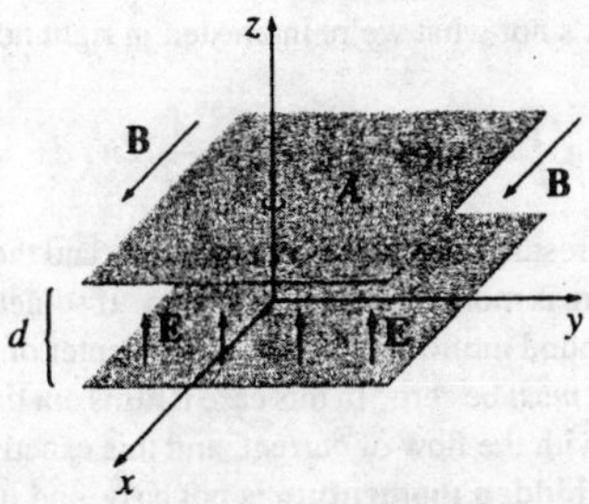

Figure 8.6

plate, and compare your answer to (b). [*Note:* This is not an *additional* force, but rather an alternative way of calculating the *same* force—in (b) we got it from the force law, and in (d) we did it by conservation of momentum.]

Problem 8.6 A charged parallel-plate capacitor (with uniform electric field $\mathbf{E} = E\,\hat{\mathbf{z}}$) is placed in a uniform magnetic field $\mathbf{B} = B\,\hat{\mathbf{x}}$, as shown in Fig. 8.6.[3]

(a) Find the electromagnetic momentum in the space between the plates.

(b) Now a resistive wire is connected between the plates, along the z axis, so that the capacitor slowly discharges. The current through the wire will experience a magnetic force; what is the total impulse delivered to the system, during the discharge?

(c) Instead of turning off the *electric* field (as in (b)), suppose we slowly reduce the *magnetic* field. This will induce a Faraday electric field, which in turn exerts a force on the plates. Show that the total impulse is (again) equal to the momentum originally stored in the fields.

8.2.4 Angular Momentum

By now the electromagnetic fields (which started out as mediators of forces between charges) have taken on a life of their own. They carry *energy* (Eq. 8.13)

$$u_{\mathrm{em}} = \frac{1}{2}\left(\epsilon_0 E^2 + \frac{1}{\mu_0} B^2\right), \tag{8.32}$$

and *momentum* (Eq. 8.30)

$$\boldsymbol{\wp}_{\mathrm{em}} = \mu_0\epsilon_0 \mathbf{S} = \epsilon_0(\mathbf{E}\times\mathbf{B}), \tag{8.33}$$

and, for that matter, *angular* momentum:

$$\boldsymbol{\ell}_{\mathrm{em}} = \mathbf{r}\times\boldsymbol{\wp}_{\mathrm{em}} = \epsilon_0\left[\mathbf{r}\times(\mathbf{E}\times\mathbf{B})\right]. \tag{8.34}$$

[3]See F. S. Johnson, B. L. Cragin, and R. R. Hodges, *Am. J. Phys.* **62**, 33 (1994).

Even perfectly static fields can harbor momentum and angular momentum, as long as $\mathbf{E} \times \mathbf{B}$ is nonzero, and it is only when these field contributions are included that the classical conservation laws hold.

Example 8.4

Imagine a very long solenoid with radius R, n turns per unit length, and current I. Coaxial with the solenoid are two long cylindrical shells of length l—one, *inside* the solenoid at radius a, carries a charge $+Q$, uniformly distributed over its surface; the other, *outside* the solenoid at radius b, carries charge $-Q$ (see Fig. 8.7; l is supposed to be much greater than b). When the current in the solenoid is gradually reduced, the cylinders begin to rotate, as we found in Ex. 7.8. *Question*: Where does the angular momentum come from?[4]

Solution: It was initially stored in the fields. Before the current was switched off, there was an electric field,

$$\mathbf{E} = \frac{Q}{2\pi\epsilon_0 l}\frac{1}{s}\,\hat{\mathbf{s}} \quad (a < s < b),$$

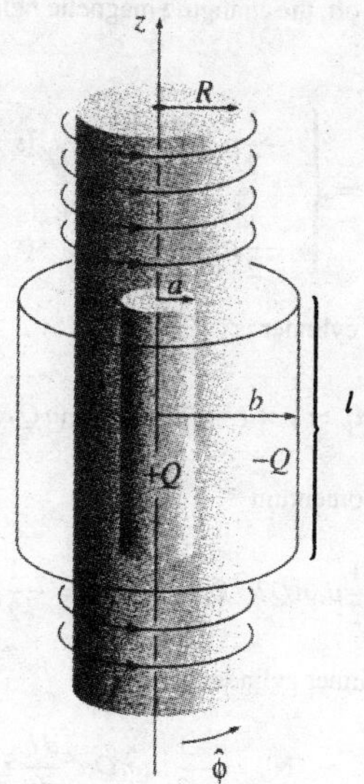

Figure 8.7

[4]This is a variation on the "Feynman disk paradox" (R. P. Feynman, R. B. Leighton, and M. Sands, *The Feynman Lectures*, vol 2, pp. 17-5 (Reading, Mass.: Addison-Wesley, 1964) suggested by F. L. Boos, Jr. (*Am. J. Phys.* **52**, 756 (1984)). A similar model was proposed earlier by R. H. Romer (*Am. J. Phys.* **34**, 772 (1966)). For further references, see T.-C. E. Ma, *Am. J. Phys.* **54**, 949 (1986).

in the region between the cylinders, and a magnetic field,

$$\mathbf{B} = \mu_0 n I \,\hat{\mathbf{z}} \; (s < R),$$

inside the solenoid. The momentum density (Eq. 8.33) was therefore

$$\boldsymbol{\wp}_{\rm em} = -\frac{\mu_0 n I Q}{2\pi l s}\,\hat{\boldsymbol{\phi}},$$

in the region $a < s < R$. The *angular* momentum density was

$$\boldsymbol{\ell}_{\rm em} = \mathbf{r} \times \boldsymbol{\wp}_{\rm em} = -\frac{\mu_0 n I Q}{2\pi l}\,\hat{\mathbf{z}},$$

which is *constant*, as it turns out; to get the *total* angular momentum in the fields, we simply multiply by the volume, $\pi(R^2 - a^2)l$:

$$\mathbf{L}_{\rm em} = -\frac{1}{2}\mu_0 n I Q (R^2 - a^2)\,\hat{\mathbf{z}}. \tag{8.35}$$

When the current is turned off, the changing magnetic field induces a circumferential electric field, given by Faraday's law:

$$\mathbf{E} = \begin{cases} -\dfrac{1}{2}\mu_0 n \dfrac{dI}{dt}\dfrac{R^2}{s}\,\hat{\boldsymbol{\phi}}, & (s > R), \\[2ex] -\dfrac{1}{2}\mu_0 n \dfrac{dI}{dt} s\,\hat{\boldsymbol{\phi}}, & (s < R). \end{cases}$$

Thus the torque on the outer cylinder is

$$\mathbf{N}_b = \mathbf{r} \times (-Q\mathbf{E}) = \frac{1}{2}\mu_0 n Q R^2 \frac{dI}{dt}\,\hat{\mathbf{z}},$$

and it picks up an angular momentum

$$\mathbf{L}_b = \frac{1}{2}\mu_0 n Q R^2\,\hat{\mathbf{z}} \int_I^0 \frac{dI}{dt}\,dt = -\frac{1}{2}\mu_0 n I Q R^2\,\hat{\mathbf{z}}.$$

Similarly, the torque on the inner cylinder is

$$\mathbf{N}_a = -\frac{1}{2}\mu_0 n Q a^2 \frac{dI}{dt}\,\hat{\mathbf{z}},$$

and its angular momentum increase is

$$\mathbf{L}_a = \frac{1}{2}\mu_0 n I Q a^2\,\hat{\mathbf{z}}.$$

So it all works out: $\mathbf{L}_{\rm em} = \mathbf{L}_a + \mathbf{L}_b$. The angular momentum *lost* by the fields is precisely equal to the angular momentum *gained* by the cylinders, and the *total* angular momentum (fields plus matter) is conserved.

Incidentally, the angular case is in some respects *cleaner* than the linear analog (Ex. 8.3), because there is no "hidden" angular momentum to compensate for the angular momentum in the fields, and the cylinders really *do* rotate when the magnetic field is turned off. If a localized system is not moving, its total *linear* momentum *has* to be zero,[5] but there is no corresponding theorem for angular momentum, and in Prob. 8.12 you will see a beautiful example in which nothing at *all* is moving—not even currents—and yet the angular momentum is nonzero.

Problem 8.7 In Ex. 8.4, suppose that instead of turning off the *magnetic* field (by reducing I) we turn off the *electric* field, by connecting a weakly[6] conducting radial spoke between the cylinders. (We'll have to cut a slot in the solenoid, so the cylinders can still rotate freely.) From the magnetic force on the current in the spoke, determine the total angular momentum delivered to the cylinders, as they discharge (they are now rigidly connected, so they rotate together). Compare the initial angular momentum stored in the fields (Eq. 8.35). (Notice that the *mechanism* by which angular momentum is transferred from the fields to the cylinders is entirely different in the two cases: in Ex. 8.4 it was Faraday's law, but here it is the Lorentz force law.)

! **Problem 8.8**[7] Imagine an iron sphere of radius R that carries a charge Q and a uniform magnetization $\mathbf{M} = M\hat{\mathbf{z}}$. The sphere is initially at rest.

(a) Compute the angular momentum stored in the electromagnetic fields.

(b) Suppose the sphere is gradually (and uniformly) demagnetized (perhaps by heating it up past the Curie point). Use Faraday's law to determine the induced electric field, find the torque this field exerts on the sphere, and calculate the total angular momentum imparted to the sphere in the course of the demagnetization.

(c) Suppose instead of *demagnetizing* the sphere we *discharge* it, by connecting a grounding wire to the north pole. Assume the current flows over the surface in such a way that the charge density remains uniform. Use the Lorentz force law to determine the torque on the sphere, and calculate the total angular momentum imparted to the sphere in the course of the discharge. (The magnetic field is discontinuous at the surface ... does this matter?) [*Answer:* $\frac{2}{9}\mu_0 MQR^2$]

More Problems on Chapter 8

Problem 8.9[8] A very long solenoid of radius a, with n turns per unit length, carries a current I_s. Coaxial with the solenoid, at radius $b \gg a$, is a circular ring of wire, with resistance R. When the current in the solenoid is (gradually) decreased, a current I_r is induced in the ring.

---

[5] S. Coleman and J. H. van Vleck, *Phys. Rev.* **171**, 1370 (1968).

[6] In Ex. 8.4 we turned the current off slowly, to keep things quasistatic; here we reduce the electric field slowly to keep the displacement current negligible.

[7] This version of the Feynman disk paradox was proposed by N. L. Sharma (*Am. J. Phys.* **56**, 420 (1988)); similar models were analyzed by E. M. Pugh and G. E. Pugh, *Am. J. Phys.* **35**, 153 (1967) and by R. H. Romer, *Am. J. Phys.* **35**, 445 (1967).

[8] For extensive discussion, see M. A. Heald, *Am. J. Phys.* **56**, 540 (1988).

(a) Calculate I_r, in terms of dI_s/dt.

(b) The power $(I_r^2 R)$ delivered to the ring must have come from the solenoid. Confirm this by calculating the Poynting vector just outside the solenoid (the *electric* field is due to the changing flux in the solenoid; the *magnetic* field is due to the current in the ring). Integrate over the entire surface of the solenoid, and check that you recover the correct total power.

Problem 8.10[9] A sphere of radius R carries a uniform polarization $\mathbf{P}$ and a uniform magnetization $\mathbf{M}$ (not necessarily in the same direction). Find the electromagnetic momentum of this configuration. [*Answer:* $(4/9)\pi\mu_0 R^3(\mathbf{M}\times\mathbf{P})$]

Problem 8.11[10] Picture the electron as a uniformly charged spherical shell, with charge e and radius R, spinning at angular velocity ω.

(a) Calculate the total energy contained in the electromagnetic fields.

(b) Calculate the total angular momentum contained in the fields.

(c) According to the Einstein formula ($E = mc^2$), the energy in the fields should contribute to the mass of the electron. Lorentz and others speculated that the *entire* mass of the electron might be accounted for in this way: $U_{\text{em}} = m_e c^2$. Suppose, moreover, that the electron's spin angular momentum is entirely attributable to the electromagnetic fields: $L_{\text{em}} = \hbar/2$. On these two assumptions, determine the radius and angular velocity of the electron. What is their product, ωR? Does this classical model make sense?

! **Problem 8.12**[11] Suppose you had an electric charge q_e and a magnetic monopole q_m. The field of the electric charge is

$$\mathbf{E} = \frac{1}{4\pi\epsilon_0}\frac{q_e}{\mathscr{r}^2}\hat{\mathscr{r}},$$

of course, and the field of the magnetic monopole is

$$\mathbf{B} = \frac{\mu_0}{4\pi}\frac{q_m}{\mathscr{r}^2}\hat{\mathscr{r}}.$$

Find the total angular momentum stored in the fields, if the two charges are separated by a distance d. [*Answer:* $(\mu_0/4\pi)q_e q_m$.][12]

Problem 8.13 Paul De Young, of Hope College, points out that because the cylinders in Ex. 8.4 are left rotating (at angular velocities ω_a and ω_b, say), there is actually a residual magnetic field, and hence angular momentum in the fields, even after the current in the solenoid has been extinguished. If the cylinders are heavy, this correction will be negligible, but it is interesting to do the problem *without* making that assumption.

---

[9] For an interesting discussion and references, see R. H. Romer, *Am. J. Phys.* **63**, 777 (1995).

[10] See J. Higbie, *Am. J. Phys.* **56**, 378 (1988).

[11] This system is known as **Thomson's dipole**. See I. Adawi, *Am. J. Phys.* **44**, 762 (1976) and *Phys. Rev.* **D31**, 3301 (1985), and K. R. Brownstein, *Am. J. Phys.* **57**, 420 (1989), for discussion and references.

[12] Note that this result is *independent of the separation distance d* (!); it points from q_e toward q_m. In quantum mechanics angular momentum comes in half-integer multiples of $\hbar$, so this result suggests that if magnetic monopoles exist, electric and magnetic charge must be quantized: $\mu_0 q_e q_m/4\pi = n\hbar/2$, for $n = 1, 2, 3, \ldots$, an idea first proposed by Dirac in 1931. If even *one* monopole exists somewhere in the universe, this would "explain" why electric charge comes in discrete units.

(a) Calculate (in terms of ω_a and ω_b) the final angular momentum in the fields.

(b) As the cylinders begin to rotate, their changing magnetic field induces an extra azimuthal electric field, which, in turn, will make an additional contribution to the torques. Find the resulting extra angular momentum, and compare it to your result in (a). [*Answer:* $\mu_0 Q^2 \omega_b (b^2 - a^2)/4\pi l$]

Problem 8.14[13] A point charge q is a distance $a > R$ from the axis of an infinite solenoid (radius R, n turns per unit length, current I). Find the linear momentum and the angular momentum in the fields. (Put q on the x axis, with the solenoid along z; treat the solenoid as a nonconductor, so you don't need to worry about induced charges on its surface.) [*Answer:* $\mathbf{p}_{\text{em}} = (\mu_0 q n I R^2/2a)\,\hat{\mathbf{y}}$; $\mathbf{L}_{\text{em}} = 0$]

Problem 8.15[14] (a) Carry through the argument in Sect. 8.1.2, starting with Eq. 8.6, but using $\mathbf{J}_f$ in place of $\mathbf{J}$. Show that the Poynting vector becomes

$$\mathbf{S} = \mathbf{E} \times \mathbf{H},$$

and the rate of change of the energy density in the fields is

$$\frac{\partial u_{\text{em}}}{\partial t} = \mathbf{E} \cdot \frac{\partial \mathbf{D}}{\partial t} + \mathbf{H} \cdot \frac{\partial \mathbf{B}}{\partial t}.$$

For *linear* media, show that

$$u_{\text{em}} = \frac{1}{2}(\mathbf{E} \cdot \mathbf{D} + \mathbf{B} \cdot \mathbf{H}).$$

(b) In the same spirit, reproduce the argument in Sect. 8.2.2, starting with Eq. 8.15, with ρ_f and $\mathbf{J}_f$ in place of ρ and $\mathbf{J}$. Don't bother to construct the Maxwell stress tensor, but do show that the momentum density is

$$\boldsymbol{\wp} = \mathbf{D} \times \mathbf{B}.$$

---

[13] See F. S. Johnson, B. L. Cragin, and R. R. Hodges, *Am. J. Phys.* **62**, 33 (1994), for a discussion of this and related problems.

[14] This problem was suggested by David Thouless of the University of Washington. Refer to Sect. 4.4.3 for the meaning of "energy" in this context.

Chapter 9

Electromagnetic Waves

9.1 Waves in One Dimension

9.1.1 The Wave Equation

What is a "wave?" I don't think I can give you an entirely satisfactory answer—the concept is intrinsically somewhat vague—but here's a start: A wave is a *disturbance of a continuous medium that propagates with a fixed shape at constant velocity.* Immediately I must add qualifiers: In the presence of absorption, the wave will diminish in size as it moves; if the medium is dispersive different frequencies travel at different speeds; in two or three dimensions, as the wave spreads out its amplitude will decrease; and of course *standing* waves don't propagate at all. But these are refinements; let's start with the simple case: fixed shape, constant speed (Fig. 9.1).

How would you represent such an object mathematically? In the figure I have drawn the wave at two different times, once at $t = 0$, and again at some later time t—each point on the wave form simply shifts to the right by an amount vt, where v is the velocity. Maybe the wave is generated by shaking one end of a taut string; $f(z, t)$ represents the displacement of the string at the point z, at time t. Given the *initial* shape of the string, $g(z) \equiv f(z, 0)$,

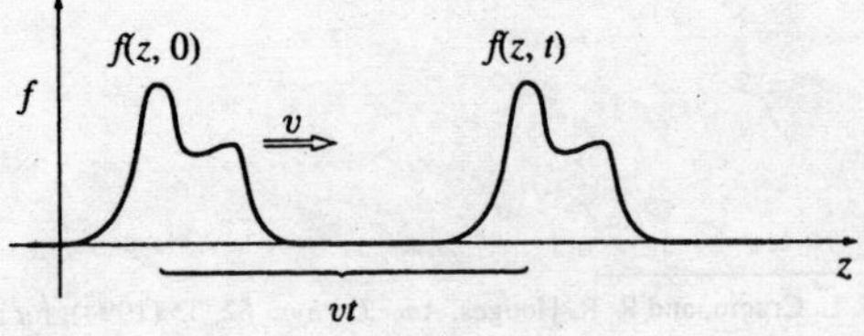

Figure 9.1

what is the subsequent form, $f(z,t)$? Evidently, the displacement at point z, at the later time t, is the same as the displacement a distance vt to the left (i.e. at $z - vt$), back at time $t = 0$:

$$f(z,t) = f(z - vt, 0) = g(z - vt). \tag{9.1}$$

That statement captures (mathematically) the essence of wave motion. It tells us that the function $f(z,t)$, which *might* have depended on z and t in *any* old way, in *fact* depends on them only in the very special combination $z - vt$; when that is true, the function $f(z,t)$ represents a wave of fixed shape traveling in the z direction at speed v. For example, if A and b are constants (with the appropriate units),

$$f_1(z,t) = Ae^{-b(z-vt)^2}, \quad f_2(z,t) = A\sin[b(z-vt)], \quad f_3(z,t) = \frac{A}{b(z-vt)^2+1}$$

all represent waves (with different shapes, of course), but

$$f_4(z,t) = Ae^{-b(bz^2+vt)}, \quad \text{and} \quad f_5(z,t) = A\sin(bz)\cos(bvt)^3,$$

do *not*.

Why does a stretched string support wave motion? Actually, it follows from Newton's second law. Imagine a very long string under tension T. If it is displaced from equilibrium, the net transverse force on the segment between z and $z + \Delta z$ (Fig. 9.2) is

$$\Delta F = T\sin\theta' - T\sin\theta,$$

where θ' is the angle the string makes with the z-direction at point $z + \Delta z$, and θ is the corresponding angle at point z. Provided that the distortion of the string is not too great, these angles are small (the figure is exaggerated, obviously), and we can replace the sine by the tangent:

$$\Delta F \cong T(\tan\theta' - \tan\theta) = T\left(\left.\frac{\partial f}{\partial z}\right|_{z+\Delta z} - \left.\frac{\partial f}{\partial z}\right|_{z}\right) \cong T\frac{\partial^2 f}{\partial z^2}\Delta z.$$

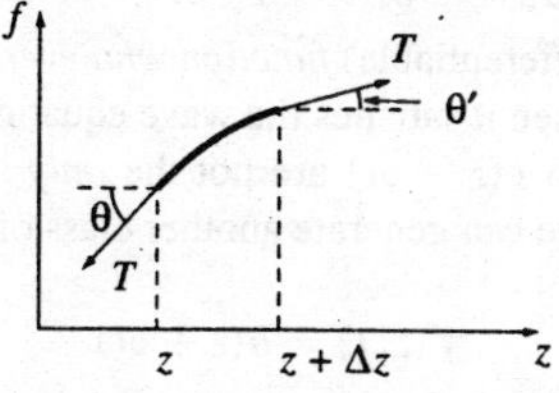

Figure 9.2

If the mass per unit length is μ, Newton's second law says

$$\Delta F = \mu(\Delta z)\frac{\partial^2 f}{\partial t^2},$$

and therefore

$$\frac{\partial^2 f}{\partial z^2} = \frac{\mu}{T}\frac{\partial^2 f}{\partial t^2}.$$

Evidently, small disturbances on the string satisfy

$$\boxed{\frac{\partial^2 f}{\partial z^2} = \frac{1}{v^2}\frac{\partial^2 f}{\partial t^2},} \tag{9.2}$$

where v (which, as we'll soon see, represents the speed of propagation) is

$$v = \sqrt{\frac{T}{\mu}}. \tag{9.3}$$

Equation 9.2 is known as the (classical) **wave equation**, because it admits as solutions all functions of the form

$$f(z, t) = g(z - vt), \tag{9.4}$$

(that is, all functions that depend on the variables z and t in the special combination $u \equiv z - vt$), and we have just learned that such functions represent waves propagating in the z direction with speed v. For Eq. 9.4 means

$$\frac{\partial f}{\partial z} = \frac{dg}{du}\frac{\partial u}{\partial z} = \frac{dg}{du}, \quad \frac{\partial f}{\partial t} = \frac{dg}{du}\frac{\partial u}{\partial t} = -v\frac{dg}{du},$$

and

$$\frac{\partial^2 f}{\partial z^2} = \frac{\partial}{\partial z}\left(\frac{dg}{du}\right) = \frac{d^2 g}{du^2}\frac{\partial u}{\partial z} = \frac{d^2 g}{du^2},$$

$$\frac{\partial^2 f}{\partial t^2} = -v\frac{\partial}{\partial t}\left(\frac{dg}{du}\right) = -v\frac{d^2 g}{du^2}\frac{\partial u}{\partial t} = v^2\frac{d^2 g}{du^2},$$

so

$$\frac{d^2 g}{du^2} = \frac{\partial^2 f}{\partial z^2} = \frac{1}{v^2}\frac{\partial^2 f}{\partial t^2}. \qquad \text{qed}$$

Note that $g(u)$ can be *any* (differentiable) *function whatever.* If the disturbance propagates without changing its shape, then it satisfies the wave equation.

But functions of the form $g(z - vt)$ are not the *only* solutions. The wave equation involves the *square* of v, so we can generate another class of solutions by simply changing the sign of the velocity:

$$f(z, t) = h(z + vt). \tag{9.5}$$

This, of course, represents a wave propagating in the *negative* z direction, and it is certainly reasonable (on physical grounds) that such solutions would be allowed. What is perhaps surprising is that the *most general* solution to the wave equation is the sum of a wave to the right and a wave to the left:

$$f(z,t) = g(z - vt) + h(z + vt). \tag{9.6}$$

(Notice that the wave equation is **linear**: The sum of any two solutions is itself a solution.) *Every* solution to the wave equation can be expressed in this form.

Like the simple harmonic oscillator equation, the wave equation is ubiquitous in physics. If something is vibrating, the oscillator equation is almost certainly responsible (at least, for small amplitudes), and if something is waving (whether the context is mechanics or acoustics, optics or oceanography), the wave equation (perhaps with some decoration) is bound to be involved.

Problem 9.1 By explicit differentiation, check that the functions f_1, f_2, and f_3 in the text satisfy the wave equation. Show that f_4 and f_5 do *not*.

Problem 9.2 Show that the **standing wave** $f(z,t) = A\sin(kz)\cos(kvt)$ satisfies the wave equation, and express it as the sum of a wave traveling to the left and a wave traveling to the right (Eq. 9.6).

9.1.2 Sinusoidal Waves

(i) Terminology. Of all possible wave forms, the sinusoidal one

$$f(z,t) = A\cos[k(z - vt) + \delta] \tag{9.7}$$

is (for good reason) the most familiar. Figure 9.3 shows this function at time $t = 0$. A is the **amplitude** of the wave (it is positive, and represents the maximum displacement from equilibrium). The argument of the cosine is called the **phase**, and δ is the **phase constant** (obviously, you can add any integer multiple of 2π to δ without changing $f(z,t)$; ordinarily, one uses a value in the range $0 \le \delta < 2\pi$). Notice that at $z = vt - \delta/k$, the phase is zero; let's call this the "central maximum." If $\delta = 0$, the central maximum passes the origin at time $t = 0$; more generally, δ/k is the distance by which the central maximum (and

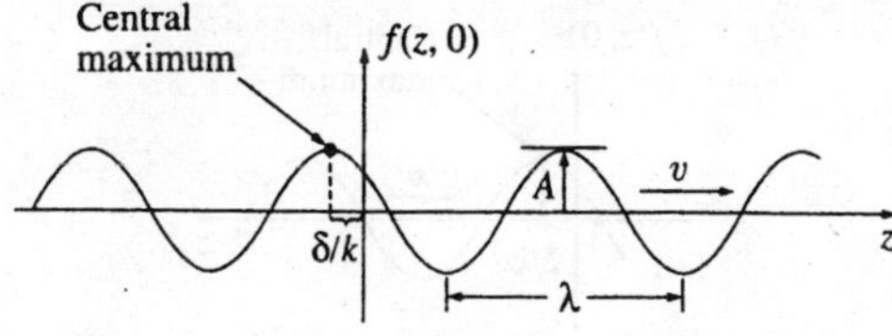

Figure 9.3

therefore the entire wave) is "delayed." Finally, k is the **wave number**; it is related to the **wavelength** λ by the equation

$$\lambda = \frac{2\pi}{k}, \tag{9.8}$$

for when z advances by $2\pi/k$, the cosine executes one complete cycle.

As time passes, the entire wave train proceeds to the right, at speed v. At any fixed point z, the string vibrates up and down, undergoing one full cycle in a **period**

$$T = \frac{2\pi}{kv}. \tag{9.9}$$

The **frequency** ν (number of oscillations per unit time) is

$$\nu = \frac{1}{T} = \frac{kv}{2\pi} = \frac{v}{\lambda}. \tag{9.10}$$

For our purposes, a more convenient unit is the **angular frequency** ω, so-called because in the analogous case of uniform circular motion it represents the number of radians swept out per unit time:

$$\omega = 2\pi\nu = kv. \tag{9.11}$$

Ordinarily, it's nicer to write sinusoidal waves (Eq. 9.7) in terms of ω, rather than v:

$$f(z,t) = A\cos(kz - \omega t + \delta). \tag{9.12}$$

A sinusoidal oscillation of wave number k and (angular) frequency ω traveling to the *left* would be written

$$f(z,t) = A\cos(kz + \omega t - \delta). \tag{9.13}$$

The sign of the phase constant is chosen for consistency with our previous convention that δ/k shall represent the distance by which the wave is "delayed" (since the wave is now moving to the *left*, a delay means a shift to the *right*). At $t = 0$, the wave looks like Fig. 9.4. Because the cosine is an *even* function, we can just as well write Eq. 9.13 thus:

$$f(z,t) = A\cos(-kz - \omega t + \delta). \tag{9.14}$$

Comparison with Eq. 9.12 reveals that, in effect, *we could simply switch the sign of k* to produce a wave with the same amplitude, phase constant, frequency, and wavelength, traveling in the opposite direction.

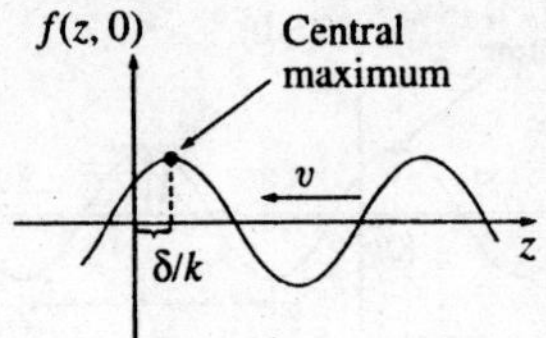

Figure 9.4

(ii) Complex notation. In view of **Euler's formula**,

$$e^{i\theta} = \cos\theta + i\sin\theta, \tag{9.15}$$

the sinusoidal wave (Eq. 9.12) can be written

$$f(z, t) = \text{Re}[Ae^{i(kz-\omega t+\delta)}], \tag{9.16}$$

where $\text{Re}(\xi)$ denotes the real part of the complex number ξ. This invites us to introduce the **complex wave function**

$$\tilde{f}(z, t) \equiv \tilde{A}e^{i(kz-\omega t)}, \tag{9.17}$$

with the **complex amplitude** $\tilde{A} \equiv Ae^{i\delta}$ absorbing the phase constant. The *actual* wave function is the real part of $\tilde{f}$:

$$f(z, t) = \text{Re}[\tilde{f}(z, t)]. \tag{9.18}$$

If you know $\tilde{f}$, it is a simple matter to find f; the *advantage* of the complex notation is that exponentials are much easier to manipulate than sines and cosines.

Example 9.1

Suppose you want to combine two sinusoidal waves:

$$f_3 = f_1 + f_2 = \text{Re}(\tilde{f}_1) + \text{Re}(\tilde{f}_2) = \text{Re}(\tilde{f}_1 + \tilde{f}_2) = \text{Re}(\tilde{f}_3),$$

with $\tilde{f}_3 = \tilde{f}_1 + \tilde{f}_2$. You simply add the corresponding *complex* wave functions, and then take the real part. In particular, if they have the same frequency and wave number,

$$\tilde{f}_3 = \tilde{A}_1 e^{i(kz-\omega t)} + \tilde{A}_2 e^{i(kz-\omega t)} = \tilde{A}_3 e^{i(kz-\omega t)},$$

where

$$\tilde{A}_3 = \tilde{A}_1 + \tilde{A}_2, \quad \text{or} \quad A_3 e^{i\delta_3} = A_1 e^{i\delta_1} + A_2 e^{i\delta_2}; \tag{9.19}$$

evidently, you just add the (complex) amplitudes. The combined wave still has the same frequency and wavelength,

$$f_3(z, t) = A_3 \cos(kz - \omega t + \delta_3),$$

and you can easily figure out A_3 and δ_3 from Eq. 9.19 (Prob. 9.3). Try doing this *without* using the complex notation—you will find yourself looking up trig identities and slogging through nasty algebra.

(iii) Linear combinations of sinusoidal waves. Although the sinusoidal function 9.17 is a very special wave form, the fact is that *any* wave can be expressed as a linear combination of sinusoidal ones:

$$\tilde{f}(z, t) = \int_{-\infty}^{\infty} \tilde{A}(k) e^{i(kz-\omega t)}\, dk. \tag{9.20}$$

Here ω is a function of k (Eq. 9.11), and I have allowed k to run through negative values in order to include waves going in both directions.[1]

[1]This does not mean that λ and ω are negative—wavelength and frequency are *always* positive. If we allow negative wave numbers, then Eqs. 9.8 and 9.11 should really be written $\lambda = 2\pi/|k|$ and $\omega = |k|v$.

The formula for $\tilde{A}(k)$, in terms of the initial conditions $f(z,0)$ and $\dot{f}(z,0)$, can be obtained from the theory of Fourier transforms (see Prob. 9.32), but the details are not relevant to my purpose here. The *point* is that any wave can be written as a linear combination of sinusoidal waves, and therefore if you know how sinusoidal waves behave, you know in principle how *any* wave behaves. So from now on we shall confine our attention to sinusoidal waves.

Problem 9.3 Use Eq. 9.19 to determine A_3 and δ_3 in terms of A_1, A_2, δ_1, and δ_2.

Problem 9.4 Obtain Eq. 9.20 directly from the wave equation, by separation of variables.

9.1.3 Boundary Conditions: Reflection and Transmission

So far I have assumed the string is infinitely long—or at any rate long enough that we don't need to worry about what happens to a wave when it reaches the end. As a matter of fact, what happens depends a lot on how the string is *attached* at the end—that is, on the specific boundary conditions to which the wave is subject. Suppose, for instance, that the string is simply tied onto a *second* string. The tension T is the same for both, but the mass per unit length μ presumably is not, and hence the wave velocities v_1 and v_2 are different (remember, $v = \sqrt{T/\mu}$). Let's say, for convenience, that the knot occurs at $z = 0$. The **incident** wave

$$\tilde{f}_I(z,t) = \tilde{A}_I e^{i(k_1 z - \omega t)}, \quad (z < 0), \tag{9.21}$$

coming in from the left, gives rise to a **reflected** wave

$$\tilde{f}_R(z,t) = \tilde{A}_R e^{i(-k_1 z - \omega t)} \quad (z < 0), \tag{9.22}$$

traveling *back* along string 1 (hence the minus sign in front of k_1), in addition to a **transmitted** wave

$$\tilde{f}_T(z,t) = \tilde{A}_T e^{i(k_2 z - \omega t)}, \quad (z > 0), \tag{9.23}$$

which continues on to the right in string 2.

The incident wave $f_I(z,t)$ is a sinusoidal oscillation that extends (in principle) all the way back to $z = -\infty$, and has been doing so for all of history. The same goes for f_R and f_T (except that the latter, of course, extends to $z = +\infty$). *All parts of the system are oscillating at the same frequency* ω (a frequency determined by the person at $z = -\infty$, who is shaking the string in the first place). Since the wave velocities are different in the two strings, however, the wavelengths and wave numbers are also different:

$$\frac{\lambda_1}{\lambda_2} = \frac{k_2}{k_1} = \frac{v_1}{v_2}. \tag{9.24}$$

Of course, this situation is pretty artificial—what's more, with incident and reflected waves of infinite extent traveling on the same piece of string, it's going to be hard for a spectator to tell them apart. You might therefore prefer to consider an incident wave of *finite* extent—say, the pulse shown in Fig. 9.5. You can work out the details for yourself, if you like (Prob. 9.5). The *trouble* with this approach is that no *finite* pulse is truly sinusoidal. The waves in Fig. 9.5 may *look* like sine functions, but they're *not*: they're little *pieces* of sines, joined onto an entirely *different* function (namely, zero). Like any other waves, they can be built up as *linear combinations* of true sinusoidal functions (Eq. 9.20), but only by putting together a whole range of frequencies and wavelengths. If you want a *single* incident frequency (as we shall in the electromagnetic case), you must let your waves extend to infinity. In practice, if you use a very *long* pulse with many oscillations, it will be *close* to the ideal of a single frequency.

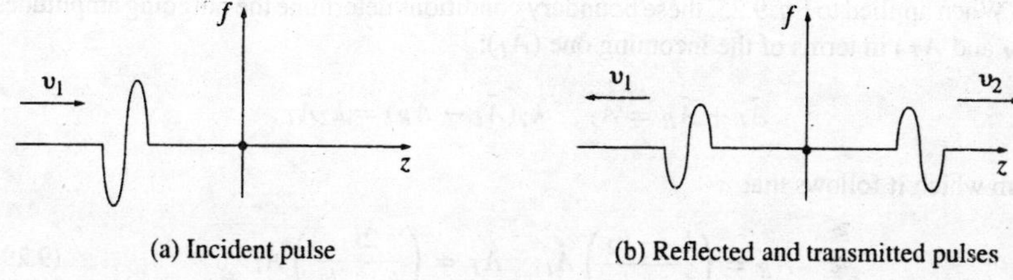

Figure 9.5

For a sinusoidal incident wave, then, the net disturbance of the string is:

$$\tilde{f}(z,t) = \begin{cases} \tilde{A}_I e^{i(k_1 z - \omega t)} + \tilde{A}_R e^{i(-k_1 z - \omega t)}, & \text{for } z < 0, \\ \tilde{A}_T e^{i(k_2 z - \omega t)}, & \text{for } z > 0. \end{cases} \tag{9.25}$$

At the join ($z = 0$), the displacement just slightly to the left ($z = 0^-$) must equal the displacement slightly to the right ($z = 0^+$), or else there would be a break between the two strings. Mathematically, $f(z, t)$ is *continuous* at $z = 0$:

$$f(0^-, t) = f(0^+, t). \tag{9.26}$$

If the knot itself is of negligible mass, then the *derivative* of f must *also* be continuous:

$$\left.\frac{\partial f}{\partial z}\right|_{0^-} = \left.\frac{\partial f}{\partial z}\right|_{0^+}. \tag{9.27}$$

Otherwise there would be a net force on the knot, and therefore an infinite acceleration (Fig. 9.6). These boundary conditions apply directly to the *real* wave function $f(z, t)$. But since the imaginary part of $\tilde{f}$ differs from the real part only in the replacement of cosine by sine (Eq. 9.15), it follows that the complex wave function $\tilde{f}(z, t)$ obeys the same rules:

$$\tilde{f}(0^-, t) = \tilde{f}(0^+, t), \qquad \left.\frac{\partial \tilde{f}}{\partial z}\right|_{0^-} = \left.\frac{\partial \tilde{f}}{\partial z}\right|_{0^+}. \tag{9.28}$$

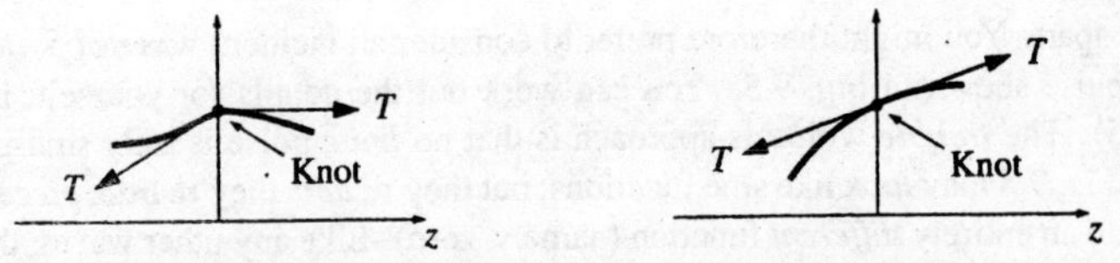

(a) Discontinuous slope; force on knot (a) Continuous slope; no force on knot

Figure 9.6

When applied to Eq. 9.25, these boundary conditions determine the outgoing amplitudes ($\tilde{A}_R$ and $\tilde{A}_T$) in terms of the incoming one ($\tilde{A}_I$):

$$\tilde{A}_I + \tilde{A}_R = \tilde{A}_T, \quad k_1(\tilde{A}_I - \tilde{A}_R) = k_2\tilde{A}_T,$$

from which it follows that

$$\tilde{A}_R = \left(\frac{k_1 - k_2}{k_1 + k_2}\right)\tilde{A}_I, \quad \tilde{A}_T = \left(\frac{2k_1}{k_1 + k_2}\right)\tilde{A}_I. \tag{9.29}$$

Or, in terms of the velocities (Eq. 9.24):

$$\tilde{A}_R = \left(\frac{v_2 - v_1}{v_2 + v_1}\right)\tilde{A}_I, \quad \tilde{A}_T = \left(\frac{2v_2}{v_2 + v_1}\right)\tilde{A}_I. \tag{9.30}$$

The *real* amplitudes and phases, then, are related by

$$A_R e^{i\delta_R} = \left(\frac{v_2 - v_1}{v_2 + v_1}\right)A_I e^{i\delta_I}, \quad A_T e^{i\delta_T} = \left(\frac{2v_2}{v_2 + v_1}\right)A_I e^{i\delta_I}. \tag{9.31}$$

If the second string is *lighter* than the first ($\mu_2 < \mu_1$, so that $v_2 > v_1$), all three waves have the same phase angle ($\delta_R = \delta_T = \delta_I$), and the outgoing amplitudes are

$$A_R = \left(\frac{v_2 - v_1}{v_2 + v_1}\right)A_I, \quad A_T = \left(\frac{2v_2}{v_2 + v_1}\right)A_I. \tag{9.32}$$

If the second string is *heavier* than the first ($v_2 < v_1$) the reflected wave is out of phase by 180° ($\delta_R + \pi = \delta_T = \delta_I$). In other words, since

$$\cos(-k_1 z - \omega t + \delta_I - \pi) = -\cos(-k_1 z - \omega t + \delta_I),$$

the reflected wave is "upside down." The amplitudes in this case are

$$A_R = \left(\frac{v_1 - v_2}{v_2 + v_1}\right)A_I \text{ and } A_T = \left(\frac{2v_2}{v_2 + v_1}\right)A_I. \tag{9.33}$$

In particular, if the second string is *infinitely* massive—or, what amounts to the same thing, if the first string is simply *nailed down* at the end—then

$$A_R = A_I \text{ and } A_T = 0.$$

Naturally, in this case there is *no* transmitted wave—*all* of it reflects back.

! **Problem 9.5** Suppose you send an incident wave of specified shape, $g_I(z - v_1 t)$, down string number 1. It gives rise to a reflected wave, $h_R(z + v_1 t)$, and a transmitted wave, $g_T(z - v_2 t)$. By imposing the boundary conditions 9.26 and 9.27, find h_R and g_T.

Problem 9.6

(a) Formulate an appropriate boundary condition, to replace Eq. 9.27, for the case of two strings under tension T joined by a knot of mass m.

(b) Find the amplitude and phase of the reflected and transmitted waves for the case where the knot has a mass m and the second string is massless.

! **Problem 9.7** Suppose string 2 is embedded in a viscous medium (such as molasses), which imposes a drag force that is proportional to its (transverse) speed:

$$\Delta F_{\text{drag}} = -\gamma \frac{\partial f}{\partial t} \Delta z.$$

(a) Derive the modified wave equation describing the motion of the string.

(b) Solve this equation, assuming the string oscillates at the incident frequency ω. That is, look for solutions of the form $\tilde{f}(z, t) = e^{i\omega t} \tilde{F}(z)$.

(c) Show that the waves are **attenuated** (that is, their amplitude decreases with increasing z). Find the characteristic penetration distance, at which the amplitude is $1/e$ of its original value, in terms of γ, T, μ, and ω.

(d) If a wave of amplitude A_I, phase $\delta_I = 0$, and frequency ω is incident from the left (string 1), find the reflected wave's amplitude and phase.

9.1.4 Polarization

The waves that travel down a string when you shake it are called **transverse**, because the displacement is perpendicular to the direction of propagation. If the string is reasonably elastic, it is also possible to stimulate *compression* waves, by giving the string little tugs. Compression waves are hard to see on a string, but if you try it with a slinky they're quite noticeable (Fig. 9.7). These waves are called **longitudinal**, because the displacement from equilibrium is along the direction of propagation. Sound waves, which are nothing but compression waves in air, are longitudinal; electromagnetic waves, as we shall see, are transverse.

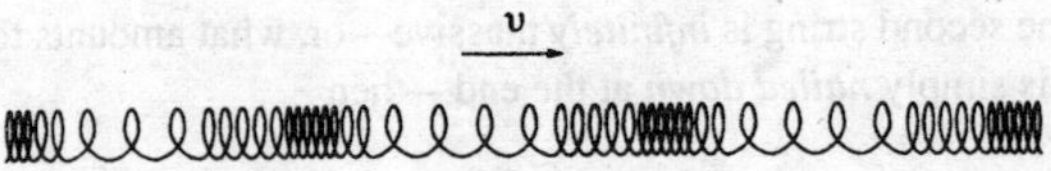

v

Figure 9.7

Now there are, of course, *two* dimensions perpendicular to any given line of propagation. Accordingly, transverse waves occur in two independent states of **polarization**: you can shake the string up-and-down ("vertical" polarization—Fig. 9.8a),

$$\tilde{\mathbf{f}}_v(z,t) = \tilde{A}e^{i(kz-\omega t)}\,\hat{\mathbf{x}}, \tag{9.34}$$

or left-and-right ("horizontal" polarization—Fig. 9.8b),

$$\tilde{\mathbf{f}}_h(z,t) = \tilde{A}e^{i(kz-\omega t)}\,\hat{\mathbf{y}}, \tag{9.35}$$

or along any other direction in the xy plane (Fig. 9.8c):

$$\tilde{\mathbf{f}}(z,t) = \tilde{A}e^{i(kz-\omega t)}\,\hat{\mathbf{n}}. \tag{9.36}$$

The **polarization vector** $\hat{\mathbf{n}}$ defines the plane of vibration.[2] Because the waves are transverse, $\hat{\mathbf{n}}$ is perpendicular to the direction of propagation:

$$\hat{\mathbf{n}}\cdot\hat{\mathbf{z}} = 0. \tag{9.37}$$

In terms of the **polarization angle** θ,

$$\hat{\mathbf{n}} = \cos\theta\,\hat{\mathbf{x}} + \sin\theta\,\hat{\mathbf{y}}. \tag{9.38}$$

Thus, the wave pictured in Fig. 9.8c can be considered a superposition of two waves—one horizontally polarized, the other vertically:

$$\tilde{\mathbf{f}}(z,t) = (\tilde{A}\cos\theta)e^{i(kz-\omega t)}\,\hat{\mathbf{x}} + (\tilde{A}\sin\theta)e^{i(kz-\omega t)}\,\hat{\mathbf{y}}. \tag{9.39}$$

Problem 9.8 Equation 9.36 describes the most general **linearly** polarized wave on a string. Linear (or "plane") polarization (so called because the displacement is parallel to a fixed vector $\hat{\mathbf{n}}$) results from the combination of horizontally and vertically polarized waves of the *same phase* (Eq. 9.39). If the two components are of equal amplitude, but *out of phase* by 90° (say, $\delta_v = 0,\ \delta_h = 90°$), the result is a *circularly* polarized wave. In that case:

(a) At a fixed point z, show that the string moves in a circle about the z axis. Does it go *clockwise* or *counterclockwise*, as you look down the axis toward the origin? How would you construct a wave circling the *other* way? (In optics, the clockwise case is called **right circular polarization**, and the counterclockwise, **left circular polarization**.)

(b) Sketch the string at time $t = 0$.

(c) How would you shake the string in order to produce a circularly polarized wave?

---

[2]Notice that you can always switch the *sign* of $\hat{\mathbf{n}}$, provided you simultaneously advance the phase constant by 180°, since both operations change the sign of the wave.

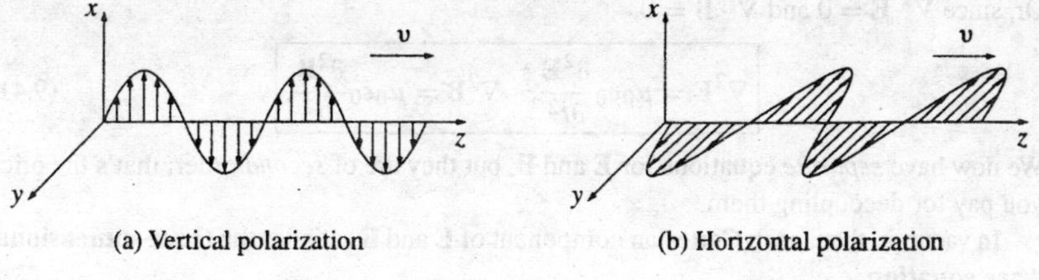

(a) Vertical polarization (b) Horizontal polarization

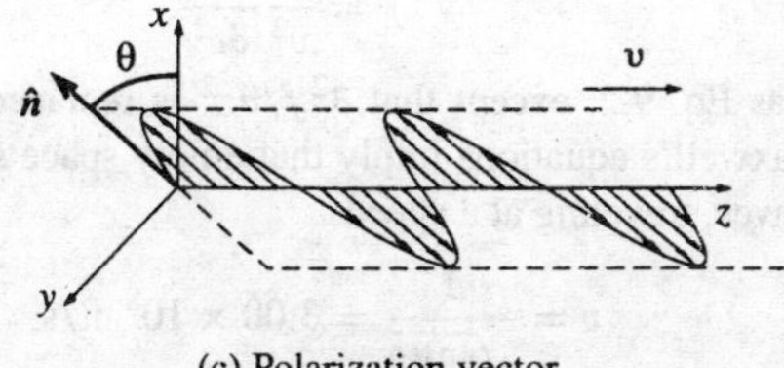

(c) Polarization vector

Figure 9.8

9.2 Electromagnetic Waves in Vacuum

9.2.1 The Wave Equation for E and B

In regions of space where there is no charge or current, Maxwell's equations read

$$\left.\begin{array}{ll}\text{(i)} \quad \nabla \cdot \mathbf{E} = 0, & \text{(iii)} \quad \nabla \times \mathbf{E} = -\dfrac{\partial \mathbf{B}}{\partial t}, \\ \text{(ii)} \quad \nabla \cdot \mathbf{B} = 0, & \text{(iv)} \quad \nabla \times \mathbf{B} = \mu_0 \epsilon_0 \dfrac{\partial \mathbf{E}}{\partial t}.\end{array}\right\} \tag{9.40}$$

They constitute a set of coupled, first-order, partial differential equations for **E** and **B**. They can be *de*coupled by applying the curl to (iii) and (iv):

$$\begin{aligned}\nabla \times (\nabla \times \mathbf{E}) &= \nabla(\nabla \cdot \mathbf{E}) - \nabla^2 \mathbf{E} = \nabla \times \left(-\frac{\partial \mathbf{B}}{\partial t}\right) \\ &= -\frac{\partial}{\partial t}(\nabla \times \mathbf{B}) = -\mu_0 \epsilon_0 \frac{\partial^2 \mathbf{E}}{\partial t^2}, \\ \nabla \times (\nabla \times \mathbf{B}) &= \nabla(\nabla \cdot \mathbf{B}) - \nabla^2 \mathbf{B} = \nabla \times \left(\mu_0 \epsilon_0 \frac{\partial \mathbf{E}}{\partial t}\right) \\ &= \mu_0 \epsilon_0 \frac{\partial}{\partial t}(\nabla \times \mathbf{E}) = -\mu_0 \epsilon_0 \frac{\partial^2 \mathbf{B}}{\partial t^2}.\end{aligned}$$

Or, since $\nabla \cdot \mathbf{E} = 0$ and $\nabla \cdot \mathbf{B} = 0$,

$$\boxed{\nabla^2 \mathbf{E} = \mu_0 \epsilon_0 \frac{\partial^2 \mathbf{E}}{\partial t^2}, \quad \nabla^2 \mathbf{B} = \mu_0 \epsilon_0 \frac{\partial^2 \mathbf{B}}{\partial t^2}.} \tag{9.41}$$

We now have *separate* equations for **E** and **B**, but they are of *second* order; that's the price you pay for decoupling them.

In vacuum, then, each Cartesian component of **E** and **B** satisfies the **three-dimensional wave equation,**

$$\nabla^2 f = \frac{1}{v^2} \frac{\partial^2 f}{\partial t^2}.$$

(This is the same as Eq. 9.2, except that $\partial^2 f/\partial z^2$ is replaced by its natural generalization, $\nabla^2 f$.) So Maxwell's equations imply that empty space supports the propagation of electromagnetic waves, traveling at a speed

$$v = \frac{1}{\sqrt{\epsilon_0 \mu_0}} = 3.00 \times 10^8 \text{ m/s}, \tag{9.42}$$

which happens to be precisely the velocity of light, c. The implication is astounding: Perhaps light *is* an electromagnetic wave.[3] Of course, this conclusion does not surprise anyone today, but imagine what a revelation it was in Maxwell's time! Remember how ϵ_0 and μ_0 came into the theory in the first place: they were constants in Coulomb's law and the Biot-Savart law, respectively. You measure them in experiments involving charged pith balls, batteries, and wires—experiments having nothing whatever to do with light. And yet, according to Maxwell's theory you can calculate c from these two numbers. Notice the crucial role played by Maxwell's contribution to Ampère's law ($\mu_0 \epsilon_0 \partial \mathbf{E}/\partial t$); without it, the wave equation would not emerge, and there would be no electromagnetic theory of light.

9.2.2 Monochromatic Plane Waves

For reasons discussed in Sect. 9.1.2, we may confine our attention to sinusoidal waves of frequency ω. Since different frequencies in the visible range correspond to different *colors*, such waves are called **monochromatic** (Table 9.1). Suppose, moreover, that the waves are traveling in the z direction and have no x or y dependence; these are called **plane waves,**[4] because the fields are uniform over every plane perpendicular to the direction of propagation (Fig. 9.9). We are interested, then, in fields of the form

$$\tilde{\mathbf{E}}(z, t) = \tilde{\mathbf{E}}_0 e^{i(kz - \omega t)}, \quad \tilde{\mathbf{B}}(z, t) = \tilde{\mathbf{B}}_0 e^{i(kz - \omega t)}, \tag{9.43}$$

---

[3] As Maxwell himself put it, "We can scarcely avoid the inference that light consists in the transverse undulations of the same medium which is the cause of electriĉ and magnetic phenomena." See Ivan Tolstoy, *James Clerk Maxwell, A Biography* (Chicago: University of Chicago Press, 1983).

[4] For a discussion of *spherical* waves, at this level, see J. R. Reitz, F. J. Milford, and R. W. Christy, *Foundations of Electromagnetic Theory*, 3rd ed., Sect. 17-5 (Reading, MA: Addison-Wesley, 1979). Or work Prob. 9.33. Of course, over small enough regions *any* wave is essentially plane, as long as the wavelength is much less than the radius of the curvature of the wave front.

The Electromagnetic Spectrum		
Frequency (Hz)	Type	Wavelength (m)
10^{22}		10^{-13}
10^{21}	gamma rays	10^{-12}
10^{20}		10^{-11}
10^{19}		10^{-10}
10^{18}	x rays	10^{-9}
10^{17}		10^{-8}
10^{16}	ultraviolet	10^{-7}
10^{15}	visible	10^{-6}
10^{14}	infrared	10^{-5}
10^{13}		10^{-4}
10^{12}		10^{-3}
10^{11}		10^{-2}
10^{10}	microwave	10^{-1}
10^{9}		1
10^{8}	TV, FM	10
10^{7}		10^{2}
10^{6}	AM	10^{3}
10^{5}		10^{4}
10^{4}	RF	10^{5}
10^{3}		10^{6}

The Visible Range		
Frequency (Hz)	Color	Wavelength (m)
1.0×10^{15}	near ultraviolet	3.0×10^{-7}
7.5×10^{14}	shortest visible blue	4.0×10^{-7}
6.5×10^{14}	blue	4.6×10^{-7}
5.6×10^{14}	green	5.4×10^{-7}
5.1×10^{14}	yellow	5.9×10^{-7}
4.9×10^{14}	orange	6.1×10^{-7}
3.9×10^{14}	longest visible red	7.6×10^{-7}
3.0×10^{14}	near infrared	1.0×10^{-6}

Table 9.1

where $\tilde{\mathbf{E}}_0$ and $\tilde{\mathbf{B}}_0$ are the (complex) amplitudes (the *physical* fields, of course, are the real parts of $\tilde{\mathbf{E}}$ and $\tilde{\mathbf{B}}$).

Now, the wave equations for **E** and **B** (Eq. 9.41) were derived from Maxwell's equations. However, whereas every solution to Maxwell's equations (in empty space) must obey the wave equation, the converse is *not* true; Maxwell's equations impose extra constraints on

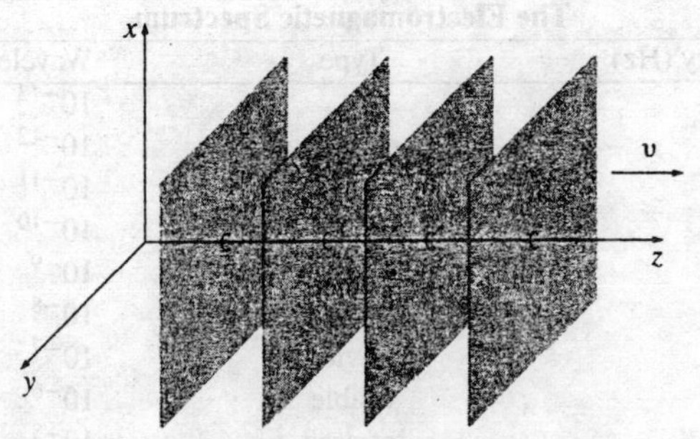

Figure 9.9

$\tilde{\mathbf{E}}_0$ and $\tilde{\mathbf{B}}_0$. In particular, since $\nabla \cdot \mathbf{E} = 0$ and $\nabla \cdot \mathbf{B} = 0$, it follows[5] that

$$(\tilde{E}_0)_z = (\tilde{B}_0)_z = 0. \tag{9.44}$$

That is, *electromagnetic waves are transverse*: the electric and magnetic fields are perpendicular to the direction of propagation. Moreover, Faraday's law, $\nabla \times \mathbf{E} = -\partial \mathbf{B}/\partial t$, implies a relation between the electric and magnetic amplitudes, to wit:

$$-k(\tilde{E}_0)_y = \omega(\tilde{B}_0)_x, \quad k(\tilde{E}_0)_x = \omega(\tilde{B}_0)_y, \tag{9.45}$$

or, more compactly:

$$\tilde{\mathbf{B}}_0 = \frac{k}{\omega}(\hat{\mathbf{z}} \times \tilde{\mathbf{E}}_0). \tag{9.46}$$

Evidently, **E** and **B** are *in phase* and *mutually perpendicular;* their (real) amplitudes are related by

$$B_0 = \frac{k}{\omega} E_0 = \frac{1}{c} E_0. \tag{9.47}$$

The fourth of Maxwell's equations, $\nabla \times \mathbf{B} = \mu_0 \epsilon_0 (\partial \mathbf{E}/\partial t)$, does not yield an independent condition; it simply reproduces Eq. 9.45.

Example 9.2

If **E** points in the x direction, then **B** points in the y direction (Eq. 9.46):

$$\tilde{\mathbf{E}}(z,t) = \tilde{E}_0 e^{i(kz-\omega t)} \hat{\mathbf{x}}, \quad \tilde{\mathbf{B}}(z,t) = \frac{1}{c} \tilde{E}_0 e^{i(kz-\omega t)} \hat{\mathbf{y}},$$

or (taking the real part)

$$\boxed{\mathbf{E}(z,t) = E_0 \cos(kz - \omega t + \delta)\, \hat{\mathbf{x}}, \quad \mathbf{B}(z,t) = \frac{1}{c} E_0 \cos(kz - \omega t + \delta)\, \hat{\mathbf{y}}.} \tag{9.48}$$

[5]Because the real part of $\tilde{\mathbf{E}}$ differs from the imaginary part only in the replacement of sine by cosine, if the former obeys Maxwell's equations, so does the latter, and hence $\tilde{\mathbf{E}}$ as well.

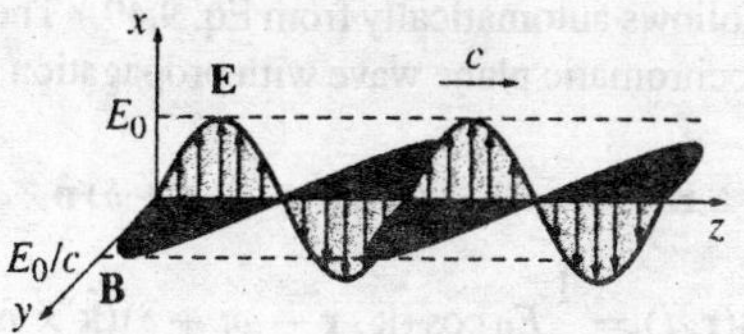

Figure 9.10

This is the paradigm for a monochromatic plane wave (see Fig. 9.10). The wave as a whole is said to be polarized in the x direction (by convention, we use the direction of **E** to specify the polarization of an electromagnetic wave).

There is nothing special about the z direction, of course—we can easily generalize to monochromatic plane waves traveling in an arbitrary direction. The notation is facilitated by the introduction of the **propagation** (or **wave**) **vector**, **k**, pointing in the direction of propagation, whose magnitude is the wave number k. The scalar product $\mathbf{k}\cdot\mathbf{r}$ is the appropriate generalization of kz (Fig. 9.11), so

$$\boxed{\begin{aligned}\tilde{\mathbf{E}}(\mathbf{r}, t) &= \tilde{E}_0 e^{i(\mathbf{k}\cdot\mathbf{r}-\omega t)}\,\hat{\mathbf{n}},\\ \tilde{\mathbf{B}}(\mathbf{r}, t) &= \frac{1}{c}\tilde{E}_0 e^{i(\mathbf{k}\cdot\mathbf{r}-\omega t)}(\hat{\mathbf{k}}\times\hat{\mathbf{n}}) = \frac{1}{c}\hat{\mathbf{k}}\times\tilde{\mathbf{E}},\end{aligned}} \tag{9.49}$$

where $\hat{\mathbf{n}}$ is the polarization vector. Because **E** is transverse,

$$\hat{\mathbf{n}}\cdot\hat{\mathbf{k}} = 0. \tag{9.50}$$

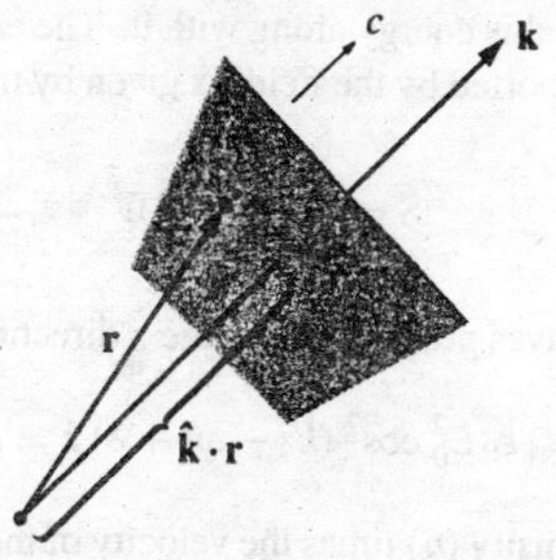

Figure 9.11

(The transversality of **B** follows automatically from Eq. 9.49.) The actual (real) electric and magnetic fields in a monochromatic plane wave with propagation vector **k** and polarization $\hat{\mathbf{n}}$ are

$$\mathbf{E}(\mathbf{r}, t) = E_0 \cos(\mathbf{k} \cdot \mathbf{r} - \omega t + \delta)\, \hat{\mathbf{n}}, \tag{9.51}$$

$$\mathbf{B}(\mathbf{r}, t) = \frac{1}{c} E_0 \cos(\mathbf{k} \cdot \mathbf{r} - \omega t + \delta)(\hat{\mathbf{k}} \times \hat{\mathbf{n}}). \tag{9.52}$$

Problem 9.9 Write down the (real) electric and magnetic fields for a monochromatic plane wave of amplitude E_0, frequency ω, and phase angle zero that is (a) traveling in the negative x direction and polarized in the z direction; (b) traveling in the direction from the origin to the point (1, 1, 1), with polarization parallel to the $x\,z$ plane. In each case, sketch the wave, and give the explicit Cartesian components of **k** and $\hat{\mathbf{n}}$.

9.2.3 Energy and Momentum in Electromagnetic Waves

According to Eq. 8.13, the energy per unit volume stored in electromagnetic fields is

$$u = \frac{1}{2}\left(\epsilon_0 E^2 + \frac{1}{\mu_0} B^2\right). \tag{9.53}$$

In the case of a monochromatic plane wave (Eq. 9.48)

$$B^2 = \frac{1}{c^2} E^2 = \mu_0 \epsilon_0 E^2, \tag{9.54}$$

so the *electric and magnetic contributions are equal*:

$$u = \epsilon_0 E^2 = \epsilon_0 E_0^2 \cos^2(kz - \omega t + \delta). \tag{9.55}$$

As the wave travels, it carries this energy along with it. The energy flux density (energy per unit area, per unit time) transported by the fields is given by the Poynting vector (Eq. 8.10):

$$\mathbf{S} = \frac{1}{\mu_0}(\mathbf{E} \times \mathbf{B}). \tag{9.56}$$

For monochromatic plane waves propagating in the z direction,

$$\mathbf{S} = c\epsilon_0 E_0^2 \cos^2(kz - \omega t + \delta)\, \hat{\mathbf{z}} = cu\, \hat{\mathbf{z}}. \tag{9.57}$$

Notice that **S** is the energy density (u) times the velocity of the waves ($c\,\hat{\mathbf{z}}$)—as it *should* be. For in a time Δt, a length $c\,\Delta t$ passes through area A (Fig. 9.12), carrying with it an energy $uAc\,\Delta t$. The energy per unit time, per unit area, transported by the wave is therefore uc.

Figure 9.12

Electromagnetic fields not only carry *energy*, they also carry *momentum*. In fact, we found in Eq. 8.30 that the momentum density stored in the fields is

$$\wp = \frac{1}{c^2}\mathbf{S}. \tag{9.58}$$

For monochromatic plane waves, then,

$$\wp = \frac{1}{c}\epsilon_0 E_0^2 \cos^2(kz - \omega t + \delta)\,\hat{\mathbf{z}} = \frac{1}{c}u\,\hat{\mathbf{z}}. \tag{9.59}$$

In the case of *light*, the wavelength is so short ($\sim 5 \times 10^{-7}$ m), and the period so brief ($\sim 10^{-15}$ s), that any macroscopic measurement will encompass many cycles. Typically, therefore, we're not interested in the fluctuating cosine-squared term in the energy and momentum densities; all we want is the *average* value. Now, the average of cosine-squared over a complete cycle[6] is $\frac{1}{2}$, so

$$\langle u \rangle = \frac{1}{2}\epsilon_0 E_0^2, \tag{9.60}$$

$$\langle \mathbf{S} \rangle = \frac{1}{2}c\epsilon_0 E_0^2\,\hat{\mathbf{z}}, \tag{9.61}$$

$$\langle \wp \rangle = \frac{1}{2c}\epsilon_0 E_0^2\,\hat{\mathbf{z}}. \tag{9.62}$$

I use brackets, $\langle\ \rangle$, to denote the (time) average over a complete cycle (or *many* cycles, if you prefer). The average power per unit area transported by an electromagnetic wave is called the **intensity**:

$$I \equiv \langle S \rangle = \frac{1}{2}c\epsilon_0 E_0^2. \tag{9.63}$$

[6]There is a cute trick for doing this in your head: $\sin^2\theta + \cos^2\theta = 1$, and over a complete cycle the average of $\sin^2\theta$ is equal to the average of $\cos^2\theta$, so $\langle\sin^2\rangle = \langle\cos^2\rangle = 1/2$. More formally,

$$\frac{1}{T}\int_0^T \cos^2(kz - 2\pi t/T + \delta)\,dt = 1/2.$$

When light falls on a perfect absorber it delivers its momentum to the surface. In a time Δt the momentum transfer is (Fig. 9.12) $\Delta \mathbf{p} = \langle \boldsymbol{\wp} \rangle Ac\,\Delta t$, so the **radiation pressure** (average force per unit area) is

$$P = \frac{1}{A}\frac{\Delta p}{\Delta t} = \frac{1}{2}\epsilon_0 E_0^2 = \frac{I}{c}. \tag{9.64}$$

(On a perfect *reflector* the pressure is *twice* as great, because the momentum switches direction, instead of simply being absorbed.) We can account for this pressure qualitatively, as follows: The electric field (Eq. 9.48) drives charges in the x direction, and the magnetic field then exerts on them a force ($q\mathbf{v} \times \mathbf{B}$) in the z direction. The net force on all the charges in the surface produces the pressure.

Problem 9.10 The intensity of sunlight hitting the earth is about 1300 $\mathrm{W/m^2}$. If sunlight strikes a perfect absorber, what pressure does it exert? How about a perfect reflector? What fraction of atmospheric pressure does this amount to?

Problem 9.11 In the complex notation there is a clever device for finding the time average of a product. Suppose $f(\mathbf{r}, t) = A\cos(\mathbf{k}\cdot\mathbf{r} - \omega t + \delta_a)$ and $g(\mathbf{r}, t) = B\cos(\mathbf{k}\cdot\mathbf{r} - \omega t + \delta_b)$. Show that $\langle fg \rangle = (1/2)\mathrm{Re}(\tilde{f}\tilde{g}^*)$, where the star denotes complex conjugation. [Note that this only works if the two waves have the same $\mathbf{k}$ and ω, but they need not have the same amplitude or phase.] For example

$$\langle u \rangle = \frac{1}{4}\mathrm{Re}(\epsilon_0 \tilde{\mathbf{E}}\cdot\tilde{\mathbf{E}}^* + \frac{1}{\mu_0}\tilde{\mathbf{B}}\cdot\tilde{\mathbf{B}}^*) \quad \text{and} \quad \langle \mathbf{S} \rangle = \frac{1}{2\mu_0}\mathrm{Re}(\tilde{\mathbf{E}}\times\tilde{\mathbf{B}}^*).$$

Problem 9.12 Find all elements of the Maxwell stress tensor for a monochromatic plane wave traveling in the z direction and linearly polarized in the x direction (Eq. 9.48). Does your answer make sense? (Remember that $\overleftrightarrow{\mathbf{T}}$ represents the momentum flux density.) How is the momentum flux density related to the energy density, in this case?

9.3 Electromagnetic Waves in Matter

9.3.1 Propagation in Linear Media

Inside matter, but in regions where there is no *free* charge or *free* current, Maxwell's equations become

$$\left.\begin{array}{llll} \text{(i)} & \nabla\cdot\mathbf{D} = 0, & \text{(iii)} & \nabla\times\mathbf{E} = -\dfrac{\partial \mathbf{B}}{\partial t}, \\ \text{(ii)} & \nabla\cdot\mathbf{B} = 0, & \text{(iv)} & \nabla\times\mathbf{H} = \dfrac{\partial \mathbf{D}}{\partial t}. \end{array}\right\} \tag{9.65}$$

If the medium is *linear*,

$$\mathbf{D} = \epsilon\mathbf{E}, \quad \mathbf{H} = \frac{1}{\mu}\mathbf{B}, \tag{9.66}$$

and *homogeneous* (so ϵ and μ do not vary from point to point), Maxwell's equations reduce to

$$\left.\begin{aligned} &\text{(i)} \quad \nabla\cdot\mathbf{E}=0, \quad \text{(iii)} \quad \nabla\times\mathbf{E}=-\frac{\partial\mathbf{B}}{\partial t}, \\ &\text{(ii)} \quad \nabla\cdot\mathbf{B}=0, \quad \text{(iv)} \quad \nabla\times\mathbf{B}=\mu\epsilon\frac{\partial\mathbf{E}}{\partial t}, \end{aligned}\right\} \tag{9.67}$$

which (remarkably) differ from the vacuum analogs (Eqs. 9.40) only in the replacement of $\mu_0\epsilon_0$ by $\mu\epsilon$.[7] Evidently electromagnetic waves propagate through a linear homogeneous medium at a speed

$$v=\frac{1}{\sqrt{\epsilon\mu}}=\frac{c}{n}, \tag{9.68}$$

where

$$n\equiv\sqrt{\frac{\epsilon\mu}{\epsilon_0\mu_0}} \tag{9.69}$$

is the **index of refraction** of the material. For most materials, μ is very close to μ_0, so

$$n\cong\sqrt{\epsilon_r}, \tag{9.70}$$

where ϵ_r is the dielectric constant (Eq. 4.34). Since ϵ_r is almost always greater than 1, light travels *more slowly* through matter—a fact that is well known from optics.

All of our previous results carry over, with the simple transcription $\epsilon_0 \rightarrow \epsilon$, $\mu_0 \rightarrow \mu$, and hence $c \rightarrow v$ (see Prob. 8.15). The energy density is[8]

$$u=\frac{1}{2}\left(\epsilon E^2+\frac{1}{\mu}B^2\right), \tag{9.71}$$

and the Poynting vector is

$$\mathbf{S}=\frac{1}{\mu}(\mathbf{E}\times\mathbf{B}). \tag{9.72}$$

For monochromatic plane waves the frequency and wave number are related by $\omega = kv$ (Eq. 9.11), the amplitude of $\mathbf{B}$ is $1/v$ times the amplitude of $\mathbf{E}$ (Eq. 9.47), and the intensity is

$$I=\frac{1}{2}\epsilon v E_0^2. \tag{9.73}$$

---

[7]This observation is mathematically pretty trivial, but the physical implications are astonishing: As the wave passes through, the fields busily polarize and magnetize all the molecules, and the resulting (oscillating) dipoles create their own electric and magnetic fields. These combine with the original fields in such a way as to create a *single* wave with the same frequency but a different speed. This extraordinary conspiracy is responsible for the phenomenon of **transparency**. It is a distinctly *non*trivial consequence of the *linearity* of the medium. For further discussion see M. B. James and D. J. Griffiths, *Am. J. Phys.* **60**, 309 (1992).

[8]Refer to Sect. 4.4.3 for the precise *meaning* of "energy density," in the context of linear media.

The interesting question is this: What happens when a wave passes from one transparent medium into another—air to water, say, or glass to plastic? As in the case of waves on a string, we expect to get a reflected wave and a transmitted wave. The details depend on the exact nature of the electrodynamic boundary conditions, which we derived in Chapter 7 (Eq. 7.64):

$$\left.\begin{array}{ll}\text{(i)}\ \epsilon_1 E_1^{\perp} = \epsilon_2 E_2^{\perp}, & \text{(iii)}\ \mathbf{E}_1^{\parallel} = \mathbf{E}_2^{\parallel}, \\ \text{(ii)}\ B_1^{\perp} = B_2^{\perp}, & \text{(iv)}\ \dfrac{1}{\mu_1}\mathbf{B}_1^{\parallel} = \dfrac{1}{\mu_2}\mathbf{B}_2^{\parallel}. \end{array}\right\} \tag{9.74}$$

These equations relate the electric and magnetic fields just to the left and just to the right of the interface between two linear media. In the following sections we use them to deduce the laws governing reflection and refraction of electromagnetic waves.

9.3.2 Reflection and Transmission at Normal Incidence

Suppose the xy plane forms the boundary between two linear media. A plane wave of frequency ω, traveling in the z direction and polarized in the x direction, approaches the interface from the left (Fig. 9.13):

$$\left.\begin{array}{l}\tilde{\mathbf{E}}_I(z,t) = \tilde{E}_{0_I} e^{i(k_1 z - \omega t)}\,\hat{\mathbf{x}}, \\ \tilde{\mathbf{B}}_I(z,t) = \dfrac{1}{v_1}\tilde{E}_{0_I} e^{i(k_1 z - \omega t)}\,\hat{\mathbf{y}}. \end{array}\right\} \tag{9.75}$$

It gives rise to a reflected wave

$$\left.\begin{array}{l}\tilde{\mathbf{E}}_R(z,t) = \tilde{E}_{0_R} e^{i(-k_1 z - \omega t)}\,\hat{\mathbf{x}}, \\ \tilde{\mathbf{B}}_R(z,t) = -\dfrac{1}{v_1}\tilde{E}_{0_R} e^{i(-k_1 z - \omega t)}\,\hat{\mathbf{y}}, \end{array}\right\} \tag{9.76}$$

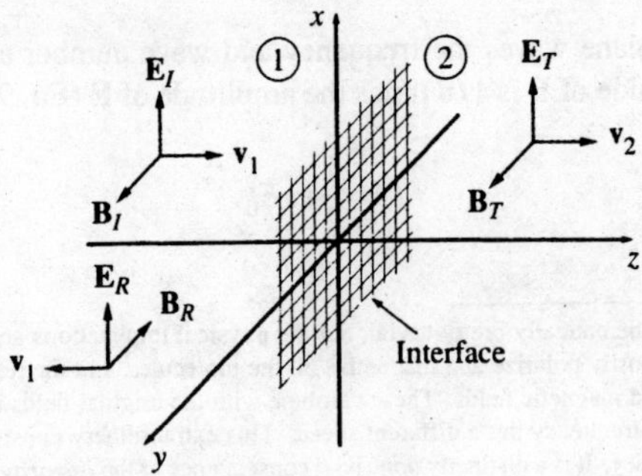

Figure 9.13

which travels back to the left in medium (1), and a transmitted wave

$$\left.\begin{aligned}\tilde{\mathbf{E}}_T(z,t) &= \tilde{E}_{0_T} e^{i(k_2 z-\omega t)}\,\hat{\mathbf{x}},\\[2ex] \tilde{\mathbf{B}}_T(z,t) &= \frac{1}{v_2}\tilde{E}_{0_T} e^{i(k_2 z-\omega t)}\,\hat{\mathbf{y}},\end{aligned}\right\}\tag{9.77}$$

which continues on the the right in medium (2). Note the minus sign in $\tilde{\mathbf{B}}_R$, as required by Eq. 9.49—or, if you prefer, by the fact that the Poynting vector aims in the direction of propagation.

At $z = 0$, the combined fields on the left, $\tilde{\mathbf{E}}_I + \tilde{\mathbf{E}}_R$ and $\tilde{\mathbf{B}}_I + \tilde{\mathbf{B}}_R$, must join the fields on the right, $\tilde{\mathbf{E}}_T$ and $\tilde{\mathbf{B}}_T$, in accordance with the boundary conditions 9.74. In this case there are no components perpendicular to the surface, so (i) and (ii) are trivial. However, (iii) requires that

$$\tilde{E}_{0_I} + \tilde{E}_{0_R} = \tilde{E}_{0_T},\tag{9.78}$$

while (iv) says

$$\frac{1}{\mu_1}\left(\frac{1}{v_1}\tilde{E}_{0_I} - \frac{1}{v_1}\tilde{E}_{0_R}\right) = \frac{1}{\mu_2}\left(\frac{1}{v_2}\tilde{E}_{0_T}\right),\tag{9.79}$$

or

$$\tilde{E}_{0_I} - \tilde{E}_{0_R} = \beta\tilde{E}_{0_T},\tag{9.80}$$

where

$$\beta \equiv \frac{\mu_1 v_1}{\mu_2 v_2} = \frac{\mu_1 n_2}{\mu_2 n_1}.\tag{9.81}$$

Equations 9.78 and 9.80 are easily solved for the outgoing amplitudes, in terms of the incident amplitude:

$$\tilde{E}_{0_R} = \left(\frac{1-\beta}{1+\beta}\right)\tilde{E}_{0_I}, \quad \tilde{E}_{0_T} = \left(\frac{2}{1+\beta}\right)\tilde{E}_{0_I}.\tag{9.82}$$

These results are strikingly similar to the ones for waves on a string. Indeed, if the permeabilities μ are close to their values in vacuum (as, remember, they *are* for most media), then $\beta = v_1/v_2$, and we have

$$\tilde{E}_{0_R} = \left(\frac{v_2 - v_1}{v_2 + v_1}\right)\tilde{E}_{0_I}, \quad \tilde{E}_{0_T} = \left(\frac{2v_2}{v_2 + v_1}\right)\tilde{E}_{0_I},\tag{9.83}$$

which are *identical* to Eqs. 9.30. In that case, as before, the reflected wave is *in phase* (right side up) if $v_2 > v_1$ and *out of phase* (upside down) if $v_2 < v_1$; the real amplitudes are related by

$$E_{0_R} = \left|\frac{v_2 - v_1}{v_2 + v_1}\right| E_{0_I}, \quad E_{0_T} = \left(\frac{2v_2}{v_2 + v_1}\right) E_{0_I},\tag{9.84}$$

or, in terms of the indices of refraction,

$$E_{0_R} = \left|\frac{n_1 - n_2}{n_1 + n_2}\right| E_{0_I}, \quad E_{0_T} = \left(\frac{2n_1}{n_1 + n_2}\right) E_{0_I}.\tag{9.85}$$

What fraction of the incident energy is reflected, and what fraction is transmitted? According to Eq. 9.73, the intensity (average power per unit area) is

$$I = \frac{1}{2}\epsilon v E_0^2.$$

If (again) $\mu_1 = \mu_2 = \mu_0$, then the ratio of the reflected intensity to the incident intensity is

$$R \equiv \frac{I_R}{I_I} = \left(\frac{E_{0_R}}{E_{0_I}}\right)^2 = \left(\frac{n_1 - n_2}{n_1 + n_2}\right)^2, \tag{9.86}$$

whereas the ratio of the transmitted intensity to the incident intensity is

$$T \equiv \frac{I_T}{I_I} = \frac{\epsilon_2 v_2}{\epsilon_1 v_1}\left(\frac{E_{0_T}}{E_{0_I}}\right)^2 = \frac{4n_1 n_2}{(n_1 + n_2)^2}. \tag{9.87}$$

R is called the **reflection coefficient** and T the **transmission coefficient**; they measure the fraction of the incident energy that is reflected and transmitted, respectively. Notice that

$$R + T = 1, \tag{9.88}$$

as conservation of energy, of course, requires. For instance, when light passes from air ($n_1 = 1$) into glass ($n_2 = 1.5$), $R = 0.04$ and $T = 0.96$. Not surprisingly, most of the light is transmitted.

Problem 9.13 Calculate the *exact* reflection and transmission coefficients, *without* assuming $\mu_1 = \mu_2 = \mu_0$. Confirm that $R + T = 1$.

Problem 9.14 In writing Eqs. 9.76 and 9.77, I tacitly assumed that the reflected and transmitted waves have the same *polarization* as the incident wave—along the x direction. Prove that this *must* be so. [*Hint:* Let the polarization vectors of the transmitted and reflected waves be

$$\hat{\mathbf{n}}_T = \cos\theta_T\,\hat{\mathbf{x}} + \sin\theta_T\,\hat{\mathbf{y}}, \quad \hat{\mathbf{n}}_R = \cos\theta_R\,\hat{\mathbf{x}} + \sin\theta_R\,\hat{\mathbf{y}},$$

and prove from the boundary conditions that $\theta_T = \theta_R = 0$.]

9.3.3 Reflection and Transmission at Oblique Incidence

In the last section I treated reflection and transmission at *normal* incidence—that is, when the incoming wave hits the interface head-on. We now turn to the more general case of *oblique* incidence, in which the incoming wave meets the boundary at an arbitrary angle θ_I (Fig. 9.14). Of course, normal incidence is really just a special case of oblique incidence, with $\theta_I = 0$, but I wanted to treat it separately, as a kind of warm-up, because the algebra is now going to get a little heavy.

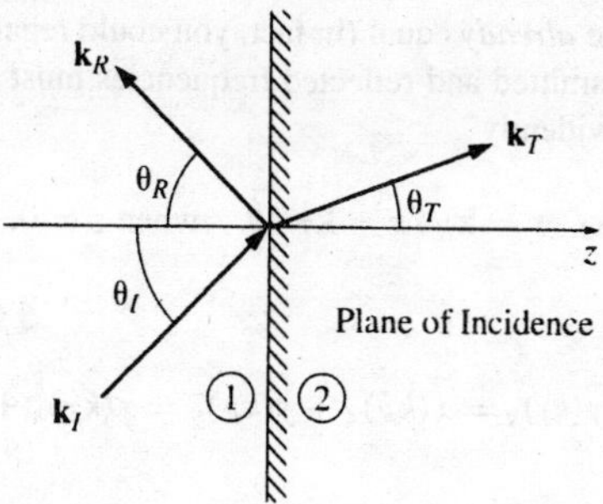

Figure 9.14

Suppose, then, that a monochromatic plane wave

$$\tilde{\mathbf{E}}_I(\mathbf{r}, t) = \tilde{\mathbf{E}}_{0_I} e^{i(\mathbf{k}_I \cdot \mathbf{r} - \omega t)}, \quad \tilde{\mathbf{B}}_I(\mathbf{r}, t) = \frac{1}{v_1}(\hat{\mathbf{k}}_I \times \tilde{\mathbf{E}}_I) \tag{9.89}$$

approaches from the left, giving rise to a reflected wave,

$$\tilde{\mathbf{E}}_R(\mathbf{r}, t) = \tilde{\mathbf{E}}_{0_R} e^{i(\mathbf{k}_R \cdot \mathbf{r} - \omega t)}, \quad \tilde{\mathbf{B}}_R(\mathbf{r}, t) = \frac{1}{v_1}(\hat{\mathbf{k}}_R \times \tilde{\mathbf{E}}_R), \tag{9.90}$$

and a transmitted wave

$$\tilde{\mathbf{E}}_T(\mathbf{r}, t) = \tilde{\mathbf{E}}_{0_T} e^{i(\mathbf{k}_T \cdot \mathbf{r} - \omega t)}, \quad \tilde{\mathbf{B}}_T(\mathbf{r}, t) = \frac{1}{v_2}(\hat{\mathbf{k}}_T \times \tilde{\mathbf{E}}_T). \tag{9.91}$$

All three waves have the same *frequency* ω—that is determined once and for all at the source (the flashlight, or whatever, that produces the incident beam). The three wave numbers are related by Eq. 9.11:

$$k_I v_1 = k_R v_1 = k_T v_2 = \omega, \quad \text{or} \quad k_I = k_R = \frac{v_2}{v_1} k_T = \frac{n_1}{n_2} k_T. \tag{9.92}$$

The combined fields in medium (1), $\tilde{\mathbf{E}}_I + \tilde{\mathbf{E}}_R$ and $\tilde{\mathbf{B}}_I + \tilde{\mathbf{B}}_R$, must now be joined to the fields $\tilde{\mathbf{E}}_T$ and $\tilde{\mathbf{B}}_T$ in medium (2), using the boundary conditions 9.74. These all share the generic structure

$$(\ )e^{i(\mathbf{k}_I \cdot \mathbf{r} - \omega t)} + (\ )e^{i(\mathbf{k}_R \cdot \mathbf{r} - \omega t)} = (\ )e^{i(\mathbf{k}_T \cdot \mathbf{r} - \omega t)}, \quad \text{at } z = 0. \tag{9.93}$$

I'll fill in the parentheses in a moment; for now, the important thing to notice is that the x, y, and t dependence is confined to the exponents. *Because the boundary conditions must hold at all points on the plane, and for all times, these exponential factors must be equal* (when $z = 0$). Otherwise, a slight change in x, say, would destroy the equality (see Prob. 9.15). Of course, the time factors are *already* equal (in fact, you could regard this as an independent confirmation that the transmitted and reflected frequencies must match the incident one). As for the spatial terms, evidently

$$\mathbf{k}_I \cdot \mathbf{r} = \mathbf{k}_R \cdot \mathbf{r} = \mathbf{k}_T \cdot \mathbf{r}, \quad \text{when } z = 0, \tag{9.94}$$

or, more explicitly,

$$x(k_I)_x + y(k_I)_y = x(k_R)_x + y(k_R)_y = x(k_T)_x + y(k_T)_y, \tag{9.95}$$

for all x and all y.

But Eq. 9.95 can *only* hold if the components are separately equal, for if $x = 0$, we get

$$(k_I)_y = (k_R)_y = (k_T)_y, \tag{9.96}$$

while $y = 0$ gives

$$(k_I)_x = (k_R)_x = (k_T)_x. \tag{9.97}$$

We may as well orient our axes so that $\mathbf{k}_I$ lies in the xz plane (i.e. $(k_I)_y = 0$); according to Eq. 9.96, so too will $\mathbf{k}_R$ and $\mathbf{k}_T$. *Conclusion:*

> **First Law:** The incident, reflected, and transmitted wave vectors form a plane (called the **plane of incidence**), which also includes the normal to the surface (here, the z axis).

Meanwhile, Eq. 9.97 implies that

$$k_I \sin\theta_I = k_R \sin\theta_R = k_T \sin\theta_T, \tag{9.98}$$

where θ_I is the **angle of incidence**, θ_R is the **angle of reflection**, and θ_T is the angle of transmission, more commonly known as the **angle of refraction**, all of them measured with respect to the normal (Fig. 9.14). In view of Eq. 9.92, then,

> **Second Law:** The angle of incidence is equal to the angle of reflection,
>
> $$\theta_I = \theta_R. \tag{9.99}$$
>
> This is the **law of reflection**.

As for the transmitted angle,

> **Third Law:**
>
> $$\frac{\sin\theta_T}{\sin\theta_I} = \frac{n_1}{n_2}. \tag{9.100}$$
>
> This is the **law of refraction**, or **Snell's law**.

These are the three fundamental laws of geometrical optics. It is remarkable how little actual *electrodynamics* went into them: we have yet to invoke any *specific* boundary conditions—all we used was their generic form (Eq. 9.93). Therefore, any *other* waves (water waves, for instance, or sound waves) can be expected to obey the same "optical" laws when they pass from one medium into another.

Now that we have taken care of the exponential factors—they cancel, given Eq. 9.94—the boundary conditions 9.74 become:

$$\left.\begin{array}{l} \text{(i)}\ \epsilon_1(\tilde{\mathbf{E}}_{0_I}+\tilde{\mathbf{E}}_{0_R})_z=\epsilon_2(\tilde{\mathbf{E}}_{0_T})_z \\ \text{(ii)}\ (\tilde{\mathbf{B}}_{0_I}+\tilde{\mathbf{B}}_{0_R})_z=(\tilde{\mathbf{B}}_{0_T})_z \\ \text{(iii)}\ (\tilde{\mathbf{E}}_{0_I}+\tilde{\mathbf{E}}_{0_R})_{x,y}=(\tilde{\mathbf{E}}_{0_T})_{x,y} \\ \text{(iv)}\ \dfrac{1}{\mu_1}(\tilde{\mathbf{B}}_{0_I}+\tilde{\mathbf{B}}_{0_R})_{x,y}=\dfrac{1}{\mu_2}(\tilde{\mathbf{B}}_{0_T})_{x,y} \end{array}\right\} \tag{9.101}$$

where $\tilde{\mathbf{B}}_0 = (1/v)\hat{\mathbf{k}} \times \tilde{\mathbf{E}}_0$ in each case. (The last two represent *pairs* of equations, one for the x-component and one for the y-component.)

Suppose that the polarization of the incident wave is *parallel* to the plane of incidence (the xz plane in Fig. 9.15); it follows (see Prob. 9.14) that the reflected and transmitted waves are also polarized in this plane. (I shall leave it for you to analyze the case of polarization *perpendicular* to the plane of incidence; see Prob. 9.16.) Then (i) reads

$$\epsilon_1(-\tilde{E}_{0_I}\sin\theta_I+\tilde{E}_{0_R}\sin\theta_R)=\epsilon_2(-\tilde{E}_{0_T}\sin\theta_T); \tag{9.102}$$

(ii) adds nothing ($0 = 0$), since the magnetic fields have no z components; (iii) becomes

$$\tilde{E}_{0_I}\cos\theta_I+\tilde{E}_{0_R}\cos\theta_R=\tilde{E}_{0_T}\cos\theta_T; \tag{9.103}$$

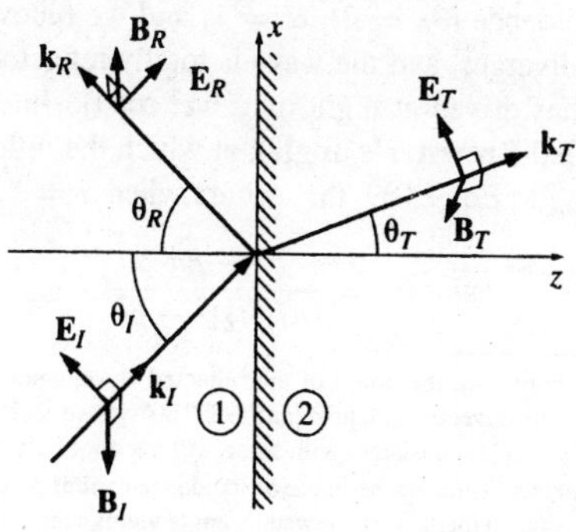

Figure 9.15

and (iv) says

$$\frac{1}{\mu_1 v_1}(\tilde{E}_{0_I} - \tilde{E}_{0_R}) = \frac{1}{\mu_2 v_2}\tilde{E}_{0_T}. \tag{9.104}$$

Given the laws of reflection and refraction, Eqs. 9.102 and 9.104 both reduce to

$$\tilde{E}_{0_I} - \tilde{E}_{0_R} = \beta \tilde{E}_{0_T}, \tag{9.105}$$

where (as before)

$$\beta \equiv \frac{\mu_1 v_1}{\mu_2 v_2} = \frac{\mu_1 n_2}{\mu_2 n_1}, \tag{9.106}$$

and Eq. 9.103 says

$$\tilde{E}_{0_I} + \tilde{E}_{0_R} = \alpha \tilde{E}_{0_T}, \tag{9.107}$$

where

$$\alpha \equiv \frac{\cos\theta_T}{\cos\theta_I}. \tag{9.108}$$

Solving Eqs. 9.105 and 9.107 for the reflected and transmitted amplitudes, we obtain

$$\boxed{\tilde{E}_{0_R} = \left(\frac{\alpha - \beta}{\alpha + \beta}\right)\tilde{E}_{0_I}, \quad \tilde{E}_{0_T} = \left(\frac{2}{\alpha + \beta}\right)\tilde{E}_{0_I}.} \tag{9.109}$$

These are known as **Fresnel's equations**, for the case of polarization in the plane of incidence. (There are two other Fresnel equations, giving the reflected and transmitted amplitudes when the polarization is *perpendicular* to the plane of incidence—see Prob. 9.16.) Notice that the transmitted wave is always *in phase* with the incident one; the reflected wave is either in phase ("right side up"), if $\alpha > \beta$, or 180° out of phase ("upside down"), if $\alpha < \beta$.[9]

The amplitudes of the transmitted and reflected waves depend on the angle of incidence, because α is a function of θ_I:

$$\alpha = \frac{\sqrt{1 - \sin^2\theta_T}}{\cos\theta_I} = \frac{\sqrt{1 - [(n_1/n_2)\sin\theta_I]^2}}{\cos\theta_I}. \tag{9.110}$$

In the case of normal incidence ($\theta_I = 0$), $\alpha = 1$, and we recover Eq. 9.82. At grazing incidence ($\theta_I = 90°$), α diverges, and the wave is totally reflected (a fact that is painfully familiar to anyone who has driven at night on a wet road). Interestingly, there is an intermediate angle, θ_B (called **Brewster's angle**), at which the reflected wave is completely extinguished.[10] According to Eq. 9.109, this occurs when $\alpha = \beta$, or

$$\sin^2\theta_B = \frac{1 - \beta^2}{(n_1/n_2)^2 - \beta^2}. \tag{9.111}$$

---

[9] There is an unavoidable ambiguity in the phase of the reflected wave, since (as I mentioned in footnote 2) changing the sign of the polarization vector is equivalent to a 180° phase shift. The convention I adopted in Fig. 9.15, with $\mathbf{E}_R$ positive "upward," is consistent with some, but not all, of the standard optics texts.

[10] Because waves polarized *perpendicular* to the plane of incidence exhibit no corresponding quenching of the reflected component, an arbitrary beam incident at Brewster's angle yields a reflected beam that is *totally* polarized parallel to the interface. That's why Polaroid glasses, with the transmission axis vertical, help to reduce glare off a horizontal surface.

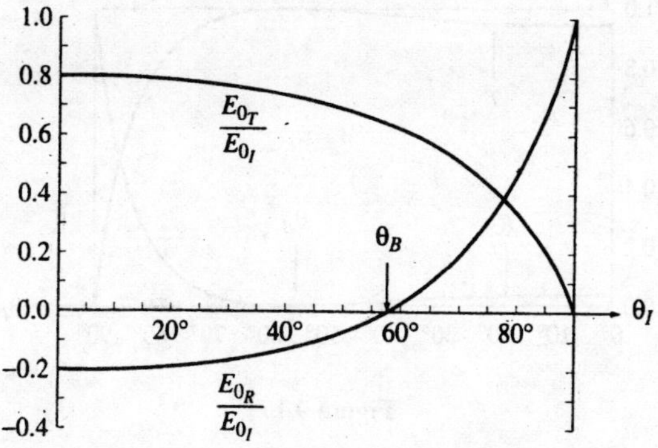

Figure 9.16

For the typical case $\mu_1 \cong \mu_2$, so $\beta \cong n_2/n_1$, $\sin^2\theta_B \cong \beta^2/(1+\beta^2)$, and hence

$$\tan\theta_B \cong \frac{n_2}{n_1}. \tag{9.112}$$

Figure 9.16 shows a plot of the transmitted and reflected amplitudes as functions of θ_I, for light incident on glass ($n_2 = 1.5$) from air ($n_1 = 1$). (On the graph, a *negative* number indicates that the wave is 180° out of phase with the incident beam—the amplitude itself is the absolute value.)

The power per unit area striking the interface is $\mathbf{S}\cdot\hat{\mathbf{z}}$. Thus the incident intensity is

$$I_I = \frac{1}{2}\epsilon_1 v_1 E_{0_I}^2\cos\theta_I, \tag{9.113}$$

while the reflected and transmitted intensities are

$$I_R = \frac{1}{2}\epsilon_1 v_1 E_{0_R}^2\cos\theta_R, \quad \text{and} \quad I_T = \frac{1}{2}\epsilon_2 v_2 E_{0_T}^2\cos\theta_T. \tag{9.114}$$

(The cosines are there because I am talking about the average power per unit area of *interface*, and the interface is at an angle to the wave front.) The reflection and transmission coefficients for waves polarized parallel to the plane of incidence are

$$R \equiv \frac{I_R}{I_I} = \left(\frac{E_{0_R}}{E_{0_I}}\right)^2 = \left(\frac{\alpha-\beta}{\alpha+\beta}\right)^2, \tag{9.115}$$

$$T \equiv \frac{I_T}{I_I} = \frac{\epsilon_2 v_2}{\epsilon_1 v_1}\left(\frac{E_{0_T}}{E_{0_I}}\right)^2\frac{\cos\theta_T}{\cos\theta_I} = \alpha\beta\left(\frac{2}{\alpha+\beta}\right)^2. \tag{9.116}$$

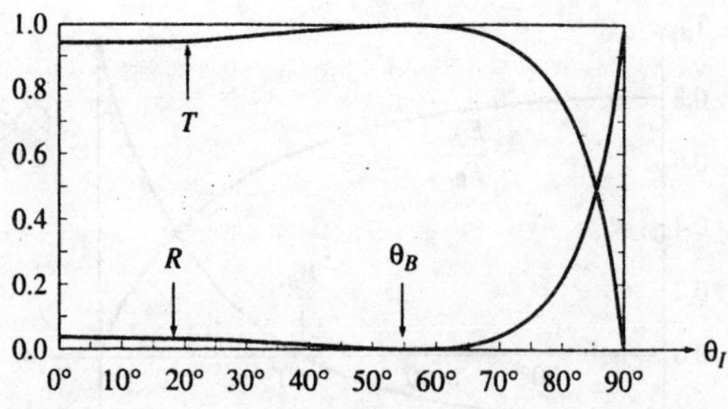

Figure 9.17

They are plotted as functions of the angle of incidence in Fig. 9.17 (for the air/glass interface). R is the fraction of the incident energy that is reflected—naturally, it goes to zero at Brewster's angle; T is the fraction transmitted—it goes to 1 at θ_B. Note that $R + T = 1$, as required by conservation of energy: the energy per unit time *reaching* a particular patch of area on the surface is equal to the energy per unit time *leaving* the patch.

Problem 9.15 Suppose $Ae^{iax} + Be^{ibx} = Ce^{icx}$, for some nonzero constants A, B, C, a, b, c, and for all x. Prove that $a = b = c$ and $A + B = C$.

! **Problem 9.16** Analyze the case of polarization *perpendicular* to the plane of incidence (i.e. electric fields in the y direction, in Fig. 9.15). Impose the boundary conditions 9.101, and obtain the Fresnel equations for $\tilde{E}_{0_R}$ and $\tilde{E}_{0_T}$. Sketch $(\tilde{E}_{0_R}/\tilde{E}_{0_I})$ and $(\tilde{E}_{0_T}/\tilde{E}_{0_I})$ as functions of θ_I, for the case $\beta = n_2/n_1 = 1.5$. (Note that for this β the reflected wave is *always* 180° out of phase.) Show that there is no Brewster's angle for *any* n_1 and n_2: $\tilde{E}_{0_R}$ is *never* zero (unless, of course, $n_1 = n_2$ and $\mu_1 = \mu_2$, in which case the two media are optically indistinguishable). Confirm that your Fresnel equations reduce to the proper forms at normal incidence. Compute the reflection and transmission coefficients, and check that they add up to 1.

Problem 9.17 The index of refraction of diamond is 2.42. Construct the graph analogous to Fig. 9.16 for the air/diamond interface. (Assume $\mu_1 = \mu_2 = \mu_0$.) In particular, calculate (a) the amplitudes at normal incidence, (b) Brewster's angle, and (c) the "crossover" angle, at which the reflected and transmitted amplitudes are equal.

9.4 Absorption and Dispersion

9.4.1 Electromagnetic Waves in Conductors

In Sect. 9.3 I stipulated that the free charge density ρ_f and the free current density $\mathbf{J}_f$ are zero, and everything that followed was predicated on that assumption. Such a restriction is perfectly reasonable when you're talking about wave propagation through a vacuum or through insulating materials such as glass or (pure) water. But in the case of conductors we do not independently control the flow of charge, and in general $\mathbf{J}_f$ is certainly *not* zero. In fact, according to Ohm's law, the (free) current density in a conductor is proportional to the electric field:

$$\mathbf{J}_f = \sigma \mathbf{E}. \tag{9.117}$$

With this, Maxwell's equations for linear media assume the form

$$\left.\begin{array}{ll} \text{(i)}\ \nabla \cdot \mathbf{E} = \dfrac{1}{\epsilon}\rho_f, & \text{(iii)}\ \nabla \times \mathbf{E} = -\dfrac{\partial \mathbf{B}}{\partial t}, \\ \text{(ii)}\ \nabla \cdot \mathbf{B} = 0, & \text{(iv)}\ \nabla \times \mathbf{B} = \mu\sigma\mathbf{E} + \mu\epsilon\dfrac{\partial \mathbf{E}}{\partial t}. \end{array}\right\} \tag{9.118}$$

Now the continuity equation for free charge,

$$\nabla \cdot \mathbf{J}_f = -\frac{\partial \rho_f}{\partial t}, \tag{9.119}$$

together with Ohm's law and Gauss's law (i), gives

$$\frac{\partial \rho_f}{\partial t} = -\sigma(\nabla \cdot \mathbf{E}) = -\frac{\sigma}{\epsilon}\rho_f$$

for a homogeneous linear medium, from which it follows that

$$\rho_f(t) = e^{-(\sigma/\epsilon)t}\rho_f(0). \tag{9.120}$$

Thus any initial free charge density $\rho_f(0)$ dissipates in a characteristic time $\tau \equiv \epsilon/\sigma$. This reflects the familiar fact that if you put some free charge on a conductor, it will flow out to the edges. The time constant τ affords a measure of how "good" a conductor is: For a "perfect" conductor $\sigma = \infty$ and $\tau = 0$; for a "good" conductor, τ is much less than the other relevant times in the problem (in oscillatory systems, that means $\tau \ll 1/\omega$); for a "poor" conductor, τ is *greater* than the characteristic times in the problem ($\tau \gg 1/\omega$).[11] At present we're not interested in this transient behavior—we'll wait for any accumulated free charge to disappear. From then on $\rho_f = 0$, and we have

$$\left.\begin{array}{ll} \text{(i)}\ \nabla \cdot \mathbf{E} = 0, & \text{(iii)}\ \nabla \times \mathbf{E} = -\dfrac{\partial \mathbf{B}}{\partial t}, \\ \text{(ii)}\ \nabla \cdot \mathbf{B} = 0, & \text{(iv)}\ \nabla \times \mathbf{B} = \mu\epsilon\dfrac{\partial \mathbf{E}}{\partial t} + \mu\sigma\mathbf{E}. \end{array}\right\} \tag{9.121}$$

[11] N. Ashby, *Am. J. Phys.* **43**, 553 (1975), points out that for good conductors τ is absurdly short (10^{-19} s, for copper, whereas the time between collisions is $\tau_c = 10^{-14}$ s). The problem is that Ohm's law itself breaks down on time scales shorter than τ_c; actually, the time it takes free charge to dissipate in a good conductor is of order τ_c, not τ. Moreover, H. C. Ohanian, *Am. J. Phys.* **51**, 1020 (1983), shows that it takes even longer for the fields and currents to equilibrate. But none of this is relevant to our present purpose; the free charge density in a conductor does *eventually* dissipate, and exactly how long the process takes is beside the point.

These differ from the corresponding equations for *non*conducting media (9.67) only in the addition of the last term in (iv).

Applying the curl to (iii) and (iv), as before, we obtain modified wave equations for **E** and **B**:

$$\nabla^2\mathbf{E} = \mu\epsilon\frac{\partial^2\mathbf{E}}{\partial t^2} + \mu\sigma\frac{\partial\mathbf{E}}{\partial t}, \quad \nabla^2\mathbf{B} = \mu\epsilon\frac{\partial^2\mathbf{B}}{\partial t^2} + \mu\sigma\frac{\partial\mathbf{B}}{\partial t}. \tag{9.122}$$

These equations still admit plane-wave solutions,

$$\tilde{\mathbf{E}}(z,t) = \tilde{\mathbf{E}}_0 e^{i(\tilde{k}z-\omega t)}, \quad \tilde{\mathbf{B}}(z,t) = \tilde{\mathbf{B}}_0 e^{i(\tilde{k}z-\omega t)}, \tag{9.123}$$

but this time the "wave number" $\tilde{k}$ is complex:

$$\tilde{k}^2 = \mu\epsilon\omega^2 + i\mu\sigma\omega, \tag{9.124}$$

as you can easily check by plugging Eq. 9.123 into Eq. 9.122. Taking the square root,

$$\tilde{k} = k + i\kappa, \tag{9.125}$$

where

$$k \equiv \omega\sqrt{\frac{\epsilon\mu}{2}}\left[\sqrt{1+\left(\frac{\sigma}{\epsilon\omega}\right)^2}+1\right]^{1/2}, \quad \kappa \equiv \omega\sqrt{\frac{\epsilon\mu}{2}}\left[\sqrt{1+\left(\frac{\sigma}{\epsilon\omega}\right)^2}-1\right]^{1/2}. \tag{9.126}$$

The imaginary part of $\tilde{k}$ results in an attenuation of the wave (decreasing amplitude with increasing z):

$$\tilde{\mathbf{E}}(z,t) = \tilde{\mathbf{E}}_0 e^{-\kappa z} e^{i(kz-\omega t)}, \quad \tilde{\mathbf{B}}(z,t) = \tilde{\mathbf{B}}_0 e^{-\kappa z} e^{i(kz-\omega t)}. \tag{9.127}$$

The distance it takes to reduce the amplitude by a factor of $1/e$ (about a third) is called the **skin depth**:

$$d \equiv \frac{1}{\kappa}; \tag{9.128}$$

it is a measure of how far the wave penetrates into the conductor. Meanwhile, the real part of $\tilde{k}$ determines the wavelength, the propagation speed, and the index of refraction, in the usual way:

$$\lambda = \frac{2\pi}{k}, \quad v = \frac{\omega}{k}, \quad n = \frac{ck}{\omega}. \tag{9.129}$$

The attenuated plane waves (Eq. 9.127) satisfy the modified wave equation (9.122) for *any* $\tilde{\mathbf{E}}_0$ and $\tilde{\mathbf{B}}_0$. But Maxwell's equations (9.121) impose further constraints, which serve to determine the relative amplitudes, phases, and polarizations of **E** and **B**. As before, (i) and (ii) rule out any z components: the fields are *transverse*. We may as well orient our axes so that **E** is polarized along the x direction:

$$\tilde{\mathbf{E}}(z,t) = \tilde{E}_0 e^{-\kappa z} e^{i(kz-\omega t)}\,\hat{\mathbf{x}}. \tag{9.130}$$

Then (iii) gives

$$\tilde{\mathbf{B}}(z,t) = \frac{\tilde{k}}{\omega}\tilde{E}_0 e^{-\kappa z} e^{i(kz-\omega t)}\,\hat{\mathbf{y}}. \tag{9.131}$$

(Equation (iv) says the same thing.) Once again, the electric and magnetic fields are mutually perpendicular.

Like any complex number, $\tilde{k}$ can be expressed in terms of its modulus and phase:

$$\tilde{k} = K e^{i\phi}, \tag{9.132}$$

where

$$K \equiv |\tilde{k}| = \sqrt{k^2+\kappa^2} = \omega\sqrt{\epsilon\mu\sqrt{1+\left(\frac{\sigma}{\epsilon\omega}\right)^2}} \tag{9.133}$$

and

$$\phi \equiv \tan^{-1}(\kappa/k). \tag{9.134}$$

According to Eq. 9.130 and 9.131, the complex amplitudes $\tilde{E}_0 = E_0 e^{i\delta_E}$ and $\tilde{B}_0 = B_0 e^{i\delta_B}$ are related by

$$B_0 e^{i\delta_B} = \frac{K e^{i\phi}}{\omega} E_0 e^{i\delta_E}. \tag{9.135}$$

Evidently the electric and magnetic fields are no longer in phase; in fact,

$$\delta_B - \delta_E = \phi; \tag{9.136}$$

the magnetic field *lags behind* the electric field. Meanwhile, the (real) amplitudes of **E** and **B** are related by

$$\frac{B_0}{E_0} = \frac{K}{\omega} = \sqrt{\epsilon\mu\sqrt{1+\left(\frac{\sigma}{\epsilon\omega}\right)^2}}. \tag{9.137}$$

The (real) electric and magnetic fields are, finally,

$$\left.\begin{aligned} \mathbf{E}(z,t) &= E_0 e^{-\kappa z}\cos(kz-\omega t+\delta_E)\,\hat{\mathbf{x}}, \\ \mathbf{B}(z,t) &= B_0 e^{-\kappa z}\cos(kz-\omega t+\delta_E+\phi)\,\hat{\mathbf{y}}. \end{aligned}\right\} \tag{9.138}$$

These fields are shown in Fig. 9.18.

Problem 9.18

(a) Suppose you imbedded some free charge in a piece of glass. About how long would it take for the charge to flow to the surface?

(b) Silver is an excellent conductor, but it's expensive. Suppose you were designing a microwave experiment to operate at a frequency of 10^{10} Hz. How thick would you make the silver coatings?

(c) Find the wavelength and propagation speed in copper for radio waves at 1 MHz. Compare the corresponding values in air (or vacuum).

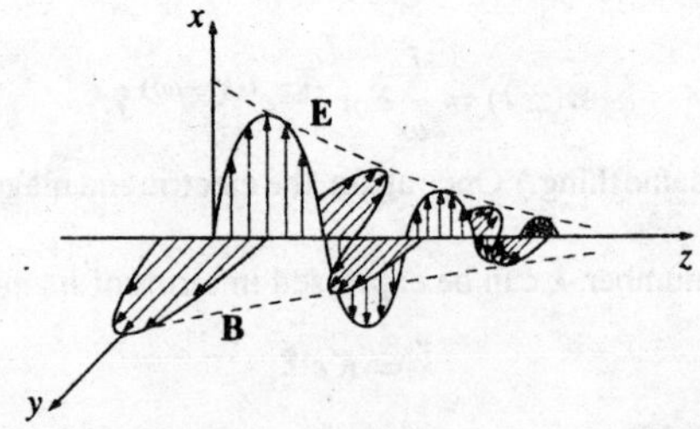

Figure 9.18

Problem 9.19

(a) Show that the skin depth in a poor conductor ($\sigma \ll \omega\epsilon$) is $(2/\sigma)\sqrt{\epsilon/\mu}$ (independent of frequency). Find the skin depth (in meters) for (pure) water.

(b) Show that the skin depth in a good conductor ($\sigma \gg \omega\epsilon$) is $\lambda/2\pi$ (where λ is the wavelength *in the conductor*). Find the skin depth (in nanometers) for a typical metal ($\sigma \approx 10^7\,(\Omega\,\text{m})^{-1}$) in the visible range ($\omega \approx 10^{15}$/s), assuming $\epsilon \approx \epsilon_0$ and $\mu \approx \mu_0$. Why are metals opaque?

(c) Show that in a good conductor the magnetic field lags the electric field by 45°, and find the ratio of their amplitudes. For a numerical example, use the "typical metal" in part (b).

Problem 9.20

(a) Calculate the (time averaged) energy density of an electromagnetic plane wave in a conducting medium (Eq. 9.138). Show that the magnetic contribution always dominates. [*Answer:* $(k^2/2\mu\omega^2)E_0^2e^{-2\kappa z}$]

(b) Show that the intensity is $(k/2\mu\omega)E_0^2e^{-2\kappa z}$.

9.4.2 Reflection at a Conducting Surface

The boundary conditions we used to analyze reflection and refraction at an interface between two dielectrics do not hold in the presence of free charges and currents. Instead, we have the more general relations (7.63):

$$\left.\begin{array}{ll}\text{(i)}\ \epsilon_1 E_1^{\perp} - \epsilon_2 E_2^{\perp} = \sigma_f, & \text{(iii)}\ \mathbf{E}_1^{\parallel} - \mathbf{E}_2^{\parallel} = 0, \\ \text{(ii)}\ B_1^{\perp} - B_2^{\perp} = 0, & \text{(iv)}\ \dfrac{1}{\mu_1}\mathbf{B}_1^{\parallel} - \dfrac{1}{\mu_2}\mathbf{B}_2^{\parallel} = \mathbf{K}_f \times \hat{\mathbf{n}},\end{array}\right\} \tag{9.139}$$

where σ_f (not to be confused with conductivity) is the free surface charge, $\mathbf{K}_f$ the free surface current, and $\hat{\mathbf{n}}$ (not to be confused with the polarization of the wave) is a unit vector perpendicular to the surface, pointing from medium (2) into medium (1). For ohmic conductors ($\mathbf{J}_f = \sigma\mathbf{E}$) there can be no free surface current, since this would require an infinite electric field at the boundary.

Suppose now that the xy plane forms the boundary between a nonconducting linear medium (1) and a conductor (2). A monochromatic plane wave, traveling in the z direction and polarized in the x direction, approaches from the left, as in Fig. 9.13:

$$\tilde{\mathbf{E}}_I(z,t) = \tilde{E}_{0_I} e^{i(k_1 z - \omega t)}\,\hat{\mathbf{x}}, \quad \tilde{\mathbf{B}}_I(z,t) = \frac{1}{v_1}\tilde{E}_{0_I} e^{i(k_1 z - \omega t)}\,\hat{\mathbf{y}}. \tag{9.140}$$

This incident wave gives rise to a reflected wave,

$$\tilde{\mathbf{E}}_R(z,t) = \tilde{E}_{0_R} e^{i(-k_1 z - \omega t)}\,\hat{\mathbf{x}}, \quad \tilde{\mathbf{B}}_R(z,t) = -\frac{1}{v_1}\tilde{E}_{0_R} e^{i(-k_1 z - \omega t)}\,\hat{\mathbf{y}}, \tag{9.141}$$

propagating back to the left in medium (1), and a transmitted wave

$$\tilde{\mathbf{E}}_T(z,t) = \tilde{E}_{0_T} e^{i(\tilde{k}_2 z - \omega t)}\,\hat{\mathbf{x}}, \quad \tilde{\mathbf{B}}_T(z,t) = \frac{\tilde{k}_2}{\omega}\tilde{E}_{0_T} e^{i(\tilde{k}_2 z - \omega t)}\,\hat{\mathbf{y}}, \tag{9.142}$$

which is attenuated as it penetrates into the conductor.

At $z = 0$, the combined wave in medium (1) must join the wave in medium (2), pursuant to the boundary conditions 9.139. Since $E^\perp = 0$ on both sides, boundary condition (i) yields $\sigma_f = 0$. Since $B^\perp = 0$, (ii) is automatically satisfied. Meanwhile, (iii) gives

$$\tilde{E}_{0_I} + \tilde{E}_{0_R} = \tilde{E}_{0_T}, \tag{9.143}$$

and (iv) (with $\mathbf{K}_f = 0$) says

$$\frac{1}{\mu_1 v_1}(\tilde{E}_{0_I} - \tilde{E}_{0_R}) - \frac{\tilde{k}_2}{\mu_2\omega}\tilde{E}_{0_T} = 0, \tag{9.144}$$

or

$$\tilde{E}_{0_I} - \tilde{E}_{0_R} = \tilde{\beta}\tilde{E}_{0_T}, \tag{9.145}$$

where

$$\tilde{\beta} \equiv \frac{\mu_1 v_1}{\mu_2\omega}\tilde{k}_2. \tag{9.146}$$

It follows that

$$\tilde{E}_{0_R} = \left(\frac{1-\tilde{\beta}}{1+\tilde{\beta}}\right)\tilde{E}_{0_I}, \quad \tilde{E}_{0_T} = \left(\frac{2}{1+\tilde{\beta}}\right)\tilde{E}_{0_I}. \tag{9.147}$$

These results are formally identical to the ones that apply at the boundary between *non*conductors (Eq. 9.82), but the resemblance is deceptive since $\tilde{\beta}$ is now a complex number.

For a *perfect* conductor ($\sigma = \infty$), $k_2 = \infty$ (Eq. 9.126), so $\tilde{\beta}$ is infinite, and

$$\tilde{E}_{0_R} = -\tilde{E}_{0_I}, \quad \tilde{E}_{0_T} = 0. \tag{9.148}$$

In this case the wave is totally reflected, with a 180° phase shift. (That's why excellent conductors make good mirrors. In practice, you paint a thin coating of silver onto the back of a pane of glass—the glass has nothing to do with the *reflection*; it's just there to support the silver and to keep it from tarnishing. Since the skin depth in silver at optical frequencies is on the order of 100 Å, you don't need a very thick layer.)

Problem 9.21 Calculate the reflection coefficient for light at an air-to-silver interface ($\mu_1 = \mu_2 = \mu_0$, $\epsilon_1 = \epsilon_0$, $\sigma = 6 \times 10^7 (\Omega \cdot \mathrm{m})^{-1}$), at optical frequencies ($\omega = 4 \times 10^{15}/\mathrm{s}$).

9.4.3 The Frequency Dependence of Permittivity

In the preceding sections, we have seen that the propagation of electromagnetic waves through matter is governed by three properties of the material, which we took to be constants: the permittivity ϵ, the permeability μ, and the conductivity σ. Actually, each of these parameters depends to some extent on the frequency of the waves you are considering. Indeed, if the permittivity were *truly* constant, then the index of refraction in a transparent medium, $n \cong \sqrt{\epsilon_r}$, would also be constant. But it is well known from optics that n is a function of wavelength (Fig. 9.19 shows the graph for a typical glass). A prism or a raindrop bends blue light more sharply than red, and spreads white light out into a rainbow of colors. This phenomenon is called **dispersion**. By extension, whenever the speed of a wave depends on its frequency, the supporting medium is called **dispersive**.[12]

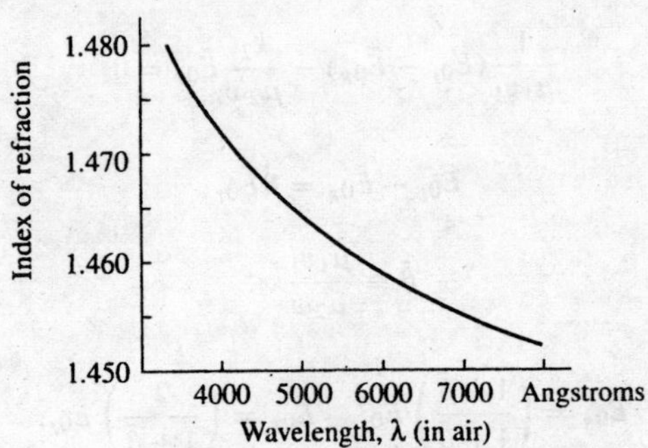

Figure 9.19

[12]Conductors, incidentally, are dispersive: see Eqs. 9.126 and 9.129.

Because waves of different frequency travel at different speeds in a dispersive medium, a wave form that incorporates a range of frequencies will change shape as it propagates. A sharply peaked wave typically flattens out, and whereas each sinusoidal component travels at the ordinary **wave (or phase) velocity**,

$$v = \frac{\omega}{k}, \tag{9.149}$$

the packet as a whole (the "envelope") travels at the so-called **group velocity**[13]

$$v_g = \frac{d\omega}{dk}. \tag{9.150}$$

[You can demonstrate this by dropping a rock into the nearest pond and watching the waves that form: While the disturbance as a whole spreads out in a circle, moving at speed v_g, the ripples that go to make it up will be seen to travel *twice* as fast ($v = 2v_g$ in this case). They appear at the back end of the packet, growing as they move forward to the center, then shrinking again and fading away at the front (Fig. 9.20).] We shall not concern ourselves with these matters—I'll stick to monochromatic waves, for which the problem does not arise. But I should just mention that the *energy* carried by a wave packet in a dispersive medium ordinarily travels at the *group* velocity, not the phase velocity. Don't be too alarmed, therefore, if in some circumstances v comes out greater than c.[14]

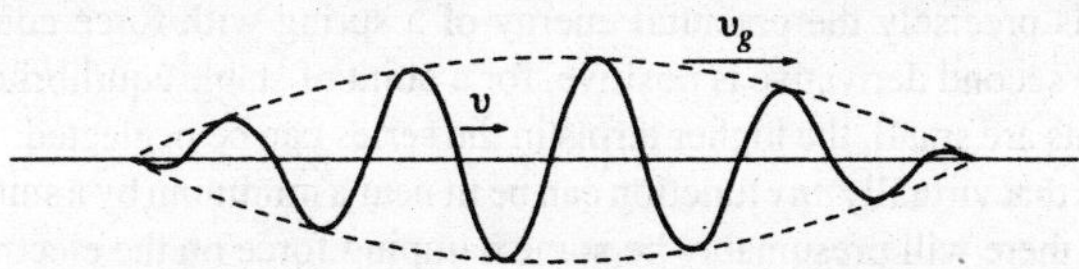

Figure 9.20

My purpose in this section is to account for the frequency dependence of ϵ in nonconductors, using a simplified model for the behavior of electrons in dielectrics. Like all classical models of atomic-scale phenomena, it is at best an approximation to the truth; nevertheless, it does yield qualitatively satisfactory results, and it provides a plausible mechanism for dispersion in transparent media.

The electrons in a nonconductor are bound to specific molecules. The actual binding forces can be quite complicated, but we shall picture each electron as attached to the end of an imaginary spring, with force constant k_{spring} (Fig. 9.21):

$$F_{\text{binding}} = -k_{\text{spring}}x = -m\omega_0^2 x, \tag{9.151}$$

[13]See A. P. French, *Vibrations and Waves*, p. 230 (New York: W. W. Norton & Co., 1971), or F. S. Crawford, Jr., *Waves*, Sect. 6.2 (New York: McGraw-Hill, 1968).

[14]Even the group velocity can exceed c in special cases—see P. C. Peters, *Am. J. Phys.* **56**, 129 (1988). Incidentally, if *two* different "speeds of light" are not enough to satisfy you, check out S. C. Bloch, *Am. J. Phys.* **45**, 538 (1977), in which no fewer than *eight* distinct velocities are identified!

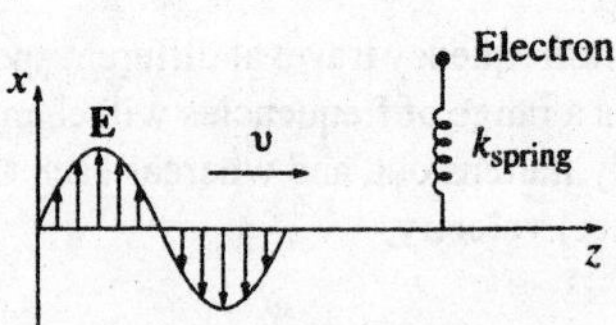

Figure 9.21

where x is displacement from equilibrium, m is the electron's mass, and ω_0 is the natural oscillation frequency, $\sqrt{k_{spring}/m}$. [If this strikes you as an implausible model, look back at Ex. 4.1, where we were led to a force of precisely this form. As a matter of fact, practically *any* binding force can be approximated this way for sufficiently small displacements from equilibrium, as you can see by expanding the potential energy in a Taylor series about the equilibrium point:

$$U(x) = U(0) + xU'(0) + \frac{1}{2}x^2U''(0) + \cdots.$$

The first term is a constant, with no dynamical significance (you can always adjust the zero of potential energy so that $U(0) = 0$). The second term automatically vanishes, since $dU/dx = -F$, and by the nature of an equilibrium the force at that point is zero. The third term is precisely the potential energy of a spring with force constant $k_{spring} = d^2U/dx^2\big|_0$ (the second derivative is positive, for a point of stable equilibrium). As long as the displacements are small, the higher terms in the series can be neglected. Geometrically, all I am saying is that virtually *any* function can be fit near a minimum by a suitable parabola.]

Meanwhile, there will presumably be some damping force on the electron:

$$F_{damping} = -m\gamma\frac{dx}{dt}. \tag{9.152}$$

[Again I have chosen the simplest possible form; the damping must be opposite in direction to the velocity, and making it *proportional* to the velocity is the easiest way to accomplish this. The *cause* of the damping does not concern us here—among other things, an oscillating charge radiates, and the radiation siphons off energy. We will calculate this "radiation damping" in Chapter 11.]

In the presence of an electromagnetic wave of frequency ω, polarized in the x direction (Fig. 9.21), the electron is subject to a driving force

$$F_{driving} = qE = qE_0\cos(\omega t), \tag{9.153}$$

where q is the charge of the electron and E_0 is the amplitude of the wave at the point z where the electron is situated. (Since we're only interested in one point, I have reset the clock so that the maximum E occurs there at $t = 0$.) Putting all this into Newton's second law gives

$$m\frac{d^2x}{dt^2} = F_{tot} = F_{binding} + F_{damping} + F_{driving},$$

or

$$m\frac{d^2x}{dt^2} + m\gamma\frac{dx}{dt} + m\omega_0^2 x = qE_0\cos(\omega t). \tag{9.154}$$

Our model, then, describes the electron as a damped harmonic oscillator, driven at frequency ω. (I assume that the much more massive nuclei remain at rest.)

Equation 9.154 is easier to handle if we regard it as the real part of a *complex* equation:

$$\frac{d^2\tilde{x}}{dt^2} + \gamma\frac{d\tilde{x}}{dt} + \omega_0^2\tilde{x} = \frac{q}{m}E_0 e^{-i\omega t}. \tag{9.155}$$

In the steady state, the system oscillates at the driving frequency:

$$\tilde{x}(t) = \tilde{x}_0 e^{-i\omega t}. \tag{9.156}$$

Inserting this into Eq. 9.155, we obtain

$$\tilde{x}_0 = \frac{q/m}{\omega_0^2 - \omega^2 - i\gamma\omega}E_0. \tag{9.157}$$

The dipole moment is the real part of

$$\tilde{p}(t) = q\tilde{x}(t) = \frac{q^2/m}{\omega_0^2 - \omega^2 - i\gamma\omega}E_0 e^{-i\omega t}. \tag{9.158}$$

The imaginary term in the denominator means that p is *out of phase* with E—lagging behind by an angle $\tan^{-1}[\gamma\omega/(\omega_0^2 - \omega^2)]$ that is very small when $\omega \ll \omega_0$ and rises to π when $\omega \gg \omega_0$.

In general, differently situated electrons within a given molecule experience different natural frequencies and damping coefficients. Let's say there are f_j electrons with frequency ω_j and damping γ_j in each molecule. If there are N molecules per unit volume, the polarization $\mathbf{P}$ is given by[15] the real part of

$$\tilde{\mathbf{P}} = \frac{Nq^2}{m}\left(\sum_j \frac{f_j}{\omega_j^2 - \omega^2 - i\gamma_j\omega}\right)\tilde{\mathbf{E}}. \tag{9.159}$$

Now, I defined the electric susceptibility as the proportionality constant between $\mathbf{P}$ and $\mathbf{E}$ (specifically, $\mathbf{P} = \epsilon_0\chi_e\mathbf{E}$). In the present case $\mathbf{P}$ is *not* proportional to $\mathbf{E}$ (this is not, strictly speaking, a linear medium) because of the difference in phase. However, the *complex* polarization $\tilde{\mathbf{P}}$ *is* proportional to the *complex* field $\tilde{\mathbf{E}}$, and this suggests that we introduce a **complex susceptibility**, $\tilde{\chi}_e$:

$$\tilde{\mathbf{P}} = \epsilon_0\tilde{\chi}_e\tilde{\mathbf{E}}. \tag{9.160}$$

[15]This applies directly to the case of a dilute gas; for denser materials the theory is modified slightly, in accordance with the Clausius-Mossotti equation (Prob. 4.38). By the way, don't confuse the "polarization" of a medium, **P**, with the "polarization" of a *wave*—same *word*, but two completely unrelated meanings.

All of the manipulations we went through before carry over, on the understanding that the physical polarization is the real part of $\tilde{\mathbf{P}}$, just as the physical field is the real part of $\tilde{\mathbf{E}}$. In particular, the proportionality between $\tilde{\mathbf{D}}$ and $\tilde{\mathbf{E}}$ is the **complex permittivity** $\tilde{\epsilon} = \epsilon_0(1+\tilde{\chi}_e)$, and the **complex dielectric constant** (in this model) is

$$\tilde{\epsilon}_r = 1 + \frac{Nq^2}{m\epsilon_0}\sum_j \frac{f_j}{\omega_j^2 - \omega^2 - i\gamma_j\omega}. \tag{9.161}$$

Ordinarily, the imaginary term is negligible; however, when ω is very close to one of the resonant frequencies (ω_j) it plays an important role, as we shall see.

In a dispersive medium the wave equation for a given frequency reads

$$\nabla^2\tilde{\mathbf{E}} = \tilde{\epsilon}\mu_0\frac{\partial^2\tilde{\mathbf{E}}}{\partial t^2}; \tag{9.162}$$

it admits plane wave solutions, as before,

$$\tilde{\mathbf{E}}(z,t) = \tilde{\mathbf{E}}_0 e^{i(\tilde{k}z-\omega t)}, \tag{9.163}$$

with the complex wave number

$$\tilde{k} \equiv \sqrt{\tilde{\epsilon}\mu_0}\,\omega. \tag{9.164}$$

Writing $\tilde{k}$ in terms of its real and imaginary parts,

$$\tilde{k} = k + i\kappa, \tag{9.165}$$

Eq. 9.163 becomes

$$\tilde{\mathbf{E}}(z,t) = \tilde{E}_0 e^{-\kappa z} e^{i(kz-\omega t)}. \tag{9.166}$$

Evidently the wave is *attenuated* (this is hardly surprising, since the damping absorbs energy). Because the intensity is proportional to E^2 (and hence to $e^{-2\kappa z}$), the quantity

$$\alpha \equiv 2\kappa \tag{9.167}$$

is called the **absorption coefficient**. Meanwhile, the wave velocity is ω/k, and the index of refraction is

$$n = \frac{ck}{\omega}. \tag{9.168}$$

I have deliberately used notation reminiscent of Sect. 9.4.1. However, in the present case k and κ have nothing to do with conductivity; rather, they are determined by the parameters of our damped harmonic oscillator. For gases, the second term in Eq. 9.161 is small, and we can approximate the square root (Eq. 9.164) by the first term in the binomial expansion, $\sqrt{1+\varepsilon} \cong 1 + \frac{1}{2}\varepsilon$. Then

$$\tilde{k} = \frac{\omega}{c}\sqrt{\tilde{\epsilon}_r} \cong \frac{\omega}{c}\left[1 + \frac{Nq^2}{2m\epsilon_0}\sum_j \frac{f_j}{\omega_j^2 - \omega^2 - i\gamma_j\omega}\right], \tag{9.169}$$

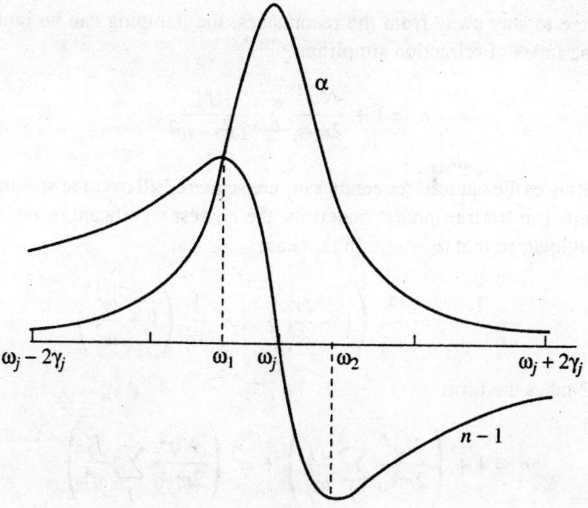

Figure 9.22

so

$$n = \frac{ck}{\omega} \cong 1 + \frac{Nq^2}{2m\epsilon_0} \sum_j \frac{f_j(\omega_j^2 - \omega^2)}{(\omega_j^2 - \omega^2)^2 + \gamma_j^2\omega^2}, \tag{9.170}$$

and

$$\alpha = 2\kappa \cong \frac{Nq^2\omega^2}{m\epsilon_0 c} \sum_j \frac{f_j\gamma_j}{(\omega_j^2 - \omega^2)^2 + \gamma_j^2\omega^2}. \tag{9.171}$$

In Fig. 9.22 I have plotted the index of refraction and the absorption coefficient in the vicinity of one of the resonances. *Most* of the time the index of refraction *rises* gradually with increasing frequency, consistent with our experience from optics (Fig. 9.19). However, in the immediate neighborhood of a resonance the index of refraction *drops* sharply. Because this behavior is atypical, it is called **anomalous dispersion**. Notice that the region of anomalous dispersion ($\omega_1 < \omega < \omega_2$, in the figure) coincides with the region of maximum absorption; in fact, the material may be practically opaque in this frequency range. The reason is that we are now driving the electrons at their "favorite" frequency; the amplitude of their oscillation is relatively large, and a correspondingly large amount of energy is dissipated by the damping mechanism.

In Fig. 9.22, n runs below 1 above the resonance, suggesting that the wave speed exceeds c. As I mentioned earlier, this is no cause for alarm, since energy does not travel at the wave velocity but rather at the *group* velocity (see Prob. 9.25). Moreover, the graph does not include the contributions of other terms in the sum, which add a relatively constant "background" that, in some cases, keeps $n > 1$ on both sides of the resonance.

If you agree to stay away from the resonances, the damping can be ignored, and the formula for the index of refraction simplifies:

$$n = 1 + \frac{Nq^2}{2m\epsilon_0} \sum_j \frac{f_j}{\omega_j^2 - \omega^2}. \tag{9.172}$$

For most substances the natural frequencies ω_j are scattered all over the spectrum in a rather chaotic fashion. But for transparent materials, the nearest significant resonances typically lie in the ultraviolet, so that $\omega < \omega_j$. In that case

$$\frac{1}{\omega_j^2 - \omega^2} = \frac{1}{\omega_j^2}\left(1 - \frac{\omega^2}{\omega_j^2}\right)^{-1} \cong \frac{1}{\omega_j^2}\left(1 + \frac{\omega^2}{\omega_j^2}\right),$$

and Eq. 9.172 takes the form

$$n = 1 + \left(\frac{Nq^2}{2m\epsilon_0} \sum_j \frac{f_j}{\omega_j^2}\right) + \omega^2 \left(\frac{Nq^2}{2m\epsilon_0} \sum_j \frac{f_j}{\omega_j^4}\right). \tag{9.173}$$

Or, in terms of the wavelength in vacuum ($\lambda = 2\pi c/\omega$):

$$n = 1 + A\left(1 + \frac{B}{\lambda^2}\right). \tag{9.174}$$

This is known as **Cauchy's formula**; the constant A is called the **coefficient of refraction** and B is called the **coefficient of dispersion**. Cauchy's equation applies reasonably well to most gases, in the optical region.

What I have described in this section is certainly not the complete story of dispersion in nonconducting media. Nevertheless, it does indicate how the damped harmonic motion of electrons can account for the frequency dependence of the index of refraction, and it explains why n is ordinarily a slowly increasing function of ω, with occasional "anomalous" regions where it precipitously drops.

Problem 9.22

(a) Shallow water is nondispersive; the waves travel at a speed that is proportional to the square root of the depth. In deep water, however, the waves can't "feel" all the way down to the bottom—they behave as though the depth were proportional to λ. (Actually, the distinction between "shallow" and "deep" itself depends on the wavelength: If the depth is less than λ the water is "shallow"; if it is substantially greater than λ the water is "deep.") Show that the wave velocity of deep water waves is *twice* the group velocity.

(b) In quantum mechanics, a free particle of mass m traveling in the x direction is described by the wave function

$$\Psi(x, t) = Ae^{i(px - Et)/\hbar},$$

where p is the momentum, and $E = p^2/2m$ is the kinetic energy. Calculate the group velocity and the wave velocity. Which one corresponds to the classical speed of the particle? Note that the wave velocity is *half* the group velocity.

Problem 9.23 If you take the model in Ex. 4.1 at face value, what natural frequency do you get? Put in the actual numbers. Where, in the electromagnetic spectrum, does this lie, assuming the radius of the atom is 0.5 Å? Find the coefficients of refraction and dispersion and compare them with those for hydrogen at 0°C and atmospheric pressure: $A = 1.36 \times 10^{-4}$, $B = 7.7 \times 10^{-15} \text{m}^2$.

Problem 9.24 Find the width of the anomalous dispersion region for the case of a single resonance at frequency ω_0. Assume $\gamma \ll \omega_0$. Show that the index of refraction assumes its maximum and minimum values at points where the absorption coefficient is at half-maximum.

Problem 9.25 Assuming negligible damping ($\gamma_j = 0$), calculate the group velocity ($v_g = d\omega/dk$) of the waves described by Eqs. 9.166 and 9.169. Show that $v_g < c$, even when $v > c$.

9.5 Guided Waves

9.5.1 Wave Guides

So far, we have dealt with plane waves of infinite extent; now we consider electromagnetic waves confined to the interior of a hollow pipe, or **wave guide** (Fig. 9.23). We'll assume the wave guide is a perfect conductor, so that $\mathbf{E} = 0$ and $\mathbf{B} = 0$ inside the material itself, and hence the boundary conditions at the inner wall are[16]

$$\left.\begin{array}{ll} \text{(i)} & \mathbf{E}^{\parallel} = 0, \\ \text{(ii)} & B^{\perp} = 0. \end{array}\right\} \tag{9.175}$$

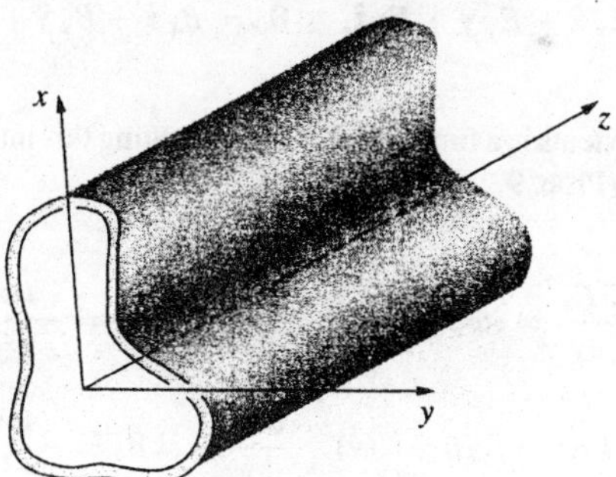

Figure 9.23

[16]See Eq. 9.139 and Prob. 7.42. In a perfect conductor $\mathbf{E} = 0$, and hence (by Faraday's law) $\partial\mathbf{B}/\partial t = 0$; assuming the magnetic field *started out* zero, then, it will *remain* so.

Free charges and currents will be induced on the surface in such a way as to enforce these constraints. We are interested in monochromatic waves that propagate down the tube, so **E** and **B** have the generic form

$$\left.\begin{aligned}&\text{(i)}\quad \tilde{\mathbf{E}}(x, y, z, t) = \tilde{\mathbf{E}}_0(x, y)e^{i(kz-\omega t)},\\ &\text{(ii)}\quad \tilde{\mathbf{B}}(x, y, z, t) = \tilde{\mathbf{B}}_0(x, y)e^{i(kz-\omega t)}.\end{aligned}\right\} \tag{9.176}$$

(For the cases of interest k is real, so I shall dispense with the tilde.) The electric and magnetic fields must, of course, satisfy Maxwell's equations, in the interior of the wave guide:

$$\left.\begin{aligned}&\text{(i)}\ \nabla\cdot\mathbf{E} = 0, \quad &&\text{(iii)}\ \nabla\times\mathbf{E} = -\frac{\partial\mathbf{B}}{\partial t},\\ &\text{(ii)}\ \nabla\cdot\mathbf{B} = 0, \quad &&\text{(iv)}\ \nabla\times\mathbf{B} = \frac{1}{c^2}\frac{\partial\mathbf{E}}{\partial t}.\end{aligned}\right\} \tag{9.177}$$

The problem, then, is to find functions $\tilde{\mathbf{E}}_0$ and $\tilde{\mathbf{B}}_0$ such that the fields (9.176) obey the differential equations (9.177), subject to boundary conditions (9.175).

As we shall soon see, *confined* waves are *not* (in general) transverse; in order to fit the boundary conditions we shall have to include longitudinal components (E_z and B_z):[17]

$$\tilde{\mathbf{E}}_0 = E_x\,\hat{\mathbf{x}} + E_y\,\hat{\mathbf{y}} + E_z\,\hat{\mathbf{z}}, \quad \tilde{\mathbf{B}}_0 = B_x\,\hat{\mathbf{x}} + B_y\,\hat{\mathbf{y}} + B_z\,\hat{\mathbf{z}}, \tag{9.178}$$

where each of the components is a function of x and y. Putting this into Maxwell's equations (iii) and (iv), we obtain (Prob. 9.26a)

$$\left.\begin{aligned}&\text{(i)}\quad \frac{\partial E_y}{\partial x} - \frac{\partial E_x}{\partial y} = i\omega B_z, \quad &&\text{(iv)}\quad \frac{\partial B_y}{\partial x} - \frac{\partial B_x}{\partial y} = -\frac{i\omega}{c^2}E_z,\\ &\text{(ii)}\quad \frac{\partial E_z}{\partial y} - ikE_y = i\omega B_x, \quad &&\text{(v)}\quad \frac{\partial B_z}{\partial y} - ikB_y = -\frac{i\omega}{c^2}E_x,\\ &\text{(iii)}\quad ikE_x - \frac{\partial E_z}{\partial x} = i\omega B_y, \quad &&\text{(vi)}\quad ikB_x - \frac{\partial B_z}{\partial x} = -\frac{i\omega}{c^2}E_y.\end{aligned}\right\} \tag{9.179}$$

[17] To avoid cumbersome notation I shall leave the subscript 0 and the tilde off the individual components.

Equations (ii), (iii), (v), and (vi) can be solved for E_x, E_y, B_x, and B_y:

$$\left.\begin{aligned}
&\text{(i)} \quad E_x = \frac{i}{(\omega/c)^2 - k^2}\left(k\frac{\partial E_z}{\partial x} + \omega\frac{\partial B_z}{\partial y}\right),\\
&\text{(ii)} \quad E_y = \frac{i}{(\omega/c)^2 - k^2}\left(k\frac{\partial E_z}{\partial y} - \omega\frac{\partial B_z}{\partial x}\right),\\
&\text{(iii)} \quad B_x = \frac{i}{(\omega/c)^2 - k^2}\left(k\frac{\partial B_z}{\partial x} - \frac{\omega}{c^2}\frac{\partial E_z}{\partial y}\right),\\
&\text{(iv)} \quad B_y = \frac{i}{(\omega/c)^2 - k^2}\left(k\frac{\partial B_z}{\partial y} + \frac{\omega}{c^2}\frac{\partial E_z}{\partial x}\right).
\end{aligned}\right\} \tag{9.180}$$

It suffices, then, to determine the longitudinal components E_z and B_z; if we knew those, we could quickly calculate all the others, just by differentiating. Inserting Eq. 9.180 into the remaining Maxwell equations (Prob. 9.26b) yields uncoupled equations for E_z and B_z:

$$\left.\begin{aligned}
&\text{(i)} \quad \left[\frac{\partial^2}{\partial x^2} + \frac{\partial^2}{\partial y^2} + (\omega/c)^2 - k^2\right] E_z = 0,\\
&\text{(ii)} \quad \left[\frac{\partial^2}{\partial x^2} + \frac{\partial^2}{\partial y^2} + (\omega/c)^2 - k^2\right] B_z = 0.
\end{aligned}\right\} \tag{9.181}$$

If $E_z = 0$ we call these **TE** ("transverse electric") **waves**; if $B_z = 0$ they are called **TM** ("transverse magnetic") **waves**; if both $E_z = 0$ and $B_z = 0$, we call them **TEM waves.**[18] It turns out that TEM waves cannot occur in a hollow wave guide.

Proof: If $E_z = 0$, Gauss's law (Eq. 9.177i) says

$$\frac{\partial E_x}{\partial x} + \frac{\partial E_y}{\partial y} = 0,$$

and if $B_z = 0$, Faraday's law (Eq. 9.177iii) says

$$\frac{\partial E_y}{\partial x} - \frac{\partial E_x}{\partial y} = 0.$$

Indeed, the vector $\tilde{\mathbf{E}}_0$ in Eq. 9.178 has zero divergence and zero curl. It can therefore be written as the gradient of a scalar potential that satisfies Laplace's equation. But the boundary condition on **E** (Eq. 9.175) requires that the surface be an equipotential, and since Laplace's equation admits no local maxima or minima (Sect. 3.1.4), this means that the potential is constant throughout, and hence the electric field is *zero*—no wave at all. qed

[18] In the case of TEM waves (including the unconfined plane waves of Sect. 9.2), $k = \omega/c$, Eqs. 9.180 are indeterminate, and you have to go back to Eqs. 9.179.

Notice that this argument applies only to a completely *empty* pipe—if you run a separate conductor down the middle, the potential at *its* surface need not be the same as on the outer wall, and hence a nontrivial potential is possible. We'll see an example of this in Sect. 9.5.3.

! **Problem 9.26**

(a) Derive Eqs. 9.179, and from these obtain Eqs. 9.180.

(b) Put Eq. 9.180 into Maxwell's equations (i) and (ii) to obtain Eq. 9.181. Check that you get the same results using (i) and (iv) of Eq. 9.179.

9.5.2 TE Waves in a Rectangular Wave Guide

Suppose we have a wave guide of rectangular shape (Fig. 9.24), with height a and width b, and we are interested in the propagation of TE waves. The problem is to solve Eq. 9.181ii, subject to the boundary condition 9.175ii. We'll do it by separation of variables. Let

$$B_z(x, y) = X(x)Y(y),$$

so that

$$Y\frac{d^2X}{dx^2} + X\frac{d^2Y}{dy^2} + [(\omega/c)^2 - k^2]XY = 0.$$

Divide by XY and note that the x- and y-dependent terms must be constant:

$$\text{(i)}\quad \frac{1}{X}\frac{d^2X}{dx^2} = -k_x^2, \quad \text{(ii)}\quad \frac{1}{Y}\frac{d^2Y}{dy^2} = -k_y^2, \tag{9.182}$$

with

$$-k_x^2 - k_y^2 + (\omega/c)^2 - k^2 = 0. \tag{9.183}$$

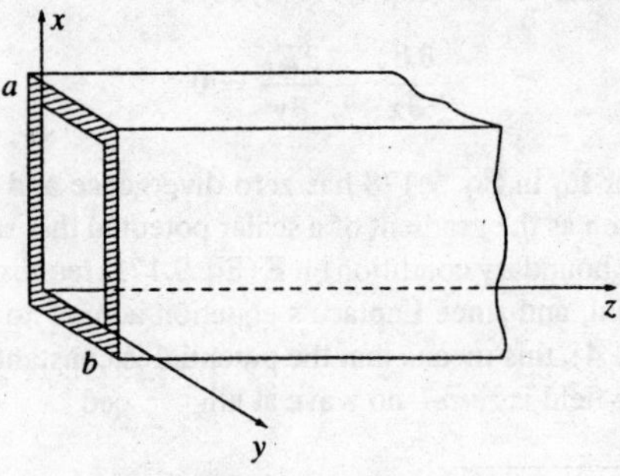

Figure 9.24

The general solution to Eq. 9.182i is

$$X(x) = A \sin(k_x x) + B \cos(k_x x).$$

But the boundary conditions require that B_x—and hence also (Eq. 9.180iii) dX/dx—vanishes at $x = 0$ and $x = a$. So $A = 0$, and

$$k_x = m\pi/a, \quad (m = 0, 1, 2, \ldots). \tag{9.184}$$

The same goes for Y, with

$$k_y = n\pi/b, \quad (n = 0, 1, 2, \ldots), \tag{9.185}$$

and we conclude that

$$B_z = B_0 \cos(m\pi x/a) \cos(n\pi y/b). \tag{9.186}$$

This solution is called the TE$_{mn}$ mode. (The first index is conventionally associated with the *larger* dimension, so we assume $a \geq b$. By the way, at least *one* of the indices must be nonzero—see Prob. 9.27.) The wave number (k) is obtained by putting Eqs. 9.184 and 9.185 into Eq. 9.183:

$$k = \sqrt{(\omega/c)^2 - \pi^2[(m/a)^2 + (n/b)^2]}. \tag{9.187}$$

If

$$\omega < c\pi\sqrt{(m/a)^2 + (n/b)^2} \equiv \omega_{mn}, \tag{9.188}$$

the wave number is imaginary, and instead of a traveling wave we have exponentially attenuated fields (Eq. 9.176). For this reason ω_{mn} is called the **cutoff frequency** for the mode in question. The *lowest* cutoff frequency for a given wave guide occurs for the mode TE$_{10}$:

$$\omega_{10} = c\pi/a. \tag{9.189}$$

Frequencies less than this will not propagate at all.

The wave number can be written more simply in terms of the cutoff frequency:

$$k = \frac{1}{c}\sqrt{\omega^2 - \omega_{mn}^2}. \tag{9.190}$$

The wave velocity is

$$v = \frac{\omega}{k} = \frac{c}{\sqrt{1 - (\omega_{mn}/\omega)^2}}, \tag{9.191}$$

which is greater than c. However (see Prob. 9.29), the energy carried by the wave travels at the *group* velocity (Eq. 9.150):

$$v_g = \frac{1}{dk/d\omega} = c\sqrt{1 - (\omega_{mn}/\omega)^2} < c. \tag{9.192}$$

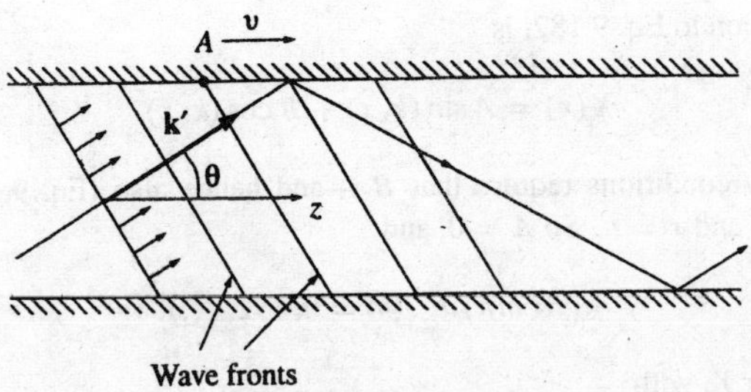

Figure 9.25

There's another way to visualize the propagation of an electromagnetic wave in a rectangular pipe, and it serves to illuminate many of these results. Consider an ordinary *plane* wave, traveling at an angle θ to the z axis, and reflecting perfectly off each conducting surface (Fig. 9.25). In the x and y directions the (multiply reflected) waves interfere to form standing wave patterns, of wavelength $\lambda_x = 2a/m$ and $\lambda_y = 2b/n$ (hence wave number $k_x = 2\pi/\lambda_x = \pi m/a$ and $k_y = \pi n/b$), respectively. Meanwhile, in the z direction there remains a traveling wave, with wave number $k_z = k$. The propagation vector for the "original" plane wave is therefore

$$\mathbf{k}' = \frac{\pi m}{a}\,\hat{\mathbf{x}} + \frac{\pi n}{b}\,\hat{\mathbf{y}} + k\,\hat{\mathbf{z}},$$

and the frequency is

$$\omega = c|\mathbf{k}'| = c\sqrt{k^2 + \pi^2[(m/a)^2 + (n/b)^2]} = \sqrt{(ck)^2 + (\omega_{mn})^2}.$$

Only certain angles will lead to one of the allowed standing wave patterns:

$$\cos\theta = \frac{k}{|\mathbf{k}'|} = \sqrt{1 - (\omega_{mn}/\omega)^2}.$$

The plane wave travels at speed c, but because it is going at an angle θ to the z axis, its net velocity down the wave guide is

$$v_g = c\cos\theta = c\sqrt{1 - (\omega_{mn}/\omega)^2}.$$

The *wave* velocity, on the other hand, is the speed of the wave fronts (A, say, in Fig. 9.25) down the pipe. Like the intersection of a line of breakers with the beach, they can move much faster than the waves themselves—in fact

$$v = \frac{c}{\cos\theta} = \frac{c}{\sqrt{1 - (\omega_{mn}/\omega)^2)}}.$$

Problem 9.27 Show that the mode TE_{00} cannot occur in a rectangular wave guide. [*Hint:* In this case $\omega/c = k$, so Eqs. 9.180 are indeterminate, and you must go back to 9.179. Show that B_z is a constant, and hence—applying Faraday's law in integral form to a cross section—that $B_z = 0$, so this would be a TEM mode.]

Problem 9.28 Consider a rectangular wave guide with dimensions 2.28 cm × 1.01 cm. What TE modes will propagate in this wave guide, if the driving frequency is 1.70×10^{10} Hz? Suppose you wanted to excite only *one* TE mode; what range of frequencies could you use? What are the corresponding wavelengths (in open space)?

Problem 9.29 Confirm that the energy in the TE_{mn} mode travels at the group velocity. [*Hint:* Find the time averaged Poynting vector $\langle \mathbf{S} \rangle$ and the energy density $\langle u \rangle$ (use Prob. 9.11 if you wish). Integrate over the cross section of the wave guide to get the energy per unit time and per unit length carried by the wave, and take their ratio.]

Problem 9.30 Work out the theory of TM modes for a rectangular wave guide. In particular, find the longitudinal electric field, the cutoff frequencies, and the wave and group velocities. Find the ratio of the lowest TM cutoff frequency to the lowest TE cutoff frequency, for a given wave guide. [*Caution*: What is the lowest TM mode?]

9.5.3 The Coaxial Transmission Line

In Sect. 9.5.1, I showed that a *hollow* wave guide cannot support TEM waves. But a coaxial transmission line, consisting of a long straight wire of radius a, surrounded by a cylindrical conducting sheath of radius b (Fig. 9.26), *does* admit modes with $E_z = 0$ and $B_z = 0$. In this case Maxwell's equations (in the form 9.179) yield

$$k = \omega/c \tag{9.193}$$

(so the waves travel at speed c, and are nondispersive),

$$cB_y = E_x \quad \text{and} \quad cB_x = -E_y \tag{9.194}$$

(so $\mathbf{E}$ and $\mathbf{B}$ are mutually perpendicular), and (together with $\nabla \cdot \mathbf{E} = 0$, $\nabla \cdot \mathbf{B} = 0$):

$$\left.\begin{array}{ll} \dfrac{\partial E_x}{\partial x} + \dfrac{\partial E_y}{\partial y} = 0, & \dfrac{\partial E_y}{\partial x} - \dfrac{\partial E_x}{\partial y} = 0, \\[2ex] \dfrac{\partial B_x}{\partial x} + \dfrac{\partial B_y}{\partial y} = 0, & \dfrac{\partial B_y}{\partial x} - \dfrac{\partial B_x}{\partial y} = 0. \end{array}\right\} \tag{9.195}$$

Figure 9.26

These are precisely the equations of *electrostatics* and *magnetostatics*, for empty space, in two dimensions; the solution with cylindrical symmetry can be borrowed directly from the case of an infinite line charge and an infinite straight current, respectively:

$$\mathbf{E}_0(s,\phi) = \frac{A}{s}\,\hat{\mathbf{s}}, \quad \mathbf{B}_0(s,\phi) = \frac{A}{cs}\,\hat{\boldsymbol{\phi}}, \tag{9.196}$$

for some constant A. Substituting these into Eq. 9.176, and taking the real part:

$$\left.\begin{aligned} \mathbf{E}(s,\phi,z,t) &= \frac{A\cos(kz-\omega t)}{s}\,\hat{\mathbf{s}}, \\ \mathbf{B}(s,\phi,z,t) &= \frac{A\cos(kz-\omega t)}{cs}\,\hat{\boldsymbol{\phi}}. \end{aligned}\right\} \tag{9.197}$$

Problem 9.31

(a) Show directly that Eqs. 9.197 satisfy Maxwell's equations (9.177) and the boundary conditions 9.175.

(b) Find the charge density, $\lambda(z,t)$, and the current, $I(z,t)$, on the inner conductor.

More Problems on Chapter 9

! **Problem 9.32** The "inversion theorem" for Fourier transforms states that

$$\tilde{\phi}(z) = \int_{-\infty}^{\infty} \tilde{\Phi}(k) e^{ikz}\,dk \iff \tilde{\Phi}(k) = \frac{1}{2\pi}\int_{-\infty}^{\infty} \tilde{\phi}(z) e^{-ikz}\,dz. \tag{9.198}$$

Use this to determine $\tilde{A}(k)$, in Eq. 9.20, in terms of $f(z,0)$ and $\dot{f}(z,0)$.

[*Answer:* $(1/2\pi)\int_{-\infty}^{\infty}[f(z,0)+(i/\omega)\dot{f}(z,0)]e^{-ikz}\,dz$]

Problem 9.33 Suppose

$$\mathbf{E}(r,\theta,\phi,t) = A\frac{\sin\theta}{r}\left[\cos(kr-\omega t) - (1/kr)\sin(kr-\omega t)\right]\hat{\boldsymbol{\phi}}, \quad \text{with } \frac{\omega}{k} = c.$$

(This is, incidentally, the simplest possible **spherical wave**. For notational convenience, let $(kr-\omega t)\equiv u$ in your calculations.)

(a) Show that **E** obeys all four of Maxwell's equations, in vacuum, and find the associated magnetic field.

(b) Calculate the Poynting vector. Average **S** over a full cycle to get the intensity vector **I**. (Does it point in the expected direction? Does it fall off like r^{-2}, as it should?)

(c) Integrate $\mathbf{I}\cdot d\mathbf{a}$ over a spherical surface to determine the total power radiated. [*Answer:* $4\pi A^2/3\mu_0 c$]

! **Problem 9.34** Light of (angular) frequency ω passes from medium 1, through a slab (thickness d) of medium 2, and into medium 3 (for instance, from water through glass into air, as in Fig. 9.27). Show that the transmission coefficient for normal incidence is given by

$$T^{-1} = \frac{1}{4n_1 n_3}\left[(n_1+n_3)^2 + \frac{(n_1^2-n_2^2)(n_3^2-n_2^2)}{n_2^2}\sin^2\left(\frac{n_2\omega d}{c}\right)\right]. \tag{9.199}$$

[*Hint:* To the *left*, there is an incident wave and a reflected wave; to the *right*, there is a transmitted wave; inside the slab there is a wave going to the right and a wave going to the left. Express each of these in terms of its complex amplitude, and relate the amplitudes by imposing suitable boundary conditions at the two interfaces. All three media are linear and homogeneous; assume $\mu_1 = \mu_2 = \mu_3 = \mu_0$.]

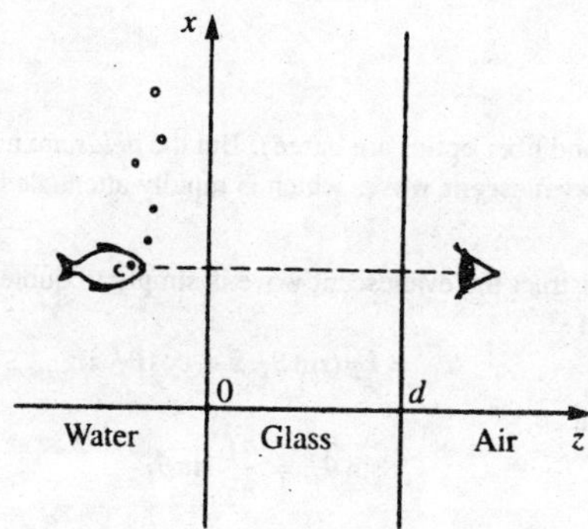

Figure 9.27

Problem 9.35 A microwave antenna radiating at 10 GHz is to be protected from the environment by a plastic shield of dielectric constant 2.5. What is the minimum thickness of this shielding that will allow perfect transmission (assuming normal incidence)? [*Hint:* use Eq. 9.199]

Problem 9.36 Light from an aquarium (Fig. 9.27) goes from water ($n = \frac{4}{3}$) through a plane of glass ($n = \frac{3}{2}$) into air ($n = 1$). Assuming it's a monochromatic plane wave and that it strikes the glass at normal incidence, find the minimum and maximum transmission coefficients (Eq. 9.199). You can see the fish clearly; how well can it see you?

! **Problem 9.37** According to Snell's law, when light passes from an optically dense medium into a less dense one ($n_1 > n_2$) the propagation vector **k** bends *away* from the normal (Fig. 9.28). In particular, if the light is incident at the **critical angle**

$$\theta_c \equiv \sin^{-1}(n_2/n_1). \tag{9.200}$$

then $\theta_T = 90°$, and the transmitted ray just grazes the surface. If θ_I *exceeds* θ_c, there is no refracted ray at all, only a reflected one (this is the phenomenon of **total internal reflection,**

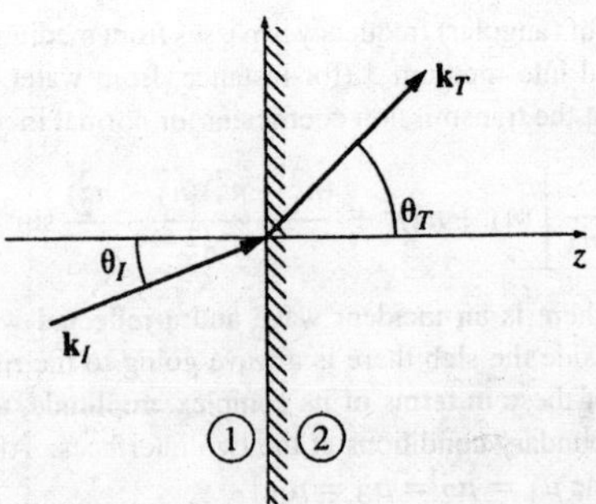

Figure 9.28

on which light pipes and fiber optics are based). But the *fields* are not zero in medium 2; what we get is a so-called **evanescent wave**, which is rapidly attenuated and transports no energy into medium 2.[19]

A quick way to construct the evanescent wave is simply to quote the results of Sect. 9.3.3, with $k_T = \omega n_2/c$ and

$$\mathbf{k}_T = k_T(\sin\theta_T\,\hat{\mathbf{x}} + \cos\theta_T\,\hat{\mathbf{z}});$$

the only change is that

$$\sin\theta_T = \frac{n_1}{n_2}\sin\theta_I$$

is now greater than 1, and

$$\cos\theta_T = \sqrt{1 - \sin^2\theta_T} = i\sqrt{\sin^2\theta_T - 1}$$

is imaginary. (Obviously, θ_T can no longer be interpreted as an *angle*!)

(a) Show that

$$\tilde{\mathbf{E}}_T(\mathbf{r}, t) = \tilde{\mathbf{E}}_{0_T} e^{-\kappa z} e^{i(kx - \omega t)} \tag{9.201}$$

where

$$\kappa \equiv \frac{\omega}{c}\sqrt{(n_1\sin\theta_I)^2 - n_2^2} \quad \text{and} \quad k \equiv \frac{\omega n_1}{c}\sin\theta_I. \tag{9.202}$$

This is a wave propagating in the x direction (*parallel* to the interface!), and attenuated in the z direction.

(b) Noting that α (Eq. 9.108) is now imaginary, use Eq. 9.109 to calculate the reflection coefficient for polarization parallel to the plane of incidence. [Notice that you get 100% reflection, which is better than at a conducting surface (see, for example, Prob. 9.21).]

(c) Do the same for polarization perpendicular to the plane of incidence (use the results of Prob. 9.16).

---

[19] The evanescent fields can be detected by placing a second interface a short distance to the right of the first; in a close analog to quantum mechanical **tunneling**, the wave crosses the gap and reassembles to the right. See F. Albiol, S. Navas, and M. V. Andres, *Am. J. Phys.* **61**, 165 (1993).

(d) In the case of polarization perpendicular to the plane of incidence, show that the (real) evanescent fields are

$$\left.\begin{aligned} \mathbf{E}(\mathbf{r}, t) &= E_0 e^{-\kappa z} \cos(kx - \omega t)\,\hat{\mathbf{y}}, \\ \mathbf{B}(\mathbf{r}, t) &= \frac{E_0}{\omega} e^{-\kappa z} \left[\kappa \sin(kx - \omega t)\,\hat{\mathbf{x}} + k \cos(kx - \omega t)\,\hat{\mathbf{z}}\right]. \end{aligned}\right\} \tag{9.203}$$

(e) Check that the fields in (d) satisfy all of Maxwell's equations (9.67).

(f) For the fields in (d), construct the Poynting vector, and show that, on average, no energy is transmitted in the z direction.

! **Problem 9.38** Consider the **resonant cavity** produced by closing off the two ends of a rectangular wave guide, at $z = 0$ and at $z = d$, making a perfectly conducting empty box. Show that the resonant frequencies for both TE and TM modes are given by

$$\omega_{lmn} = c\pi\sqrt{(l/d)^2 + (m/a)^2 + (n/b)^2}, \tag{9.204}$$

for integers l, m, and n. Find the associated electric and magnetic fields.

Chapter 10

Potentials and Fields

10.1 The Potential Formulation

10.1.1 Scalar and Vector Potentials

In this chapter we ask how the sources (ρ and $\mathbf{J}$) generate electric and magnetic fields; in other words, we seek the *general* solution to Maxwell's equations,

$$\left.\begin{array}{llll}\text{(i)} & \nabla\cdot\mathbf{E}=\dfrac{1}{\epsilon_0}\rho, & \text{(iii)} & \nabla\times\mathbf{E}=-\dfrac{\partial\mathbf{B}}{\partial t}, \\ \text{(ii)} & \nabla\cdot\mathbf{B}=0, & \text{(iv)} & \nabla\times\mathbf{B}=\mu_0\mathbf{J}+\mu_0\epsilon_0\dfrac{\partial\mathbf{E}}{\partial t}.\end{array}\right\} \tag{10.1}$$

Given $\rho(\mathbf{r},t)$ and $\mathbf{J}(\mathbf{r},t)$, what are the fields $\mathbf{E}(\mathbf{r},t)$ and $\mathbf{B}(\mathbf{r},t)$? In the static case Coulomb's law and the Biot-Savart law provide the answer. What we're looking for, then, is the generalization of those laws to time-dependent configurations.

This is not an easy problem, and it pays to begin by representing the fields in terms of potentials. In electrostatics $\nabla\times\mathbf{E}=0$ allowed us to write $\mathbf{E}$ as the gradient of a scalar potential: $\mathbf{E}=-\nabla V$. In electro*dynamics* this is no longer possible, because the curl of $\mathbf{E}$ is nonzero. But $\mathbf{B}$ remains divergenceless, so we can still write

$$\boxed{\mathbf{B}=\nabla\times\mathbf{A},} \tag{10.2}$$

as in magnetostatics. Putting this into Faraday's law (iii) yields

$$\nabla\times\mathbf{E}=-\frac{\partial}{\partial t}(\nabla\times\mathbf{A}),$$

or

$$\nabla\times\left(\mathbf{E}+\frac{\partial\mathbf{A}}{\partial t}\right)=0.$$

Here is a quantity, unlike **E** alone, whose curl *does* vanish; it can therefore be written as the gradient of a scalar:

$$\mathbf{E}+\frac{\partial \mathbf{A}}{\partial t}=-\nabla V.$$

In terms of V and $\mathbf{A}$, then,

$$\boxed{\mathbf{E}=-\nabla V-\frac{\partial \mathbf{A}}{\partial t}.} \tag{10.3}$$

This reduces to the old form, of course, when **A** is constant.

The potential representation (Eqs. 10.2 and 10.3) automatically fulfills the two homogeneous Maxwell equations, (ii) and (iii). How about Gauss's law (i) and the Ampère/Maxwell law (iv)? Putting Eq. 10.3 into (i), we find that

$$\nabla^2 V+\frac{\partial}{\partial t}(\nabla\cdot\mathbf{A})=-\frac{1}{\epsilon_0}\rho; \tag{10.4}$$

this replaces Poisson's equation (to which it reduces in the static case). Putting Eqs. 10.2 and 10.3 into (iv) yields

$$\nabla\times(\nabla\times\mathbf{A})=\mu_0\mathbf{J}-\mu_0\epsilon_0\nabla\left(\frac{\partial V}{\partial t}\right)-\mu_0\epsilon_0\frac{\partial^2\mathbf{A}}{\partial t^2},$$

or, using the vector identity $\nabla\times(\nabla\times\mathbf{A})=\nabla(\nabla\cdot\mathbf{A})-\nabla^2\mathbf{A}$, and rearranging the terms a bit:

$$\left(\nabla^2\mathbf{A}-\mu_0\epsilon_0\frac{\partial^2\mathbf{A}}{\partial t^2}\right)-\nabla\left(\nabla\cdot\mathbf{A}+\mu_0\epsilon_0\frac{\partial V}{\partial t}\right)=-\mu_0\mathbf{J}. \tag{10.5}$$

Equations 10.4 and 10.5 contain all the information in Maxwell's equations.

Example 10.1

Find the charge and current distributions that would give rise to the potentials

$$V=0,\quad \mathbf{A}=\begin{cases}\dfrac{\mu_0 k}{4c}(ct-|x|)^2\,\hat{\mathbf{z}}, & \text{for } |x|<ct,\\[2ex] 0, & \text{for } |x|>ct,\end{cases}$$

where k is a constant, and $c=1/\sqrt{\epsilon_0\mu_0}$.

Solution: First we'll determine the electric and magnetic fields, using Eqs. 10.2 and 10.3:

$$\mathbf{E}=-\frac{\partial \mathbf{A}}{\partial t}=-\frac{\mu_0 k}{2}(ct-|x|)\,\hat{\mathbf{z}},$$

$$\mathbf{B}=\nabla\times\mathbf{A}=-\frac{\mu_0 k}{4c}\frac{\partial}{\partial x}(ct-|x|)^2\,\hat{\mathbf{y}}=\pm\frac{\mu_0 k}{2c}(ct-|x|)\,\hat{\mathbf{y}},$$

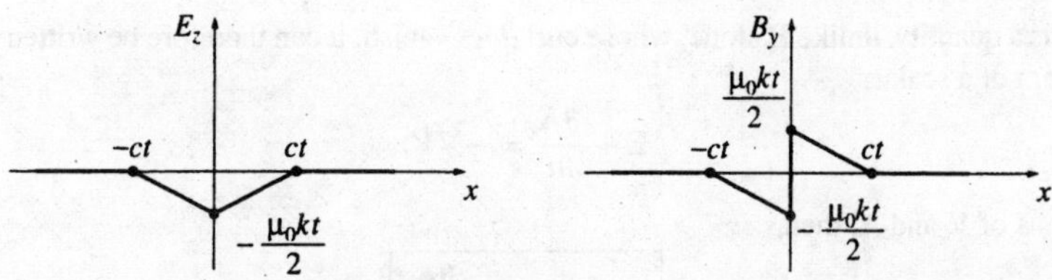

Figure 10.1

(plus for $x > 0$, minus for $x < 0$). These are for $|x| < ct$; when $|x| > ct$, $\mathbf{E} = \mathbf{B} = 0$ (Fig.10.1). Calculating every derivative in sight, I find

$$\nabla \cdot \mathbf{E} = 0; \quad \nabla \cdot \mathbf{B} = 0; \quad \nabla \times \mathbf{E} = \mp \frac{\mu_0 k}{2} \hat{\mathbf{y}}; \quad \nabla \times \mathbf{B} = -\frac{\mu_0 k}{2c} \hat{\mathbf{z}};$$

$$\frac{\partial \mathbf{E}}{\partial t} = -\frac{\mu_0 kc}{2} \hat{\mathbf{z}}; \quad \frac{\partial \mathbf{B}}{\partial t} = \pm \frac{\mu_0 k}{2} \hat{\mathbf{y}}.$$

As you can easily check, Maxwell's equations are all satisfied, with ρ and $\mathbf{J}$ both *zero*. Notice, however, that $\mathbf{B}$ has a discontinuity at $x = 0$, and this signals the presence of a surface current $\mathbf{K}$ in the yz plane; boundary condition (iv) in Eq. 7.63 gives

$$kt\,\hat{\mathbf{y}} = \mathbf{K} \times \hat{\mathbf{x}},$$

and hence

$$\mathbf{K} = kt\,\hat{\mathbf{z}}.$$

Evidently we have here a uniform surface current flowing in the z direction over the plane $x = 0$, which starts up at $t = 0$, and increases in proportion to t. Notice that the news travels out (in both directions) at the speed of light: for points $|x| > ct$ the message (that current is now flowing) has not yet arrived, so the fields are zero.

Problem 10.1 Show that the differential equations for V and $\mathbf{A}$ (Eqs. 10.4 and 10.5) can be written in the more symmetrical form

$$\left.\begin{aligned} \Box^2 V + \frac{\partial L}{\partial t} &= -\frac{1}{\epsilon_0}\rho, \\ \Box^2 \mathbf{A} - \nabla L &= -\mu_0 \mathbf{J}, \end{aligned}\right\} \tag{10.6}$$

where

$$\Box^2 \equiv \nabla^2 - \mu_0\epsilon_0 \frac{\partial^2}{\partial t^2} \quad \text{and} \quad L \equiv \nabla \cdot \mathbf{A} + \mu_0\epsilon_0 \frac{\partial V}{\partial t}.$$

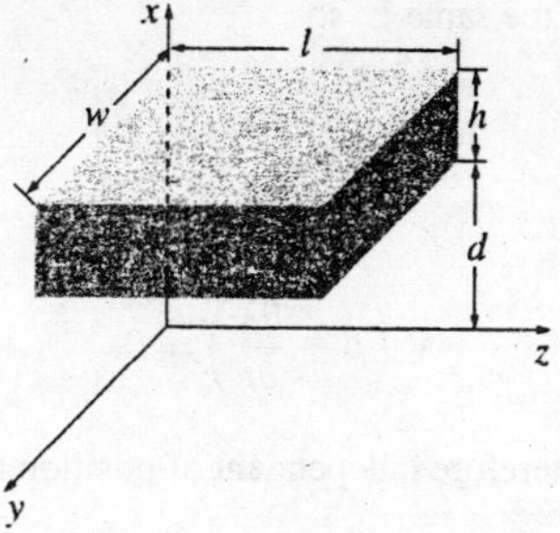

Figure 10.2

Problem 10.2 For the configuration in Ex. 10.1, consider a rectangular box of length l, width w, and height h, situated a distance d above the yz plane (Fig. 10.2).

(a) Find the energy in the box at time $t_1 = d/c$, and at $t_2 = (d+h)/c$.

(b) Find the Poynting vector, and determine the energy per unit time flowing into the box during the interval $t_1 < t < t_2$.

(c) Integrate the result in (b) from t_1 to t_2 and confirm that the increase in energy (part (a)) equals the net influx.

10.1.2 Gauge Transformations

Equations 10.4 and 10.5 are *ugly*, and you might be inclined at this stage to abandon the potential formulation altogether. However, we *have* succeeded in reducing six problems—finding **E** and **B** (three components each)—down to four: V (one component) and **A** (three more). Moreover, Eqs. 10.2 and 10.3 do not uniquely define the potentials; we are free to impose extra conditions on V and **A**, as long as nothing happens to **E** and **B**. Let's work out precisely what this **gauge freedom** entails. Suppose we have two sets of potentials, $(V, \mathbf{A})$ and $(V', \mathbf{A}')$, which correspond to the *same* electric and magnetic fields. By how much can they differ? Write

$$\mathbf{A}' = \mathbf{A} + \boldsymbol{\alpha} \quad \text{and} \quad V' = V + \beta.$$

Since the two **A**'s give the same **B**, their curls must be equal, and hence

$$\nabla \times \boldsymbol{\alpha} = 0.$$

We can therefore write $\boldsymbol{\alpha}$ as the gradient of some scalar:

$$\boldsymbol{\alpha} = \nabla\lambda.$$

The two potentials also give the same **E**, so

$$\nabla \beta + \frac{\partial \boldsymbol{\alpha}}{\partial t} = 0,$$

or

$$\nabla \left(\beta + \frac{\partial \lambda}{\partial t} \right) = 0.$$

The term in parentheses is therefore independent of position (it could, however, depend on time); call it $k(t)$:

$$\beta = -\frac{\partial \lambda}{\partial t} + k(t).$$

Actually, we might as well absorb $k(t)$ into λ, defining a new λ by adding $\int_0^t k(t')dt'$ to the old one. This will not affect the gradient of λ; it just adds $k(t)$ to $\partial\lambda/\partial t$. It follows that

$$\left.\begin{aligned} \mathbf{A}' &= \mathbf{A} + \nabla\lambda, \\ V' &= V - \frac{\partial \lambda}{\partial t}. \end{aligned}\right\} \tag{10.7}$$

Conclusion: For any old scalar function λ, we can with impunity add $\nabla\lambda$ to **A**, provided we simultaneously subtract $\partial\lambda/\partial t$ from V. None of this will affect the physical quantities **E** and **B**. Such changes in V and **A** are called **gauge transformations**. They can be exploited to adjust the divergence of **A**, with a view to simplifying the "ugly" equations 10.4 and 10.5. In magnetostatics, it was best to choose $\nabla \cdot \mathbf{A} = 0$ (Eq. 5.61); in electrodynamics the situation is not so clear cut, and the most convenient gauge depends to some extent on the problem at hand. There are many famous gauges in the literature; I'll show you the two most popular ones.

Problem 10.3 Find the fields, and the charge and current distributions, corresponding to

$$V(\mathbf{r}, t) = 0, \quad \mathbf{A}(\mathbf{r}, t) = -\frac{1}{4\pi\epsilon_0}\frac{qt}{r^2}\,\hat{\mathbf{r}}.$$

Problem 10.4 Suppose $V = 0$ and $\mathbf{A} = A_0 \sin(kx - \omega t)\,\hat{\mathbf{y}}$, where A_0, ω, and k are constants. Find **E** and **B**, and check that they satisfy Maxwell's equations in vacuum. What condition must you impose on ω and k?

Problem 10.5 Use the gauge function $\lambda = -(1/4\pi\epsilon_0)(qt/r)$ to transform the potentials in Prob. 10.3, and comment on the result.

10.1.3 Coulomb Gauge and Lorentz* Gauge

The Coulomb Gauge. As in magnetostatics, we pick

$$\nabla \cdot \mathbf{A} = 0. \tag{10.8}$$

With this, Eq. 10.4 becomes

$$\nabla^2 V = -\frac{1}{\epsilon_0}\rho. \tag{10.9}$$

This is Poisson's equation, and we already know how to solve it: setting $V = 0$ at infinity,

$$V(\mathbf{r}, t) = \frac{1}{4\pi\epsilon_0}\int \frac{\rho(\mathbf{r}', t)}{\mathscr{r}}\, d\tau'. \tag{10.10}$$

Don't be fooled, though—unlike electrostatics, V by itself doesn't tell you **E**; you have to know **A** as well (Eq. 10.3).

There is a peculiar thing about the scalar potential in the Coulomb gauge: it is determined by the distribution of charge *right now*. If I move an electron in my laboratory, the potential V on the moon immediately records this change. That sounds particularly odd in the light of special relativity, which allows no message to travel faster than the speed of light. The point is that V *by itself* is not a physically measurable quantity—all the man in the moon can measure is **E**, and that involves **A** as well. Somehow it is built into the vector potential, in the Coulomb gauge, that whereas V instantaneously reflects all changes in ρ, the combination $-\nabla V - (\partial \mathbf{A}/\partial t)$ does *not*; **E** will change only after sufficient time has elapsed for the "news" to arrive.[1]

The *advantage* of the Coulomb gauge is that the *scalar* potential is particularly simple to calculate; the *disadvantage* (apart from the acausal appearance of V) is that **A** is particularly *difficult* to calculate. The differential equation for **A** (10.5) in the Coulomb gauge reads

$$\nabla^2\mathbf{A} - \mu_0\epsilon_0\frac{\partial^2\mathbf{A}}{\partial t^2} = -\mu_0\mathbf{J} + \mu_0\epsilon_0\nabla\left(\frac{\partial V}{\partial t}\right). \tag{10.11}$$

The Lorentz gauge. In the Lorentz gauge we pick

$$\boxed{\nabla \cdot \mathbf{A} = -\mu_0\epsilon_0\frac{\partial V}{\partial t}.} \tag{10.12}$$

This is designed to eliminate the middle term in Eq. 10.5 (in the language of Prob. 10.1, it sets $L = 0$). With this

$$\nabla^2\mathbf{A} - \mu_0\epsilon_0\frac{\partial^2\mathbf{A}}{\partial t^2} = -\mu_0\mathbf{J}. \tag{10.13}$$

Meanwhile, the differential equation for V, (10.4), becomes

$$\nabla^2 V - \mu_0\epsilon_0\frac{\partial^2 V}{\partial t^2} = -\frac{1}{\epsilon_0}\rho. \tag{10.14}$$

---

*There is some question whether this should be attibuted to H. A. Lorentz or to L. V. Lorenz (see J. Van Bladel, *IEEE Antennas and Propagation Magazine* **33**(2), 69 (1991)). But all the standard textbooks include the t, and to avoid possible confusion I shall adhere to that practice.

[1] See O. L. Brill and B. Goodman. *Am. J. Phys.* **35**, 832 (1967).

The virtue of the Lorentz gauge is that it treats V and $\mathbf{A}$ on an equal footing: the same differential operator

$$\boxed{\nabla^2 - \mu_0\epsilon_0\frac{\partial^2}{\partial t^2} \equiv \Box^2,} \tag{10.15}$$

(called the **d'Alembertian**) occurs in both equations:

$$\boxed{\begin{aligned} &\text{(i)} \quad \Box^2 V = -\frac{1}{\epsilon_0}\rho, \\ &\text{(ii)} \quad \Box^2 \mathbf{A} = -\mu_0 \mathbf{J}. \end{aligned}} \tag{10.16}$$

This democratic treatment of V and $\mathbf{A}$ is particularly nice in the context of special relativity, where the d'Alembertian is the natural generalization of the Laplacian, and Eqs. 10.16 can be regarded as four-dimensional versions of Poisson's equation. (In this same spirit the wave equation, for propagation speed c, $\Box^2 f = 0$, might be regarded as the four-dimensional version of Laplace's equation.) In the Lorentz gauge V and $\mathbf{A}$ satisfy the **inhomogeneous wave equation**, with a "source" term (in place of zero) on the right. From now on I shall use the Lorentz gauge exclusively, and the whole of electrodynamics reduces to the problem of *solving the inhomogeneous wave equation for specified sources*. That's my project for the next section.

Problem 10.6 Which of the potentials in Ex. 10.1, Prob. 10.3, and Prob. 10.4 are in the Coulomb gauge? Which are in the Lorentz gauge? (Notice that these gauges are not mutually exclusive.)

Problem 10.7 In Chapter 5, I showed that it is always possible to pick a vector potential whose divergence is zero (Coulomb gauge). Show that it is always possible to choose $\nabla \cdot \mathbf{A} = -\mu_0\epsilon_0(\partial V/\partial t)$, as required for the Lorentz gauge, assuming you know how to solve equations of the form 10.16. Is it always possible to pick $V = 0$? How about $\mathbf{A} = 0$?

10.2 Continuous Distributions

10.2.1 Retarded Potentials

In the static case, Eqs. 10.16 reduce to (four copies of) Poisson's equation,

$$\nabla^2 V = -\frac{1}{\epsilon_0}\rho, \quad \nabla^2 \mathbf{A} = -\mu_0 \mathbf{J},$$

with the familiar solutions

$$V(\mathbf{r}) = \frac{1}{4\pi\epsilon_0}\int \frac{\rho(\mathbf{r}')}{\imath}\,d\tau', \quad \mathbf{A}(\mathbf{r}) = \frac{\mu_0}{4\pi}\int \frac{\mathbf{J}(\mathbf{r}')}{\imath}\,d\tau', \tag{10.17}$$

Figure 10.3

where $\mathscr{r}$, as always, is the distance from the source point $\mathbf{r}'$ to the field point $\mathbf{r}$ (Fig. 10.3). Now, electromagnetic "news" travels at the speed of light. In the *nonstatic* case, therefore, it's not the status of the source *right now* that matters, but rather its condition at some earlier time t_r (called the **retarded time**) when the "message" left. Since this message must travel a distance $\mathscr{r}$, the delay is $\mathscr{r}/c$:

$$\boxed{t_r \equiv t - \frac{\mathscr{r}}{c}.} \tag{10.18}$$

The natural generalization of Eq. 10.17 for nonstatic sources is therefore

$$\boxed{V(\mathbf{r}, t) = \frac{1}{4\pi\epsilon_0}\int \frac{\rho(\mathbf{r}', t_r)}{\mathscr{r}}\, d\tau', \quad \mathbf{A}(\mathbf{r}, t) = \frac{\mu_0}{4\pi}\int \frac{\mathbf{J}(\mathbf{r}', t_r)}{\mathscr{r}}\, d\tau'.} \tag{10.19}$$

Here $\rho(\mathbf{r}', t_r)$ is the charge density that prevailed at point $\mathbf{r}'$ at the retarded time t_r. Because the integrands are evaluated at the retarded time, these are called **retarded potentials**. (I speak of "the" retarded time, but of course the most distant parts of the charge distribution have earlier retarded times than nearby ones. It's just like the night sky: The light we see now left each star at the retarded time corresponding to that star's distance from the earth.) Note that the retarded potentials reduce properly to Eq. 10.17 in the static case, for which ρ and $\mathbf{J}$ are independent of time.

Well, that all sounds *reasonable*—and surprisingly simple. But are we sure it's *right*? I didn't actually *derive* these formulas for V and $\mathbf{A}$; all I did was invoke a heuristic argument ("electromagnetic news travels at the speed of light") to make them seem *plausible*. To *prove* them, I must show that they satisfy the inhomogeneous wave equation (10.16) and meet the Lorentz condition (10.12). In case you think I'm being fussy, let me warn you that if you apply the same argument to the *fields* you'll get entirely the *wrong* answer:

$$\mathbf{E}(\mathbf{r}, t) \neq \frac{1}{4\pi\epsilon_0}\int \frac{\rho(\mathbf{r}', t_r)}{\mathscr{r}^2}\hat{\boldsymbol{\mathscr{r}}}\, d\tau', \quad \mathbf{B}(\mathbf{r}, t) \neq \frac{\mu_0}{4\pi}\int \frac{\mathbf{J}(\mathbf{r}', t_r)\times\hat{\boldsymbol{\mathscr{r}}}}{\mathscr{r}^2}\, d\tau',$$

as you would expect if the same "logic" worked for Coulomb's law and the Biot-Savart law. Let's stop and check, then, that the retarded scalar potential satisfies Eq. 10.16; essentially the same argument would serve for the vector potential.[2] I shall leave it for you (Prob. 10.8) to check that the retarded potentials obey the Lorentz condition.

In calculating the Laplacian of $V(\mathbf{r}, t)$, the crucial point to notice is that the integrand (in Eq. 10.19) depends on $\mathbf{r}$ in *two* places: *explicitly*, in the denominator ($\imath = |\mathbf{r} - \mathbf{r}'|$), and *implicitly*, through $t_r = t - \imath/c$, in the numerator. Thus

$$\nabla V = \frac{1}{4\pi\epsilon_0}\int\left[(\nabla\rho)\frac{1}{\imath} + \rho\nabla\left(\frac{1}{\imath}\right)\right]d\tau', \tag{10.20}$$

and

$$\nabla\rho = \dot{\rho}\nabla t_r = -\frac{1}{c}\dot{\rho}\nabla\imath \tag{10.21}$$

(the dot denotes differentiation with respect to time).[3] Now $\nabla\imath = \hat{\boldsymbol{\imath}}$ and $\nabla(1/\imath) = -\hat{\boldsymbol{\imath}}/\imath^2$ (Prob. 1.13), so

$$\nabla V = \frac{1}{4\pi\epsilon_0}\int\left[-\frac{\dot{\rho}}{c}\frac{\hat{\boldsymbol{\imath}}}{\imath} - \rho\frac{\hat{\boldsymbol{\imath}}}{\imath^2}\right]d\tau'. \tag{10.22}$$

Taking the divergence,

$$\nabla^2 V = \frac{1}{4\pi\epsilon_0}\int\left\{-\frac{1}{c}\left[\frac{\hat{\boldsymbol{\imath}}}{\imath}\cdot(\nabla\dot{\rho}) + \dot{\rho}\nabla\cdot\left(\frac{\hat{\boldsymbol{\imath}}}{\imath}\right)\right] - \left[\frac{\hat{\boldsymbol{\imath}}}{\imath^2}\cdot(\nabla\rho) + \rho\nabla\cdot\left(\frac{\hat{\boldsymbol{\imath}}}{\imath^2}\right)\right]\right\}d\tau'.$$

But

$$\nabla\dot{\rho} = -\frac{1}{c}\ddot{\rho}\nabla\imath = -\frac{1}{c}\ddot{\rho}\hat{\boldsymbol{\imath}},$$

as in Eq. 10.21, and

$$\nabla\cdot\left(\frac{\hat{\boldsymbol{\imath}}}{\imath}\right) = \frac{1}{\imath^2}$$

(Prob. 1.62), whereas

$$\nabla\cdot\left(\frac{\hat{\boldsymbol{\imath}}}{\imath^2}\right) = 4\pi\delta^3(\boldsymbol{\imath})$$

(Eq. 1.100). So

$$\nabla^2 V = \frac{1}{4\pi\epsilon_0}\int\left[\frac{1}{c^2}\frac{\ddot{\rho}}{\imath} - 4\pi\rho\delta^3(\boldsymbol{\imath})\right]d\tau' = \frac{1}{c^2}\frac{\partial^2 V}{\partial t^2} - \frac{1}{\epsilon_0}\rho(\mathbf{r}, t),$$

confirming that the retarded potential (10.19) satisfies the inhomogeneous wave equation (10.16). qed

---

[2] I'll give you the straightforward but cumbersome proof; for a clever indirect argument see M. A. Heald and J. B. Marion, *Classical Electromagnetic Radiation*, 3d ed., Sect. 8.1 (Orlando, FL: Saunders (1995)).

[3] Note that $\partial/\partial t_r = \partial/\partial t$, since $t_r = t - \imath/c$ and $\imath$ is independent of t.

Incidentally, this proof applies equally well to the **advanced potentials,**

$$V_a(\mathbf{r}, t) = \frac{1}{4\pi\epsilon_0}\int \frac{\rho(\mathbf{r}', t_a)}{\eta}\, d\tau', \quad \mathbf{A}_a(\mathbf{r}, t) = \frac{\mu_0}{4\pi}\int \frac{\mathbf{J}(\mathbf{r}', t_a)}{\eta}\, d\tau', \tag{10.23}$$

in which the charge and the current densities are evaluated at the **advanced time**

$$t_a \equiv t + \frac{\eta}{c}. \tag{10.24}$$

A few signs are changed, but the final result is unaffected. Although the advanced potentials are entirely consistent with Maxwell's equations, they violate the most sacred tenet in all of physics: the principle of **causality**. They suggest that the potentials *now* depend on what the charge and the current distribution *will* be at some time in the future—the effect, in other words, precedes the cause. Although the advanced potentials are of some theoretical interest, they have no direct physical significance.[4]

Example 10.2

An infinite straight wire carries the current

$$I(t) = \begin{cases} 0, & \text{for } t \le 0, \\ I_0, & \text{for } t > 0. \end{cases}$$

That is, a constant current I_0 is turned on abruptly at $t = 0$. Find the resulting electric and magnetic fields.

Solution: The wire is presumably electrically neutral, so the scalar potential is zero. Let the wire lie along the z axis (Fig. 10.4); the retarded vector potential at point P is

$$\mathbf{A}(s, t) = \frac{\mu_0}{4\pi}\,\hat{\mathbf{z}}\int_{-\infty}^{\infty} \frac{I(t_r)}{\eta}\, dz.$$

For $t < s/c$, the "news" has not yet reached P, and the potential is zero. For $t > s/c$, only the segment

$$|z| \le \sqrt{(ct)^2 - s^2} \tag{10.25}$$

contributes (outside this range t_r is negative, so $I(t_r) = 0$); thus

$$\begin{aligned} \mathbf{A}(s, t) &= \left(\frac{\mu_0 I_0}{4\pi}\,\hat{\mathbf{z}}\right) 2\int_0^{\sqrt{(ct)^2 - s^2}} \frac{dz}{\sqrt{s^2 + z^2}} \\ &= \frac{\mu_0 I_0}{2\pi}\,\hat{\mathbf{z}}\ln\left(\sqrt{s^2 + z^2} + z\right)\Big|_0^{\sqrt{(ct)^2 - s^2}} = \frac{\mu_0 I_0}{2\pi}\ln\left(\frac{ct + \sqrt{(ct)^2 - s^2}}{s}\right)\hat{\mathbf{z}}. \end{aligned}$$

---

[4] Because the d'Alembertian involves t^2 (as opposed to t), the theory itself is **time-reversal invariant,** and does not distinguish "past" from "future." Time asymmetry is introduced when we select the retarded potentials in preference to the advanced ones, reflecting the (not unreasonable!) belief that electromagnetic influences propagate forward, not backward, in time.

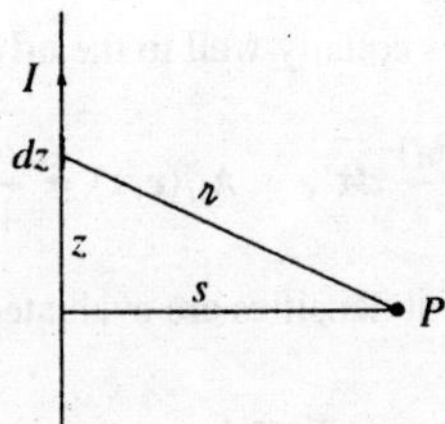

Figure 10.4

The electric field is

$$\mathbf{E}(s,t) = -\frac{\partial \mathbf{A}}{\partial t} = -\frac{\mu_0 I_0 c}{2\pi\sqrt{(ct)^2 - s^2}}\,\hat{\mathbf{z}},$$

and the magnetic field is

$$\mathbf{B}(s,t) = \nabla \times \mathbf{A} = -\frac{\partial A_z}{\partial s}\,\hat{\boldsymbol{\phi}} = \frac{\mu_0 I_0}{2\pi s}\frac{ct}{\sqrt{(ct)^2 - s^2}}\,\hat{\boldsymbol{\phi}}.$$

Notice that as $t \to \infty$ we recover the static case: $\mathbf{E} = 0$, $\mathbf{B} = (\mu_0 I_0/2\pi s)\,\hat{\boldsymbol{\phi}}$.

! **Problem 10.8** Confirm that the retarded potentials satisfy the Lorentz gauge condition. [*Hint:* First show that

$$\nabla \cdot \left(\frac{\mathbf{J}}{\mathscr{r}}\right) = \frac{1}{\mathscr{r}}(\nabla \cdot \mathbf{J}) + \frac{1}{\mathscr{r}}(\nabla' \cdot \mathbf{J}) - \nabla' \cdot \left(\frac{\mathbf{J}}{\mathscr{r}}\right),$$

where ∇ denotes derivatives with respect to $\mathbf{r}$, and ∇' denotes derivatives with respect to $\mathbf{r}'$. Next, noting that $\mathbf{J}(\mathbf{r}', t - \mathscr{r}/c)$ depends on $\mathbf{r}'$ both explicitly and through $\mathscr{r}$, whereas it depends on $\mathbf{r}$ only through $\mathscr{r}$, confirm that

$$\nabla \cdot \mathbf{J} = -\frac{1}{c}\dot{\mathbf{J}} \cdot (\nabla \mathscr{r}), \quad \nabla' \cdot \mathbf{J} = -\dot{\rho} - \frac{1}{c}\dot{\mathbf{J}} \cdot (\nabla' \mathscr{r}).$$

Use this to calculate the divergence of $\mathbf{A}$ (Eq. 10.19).]

! **Problem 10.9**

(a) Suppose the wire in Ex. 10.2 carries a linearly increasing current

$$I(t) = kt,$$

for $t > 0$. Find the electric and magnetic fields generated.

(b) Do the same for the case of a sudden burst of current:

$$I(t) = q_0\delta(t).$$

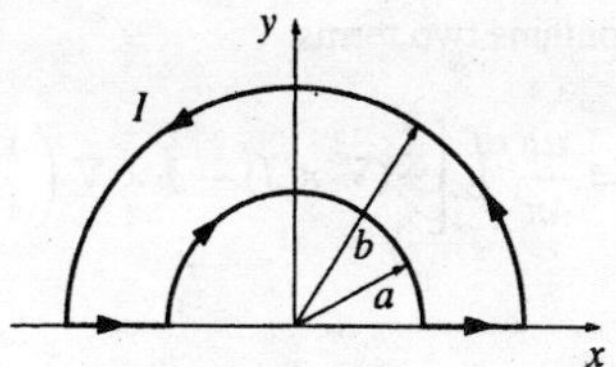

Figure 10.5

Problem 10.10 A piece of wire bent into a loop, as shown in Fig. 10.5, carries a current that increases linearly with time:

$$I(t) = kt.$$

Calculate the retarded vector potential **A** at the center. Find the electric field at the center. Why does this (neutral) wire produce an *electric* field? (Why can't you determine the *magnetic* field from this expression for **A**?)

10.2.2 Jefimenko's Equations

Given the retarded potentials

$$V(\mathbf{r}, t) = \frac{1}{4\pi\epsilon_0}\int \frac{\rho(\mathbf{r}', t_r)}{\imath}\, d\tau', \quad \mathbf{A}(\mathbf{r}, t) = \frac{\mu_0}{4\pi}\int \frac{\mathbf{J}(\mathbf{r}', t_r)}{\imath}\, d\tau', \tag{10.26}$$

it is, in principle, a straightforward matter to determine the fields:

$$\mathbf{E} = -\nabla V - \frac{\partial \mathbf{A}}{\partial t}, \qquad \mathbf{B} = \nabla \times \mathbf{A}. \tag{10.27}$$

But the details are not entirely trivial because, as I mentioned earlier, the integrands depend on **r** both explicitly, through $\imath = |\mathbf{r} - \mathbf{r}'|$ in the denominator, and implicitly, through the retarded time $t_r = t - \imath/c$ in the argument of the numerator.

I already calculated the gradient of V (Eq. 10.22); the time derivative of **A** is easy:

$$\frac{\partial \mathbf{A}}{\partial t} = \frac{\mu_0}{4\pi}\int \frac{\dot{\mathbf{J}}}{\imath}\, d\tau'. \tag{10.28}$$

Putting them together (and using $c^2 = 1/\mu_0\epsilon_0$):

$$\boxed{\mathbf{E}(\mathbf{r}, t) = \frac{1}{4\pi\epsilon_0}\int \left[\frac{\rho(\mathbf{r}', t_r)}{\imath^2}\hat{\boldsymbol{\imath}} + \frac{\dot{\rho}(\mathbf{r}', t_r)}{c\imath}\hat{\boldsymbol{\imath}} - \frac{\dot{\mathbf{J}}(\mathbf{r}', t_r)}{c^2\imath}\right] d\tau'.} \tag{10.29}$$

This is the time-dependent generalization of Coulomb's law, to which it reduces in the static case (where the second and third terms drop out and the first term loses its dependence on t_r).

As for **B**, the curl of **A** contains two terms:

$$\nabla \times \mathbf{A} = \frac{\mu_0}{4\pi} \int \left[\frac{1}{\imath} (\nabla \times \mathbf{J}) - \mathbf{J} \times \nabla \left(\frac{1}{\imath} \right) \right] d\tau'.$$

Now

$$(\nabla \times \mathbf{J})_x = \frac{\partial J_z}{\partial y} - \frac{\partial J_y}{\partial z},$$

and

$$\frac{\partial J_z}{\partial y} = \dot{J}_z \frac{\partial t_r}{\partial y} = -\frac{1}{c} \dot{J}_z \frac{\partial \imath}{\partial y},$$

so

$$(\nabla \times \mathbf{J})_x = -\frac{1}{c} \left(\dot{J}_z \frac{\partial \imath}{\partial y} - \dot{J}_y \frac{\partial \imath}{\partial z} \right) = \frac{1}{c} \left[\dot{\mathbf{J}} \times (\nabla \imath) \right]_x .$$

But $\nabla \imath = \hat{\imath}$ (Prob. 1.13), so

$$\nabla \times \mathbf{J} = \frac{1}{c} \dot{\mathbf{J}} \times \hat{\imath}. \tag{10.30}$$

Meanwhile $\nabla(1/\imath) = -\hat{\imath}/\imath^2$ (again, Prob. 1.13), and hence

$$\boxed{\mathbf{B}(\mathbf{r}, t) = \frac{\mu_0}{4\pi} \int \left[\frac{\mathbf{J}(\mathbf{r}', t_r)}{\imath^2} + \frac{\dot{\mathbf{J}}(\mathbf{r}', t_r)}{c\imath} \right] \times \hat{\imath}\, d\tau'.} \tag{10.31}$$

This is the time-dependent generalization of the Biot-Savart law, to which it reduces in the static case.

Equations 10.29 and 10.31 are the (causal) solutions to Maxwell's equations. For some reason they do not seem to have been published until quite recently—the earliest explicit statement of which I am aware was by Oleg Jefimenko, in 1966.[5] In practice **Jefimenko's equations** are of limited utility, since it is typically easier to calculate the retarded potentials and differentiate them, rather than going directly to the fields. Nevertheless, they provide a satisfying sense of closure to the theory. They also help to clarify an observation I made in the previous section: To get to the retarded *potentials*, all you do is replace t by t_r in the electrostatic and magnetostatic formulas, but in the case of the *fields* not only is time replaced by retarded time, but completely new terms (involving derivatives of ρ and **J**) appear. And they provide surprisingly strong support for the quasistatic approximation (see Prob. 10.12).

---

[5] O. D. Jefimenko, *Electricity and Magnetism*, Sect. 15.7 (New York: Appleton-Century-Crofts, 1996). Closely related expressions appear in W. K. H. Panofsky and M. Phillips, *Classical Electricity and Magnetism*, Sect. 14.3 (Reading, MA: Addison-Wesley, 1962). See K. T. McDonald, *Am. J. Phys.* **65**, 1074 (1997) for illuminating commentary and references.

Problem 10.11 Suppose $\mathbf{J}(\mathbf{r})$ is constant in time, so (Prob. 7.55) $\rho(\mathbf{r}, t) = \rho(\mathbf{r}, 0) + \dot{\rho}(\mathbf{r}, 0)t$. Show that

$$\mathbf{E}(\mathbf{r}, t) = \frac{1}{4\pi\epsilon_0}\int \frac{\rho(\mathbf{r}', t)}{\mathscr{r}^2}\hat{\mathscr{r}}\,d\tau';$$

that is, Coulomb's law holds, with the charge density evaluated at the *non-retarded* time.

Problem 10.12 Suppose the current density changes slowly enough that we can (to good approximation) ignore all higher derivatives in the Taylor expansion

$$\mathbf{J}(t_r) = \mathbf{J}(t) + (t_r - t)\dot{\mathbf{J}}(t) + \ldots$$

(for clarity, I suppress the $\mathbf{r}$-dependence, which is not at issue). Show that a fortuitous cancellation in Eq. 10.31 yields

$$\mathbf{B}(\mathbf{r}, t) = \frac{\mu_0}{4\pi}\int \frac{\mathbf{J}(\mathbf{r}', t)\times\hat{\mathscr{r}}}{\mathscr{r}^2}\,d\tau'.$$

That is: the Biot-Savart law holds, with $\mathbf{J}$ evaluated at the *non-retarded* time. This means that the quasistatic approximation is actually much *better* than we had any right to expect: the *two* errors involved (neglecting retardation and dropping the second term in Eq. 10.31) *cancel* one another, to first order.

10.3 Point Charges

10.3.1 Liénard-Wiechert Potentials

My next project is to calculate the (retarded) potentials, $V(\mathbf{r}, t)$ and $\mathbf{A}(\mathbf{r}, t)$, of a point charge q that is moving on a specified trajectory

$$\mathbf{w}(t) \equiv \text{position of } q \text{ at time } t. \tag{10.32}$$

The retarded time is determined implicitly by the equation

$$|\mathbf{r} - \mathbf{w}(t_r)| = c(t - t_r), \tag{10.33}$$

for the left side is the distance the "news" must travel, and $(t - t_r)$ is the time it takes to make the trip (Fig. 10.6). I shall call $\mathbf{w}(t_r)$ the **retarded position** of the charge; $\boldsymbol{\mathscr{r}}$ is the vector from the retarded position to the field point $\mathbf{r}$:

$$\boldsymbol{\mathscr{r}} = \mathbf{r} - \mathbf{w}(t_r). \tag{10.34}$$

It is important to note that at most *one* point on the trajectory is "in communication" with $\mathbf{r}$ at any particular time t. For suppose there were *two* such points, with retarded times t_1 and t_2:

$$\mathscr{r}_1 = c(t - t_1) \quad\text{and}\quad \mathscr{r}_2 = c(t - t_2).$$

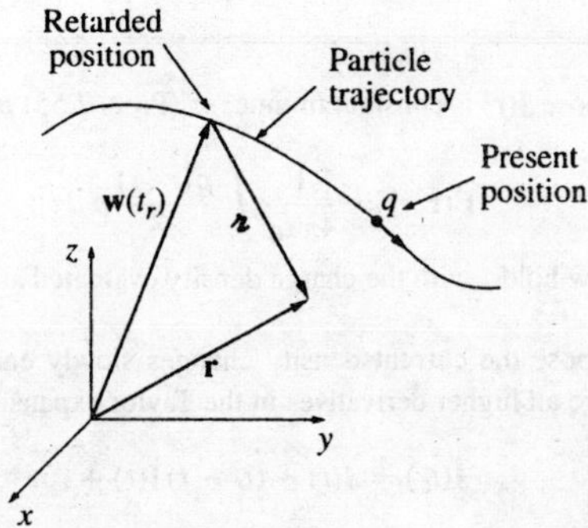

Figure 10.6

Then $\mathscr{r}_1 - \mathscr{r}_2 = c(t_2 - t_1)$, so the average velocity of the particle in the direction of **r** would have to be c—and that's not counting whatever velocity the charge might have in *other* directions. Since no charged particle can travel at the speed of light, it follows that only *one retarded point contributes to the potentials, at any given moment.*[6]

Now, a naïve reading of the formula

$$V(\mathbf{r}, t) = \frac{1}{4\pi\epsilon_0}\int \frac{\rho(\mathbf{r}', t_r)}{\mathscr{r}}\, d\tau' \tag{10.35}$$

might suggest to you that the retarded potential of a point charge is simply

$$\frac{1}{4\pi\epsilon_0}\frac{q}{\mathscr{r}}$$

(the same as in the static case, only with the understanding that $\mathscr{r}$ is the distance to the *retarded* position of the charge). But this is wrong, for a very subtle reason: It is true that for a point source the denominator $\mathscr{r}$ comes outside the integral,[7] but what remains,

$$\int \rho(\mathbf{r}', t_r)\, d\tau', \tag{10.36}$$

is *not* equal to the charge of the particle. To calculate the total charge of a configuration you must integrate ρ over the entire distribution at *one instant of time*, but here the retardation, $t_r = t - \mathscr{r}/c$, obliges us to evaluate ρ at *different times* for different parts of the configuration. If the source is moving, this will give a distorted picture of the total charge. You might

[6]For the same reason, an observer at **r** *sees* the particle in only one place at a time. By contrast, it is possible to *hear* an object in two places at once. Consider a bear who growls at you and then runs toward you at the speed of sound and growls again; you hear both growls at the same time, coming from two different locations, but there's only one bear.

[7]There is, however, an implicit change in its functional dependence: *Before* the integration, $\mathscr{r} = |\mathbf{r} - \mathbf{r}'|$ is a function of **r** and $\mathbf{r}'$; *after* the integration (which fixes $\mathbf{r}' = \mathbf{w}(t_r)$) $\mathscr{r} = |\mathbf{r} - \mathbf{w}(t_r)|$ is (like t_r) a function of **r** and t.

think that this problem would disappear for *point* charges, but it doesn't. In Maxwell's electrodynamics, formulated as it is in terms of charge and current *densities*, a point charge must be regarded as the limit of an extended charge, when the size goes to zero. And for an extended particle, no matter how small, the retardation in Eq. 10.36 throws in a factor $(1 - \hat{\boldsymbol{\imath}} \cdot \mathbf{v}/c)^{-1}$, where $\mathbf{v}$ is the velocity of the charge at the retarded time:

$$\int \rho(\mathbf{r}', t_r)\, d\tau' = \frac{q}{1 - \hat{\boldsymbol{\imath}} \cdot \mathbf{v}/c}. \tag{10.37}$$

Proof: This is a purely *geometrical* effect, and it may help to tell the story in a less abstract context. You will not have noticed it, for obvious reasons, but the fact is that a train coming towards you looks a little longer than it really is, because the light you receive from the caboose left earlier than the light you receive simultaneously from the engine, and at that earlier time the train was farther away (Fig. 10.7). In the interval it takes light from the caboose to travel the extra distance L', the train itself moves a distance $L' - L$:

$$\frac{L'}{c} = \frac{L' - L}{v}, \quad \text{or} \quad L' = \frac{L}{1 - v/c}.$$

So approaching trains appear *longer*, by a factor $(1 - v/c)^{-1}$. By contrast, a train going *away* from you looks *shorter*,[8] by a factor $(1 + v/c)^{-1}$. In general, if the train's velocity makes an angle θ with your line of sight,[9] the extra distance light from the caboose must cover is $L' \cos\theta$ (Fig. 10.8). In the time $L' \cos\theta / c$, then, the train moves a distance $(L' - L)$:

$$\frac{L' \cos\theta}{c} = \frac{L' - L}{v}, \quad \text{or} \quad L' = \frac{L}{1 - v\cos\theta/c}.$$

Figure 10.7

[8]Please note that this has nothing whatever to do with special relativity or Lorentz contraction—L is the length of the *moving* train, and its *rest* length is not at issue. The argument *is* somewhat reminiscent of the Doppler effect.

[9]I assume the train is far enough away or (more to the point) *short* enough so that rays from the caboose and engine can be considered parallel.

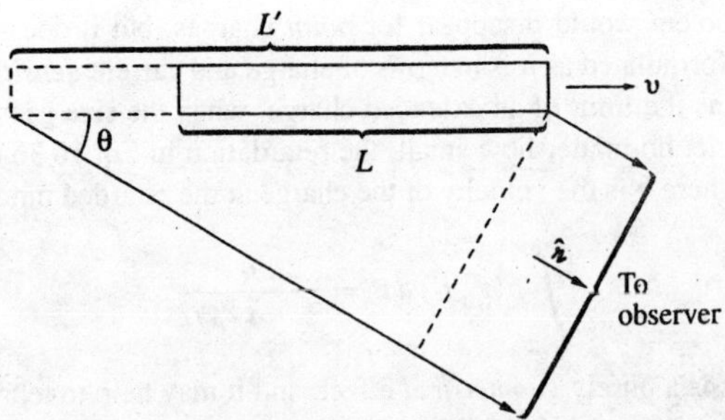

Figure 10.8

Notice that this effect does *not* distort the dimensions perpendicular to the motion (the height and width of the train). Never mind that the light from the far side is delayed in reaching you (relative to light from the near side)—since there's no *motion* in that direction, they'll still look the same distance apart. The apparent *volume* τ' of the train, then, is related to the *actual* volume τ by

$$\tau' = \frac{\tau}{1 - \hat{\boldsymbol{\imath}} \cdot \mathbf{v}/c}, \tag{10.38}$$

where $\hat{\boldsymbol{\imath}}$ is a unit vector from the train to the observer.

In case the connection between moving trains and retarded potentials escapes you, the point is this: Whenever you do an integral of the type 10.37, in which the integrand is evaluated at the retarded time, the effective volume is modified by the factor in Eq. 10.38, just as the apparent volume of the train was—and for the same reason. Because this correction factor makes no reference to the size of the particle, it is every bit as significant for a point charge as for an extended charge. qed

It follows, then, that

$$\boxed{V(\mathbf{r}, t) = \frac{1}{4\pi\epsilon_0} \frac{qc}{(\imath c - \boldsymbol{\imath} \cdot \mathbf{v})},} \tag{10.39}$$

where $\mathbf{v}$ is the velocity of the charge at the retarded time, and $\boldsymbol{\imath}$ is the vector from the retarded position to the field point $\mathbf{r}$. Meanwhile, since the current density of a rigid object is $\rho\mathbf{v}$ (Eq. 5.26), we also have

$$\mathbf{A}(\mathbf{r}, t) = \frac{\mu_0}{4\pi} \int \frac{\rho(\mathbf{r}', t_r)\mathbf{v}(t_r)}{\imath} \, d\tau' = \frac{\mu_0}{4\pi} \frac{\mathbf{v}}{\imath} \int \rho(\mathbf{r}', t_r) \, d\tau',$$

or

$$\boxed{\mathbf{A}(\mathbf{r},t) = \frac{\mu_0}{4\pi}\frac{qc\mathbf{v}}{(\mathscr{r}c - \boldsymbol{\mathscr{r}}\cdot\mathbf{v})} = \frac{\mathbf{v}}{c^2}V(\mathbf{r},t).} \tag{10.40}$$

Equations 10.39 and 10.40 are the famous **Liénard-Wiechert potentials** for a moving point charge.[10]

Example 10.3

Find the potentials of a point charge moving with constant velocity.

Solution: For convenience, let's say the particle passes through the origin at time $t = 0$, so that

$$\mathbf{w}(t) = \mathbf{v}t.$$

We first compute the retarded time, using Eq. 10.33:

$$|\mathbf{r} - \mathbf{v}t_r| = c(t - t_r),$$

or, squaring:

$$r^2 - 2\mathbf{r}\cdot\mathbf{v}t_r + v^2t_r^2 = c^2(t^2 - 2tt_r + t_r^2).$$

Solving for t_r by the quadratic formula, I find that

$$t_r = \frac{(c^2t - \mathbf{r}\cdot\mathbf{v}) \pm \sqrt{(c^2t - \mathbf{r}\cdot\mathbf{v})^2 + (c^2 - v^2)(r^2 - c^2t^2)}}{c^2 - v^2}. \tag{10.41}$$

To fix the sign, consider the limit $v = 0$:

$$t_r = t \pm \frac{r}{c}.$$

In this case the charge is at rest at the origin, and the retarded time should be $(t - r/c)$; evidently we want the *minus* sign.

Now, from Eqs. 10.33 and 10.34,

$$\mathscr{r} = c(t - t_r), \quad \text{and} \quad \hat{\boldsymbol{\mathscr{r}}} = \frac{\mathbf{r} - \mathbf{v}t_r}{c(t - t_r)},$$

so

$$\begin{aligned}
\mathscr{r}(1 - \hat{\boldsymbol{\mathscr{r}}}\cdot\mathbf{v}/c) &= c(t - t_r)\left[1 - \frac{\mathbf{v}}{c}\cdot\frac{(\mathbf{r} - \mathbf{v}t_r)}{c(t - t_r)}\right] = c(t - t_r) - \frac{\mathbf{v}\cdot\mathbf{r}}{c} - \frac{v^2}{c}t_r \\
&= \frac{1}{c}[(c^2t - \mathbf{r}\cdot\mathbf{v}) - (c^2 - v^2)t_r] \\
&= \frac{1}{c}\sqrt{(c^2t - \mathbf{r}\cdot\mathbf{v})^2 + (c^2 - v^2)(r^2 - c^2t^2)}
\end{aligned}$$

---

[10] There are many ways to obtain the Liénard-Wiechert potentials. I have tried to emphasize the *geometrical* origin of the factor $(1 - \hat{\boldsymbol{\mathscr{r}}}\cdot\mathbf{v}/c)^{-1}$; for illuminating commentary see W. K. H. Panofsky and M. Phillips, *Classical Electricity and Magnetism*, 2d ed., pp. 342-3 (Reading, MA: Addison-Wesley, 1962). A more rigorous derivation is provided by J. R. Reitz, F. J. Milford, and R. W. Christy, *Foundations of Electromagnetic Theory*, 3d ed., Sect. 21.1 (Reading, MA: Addison-Wesley, 1979), or M. A. Heald and J. B. Marion, *Classical Electromagnetic Radiation*, 3d ed., Sect. 8.3 (Orlando, FL: Saunders, 1995).

(I used Eq. 10.41, with the minus sign, in the last step). Therefore,

$$V(\mathbf{r}, t) = \frac{1}{4\pi\epsilon_0} \frac{qc}{\sqrt{(c^2 t - \mathbf{r}\cdot\mathbf{v})^2 + (c^2 - v^2)(r^2 - c^2 t^2)}}, \tag{10.42}$$

and (Eq. 10.40)

$$\mathbf{A}(\mathbf{r}, t) = \frac{\mu_0}{4\pi} \frac{qc\mathbf{v}}{\sqrt{(c^2 t - \mathbf{r}\cdot\mathbf{v})^2 + (c^2 - v^2)(r^2 - c^2 t^2)}}. \tag{10.43}$$

Problem 10.13 A particle of charge q moves in a circle of radius a at constant angular velocity ω. (Assume that the circle lies in the xy plane, centered at the origin, and at time $t = 0$ the charge is at $(a, 0)$, on the positive x axis.) Find the Liénard-Wiechert potentials for points on the z axis.

Problem 10.14 Show that the scalar potential of a point charge moving with constant velocity (Eq. 10.42) can be written equivalently as

$$V(\mathbf{r}, t) = \frac{1}{4\pi\epsilon_0} \frac{q}{R\sqrt{1 - v^2 \sin^2\theta/c^2}}, \tag{10.44}$$

where $\mathbf{R} \equiv \mathbf{r} - \mathbf{v}t$ is the vector from the *present (!)* position of the particle to the field point $\mathbf{r}$, and θ is the angle between $\mathbf{R}$ and $\mathbf{v}$ (Fig. 10.9). Evidently for nonrelativistic velocities ($v^2 \ll c^2$),

$$V(\mathbf{r}, t) \cong \frac{1}{4\pi\epsilon_0} \frac{q}{R}.$$

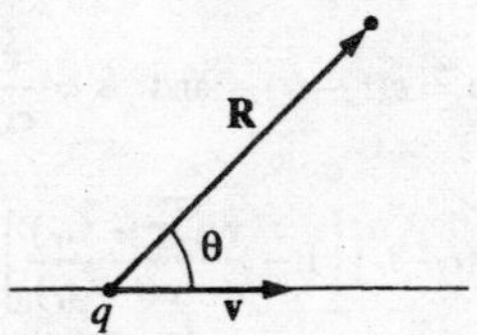

Figure 10.9

Problem 10.15 I showed that *at most one* point on the particle trajectory communicates with $\mathbf{r}$ at any given time. In some cases there may be *no* such point (an observer at $\mathbf{r}$ would not see the particle—in the colorful language of General Relativity it is "beyond the **horizon**"). As an example, consider a particle in **hyperbolic motion** along the x axis:

$$\mathbf{w}(t) = \sqrt{b^2 + (ct)^2}\,\hat{\mathbf{x}} \qquad (-\infty < t < \infty). \tag{10.45}$$

(In Special Relativity this is the trajectory of a particle subject to a constant force $F = mc^2/b$.) Sketch the graph of w versus t. At four or five representative points on the curve, draw the trajectory of a light signal emitted by the particle at that point—both in the plus x direction and in the minus x direction. What region on your graph corresponds to points and times (x, t) from which the particle cannot be seen? At what time does someone at point x first see the particle? (Prior to this the potential at x is evidently zero.) Is it possible for a particle, once seen, to *disappear* from view?

! **Problem 10.16** Determine the Liénard-Wiechert potentials for a charge in hyperbolic motion (Eq. 10.45). Assume the point $\mathbf{r}$ is on the x axis and to the right of the charge.

10.3.2 The Fields of a Moving Point Charge

We are now in a position to calculate the electric and magnetic fields of a point charge in arbitrary motion, using the Liénard-Wiechert potentials:[11]

$$V(\mathbf{r}, t) = \frac{1}{4\pi\epsilon_0} \frac{qc}{(\mathscr{r}c - \boldsymbol{\mathscr{r}} \cdot \mathbf{v})}, \qquad \mathbf{A}(\mathbf{r}, t) = \frac{\mathbf{v}}{c^2} V(\mathbf{r}, t), \tag{10.46}$$

and the equations for $\mathbf{E}$ and $\mathbf{B}$:

$$\mathbf{E} = -\nabla V - \frac{\partial \mathbf{A}}{\partial t}, \qquad \mathbf{B} = \nabla \times \mathbf{A}.$$

The differentiation is tricky, however, because

$$\boldsymbol{\mathscr{r}} = \mathbf{r} - \mathbf{w}(t_r) \quad \text{and} \quad \mathbf{v} = \dot{\mathbf{w}}(t_r) \tag{10.47}$$

are both evaluated at the retarded time, and t_r—defined implicitly by the equation

$$|\mathbf{r} - \mathbf{w}(t_r)| = c(t - t_r) \tag{10.48}$$

—is *itself* a function of $\mathbf{r}$ and t.[12] So hang on: the next two pages are rough going ... but the answer is worth the effort.

Let's begin with the gradient of V:

$$\nabla V = \frac{qc}{4\pi\epsilon_0} \frac{-1}{(\mathscr{r}c - \boldsymbol{\mathscr{r}} \cdot \mathbf{v})^2} \nabla(\mathscr{r}c - \boldsymbol{\mathscr{r}} \cdot \mathbf{v}). \tag{10.49}$$

---

[11] You can get the fields directly from Jefimenko's equations, but it's not easy. See, for example, M. A. Heald and J. B. Marion, *Classical Electromagnetic Radiation*, 3d ed., Sect. 8.4 (Orlando, FL: Saunders, 1995).

[12] The following calculation is done by the most direct, "brute force" method. For a more clever and efficient approach see J. D. Jackson, *Classical Electrodynamics*, 3d ed., Sect. 14.1 (New York: John Wiley, 1999).

Since $\mathscr{r} = c(t - t_r)$,

$$\nabla \mathscr{r} = -c\nabla t_r. \tag{10.50}$$

As for the second term, product rule 4 gives

$$\nabla(\boldsymbol{\mathscr{r}} \cdot \mathbf{v}) = (\boldsymbol{\mathscr{r}} \cdot \nabla)\mathbf{v} + (\mathbf{v} \cdot \nabla)\boldsymbol{\mathscr{r}} + \boldsymbol{\mathscr{r}} \times (\nabla \times \mathbf{v}) + \mathbf{v} \times (\nabla \times \boldsymbol{\mathscr{r}}). \tag{10.51}$$

Evaluating these terms one at a time:

$$\begin{aligned}(\boldsymbol{\mathscr{r}} \cdot \nabla)\mathbf{v} &= \left(\mathscr{r}_x \frac{\partial}{\partial x} + \mathscr{r}_y \frac{\partial}{\partial y} + \mathscr{r}_z \frac{\partial}{\partial z}\right)\mathbf{v}(t_r) \\ &= \mathscr{r}_x \frac{d\mathbf{v}}{dt_r}\frac{\partial t_r}{\partial x} + \mathscr{r}_y \frac{d\mathbf{v}}{dt_r}\frac{\partial t_r}{\partial y} + \mathscr{r}_z \frac{d\mathbf{v}}{dt_r}\frac{\partial t_r}{\partial z} \\ &= \mathbf{a}(\boldsymbol{\mathscr{r}} \cdot \nabla t_r),\end{aligned} \tag{10.52}$$

where $\mathbf{a} \equiv \dot{\mathbf{v}}$ is the *acceleration* of the particle at the retarded time. Now

$$(\mathbf{v} \cdot \nabla)\boldsymbol{\mathscr{r}} = (\mathbf{v} \cdot \nabla)\mathbf{r} - (\mathbf{v} \cdot \nabla)\mathbf{w}, \tag{10.53}$$

and

$$\begin{aligned}(\mathbf{v} \cdot \nabla)\mathbf{r} &= \left(v_x \frac{\partial}{\partial x} + v_y \frac{\partial}{\partial y} + v_z \frac{\partial}{\partial z}\right)(x\,\hat{\mathbf{x}} + y\,\hat{\mathbf{y}} + z\,\hat{\mathbf{z}}) \\ &= v_x\,\hat{\mathbf{x}} + v_y\,\hat{\mathbf{y}} + v_z\,\hat{\mathbf{z}} = \mathbf{v},\end{aligned} \tag{10.54}$$

while

$$(\mathbf{v} \cdot \nabla)\mathbf{w} = \mathbf{v}(\mathbf{v} \cdot \nabla t_r)$$

(same reasoning as Eq. 10.52). Moving on to the third term in Eq. 10.51,

$$\nabla \times \mathbf{v} = \left(\frac{\partial v_z}{\partial y} - \frac{\partial v_y}{\partial z}\right)\hat{\mathbf{x}} + \left(\frac{\partial v_x}{\partial z} - \frac{\partial v_z}{\partial x}\right)\hat{\mathbf{y}} + \left(\frac{\partial v_y}{\partial x} - \frac{\partial v_x}{\partial y}\right)\hat{\mathbf{z}}$$

$$= \left(\frac{dv_z}{dt_r}\frac{\partial t_r}{\partial y} - \frac{dv_y}{dt_r}\frac{\partial t_r}{\partial z}\right)\hat{\mathbf{x}} + \left(\frac{dv_x}{dt_r}\frac{\partial t_r}{\partial z} - \frac{dv_z}{dt_r}\frac{\partial t_r}{\partial x}\right)\hat{\mathbf{y}} + \left(\frac{dv_y}{dt_r}\frac{\partial t_r}{\partial x} - \frac{dv_x}{dt_r}\frac{\partial t_r}{\partial y}\right)\hat{\mathbf{z}}$$

$$= -\mathbf{a} \times \nabla t_r. \tag{10.55}$$

Finally,

$$\nabla \times \boldsymbol{\mathscr{r}} = \nabla \times \mathbf{r} - \nabla \times \mathbf{w}, \tag{10.56}$$

but $\nabla \times \mathbf{r} = 0$, while, by the same argument as Eq. 10.55,

$$\nabla \times \mathbf{w} = -\mathbf{v} \times \nabla t_r. \tag{10.57}$$

Putting all this back into Eq. 10.51, and using the "BAC-CAB" rule to reduce the triple cross products,

$$\begin{aligned}\nabla(\boldsymbol{\mathscr{r}}\cdot\mathbf{v}) &= \mathbf{a}(\boldsymbol{\mathscr{r}}\cdot\nabla t_r)+\mathbf{v}-\mathbf{v}(\mathbf{v}\cdot\nabla t_r)-\boldsymbol{\mathscr{r}}\times(\mathbf{a}\times\nabla t_r)+\mathbf{v}\times(\mathbf{v}\times\nabla t_r)\\ &= \mathbf{v}+(\boldsymbol{\mathscr{r}}\cdot\mathbf{a}-v^2)\nabla t_r.\end{aligned} \tag{10.58}$$

Collecting Eqs. 10.50 and 10.58 together, we have

$$\nabla V=\frac{qc}{4\pi\epsilon_0}\frac{1}{(\mathscr{r}c-\boldsymbol{\mathscr{r}}\cdot\mathbf{v})^2}\left[\mathbf{v}+(c^2-v^2+\boldsymbol{\mathscr{r}}\cdot\mathbf{a})\nabla t_r\right]. \tag{10.59}$$

To complete the calculation, we need to know ∇t_r. This can be found by taking the gradient of the defining equation (10.48)—which we have already done in Eq. 10.50—and expanding out $\nabla\mathscr{r}$:

$$\begin{aligned}-c\nabla t_r &= \nabla\mathscr{r}=\nabla\sqrt{\boldsymbol{\mathscr{r}}\cdot\boldsymbol{\mathscr{r}}}=\frac{1}{2\sqrt{\boldsymbol{\mathscr{r}}\cdot\boldsymbol{\mathscr{r}}}}\nabla(\boldsymbol{\mathscr{r}}\cdot\boldsymbol{\mathscr{r}})\\ &= \frac{1}{\mathscr{r}}[(\boldsymbol{\mathscr{r}}\cdot\nabla)\boldsymbol{\mathscr{r}}+\boldsymbol{\mathscr{r}}\times(\nabla\times\boldsymbol{\mathscr{r}})].\end{aligned} \tag{10.60}$$

But

$$(\boldsymbol{\mathscr{r}}\cdot\nabla)\boldsymbol{\mathscr{r}}=\boldsymbol{\mathscr{r}}-\mathbf{v}(\boldsymbol{\mathscr{r}}\cdot\nabla t_r)$$

(same idea as Eq. 10.53), while (from Eq. 10.56 and 10.57)

$$\nabla\times\boldsymbol{\mathscr{r}}=(\mathbf{v}\times\nabla t_r).$$

Thus

$$-c\nabla t_r=\frac{1}{\mathscr{r}}[\boldsymbol{\mathscr{r}}-\mathbf{v}(\boldsymbol{\mathscr{r}}\cdot\nabla t_r)+\boldsymbol{\mathscr{r}}\times(\mathbf{v}\times\nabla t_r)]=\frac{1}{\mathscr{r}}[\boldsymbol{\mathscr{r}}-(\boldsymbol{\mathscr{r}}\cdot\mathbf{v})\nabla t_r],$$

and hence

$$\nabla t_r=\frac{-\boldsymbol{\mathscr{r}}}{\mathscr{r}c-\boldsymbol{\mathscr{r}}\cdot\mathbf{v}}. \tag{10.61}$$

Incorporating this result into Eq. 10.59, I conclude that

$$\nabla V=\frac{1}{4\pi\epsilon_0}\frac{qc}{(\mathscr{r}c-\boldsymbol{\mathscr{r}}\cdot\mathbf{v})^3}\left[(\mathscr{r}c-\boldsymbol{\mathscr{r}}\cdot\mathbf{v})\mathbf{v}-(c^2-v^2+\boldsymbol{\mathscr{r}}\cdot\mathbf{a})\boldsymbol{\mathscr{r}}\right]. \tag{10.62}$$

A similar calculation, which I shall leave for you (Prob. 10.17), yields

$$\begin{aligned}\frac{\partial\mathbf{A}}{\partial t} &= \frac{1}{4\pi\epsilon_0}\frac{qc}{(\mathscr{r}c-\boldsymbol{\mathscr{r}}\cdot\mathbf{v})^3}\Big[(\mathscr{r}c-\boldsymbol{\mathscr{r}}\cdot\mathbf{v})(-\mathbf{v}+\mathscr{r}\mathbf{a}/c)\\ &\quad+\frac{\mathscr{r}}{c}(c^2-v^2+\boldsymbol{\mathscr{r}}\cdot\mathbf{a})\mathbf{v}\Big].\end{aligned} \tag{10.63}$$

Combining these results, and introducing the vector

$$\mathbf{u} \equiv c\hat{\boldsymbol{\mathscr{r}}} - \mathbf{v}, \tag{10.64}$$

I find

$$\boxed{\mathbf{E}(\mathbf{r}, t) = \frac{q}{4\pi\epsilon_0} \frac{\mathscr{r}}{(\boldsymbol{\mathscr{r}} \cdot \mathbf{u})^3}[(c^2 - v^2)\mathbf{u} + \boldsymbol{\mathscr{r}} \times (\mathbf{u} \times \mathbf{a})].} \tag{10.65}$$

Meanwhile,

$$\nabla \times \mathbf{A} = \frac{1}{c^2}\nabla \times (V\mathbf{v}) = \frac{1}{c^2}[V(\nabla \times \mathbf{v}) - \mathbf{v} \times (\nabla V)].$$

We have already calculated $\nabla \times \mathbf{v}$ (Eq. 10.55) and ∇V (Eq. 10.62). Putting these together,

$$\nabla \times \mathbf{A} = -\frac{1}{c}\frac{q}{4\pi\epsilon_0}\frac{1}{(\mathbf{u} \cdot \boldsymbol{\mathscr{r}})^3}\boldsymbol{\mathscr{r}} \times [(c^2 - v^2)\mathbf{v} + (\boldsymbol{\mathscr{r}} \cdot \mathbf{a})\mathbf{v} + (\boldsymbol{\mathscr{r}} \cdot \mathbf{u})\mathbf{a}].$$

The quantity in brackets is strikingly similar to the one in Eq. 10.65, which can be written, using the BAC-CAB rule, as $[(c^2 - v^2)\mathbf{u} + (\boldsymbol{\mathscr{r}} \cdot \mathbf{a})\mathbf{u} - (\boldsymbol{\mathscr{r}} \cdot \mathbf{u})\mathbf{a}]$; the main difference is that we have $\mathbf{v}$'s instead of $\mathbf{u}$'s in the first two terms. In fact, since it's all crossed into $\boldsymbol{\mathscr{r}}$ *anyway*, we can with impunity *change* these $\mathbf{v}$'s into $-\mathbf{u}$'s; the extra term proportional to $\hat{\boldsymbol{\mathscr{r}}}$ disappears in the cross product. It follows that

$$\boxed{\mathbf{B}(\mathbf{r}, t) = \frac{1}{c}\hat{\boldsymbol{\mathscr{r}}} \times \mathbf{E}(\mathbf{r}, t).} \tag{10.66}$$

Evidently *the magnetic field of a point charge is always perpendicular to the electric field, and to the vector from the retarded point.*

The first term in $\mathbf{E}$ (the one involving $(c^2 - v^2)\mathbf{u}$) falls off as the inverse *square* of the distance from the particle. If the velocity and acceleration are both zero, this term alone survives and reduces to the old electrostatic result

$$\mathbf{E} = \frac{1}{4\pi\epsilon_0}\frac{q}{\mathscr{r}^2}\hat{\boldsymbol{\mathscr{r}}}.$$

For this reason, the first term in $\mathbf{E}$ is sometimes called the **generalized Coulomb field**. (Because it does not depend on the acceleration, it is also known as the **velocity field**.) The second term (the one involving $\boldsymbol{\mathscr{r}} \times (\mathbf{u} \times \mathbf{a})$) falls off as the inverse *first* power of $\mathscr{r}$ and is therefore dominant at large distances. As we shall see in Chapter 11, it is this term that is responsible for electromagnetic radiation; accordingly, it is called the **radiation field**—or, since it is proportional to a, the **acceleration field**. The same terminology applies to the magnetic field.

Back in Chapter 2, I commented that if we could only write down the formula for the force one charge exerts on another, we would be done with electrodynamics, in principle. That, together with the superposition principle, would tell us the force exerted on a test charge Q by any configuration whatsoever. Well ... here we are: Eqs. 10.65 and 10.66 give us the fields, and the Lorentz force law determines the resulting force:

$$\begin{aligned} \mathbf{F} &= \frac{qQ}{4\pi\epsilon_0}\frac{\mathscr{r}}{(\boldsymbol{\mathscr{r}}\cdot\mathbf{u})^3}\Big\{[(c^2-v^2)\mathbf{u}+\boldsymbol{\mathscr{r}}\times(\mathbf{u}\times\mathbf{a})] \\ &+ \frac{\mathbf{V}}{c}\times\Big[\hat{\boldsymbol{\mathscr{r}}}\times[(c^2-v^2)\mathbf{u}+\boldsymbol{\mathscr{r}}\times(\mathbf{u}\times\mathbf{a})]\Big]\Big\}, \end{aligned} \tag{10.67}$$

where $\mathbf{V}$ is the velocity of Q, and $\boldsymbol{\mathscr{r}}$, $\mathbf{u}$, $\mathbf{v}$, and $\mathbf{a}$ are all evaluated at the retarded time. The entire theory of classical electrodynamics is contained in that equation ... but you see why I preferred to start out with Coulomb's law.

Example 10.4

Calculate the electric and magnetic fields of a point charge moving with constant velocity.

Solution: Putting $\mathbf{a} = 0$ in Eq. 10.65,

$$\mathbf{E} = \frac{q}{4\pi\epsilon_0}\frac{(c^2-v^2)\mathscr{r}}{(\boldsymbol{\mathscr{r}}\cdot\mathbf{u})^3}\mathbf{u}.$$

In this case, using $\mathbf{w} = \mathbf{v}t$,

$$\mathscr{r}\mathbf{u} = c\boldsymbol{\mathscr{r}} - \mathscr{r}\mathbf{v} = c(\mathbf{r}-\mathbf{v}t_r) - c(t-t_r)\mathbf{v} = c(\mathbf{r}-\mathbf{v}t).$$

In Ex. 10.3 we found that

$$\mathscr{r}c - \boldsymbol{\mathscr{r}}\cdot\mathbf{v} = \boldsymbol{\mathscr{r}}\cdot\mathbf{u} = \sqrt{(c^2t-\mathbf{r}\cdot\mathbf{v})^2+(c^2-v^2)(r^2-c^2t^2)}.$$

In Prob. 10.14, you showed that this radical could be written as

$$Rc\sqrt{1-v^2\sin^2\theta/c^2},$$

where

$$\mathbf{R} \equiv \mathbf{r} - \mathbf{v}t$$

is the vector from the *present* location of the particle to $\mathbf{r}$, and θ is the angle between $\mathbf{R}$ and $\mathbf{v}$ (Fig. 10.9). Thus

$$\boxed{\mathbf{E}(\mathbf{r},t) = \frac{q}{4\pi\epsilon_0}\frac{1-v^2/c^2}{\left(1-v^2\sin^2\theta/c^2\right)^{3/2}}\frac{\hat{\mathbf{R}}}{R^2}.} \tag{10.68}$$

Notice that $\mathbf{E}$ points along the line from the *present* position of the particle. This is an *extraordinary* coincidence, since the "message" came from the *retarded* position. Because of the $\sin^2\theta$ in the denominator, the field of a fast-moving charge is flattened out like a pancake in the direction perpendicular to the motion (Fig. 10.10). In the forward and backward directions $\mathbf{E}$ is *reduced* by a factor $(1-v^2/c^2)$ relative to the field of a charge at rest; in the perpendicular direction it is *enhanced* by a factor $1/\sqrt{1-v^2/c^2}$.

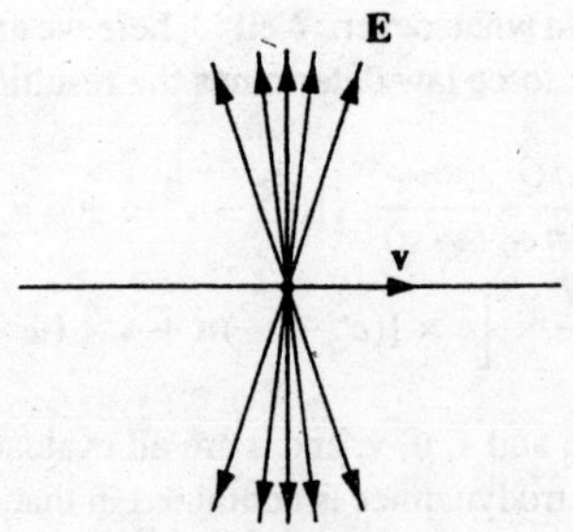

Figure 10.10

As for **B**, we have

$$\hat{\boldsymbol{\mathscr{r}}} = \frac{\mathbf{r} - \mathbf{v}t_r}{\mathscr{r}} = \frac{(\mathbf{r} - \mathbf{v}t) + (t - t_r)\mathbf{v}}{\mathscr{r}} = \frac{\mathbf{R}}{\mathscr{r}} + \frac{\mathbf{v}}{c},$$

and therefore

$$\boxed{\mathbf{B} = \frac{1}{c}(\hat{\boldsymbol{\mathscr{r}}} \times \mathbf{E}) = \frac{1}{c^2}(\mathbf{v} \times \mathbf{E}).} \tag{10.69}$$

Lines of **B** *circle around* the charge, as shown in Fig. 10.11.

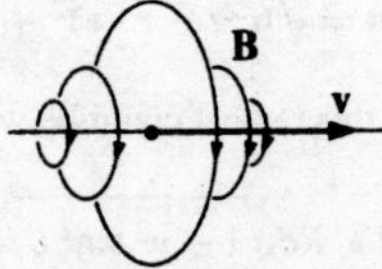

Figure 10.11

The fields of a point charge moving at constant velocity (Eqs. 10.68 and 10.69) were first obtained by Oliver Heaviside in 1888.[13] When $v^2 \ll c^2$ they reduce to

$$\mathbf{E}(\mathbf{r}, t) = \frac{1}{4\pi\epsilon_0}\frac{q}{R^2}\hat{\mathbf{R}}; \qquad \mathbf{B}(\mathbf{r}, t) = \frac{\mu_0}{4\pi}\frac{q}{R^2}(\mathbf{v} \times \hat{\mathbf{R}}). \tag{10.70}$$

The first is essentially Coulomb's law, and the latter is the "Biot-Savart law for a point charge" I warned you about in Chapter 5 (Eq. 5.40).

---

[13]For history and references, see O. J. Jefimenko, *Am. J. Phys.* **62**, 79 (1994).

Problem 10.17 Derive Eq. 10.63. First show that

$$\frac{\partial t_r}{\partial t} = \frac{\mathscr{r} c}{\boldsymbol{\mathscr{r}} \cdot \mathbf{u}}. \tag{10.71}$$

Problem 10.18 Suppose a point charge q is constrained to move along the x axis. Show that the fields at points on the axis to the *right* of the charge are given by

$$\mathbf{E} = \frac{q}{4\pi\epsilon_0}\frac{1}{\mathscr{r}^2}\left(\frac{c+v}{c-v}\right)\hat{\mathbf{x}}, \quad \mathbf{B} = 0.$$

What are the fields on the axis to the *left* of the charge?

Problem 10.19
(a) Use Eq. 10.68 to calculate the electric field a distance d from an infinite straight wire carrying a uniform line charge λ, moving at a constant speed v down the wire.
(b) Use Eq. 10.69 to find the *magnetic* field of this wire.

Problem 10.20 For the configuration in Prob. 10.13, find the electric and magnetic fields at the center. From your formula for $\mathbf{B}$, determine the magnetic field at the center of a circular loop carrying a steady current I, and compare your answer with the result of Ex. 5.6

More Problems on Chapter 10

Problem 10.21 Suppose you take a plastic ring of radius a and glue charge on it, so that the line charge density is $\lambda_0|\sin(\theta/2)|$. Then you spin the loop about its axis at an angular velocity ω. Find the (exact) scalar and vector potentials at the center of the ring. [*Answer:* $\mathbf{A} = (\mu_0\lambda_0\omega a/3\pi)\left\{\sin[\omega(t-a/c)]\,\hat{\mathbf{x}} - \cos[\omega(t-a/c)]\,\hat{\mathbf{y}}\right\}$]

Problem 10.22 Figure 2.35 summarizes the laws of *electrostatics* in a "triangle diagram" relating the *source* (ρ), the *field* ($\mathbf{E}$), and the *potential* (V). Figure 5.48 does the same for *magnetostatics*, where the source is $\mathbf{J}$, the field is $\mathbf{B}$, and the potential is $\mathbf{A}$. Construct the analogous diagram for *electrodynamics*, with sources ρ and $\mathbf{J}$ (constrained by the continuity equation), fields $\mathbf{E}$ and $\mathbf{B}$, and potentials V and $\mathbf{A}$ (constrained by the Lorentz gauge condition). Do not include formulas for V and $\mathbf{A}$ in terms of $\mathbf{E}$ and $\mathbf{B}$.

Problem 10.23 Check that the potentials of a point charge moving at constant velocity (Eqs. 10.42 and 10.43) satisfy the Lorentz gauge condition (Eq. 10.12).

Problem 10.24 One particle, of charge q_1, is held at rest at the origin. Another particle, of charge q_2, approaches along the x axis, in hyperbolic motion:

$$x(t) = \sqrt{b^2 + (ct)^2};$$

it reaches the closest point, b, at time $t = 0$, and then returns out to infinity.

(a) What is the force F_2 on q_2 (due to q_1) at time t?

(b) What total impulse ($I_2 = \int_{-\infty}^{\infty} F_2 dt$) is delivered by q_2 to q_1?

(c) What is the force F_1 on q_1 (due to q_2) at time t?

(d) What total impulse ($I_1 = \int_{-\infty}^{\infty} F_1 dt$) is delivered to q_1 by q_2? [*Hint:* It might help to review Prob. 10.15 before doing this integral. *Answer:* $I_2 = -I_1 = q_1 q_2 / 4\epsilon_0 bc$]

Problem 10.25 A particle of charge q is traveling at constant speed v along the x axis. Calculate the total power passing through the plane $x = a$, at the moment the particle itself is at the origin. [*Answer:* $q^2 v / 32\pi\epsilon_0 a^2$]

Problem 10.26[14] A particle of charge q_1 is at rest at the origin. A second particle, of charge q_2, moves along the z axis at constant velocity v.

(a) Find the force $\mathbf{F}_{12}(t)$ of q_1 on q_2, at time t (when q_2 is at $z = vt$).

(b) Find the force $\mathbf{F}_{21}(t)$ of q_2 on q_1, at time t. Does Newton's third law hold, in this case?

! (c) Calculate the linear momentum $\mathbf{p}(t)$ in the electromagnetic fields, at time t. (Don't bother with any terms that are constant in time, since you won't need them in part (d)). [*Answer:* $(\mu_0 q_1 q_2 / 4\pi t)\,\hat{\mathbf{z}}$]

(d) Show that the sum of the forces is equal to minus the rate of change of the momentum in the fields, and interpret this result physically.

---

[14] See J. J. G. Scanio, *Am. J. Phys.* **43**, 258 (1975).

Chapter 11

Radiation

11.1 Dipole Radiation

11.1.1 What is Radiation?

In Chapter 9 we discussed the propagation of plane electromagnetic waves through various media, but I did not tell you how the waves got started in the first place. Like all electromagnetic fields, their source is some arrangement of electric charge. But a charge at rest does not generate electromagnetic waves; nor does a steady current. It takes *accelerating* charges, and *changing* currents, as we shall see. My purpose in this chapter is to show you how such configurations produce electromagnetic waves—that is, how they **radiate**.

Once established, electromagnetic waves in vacuum propagate out "to infinity," carrying energy with them; the *signature* of radiation is this irreversible flow of energy away from the source. Throughout this chapter I shall assume the source is *localized*[1] near the origin. Imagine a gigantic spherical shell, out at radius r (Fig. 11.1); the total power passing out through this surface is the integral of the Poynting vector:

$$P(r) = \oint \mathbf{S} \cdot d\mathbf{a} = \frac{1}{\mu_0} \oint (\mathbf{E} \times \mathbf{B}) \cdot d\mathbf{a}. \tag{11.1}$$

The power *radiated* is the limit of this quantity as r goes to infinity:

$$P_{\text{rad}} \equiv \lim_{r \to \infty} P(r). \tag{11.2}$$

This is the energy (per unit time) that is transported out to infinity, and never comes back.

Now, the area of the sphere is $4\pi r^2$, so for radiation to occur the Poynting vector must decrease (at large r) no faster than $1/r^2$ (if it went like $1/r^3$, for example, then $P(r)$ would go like $1/r$, and P_{rad} would be zero). According to Coulomb's law, electro*static* fields fall off like $1/r^2$ (or even faster, if the total charge is zero), and the Biot-Savart law says

[1]For *non*localized sources, such as infinite planes, wires, or solenoids, the whole concept of "radiation" must be reformulated—see Prob. 11.24.

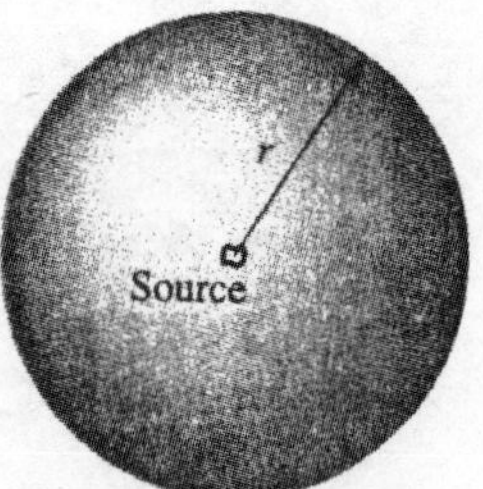

Figure 11.1

that magneto*static* fields go like $1/r^2$ (or faster), which means that $S \sim 1/r^4$, for static configurations. So *static* sources do not radiate. But Jefimenko's equations (10.29 and 10.31) indicate that *time-dependent* fields include terms (involving $\dot{\rho}$ and $\dot{\mathbf{J}}$) that go like $1/r$; it is *these* terms that are responsible for electromagnetic radiation.

The study of radiation, then, involves picking out the parts of **E** and **B** that go like $1/r$ at large distances from the source, constructing from them the $1/r^2$ term in **S**, integrating over a large spherical[2] surface, and taking the limit as $r \to \infty$. I'll carry through this procedure first for oscillating electric and magnetic dipoles; then, in Sect. 11.2, we'll consider the more difficult case of radiation from an accelerating point charge.

11.1.2 Electric Dipole Radiation

Picture two tiny metal spheres separated by a distance d and connected by a fine wire (Fig. 11.2); at time t the charge on the upper sphere is $q(t)$, and the charge on the lower sphere is $-q(t)$. Suppose that we drive the charge back and forth through the wire, from one end to the other, at an angular frequency ω:

$$q(t) = q_0 \cos(\omega t). \tag{11.3}$$

The result is an oscillating electric dipole:[3]

$$\mathbf{p}(t) = p_0 \cos(\omega t)\,\hat{\mathbf{z}}, \tag{11.4}$$

where

$$p_0 \equiv q_0 d$$

is the the maximum value of the dipole moment.

---

[2]It doesn't have to be a sphere, of course, but this makes the calculations a lot easier.

[3]It might occur to you that a more natural model would consist of equal and opposite charges mounted on a spring, say, so that q is constant while d oscillates, instead of the other way around. Such a model would lead to the same result, but there is a subtle problem in calculating the retarded potentials of a moving point charge, which I would prefer to save for Sect. 11.2.

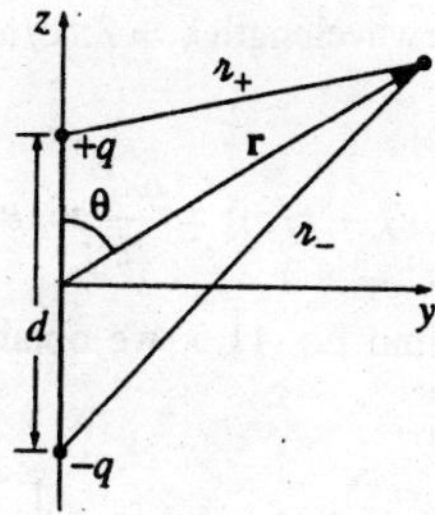

Figure 11.2

The retarded potential (Eq. 10.19) is

$$V(\mathbf{r},t)=\frac{1}{4\pi\epsilon_0}\left\{\frac{q_0\cos[\omega(t-\mathscr{r}_+/c)]}{\mathscr{r}_+}-\frac{q_0\cos[\omega(t-\mathscr{r}_-/c)]}{\mathscr{r}_-}\right\}, \tag{11.5}$$

where, by the law of cosines,

$$\mathscr{r}_\pm=\sqrt{r^2\mp rd\cos\theta+(d/2)^2}. \tag{11.6}$$

Now, to make this *physical* dipole into a *perfect* dipole, we want the separation distance to be extremely small:

$$\textbf{approximation 1}: \; d \ll r. \tag{11.7}$$

Of course, if d is *zero* we get no potential at all; what we want is an expansion carried to *first order* in d. Thus

$$\mathscr{r}_\pm \cong r\left(1\mp\frac{d}{2r}\cos\theta\right). \tag{11.8}$$

It follows that

$$\frac{1}{\mathscr{r}_\pm}\cong\frac{1}{r}\left(1\pm\frac{d}{2r}\cos\theta\right), \tag{11.9}$$

and

$$\cos[\omega(t-\mathscr{r}_\pm/c)]\cong\cos\left[\omega(t-r/c)\pm\frac{\omega d}{2c}\cos\theta\right]$$

$$=\cos[\omega(t-r/c)]\cos\left(\frac{\omega d}{2c}\cos\theta\right)\mp\sin[\omega(t-r/c)]\sin\left(\frac{\omega d}{2c}\cos\theta\right).$$

In the perfect dipole limit we have, further,

$$\textbf{approximation 2}: \; d \ll \frac{c}{\omega}. \tag{11.10}$$

(Since waves of frequency ω have a wavelength $\lambda = 2\pi c/\omega$, this amounts to the requirement $d \ll \lambda$.) Under these conditions

$$\cos[\omega(t - \imath_{\pm}/c)] \cong \cos[\omega(t - r/c)] \mp \frac{\omega d}{2c}\cos\theta \sin[\omega(t - r/c)]. \tag{11.11}$$

Putting Eqs. 11.9 and 11.11 into Eq. 11.5, we obtain the potential of an oscillating perfect dipole:

$$V(r, \theta, t) = \frac{p_0 \cos\theta}{4\pi\epsilon_0 r}\left\{-\frac{\omega}{c}\sin[\omega(t - r/c)] + \frac{1}{r}\cos[\omega(t - r/c)]\right\}. \tag{11.12}$$

In the static limit ($\omega \to 0$) the second term reproduces the old formula for the potential of a stationary dipole (Eq. 3.99):

$$V = \frac{p_0 \cos\theta}{4\pi\epsilon_0 r^2}.$$

This is not, however, the term that concerns us now; we are interested in the fields that survive at *large distances from the source*, in the so-called **radiation zone**:[4]

$$\textbf{approximation 3}: \quad r \gg \frac{c}{\omega}. \tag{11.13}$$

(Or, in terms of the wavelength, $r \gg \lambda$.) In this region the potential reduces to

$$\boxed{V(r, \theta, t) = -\frac{p_0\omega}{4\pi\epsilon_0 c}\left(\frac{\cos\theta}{r}\right)\sin[\omega(t - r/c)].} \tag{11.14}$$

Meanwhile, the *vector* potential is determined by the current flowing in the wire:

$$\mathbf{I}(t) = \frac{dq}{dt}\,\hat{\mathbf{z}} = -q_0\omega\sin(\omega t)\,\hat{\mathbf{z}}. \tag{11.15}$$

Referring to Fig. 11.3,

$$\mathbf{A}(\mathbf{r}, t) = \frac{\mu_0}{4\pi}\int_{-d/2}^{d/2}\frac{-q_0\omega\sin[\omega(t - \imath/c)]\,\hat{\mathbf{z}}}{\imath}\,dz. \tag{11.16}$$

Because the integration itself introduces a factor of d, we can, to first order, replace the integrand by its value at the center:

$$\boxed{\mathbf{A}(r, \theta, t) = -\frac{\mu_0 p_0 \omega}{4\pi r}\sin[\omega(t - r/c)]\,\hat{\mathbf{z}}.} \tag{11.17}$$

(Notice that whereas I implicitly used approximations 1 and 2, in keeping only the first order in d, Eq. 11.17 is *not* subject to approximation 3.)

---

[4]Note that approximations 2 and 3 subsume approximation 1; all together, we have $d \ll \lambda \ll r$.

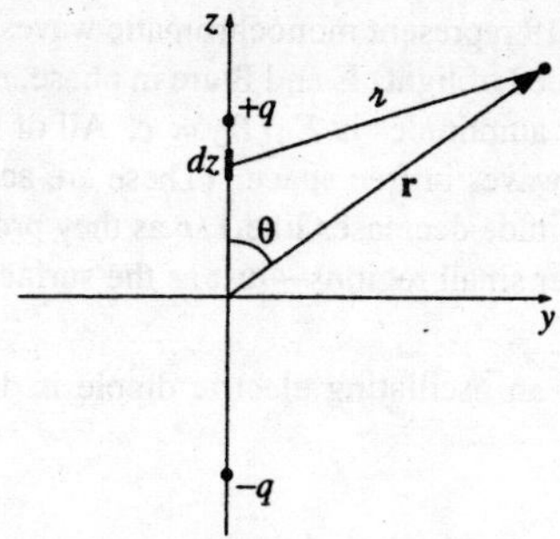

Figure 11.3

From the potentials, it is a straightforward matter to compute the fields.

$$\nabla V = \frac{\partial V}{\partial r}\hat{\mathbf{r}} + \frac{1}{r}\frac{\partial V}{\partial \theta}\hat{\boldsymbol{\theta}}$$

$$= -\frac{p_0\omega}{4\pi\epsilon_0 c}\left\{\cos\theta\left(-\frac{1}{r^2}\sin[\omega(t-r/c)] - \frac{\omega}{rc}\cos[\omega(t-r/c)]\right)\hat{\mathbf{r}} - \frac{\sin\theta}{r^2}\sin[\omega(t-r/c)]\hat{\boldsymbol{\theta}}\right\}$$

$$\cong \frac{p_0\omega^2}{4\pi\epsilon_0 c^2}\left(\frac{\cos\theta}{r}\right)\cos[\omega(t-r/c)]\,\hat{\mathbf{r}}.$$

(I dropped the first and last terms, in accordance with approximation 3.) Likewise,

$$\frac{\partial \mathbf{A}}{\partial t} = -\frac{\mu_0 p_0\omega^2}{4\pi r}\cos[\omega(t-r/c)](\cos\theta\,\hat{\mathbf{r}} - \sin\theta\,\hat{\boldsymbol{\theta}}),$$

and therefore

$$\boxed{\mathbf{E} = -\nabla V - \frac{\partial \mathbf{A}}{\partial t} = -\frac{\mu_0 p_0\omega^2}{4\pi}\left(\frac{\sin\theta}{r}\right)\cos[\omega(t-r/c)]\,\hat{\boldsymbol{\theta}}.} \quad (11.18)$$

Meanwhile

$$\nabla \times \mathbf{A} = \frac{1}{r}\left[\frac{\partial}{\partial r}(rA_\theta) - \frac{\partial A_r}{\partial \theta}\right]\hat{\boldsymbol{\phi}}$$

$$= -\frac{\mu_0 p_0\omega}{4\pi r}\left\{\frac{\omega}{c}\sin\theta\cos[\omega(t-r/c)] + \frac{\sin\theta}{r}\sin[\omega(t-r/c)]\right\}\hat{\boldsymbol{\phi}}.$$

The second term is again eliminated by approximation 3, so

$$\boxed{\mathbf{B} = \nabla \times \mathbf{A} = -\frac{\mu_0 p_0\omega^2}{4\pi c}\left(\frac{\sin\theta}{r}\right)\cos[\omega(t-r/c)]\,\hat{\boldsymbol{\phi}}.} \quad (11.19)$$

Equations 11.18 and 11.19 represent monochromatic waves of frequency ω traveling in the radial direction at the speed of light. **E** and **B** are in phase, mutually perpendicular, and transverse; the ratio of their amplitudes is $E_0/B_0 = c$. All of which is precisely what we expect for electromagnetic waves in free space. (These are actually *spherical* waves, not plane waves, and their amplitude decreases like $1/r$ as they progress. But for large r, they are approximately plane over small regions—just as the surface of the earth is reasonably flat, locally.)

The energy radiated by an oscillating electric dipole is determined by the Poynting vector:

$$\mathbf{S} = \frac{1}{\mu_0}(\mathbf{E} \times \mathbf{B}) = \frac{\mu_0}{c}\left\{\frac{p_0\omega^2}{4\pi}\left(\frac{\sin\theta}{r}\right)\cos[\omega(t - r/c)]\right\}^2 \hat{\mathbf{r}}. \tag{11.20}$$

The intensity is obtained by averaging (in time) over a complete cycle:

$$\langle \mathbf{S} \rangle = \left(\frac{\mu_0 p_0^2 \omega^4}{32\pi^2 c}\right)\frac{\sin^2\theta}{r^2}\,\hat{\mathbf{r}}. \tag{11.21}$$

Notice that there is no radiation along the *axis* of the dipole (here $\sin\theta = 0$); the intensity profile[5] takes the form of a donut, with its maximum in the equatorial plane (Fig. 11.4). The total power radiated is found by integrating $\langle \mathbf{S} \rangle$ over a sphere of radius r:

$$\langle P \rangle = \int \langle \mathbf{S} \rangle \cdot d\mathbf{a} = \frac{\mu_0 p_0^2 \omega^4}{32\pi^2 c}\int \frac{\sin^2\theta}{r^2} r^2 \sin\theta \, d\theta \, d\phi = \frac{\mu_0 p_0^2 \omega^4}{12\pi c}. \tag{11.22}$$

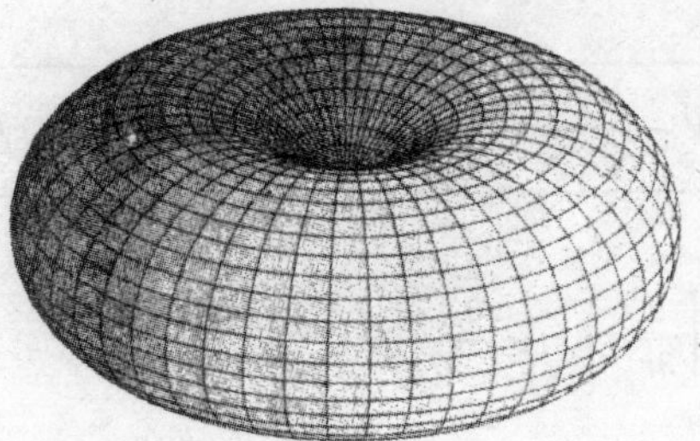

Figure 11.4

[5] The "radial" coordinate in Fig. 11.4 represents the magnitude of $\langle \mathbf{S} \rangle$ (at fixed r), as a function of θ and ϕ.

It is independent of the radius of the sphere, as one would expect from conservation of energy (with approximation 3 we were anticipating the limit $r \to \infty$).

Example 11.1

The sharp frequency dependence of the power formula is what accounts for the blueness of the sky. Sunlight passing through the atmosphere stimulates atoms to oscillate as tiny dipoles. The incident solar radiation covers a broad range of frequencies (white light), but the energy absorbed and reradiated by the atmospheric dipoles is stronger at the higher frequencies because of the ω^4 in Eq. 11.22. It is more intense in the blue, then, than in the red. It is this reradiated light that you see when you look up in the sky—unless, of course, you're staring directly at the sun.

Because electromagnetic waves are transverse, the dipoles oscillate in a plane orthogonal to the sun's rays. In the celestial arc perpendicular to these rays, where the blueness is most pronounced, the dipoles oscillating along the line of sight send no radiation to the observer (because of the $\sin^2\theta$ in equation Eq. 11.21); light received at this angle is therefore polarized perpendicular to the sun's rays (Fig. 11.5).

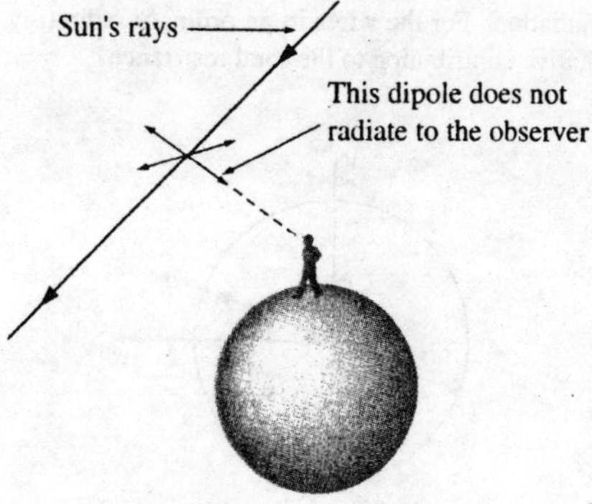

Figure 11.5

The redness of sunset is the other side of the same coin: Sunlight coming in at a tangent to the earth's surface must pass through a much longer stretch of atmosphere than sunlight coming from overhead (Fig. 11.6). Accordingly, much of the blue has been *removed* by scattering and what's left is red.

Problem 11.1 Check that the retarded potentials of an oscillating dipole (Eqs. 11.12 and 11.17) satisfy the Lorentz gauge condition. Do *not* use approximation 3.

Problem 11.2 Equation 11.14 can be expressed in "coordinate-free" form by writing $p_0\cos\theta = \mathbf{p}_0\cdot\hat{\mathbf{r}}$. Do so, and likewise for Eqs. 11.17, 11.18, 11.19, and 11.21.

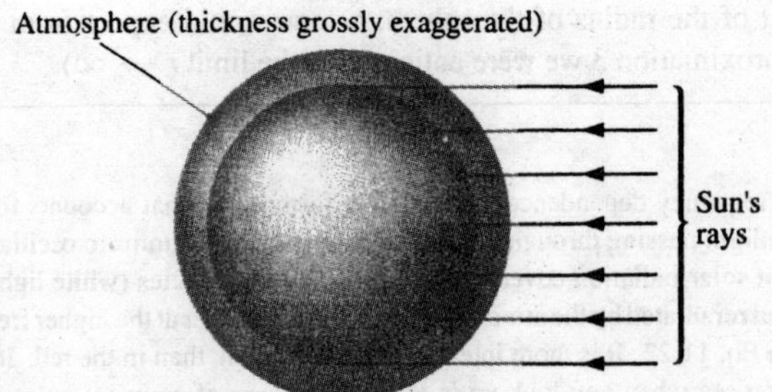

Figure 11.6

Problem 11.3 Find the **radiation resistance** of the wire joining the two ends of the dipole. (This is the resistance that would give the same average power loss—to heat—as the oscillating dipole in *fact* puts out in the form of radiation.) Show that $R = 790\,(d/\lambda)^2\ \Omega$, where λ is the wavelength of the radiation. For the wires in an ordinary radio (say, $d = 5$ cm), should you worry about the radiative contribution to the total resistance?

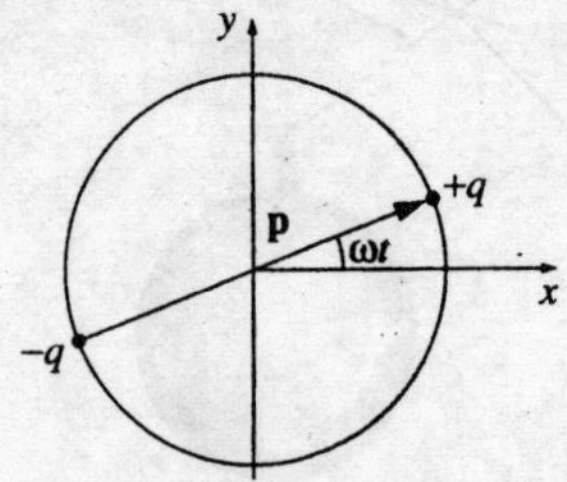

Figure 11.7

! **Problem 11.4** A *rotating* electric dipole can be thought of as the superposition of two *oscillating* dipoles, one along the x axis, and the other along the y axis (Fig. 11.7), with the latter out of phase by 90°:

$$\mathbf{p} = p_0[\cos(\omega t)\,\hat{\mathbf{x}} + \sin(\omega t)\,\hat{\mathbf{y}}].$$

Using the principle of superposition and Eqs. 11.18 and 11.19 (perhaps in the form suggested by Prob. 11.2), find the fields of a rotating dipole. Also find the Poynting vector and the intensity of the radiation. Sketch the intensity profile as a function of the polar angle θ, and calculate the total power radiated. Does the answer seem reasonable? (Note that power, being *quadratic* in the fields, does *not* satisfy the superposition principle. In this instance, however, it *seems* to. Can you account for this?)

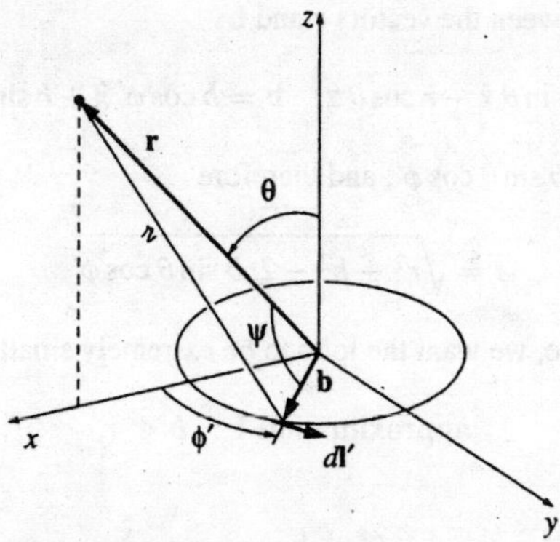

Figure 11.8

11.1.3 Magnetic Dipole Radiation

Suppose now that we have a wire loop of radius b (Fig. 11.8), around which we drive an alternating current:

$$I(t) = I_0 \cos(\omega t). \tag{11.23}$$

This is a model for an oscillating *magnetic* dipole,

$$\mathbf{m}(t) = \pi b^2 I(t)\,\hat{\mathbf{z}} = m_0 \cos(\omega t)\,\hat{\mathbf{z}}, \tag{11.24}$$

where

$$m_0 \equiv \pi b^2 I_0 \tag{11.25}$$

is the maximum value of the magnetic dipole moment.

The loop is uncharged, so the scalar potential is zero. The retarded vector potential is

$$\mathbf{A}(\mathbf{r}, t) = \frac{\mu_0}{4\pi} \int \frac{I_0 \cos[\omega(t - \mathscr{r}/c)]}{\mathscr{r}}\, d\mathbf{l}'. \tag{11.26}$$

For a point $\mathbf{r}$ directly above the x axis (Fig. 11.8), $\mathbf{A}$ must aim in the y direction, since the x components from symmetrically placed points on either side of the x axis will cancel. Thus

$$\mathbf{A}(\mathbf{r}, t) = \frac{\mu_0 I_0 b}{4\pi}\,\hat{\mathbf{y}} \int_0^{2\pi} \frac{\cos[\omega(t - \mathscr{r}/c)]}{\mathscr{r}} \cos\phi'\, d\phi' \tag{11.27}$$

($\cos\phi'$ serves to pick out the y-component of $d\mathbf{l}'$). By the law of cosines,

$$\mathscr{r} = \sqrt{r^2 + b^2 - 2rb\cos\psi}\,,$$

where ψ is the angle between the vectors $\mathbf{r}$ and $\mathbf{b}$:

$$\mathbf{r} = r\sin\theta\,\hat{\mathbf{x}} + r\cos\theta\,\hat{\mathbf{z}}, \quad \mathbf{b} = b\cos\phi'\,\hat{\mathbf{x}} + b\sin\phi'\,\hat{\mathbf{y}}.$$

So $rb\cos\psi = \mathbf{r}\cdot\mathbf{b} = rb\sin\theta\cos\phi'$, and therefore

$$\mathscr{r} = \sqrt{r^2 + b^2 - 2rb\sin\theta\cos\phi'}. \tag{11.28}$$

For a "perfect" dipole, we want the loop to be extremely small:

$$\textbf{approximation 1}: \quad b \ll r. \tag{11.29}$$

To first order in b, then,

$$\mathscr{r} \cong r\left(1 - \frac{b}{r}\sin\theta\cos\phi'\right),$$

so

$$\frac{1}{\mathscr{r}} \cong \frac{1}{r}\left(1 + \frac{b}{r}\sin\theta\cos\phi'\right) \tag{11.30}$$

and

$$\cos[\omega(t - \mathscr{r}/c)] \cong \cos\left[\omega(t - r/c) + \frac{\omega b}{c}\sin\theta\cos\phi'\right]$$

$$= \cos[\omega(t - r/c)]\cos\left(\frac{\omega b}{c}\sin\theta\cos\phi'\right) - \sin[\omega(t - r/c)]\sin\left(\frac{\omega b}{c}\sin\theta\cos\phi'\right).$$

As before, we also assume the size of the dipole is small compared to the wavelength radiated:

$$\textbf{approximation 2}: \quad b \ll \frac{c}{\omega}. \tag{11.31}$$

In that case,

$$\cos[\omega(t - \mathscr{r}/c)] \cong \cos[\omega(t - r/c)] - \frac{\omega b}{c}\sin\theta\cos\phi'\sin[\omega(t - r/c)]. \tag{11.32}$$

Inserting Eqs. 11.30 and 11.32 into Eq. 11.27, and dropping the second-order term:

$$\mathbf{A}(\mathbf{r}, t) \cong \frac{\mu_0 I_0 b}{4\pi r}\,\hat{\mathbf{y}}\int_0^{2\pi}\Bigg\{\cos[\omega(t - r/c)]$$

$$+ b\sin\theta\cos\phi'\left(\frac{1}{r}\cos[\omega(t - r/c)] - \frac{\omega}{c}\sin[\omega(t - r/c)]\right)\Bigg\}\cos\phi'\,d\phi'.$$

The first term integrates to zero:

$$\int_0^{2\pi}\cos\phi'\,d\phi' = 0.$$

The second term involves the integral of cosine squared:

$$\int_0^{2\pi} \cos^2\phi'\,d\phi' = \pi.$$

Putting this in, and noting that in general **A** points in the $\hat{\boldsymbol{\phi}}$-direction, I conclude that the vector potential of an oscillating perfect magnetic dipole is

$$\mathbf{A}(r,\theta,t) = \frac{\mu_0 m_0}{4\pi}\left(\frac{\sin\theta}{r}\right)\left\{\frac{1}{r}\cos[\omega(t-r/c)] - \frac{\omega}{c}\sin[\omega(t-r/c)]\right\}\hat{\boldsymbol{\phi}}. \tag{11.33}$$

In the static limit ($\omega = 0$) we recover the familiar formula for the potential of a magnetic dipole (Eq. 5.85),

$$\mathbf{A}(r,\theta) = \frac{\mu_0}{4\pi}\frac{m_0\sin\theta}{r^2}\hat{\boldsymbol{\phi}}.$$

In the radiation zone,

$$\textbf{approximation 3}: \; r \gg \frac{c}{\omega}, \tag{11.34}$$

the first term in **A** is negligible, so

$$\boxed{\mathbf{A}(r,\theta,t) = -\frac{\mu_0 m_0 \omega}{4\pi c}\left(\frac{\sin\theta}{r}\right)\sin[\omega(t-r/c)]\hat{\boldsymbol{\phi}}.} \tag{11.35}$$

From **A** we obtain the fields at large r:

$$\boxed{\mathbf{E} = -\frac{\partial \mathbf{A}}{\partial t} = \frac{\mu_0 m_0 \omega^2}{4\pi c}\left(\frac{\sin\theta}{r}\right)\cos[\omega(t-r/c)]\hat{\boldsymbol{\phi}},} \tag{11.36}$$

and

$$\boxed{\mathbf{B} = \nabla\times\mathbf{A} = -\frac{\mu_0 m_0 \omega^2}{4\pi c^2}\left(\frac{\sin\theta}{r}\right)\cos[\omega(t-r/c)]\hat{\boldsymbol{\theta}}.} \tag{11.37}$$

(I used approximation 3 in calculating **B**.) These fields are in phase, mutually perpendicular, and transverse to the direction of propagation ($\hat{\mathbf{r}}$), and the ratio of their amplitudes is $E_0/B_0 = c$, all of which is as expected for electromagnetic waves. They are, in fact, remarkably similar in structure to the fields of an oscillating *electric* dipole (Eqs. 11.18 and 11.19), only this time it is **B** that points in the $\hat{\boldsymbol{\theta}}$ direction and **E** in the $\hat{\boldsymbol{\phi}}$ direction, whereas for electric dipoles it's the other way around.

The energy flux for magnetic dipole radiation is

$$\mathbf{S} = \frac{1}{\mu_0}(\mathbf{E}\times\mathbf{B}) = \frac{\mu_0}{c}\left\{\frac{m_0\omega^2}{4\pi c}\left(\frac{\sin\theta}{r}\right)\cos[\omega(t-r/c)]\right\}^2\hat{\mathbf{r}}, \tag{11.38}$$

the intensity is

$$\langle\mathbf{S}\rangle = \left(\frac{\mu_0 m_0^2\omega^4}{32\pi^2 c^3}\right)\frac{\sin^2\theta}{r^2}\hat{\mathbf{r}}, \tag{11.39}$$

and the total radiated power is

$$\langle P\rangle = \frac{\mu_0 m_0^2 \omega^4}{12\pi c^3}. \tag{11.40}$$

Once again, the intensity profile has the shape of a donut (Fig. 11.4), and the power radiated goes like ω^4. There is, however, one important difference between electric and magnetic dipole radiation: For configurations with comparable dimensions, the power radiated electrically is enormously greater. Comparing Eqs. 11.22 and 11.40,

$$\frac{P_{\text{magnetic}}}{P_{\text{electric}}} = \left(\frac{m_0}{p_0 c}\right)^2, \tag{11.41}$$

where (remember) $m_0 = \pi b^2 I_0$, and $p_0 = q_0 d$. The amplitude of the current in the electrical case was $I_0 = q_0\omega$ (Eq. 11.15). Setting $d = \pi b$, for the sake of comparison, I get

$$\frac{P_{\text{magnetic}}}{P_{\text{electric}}} = \left(\frac{\omega b}{c}\right)^2. \tag{11.42}$$

But $\omega b/c$ is precisely the quantity we assumed was very small (approximation 2), and here it appears *squared*. Ordinarily, then, one should expect electric dipole radiation to dominate. Only when the system is carefully contrived to exclude any electric contribution (as in the case just treated) will the magnetic dipole radiation reveal itself.

Problem 11.5 Calculate the electric and magnetic fields of an oscillating magnetic dipole *without* using approximation 3. [Do they look familiar? Compare Prob. 9.33.] Find the Poynting vector, and show that the intensity of the radiation is exactly the same as we got using approximation 3.

Problem 11.6 Find the radiation resistance (Prob. 11.3) for the oscillating magnetic dipole in Fig. 11.8. Express your answer in terms of λ and b, and compare the radiation resistance of the *electric* dipole. [*Answer:* $3 \times 10^5\ (b/\lambda)^4\ \Omega$]

Problem 11.7 Use the "duality" transformation of Prob. 7.60, together with the fields of an oscillating *electric* dipole (Eqs. 11.18 and 11.19), to determine the fields that would be produced by an oscillating "Gilbert" *magnetic* dipole (composed of equal and opposite magnetic charges, instead of an electric current loop). Compare Eqs. 11.36 and 11.37, and comment on the result.

11.1.4 Radiation from an Arbitrary Source

In the previous sections we studied the radiation produced by two specific systems: oscillating electric dipoles and oscillating magnetic dipoles. Now I want to apply the same procedures to a configuration of charge and current that is entirely arbitrary, except that it is localized within some finite volume near the origin (Fig. 11.9). The retarded scalar potential is

$$V(\mathbf{r}, t) = \frac{1}{4\pi\epsilon_0}\int \frac{\rho(\mathbf{r}', t - \mathscr{r}/c)}{\mathscr{r}}\, d\tau', \tag{11.43}$$

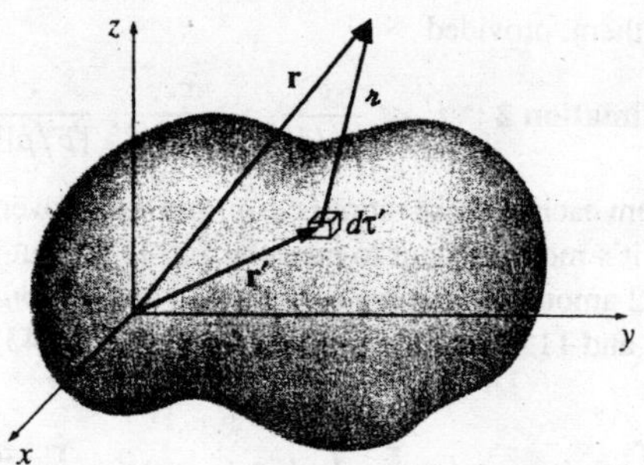

Figure 11.9

where

$$\mathscr{r} = \sqrt{r^2 + r'^2 - 2\mathbf{r} \cdot \mathbf{r}'}. \tag{11.44}$$

As before, we shall assume that the field point **r** is far away, in comparison to the dimensions of the source:

$$\textbf{approximation 1}: \; r' \ll r. \tag{11.45}$$

(Actually, r' is a variable of integration; approximation 1 means that the *maximum* value of r', as it ranges over the source, is much less than r.) On this assumption,

$$\mathscr{r} \cong r\left(1 - \frac{\mathbf{r} \cdot \mathbf{r}'}{r^2}\right), \tag{11.46}$$

so

$$\frac{1}{\mathscr{r}} \cong \frac{1}{r}\left(1 + \frac{\mathbf{r} \cdot \mathbf{r}'}{r^2}\right) \tag{11.47}$$

and

$$\rho(\mathbf{r}', t - \mathscr{r}/c) \cong \rho\left(\mathbf{r}', t - \frac{r}{c} + \frac{\hat{\mathbf{r}} \cdot \mathbf{r}'}{c}\right).$$

Expanding ρ as a Taylor series in t about the retarded time at the origin,

$$t_0 \equiv t - \frac{r}{c}, \tag{11.48}$$

we have

$$\rho(\mathbf{r}', t - \mathscr{r}/c) \cong \rho(\mathbf{r}', t_0) + \dot{\rho}(\mathbf{r}', t_0)\left(\frac{\hat{\mathbf{r}} \cdot \mathbf{r}'}{c}\right) + \ldots \tag{11.49}$$

where the dot signifies differentiation with respect to time. The next terms in the series would be

$$\frac{1}{2}\ddot{\rho}\left(\frac{\hat{\mathbf{r}} \cdot \mathbf{r}'}{c}\right)^2, \quad \frac{1}{3!}\dddot{\rho}\left(\frac{\hat{\mathbf{r}} \cdot \mathbf{r}'}{c}\right)^3, \quad \ldots .$$

We can afford to drop them, provided

$$\textbf{approximation 2}: \quad r' \ll \frac{c}{|\ddot{\rho}/\dot{\rho}|}, \frac{c}{|\dddot{\rho}/\dot{\rho}|^{1/2}}, \frac{c}{|\ddddot{\rho}/\dot{\rho}|^{1/3}}, \dots \tag{11.50}$$

For an oscillating system each of these ratios is c/ω, and we recover the old approximation 2. In the general case it's more difficult to interpret Eq. 11.50, but as a *procedural* matter approximations 1 and 2 amount to *keeping only the first-order terms in* r'.

Putting Eqs. 11.47 and 11.49 into the formula for V (Eq. 11.43), and again discarding the second-order term:

$$V(\mathbf{r}, t) \cong \frac{1}{4\pi\epsilon_0 r}\left[\int \rho(\mathbf{r}', t_0)\, d\tau' + \frac{\hat{\mathbf{r}}}{r} \cdot \int \mathbf{r}'\rho(\mathbf{r}', t_0)\, d\tau' + \frac{\hat{\mathbf{r}}}{c} \cdot \frac{d}{dt}\int \mathbf{r}'\rho(\mathbf{r}', t_0)\, d\tau'\right].$$

The first integral is simply the total charge, Q, at time t_0. Because charge is conserved, however, Q is actually *independent* of time. The other two integrals represent the electric dipole moment at time t_0. Thus

$$V(\mathbf{r}, t) \cong \frac{1}{4\pi\epsilon_0}\left[\frac{Q}{r} + \frac{\hat{\mathbf{r}} \cdot \mathbf{p}(t_0)}{r^2} + \frac{\hat{\mathbf{r}} \cdot \dot{\mathbf{p}}(t_0)}{rc}\right]. \tag{11.51}$$

In the static case, the first two terms are the monopole and dipole contributions to the multipole expansion for V; the third term, of course, would not be present.

Meanwhile, the vector potential is

$$\mathbf{A}(\mathbf{r}, t) = \frac{\mu_0}{4\pi}\int \frac{\mathbf{J}(\mathbf{r}', t - \mathscr{r}/c)}{\mathscr{r}}\, d\tau'. \tag{11.52}$$

As you'll see in a moment, to first order in r' it suffices to replace $\mathscr{r}$ by r in the integrand:

$$\mathbf{A}(\mathbf{r}, t) \cong \frac{\mu_0}{4\pi r}\int \mathbf{J}(\mathbf{r}', t_0)\, d\tau'. \tag{11.53}$$

According to Prob. 5.7, the integral of $\mathbf{J}$ is the time derivative of the dipole moment, so

$$\mathbf{A}(\mathbf{r}, t) \cong \frac{\mu_0}{4\pi}\frac{\dot{\mathbf{p}}(t_0)}{r}. \tag{11.54}$$

Now you see why it was unnecessary to carry the approximation of $\mathscr{r}$ beyond the zeroth order ($\mathscr{r} \cong r$): $\mathbf{p}$ is *already* first order in r', and any refinements would be corrections of *second* order.

Next we must calculate the fields. Once again, we are interested in the radiation zone (that is, in the fields that survive at large distances from the source), so we keep only those terms that go like $1/r$:

$$\textbf{approximation 3}: \quad \text{discard } 1/r^2 \text{ terms in } \mathbf{E} \text{ and } \mathbf{B}. \tag{11.55}$$

For instance, the Coulomb field,

$$\mathbf{E} = \frac{1}{4\pi\epsilon_0}\frac{Q}{r^2}\hat{\mathbf{r}},$$

coming from the first term in Eq. 11.51, does not contribute to the electromagnetic radiation. In fact, the radiation comes entirely from those terms in which we differentiate the argument t_0. From Eq. 11.48 it follows that

$$\nabla t_0 = -\frac{1}{c}\nabla r = -\frac{1}{c}\hat{\mathbf{r}},$$

and hence

$$\nabla V \cong \nabla\left[\frac{1}{4\pi\epsilon_0}\frac{\hat{\mathbf{r}}\cdot\dot{\mathbf{p}}(t_0)}{rc}\right] \cong \frac{1}{4\pi\epsilon_0}\left[\frac{\hat{\mathbf{r}}\cdot\ddot{\mathbf{p}}(t_0)}{rc}\right]\nabla t_0 = -\frac{1}{4\pi\epsilon_0 c^2}\frac{[\hat{\mathbf{r}}\cdot\ddot{\mathbf{p}}(t_0)]}{r}\hat{\mathbf{r}}.$$

Similarly,

$$\nabla\times\mathbf{A} \cong \frac{\mu_0}{4\pi r}[\nabla\times\dot{\mathbf{p}}(t_0)] = \frac{\mu_0}{4\pi r}[(\nabla t_0)\times\ddot{\mathbf{p}}(t_0)] = -\frac{\mu_0}{4\pi rc}[\hat{\mathbf{r}}\times\ddot{\mathbf{p}}(t_0)],$$

while

$$\frac{\partial\mathbf{A}}{\partial t} \cong \frac{\mu_0}{4\pi}\frac{\ddot{\mathbf{p}}(t_0)}{r}.$$

So

$$\boxed{\mathbf{E}(\mathbf{r},t) \cong \frac{\mu_0}{4\pi r}[(\hat{\mathbf{r}}\cdot\ddot{\mathbf{p}})\hat{\mathbf{r}}-\ddot{\mathbf{p}}] = \frac{\mu_0}{4\pi r}[\hat{\mathbf{r}}\times(\hat{\mathbf{r}}\times\ddot{\mathbf{p}})],} \tag{11.56}$$

where $\ddot{\mathbf{p}}$ is evaluated at time $t_0 = t - r/c$, and

$$\boxed{\mathbf{B}(\mathbf{r},t) \cong -\frac{\mu_0}{4\pi rc}[\hat{\mathbf{r}}\times\ddot{\mathbf{p}}].} \tag{11.57}$$

In particular, if we use spherical polar coordinates, with the z axis in the direction of $\ddot{\mathbf{p}}(t_0)$, then

$$\left.\begin{aligned} \mathbf{E}(r,\theta,t) &\cong \frac{\mu_0\ddot{p}(t_0)}{4\pi}\left(\frac{\sin\theta}{r}\right)\hat{\boldsymbol{\theta}}, \\ \mathbf{B}(r,\theta,t) &\cong \frac{\mu_0\ddot{p}(t_0)}{4\pi c}\left(\frac{\sin\theta}{r}\right)\hat{\boldsymbol{\phi}}. \end{aligned}\right\} \tag{11.58}$$

The Poynting vector is

$$\mathbf{S} \cong \frac{1}{\mu_0}(\mathbf{E}\times\mathbf{B}) = \frac{\mu_0}{16\pi^2 c}[\ddot{p}(t_0)]^2\left(\frac{\sin^2\theta}{r^2}\right)\hat{\mathbf{r}}, \tag{11.59}$$

and the total radiated power is

$$P \cong \int \mathbf{S}\cdot d\mathbf{a} = \frac{\mu_0\ddot{p}^2}{6\pi c}. \tag{11.60}$$

Notice that $\mathbf{E}$ and $\mathbf{B}$ are mutually perpendicular, transverse to the direction of propagation ($\hat{\mathbf{r}}$), and in the ratio $E/B = c$, as always for radiation fields.

Example 11.2

(a) In the case of an oscillating electric dipole,

$$p(t) = p_0 \cos(\omega t), \quad \ddot{p}(t) = -\omega^2 p_0 \cos(\omega t),$$

and we recover the results of Sect. 11.1.2.

(b) For a single point charge q, the dipole moment is

$$\mathbf{p}(t) = q\mathbf{d}(t),$$

where $\mathbf{d}$ is the position of q with respect to the origin. Accordingly,

$$\ddot{\mathbf{p}}(t) = q\mathbf{a}(t),$$

where $\mathbf{a}$ is the acceleration of the charge. In this case the power radiated (Eq. 11.60) is

$$P = \frac{\mu_0 q^2 a^2}{6\pi c}. \tag{11.61}$$

This is the famous **Larmor formula**; I'll derive it again, by rather different means, in the next section. Notice that the power radiated by a point charge is proportional to the *square* of its *acceleration.*

What I have done in this section amounts to a multipole expansion of the retarded potentials, carried to the lowest order in r' that is capable of producing electromagnetic radiation (fields that go like $1/r$). This turns out to be the electric dipole term. Because charge is conserved, an electric *monopole* does not radiate—if charge were *not* conserved, the first term in Eq. 11.51 would read

$$V_{\text{mono}} = \frac{1}{4\pi\epsilon_0}\frac{Q(t_0)}{r},$$

and we would get a monopole field proportional to $1/r$:

$$\mathbf{E}_{\text{mono}} = \frac{1}{4\pi\epsilon_0 c}\frac{\dot{Q}(t_0)}{r}\hat{\mathbf{r}}.$$

You might think that a charged sphere whose radius oscillates in and out would radiate, but it *doesn't*—the field outside, according to Gauss's law, is exactly $(Q/4\pi\epsilon_0 r^2)\hat{\mathbf{r}}$, regardless of the fluctuations in size. (In the acoustical analog, by the way, monopoles *do* radiate: witness the croak of a bullfrog.)

If the electric dipole moment should happen to vanish (or, at any rate, if its second time derivative is zero), then there is no electric dipole radiation, and one must look to the next term: the one of *second* order in r'. As it happens, this term can be separated into two parts, one of which is related to the *magnetic* dipole moment of the source, the other to its electric *quadrupole* moment. (The former is a generalization of the magnetic dipole radiation we considered in Sect. 11.1.3.) If the magnetic dipole and electric quadrupole contributions vanish, the $(r')^3$ term must be considered. This yields magnetic quadrupole and electric octopole radiation . . . and so it goes.

Problem 11.8 Apply Eqs. 11.59 and 11.60 to the rotating dipole of Prob. 11.4. Explain any apparent discrepancies with your previous answer.

Problem 11.9 An insulating circular ring (radius b) lies in the xy plane, centered at the origin. It carries a linear charge density $\lambda = \lambda_0 \sin\phi$, where λ_0 is constant and ϕ is the usual azimuthal angle. The ring is now set spinning at a constant angular velocity ω about the z axis. Calculate the power radiated.

Problem 11.10 An electron is released from rest and falls under the influence of gravity. In the first centimeter, what fraction of the potential energy lost is radiated away?

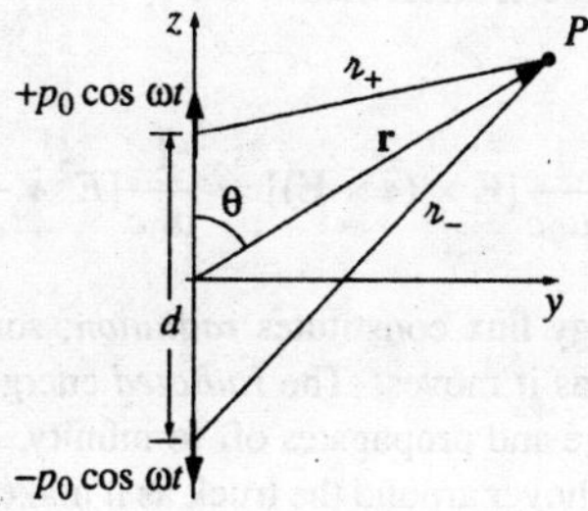

Figure 11.10

! **Problem 11.11** As a model for electric quadrupole radiation, consider two oppositely oriented oscillating electric dipoles, separated by a distance d, as shown in Fig. 11.10. Use the results of Sect. 11.1.2 for the potentials of each dipole, but note that they are *not* located at the origin. Keeping only the terms of first order in d:

(a) Find the scalar and vector potentials.

(b) Find the electric and magnetic fields.

(c) Find the Poynting vector and the power radiated. Sketch the intensity profile as a function of θ.

! **Problem 11.12** A current $I(t)$ flows around the circular ring in Fig. 11.8. Derive the general formula for the power radiated (analogous to Eq. 11.60), expressing your answer in terms of the magnetic dipole moment ($m(t)$) of the loop. [*Answer:* $P = \mu_0 \ddot{m}^2/6\pi c^3$]

11.2 Point Charges

11.2.1 Power Radiated by a Point Charge

In Chapter 10 we derived the fields of a point charge q in arbitrary motion (Eqs. 10.65 and 10.66):

$$\mathbf{E}(\mathbf{r}, t) = \frac{q}{4\pi\epsilon_0} \frac{\mathscr{r}}{(\boldsymbol{\mathscr{r}} \cdot \mathbf{u})^3} [(c^2 - v^2)\mathbf{u} + \boldsymbol{\mathscr{r}} \times (\mathbf{u} \times \mathbf{a})], \tag{11.62}$$

where $\mathbf{u} = c\hat{\boldsymbol{\mathscr{r}}} - \mathbf{v}$, and

$$\mathbf{B}(\mathbf{r}, t) = \frac{1}{c}\hat{\boldsymbol{\mathscr{r}}} \times \mathbf{E}(\mathbf{r}, t). \tag{11.63}$$

The first term in Eq. 11.62 is called the **velocity field**, and the second one (with the triple cross-product) is called the **acceleration field**.

The Poynting vector is

$$\mathbf{S} = \frac{1}{\mu_0}(\mathbf{E} \times \mathbf{B}) = \frac{1}{\mu_0 c}[\mathbf{E} \times (\hat{\boldsymbol{\mathscr{r}}} \times \mathbf{E})] = \frac{1}{\mu_0 c}[E^2\hat{\boldsymbol{\mathscr{r}}} - (\hat{\boldsymbol{\mathscr{r}}} \cdot \mathbf{E})\mathbf{E}]. \tag{11.64}$$

However, not all of this energy flux constitutes *radiation*; some of it is just field energy carried along by the particle as it moves. The *radiated* energy is the stuff that, in effect, *detaches* itself from the charge and propagates off to infinity. (It's like flies breeding on a garbage truck: Some of them hover around the truck as it makes its rounds; others fly away and never come back.) To calculate the total power radiated by the particle at time t_r, we draw a huge sphere of radius $\mathscr{r}$ (Fig. 11.11), centered at the position of the particle (at time t_r), wait the appropriate interval

$$t - t_r = \frac{\mathscr{r}}{c} \tag{11.65}$$

for the radiation to reach the sphere, and at that moment integrate the Poynting vector over the surface.[6] I have used the notation t_r because, in fact, this *is* the retarded time for all points on the sphere at time t.

Now, the area of the sphere is proportional to $\mathscr{r}^2$, so any term in $\mathbf{S}$ that goes like $1/\mathscr{r}^2$ will yield a finite answer, but terms like $1/\mathscr{r}^3$ or $1/\mathscr{r}^4$ will contribute nothing in the limit $\mathscr{r} \to \infty$. For this reason only the *acceleration* fields represent true radiation (hence their other name, **radiation fields**):

$$\mathbf{E}_{\text{rad}} = \frac{q}{4\pi\epsilon_0} \frac{\mathscr{r}}{(\boldsymbol{\mathscr{r}} \cdot \mathbf{u})^3} [\boldsymbol{\mathscr{r}} \times (\mathbf{u} \times \mathbf{a})]. \tag{11.66}$$

---

[6]Note the subtle change in strategy here: In Sect. 11.1 we worked from a fixed point (the origin), but here it is more appropriate to use the (moving) location of the charge. The implications of this change in perspective will become clearer in a moment.

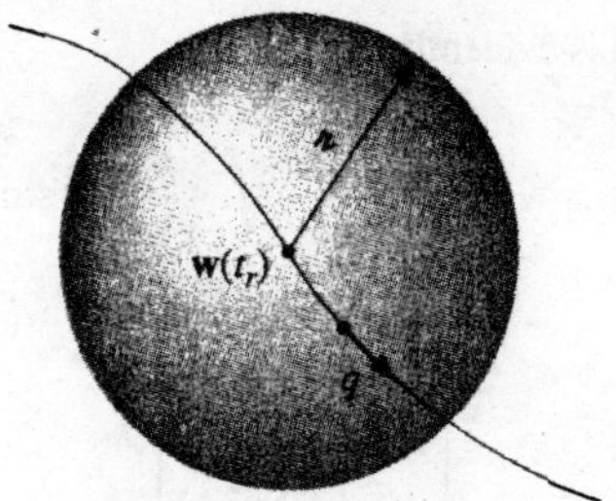

Figure 11.11

The velocity fields carry energy, to be sure, and as the charge moves this energy is dragged along—but it's not *radiation*. (It's like the flies that stay with the garbage truck.) Now $\mathbf{E}_{\text{rad}}$ is perpendicular to $\hat{\boldsymbol{\mathscr{r}}}$, so the second term in Eq. 11.64 vanishes:

$$\mathbf{S}_{\text{rad}} = \frac{1}{\mu_0 c} E_{\text{rad}}^2 \,\hat{\boldsymbol{\mathscr{r}}}. \tag{11.67}$$

If the charge is instantaneously at *rest* (at time t_r), then $\mathbf{u} = c\hat{\boldsymbol{\mathscr{r}}}$, and

$$\mathbf{E}_{\text{rad}} = \frac{q}{4\pi\epsilon_0 c^2 \mathscr{r}}[\hat{\boldsymbol{\mathscr{r}}} \times (\hat{\boldsymbol{\mathscr{r}}} \times \mathbf{a})] = \frac{\mu_0 q}{4\pi \mathscr{r}}[(\hat{\boldsymbol{\mathscr{r}}} \cdot \mathbf{a})\,\hat{\boldsymbol{\mathscr{r}}} - \mathbf{a}]. \tag{11.68}$$

In that case

$$\mathbf{S}_{\text{rad}} = \frac{1}{\mu_0 c}\left(\frac{\mu_0 q}{4\pi \mathscr{r}}\right)^2 [a^2 - (\hat{\boldsymbol{\mathscr{r}}} \cdot \mathbf{a})^2]\,\hat{\boldsymbol{\mathscr{r}}} = \frac{\mu_0 q^2 a^2}{16\pi^2 c}\left(\frac{\sin^2\theta}{\mathscr{r}^2}\right)\hat{\boldsymbol{\mathscr{r}}}, \tag{11.69}$$

where θ is the angle between $\hat{\boldsymbol{\mathscr{r}}}$ and $\mathbf{a}$. No power is radiated in the forward or backward direction—rather, it is emitted in a donut about the direction of instantaneous acceleration (Fig. 11.12).

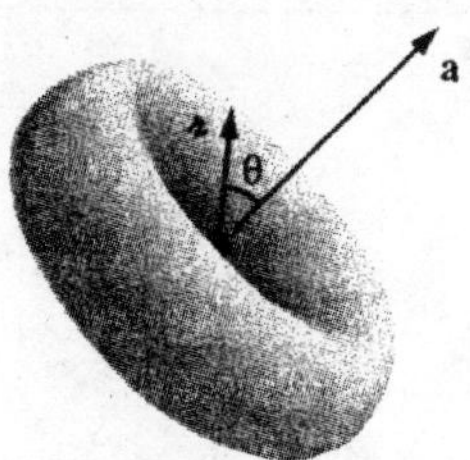

Figure 11.12

The total power radiated is evidently

$$P = \oint \mathbf{S}_{\text{rad}} \cdot d\mathbf{a} = \frac{\mu_0 q^2 a^2}{16\pi^2 c} \int \frac{\sin^2\theta}{\mathscr{r}^2} \mathscr{r}^2 \sin\theta \, d\theta \, d\phi,$$

or

$$\boxed{P = \frac{\mu_0 q^2 a^2}{6\pi c}.} \tag{11.70}$$

This, again, is the **Larmor formula**, which we obtained earlier by another route (Eq. 11.61).

Although I *derived* them on the assumption that $v = 0$, Eqs. 11.69 and 11.70 actually hold to good approximation as long as $v \ll c$. An exact treatment of the case $v \neq 0$ is more difficult,[7] both for the obvious reason that $\mathbf{E}_{\text{rad}}$ is more complicated, and also for the more subtle reason that $\mathbf{S}_{\text{rad}}$, the rate at which energy passes through the sphere, is *not* the same as the rate at which energy left the particle. Suppose someone is firing a stream of bullets out the window of a moving car (Fig. 11.13). The rate N_t at which the bullets strike a stationary target is not the same as the rate N_g at which they left the gun, because of the motion of the car. In fact, you can easily check that $N_g = (1 - v/c)N_t$, if the car is moving towards the target, and

$$N_g = \left(1 - \frac{\hat{\mathscr{r}} \cdot \mathbf{v}}{c}\right) N_t$$

for arbitrary directions (here $\mathbf{v}$ is the velocity of the car, c is that of the bullets—relative to the ground—and $\hat{\mathscr{r}}$ is a unit vector from car to target). In our case, if dW/dt is the rate at which energy passes through the sphere at radius $\mathscr{r}$, then the rate at which energy left the charge was

$$\frac{dW}{dt_r} = \frac{dW/dt}{\partial t_r/\partial t} = \left(\frac{\boldsymbol{\mathscr{r}} \cdot \mathbf{u}}{\mathscr{r} c}\right) \frac{dW}{dt}. \tag{11.71}$$

Figure 11.13

[7] In the context of special relativity, the condition $v = 0$ simply represents an astute choice of reference system, with no essential loss of generality. If you can decide how P transforms, you can *deduce* the general (Liénard) result from the $v = 0$ (Larmor) formula (see Prob. 12.69).

(I used Eq. 10.71 to express $\partial t_r/\partial t$.) But

$$\frac{\boldsymbol{\mathscr{r}}\cdot\mathbf{u}}{\mathscr{r}c} = 1 - \frac{\hat{\boldsymbol{\mathscr{r}}}\cdot\mathbf{v}}{c},$$

which is precisely the ratio of N_g to N_t; it's a purely geometrical factor (the same as in the Doppler effect).

The power radiated by the particle into a patch of area $\mathscr{r}^2 \sin\theta\, d\theta\, d\phi = \mathscr{r}^2\, d\Omega$ on the sphere is therefore given by

$$\frac{dP}{d\Omega} = \left(\frac{\boldsymbol{\mathscr{r}}\cdot\mathbf{u}}{\mathscr{r}c}\right)\frac{1}{\mu_0 c}E_{\text{rad}}^2\mathscr{r}^2 = \frac{q^2}{16\pi^2\epsilon_0}\frac{|\hat{\boldsymbol{\mathscr{r}}}\times(\mathbf{u}\times\mathbf{a})|^2}{(\hat{\boldsymbol{\mathscr{r}}}\cdot\mathbf{u})^5}, \tag{11.72}$$

where $d\Omega = \sin\theta\, d\theta\, d\phi$ is the **solid angle** into which this power is radiated. Integrating over θ and ϕ to get the total power radiated is no picnic, and for once I shall simply quote the answer:

$$P = \frac{\mu_0 q^2\gamma^6}{6\pi c}\left(a^2 - \left|\frac{\mathbf{v}\times\mathbf{a}}{c}\right|^2\right), \tag{11.73}$$

where $\gamma \equiv 1/\sqrt{1-v^2/c^2}$. This is **Liénard's generalization** of the Larmor formula (to which it reduces when $v \ll c$). The factor γ^6 means that the radiated power increases enormously as the particle velocity approaches the speed of light.

Example 11.3

Suppose $\mathbf{v}$ and $\mathbf{a}$ are instantaneously collinear (at time t_r), as, for example, in straight-line motion. Find the angular distribution of the radiation (Eq. 11.72) and the total power emitted.

Solution: In this case $(\mathbf{u}\times\mathbf{a}) = c(\hat{\boldsymbol{\mathscr{r}}}\times\mathbf{a})$, so

$$\frac{dP}{d\Omega} = \frac{q^2c^2}{16\pi^2\epsilon_0}\frac{|\hat{\boldsymbol{\mathscr{r}}}\times(\hat{\boldsymbol{\mathscr{r}}}\times\mathbf{a})|^2}{(c-\hat{\boldsymbol{\mathscr{r}}}\cdot\mathbf{v})^5}.$$

Now

$$\hat{\boldsymbol{\mathscr{r}}}\times(\hat{\boldsymbol{\mathscr{r}}}\times\mathbf{a}) = (\hat{\boldsymbol{\mathscr{r}}}\cdot\mathbf{a})\hat{\boldsymbol{\mathscr{r}}} - \mathbf{a}, \quad \text{so } |\hat{\boldsymbol{\mathscr{r}}}\times(\hat{\boldsymbol{\mathscr{r}}}\times\mathbf{a})|^2 = a^2 - (\hat{\boldsymbol{\mathscr{r}}}\cdot\mathbf{a})^2.$$

In particular, if we let the z axis point along $\mathbf{v}$, then

$$\frac{dP}{d\Omega} = \frac{\mu_0 q^2 a^2}{16\pi^2 c}\frac{\sin^2\theta}{(1-\beta\cos\theta)^5}, \tag{11.74}$$

where $\beta \equiv v/c$. This is consistent, of course, with Eq. 11.69, in the case $v = 0$. However, for very large v ($\beta \approx 1$) the donut of radiation (Fig. 11.12) is stretched out and pushed forward by the factor $(1-\beta\cos\theta)^{-5}$, as indicated in Fig. 11.14. Although there is still no radiation in *precisely* the forward direction, most of it is concentrated within an increasingly narrow cone *about* the forward direction (see Prob. 11.15).

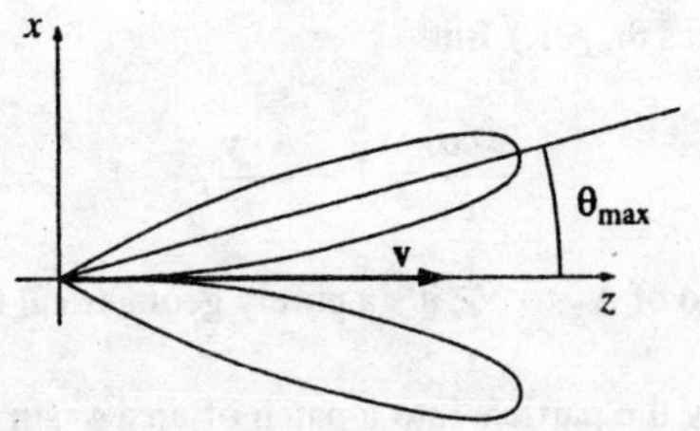

Figure 11.14

The *total* power emitted is found by integrating Eq. 11.74 over all angles:

$$P = \int \frac{dP}{d\Omega}\,d\Omega = \frac{\mu_0 q^2 a^2}{16\pi^2 c}\int \frac{\sin^2\theta}{(1-\beta\cos\theta)^5}\sin\theta\,d\theta\,d\phi.$$

The ϕ integral is 2π; the θ integral is simplified by the substitution $x \equiv \cos\theta$:

$$P = \frac{\mu_0 q^2 a^2}{8\pi c}\int_{-1}^{+1}\frac{(1-x^2)}{(1-\beta x)^5}\,dx.$$

Integration by parts yields $\frac{4}{3}(1-\beta^2)^{-3}$, and I conclude that

$$P = \frac{\mu_0 q^2 a^2 \gamma^6}{6\pi c}. \tag{11.75}$$

This result is consistent with the Liénard formula (Eq. 11.73), for the case of collinear **v** and **a**. Notice that the angular distribution of the radiation is the same whether the particle is *accelerating* or *decelerating*; it only depends on the *square* of a, and is concentrated in the forward direction (with respect to the velocity) in either case. When a high speed electron hits a metal target it rapidly decelerates, giving off what is called **bremsstrahlung**, or "braking radiation." What I have described in this example is essentially the classical theory of bremsstrahlung.

Problem 11.13

(a) Suppose an electron decelerated at a constant rate a from some initial velocity v_0 down to zero. What fraction of its initial kinetic energy is lost to radiation? (The rest is absorbed by whatever mechanism keeps the acceleration constant.) Assume $v_0 \ll c$ so that the Larmor formula can be used.

(b) To get a sense of the numbers involved, suppose the initial velocity is thermal (around 10^5 m/s) and the distance the electron goes is 30 Å. What can you conclude about radiation losses for the electrons in an ordinary conductor?

Problem 11.14 In Bohr's theory of hydrogen, the electron in its ground state was supposed to travel in a circle of radius 5×10^{-11}m, held in orbit by the Coulomb attraction of the proton.

According to classical electrodynamics, this electron should radiate, and hence spiral in to the nucleus. Show that $v \ll c$ for most of the trip (so you can use the Larmor formula), and calculate the lifespan of Bohr's atom. (Assume each revolution is essentially circular.)

Problem 11.15 Find the angle θ_{max} at which the maximum radiation is emitted, in Ex. 11.3 (see Fig. 11.14). Show that for ultrarelativistic speeds (v close to c), $\theta_{max} \cong \sqrt{(1-\beta)/2}$. What is the intensity of the radiation in this maximal direction (in the ultrarelativistic case), in proportion to the same quantity for a particle instantaneously at rest? Give your answer in terms of γ.

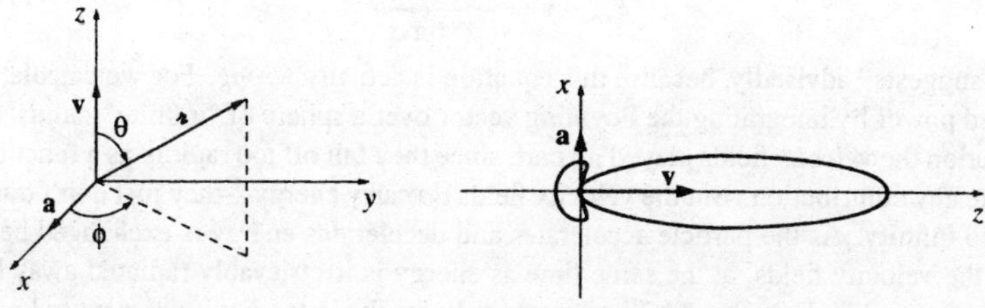

Figure 11.15 Figure 11.16

! **Problem 11.16** In Ex. 11.3 we assumed the velocity and acceleration were (instantaneously, at least) *collinear.* Carry out the same analysis for the case where they are *perpendicular.* Choose your axes so that $\mathbf{v}$ lies along the z axis and $\mathbf{a}$ along the x axis (Fig. 11.15), so that $\mathbf{v} = v\,\hat{\mathbf{z}}$, $\mathbf{a} = a\,\hat{\mathbf{x}}$, and $\hat{\boldsymbol{\imath}} = \sin\theta\cos\phi\,\hat{\mathbf{x}} + \sin\theta\sin\phi\,\hat{\mathbf{y}} + \cos\theta\,\hat{\mathbf{z}}$. Check that P is consistent with the Liénard formula. [*Answer:*

$$\frac{dP}{d\Omega} = \frac{\mu_0 q^2 a^2}{16\pi^2 c}\frac{[(1-\beta\cos\theta)^2 - (1-\beta^2)\sin^2\theta\cos^2\phi]}{(1-\beta\cos\theta)^5}, \quad P = \frac{\mu_0 q^2 a^2 \gamma^4}{6\pi c}.$$

For relativistic velocities ($\beta \approx 1$) the radiation is again sharply peaked in the forward direction (Fig. 11.16). The most important application of these formulas is to *circular* motion—in this case the radiation is called **synchrotron radiation.** For a relativistic electron the radiation sweeps around like a locomotive's headlight as the particle moves.]

11.2.2 Radiation Reaction

According to the laws of classical electrodynamics, an accelerating charge radiates. This radiation carries off energy, which must come at the expense of the particle's kinetic energy. Under the influence of a given force, therefore, a charged particle accelerates *less* than a neutral one of the same mass. The radiation evidently exerts a force ($\mathbf{F}_{rad}$) back on the charge—a *recoil* force, rather like that of a bullet on a gun. In this section we'll derive the

radiation reaction force from conservation of energy. Then in the next section I'll show you the actual *mechanism* responsible, and derive the reaction force again in the context of a simple model.

For a nonrelativistic particle ($v \ll c$) the total power radiated is given by the Larmor formula (Eq. 11.70):

$$P = \frac{\mu_0 q^2 a^2}{6\pi c}. \tag{11.76}$$

Conservation of energy suggests that this is also the rate at which the particle *loses* energy, under the influence of the radiation reaction force $\mathbf{F}_{\text{rad}}$:

$$\mathbf{F}_{\text{rad}} \cdot \mathbf{v} = -\frac{\mu_0 q^2 a^2}{6\pi c}. \tag{11.77}$$

I say "suggests" advisedly, because this equation is actually *wrong*. For we calculated the radiated power by integrating the Poynting vector over a sphere of "infinite" radius; in this calculation the *velocity* fields played no part, since they fall off too rapidly as a function of $\mathscr{r}$ to make any contribution. But the velocity fields *do* carry energy—they just don't transport it out to infinity. As the particle accelerates and decelerates energy is exchanged between it and the velocity fields, at the same time as energy is irretrievably radiated away by the acceleration fields. Equation 11.77 accounts only for the latter, but if we want to know the recoil force exerted by the fields on the charge, we need to consider the *total* power lost at any instant, not just the portion that eventually escapes in the form of radiation. (The term "radiation reaction" is a misnomer. We should really call it the *field reaction.* In fact, we'll soon see that $\mathbf{F}_{\text{rad}}$ is determined by the *time derivative* of the acceleration and can be nonzero even when the acceleration itself is instantaneously zero, so that the particle is not radiating.)

The energy lost by the particle in any given time interval, then, must equal the energy carried away by the radiation *plus* whatever extra energy has been pumped into the velocity fields.[8] However, if we agree to consider only intervals over which the system returns to its initial state, then the energy in the velocity fields is the same at both ends, and the only *net* loss is in the form of radiation. Thus Eq. 11.77, while incorrect *instantaneously*, is valid on the *average*:

$$\int_{t_1}^{t_2} \mathbf{F}_{\text{rad}} \cdot \mathbf{v}\, dt = -\frac{\mu_0 q^2}{6\pi c} \int_{t_1}^{t_2} a^2\, dt, \tag{11.78}$$

with the stipulation that the *state of the system is identical at* t_1 *and* t_2. In the case of periodic motion, for instance, we must integrate over an integral number of full cycles.[9] Now, the

[8] Actually, while the total field is the sum of velocity and acceleration fields, $\mathbf{E} = \mathbf{E}_v + \mathbf{E}_a$, the *energy* is proportional to $E^2 = E_v^2 + 2\mathbf{E}_v \cdot \mathbf{E}_a + E_a^2$ and contains *three* terms: energy stored in the velocity fields alone (E_v^2), energy radiated away (E_a^2), and a cross term $\mathbf{E}_v \cdot \mathbf{E}_a$. For the sake of simplicity, I'm referring to the *combination* ($E_v^2 + 2\mathbf{E}_v \cdot \mathbf{E}_a$) as "energy stored in the velocity fields." These terms go like $1/\mathscr{r}^4$ and $1/\mathscr{r}^3$, respectively, so neither one contributes to the radiation.

[9] For *non*periodic motion the condition that the energy in the velocity fields be the same at t_1 and t_2 is more difficult to achieve. It is not enough that the instantaneous velocities and accelerations be equal, since the fields farther out depend on v and a at *earlier times*. In *principle*, then, v and *a and all higher derivatives* must be identical at t_1 and t_2. In *practice*, since the velocity fields fall off rapidly with $\mathscr{r}$, it is sufficient that v and a be the same over a brief interval prior to t_1 and t_2.

right side of Eq. 11.78 can be integrated by parts:

$$\int_{t_1}^{t_2} a^2\,dt = \int_{t_1}^{t_2}\left(\frac{d\mathbf{v}}{dt}\right)\cdot\left(\frac{d\mathbf{v}}{dt}\right)dt = \left(\mathbf{v}\cdot\frac{d\mathbf{v}}{dt}\right)\Bigg|_{t_1}^{t_2} - \int_{t_1}^{t_2}\frac{d^2\mathbf{v}}{dt^2}\cdot\mathbf{v}\,dt.$$

The boundary term drops out, since the velocities and accelerations are identical at t_1 and t_2, so Eq. 11.78 can be written equivalently as

$$\int_{t_1}^{t_2}\left(\mathbf{F}_{\text{rad}} - \frac{\mu_0 q^2}{6\pi c}\dot{\mathbf{a}}\right)\cdot\mathbf{v}\,dt = 0. \tag{11.79}$$

Equation 11.79 will certainly be satisfied if

$$\boxed{\mathbf{F}_{\text{rad}} = \frac{\mu_0 q^2}{6\pi c}\dot{\mathbf{a}}.} \tag{11.80}$$

This is the **Abraham-Lorentz formula** for the radiation reaction force.

Of course, Eq. 11.79 doesn't *prove* Eq. 11.80. It tells you nothing whatever about the component of $\mathbf{F}_{\text{rad}}$ *perpendicular* to $\mathbf{v}$; and it only tells you the *time average* of the parallel component—the average, moreover, over very special time intervals. As we'll see in the next section, there are other reasons for believing in the Abraham-Lorentz formula, but for now the best that can be said is that it represents the *simplest* form the radiation reaction force could take, consistent with conservation of energy.

The Abraham-Lorentz formula has disturbing implications, which are not entirely understood nearly a century after the law was first proposed. For suppose a particle is subject to no *external* forces; then Newton's second law says

$$F_{\text{rad}} = \frac{\mu_0 q^2}{6\pi c}\dot{a} = ma,$$

from which it follows that

$$a(t) = a_0 e^{t/\tau}, \tag{11.81}$$

where

$$\tau \equiv \frac{\mu_0 q^2}{6\pi mc}. \tag{11.82}$$

(In the case of the electron, $\tau = 6\times10^{-24}$s.) The acceleration spontaneously *increases* exponentially with time! This absurd conclusion can be avoided if we insist that $a_0 = 0$, but it turns out that the systematic exclusion of such **runaway solutions** has an even more unpleasant consequence: If you *do* apply an external force, the particle starts to respond *before the force acts!* (See Prob. 11.19.) This **acausal preacceleration** jumps the gun by only a short time τ; nevertheless, it is (to my mind) philosophically repugnant that the theory should countenance it at all.[10]

[10]These difficulties persist in the relativistic version of the Abraham-Lorentz equation, which can be derived by starting with Liénard's formula instead of Larmor's (see Prob. 12.70). Perhaps they are telling us that there can be no such thing as a point charge in classical electrodynamics, or maybe they presage the onset of quantum mechanics. For guides to the literature see Philip Pearle's chapter in D. Teplitz, ed., *Electromagnetism: Paths to Research* (New York: Plenum, 1982) and F. Rohrlich, *Am. J. Phys.* **65**, 1051 (1997).

Example 11.4

Calculate the **radiation damping** of a charged particle attached to a spring of natural frequency ω_0, driven at frequency ω.

Solution: The equation of motion is

$$m\ddot{x} = F_{\text{spring}} + F_{\text{rad}} + F_{\text{driving}} = -m\omega_0^2 x + m\tau\dddot{x} + F_{\text{driving}}.$$

With the system oscillating at frequency ω,

$$x(t) = x_0 \cos(\omega t + \delta),$$

so

$$\dddot{x} = -\omega^2 \dot{x}.$$

Therefore

$$m\ddot{x} + m\gamma\dot{x} + m\omega_0^2 x = F_{\text{driving}}, \tag{11.83}$$

and the damping factor γ is given by

$$\gamma = \omega^2 \tau. \tag{11.84}$$

[When I wrote $F_{\text{damping}} = -\gamma m v$, back in Chap. 9 (Eq. 9.152), I assumed for simplicity that the damping was proportional to the velocity. We now know that *radiation* damping, at least, is proportional to $\ddot{v}$. But it hardly matters: for sinusoidal oscillations *any* even number of derivatives of v would do, since they're all proportional to v.]

Problem 11.17

(a) A particle of charge q moves in a circle of radius R at a constant speed v. To sustain the motion, you must, of course, provide a centripetal force mv^2/R; what *additional* force ($\mathbf{F}_e$) must you exert, in order to counteract the radiation reaction? [It's easiest to express the answer in terms of the instantaneous velocity $\mathbf{v}$.] What power (P_e) does this extra force deliver? Compare P_e with the power radiated (use the Larmor formula).

(b) Repeat part (a) for a particle in simple harmonic motion with amplitude A and angular frequency ω ($\mathbf{w}(t) = A\cos(\omega t)\,\hat{\mathbf{z}}$). Explain the discrepancy.

(c) Consider the case of a particle in free fall (constant acceleration g). What is the radiation reaction force? What is the power radiated? Comment on these results.

Problem 11.18

(a) Assuming (implausibly) that γ is entirely attributable to radiation damping (Eq. 11.84), show that for optical dispersion the damping is "small" ($\gamma \ll \omega_0$). Assume that the relevant resonances lie in or near the optical frequency range.

(b) Using your results from Prob. 9.24, estimate the width of the anomalous dispersion region, for the model in Prob. 9.23.

! **Problem 11.19** With the inclusion of the radiation reaction force (Eq. 11.80), Newton's second law for a charged particle becomes

$$a = \tau \dot{a} + \frac{F}{m},$$

where F is the external force acting on the particle.

(a) In contrast to the case of an *uncharged* particle ($a = F/m$), acceleration (like position and velocity) must now be a *continuous* function of time, even if the force changes abruptly. (Physically, the radiation reaction damps out any rapid change in a.) *Prove* that a is continuous at any time t, by integrating the equation of motion above from $(t - \epsilon)$ to $(t + \epsilon)$ and taking the limit $\epsilon \to 0$.

(b) A particle is subjected to a constant force F, beginning at time $t = 0$ and lasting until time T. Find the most general solution $a(t)$ to the equation of motion in each of the three periods: (i) $t < 0$; (ii) $0 < t < T$; (iii) $t > T$.

(c) Impose the continuity condition (a) at $t = 0$ and $t = T$. Show that you can *either* eliminate the runaway in region (iii) *or* avoid preacceleration in region (i), *but not both*.

(d) If you choose to eliminate the runaway, what is the acceleration as a function of time, in each interval? How about the velocity? (The latter must, of course, be continuous at $t = 0$ and $t = T$.) Assume the particle was originally at rest: $v(-\infty) = 0$.

(e) Plot $a(t)$ and $v(t)$, both for an *uncharged* particle and for a (nonrunaway) charged particle, subject to this force.

11.2.3 The Physical Basis of the Radiation Reaction

In the last section I derived the Abraham-Lorentz formula for the radiation reaction, using conservation of energy. I made no attempt to identify the actual *mechanism* responsible for this force, except to point out that it must be a recoil effect of the particle's own fields acting back on the charge. Unfortunately, the fields of a point charge blow up right at the particle, so it's hard to see how one can calculate the force they exert.[11] Let's avoid this problem by considering an *extended* charge distribution, for which the field is finite everywhere; at the end, we'll take the limit as the size of the charge goes to zero. In general, the electromagnetic force of one part (A) on another part (B) is *not* equal and opposite to the force of B on A (Fig. 11.17). If the distribution is divided up into infinitesimal chunks, and the imbalances are added up for all such pairs, the result is a *net force of the charge* on itself. It is this **self-force**, resulting from the breakdown of Newton's third law within the structure of the particle, that accounts for the radiation reaction.

Lorentz originally calculated the electromagnetic self-force using a spherical charge distribution, which seems reasonable but makes the mathematics rather cumbersome.[12] Because I am only trying to elucidate the *mechanism* involved, I shall use a less realistic model: a "dumbbell" in which the total charge q is divided into two halves separated by

[11] It can be done by a suitable averaging of the field, but it's not easy. See T. H. Boyer, *Am. J. Phys.* **40**, 1843 (1972), and references cited there.

[12] See J. D. Jackson, *Classical Electrodynamics*, 3rd ed., Sect. 16.3 (New York: John Wiley, 1999).

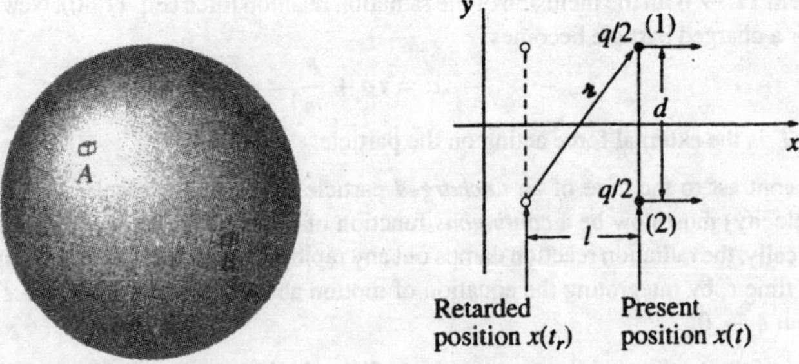

Figure 11.17 Figure 11.18

a fixed distance d (Fig. 11.18). This is the simplest possible arrangement of the charge that permits the essential mechanism (imbalance of internal electromagnetic forces) to function. Never mind that it's an unlikely model for an elementary particle: in the point limit ($d \to 0$) *any* model must yield the Abraham-Lorentz formula, to the extent that conservation of energy alone dictates that answer.

Let's assume the dumbbell moves in the x direction, and is (instantaneously) at rest at the retarded time. The electric field at (1) due to (2) is

$$\mathbf{E}_1 = \frac{(q/2)}{4\pi\epsilon_0}\frac{\mathscr{r}}{(\boldsymbol{\mathscr{r}}\cdot\mathbf{u})^3}[(c^2+\boldsymbol{\mathscr{r}}\cdot\mathbf{a})\mathbf{u}-(\boldsymbol{\mathscr{r}}\cdot\mathbf{u})\mathbf{a}] \tag{11.85}$$

(Eq. 10.65), where

$$\mathbf{u} = c\,\hat{\boldsymbol{\mathscr{r}}} \quad \text{and} \quad \boldsymbol{\mathscr{r}} = l\,\hat{\mathbf{x}} + d\,\hat{\mathbf{y}}, \tag{11.86}$$

so that

$$\boldsymbol{\mathscr{r}}\cdot\mathbf{u} = c\mathscr{r}, \quad \boldsymbol{\mathscr{r}}\cdot\mathbf{a} = la, \quad \text{and} \quad \mathscr{r} = \sqrt{l^2+d^2}. \tag{11.87}$$

Actually, we're only interested in the x component of $\mathbf{E}_1$, since the y components will cancel when we add the forces on the two ends (for the same reason, we don't need to worry about the magnetic forces). Now

$$u_x = \frac{cl}{\mathscr{r}}, \tag{11.88}$$

and hence

$$E_{1_x} = \frac{q}{8\pi\epsilon_0 c^2}\frac{(lc^2-ad^2)}{(l^2+d^2)^{3/2}}. \tag{11.89}$$

By symmetry, $E_{2_x} = E_{1_x}$, so the net force on the dumbbell is

$$\mathbf{F}_{\text{self}} = \frac{q}{2}(\mathbf{E}_1+\mathbf{E}_2) = \frac{q^2}{8\pi\epsilon_0 c^2}\frac{(lc^2-ad^2)}{(l^2+d^2)^{3/2}}\,\hat{\mathbf{x}}. \tag{11.90}$$

So far everything is exact. The idea now is to expand in powers of d; when the size of the particle goes to zero, all *positive* powers will disappear. Using Taylor's theorem

$$x(t) = x(t_r) + \dot{x}(t_r)(t - t_r) + \frac{1}{2}\ddot{x}(t_r)(t - t_r)^2 + \frac{1}{3!}\dddot{x}(t_r)(t - t_r)^3 + \cdots,$$

we have,

$$l = x(t) - x(t_r) = \frac{1}{2}aT^2 + \frac{1}{6}\dot{a}T^3 + \cdots, \tag{11.91}$$

where $T \equiv t - t_r$, for short. Now T is determined by the retarded time condition

$$(cT)^2 = l^2 + d^2, \tag{11.92}$$

so

$$d = \sqrt{(cT)^2 - l^2} = cT\sqrt{1 - \left(\frac{aT}{2c} + \frac{\dot{a}T^2}{6c} + \cdots\right)^2} = cT - \frac{a^2}{8c}T^3 + (\;)T^4 + \cdots.$$

This equation tells us d, in terms of T; we need to "solve" it for T as a function of d. There's a systematic procedure for doing this, known as **reversion of series**,[13] but we can get the first couple of terms more informally as follows: Ignoring all higher powers of T,

$$d \cong cT \quad\Rightarrow\quad T \cong \frac{d}{c};$$

using this as an approximation for the cubic term,

$$d \cong cT - \frac{a^2}{8c}\frac{d^3}{c^3} \quad\Rightarrow\quad T \cong \frac{d}{c} + \frac{a^2 d^3}{8c^5},$$

and so on. Evidently

$$T = \frac{1}{c}d + \frac{a^2}{8c^5}d^3 + (\;)d^4 + \cdots. \tag{11.93}$$

Returning to Eq. 11.91, we construct the power series for l in terms of d:

$$l = \frac{a}{2c^2}d^2 + \frac{\dot{a}}{6c^3}d^3 + (\;)d^4 + \cdots. \tag{11.94}$$

Putting this into Eq. 11.90, I conclude that

$$\mathbf{F}_{\text{self}} = \frac{q^2}{4\pi\epsilon_0}\left[-\frac{a}{4c^2 d} + \frac{\dot{a}}{12c^3} + (\;)d + \cdots\right]\hat{\mathbf{x}}. \tag{11.95}$$

Here a and $\dot{a}$ are evaluated at the *retarded* time (t_r), but it's easy to rewrite the result in terms of the *present* time t:

$$a(t_r) = a(t) + \dot{a}(t)(t - t_r) + \cdots = a(t) - \dot{a}(t)T + \cdots = a(t) - \dot{a}(t)\frac{d}{c} + \cdots,$$

[13] See, for example, the *CRC Standard Mathematical Tables* (Cleveland: CRC Press).

and it follows that

$$\mathbf{F}_{\text{self}} = \frac{q^2}{4\pi\epsilon_0}\left[-\frac{a(t)}{4c^2d} + \frac{\dot{a}(t)}{3c^3} + (\ )d + \cdots\right]\hat{\mathbf{x}}. \tag{11.96}$$

The first term on the right is proportional to the acceleration of the charge; if we pull it over to the other side of Newton's second law, it simply adds to the dumbbell's mass. In effect, the total inertia of the charged dumbbell is

$$m = 2m_0 + \frac{1}{4\pi\epsilon_0}\frac{q^2}{4dc^2}, \tag{11.97}$$

where m_0 is the mass of either end alone. In the context of special relativity it is not surprising that the electrical repulsion of the charges should enhance the mass of the dumbbell. For the potential energy of this configuration (in the static case) is

$$\frac{1}{4\pi\epsilon_0}\frac{(q/2)^2}{d}, \tag{11.98}$$

and according to Einstein's formula $E = mc^2$, this energy contributes to the inertia of the object.[14]

The second term in Eq. 11.96 is the radiation reaction:

$$F_{\text{rad}}^{\text{int}} = \frac{\mu_0 q^2 \dot{a}}{12\pi c}. \tag{11.99}$$

It alone (apart from the mass correction[15]) survives in the "point dumbbell" limit $d \to 0$. Unfortunately, it differs from the Abraham-Lorentz formula by a factor of 2. But then, this is only the self-force associated with the *interaction* between 1 and 2—hence, the superscript "int." There remains the force of *each end on itself.* When the latter is included (see Prob. 11.20) the result is

$$F_{\text{rad}} = \frac{\mu_0 q^2 \dot{a}}{6\pi c}, \tag{11.100}$$

reproducing the Abraham-Lorentz formula exactly. *Conclusion: The radiation reaction is due to the force of the charge on itself*—or, more elaborately, the net force exerted by the fields generated by different parts of the charge distribution acting on one another.

---

[14]The fact that the *numbers* work out perfectly is a lucky feature of this configuration. If you do the same calculation for the dumbbell in *longitudinal* motion, the mass correction is only *half* of what it "should" be (there's a 2, instead of a 4, in Eq. 11.97), and for a sphere it's off by a factor of 3/4. This notorious paradox has been the subject of much debate over the years. See D. J. Griffiths and R. E. Owen, *Am. J. Phys.* **51**, 1120 (1983).

[15]Of course, the limit $d \to 0$ has an embarrassing effect on the mass term. In a sense, it doesn't matter, since only the *total* mass m is observable; maybe m_0 somehow has a compensating (negative!) infinity, so that m comes out finite. This awkward problem persists in *quantum* electrodynamics, where it is "swept under the rug" in a process known as **mass renormalization.**

Problem 11.20 Deduce Eq. 11.100 from Eq. 11.99, as follows:

(a) Use the Abraham-Lorentz formula to determine the radiation reaction on each end of the dumbbell; add this to the interaction term (Eq. 11.99).

(b) Method (a) has the defect that it *uses* the Abraham-Lorentz formula—the very thing that we were trying to *derive*. To avoid this, smear out the charge along a strip of length L oriented perpendicular to the motion (the charge density, then, is $\lambda = q/L$); find the cumulative interaction force for all pairs of segments, using Eq. 11.99 (with the correspondence $q/2 \rightarrow \lambda\, dy_1$, at one end and $q/2 \rightarrow \lambda\, dy_2$ at the other). Make sure you don't count the same pair twice.

More Problems on Chapter 11

Problem 11.21 A particle of mass m and charge q is attached to a spring with force constant k, hanging from the ceiling (Fig. 11.19). Its equilibrium position is a distance h above the floor. It is pulled down a distance d below equilibrium and released, at time $t = 0$.

(a) Under the usual assumptions ($d \ll \lambda \ll h$), calculate the intensity of the radiation hitting the floor, as a function of the distance R from the point directly below q. [*Note:* The intensity here is the average power per unit area of *floor*.] At what R is the radiation most intense? Neglect the radiative damping of the oscillator. [*Answer:* $\mu_0 q^2 d^2 \omega^4 R^2 h / 32\pi^2 c(R^2 + h^2)^{5/2}$]

(b) As a check on your formula, assume the floor is of infinite extent, and calculate the average energy per unit time striking the entire floor. Is it what you'd expect?

(c) Because it is losing energy in the form of radiation, the amplitude of the oscillation will gradually decrease. After what time τ has the amplitude been reduced to d/e? (Assume the fraction of the total energy lost in one cycle is very small.)

Problem 11.22 A radio tower rises to height h above flat horizontal ground. At the top is a magnetic dipole antenna, of radius b, with its axis vertical. FM station KRUD broadcasts from this antenna at angular frequency ω, with a total radiated power P (that's averaged, of course, over a full cycle). Neighbors have complained about problems they attribute to excessive

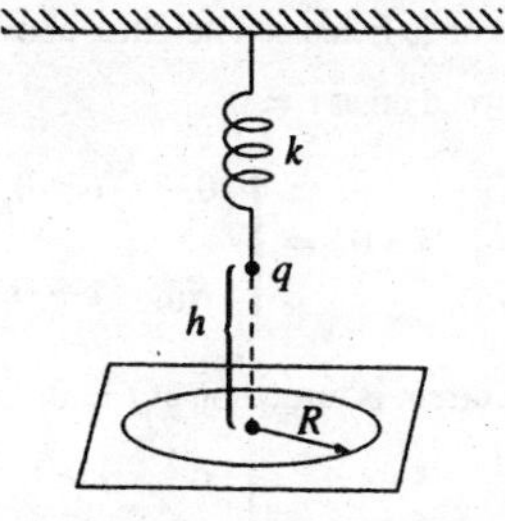

Figure 11.19

radiation from the tower—interference with their stereo systems, mechanical garage doors opening and closing mysteriously, and a variety of suspicious medical problems. But the city engineer who measured the radiation level at the base of the tower found it to be well below the accepted standard. You have been hired by the Neighborhood Association to assess the engineer's report.

(a) In terms of the variables given (not all of which may be relevant, of course), find the formula for the intensity of the radiation at ground level, a distance R from the base of the tower. You may assume that $b \ll c/\omega \ll h$. [*Note:* we are interested only in the *magnitude* of the radiation, not in its *direction*—when measurements are taken the detector will be aimed directly at the antenna.]

(b) How far from the base of the tower *should* the engineer have made the measurement? What is the formula for the intensity at this location?

(c) KRUD's actual power output is 35 kilowatts, its frequency is 90 MHz, the antenna's radius is 6 cm, and the height of the tower is 200 m. The city's radio-emission limit is 200 microwatts/cm^2. Is KRUD in compliance?

Problem 11.23 As you know, the magnetic north pole of the earth does not coincide with the geographic north pole—in fact, it's off by about 11°. Relative to the fixed axis of rotation, therefore, the magnetic dipole moment vector of the earth is changing with time, and the earth must be giving off magnetic dipole radiation.

(a) Find the formula for the total power radiated, in terms of the following parameters: Ψ (the angle between the geographic and magnetic north poles), M (the magnitude of the earth's magnetic dipole moment), and ω (the angular velocity of rotation of the earth). [*Hint:* refer to Prob. 11.4 or Prob. 11.12.]

(b) Using the fact that the earth's magnetic field is about half a gauss at the equator, estimate the magnetic dipole moment M of the earth.

(c) Find the power radiated. [*Answer:* 4×10^{-5} W]

(d) Pulsars are thought to be rotating neutron stars, with a typical radius of 10 km, a rotational period of 10^{-3} s, and a surface magnetic field of 10^8 T. What sort of radiated power would you expect from such a star? [See J. P. Ostriker and J. E. Gunn, *Astrophys. J.* **157**, 1395 (1969).] [*Answer:* 2×10^{36} W]

Problem 11.24 Suppose the (electrically neutral) yz plane carries a time-dependent but uniform surface current $K(t)\,\hat{\mathbf{z}}$.

(a) Find the electric and magnetic fields at a height x above the plane if

(i) a constant current is turned on at $t = 0$:

$$K(t) = \begin{cases} 0, & t \le 0, \\ K_0, & t > 0. \end{cases}$$

(ii) a linearly increasing current is turned on at $t = 0$:

$$K(t) = \begin{cases} 0, & t \le 0, \\ \alpha t, & t > 0. \end{cases}$$

(b) Show that the retarded vector potential can be written in the form

$$\mathbf{A}(x,t) = \frac{\mu_0 c}{2}\,\hat{\mathbf{z}} \int_0^\infty K\left(t - \frac{x}{c} - u\right) du,$$

and from this determine **E** and **B**.

(c) Show that the total power radiated per unit area of surface is

$$\frac{\mu_0 c}{2}[K(t)]^2.$$

Explain what you mean by "radiation," in this case, given that the source is not localized. [For discussion and related problems, see B. R. Holstein, *Am. J. Phys.* **63**, 217 (1995), T. A. Abbott and D. J. Griffiths, *Am. J. Phys.* **53**, 1203 (1985).]

Problem 11.25 When a charged particle approaches (or leaves) a conducting surface, radiation is emitted, associated with the changing electric dipole moment of the charge and its image. If the particle has mass m and charge q, find the total radiated power, as a function of its height z above the plane. [*Answer:* $(\mu_0 c q^2/4\pi)^3/6m^2z^4$]

Problem 11.26 Use the duality transformation (Prob. 7.60) to construct the electric and magnetic fields of a magnetic monopole q_m in arbitrary motion, and find the "Larmor formula" for the power radiated. [For related applications see J. A. Heras, *Am. J. Phys.* **63**, 242 (1995).]

Problem 11.27 Assuming you exclude the runaway solution in Prob. 11.19, calculate

(a) the work done by the external force,

(b) the final kinetic energy (assume the initial kinetic energy was zero),

(c) the total energy radiated.

Check that energy is conserved in this process.[16]

Problem 11.28

(a) Repeat Prob. 11.19, but this time let the external force be a Dirac delta function: $F(t) = k\delta(t)$ (for some constant k).[17] [Note that the acceleration is now *discontinuous* at $t = 0$ (though the *velocity* must still be continuous); use the method of Prob. 11.19 (a) to show that $\Delta a = -k/m\tau$. In this problem there are only *two* intervals to consider: (i) $t < 0$, and (ii) $t > 0$.]

(b) As in Prob. 11.27, check that energy is conserved in this process.

Problem 11.29 A charged particle, traveling in from $-\infty$ along the x axis, encounters a rectangular potential energy barrier

$$U(x) = \begin{cases} U_0, & \text{if } 0 < x < L, \\ 0, & \text{otherwise.} \end{cases}$$

Show that, because of the radiation reaction, it is possible for the particle to **tunnel** through the barrier—that is: even if the incident kinetic energy is less than U_0, the particle can pass

[16] Problems 11.27 and 11.28 were suggested by G. L. Pollack.

[17] This example was first analyzed by P. A. M. Dirac, *Proc. Roy. Soc.* **A167**, 148 (1938).

through. (See F. Denef *et al.*, *Phys. Rev. E* **56**, 3624 (1997).) [*Hint:* Your task is to solve the equation

$$a = \tau \dot{a} + \frac{F}{m},$$

subject to the force

$$F(x) = U_0[-\delta(x) + \delta(x - L)].$$

Refer to Probs. 11.19 and 11.28, but notice that this time the force is a specified function of x, not t. There are three regions to consider: (i) $x < 0$, (ii) $0 < x < L$, (iii) $x > L$. Find the general solution (for $a(t)$, $v(t)$, and $x(t)$) in each region, exclude the runaway in region (iii), and impose the appropriate boundary conditions at $x = 0$ and $x = L$. Show that the final velocity (v_f) is related to the time T spent traversing the barrier by the equation

$$L = v_f T - \frac{U_0}{m v_f}\left(\tau e^{-T/\tau} + T - \tau\right),$$

and the initial velocity (at $x = -\infty$) is

$$v_i = v_f - \frac{U_0}{m v_f}\left[1 - \frac{1}{1 + \frac{U_0}{m v_f^2}\left(e^{-T/\tau} - 1\right)}\right].$$

To simplify these results (since all we're looking for is a specific example), suppose the final kinetic energy is half the barrier height. Show that in this case

$$v_i = \frac{v_f}{1 - (L/v_f \tau)}.$$

In particular, if you choose $L = v_f \tau/4$, then $v_i = (4/3)v_f$, the initial kinetic energy is $(8/9)U_0$, and the particle makes it through, even though it didn't have sufficient energy to get over the barrier!]

! **Problem 11.30**

(a) Find the radiation reaction force on a particle moving with arbitrary velocity in a straight line, by reconstructing the argument in Sect. 11.2.3 *without* assuming $v(t_r) = 0$. [*Answer:* $(\mu_0 q^2 \gamma^4 / 6\pi c)(\dot{a} + 3\gamma^2 a^2 v/c^2)$]

(b) Show that this result is consistent (in the sense of Eq. 11.78) with the power radiated by such a particle (Eq. 11.75).

Problem 11.31

(a) Does a particle in hyperbolic motion (Eq. 10.45) radiate? (Use the exact formula (Eq. 11.75) to calculate the power radiated.)

(b) Does a particle in hyperbolic motion experience a radiation reaction? (Use the exact formula (Prob. 11.30) to determine the reaction force.)

[*Comment:* These famous questions carry important implications for the **principle of equivalence**. See T. Fulton and F. Rohrlich, *Annals of Physics* **9**, 499 (1960); J. Cohn, *Am. J. Phys.* **46**, 225 (1978); Chapter 8 of R. Peierls, *Surprises in Theoretical Physics* (Princeton: Princeton University Press, 1979); and the article by P. Pearle in *Electromagnetism: Paths to Research*, ed. D. Teplitz (New York: Plenum Press, 1982).]

Chapter 12

Electrodynamics and Relativity

12.1 The Special Theory of Relativity

12.1.1 Einstein's Postulates

Classical mechanics obeys the **principle of relativity**: the same laws apply in any **inertial reference frame**. By "inertial" I mean that the system is at rest or moving with constant velocity.[1] Imagine, for example, that you have loaded a billiard table onto a railroad car, and the train is going at constant speed down a smooth straight track. The game would proceed exactly the same as it would if the train were parked in the station; you don't have to "correct" your shots for the fact that the train is moving—indeed, if you pulled all the curtains you would have no way of knowing whether the train was moving or not. Notice by contrast that you would know it *immediately* if the train sped up, or slowed down, or turned a corner, or went over a bump—the billiard balls would roll in weird curved trajectories, and you yourself would feel a lurch. The laws of mechanics, then, are certainly *not* the same in *accelerating* reference frames.

In its application to classical mechanics, the principle of relativity is hardly new; it was stated clearly by Galileo. *Question:* does it also apply to the laws of electrodynamics? At first glance the answer would seem to be *no*. After all, a charge in motion produces a magnetic field, whereas a charge at rest does not. A charge carried along by the train would generate a magnetic field, but someone on the train, applying the laws of electrodynamics

[1] This raises an awkward problem: If the laws of physics hold just as well in a uniformly moving frame, then we have no way of identifying the "rest" frame in the first place, and hence no way of checking that some other frame is moving at constant velocity. To avoid this trap we define an inertial frame formally as *one in which Newton's first law holds*. If you want to know whether you're in an inertial frame, throw some rocks around—if they travel in straight lines at constant speed, you've got yourself an inertial frame, and any frame moving at constant velocity with respect to you will be another inertial frame (see Prob. 12.1).

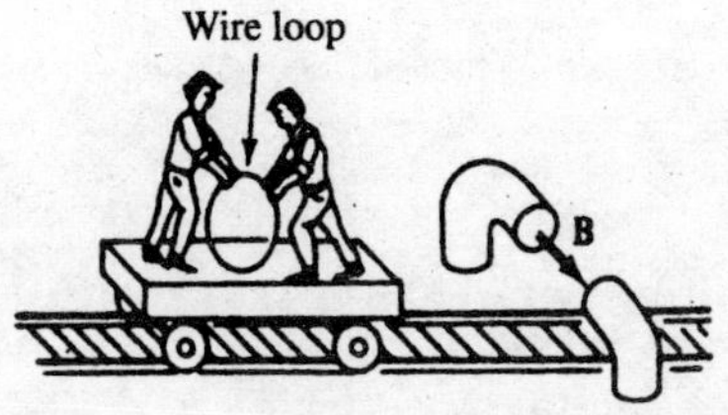

Figure 12.1

in that system, would predict no magnetic field. In fact, many of the equations of electrodynamics, starting with the Lorentz force law, make explicit reference to "the" velocity of the charge. It certainly appears, therefore, that electromagnetic theory presupposes the existence of a unique stationary reference frame, with respect to which all velocities are to be measured.

And yet there is an extraordinary coincidence that gives us pause. Suppose we mount a wire loop on a freight car, and have the train pass between the poles of a giant magnet (Fig. 12.1). As the loop rides through the magnetic field, a motional emf is established; according to the flux rule (Eq. 7.13),

$$\mathcal{E} = -\frac{d\Phi}{dt}.$$

This emf, remember, is due to the magnetic force on charges in the wire loop, which are moving along with the train. On the other hand, if someone on the train naïvely applied the laws of electrodynamics in *that* system, what would the prediction be? No *magnetic* force, because the loop is at rest. But as the magnet flies by, the magnetic field in the freight car will change, and a changing magnetic field induces an electric field, by Faraday's law. The resulting *electric* force would generate an emf in the loop given by Eq. 7.14:

$$\mathcal{E} = -\frac{d\Phi}{dt}.$$

Because Faraday's law and the flux rule predict exactly the same emf, people on the train will get the right answer, *even though their physical interpretation of the process is completely wrong.*

Or *is* it? Einstein could not believe this was a mere coincidence; he took it, rather, as a clue that electromagnetic phenomena, like mechanical ones, obey the principle of relativity. In his view the analysis by the observer on the train is just as valid as that of the observer on the ground. If their *interpretations* differ (one calling the process electric, the other magnetic), so be it; their actual *predictions* are in agreement. Here's what he wrote on the first page of his 1905 paper introducing the **special theory of relativity**:

> It is known that Maxwell's electrodynamics—as usually understood at the present time—when applied to moving bodies, leads to asymmetries which do not appear to be inherent in the phenomena. Take, for example, the reciprocal electrodynamic action of a magnet and a conductor. The observable phenomenon here depends only on the relative motion of the conductor and the magnet, whereas the customary view draws a sharp distinction between the two cases in which either one or the other of these bodies is in motion. For if the magnet is in motion and the conductor at rest, there arises in the neighborhood of the magnet an electric field . . . producing a current at the places where parts of the conductor are situated. But if the magnet is stationary and the conductor in motion, no electric field arises in the neighborhood of the magnet. In the conductor, however, we find an electromotive force . . . which gives rise—assuming equality of relative motion in the two cases discussed—to electric currents of the same path and intensity as those produced by the electric forces in the former case.
>
> Examples of this sort, together with unsuccessful attempts to discover any motion of the earth relative to the "light medium," suggest that the phenomena of electrodynamics as well as of mechanics possess no properties corresponding to the idea of absolute rest.[2]

But I'm getting ahead of the story. To Einstein's predecessors the equality of the two emf's was just a lucky accident; they had no doubt that one observer was right and the other was wrong. They thought of electric and magnetic fields as strains in an invisible jellylike medium called **ether**, which permeated all of space. The speed of the charge was to be measured *with respect to the ether*—only then would the laws of electrodynamics be valid. The train observer is wrong, because that frame is *moving* relative to the ether.

But wait a minute! How do we know the *ground* observer isn't moving relative to the ether, too? After all, the earth rotates on its axis once a day and revolves around the sun once a year; the solar system circulates around the galaxy, and for all I know the galaxy itself may be moving at a high speed through the cosmos. All told, we should be traveling at well over 50 km/s with respect to the ether. Like a motorcycle rider on the open road, we face an "ether wind" of high velocity—unless by some miraculous coincidence we just happen to find ourselves in a tailwind of precisely the right strength, or the earth has some sort of "windshield" and drags its local supply of ether along with it. Suddenly it becomes a matter of crucial importance to *find* the ether frame, experimentally, or else *all* our calculations will be invalid.

The problem, then, is to determine our motion through the ether—to measure the speed and direction of the "ether wind." How shall we do it? At first glance you might suppose that practically *any* electromagnetic experiment would suffice: If Maxwell's equations are valid only with respect to the ether frame, any discrepancy between the experimental result and the theoretical prediction would be ascribable to the ether wind. Unfortunately, as nineteenth century physicists soon realized, the anticipated error in a typical experiment is

[2]A translation of Einstein's first relativity paper, "On the Electrodynamics of Moving Bodies," is reprinted in *The Principle of Relativity*, by H. A. Lorentz *et al.* (New York: Dover, 1923).

extremely small; as in the example above, "coincidences" always seem to conspire to hide the fact that we are using the "wrong" reference frame. So it takes an uncommonly delicate experiment to do the job.

Now, among the results of classical electrodynamics is the prediction that electromagnetic waves travel through the vacuum at a speed

$$\frac{1}{\sqrt{\epsilon_0\mu_0}} = 3.00 \times 10^8 \mathrm{m/s},$$

relative (presumably) *to the ether*. In principle, then, one should be able to detect the ether wind by simply measuring the speed of light in various directions. Like a motorboat on a river, the net speed "downstream" should be a maximum, for here the light is swept along by the ether; in the opposite direction, where it is bucking the current, the speed should be a minimum (Fig. 12.2). While the *idea* of this experiment could not be simpler, its *execution* is another matter, because light travels so inconveniently fast. If it weren't for that "technical detail" you could do it all with a flashlight and a stopwatch. As it happened, an elaborate and lovely experiment was devised by Michelson and Morley, using an optical interferometer of fantastic precision. I shall not go into the details here, because I do not want to distract your attention from the two essential points: (1) all Michelson and Morley were trying to do was compare the speed of light in different directions, and (2) what they in fact *discovered* was that this speed is *exactly the same in all directions*.

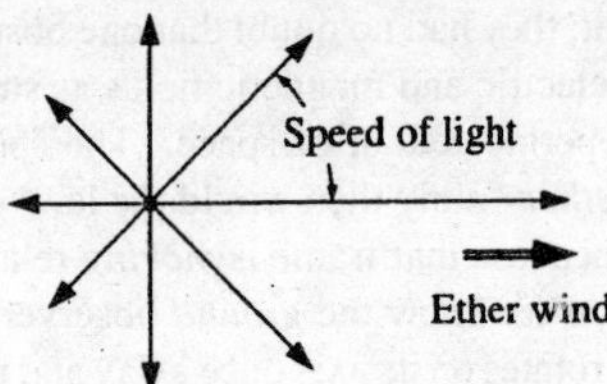

Figure 12.2

Nowadays, when students are taught in high school to snicker at the naïveté of the ether model, it takes some imagination to comprehend how utterly perplexing this result must have been at the time. All other waves (water waves, sound waves, waves on a string) travel at a prescribed speed *relative to the propagating medium* (the stuff that does the waving), and if this medium is in motion with respect to the observer, the net speed is always greater "downstream" than "upstream." Over the next 20 years a series of improbable schemes were concocted in an effort to explain why this does *not* occur with light. Michelson and Morley themselves interpreted their experiment as confirmation of the "ether drag" hypothesis, which held that the earth somehow pulls the ether along with it. But this was found to be inconsistent with other observations, notably the aberration of starlight.[3] Various so-

[3] A discussion of the Michelson-Morley experiment and related matters is to be found in R. Resnick's *Introduction to Special Relativity*, Chap. 1 (New York: John Wiley, 1968).

called "emission" theories were proposed, according to which the speed of electromagnetic waves is governed by the motion of the *source*—as it would be in a corpuscular theory (conceiving of light as a stream of particles). Such theories called for implausible modifications in Maxwell's equations, but in any event they were discredited by experiments using extraterrestrial light sources. Meanwhile, Fitzgerald and Lorentz suggested that the ether wind physically compresses all matter (including the Michelson-Morley apparatus itself) in just the right way to compensate for, and thereby conceal, the variation in speed with direction. As it turns out, there is a grain of truth in this, although their idea of the reason for the contraction was quite wrong.

At any rate, it was not until Einstein that anyone took the Michelson-Morley result at face value and suggested that the speed of light is a universal constant, the same in all directions, regardless of the motion of the observer or the source. There *is* no ether wind because there is no ether. *Any* inertial system is a suitable reference frame for the application of Maxwell's equations, and the velocity of a charge is to be measured *not* with respect to a (nonexistent) absolute rest frame, nor with respect to a (nonexistent) ether, but simply with respect to the particular reference system you happen to have chosen.

Inspired, then, both by internal theoretical hints (the fact that the laws of electrodynamics are such as to give the right answer even when applied in the "wrong" system) and by external empirical evidence (the Michelson-Morley experiment[4]), Einstein proposed his two famous postulates:

1. **The principle of relativity.** The laws of physics apply in all inertial reference systems.
2. **The universal speed of light.** The speed of light in vacuum is the same for all inertial observers, regardless of the motion of the source.

The special theory of relativity derives from these two postulates. The first elevates Galileo's observation about classical mechanics to the status of a general law, applying to *all* of physics. It states that there is no absolute rest system. The second might be considered Einstein's response to the Michelson-Morley experiment. It means that there is no ether. (Some authors consider Einstein's second postulate redundant—no more than a special case of the first. They maintain that the very existence of ether would violate the principle of relativity, in the sense that it would define a unique stationary reference frame. I think this is nonsense. The existence of air as a medium for sound does not invalidate the theory of relativity. Ether is no more an absolute rest system than the water in a goldfish bowl—which is a *special* system, if you happen to be the goldfish, but scarcely "absolute.")[5]

Unlike the principle of relativity, which had roots going back several centuries, the universal speed of light was radically new—and, on the face of it, preposterous. For if I walk 5 mi/h down the corridor of a train going 60 mi/h, my net speed relative to the ground

[4]Actually, Einstein appears to have been only dimly aware of the Michelson-Morley experiment at the time. For him, the theoretical argument alone was decisive.

[5]I put it this way in an effort to dispel some misunderstanding as to what constitutes an absolute rest frame. In 1977, it became possible to measure the speed of the earth through the 3 K background radiation left over from the "big bang." Does this mean we have found an absolute rest system, and relativity is out the window? Of course not.

is "obviously" 65 mi/h—the speed of A (me) with respect to C (ground) is equal to the speed of A relative to B (train) plus the speed of B relative to C:

$$v_{AC} = v_{AB} + v_{BC}. \tag{12.1}$$

And yet, if A is a *light* signal (whether is comes from a flashlight on the train or a lamp on the ground or a star in the sky) Einstein would have us believe that its speed is c relative to the train *and* c relative to the ground:

$$v_{AC} = v_{AB} = c. \tag{12.2}$$

Evidently, Eq. 12.1, which we now call **Galileo's velocity addition rule** (no one before Einstein would have bothered to give it a name at all) is incompatible with the second postulate. In special relativity, as we shall see, it is replaced by **Einstein's velocity addition rule**:

$$\boxed{v_{AC} = \frac{v_{AB} + v_{BC}}{1 + (v_{AB}v_{BC}/c^2)}}. \tag{12.3}$$

For "ordinary" speeds ($v_{AB} \ll c$, $v_{BC} \ll c$), the denominator is so close to 1 that the discrepancy between Galileo's formula and Einstein's formula is negligible. On the other hand, Einstein's formula has the desired property that if $v_{AB} = c$, then *automatically* $v_{AC} = c$:

$$v_{AC} = \frac{c + v_{BC}}{1 + (cv_{BC}/c^2)} = c.$$

But how can Galileo's rule, which seems to rely on nothing but common sense, possibly be wrong? And if it *is* wrong, what does this do to all of classical physics? The answer is that special relativity compels us to alter our notions of space and time themselves, and therefore also of such derived quantities as velocity, momentum, and energy. Although it developed historically out of Einstein's contemplation of electrodynamics, the special theory is not limited to any particular class of phenomena—rather, it is a description of the space-time "arena" in which *all* physical phenomena take place. And in spite of the reference to the speed of light in the second postulate, relativity has nothing to do with light: c is evidently a fundamental velocity, and it happens that light travels at that speed, but it is perfectly possible to conceive of a universe in which there are no electric charges, and hence no electromagnetic fields or waves, and yet relativity would still prevail. Because relativity defines the structure of space and time, it claims authority not merely over all presently known phenomena, but over those not yet discovered. It is, as Kant would say, a "prolegomenon to any future physics."

Problem 12.1 Use Galileo's velocity addition rule. Let S be an inertial reference system.

(a) Suppose that $\bar{S}$ moves with constant velocity relative to S. Show that $\bar{S}$ is also an inertial reference system. [*Hint:* use the definition in footnote 1.]

(b) Conversely, show that if $\bar{S}$ is an inertial system, then it moves with respect to S at constant velocity.

Problem 12.2 As an illustration of the principle of relativity in classical mechanics, consider the following generic collision: In inertial frame $\mathcal{S}$, particle A (mass m_A, velocity $\mathbf{u}_A$) hits particle B (mass m_B, velocity $\mathbf{u}_B$). In the course of the collision some mass rubs off A and onto B, and we are left with particles C (mass m_C, velocity $\mathbf{u}_C$) and D (mass m_D, velocity $\mathbf{u}_D$). Assume that momentum ($\mathbf{p} \equiv m\mathbf{u}$) is conserved in $\mathcal{S}$.

(a) Prove that momentum is also conserved in inertial frame $\bar{\mathcal{S}}$, which moves with velocity $\mathbf{v}$ relative to $\mathcal{S}$. [Use Galileo's velocity addition rule—this is an entirely classical calculation. What must you assume about mass?]

(b) Suppose the collision is elastic in $\mathcal{S}$; show that it is also elastic in $\bar{\mathcal{S}}$.

Problem 12.3

(a) What's the percent error introduced when you use Galileo's rule, instead of Einstein's, with $v_{AB} = 5$ mi/h and $v_{BC} = 60$ mi/h?

(b) Suppose you could run at half the speed of light down the corridor of a train going three-quarters the speed of light. What would your speed be relative to the ground?

(c) Prove, using Eq. 12.3, that if $v_{AB} < c$ and $v_{BC} < c$ then $v_{AC} < c$. Interpret this result.

Figure 12.3

Problem 12.4 As the outlaws escape in their getaway car, which goes $\frac{3}{4}c$, the police officer fires a bullet from the pursuit car, which only goes $\frac{1}{2}c$ (Fig. 12.3). The muzzle velocity of the bullet (relative to the gun) is $\frac{1}{3}c$. Does the bullet reach its target (a) according to Galileo, (b) according to Einstein?

12.1.2 The Geometry of Relativity

In this section I present a series of *gedanken* (thought) experiments that serve to introduce the three most striking geometrical consequences of Einstein's postulates: time dilation, Lorentz contraction, and the relativity of simultaneity. In Sect. 12.1.3 the same results will be derived more systematically, using Lorentz transformations.

(i) The relativity of simultaneity. Imagine a freight car, traveling at constant speed along a smooth, straight track (Fig. 12.4). In the very center of the car there hangs a light bulb. When someone switches it on, the light spreads out in all directions at speed c. Because the lamp is equidistant from the two ends, an observer on the train will find that the light reaches the front end at the same instant as it reaches the back end: The two events in question—(a) light reaches the front end and (b) light reaches the back end—occur *simultaneously*. However, to an observer on the *ground* these same two events are *not*

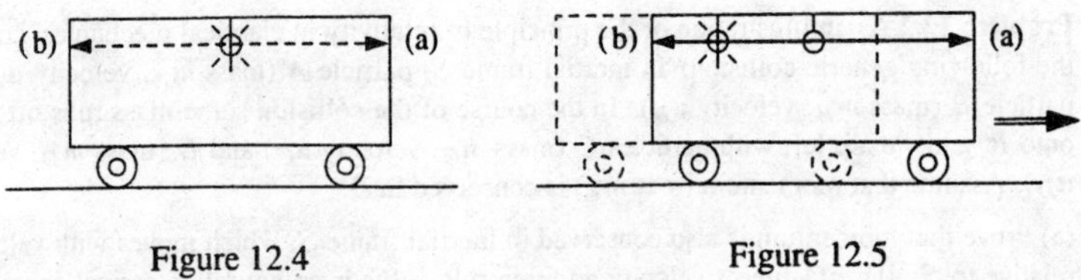

Figure 12.4 Figure 12.5

simultaneous. For as the light travels out from the bulb, the train itself moves forward, so the beam going to the back end has a shorter distance to travel than the one going forward (Fig. 12.5). According to this observer, therefore, event (b) happens *before* event (a). An observer passing by on an express train, meanwhile, would report that (a) preceded (b). *Conclusion*:

> **Two events that are simultaneous in one inertial system are not, in general, simultaneous in another.**

Naturally, the train has to be going awfully fast before the discrepancy becomes detectable—that's why you don't notice it all the time.

Of course, it is *always* possible for a naïve witness to be ***mistaken*** about simultaneity: you hear the thunder *after* you see the lightning, and a child might infer that the source of the light was not simultaneous with the source of the sound. But this is a trivial error, having nothing to do with moving observers or relativity—*obviously*, you must correct for the time the signal (sound, light, carrier pigeon, or whatever) takes to reach you. When I speak of an **observer**, I mean someone having the sense to make this correction, and an **observation** is what an observer records *after* doing so. What you *see*, therefore, is not the same as what you *observe*. An observation cannot be made with a camera—it is an artificial reconstruction after the fact, when all the data are in. In fact, a wise observer will avoid the whole problem, by stationing assistants at strategic locations, each equipped with a watch synchronized to a master clock, so that time measurements can be made right at the scene. I belabor this point in order to emphasize that the relativity of simultaneity is a genuine discrepancy between measurements made by competent observers in relative motion, not a simple mistake arising from a failure to account for the travel time of light signals.

Problem 12.5 Synchronized clocks are stationed at regular intervals, a million km apart, along a straight line. When the clock next to you reads 12 noon:

(a) What time do you *see* on the 90th clock down the line?

(b) What time do you *observe* on that clock?

Problem 12.6 Every 2 years, more or less, *The New York Times* publishes an article in which some astronomer claims to have found an object traveling faster than the speed of light. Many of these reports result from a failure to distinguish what is *seen* from what is *observed*—that is, from a failure to account for light travel time. Here's an example: A star is traveling with speed v at an angle θ to the line of sight (Fig. 12.6). What is its apparent speed across the sky?

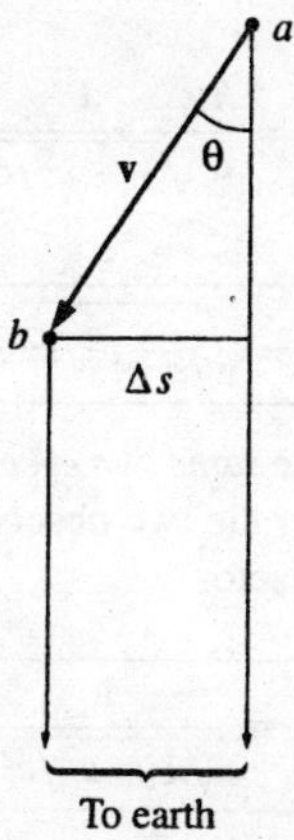

Figure 12.6

(Suppose the light signal from b reaches the earth at a time Δt after the signal from a, and the star has meanwhile advanced a distance Δs across the celestial sphere; by "apparent speed" I mean $\Delta s/\Delta t$.) What angle θ gives the maximum apparent speed? Show that the apparent speed can be much greater than c, even if v itself is less than c.

(ii) Time dilation. Now let's consider a light ray that leaves the bulb and strikes the floor of the car directly below. *Question*: How long does it take the light to make this trip? From the point of view of an observer on the train, the answer is easy: If the height of the car is h, the time is

$$\Delta \bar{t} = \frac{h}{c}. \tag{12.4}$$

(I'll use an overbar to denote measurements made on the train.) On the other hand, as observed from the ground this same ray must travel farther, because the train itself is moving. From Fig. 12.7 I see that this distance is $\sqrt{h^2 + (v\Delta t)^2}$, so

$$\Delta t = \frac{\sqrt{h^2 + (v\Delta t)^2}}{c}.$$

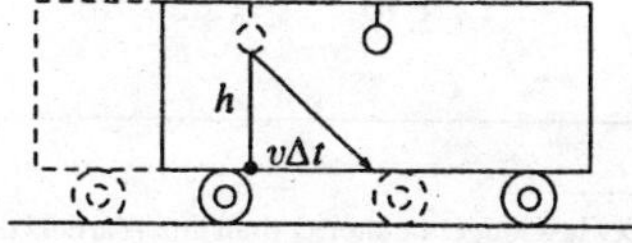

Figure 12.7

Solving for Δt, we have

$$\Delta t = \frac{h}{c}\frac{1}{\sqrt{1 - v^2/c^2}},$$

and therefore

$$\boxed{\Delta \bar{t} = \sqrt{1 - v^2/c^2}\,\Delta t.} \tag{12.5}$$

Evidently the time elapsed between the *same two events*—(a) light leaves bulb, and (b) light strikes center of floor—is different for the two observers. In fact, the interval recorded on the train clock, $\Delta \bar{t}$, is *shorter* by the factor

$$\boxed{\gamma \equiv \frac{1}{\sqrt{1 - v^2/c^2}}.} \tag{12.6}$$

Conclusion:

Moving clocks run slow.

This is called **time dilation**. It doesn't have anything to do with the mechanics of clocks; it's a statement about the nature of time, which applies to *all* properly functioning timepieces.

Of all Einstein's predictions, none has received more spectacular and persuasive confirmation than time dilation. Most elementary particles are unstable: they disintegrate after a characteristic lifetime[6] that varies from one species to the next. The lifetime of a neutron is 15 min, of a muon, 2×10^{-6} s, of a neutral pion, 9×10^{-17} s. But these are lifetimes of particles at *rest*. When particles are moving at speeds close to c they last much longer, for their internal clocks (whatever it is that tells them when their time is up) are running slow, in accordance with Einstein's time dilation formula.

Example 12.1

A muon is traveling through the laboratory at three-fifths the speed of light. How long does it last?

Solution: In this case,

$$\gamma = \frac{1}{\sqrt{1 - (3/5)^2}} = \frac{5}{4},$$

so it lives longer (than at rest) by a factor of $\frac{5}{4}$:

$$\frac{5}{4} \times (2 \times 10^{-6})\,\text{s} = 2.5 \times 10^{-6}\,\text{s}.$$

---

[6] Actually, an individual particle may last longer or shorter than this. Particle disintegration is a random process, and I should really speak of the *average* lifetime for the species. But to avoid irrelevant complication I shall pretend that every particle disintegrates after precisely the average lifetime.

It may strike you that time dilation is inconsistent with the principle of relativity. For if the ground observer says the train clock runs slow, the train observer can with equal justice claim that the *ground* clock runs slow—after all, from the train's point of view it is the ground that is in motion. Who's right? *Answer:* They're *both* right! On closer inspection the "contradiction," which seems so stark, evaporates. Let me explain: In order to check the rate of the train clock, the ground observer uses *two* of his own clocks (Fig. 12.8): one to compare times at the beginning of the interval, when the train clock passes point A, the other to compare times at the end of the interval, when the train clock passes point B. Of course, he must be careful to synchronize his clocks before the experiment. What he finds is that while the train clock ticked off, say, 3 minutes, the interval between his own two clock readings was 5 minutes. He concludes that the *train* clock runs slow.

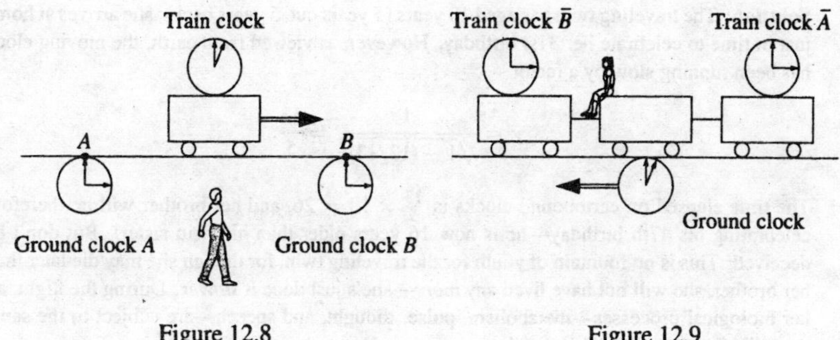

Figure 12.8

Figure 12.9

Meanwhile, the observer on the train is checking the rate of the ground clock by the same procedure: She uses two carefully synchronized train clocks, and compares times with a single ground clock as it passes by each of them in turn (Fig. 12.9). She finds that while the ground clock ticks off 3 minutes, the interval between her train clocks is 5 minutes, and concludes that the *ground* clock runs slow. Is there a contradiction? *No*, for the two observers have measured *different things*. The ground observer compared *one* train clock with *two* ground clocks; the train observer compared one *ground* clock with two *train* clocks. Each followed a sensible and correct procedure, comparing a single moving clock with two stationary ones. "So what," you say, "the stationary clocks were synchronized in each instance, so it cannot matter that they used two different ones." But there's the rub: *Clocks that are properly synchronized in one system will not be synchronized when observed from another system.* They *can't* be, for to say that two clocks are synchronized is to say that they read 12 noon *simultaneously*, and we have already learned that what's simultaneous to one observer is *not* simultaneous to another. So whereas each observer conducted a perfectly sound measurement, from his/her own point of view, the *other* observer, watching the process, considers that she/he made the most elementary blunder, by using two unsynchronized clocks. That's how, in spite of the fact that *his* clocks "actually" run slow, he manages to conclude that *hers* are running slow (and vice versa).

Because moving clocks are not synchronized, it is essential when checking time dilation to focus attention on a *single* moving clock. *All* moving clocks run slow by the same factor, but you can't start timing on one clock and then switch to another because they weren't in step to begin with. But you can use as many *stationary* clocks (stationary with respect to you, the observer) as you please, for they *are* properly synchronized (moving observers would dispute this, but that's *their* problem).

Example 12.2

The twin paradox. On her 21st birthday, an astronaut takes off in a rocket ship at a speed of $\frac{12}{13}c$. After 5 years have elapsed on her watch, she turns around and heads back at the same speed to rejoin her twin brother, who stayed at home. *Question*: How old is each twin at their reunion?

Solution: The traveling twin has aged 10 years (5 years out, 5 years back); she arrives at home just in time to celebrate her 31st birthday. However, as viewed from earth, the moving clock has been running slow by a factor

$$\gamma = \frac{1}{\sqrt{1-(12/13)^2}} = \frac{13}{5}.$$

The time elapsed on earthbound clocks is $\frac{13}{5} \times 10 = 26$, and her brother will be therefore celebrating his 47th birthday—he is now 16 years older than his twin sister! But don't be deceived: This is no fountain of youth for the traveling twin, for though she may die later than her brother, she will not have lived any *more*—she's just done it *slower*. During the flight, all her biological processes—metabolism, pulse, thought, and speech—are subject to the same time dilation that affects her watch.

The so-called **twin paradox** arises when you try to tell this story from the point of view of the *traveling* twin. She sees the *earth* fly off at $\frac{12}{13}c$, turn around after 5 years, and return. From her point of view, it would seem, *she's* at rest, whereas her *brother* is in motion, and hence it is *he* who should be younger at the reunion. An enormous amount has been written about the twin paradox, but the truth is there's really no paradox here at all: this second analysis is simply *wrong*. The two twins are *not* equivalent. The traveling twin experiences *acceleration* when she turns around to head home, but her brother does *not*. To put it in fancier language, the traveling twin is not in an inertial system—more precisely, she's in *one* inertial system on the way out and a completely different one on the way back. You'll see in Prob. 12.16 how to analyze this problem *correctly* from her point of view, but as far as the resolution of the "paradox" is concerned, it is enough to note that the *traveling twin cannot claim to be a stationary observer* because you can't undergo acceleration and remain stationary.

Problem 12.7 In a laboratory experiment a muon is observed to travel 800 m before disintegrating. A graduate student looks up the lifetime of a muon (2×10^{-6} s) and concludes that its speed was

$$v = \frac{800\,\text{m}}{2 \times 10^{-6}\,\text{s}} = 4 \times 10^8\,\text{m/s}.$$

Faster than light! Identify the student's error, and find the *actual* speed of this muon.

Problem 12.8 A rocket ship leaves earth at a speed of $\frac{3}{5}c$. When a clock on the rocket says 1 hour has elapsed, the rocket ship sends a light signal back to earth.

(a) According to *earth* clocks, when was the signal sent?

(b) According to *earth* clocks, how long after the rocket left did the signal arrive back on earth?

(c) According to the *rocket* observer, how long after the rocket left did the signal arrive back on earth?

(iii) Lorentz contraction. For the third gedanken experiment you must imagine that we have set up a lamp at one end of a boxcar and a mirror at the other, so that a light signal can be sent down and back (Fig. 12.10). *Question*: How long does the signal take to complete the round trip? To an observer on the train, the answer is

$$\Delta\bar{t} = 2\frac{\Delta\bar{x}}{c}, \tag{12.7}$$

where $\Delta\bar{x}$ is the length of the car (the overbar, as before, denotes measurements made on the train). To an observer on the ground the process is more complicated because of the motion of the train. If Δt_1 is the time for the light signal to reach the front end, and Δt_2 is the return time, then (see Fig. 12.11):

$$\Delta t_1 = \frac{\Delta x + v\Delta t_1}{c}, \quad \Delta t_2 = \frac{\Delta x - v\Delta t_2}{c},$$

or, solving for Δt_1 and Δt_2:

$$\Delta t_1 = \frac{\Delta x}{c - v}, \quad \Delta t_2 = \frac{\Delta x}{c + v}.$$

So the round-trip time is

$$\Delta t = \Delta t_1 + \Delta t_2 = 2\frac{\Delta x}{c}\frac{1}{(1 - v^2/c^2)}. \tag{12.8}$$

Meanwhile, these same intervals are related by the time dilation formula, Eq. 12.5:

$$\Delta\bar{t} = \sqrt{1 - v^2/c^2}\,\Delta t.$$

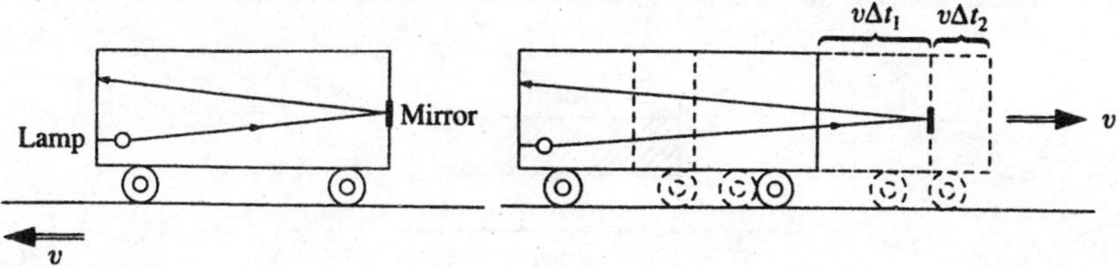

Figure 12.10

Figure 12.11

Applying this to Eqs. 12.7 and 12.8, I conclude that

$$\boxed{\Delta\bar{x} = \frac{1}{\sqrt{1 - v^2/c^2}}\Delta x.} \tag{12.9}$$

The length of the boxcar is not the same when measured by an observer on the ground, as it is when measured by an observer on the train—from the ground point of view it is somewhat *shorter*. *Conclusion*:

Moving objects are shortened.

We call this **Lorentz contraction.** Notice that the same factor,

$$\gamma \equiv \frac{1}{\sqrt{1 - v^2/c^2}},$$

appears in both the time dilation formula and the Lorentz contraction formula. This makes it all very easy to remember: Moving clocks run slow, moving sticks are shortened, and the factor is always γ.

Of course, the observer on the train doesn't think her car is shortened—her meter sticks are contracted by that same factor, so all her measurements come out the same as when the train was standing in the station. In fact, from *her* point of view it is objects on the *ground* that are shortened. This raises again a paradoxical problem: If A says B's sticks are short, and B says A's sticks are short, who is right? *Answer:* They *both* are! But to reconcile the rival claims we must study carefully the actual process by which length is measured.

Suppose you want to find the length of a board. If it's at rest (with respect to you) you simply lay your ruler down next to the board, record the readings at each end, and subtract them to get the length of the board (Fig. 12.12). (If you're really clever, you'll line up the left end of the ruler against the left end of the board—then you only have to read *one* number.)

But what if the board is *moving*? Same story, only this time, of course, you must be careful to read the two ends *at the same instant of time.* If you don't, the board will move in the course of measurement, and obviously you'll get the wrong answer. But therein lies the problem: Because of the relativity of simultaneity the two observers disagree on what constitutes "the same instant of time." When the person on the ground measures the length of the boxcar, he reads the position of the two ends at the same instant *in his system.* But

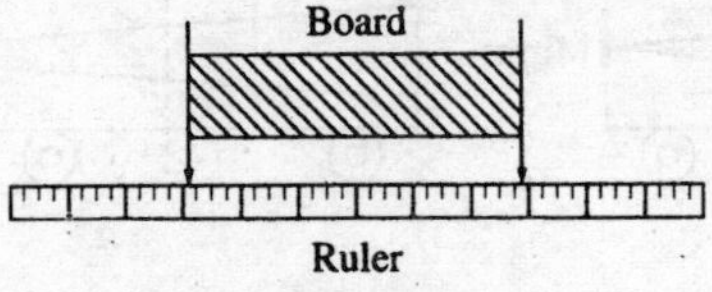

Figure 12.12

the person on the train, watching him do it, complains that he read the front end first, then waited a moment before reading the back end. *Naturally*, he came out short, in spite of the fact that (to her) he was using an undersized meter stick, which would otherwise have yielded a number too *large*. Both observers measure lengths correctly (from the point of view of their respective inertial frames) and each finds the other's sticks to be shortened. Yet there is no inconsistency, for they are measuring different things, and each considers the other's method improper.

Example 12.3

The barn and ladder paradox. Unlike time dilation, there is no direct experimental confirmation of Lorentz contraction, simply because it's too difficult to get an object of measurable size going anywhere near the speed of light. The following parable illustrates how bizarre the world would be if the speed of light were more accessible.

There once was a farmer who had a ladder too long to store in his barn (Fig. 12.13a). He chanced one day to read some relativity, and a solution to his problem suggested itself. He instructed his daughter to run with the ladder as fast as she could—the moving ladder having Lorentz-contracted to a size the barn could easily accommodate, she was to rush through the door, whereupon the farmer would slam it behind her, capturing the ladder inside (Fig. 12.13b). The daughter, however, has read somewhat farther in the relativity book; she points out that in *her* reference frame the *barn*, not the ladder, will contract, and the fit will be even worse than it was with the two at rest (Fig. 12.13c). *Question*: Who's right? Will the ladder fit inside the barn, or won't it?

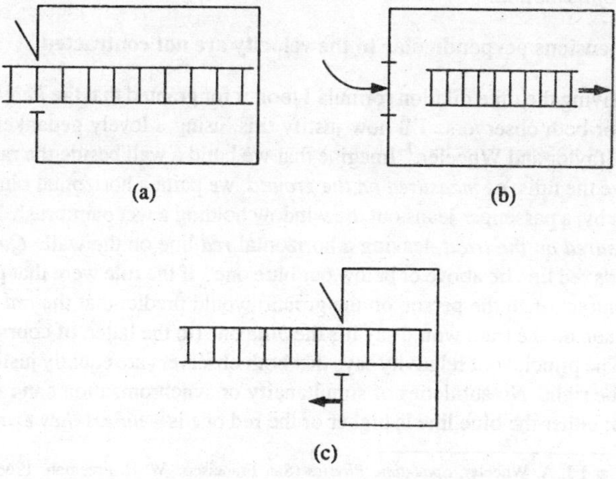

Figure 12.13

Solution: They're *both* right! When you say "the ladder is in the barn," you mean that all parts of it are inside *at one instant of time*, but in view of the relativity of simultaneity, that's a condition that depends on the observer. There are really *two* relevant events here:

a. Back end of ladder makes it in the door.
b. Front end of ladder hits far wall of barn.

The farmer says *a* occurs before *b*, so there *is* a time when the ladder is entirely within the barn; his daughter says *b* precedes *a*, so there is *not*. *Contradiction?* Nope—just a difference in perspective.

"But *come* now," I hear you protest, "when it's all over and the dust clears, either the ladder is inside the barn, or it isn't. There can be no dispute about *that*." Quite so, but now you're introducing a new element into the story: What happens *as the ladder is brought to a stop?* Suppose the farmer grabs the last rung of the ladder firmly with one hand, while he slams the door with the other. Assuming it remains intact, the ladder must now stretch out to its rest length. Evidently, the front end keeps going, even after the rear end has been stopped! Expanding like an accordian, the front end of the ladder smashes into the far side of the barn. In truth, the whole notion of a "rigid" object loses its meaning in relativity, for when it changes its speed, different parts do not in general accelerate simultaneously—in this way the material stretches or shrinks to reach the length appropriate to its new velocity.

But to return to the question at hand: When the ladder finally comes to a stop, is it inside the barn or not? The answer is indeterminate. When the front end of the ladder hits the far side of the barn, something has to give, and the farmer is left either with a broken ladder inside the barn or with the ladder intact poking through a hole in the wall. In any event, he is unlikely to be pleased with the outcome.

One final comment on Lorentz contraction. A moving object is shortened *only along the direction of its motion:*

Dimensions perpendicular to the velocity are not contracted.

Indeed, in deriving the time dilation formula I took it for granted that the *height* of the train is the same for both observers. I'll now justify this, using a lovely gedanken experiment suggested by Taylor and Wheeler.[7] Imagine that we build a wall beside the railroad tracks, and 1 m above the rails, *as measured on the ground*, we paint a horizontal blue line. When the train goes by, a passenger leans out the window holding a wet paintbrush 1 m above the rails, *as measured on the train*, leaving a horizontal *red* line on the wall. *Question*: Does the passenger's red line lie above or below our blue one? If the rule were that perpendicular directions contract, then the person on the ground would predict that the *red* line is lower, while the person on the train would say it's the *blue* one (to the latter, of course, the *ground* is moving). The principle of relativity says that both observers are equally justified, but they cannot both be right. No subtleties of simultaneity or synchronization can rationalize this contradiction; either the blue line is higher or the red one is—*unless they exactly coincide*,

---

[7]E. F. Taylor and J. A. Wheeler, *Spacetime Physics* (San Francisco: W. H. Freeman, 1966). A somewhat different version of the same argument is given in J. H. Smith, *Introduction to Special Relativity* (Champaign, IL: Stipes, 1965).

which is the inescapable conclusion. There *cannot* be a law of contraction (or expansion) of perpendicular dimensions, for it would lead to irreconcilably inconsistent predictions.

Problem 12.9 A Lincoln Continental is twice as long as a VW Beetle, when they are at rest. As the Continental overtakes the VW, going through a speed trap, a (stationary) policeman observes that they both have the same length. The VW is going at half the speed of light. How fast is the Lincoln going? (Leave your answer as a multiple of c.)

Problem 12.10 A sailboat is manufactured so that the mast leans at an angle $\bar{\theta}$ with respect to the deck. An observer standing on a dock sees the boat go by at speed v (Fig. 12.14). What angle does this *observer* say the mast makes?

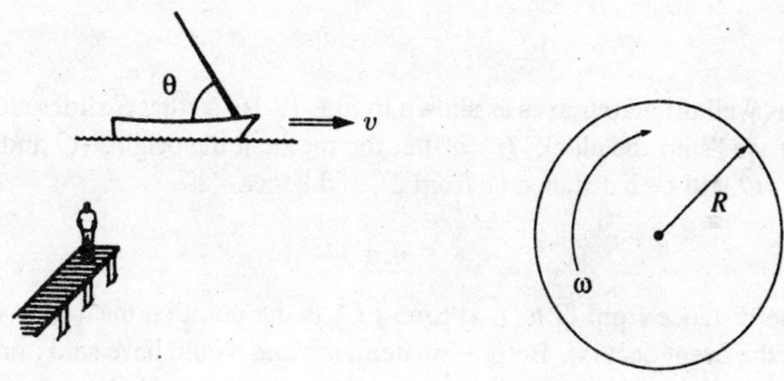

Figure 12.14 Figure 12.15

! **Problem 12.11** A record turntable of radius R rotates at angular velocity ω (Fig. 12.15). The circumference is presumably Lorentz-contracted, but the radius (being perpendicular to the velocity) is *not*. What's the ratio of the circumference to the diameter, in terms of ω and R? According to the rules of ordinary geometry, that has to be π. What's going on here? [This is known as **Ehrenfest's paradox**; for discussion and references see H. Arzelies, *Relativistic Kinematics*, Chap. IX (Elmsford, NY: Pergamon Press, 1966) and T. A. Weber, *Am. J. Phys.* **65**, 486 (1997).]

12.1.3 The Lorentz Transformations

Any physical process consists of one or more **events**. An "event" is something that takes place at a specific location (x, y, z), at a precise time (t). The explosion of a firecracker, for example, is an event; a tour of Europe is not. Suppose that we know the coordinates (x, y, z, t) of a particular event E in *one* inertial system $\mathcal{S}$, and we would like to calculate the coordinates $(\bar{x}, \bar{y}, \bar{z}, \bar{t})$ of that *same event* in some other inertial system $\bar{\mathcal{S}}$. What we need is a "dictionary" for translating from the language of $\mathcal{S}$ to the language of $\bar{\mathcal{S}}$.

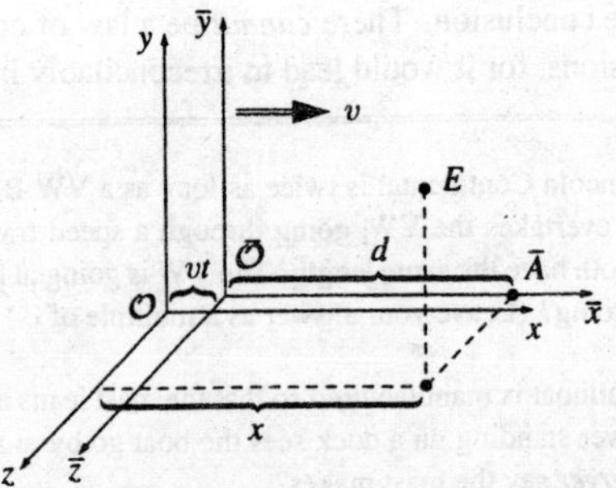

Figure 12.16

We may as well orient our axes as shown in Fig. 12.16, so that $\bar{S}$ slides along the x axis at speed v. If we "start the clock" ($t = 0$) at the moment the origins ($\mathcal{O}$ and $\bar{\mathcal{O}}$) coincide, then at time t, $\bar{\mathcal{O}}$ will be a distance vt from $\mathcal{O}$, and hence

$$x = d + vt, \tag{12.10}$$

where d is the distance from $\bar{\mathcal{O}}$ to $\bar{A}$ at time t ($\bar{A}$ is the point on the $\bar{x}$ axis which is even with E when the event occurs). Before Einstein, anyone would have said immediately that

$$d = \bar{x}, \tag{12.11}$$

and thus constructed the "dictionary"

$$\left.\begin{array}{ll} \text{(i)} & \bar{x} = x - vt, \\ \text{(ii)} & \bar{y} = y, \\ \text{(iii)} & \bar{z} = z, \\ \text{(iv)} & \bar{t} = t. \end{array}\right\} \tag{12.12}$$

These are now called the **Galilean transformations**, though they scarcely deserve so fine a title—the last one, in particular, went without saying, since everyone assumed the flow of time was the same for all observers. In the context of special relativity, however, we must expect (iv) to be replaced by a rule that incorporates time dilation, the relativity of simultaneity, and the nonsynchronization of moving clocks. Likewise, there will be a modification in (i) to account for Lorentz contraction. As for (ii) and (iii), they, at least, remain unchanged, for we have already seen that there can be no modification of lengths perpendicular to the motion.

But where does the classical derivation of (i) break down? *Answer:* In Eq. 12.11. For d is the distance from $\bar{\mathcal{O}}$ to $\bar{A}$ *as measured in* S, whereas $\bar{x}$ is the distance from $\bar{\mathcal{O}}$ to $\bar{A}$ *as measured in* $\bar{\mathcal{S}}$. Because $\bar{\mathcal{O}}$ and $\bar{A}$ are at rest in $\bar{\mathcal{S}}$, $\bar{x}$ is the "moving stick," which appears contracted to $\mathcal{S}$:

$$d = \frac{1}{\gamma}\bar{x}. \tag{12.13}$$

When this is inserted in Eq. 12.10 we obtain the relativistic version of (i):

$$\bar{x} = \gamma(x - vt). \tag{12.14}$$

Of course, we could have run the same argument from the point of view of $\bar{\mathcal{S}}$. The diagram (Fig. 12.17) looks similar, but in this case it depicts the scene *at time* $\bar{t}$, whereas Fig. 12.16 showed the scene *at time* t. (Note that t and $\bar{t}$ represent the same physical instant *at* E, but not elsewhere, because of the relativity of simultaneity.) If we assume that $\bar{\mathcal{S}}$ also starts the clock when the origins coincide, then at time $\bar{t}$, $\mathcal{O}$ will be a distance $v\bar{t}$ from $\bar{\mathcal{O}}$, and therefore

$$\bar{x} = \bar{d} - v\bar{t}, \tag{12.15}$$

where $\bar{d}$ is the distance from $\mathcal{O}$ to A at time $\bar{t}$, and A is that point on the x axis which is even with E when the event occurs. The classical physicist would have said that $x = \bar{d}$, and, using (iv), recovered (i). But, as before, relativity demands that we observe a subtle distinction: x is the distance from $\mathcal{O}$ to A *in* $\mathcal{S}$, whereas $\bar{d}$ is the distance from $\mathcal{O}$ to A *in* $\bar{\mathcal{S}}$. Because $\mathcal{O}$ and A are at rest in $\mathcal{S}$, x is the "moving stick," and

$$\bar{d} = \frac{1}{\gamma}x. \tag{12.16}$$

It follows that

$$x = \gamma(\bar{x} + v\bar{t}). \tag{12.17}$$

This last equation comes as no surprise, for the symmetry of the situation dictates that the formula for x, in terms of $\bar{x}$ and $\bar{t}$, should be identical to the formula for $\bar{x}$ in terms of x and t (Eq. 12.14), except for a switch in the sign of v. (If $\bar{\mathcal{S}}$ is going to the *right* at speed

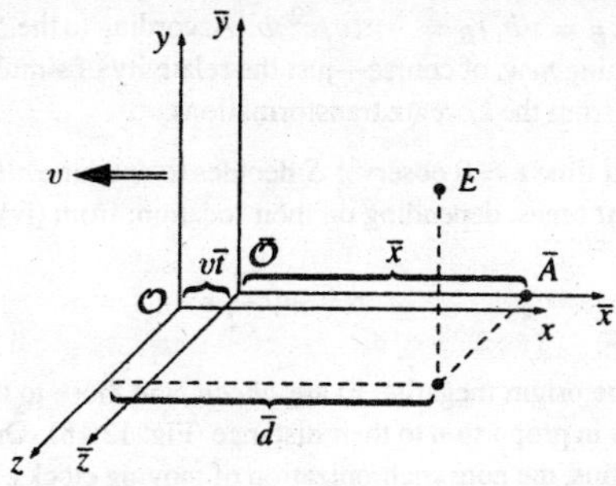

Figure 12.17

v, with respect to $\mathcal{S}$, then $\mathcal{S}$ is going to the *left* at speed v, with respect to $\bar{\mathcal{S}}$.) Nevertheless, this is a useful result, for if we substitute $\bar{x}$ from Eq. 12.14, and solve for $\bar{t}$, we complete the relativistic "dictionary":

$$\boxed{\begin{aligned} &\text{(i) } \bar{x} = \gamma(x - vt), \\ &\text{(ii) } \bar{y} = y, \\ &\text{(iii) } \bar{z} = z, \\ &\text{(iv) } \bar{t} = \gamma\left(t - \frac{v}{c^2}x\right). \end{aligned}} \tag{12.18}$$

These are the famous **Lorentz transformations**, with which Einstein replaced the Galilean ones. They contain all the geometrical information in the special theory, as the following examples illustrate. The reverse dictionary, which carries you from $\bar{\mathcal{S}}$ back to $\mathcal{S}$, can be obtained algebraically by solving (i) and (iv) for x and t, or, more simply, by switching the sign of v:

$$\left.\begin{aligned} &\text{(i}'\text{) } x = \gamma(\bar{x} + v\bar{t}), \\ &\text{(ii}'\text{) } y = \bar{y}, \\ &\text{(iii}'\text{) } z = \bar{z}, \\ &\text{(iv}'\text{) } t = \gamma\left(\bar{t} + \frac{v}{c^2}\bar{x}\right). \end{aligned}\right\} \tag{12.19}$$

Example 12.4

Simultaneity, synchronization, and time dilation. Suppose event A occurs at $x_A = 0$, $t_A = 0$, and event B occurs at $x_B = b$, $t_B = 0$. The two events are simultaneous in $\mathcal{S}$ (they both take place at $t = 0$). But they are *not* simultaneous in $\bar{\mathcal{S}}$, for the Lorentz transformations give $\bar{x}_A = 0$, $\bar{t}_A = 0$ and $\bar{x}_B = \gamma b$, $\bar{t}_B = -\gamma(v/c^2)b$. According to the $\bar{\mathcal{S}}$ clocks, then, B occurred *before* A. This is nothing *new*, of course—just the relativity of simultaneity. But I wanted you to see how it follows from the Lorentz transformations.

Now suppose that at time $t = 0$ observer $\mathcal{S}$ decides to examine *all* the clocks in $\bar{\mathcal{S}}$. He finds that they read *different* times, depending on their location; from (iv):

$$\bar{t} = -\gamma\frac{v}{c^2}x.$$

Those to the left of the origin (negative x) are *ahead*, and those to the right are *behind*, by an amount that increases in proportion to their distance (Fig. 12.18). Only the master clock at the origin reads $\bar{t} = 0$. Thus, the nonsynchronization of moving clocks, too, follows directly from the Lorentz transformations. Of course, from the $\bar{\mathcal{S}}$ viewpoint it is the $\mathcal{S}$ clocks that are out of synchronization, as you can check by putting $\bar{t} = 0$ into equation (iv$'$).

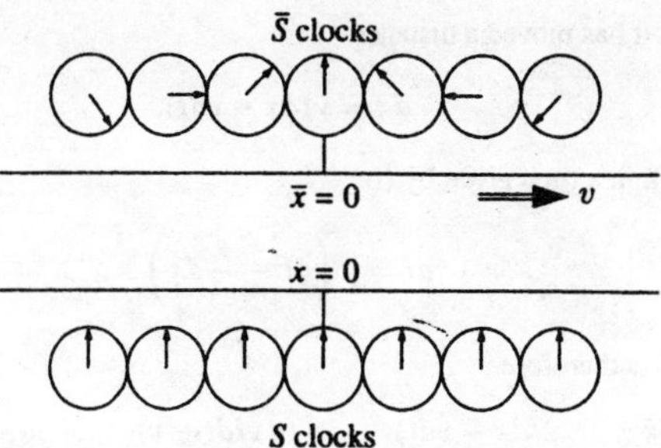

Figure 12.18

Finally, suppose S focuses his attention on a single clock in the $\bar{S}$ frame (say, the one at $\bar{x} = a$), and watches it over some interval Δt. How much time elapses on the moving clock? Because $\bar{x}$ is fixed, (iv′) gives $\Delta t = \gamma \Delta \bar{t}$, or

$$\Delta \bar{t} = \frac{1}{\gamma} \Delta t.$$

That's the old time dilation formula, derived now from the Lorentz transformations. Please note that it's $\bar{x}$ we hold fixed, here, because we're watching *one moving clock*. If you hold x fixed, then you're watching a whole series of different $\bar{S}$ clocks as they pass by, and that won't tell you whether any one of them is running slow.

Example 12.5

Lorentz contraction. Imagine a stick moving to the right at speed v. Its rest length (that is, its length as measured in $\bar{S}$) is $\Delta \bar{x} = \bar{x}_r - \bar{x}_l$, where the subscripts denote the right and left ends of the stick. If an observer in S were to measure the stick, he would subtract the positions of the two ends at one instant of *his* time t: $\Delta x = x_r - x_l$. According to (i), then,

$$\Delta x = \frac{1}{\gamma} \Delta \bar{x}.$$

This is the old Lorentz contraction formula. Note that it's t we hold fixed, here, because we're talking about a measurement made by S, and he marks off the two ends at the same instant of his time. ($\bar{S}$ doesn't have to be so fussy, since the stick is at rest in her frame.)

Example 12.6

Einstein's velocity addition rule. Suppose a particle moves a distance dx (in S) in a time dt. Its velocity u is then

$$u = \frac{dx}{dt},$$

In $\bar{S}$, meanwhile, it has moved a distance

$$d\bar{x} = \gamma(dx - v\,dt),$$

as we see from (i), in a time given by (iv):

$$d\bar{t} = \gamma\left(dt - \frac{v}{c^2}dx\right).$$

The velocity in $\bar{S}$ is therefore

$$\bar{u} = \frac{d\bar{x}}{d\bar{t}} = \frac{\gamma(dx - v\,dt)}{\gamma\left(dt - v/c^2 dx\right)} = \frac{(dx/dt - v)}{1 - v/c^2 dx/dt} = \frac{u - v}{1 - uv/c^2}. \tag{12.20}$$

This is **Einstein's velocity addition rule**. To recover the more transparent notation of Eq. 12.3, let A be the particle, B be S, and C be $\bar{S}$; then $u = v_{AB}, \bar{u} = v_{AC}$, and $v = v_{CB} = -v_{BC}$, so Eq. 12.20 becomes

$$v_{AC} = \frac{v_{AB} + v_{BC}}{1 + (v_{AB}v_{BC}/c^2)},$$

as before.

Problem 12.12 Solve Eqs. 12.18 for x, y, z, t in terms of $\bar{x}, \bar{y}, \bar{z}, \bar{t}$, and check that you recover Eqs. 12.19.

Problem 12.13 Sophie Zabar, clairvoyante, cried out in pain at precisely the instant her twin brother, 500 km away, hit his thumb with a hammer. A skeptical scientist observed both events (brother's accident, Sophie's cry) from an airplane traveling at $\frac{12}{13}c$ to the right (see Fig. 12.19). Which event occurred first, according to the scientist? How *much* earlier was it, in seconds?

Problem 12.14

(a) In Ex. 12.6 we found how velocities *in the x direction* transform when you go from S to $\bar{S}$. Derive the analogous formulas for velocities in the y and z directions.

(b) A spotlight is mounted on a boat so that its beam makes an angle $\bar{\theta}$ with the deck (Fig. 12.20). If this boat is then set in motion at speed v, what angle θ does an observer on the *dock* say the beam makes with the deck? Compare Prob. 12.10, and explain the difference.

Problem 12.15 You probably did Prob. 12.4 from the point of view of an observer on the *ground*. Now do it from the point of view of the police car, the outlaws, and the bullet. That is, fill in the gaps in the following table:

speed of → / relative to ↓	Ground	Police	Outlaws	Bullet	Do they escape?
Ground	0	$\frac{1}{2}c$	$\frac{3}{4}c$		
Police				$\frac{1}{3}c$	
Outlaws					
Bullet					

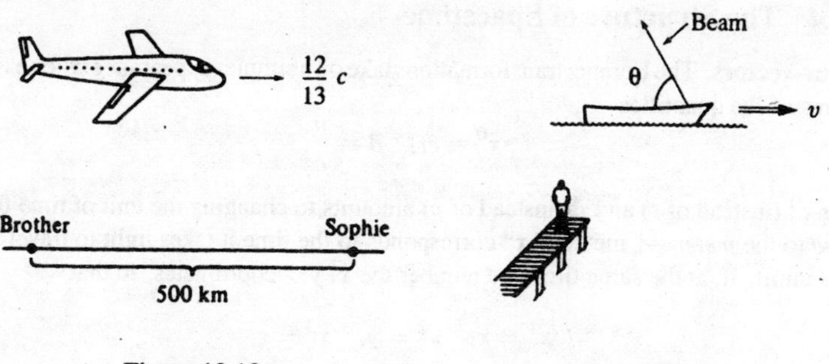

Figure 12.19

Figure 12.20

! **Problem 12.16 The twin paradox revisited.** On their 21st birthday, one twin gets on a moving sidewalk, which carries her out to star X at speed $\frac{4}{5}c$; her twin brother stays home. When the traveling twin gets to star X, she immediately jumps onto the returning moving sidewalk and comes back to earth, again at speed $\frac{4}{5}c$. She arrives on her 39th birthday (as determined by *her* watch).

(a) How old is her twin brother (who stayed at home)?

(b) How far away is star X? (Give your answer in light years.)

Call the outbound sidewalk system $\bar{S}$ and the inbound one $\tilde{S}$ (the earth system is S). All three systems set their master clocks, and choose their origins, so that $x = \bar{x} = \tilde{x} = 0$, $t = \bar{t} = \tilde{t} = 0$ at the moment of departure.

(c) What are the coordinates (x, t) of the jump (from outbound to inbound sidewalk) in S?

(d) What are the coordinates $(\bar{x}, \bar{t})$ of the jump in $\bar{S}$?

(e) What are the coordinates $(\tilde{x}, \tilde{t})$ of the jump in $\tilde{S}$?

(f) If the traveling twin wanted her watch to agree with the clock in $\tilde{S}$, how would she have to reset it immediately after the jump? If she *did* this, what would her watch read when she got home? (This wouldn't change her *age*, of course—she's still 39—it would just make her watch agree with the standard synchronization in $\tilde{S}$.)

(g) If the traveling twin is asked the question, "How old is your brother *right now*?", what is the correct reply (i) just *before* she makes the jump, (ii) just *after* she makes the jump? (Nothing dramatic happens to her brother during the split second between (i) and (ii), of course; what *does* change abruptly is his sister's notion of what "right now, back home" *means*.)

(h) How many earth years does the return trip take? Add this to (ii) from (g) to determine how old *she* expects him to be at their reunion. Compare your answer to (a).

12.1.4 The Structure of Spacetime

(i) Four-vectors. The Lorentz transformations take on a simpler appearance when expressed in terms of the quantities

$$x^0 \equiv ct, \quad \beta \equiv \frac{v}{c}. \tag{12.21}$$

Using x^0 (instead of t) and β (instead of v) amounts to changing the unit of time from the *second* to the *meter*—1 meter of x^0 corresponds to the time it takes light to travel 1 meter (in vacuum). If, at the same time, we number the x, y, z coordinates, so that

$$x^1 = x, \quad x^2 = y, \quad x^3 = z, \tag{12.22}$$

then the Lorentz transformations read

$$\left.\begin{aligned} \bar{x}^0 &= \gamma(x^0 - \beta x^1), \\ \bar{x}^1 &= \gamma(x^1 - \beta x^0), \\ \bar{x}^2 &= x^2, \\ \bar{x}^3 &= x^3. \end{aligned}\right\} \tag{12.23}$$

Or, in matrix form:

$$\begin{pmatrix} \bar{x}^0 \\ \bar{x}^1 \\ \bar{x}^2 \\ \bar{x}^3 \end{pmatrix} = \begin{pmatrix} \gamma & -\gamma\beta & 0 & 0 \\ -\gamma\beta & \gamma & 0 & 0 \\ 0 & 0 & 1 & 0 \\ 0 & 0 & 0 & 1 \end{pmatrix} \begin{pmatrix} x^0 \\ x^1 \\ x^2 \\ x^3 \end{pmatrix}. \tag{12.24}$$

Letting Greek indices run from 0 to 3, this can be distilled into a single equation:

$$\bar{x}^\mu = \sum_{\nu=0}^{3} (\Lambda^\mu_\nu) x^\nu, \tag{12.25}$$

where Λ is the **Lorentz transformation matrix** in Eq. 12.24 (the superscript μ labels the row, the subscript ν labels the column). One virtue of writing things in this abstract manner is that we can handle in the same format a more general transformation, in which the relative motion is *not* along a common $x\,\bar{x}$ axis; the matrix Λ would be more complicated, but the structure of Eq. 12.25 is unchanged.

If this reminds you of the *rotations* we studied in Sect. 1.1.5, it's no accident. There we were concerned with the change in components when you switch to a *rotated* coordinate system; here we are interested in the change of components when you go to a *moving* system. In Chapter 1 we defined a (3-) vector as any set of three components that transform under rotations the same way (x, y, z) do; by extension, we now define a **4-vector** as any set of *four* components that transform in the same manner as (x^0, x^1, x^2, x^3) under Lorentz transformations:

$$\bar{a}^\mu = \sum_{\nu=0}^{3} \Lambda^\mu_\nu a^\nu. \tag{12.26}$$

For the particular case of a transformation along the x axis:

$$\left.\begin{aligned} \bar{a}^0 &= \gamma(a^0 - \beta a^1), \\ \bar{a}^1 &= \gamma(a^1 - \beta a^0), \\ \bar{a}^2 &= a^2, \\ \bar{a}^3 &= a^3 \end{aligned}\right\} \tag{12.27}$$

There is a 4-vector analog to the dot product ($\mathbf{A}\cdot\mathbf{B} \equiv A_x B_x + A_y B_y + A_z B_z$), but it's not just the sum of the products of like components; rather, the zeroth components have a minus sign:

$$-a^0 b^0 + a^1 b^1 + a^2 b^2 + a^3 b^3. \tag{12.28}$$

This is the **four-dimensional scalar product**; you should check for yourself (Prob. 12.17) that it has the same value in all inertial systems:

$$-\bar{a}^0\bar{b}^0 + \bar{a}^1\bar{b}^1 + \bar{a}^2\bar{b}^2 + \bar{a}^3\bar{b}^3 = -a^0 b^0 + a^1 b^1 + a^2 b^2 + a^3 b^3. \tag{12.29}$$

Just as the ordinary dot product is **invariant** (unchanged) under rotations, this combination is invariant under Lorentz transformations.

To keep track of the minus sign it is convenient to introduce the **covariant** vector a_μ, which differs from the **contravariant** a^μ only in the sign of the zeroth component:

$$a_\mu = (a_0, a_1, a_2, a_3) \equiv (-a^0, a^1, a^2, a^3). \tag{12.30}$$

You must be scrupulously careful about the placement of indices in this business: *upper* indices designate *contravariant* vectors; *lower* indices are for *covariant* vectors. Raising or lowering the temporal index costs a minus sign ($a_0 = -a^0$); raising or lowering a spatial index changes nothing ($a_1 = a^1, a_2 = a^2, a_3 = a^3$). The scalar product can now be written with the summation symbol,

$$\sum_{\mu=0}^{3} a_\mu b^\mu, \tag{12.31}$$

or, more compactly still,

$$a_\mu b^\mu. \tag{12.32}$$

Summation is *implied* whenever a Greek index is repeated in a product—once as a covariant index and once as contravariant. This is called the **Einstein summation convention**, after its inventor, who regarded it as one of his most important contributions. Of course, we could as well take care of the minus sign by switching to covariant b:

$$a_\mu b^\mu = a^\mu b_\mu = -a^0 b^0 + a^1 b^1 + a^2 b^2 + a^3 b^3. \tag{12.33}$$

• **Problem 12.17** Check Eq. 12.29, using Eq. 12.27. [This only proves the invariance of the scalar product for transformations along the x direction. But the scalar product is also invariant under *rotations*, since the first term is not affected at all, and the last three constitute the three-dimensional dot product $\mathbf{a}\cdot\mathbf{b}$. By a suitable rotation, the x direction can be aimed any way you please, so the four-dimensional scalar product is actually invariant under *arbitrary* Lorentz transformations.]

Problem 12.18

(a) Write out the matrix that describes a *Galilean* transformation (Eq. 12.12).

(b) Write out the matrix describing a Lorentz transformation along the y axis.

(c) Find the matrix describing a Lorentz transformation with velocity v along the x axis followed by a Lorentz transformation with velocity $\bar{v}$ along the y axis. Does it matter in what order the transformations are carried out?

Problem 12.19 The parallel between rotations and Lorentz transformations is even more striking if we introduce the **rapidity**:

$$\theta \equiv \tanh^{-1}(v/c). \tag{12.34}$$

(a) Express the Lorentz transformation matrix Λ (Eq. 12.24) in terms of θ, and compare it to the rotation matrix (Eq. 1.29).

In some respects rapidity is a more natural way to describe motion than velocity. [See E. F. Taylor and J. A. Wheeler, *Spacetime Physics* (San Francisco: W. H. Freeman, 1966).] For one thing, it ranges from $-\infty$ to $+\infty$, instead of $-c$ to $+c$. More significantly, rapidities add, whereas velocities do not.

(b) Express the Einstein velocity addition law in terms of rapidity.

(ii) The invariant interval. Suppose event A occurs at $(x_A^0, x_A^1, x_A^2, x_A^3)$, and event B at $(x_B^0, x_B^1, x_B^2, x_B^3)$. The difference,

$$\Delta x^\mu \equiv x_A^\mu - x_B^\mu, \tag{12.35}$$

is the **displacement 4-vector**. The scalar product of Δx^μ with itself is a quantity of special importance; we call it the **interval** between two events:

$$I \equiv (\Delta x)_\mu (\Delta x)^\mu = -(\Delta x^0)^2 + (\Delta x^1)^2 + (\Delta x^2)^2 + (\Delta x^3)^2 = -c^2 t^2 + d^2, \tag{12.36}$$

where t is the time difference between the two events and d is their spatial separation. When you transform to a moving system, the *time* between A and B is altered ($\bar{t} \neq t$), and so is the *spatial separation* ($\bar{d} \neq d$), but the interval I remains the same.

Depending on the two events in question, the interval can be positive, negative, or zero:

1. If $I < 0$ we call the interval **timelike**, for this is the sign we get when the two occur at the *same place* ($d = 0$), and are separated only temporally.
2. If $I > 0$ we call the interval **spacelike**, for this is the sign we get when the two occur at the *same time* ($t = 0$), and are separated only spatially.
3. If $I = 0$ we call the interval **lightlike**, for this is the relation that holds when the two events are connected by a signal traveling at the speed of light.

If the interval between the two events is timelike, there exists an inertial system (accessible by Lorentz transformation) in which they occur at the same point. For if I hop on a train going from (A) to (B) at the speed $v = d/t$, leaving event A when it occurs, I shall be just in time to pass B when *it* occurs; in the train system, A and B take place at the same point. You cannot do this for a *spacelike* interval, of course, because v would have to be greater than c, and no observer can exceed the speed of light (γ would be imaginary and the Lorentz transformations would be nonsense). On the other hand, if the interval is spacelike, then there exists a system in which the two events occur at the same time (see Prob. 12.21).

Problem 12.20

(a) Event A happens at point ($x_A = 5$, $y_A = 3$, $z_A = 0$) and at time t_A given by $ct_A = 15$; event B occurs at (10, 8, 0) and $ct_B = 5$, both in system S.

(i) What is the invariant interval between A and B?

(ii) Is there an inertial system in which they occur *simultaneously*? If so, find its velocity (magnitude and direction) relative to S.

(iii) Is there an inertial system in which they occur at the same point? If so, find its velocity relative to S

(b) Repeat part (a) for $A = (2, 0, 0)$, $ct = 1$; and $B = (5, 0, 0)$, $ct = 3$.

Problem 12.21 The coordinates of event A are $(x_A, 0, 0)$, t_A, and the coordinates of event B are $(x_B, 0, 0)$, t_B. Assuming the interval between them is spacelike, find the velocity of the system in which they are simultaneous.

(iii) Space-time diagrams. If you want to represent the motion of a particle graphically, the normal practice is to plot the position versus time (that is, x runs vertically and t horizontally). On such a graph, the velocity can be read off as the slope of the curve. For some reason the convention is reversed in relativity: everyone plots position horizontally and time (or, better, $x^0 = ct$) vertically. Velocity is then given by the *reciprocal* of the slope. A particle at rest is represented by a vertical line; a photon, traveling at the speed of light, is described by a 45° line; and a rocket going at some intermediate speed follows a line of slope $c/v = 1/\beta$ (Fig. 12.21). We call such plots **Minkowski diagrams.**

The trajectory of a particle on a Minkowski diagram is called a **world line**. Suppose you set out from the origin at time $t = 0$. Because no material object can travel faster than light, your world line can never have a slope less than 1. Accordingly, your motion is

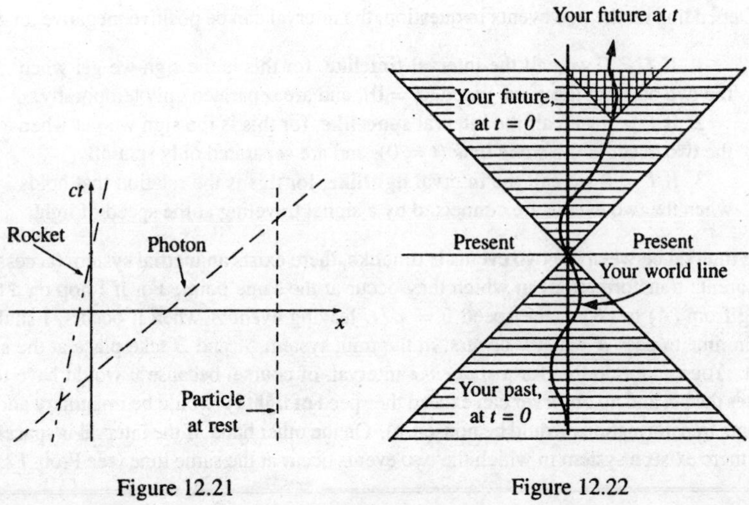

Figure 12.21

Figure 12.22

restricted to the wedge-shaped region bounded by the two 45° lines (Fig. 12.22). We call this your "future," in the sense that it is the locus of all points accessible to you. Of course, as time goes on, and you move along your chosen world line, your options progressively narrow: your "future" at any moment is the forward "wedge" constructed at whatever point you find yourself. Meanwhile, the *backward* wedge represents your "past," in the sense that it is the locus of all points from which you might have come. As for the rest (the region outside the forward and backward wedges) this is the generalized "present." You can't *get* there, and you didn't *come* from there. In fact, there's no way you can influence any event in the present (the message would have to travel faster than light); it's a vast expanse of spacetime that is absolutely inaccessible to you.

I've been ignoring the y and z directions. If we include a y axis coming out of the page, the "wedges" become cones—and, with an undrawable z axis, hypercones. Because their boundaries are the trajectories of light rays, we call them the **forward light cone** and the **backward light cone**. Your future, in other words, lies within your forward light cone, your past within your backward light cone.

Notice that the slope of the line connecting two events on a space-time diagram tells you at a glance whether the invariant interval between them is timelike (slope greater than 1), spacelike (slope less than 1), or lightlike (slope 1). For example, all points in the past and future are timelike with respect to your present location, whereas points in the present are spacelike, and points on the light cone are lightlike.

Hermann Minkowski, who was the first to recognize the full geometrical significance of special relativity, began a classic paper with the words, "Henceforth space by itself, and time by itself, are doomed to fade away into mere shadows, and only a kind of union of the two will preserve an independent reality." It is a lovely thought, but you must be careful

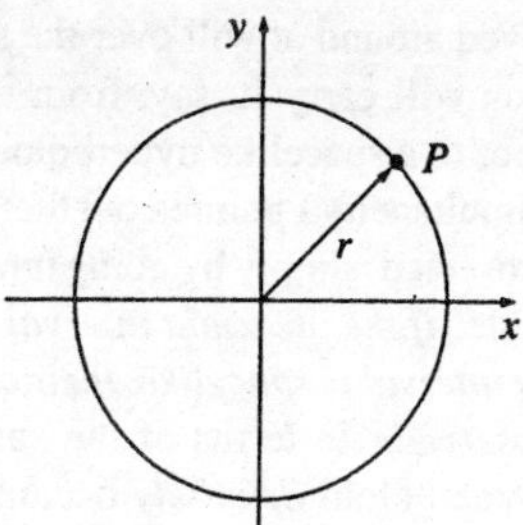

Figure 12.23

not to read too much into it. For it is not at all the case that time is "just another coordinate, on the same footing with x, y, and z" (except that for obscure reasons we measure it on clocks instead of rulers). *No:* Time is *utterly different* from the others, and the mark of its distinction is the minus sign in the invariant interval. That minus sign imparts to spacetime a hyperbolic geometry that is much richer than the circular geometry of 3-space.

Under rotations about the z axis, a point P in the xy plane describes a *circle:* the locus of all points a fixed distance $r = \sqrt{x^2 + y^2}$ from the origin (Fig. 12.23). Under Lorentz transformations, however, it is the interval $I = (x^2 - c^2t^2)$ that is preserved, and the locus of all points with a given value of I is a *hyperbola*—or, if we include the y axis, a *hyperboloid of revolution.* When the interval is *timelike*, it's a "hyperboloid of two sheets" (Fig. 12.24a); when the interval is *spacelike*, it's a "hyperboloid of one sheet" (Fig. 12.24b). When you perform a Lorentz transformation (that is, when you go into a moving inertial system), the coordinates (x, t) of a given event will change to $(\bar{x}, \bar{t})$, but these new coordinates *will lie on the same hyperbola* as (x, t). By appropriate combinations of Lorentz transformations

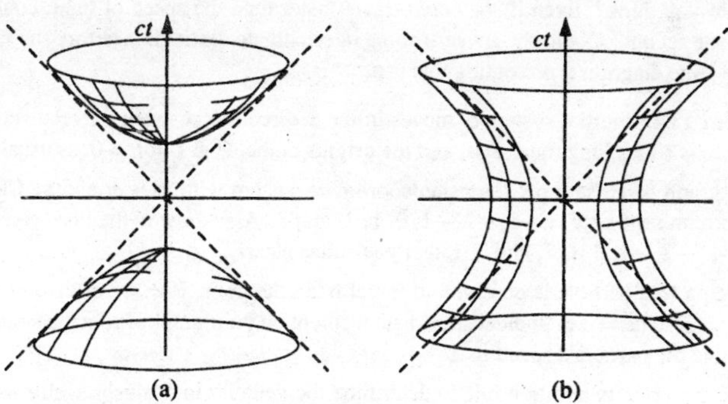

Figure 12.24

and rotations, a spot can be moved around at will over the surface of a given hyperboloid, but no amount of transformation will carry it, say, from the upper sheet of the timelike hyperboloid to the lower sheet, or to a spacelike hyperboloid.

When we were discussing simultaneity I pointed out that the time ordering of two events can, at least in certain cases, be reversed, simply by going into a moving system. But we now see that this is not *always* possible: *If the invariant interval between two events is timelike, their ordering is absolute; if the interval is spacelike, their ordering depends on the inertial system from which they are observed.* In terms of the space-time diagram, an event on the upper sheet of a timelike hyperboloid *definitely* occurred *after* (0, 0), and one on the lower sheet certainly occurred *before;* but an event on a spacelike hyperboloid occurred at positive t, or negative t, depending on your reference frame. This is not an idle curiosity, for it rescues the notion of **causality**, on which all physics is based. If it were *always* possible to reverse the order of two events, then we could never say "A caused B," since a rival observer would retort that *B preceded A*. This embarrassment is avoided, provided the two events are timelike-separated. And causally related events *are* timelike-separated—otherwise no influence could travel from one to the other. *Conclusion:* The invariant interval between causally related events is always timelike, and their temporal ordering is the same for all inertial observers.

Problem 12.22

(a) Draw a space-time diagram representing a game of catch (or a conversation) between two people at rest, 10 ft apart. How is it possible for them to communicate, given that their separation is spacelike?

(b) There's an old limerick that runs as follows:

> There once was a girl named Ms. Bright,
> Who could travel much faster than light.
> She departed one day,
> The Einsteinian way,
> And returned on the previous night.

What do you think? Even if she *could* travel faster than the speed of light, could she return before she set out? Could she arrive at some intermediate destination before she set out? Draw a space-time diagram representing this trip.

Problem 12.23 Inertial system $\bar{S}$ moves in the x direction at speed $\frac{3}{5}c$ relative to system S. (The $\bar{x}$ axis slides long the x axis, and the origins coincide at $t = \bar{t} = 0$, as usual.)

(a) On graph paper set up a Cartesian coordinate system with axes ct and x. Carefully draw in lines representing $\bar{x} = -3, -2, -1, 0, 1, 2,$ and 3. Also draw in the lines corresponding to $c\bar{t} = -3, -2, -1, 0, 1, 2,$ and 3. Label your lines clearly.

(b) In $\bar{S}$, a free particle is observed to travel from the point $\bar{x} = -2$ at time $c\bar{t} = -2$ to the point $\bar{x} = 2$ at $c\bar{t} = +3$. Indicate this displacement on your graph. From the slope of this line, determine the particle's speed in S.

(c) Use the velocity addition rule to determine the velocity in S algebraically, and check that your answer is consistent with the graphical solution in (b).

12.2 Relativistic Mechanics

12.2.1 Proper Time and Proper Velocity

As you progress along your world line, your watch runs slow; while the clock on the wall ticks off an interval dt, your watch only advances $d\tau$:

$$d\tau = \sqrt{1 - u^2/c^2}\, dt. \tag{12.37}$$

(I'll use u for the velocity of a particular object—you, in this instance—and reserve v for the relative velocity of two inertial systems.) The time τ your watch registers (or, more generally, the time associated with the moving object) is called **proper time**. (The word suggests a mistranslation of the French *propre*, meaning "own.") In some cases τ may be a more relevant or useful quantity than t. For one thing, proper time is invariant, whereas "ordinary" time depends on the particular reference frame you have in mind.

Now, imagine you're on a flight to Los Angeles, and the pilot announces that the plane's velocity is $\frac{4}{5}c$, due South. What precisely does he mean by "velocity"? Well, of course, he means the displacement divided by the time:

$$\mathbf{u} = \frac{d\mathbf{l}}{dt}, \tag{12.38}$$

and, since he is presumably talking about the velocity relative to ground, both $d\mathbf{l}$ and dt are to be measured by the ground observer. That's the important number to know, if you're concerned about being on time for an appointment in Los Angeles, but if you're wondering whether you'll be hungry on arrival, you might be more interested in the distance covered per unit *proper* time:

$$\boldsymbol{\eta} \equiv \frac{d\mathbf{l}}{d\tau}. \tag{12.39}$$

This hybrid quantity—distance measured on the ground, over time measured in the airplane—is called **proper velocity**; for contrast, I'll call $\mathbf{u}$ the **ordinary velocity**. The two are related by Eq. 12.37:

$$\boldsymbol{\eta} = \frac{1}{\sqrt{1 - u^2/c^2}}\mathbf{u}. \tag{12.40}$$

For speeds much less than c, of course, the difference between ordinary and proper velocity is negligible.

From a theoretical standpoint, however, proper velocity has an enormous advantage over ordinary velocity: it transforms simply, when you go from one inertial system to another. In fact, $\boldsymbol{\eta}$ is the spatial part of a 4-vector,

$$\eta^\mu \equiv \frac{dx^\mu}{d\tau}, \tag{12.41}$$

whose zeroth component is

$$\eta^0 = \frac{dx^0}{d\tau} = c\frac{dt}{d\tau} = \frac{c}{\sqrt{1 - u^2/c^2}}. \tag{12.42}$$

For the numerator, dx^μ, is a displacement 4-vector, while the denominator, $d\tau$, is invariant. Thus, for instance, when you go from system $\mathcal{S}$ to system $\bar{\mathcal{S}}$, moving at speed v along the common $x\,\bar{x}$ axis,

$$\left.\begin{aligned} \bar{\eta}^0 &= \gamma(\eta^0 - \beta\eta^1), \\ \bar{\eta}^1 &= \gamma(\eta^1 - \beta\eta^0), \\ \bar{\eta}^2 &= \eta^2, \\ \bar{\eta}^3 &= \eta^3. \end{aligned}\right\} \tag{12.43}$$

More generally,

$$\bar{\eta}^\mu = \Lambda^\mu_\nu \eta^\nu; \tag{12.44}$$

η^μ is called the **proper velocity 4-vector**, or simply the **4-velocity**.

By contrast, the transformation rule for *ordinary* velocities is extremely cumbersome, as we found in Ex. 12.6 and Prob. 12.14:

$$\left.\begin{aligned} \bar{u}_x &= \frac{d\bar{x}}{d\bar{t}} = \frac{u_x - v}{(1 - vu_x/c^2)}, \\ \bar{u}_y &= \frac{d\bar{y}}{d\bar{t}} = \frac{u_y}{\gamma(1 - vu_x/c^2)}, \\ \bar{u}_z &= \frac{d\bar{z}}{d\bar{t}} = \frac{u_z}{\gamma(1 - vu_x/c^2)}. \end{aligned}\right\} \tag{12.45}$$

The *reason* for the added complexity is plain: we're obliged to transform both the numerator $d\mathbf{l}$ *and the denominator* dt, whereas for *proper* velocity the denominator $d\tau$ is invariant, so the ratio inherits the transformation rule of the numerator alone.

Problem 12.24

(a) Equation 12.40 defines proper velocity in terms of ordinary velocity. Invert that equation to get the formula for $\mathbf{u}$ in terms of $\boldsymbol{\eta}$.

(b) What is the relation between proper velocity and *rapidity* (Eq. 12.34)? Assume the velocity is along the x direction, and find η as a function of θ.

Problem 12.25 A car is traveling along the 45° line in $\mathcal{S}$ (Fig. 12.25), at (ordinary) speed $(2/\sqrt{5})c$.

(a) Find the components u_x and u_y of the (ordinary) velocity.

(b) Find the components η_x and η_y of the proper velocity.

(c) Find the zeroth component of the 4-velocity, η^0.

System $\bar{\mathcal{S}}$ is moving in the x direction with (ordinary) speed $\sqrt{2/5}\,c$, relative to $\mathcal{S}$. By using the appropriate transformation laws:

(d) Find the (ordinary) velocity components $\bar{u}_x$ and $\bar{u}_y$ in $\bar{\mathcal{S}}$.

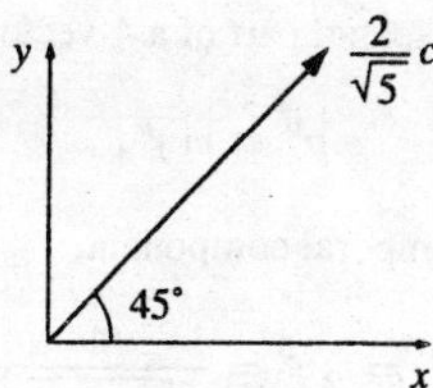

Figure 12.25

(e) Find the proper velocity components $\bar{\eta}_x$ and $\bar{\eta}_y$ in $\bar{S}$.

(f) As a consistency check, verify that

$$\bar{\boldsymbol{\eta}} = \frac{\bar{\mathbf{u}}}{\sqrt{1-\bar{u}^2/c^2}}.$$

• **Problem 12.26** Find the invariant product of the 4-velocity with itself, $\eta^\mu \eta_\mu$.

Problem 12.27 Consider a particle in hyperbolic motion,

$$x(t) = \sqrt{b^2 + (ct)^2}, \quad y = z = 0.$$

(a) Find the proper time τ as a function of t, assuming the clocks are set so that $\tau = 0$ when $t = 0$. [*Hint:* Integrate Eq. 12.37.]

(b) Find x and v (ordinary velocity) as functions of τ.

(c) Find η^μ (proper velocity) as a function of t.

12.2.2 Relativistic Energy and Momentum

In classical mechanics momentum is mass times velocity. I would like to extend this definition to the relativistic domain, but immediately a question arises: Should I use *ordinary* velocity or *proper* velocity? In classical physics $\boldsymbol{\eta}$ and $\mathbf{u}$ are identical, so there is no *a priori* reason to favor one over the other. However, in the context of relativity it is essential that we use *proper* velocity, for the law of conservation of momentum would be inconsistent with the principle of relativity if we were to define momentum as $m\mathbf{u}$ (see Prob. 12.28). Thus

$$\boxed{\mathbf{p} \equiv m\boldsymbol{\eta} = \frac{m\mathbf{u}}{\sqrt{1-u^2/c^2}};} \tag{12.46}$$

this is the **relativistic momentum.**

Relativistic momentum is the spatial part of a 4-vector,

$$p^{\mu} \equiv m\eta^{\mu}, \tag{12.47}$$

and it is natural to ask what the temporal component,

$$p^0 = m\eta^0 = \frac{mc}{\sqrt{1-u^2/c^2}} \tag{12.48}$$

represents. Einstein called

$$m_{\text{rel}} \equiv \frac{m}{\sqrt{1-u^2/c^2}} \tag{12.49}$$

the **relativistic mass** (so that $p^0 = m_{\text{rel}}c$ and $\mathbf{p} = m_{\text{rel}}\mathbf{u}$; m itself was then called the **rest mass**), but modern usage has abandoned this terminology in favor of **relativistic energy**:

$$\boxed{E \equiv \frac{mc^2}{\sqrt{1-u^2/c^2}}} \tag{12.50}$$

(so $p^0 = E/c$).[8] Because p^0 is (apart from the factor $1/c$) the relativistic energy, p^{μ} is called the **energy-momentum 4-vector** (or the **momentum 4-vector**, for short).

Notice that the relativistic energy is nonzero *even when the object is stationary*; we call this **rest energy**:

$$E_{\text{rest}} \equiv mc^2. \tag{12.51}$$

The remainder, which is attributable to the *motion*, we call **kinetic energy**

$$E_{\text{kin}} \equiv E - mc^2 = mc^2\left(\frac{1}{\sqrt{1-u^2/c^2}} - 1\right). \tag{12.52}$$

In the nonrelativistic régime ($u \ll c$) the square root can be expanded in powers of u^2/c^2, giving

$$E_{\text{kin}} = \frac{1}{2}mu^2 + \frac{3}{8}\frac{mu^4}{c^2} + \cdots; \tag{12.53}$$

the leading term reproduces the classical formula.

So far, this is all just *notation*. The *physics* resides in the experimental fact that E and $\mathbf{p}$, as defined by Eqs. 12.46 and 12.50, are *conserved:*

> **In every closed[9] system, the total relativistic energy and momentum are conserved.**

---

[8] Since E and m_{rel} differ only by a constant factor (c^2), there's nothing to be gained by keeping both terms in circulation, and m_{rel} has gone the way of the two dollar bill.

[9] If there are *external* forces at work, then (just as in the classical case) the energy and momentum of the system itself will *not*, in general, be conserved.

"Relativistic mass" (if you care to use that term) is *also* conserved—but this is equivalent to conservation of energy. *Rest* mass is *not* conserved—a fact that has been painfully familiar to everyone since 1945 (though the so-called "conversion of mass into energy" is really a conversion of *rest* energy into *kinetic* energy). Note the distinction between an **invariant** quantity (same value in all inertial systems) and a **conserved** quantity (same value before and after some process). Mass is invariant, but not conserved; energy is conserved but not invariant; electric charge (as we shall see) is both conserved *and* invariant; velocity is neither conserved *nor* invariant.

The scalar product of p^μ with itself is

$$p^\mu p_\mu = -(p^0)^2 + (\mathbf{p} \cdot \mathbf{p}) = -m^2c^2, \tag{12.54}$$

as you can quickly check using the result of Prob. 12.26. In terms of the relativistic energy,

$$\boxed{E^2 - p^2c^2 = m^2c^4.} \tag{12.55}$$

This result is extremely useful, for it enables you to calculate E (if you know p), or p (knowing E), without ever having to determine the velocity.

Problem 12.28

(a) Repeat Prob. 12.2 using the (incorrect) definition $\mathbf{p} = m\mathbf{u}$, but with the (correct) Einstein velocity addition rule. Notice that if momentum (so defined) is conserved in S, it is *not* conserved in $\bar{S}$. Assume all motion is along the x axis.

(b) Now do the same using the correct definition, $\mathbf{p} = m\boldsymbol{\eta}$. Notice that if momentum (so defined) is conserved in S it is automatically also conserved in $\bar{S}$. [*Hint:* Use Eq. 12.43 to transform the proper velocity.] What must you assume about relativistic energy?

Problem 12.29 If a particle's kinetic energy is n times its rest energy, what is its speed?

Problem 12.30 Suppose you have a collection of particles, all moving in the x direction, with energies $E_1, E_2, E_3, \ldots$ and momenta $p_1, p_2, p_3, \ldots$. Find the velocity of the **center of momentum** frame, in which the total momentum is zero.

12.2.3 Relativistic Kinematics

In this section we'll explore some applications of the conservation laws to particle decays and collisions.

Example 12.7

Two lumps of clay, each of (rest) mass m, collide head-on at $\frac{3}{5}c$ (Fig. 12.26). They stick together. *Question*: what is the mass (M) of the composite lump?

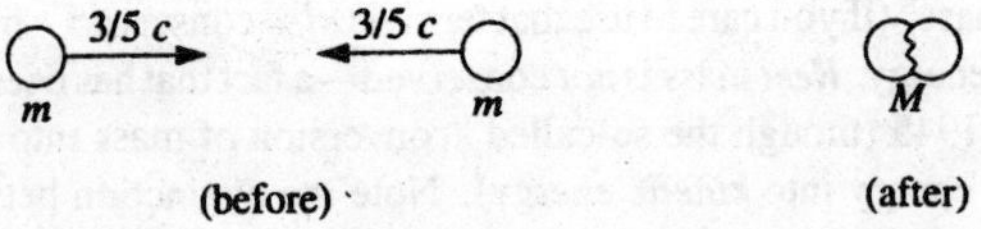

3/5 c 3/5 c
m m M
(before) (after)

Figure 12.26

Solution: In this case conservation of momentum is trivial: zero before, zero after. The energy of each lump prior to the collision is

$$\frac{mc^2}{\sqrt{1-(3/5)^2}} = \frac{5}{4}mc^2,$$

and the energy of the composite lump after the collision is Mc^2 (since it's at rest). So conservation of energy says

$$\frac{5}{4}mc^2 + \frac{5}{4}mc^2 = Mc^2,$$

and hence

$$M = \frac{5}{2}m.$$

Notice that this is *greater* than the sum of the initial masses! Mass was not conserved in this collision; kinetic energy was converted into rest energy, so the mass increased.

In the *classical* analysis of such a collision, we say that kinetic energy was converted into *thermal* energy—the composite lump is *hotter* than the two colliding pieces. This is, of course, true in the relativistic picture too. But what *is* thermal energy? It's the sum total of the random kinetic and potential energies of all the atoms and molecules in the substance. Relativity tells us that these microscopic energies are represented in the *mass* of the object: a hot potato is *heavier* than a cold potato, and a compressed spring is *heavier* than a relaxed spring. Not by *much*, it's true—internal energy (U) contributes an amount U/c^2 to the mass, and c^2 is a very large number by everyday standards. You could never get two lumps of clay going anywhere *near* fast enough to detect the nonconservation of mass in their collision. But in the realm of elementary particles, the effect can be very striking. For example, when the neutral pi meson (mass 2.4×10^{-28} kg) decays into an electron and a positron (each of mass 9.11×10^{-31} kg), the rest energy is converted *almost entirely* into kinetic energy—less than 1% of the original mass remains.

In classical mechanics there's no such thing as a massless particle—its kinetic energy ($\frac{1}{2}mu^2$) and its momentum ($m\mathbf{u}$) would be zero, you couldn't apply a force to it ($\mathbf{F} = m\mathbf{a}$), and hence (by Newton's third law) *it* couldn't apply a force on anything else—it's a cipher, as far as physics is concerned. You might at first assume that the same is true in relativity; after all, $\mathbf{p}$ and E are still proportional to m. However, a closer inspection of Eqs. 12.46 and 12.50 reveals a loophole worthy of a congressman: If $u = c$, then the zero in the numerator is balanced by a zero in the denominator, leaving $\mathbf{p}$ and E indeterminate (zero over zero). It is conceivable, therefore, that a massless particle could carry energy and momentum, *provided it always travels at the speed of light.* Although Eqs. 12.46 and 12.50 would no longer suffice to determine E and $\mathbf{p}$, Eq. 12.55 suggests that the two should be related by

$$E = pc. \tag{12.56}$$

Personally, I would regard this argument as a joke, were it not for the fact that at least one massless particle is known to exist in nature: the photon.[10] Photons *do* travel at the speed of light, and they obey Eq. 12.56.[11] They force us to take the "loophole" seriously. (By the way, you might ask what distinguishes a photon with a *lot* of energy from one with very little—after all, they have the same mass (zero) and the same speed (c). Relativity offers no answer to this question; curiously, quantum mechanics *does:* According to the Planck formula, $E = h\nu$, where h is **Planck's constant** and ν is the *frequency.* A *blue* photon is more energetic than a *red* one!)

Example 12.8

A pion at rest decays into a muon and a neutrino (Fig. 12.27). Find the energy of the outgoing muon, in terms of the two masses, m_π and m_μ (assume $m_\nu = 0$).

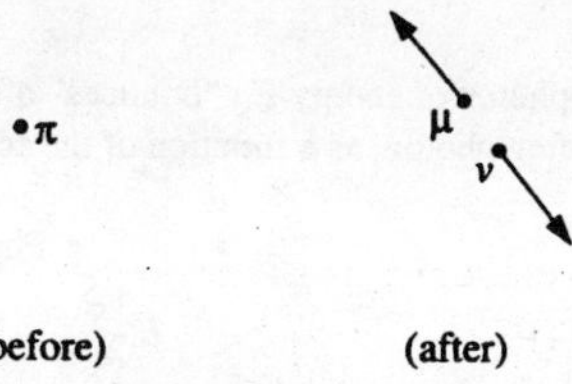

Figure 12.27

Solution: In this case

$$E_{\text{before}} = m_\pi c^2, \qquad \mathbf{p}_{\text{before}} = 0,$$

$$E_{\text{after}} = E_\mu + E_\nu, \qquad \mathbf{p}_{\text{after}} = \mathbf{p}_\mu + \mathbf{p}_\nu.$$

Conservation of momentum requires that $\mathbf{p}_\nu = -\mathbf{p}_\mu$. Conservation of energy says that

$$E_\mu + E_\nu = m_\pi c^2.$$

---

[10]Until recently neutrinos were also generally assumed to be massless, but experiments in 1998 indicate that they may in fact carry a (very small) mass.

[11]The photon is the **quantum** of the electromagnetic field, and it is no accident that the same ratio between energy and momentum holds for electromagnetic waves (see Eqs. 9.60 and 9.62).

Now, $E_\nu = |\mathbf{p}_\nu|c$, by Eq. 12.56, whereas $|\mathbf{p}_\mu| = \sqrt{E_\mu^2 - m_\mu^2 c^4}/c$, by Eq. 12.55, so

$$E_\mu + \sqrt{E_\mu^2 - m_\mu^2 c^4} = m_\pi c^2,$$

from which it follows that

$$E_\mu = \frac{(m_\pi^2 + m_\mu^2)c^2}{2m_\pi}.$$

In a classical collision, momentum and mass are always conserved, whereas kinetic energy, in general, is not. A "sticky" collision generates heat at the expense of kinetic energy; an "explosive" collision generates kinetic energy at the expense of chemical energy (or some other kind). If the kinetic energy *is* conserved, as in the ideal collision of the two billiard balls, we call the process *elastic*. In the relativistic case, momentum and total energy are always conserved but mass and kinetic energy, in general, are not. Once again, we call the process **elastic** if kinetic energy is conserved. In such a case the rest energy (being the total minus the kinetic) is *also* conserved, and therefore so too is the mass. In practice this means that the same particles come out as went in. Examples 12.7 and 12.8 were *inelastic* processes; the next one is *elastic*.

Example 12.9

Compton scattering. A photon of energy E_0 "bounces" off an electron, initially at rest. Find the energy E of the outgoing photon, as a function of the **scattering angle** θ (see Fig. 12.28).

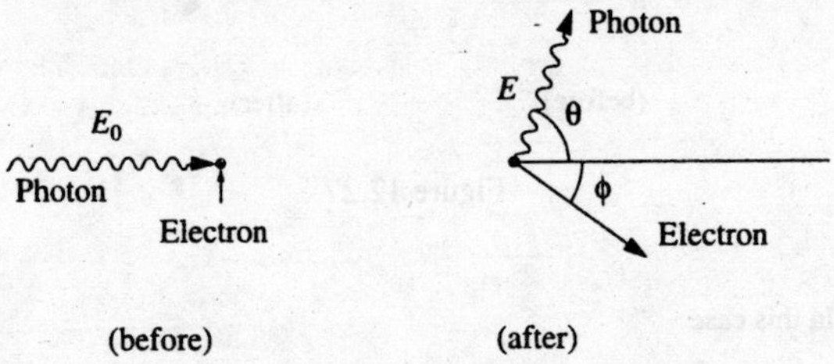

Figure 12.28

Solution: Conservation of momentum in the "vertical" direction gives $p_e \sin\phi = p_p \sin\theta$, or, since $p_p = E/c$,

$$\sin\phi = \frac{E}{p_e c}\sin\theta.$$

Conservation of momentum in the "horizontal" direction gives

$$\frac{E_0}{c} = p_p\cos\theta + p_e\cos\phi = \frac{E}{c}\cos\theta + p_e\sqrt{1 - \left(\frac{E}{p_e c}\sin\theta\right)^2},$$

or

$$p_e^2 c^2 = (E_0 - E\cos\theta)^2 + E^2\sin^2\theta = E_0^2 - 2E_0 E\cos\theta + E^2.$$

Finally, conservation of energy says that

$$E_0 + mc^2 = E + E_e = E + \sqrt{m^2c^4 + p_e^2c^2}$$

$$= E + \sqrt{m^2c^4 + E_0^2 - 2E_0E\cos\theta + E^2}.$$

Solving for E, I find that

$$E = \frac{1}{(1-\cos\theta)/mc^2 + (1/E_0)}. \quad (12.57)$$

The answer looks nicer when expressed in terms of photon *wavelength:*

$$E = h\nu = \frac{hc}{\lambda},$$

so

$$\lambda = \lambda_0 + \frac{h}{mc}(1-\cos\theta). \quad (12.58)$$

The quantity (h/mc) is called the **Compton wavelength** of the electron.

Problem 12.31 Find the velocity of the muon in Ex. 12.8.

Problem 12.32 A particle of mass m whose total energy is twice its rest energy collides with an identical particle at rest. If they stick together, what is the mass of the resulting composite particle? What is its velocity?

Problem 12.33 A neutral pion of (rest) mass m and (relativistic) momentum $p = \frac{3}{4}mc$ decays into two photons. One of the photons is emitted in the same direction as the original pion, and the other in the opposite direction. Find the (relativistic) energy of each photon.

Problem 12.34 In the past, most experiments in particle physics involved stationary targets: one particle (usually a proton or an electron) was accelerated to a high energy E, and collided with a target particle at rest (Fig. 12.29a). Far higher *relative* energies are obtainable (with the same accelerator) if you accelerate *both* particles to energy E, and fire them at each other (Fig. 12.29b). *Classically,* the energy $\bar{E}$ of one particle, relative to the other, is just 4E (why?)—not much of a gain (only a factor of 4). But *relativistically* the gain can be *enormous.* Assuming the two particles have the same mass, m, show that

$$\bar{E} = \frac{2E^2}{mc^2} - mc^2. \quad (12.59)$$

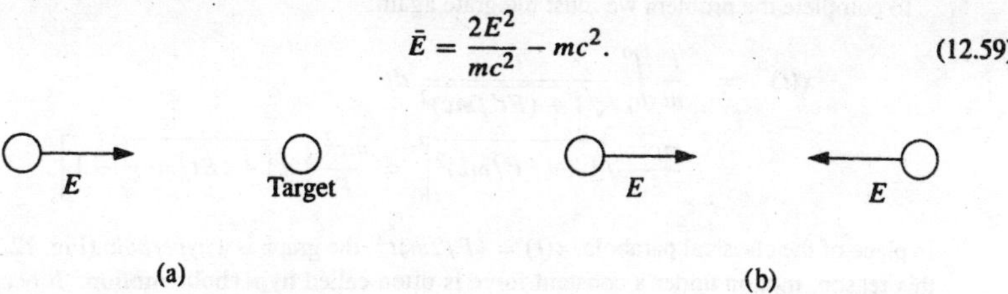

Figure 12.29

Suppose you use protons ($mc^2 = 1$ GeV) with $E = 30$ GeV. What $\bar{E}$ do you get? What multiple of E does this amount to? (1 GeV=10^9 electron volts.) [Because of this relativistic enhancement, most modern elementary particle experiments involve **colliding beams**, instead of fixed targets.]

Problem 12.35 In a **pair annihilation** experiment, an electron (mass m) with momentum p_e hits a positron (same mass, but opposite charge) at rest. They annihilate, producing two photons. (Why couldn't they produce just *one* photon?) If one of the photons emerges at 60° to the incident electron direction, what is its energy?

12.2.4 Relativistic Dynamics

Newton's *first* law is built into the principle of relativity. His second law, in the form

$$\boxed{\mathbf{F} = \frac{d\mathbf{p}}{dt}}, \tag{12.60}$$

retains its validity in relativistic mechanics, *provided we use the relativistic momentum.*

Example 12.10

Motion under a constant force. A particle of mass m is subject to a constant force F. If it starts from rest at the origin, at time $t = 0$, find its position (x), as a function of time.

Solution:

$$\frac{dp}{dt} = F \Rightarrow p = Ft + \text{constant},$$

but since $p = 0$ at $t = 0$, the constant must be zero, and hence

$$p = \frac{mu}{\sqrt{1-u^2/c^2}} = Ft.$$

Solving for u, we obtain

$$u = \frac{(F/m)t}{\sqrt{1+(Ft/mc)^2}}. \tag{12.61}$$

The numerator, of course, is the classical answer—it's approximately right, if $(F/m)t \ll c$. But the relativistic denominator ensures that u never exceeds c; in fact, as $t \to \infty$, $u \to c$.

To complete the problem we must integrate again:

$$\begin{aligned} x(t) &= \frac{F}{m}\int_0^t \frac{t'}{\sqrt{1+(Ft'/mc)^2}}\,dt' \\ &= \frac{mc^2}{F}\sqrt{1+(Ft'/mc)^2}\bigg|_0^t = \frac{mc^2}{F}\left[\sqrt{1+(Ft/mc)^2}-1\right]. \end{aligned} \tag{12.62}$$

In place of the classical parabola, $x(t) = (F/2m)t^2$, the graph is a *hyperbola* (Fig. 12.30); for this reason, motion under a constant force is often called **hyperbolic motion.** It occurs, for example, when a charged particle is placed in a uniform electric field.

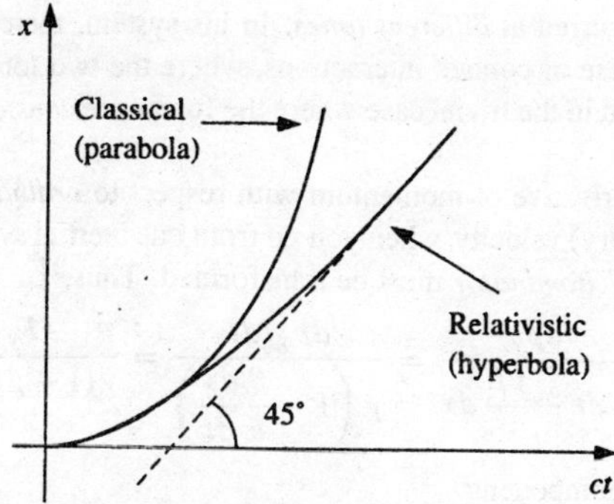

Figure 12.30

Work, as always, is the line integral of the force:

$$W \equiv \int \mathbf{F} \cdot d\mathbf{l}. \tag{12.63}$$

The **work-energy theorem** ("the net work done on a particle equals the increase in its kinetic energy") holds relativistically:

$$W = \int \frac{d\mathbf{p}}{dt} \cdot d\mathbf{l} = \int \frac{d\mathbf{p}}{dt} \cdot \frac{d\mathbf{l}}{dt}\, dt = \int \frac{d\mathbf{p}}{dt} \cdot \mathbf{u}\, dt,$$

while

$$\begin{aligned} \frac{d\mathbf{p}}{dt} \cdot \mathbf{u} &= \frac{d}{dt}\left(\frac{m\mathbf{u}}{\sqrt{1-u^2/c^2}}\right) \cdot \mathbf{u} \\ &= \frac{m\mathbf{u}}{(1-u^2/c^2)^{3/2}} \cdot \frac{d\mathbf{u}}{dt} = \frac{d}{dt}\left(\frac{mc^2}{\sqrt{1-u^2/c^2}}\right) = \frac{dE}{dt}, \end{aligned} \tag{12.64}$$

so

$$W = \int \frac{dE}{dt}\, dt = E_{\text{final}} - E_{\text{initial}}. \tag{12.65}$$

(Since the *rest* energy is constant, it doesn't matter whether we use the total energy, here, or the kinetic energy.)

Unlike to the first two, Newton's *third* law does *not*, in general, extend to the relativistic domain. Indeed, if the two objects in question are separated in space, the third law is incompatible with the relativity of simultaneity. For suppose the force of A on B at some instant t is $\mathbf{F}(t)$, and the force of B on A at the same instant is $-\mathbf{F}(t)$; then the third law applies, *in this reference frame*. But a moving observer will report that these equal and opposite forces occurred at *different times;* in his system, therefore, the third law is *violated.* Only in the case of contact interactions, where the two forces are applied at the *same physical point* (and in the trivial case where the forces are *constant*), can the third law be retained.

Because **F** is the derivative of momentum with respect to *ordinary* time, it shares the ugly behavior of (ordinary) velocity, when you go from one inertial system to another: both the numerator *and the denominator* must be transformed. Thus,[12]

$$\bar{F}_y = \frac{d\bar{p}_y}{d\bar{t}} = \frac{dp_y}{\gamma\, dt - \frac{\gamma\beta}{c}\, dx} = \frac{dp_y/dt}{\gamma\left(1 - \frac{\beta}{c}\frac{dx}{dt}\right)} = \frac{F_y}{\gamma(1 - \beta u_x/c)}, \tag{12.66}$$

and similarly for the z component:

$$\bar{F}_z = \frac{F_z}{\gamma(1 - \beta u_x/c)}.$$

The x component is even worse:

$$\bar{F}_x = \frac{d\bar{p}_x}{d\bar{t}} = \frac{\gamma\, dp_x - \gamma\beta\, dp^0}{\gamma\, dt - \frac{\gamma\beta}{c}\, dx} = \frac{\frac{dp_x}{dt} - \beta\frac{dp^0}{dt}}{1 - \frac{\beta}{c}\frac{dx}{dt}} = \frac{F_x - \frac{\beta}{c}\left(\frac{dE}{dt}\right)}{1 - \beta u_x/c}.$$

We calculated dE/dt in Eq. 12.64; putting that in,

$$\bar{F}_x = \frac{F_x - \beta(\mathbf{u}\cdot\mathbf{F})/c}{1 - \beta u_x/c}. \tag{12.67}$$

Only in one special case are these equations reasonably tractable: *If the particle is (instantaneously) at rest in* $\mathcal{S}$, so that $\mathbf{u} = 0$, then

$$\bar{\mathbf{F}}_\perp = \frac{1}{\gamma}\mathbf{F}_\perp, \quad \bar{F}_\parallel = F_\parallel. \tag{12.68}$$

That is, the component of **F** *parallel* to the motion of $\bar{\mathcal{S}}$ is unchanged, whereas components perpendicular are divided by γ.

It has perhaps occurred to you that we could avoid the bad transformation behavior of **F** by introducing a "proper" force, analogous to proper velocity, which would be the derivative of momentum with respect to *proper* time:

$$K^\mu \equiv \frac{dp^\mu}{d\tau}. \tag{12.69}$$

This is called the **Minkowski force**; it is plainly a 4-vector, since p^μ is a 4-vector and proper time is invariant. The spatial components of K^μ are related to the "ordinary" force by

$$\mathbf{K} = \left(\frac{dt}{d\tau}\right)\frac{d\mathbf{p}}{dt} = \frac{1}{\sqrt{1 - u^2/c^2}}\mathbf{F}, \tag{12.70}$$

[12]Remember: γ and β pertain to the motion of $\bar{\mathcal{S}}$ with respect $\mathcal{S}$—they are *constants;* **u** is the velocity of the *particle* with respect to $\mathcal{S}$.

while the zeroth component,

$$K^0 = \frac{dp^0}{d\tau} = \frac{1}{c}\frac{dE}{d\tau}, \tag{12.71}$$

is, apart from the $1/c$, the (proper) rate at which the energy of the particle increases—in other words, the (proper) *power* delivered to the particle.

Relativistic dynamics can be formulated in terms of the ordinary force *or* in terms of the Minkowski force. The latter is generally much *neater*, but since in the long run we are interested in the particle's trajectory as a function of *ordinary* time, the former is often more useful. When we wish to generalize some classical force law, such as Lorentz's, to the relativistic domain, the question arises: Does the classical formula correspond to the *ordinary* force or to the Minkowski force? In other words, should we write

$$\mathbf{F} = q(\mathbf{E} + \mathbf{u} \times \mathbf{B}),$$

or should it rather be

$$\mathbf{K} = q(\mathbf{E} + \mathbf{u} \times \mathbf{B})?$$

Since proper time and ordinary time are identical in classical physics, there is no way at this stage to decide the issue. The Lorentz force law, as it turns out, is an *ordinary* force—later on I'll explain why this is so, and show you how to construct the electromagnetic Minkowski force.

Example 12.11

The typical trajectory of a charged particle in a uniform *magnetic* field is **cyclotron motion** (Fig. 12.31). The magnetic force pointing toward the center,

$$F = QuB,$$

provides the centripetal acceleration necessary to sustain circular motion. Beware, however—in special relativity the centripetal force is *not* mu^2/R, as in classical mechanics. Rather, as you can see from Fig. 12.32, $dp = p\,d\theta$, so

$$F = \frac{dp}{dt} = p\frac{d\theta}{dt} = p\frac{u}{R}.$$

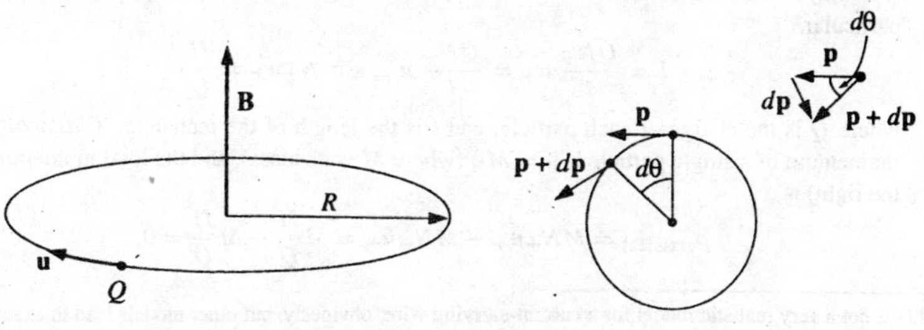

Figure 12.31

Figure 12.32

(Classically, of course, $p = mu$, so $F = mu^2/R$.) Thus,

$$QuB = p\frac{u}{R},$$

or

$$p = QBR. \tag{12.72}$$

In this form the relativistic cyclotron formula is identical to the nonrelativistic one, Eq. 5.3—the only difference is that p is now the relativistic momentum.

Example 12.12

Hidden momentum. As a model for a magnetic dipole **m**, consider a rectangular loop of wire carrying a steady current. Picture the current as a stream of noninteracting positive charges that move freely within the wire. When a uniform electric field **E** is applied (Fig. 12.33), the charges accelerate in the left segment and decelerate in the right one.[13] Find the total momentum of all the charges in the loop.

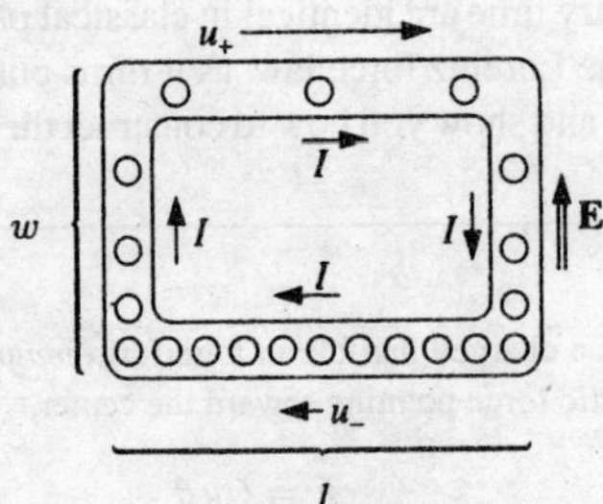

Figure 12.33

Solution: The momenta of the left and right segments cancel, so we need only consider the top and the bottom. Say there are N_+ charges in the top segment, going at speed u_+ to the right, and N_- charges in the lower segment, going at (slower) speed u_- to the left. The current ($I = \lambda u$) is the same in all four segments (or else charge would be piling up somewhere); in particular,

$$I = \frac{QN_+}{l}u_+ = \frac{QN_-}{l}u_-, \quad \text{so} \quad N_\pm u_\pm = \frac{Il}{Q},$$

where Q is the charge of each particle, and l is the length of the rectangle. *Classically,* the momentum of a single particle is $\mathbf{p} = M\mathbf{u}$ (where M is its mass), and the total momentum (to the right) is

$$p_{\text{classical}} = MN_+u_+ - MN_-u_- = M\frac{Il}{Q} - M\frac{Il}{Q} = 0,$$

[13] This is not a very realistic model for a current-carrying wire, obviously, but other models lead to exactly the same result. See V. Hnizdo, *Am. J. Phys.* **65**, 92 (1997).

as one would certainly expect (after all, the loop as a whole is not moving). But relativistically $\mathbf{p} = \gamma M \mathbf{u}$, and we get

$$p = \gamma_+ M N_+ u_+ - \gamma_- M N_- u_- = \frac{MIl}{Q}(\gamma_+ - \gamma_-),$$

which is *not* zero, because the particles in the upper segment are moving faster.

In fact, the gain in energy (γMc^2), as a particle goes up the left segment, is equal to the work done by the electric force, QEw, where w is the height of the rectangle, so

$$\gamma_+ - \gamma_- = \frac{QEw}{Mc^2},$$

and hence

$$p = \frac{IlEw}{c^2}.$$

But Ilw is the magnetic dipole moment of the loop; as vectors, $\mathbf{m}$ points into the page and $\mathbf{p}$ is to the right, so

$$\mathbf{p} = \frac{1}{c^2}(\mathbf{m} \times \mathbf{E}).$$

Thus a magnetic dipole in an electric field carries linear momentum, *even though it is not moving!* This so-called **hidden momentum** is strictly relativistic, and purely mechanical; it precisely cancels the electromagnetic momentum stored in the fields (see Ex. 8.3; note that both results can be expressed in the form $p = IlV/c^2$).

Problem 12.36 In classical mechanics Newton's law can be written in the more familiar form $\mathbf{F} = m\mathbf{a}$. The relativistic equation, $\mathbf{F} = d\mathbf{p}/dt$, *cannot* be so simply expressed. Show, rather, that

$$\mathbf{F} = \frac{m}{\sqrt{1 - u^2/c^2}}\left[\mathbf{a} + \frac{\mathbf{u}(\mathbf{u}\cdot\mathbf{a})}{c^2 - u^2}\right], \tag{12.73}$$

where $\mathbf{a} \equiv d\mathbf{u}/dt$ is the **ordinary acceleration**.

Problem 12.37 Show that it is possible to outrun a light ray, if you're given a sufficient head start, and your feet generate a constant force.

Problem 12.38 Define **proper acceleration** in the obvious way:

$$\alpha^\mu \equiv \frac{d\eta^\mu}{d\tau} = \frac{d^2x^\mu}{d\tau^2}. \tag{12.74}$$

(a) Find α^0 and $\boldsymbol{\alpha}$ in terms of $\mathbf{u}$ and $\mathbf{a}$ (the ordinary acceleration).

(b) Express $\alpha_\mu \alpha^\mu$ in terms of $\mathbf{u}$ and $\mathbf{a}$.

(c) Show that $\eta^\mu \alpha_\mu = 0$.

(d) Write the Minkowski version of Newton's second law, Eq. 12.70, in terms of α^μ. Evaluate the invariant product $K^\mu \eta_\mu$.

Problem 12.39 Show that

$$K_\mu K^\mu = \frac{1 - (u^2/c^2)\cos^2\theta}{1 - u^2/c^2} F^2,$$

where θ is the angle between $\mathbf{u}$ and $\mathbf{F}$.

Problem 12.40 Show that the (ordinary) acceleration of a particle of mass m and charge q, moving at velocity $\mathbf{u}$ under the influence of electromagnetic fields $\mathbf{E}$ and $\mathbf{B}$, is given by

$$\mathbf{a} = \frac{q}{m}\sqrt{1 - u^2/c^2}\left[\mathbf{E} + \mathbf{u} \times \mathbf{B} - \frac{1}{c^2}\mathbf{u}(\mathbf{u} \cdot \mathbf{E})\right].$$

[*Hint:* Use Eq. 12.73.]

12.3 Relativistic Electrodynamics

12.3.1 Magnetism as a Relativistic Phenomenon

Unlike Newtonian mechanics, classical electrodynamics is *already* consistent with special relativity. Maxwell's equations and the Lorentz force law can be applied legitimately in any inertial system. Of course, what one observer interprets as an electrical process another may regard as magnetic, but the actual particle motions they predict will be identical. To the extent that this did *not* work out for Lorentz and others, who studied the question in the late nineteenth century, the fault lay with the nonrelativistic mechanics they used, not with the electrodynamics. Having corrected Newtonian mechanics, we are now in a position to develop a complete and consistent formulation of relativistic electrodynamics. But I emphasize that we will not be changing the rules of electrodynamics in the slightest—rather, we will be *expressing* these rules in a notation that exposes and illuminates their relativistic character. As we go along, I shall pause now and then to rederive, using the Lorentz transformations, results obtained earlier by more laborious means. But the main purpose of this section is to provide you with a deeper understanding of the structure of electrodynamics—laws that had seemed arbitrary and unrelated before take on a kind of coherence and inevitability when approached from the point of view of relativity.

To begin with I'd like to show you why there *had* to be such a thing as magnetism, given electrostatics and relativity, and how, in particular, you can calculate the magnetic force between a current-carrying wire and a moving charge without ever invoking the laws of magnetism.[14] Suppose you had a string of positive charges moving along to the right at speed v. I'll assume the charges are close enough together so that we may regard them as a continuous line charge λ. Superimposed on this positive string is a negative one, $-\lambda$ proceeding to the left at the same speed v. We have, then, a net current to the right, of magnitude

$$I = 2\lambda v. \tag{12.75}$$

[14]This and several other arguments in this section are adapted from E. M. Purcell's *Electricity and Magnetism*, 2d ed. (New York: McGraw-Hill, 1985).

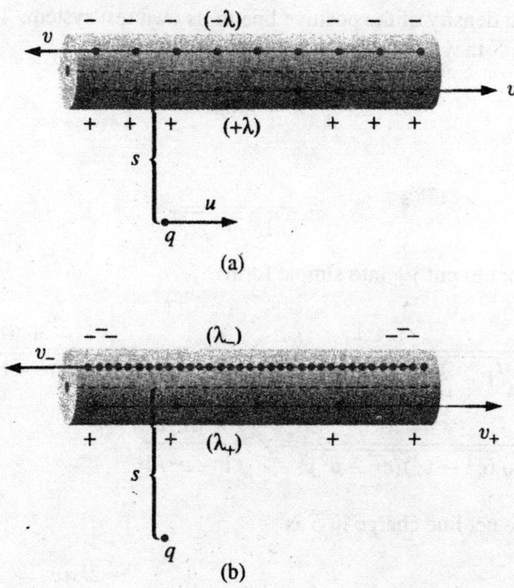

Figure 12.34

Meanwhile, a distance s away there is a point charge q traveling to the right at speed $u < v$ (Fig. 12.34a). Because the two line charges cancel, there is *no electrical force on q* in this system ($\mathcal{S}$).

However, let's examine the same situation from the point of view of system $\bar{\mathcal{S}}$, which moves to the right with speed u (Fig. 12.34b). In this reference frame q is at rest. By the Einstein velocity addition rule, the velocities of the positive and negative lines are now

$$v_{\pm} = \frac{v \mp u}{1 \mp vu/c^2}. \tag{12.76}$$

Because v_- is greater than v_+, the Lorentz contraction of the spacing between negative charges is more severe than that between positive charges; *in this frame,* therefore, *the wire carries a net negative charge!* In fact,

$$\lambda_{\pm} = \pm(\gamma_{\pm})\lambda_0, \tag{12.77}$$

where

$$\gamma_{\pm} = \frac{1}{\sqrt{1 - v_{\pm}^2/c^2}}, \tag{12.78}$$

and λ_0 is the charge density of the positive line in its own rest system. That's not the same as λ, of course—in $\mathcal{S}$ they're already moving at speed v, so

$$\lambda = \gamma \lambda_0, \tag{12.79}$$

where

$$\gamma = \frac{1}{\sqrt{1 - v^2/c^2}}. \tag{12.80}$$

It takes some algebra to put $\gamma_\pm$ into simple form:

$$\begin{aligned}
\gamma_\pm &= \frac{1}{\sqrt{1 - \frac{1}{c^2}(v \mp u)^2 (1 \mp vu/c^2)^{-2}}} = \frac{c^2 \mp uv}{\sqrt{(c^2 \mp uv)^2 - c^2(v \mp u)^2}} \\
&= \frac{c^2 \mp uv}{\sqrt{(c^2 - v^2)(c^2 - u^2)}} = \gamma \frac{1 \mp uv/c^2}{\sqrt{1 - u^2/c^2}}.
\end{aligned} \tag{12.81}$$

Evidently, then, the net line charge in $\bar{\mathcal{S}}$ is

$$\lambda_{\text{tot}} = \lambda_+ + \lambda_- = \lambda_0(\gamma_+ - \gamma_-) = \frac{-2\lambda u v}{c^2 \sqrt{1 - u^2/c^2}}. \tag{12.82}$$

Conclusion: As a result of unequal Lorentz contraction of the positive and negative lines, a current-carrying wire that is electrically neutral in one inertial system will be charged in another.

Now, a line charge λ_{tot} sets up an *electric* field

$$E = \frac{\lambda_{\text{tot}}}{2\pi\epsilon_0 s},$$

so *there is an electrical force on* q *in* $\bar{\mathcal{S}}$, to wit:

$$\bar{F} = qE = -\frac{\lambda v}{\pi\epsilon_0 c^2 s} \frac{qu}{\sqrt{1 - u^2/c^2}}. \tag{12.83}$$

But if there's a force on q in $\bar{\mathcal{S}}$, there must be one in $\mathcal{S}$; in fact, we can *calculate* it by using the transformation rules for forces. Since q is at rest $\bar{\mathcal{S}}$, and $\bar{F}$ is perpendicular to u, the force in $\mathcal{S}$ is given by Eq. 12.68:

$$F = \sqrt{1 - u^2/c^2}\, \bar{F} = -\frac{\lambda v}{\pi\epsilon_0 c^2} \frac{qu}{s}. \tag{12.84}$$

The charge is attracted toward the wire by a force that is purely electrical in $\bar{\mathcal{S}}$ (where the wire is charged, and q is at rest), but distinctly *non*electrical in $\mathcal{S}$ (where the wire is neutral). Taken together, then, electrostatics and relativity imply the existence of another force. This "other force" is, of course, *magnetic.* In fact, we can cast Eq. 12.84 into more familiar form by using $c^2 = (\epsilon_0\mu_0)^{-1}$ and expressing λv in terms of the current (Eq. 12.75):

$$F = -qu\left(\frac{\mu_0 I}{2\pi s}\right). \tag{12.85}$$

The term in parentheses is the magnetic field of a long, straight wire, and the force is precisely what we would have obtained by using the Lorentz force law in system $\mathcal{S}$.

12.3.2 How the Fields Transform

We have learned, in various special cases, that one observer's electric field is another's magnetic field. It would be nice to know the *general* transformation rules for electromagnetic fields: Given the fields in $\mathcal{S}$, what are the fields in $\bar{\mathcal{S}}$? Your first guess might be that $\mathbf{E}$ is the spatial part of one 4-vector and $\mathbf{B}$ the spatial part of another. If so, your intuition is wrong—it's more complicated than that. Let me begin by making explicit an assumption that was already used implicitly in Sect. 12.3.1: *Charge is invariant.* Like mass, but unlike energy, the charge of a particle is a fixed number, independent of how fast it happens to be moving. We shall assume also that the transformation rules are the same no matter how the fields were produced—electric fields generated by changing magnetic fields transform the same way as those set up by stationary charges. Were this not the case we'd have to abandon the field formulation altogether, for it is the essence of a field theory that the fields at a given point tell you *all there is to know,* electromagnetically, about that point; you do *not* have to append extra information regarding their source.

With this in mind, consider the *simplest possible* electric field: the uniform field in the region between the plates of a large parallel-plate capacitor (Fig. 12.35a). Say the capacitor is at rest in $\mathcal{S}_0$ and carries surface charges $\pm\sigma_0$. Then

$$\mathbf{E}_0 = \frac{\sigma_0}{\epsilon_0}\hat{\mathbf{y}}. \tag{12.86}$$

But what if we examine this same capacitor from system $\mathcal{S}$, moving to the right at speed v_0 (Fig. 12.35b)? In this system the plates are moving to the left, but the field still takes the form

$$\mathbf{E} = \frac{\sigma}{\epsilon_0}\hat{\mathbf{y}}; \tag{12.87}$$

the only difference is the value of the surface charge σ. [Wait a minute! *Is* that the only difference? The formula $E = \sigma/\epsilon_0$ for a parallel plate capacitor came from Gauss's law, and whereas Gauss's law is perfectly valid for moving charges, this particular application also relies on symmetry. Are we sure that the field is still perpendicular to the plates? What if the field of a moving plane *tilts,* say, in the direction of motion, as in Fig. 12.35c? Well, *even if it did* (it *doesn't*), the field between the plates, being the superposition of the $+\sigma$ field and the $-\sigma$ field, would nevertheless run perpendicular to the plates. For the $-\sigma$ field would aim as indicated in Fig. 12.35c (changing the sign of the charges reverses the direction of the field), and the vector sum kills off the parallel components.]

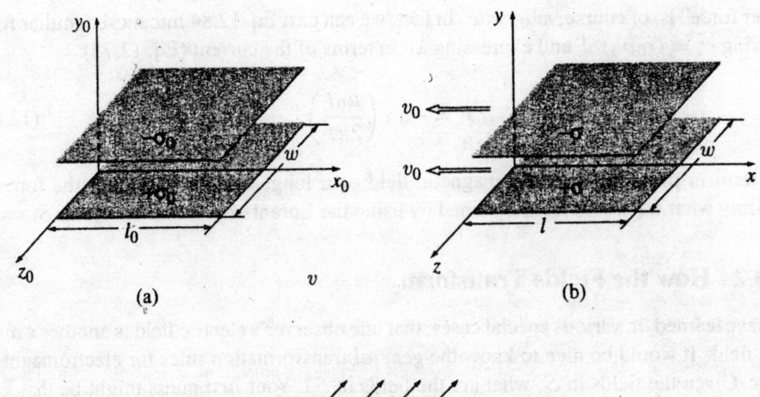

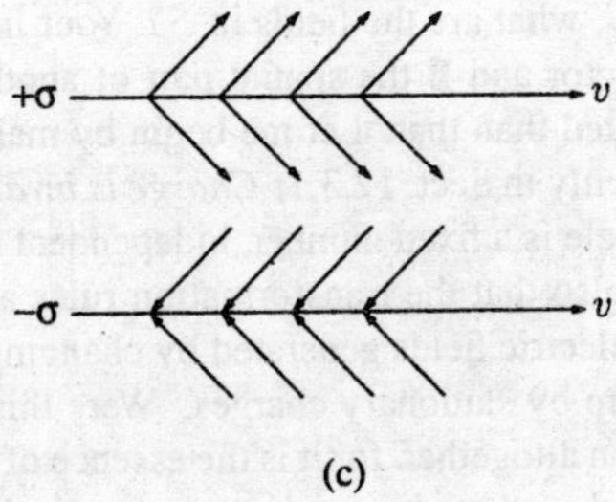

Figure 12.35

Now, the total charge on each plate is invariant, and the *width* (w) is unchanged, but the *length* (l) is Lorentz-contracted by a factor

$$\frac{1}{\gamma_0} = \sqrt{1 - v_0^2/c^2}, \tag{12.88}$$

so the charge per unit area is *increased* by a factor γ_0:

$$\sigma = \gamma_0 \sigma_0. \tag{12.89}$$

Accordingly,

$$\mathbf{E}^{\perp} = \gamma_0 \mathbf{E}_0{}^{\perp}. \tag{12.90}$$

I have put in the superscript $\perp$ to make it clear that this rule pertains to components of $\mathbf{E}$ that are *perpendicular* to the direction of motion of $\mathcal{S}$. To get the rule for *parallel* components, consider the capacitor lined up with the yz plane (Fig. 12.36). This time it is the plate separation (d) that is Lorentz-contracted, whereas l and w (and hence also σ) are the same in both frames. Since the field does not depend on d, it follows that

$$E^{\parallel} = E_0{}^{\parallel}. \tag{12.91}$$

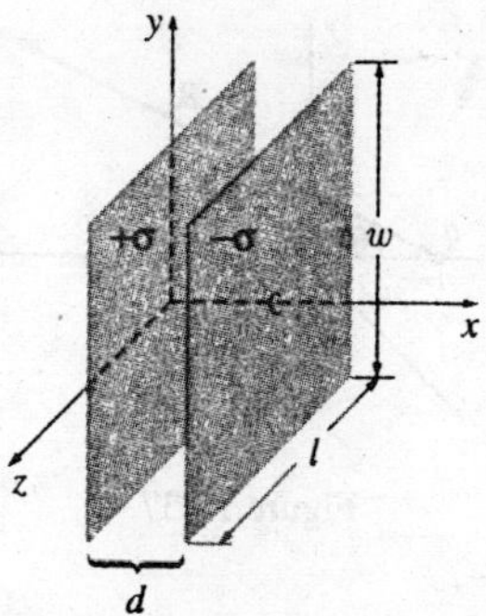

Figure 12.36

Example 12.13

Electric field of a point charge in uniform motion. A point charge q is at rest at the origin in system S_0. *Question*: What is the electric field of this same charge in system S, which moves to the right at speed v_0 relative to S_0?

Solution: In S_0 the field is

$$\mathbf{E}_0 = \frac{1}{4\pi\epsilon_0}\frac{q}{r_0^2}\hat{\mathbf{r}}_0,$$

or

$$\left\{\begin{array}{lcl} E_{x0} & = & \dfrac{1}{4\pi\epsilon_0}\dfrac{qx_0}{(x_0^2+y_0^2+z_0^2)^{3/2}}, \\ E_{y0} & = & \dfrac{1}{4\pi\epsilon_0}\dfrac{qy_0}{(x_0^2+y_0^2+z_0^2)^{3/2}}, \\ E_{z0} & = & \dfrac{1}{4\pi\epsilon_0}\dfrac{qz_0}{(x_0^2+y_0^2+z_0^2)^{3/2}}. \end{array}\right.$$

From the transformation rules (Eqs. 12.90 and 12.91), we have

$$\left\{\begin{array}{lclcl} E_x & = & E_{x0} & = & \dfrac{1}{4\pi\epsilon_0}\dfrac{qx_0}{(x_0^2+y_0^2+z_0^2)^{3/2}}, \\ E_y & = & \gamma_0 E_{y0} & = & \dfrac{1}{4\pi\epsilon_0}\dfrac{\gamma_0 q y_0}{(x_0^2+y_0^2+z_0^2)^{3/2}}, \\ E_z & = & \gamma_0 E_{z0} & = & \dfrac{1}{4\pi\epsilon_0}\dfrac{\gamma_0 q z_0}{(x_0^2+y_0^2+z_0^2)^{3/2}}. \end{array}\right.$$

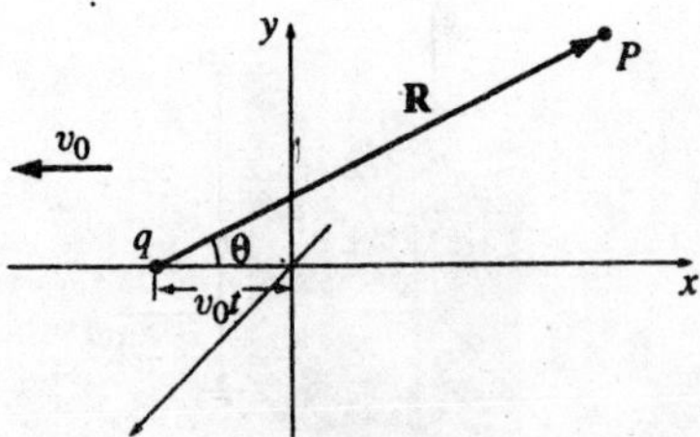

Figure 12.37

These are still expressed in terms of the S_0 coordinates (x_0, y_0, z_0) of the field point (P); I'd prefer to write them in terms of the S coordinates of P. From the Lorentz transformations (or, actually, the inverse transformations),

$$\left\{\begin{array}{lllll} x_0 & = & \gamma_0(x+v_0t) & = & \gamma_0 R_x, \\ y_0 & = & y & = & R_y, \\ z_0 & = & z & = & R_z, \end{array}\right.$$

where **R** is the vector from q to P (Fig. 12.37). Thus

$$\begin{aligned} \mathbf{E} &= \frac{1}{4\pi\epsilon_0}\frac{\gamma_0 q\mathbf{R}}{(\gamma_0^2 R^2\cos^2\theta + R^2\sin^2\theta)^{3/2}} \\ &= \frac{1}{4\pi\epsilon_0}\frac{q(1-v_0^2/c^2)}{[1-(v_0^2/c^2)\sin^2\theta]^{3/2}}\frac{\hat{\mathbf{R}}}{R^2}. \end{aligned} \tag{12.92}$$

This, then, is the field of a charge in uniform motion; we got the same result in Chapter 10 using the retarded potentials (Eq. 10.68). The present derivation is far more efficient, and sheds some light on the remarkable fact that the field points away from the instantaneous (as opposed to the retarded) position of the charge: E_x gets a factor of γ_0 from the Lorentz transformation of the *coordinates;* E_y and E_z pick up theirs from the transformation of the *field.* It's the balancing of these two γ_0's that leaves **E** parallel to **R**.

But Eqs. 12.90 and 12.91 are not the most general transformation laws, for we began with a system S_0 in which the charges were at rest and where, consequently, there was no magnetic field. To derive the *general* rule we must start out in a system with both electric and magnetic fields. For this purpose S itself will serve nicely. In addition to the electric field

$$E_y = \frac{\sigma}{\epsilon_0}, \tag{12.93}$$

there is a *magnetic* field due to the surface currents (Fig. 12.35b):

$$\mathbf{K}_\pm = \mp\sigma v_0\,\hat{\mathbf{x}}. \tag{12.94}$$

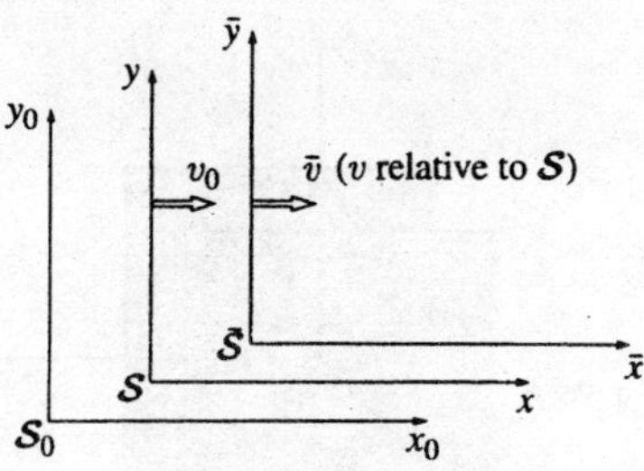

Figure 12.38

By the right-hand rule, this field points in the negative z direction; its magnitude is given by Ampère's law:

$$B_z = -\mu_0 \sigma v_0. \tag{12.95}$$

In a *third* system, $\bar{\mathcal{S}}$, traveling to the right with speed v relative to $\mathcal{S}$ (Fig. 12.38), the fields would be

$$\bar{E}_y = \frac{\bar{\sigma}}{\epsilon_0}, \quad \bar{B}_z = -\mu_0 \bar{\sigma} \bar{v}, \tag{12.96}$$

where $\bar{v}$ is the velocity of $\bar{\mathcal{S}}$ relative to $\mathcal{S}_0$:

$$\bar{v} = \frac{v + v_0}{1 + v v_0 / c^2}, \quad \bar{\gamma} = \frac{1}{\sqrt{1 - \bar{v}^2/c^2}}, \tag{12.97}$$

and

$$\bar{\sigma} = \bar{\gamma} \sigma_0. \tag{12.98}$$

It remains only to express $\bar{\mathbf{E}}$ and $\bar{\mathbf{B}}$ (Eq. 12.96), in terms of $\mathbf{E}$ and $\mathbf{B}$ (Eqs. 12.93 and 12.95). In view of Eqs. 12.89 and 12.98, we have

$$\bar{E}_y = \left(\frac{\bar{\gamma}}{\gamma_0}\right) \frac{\sigma}{\epsilon_0}, \quad \bar{B}_z = -\left(\frac{\bar{\gamma}}{\gamma_0}\right) \mu_0 \sigma \bar{v}. \tag{12.99}$$

With a little algebra, you will find that

$$\frac{\bar{\gamma}}{\gamma_0} = \frac{\sqrt{1 - v_0^2/c^2}}{\sqrt{1 - \bar{v}^2/c^2}} = \frac{1 + v v_0/c^2}{\sqrt{1 - v^2/c^2}} = \gamma \left(1 + \frac{v v_0}{c^2}\right), \tag{12.100}$$

where

$$\gamma = \frac{1}{\sqrt{1 - v^2/c^2}}, \tag{12.101}$$

as always. Thus,

$$\bar{E}_y = \gamma \left(1 + \frac{v v_0}{c^2}\right) \frac{\sigma}{\epsilon_0} = \gamma \left(E_y - \frac{v}{c^2 \epsilon_0 \mu_0} B_z\right),$$

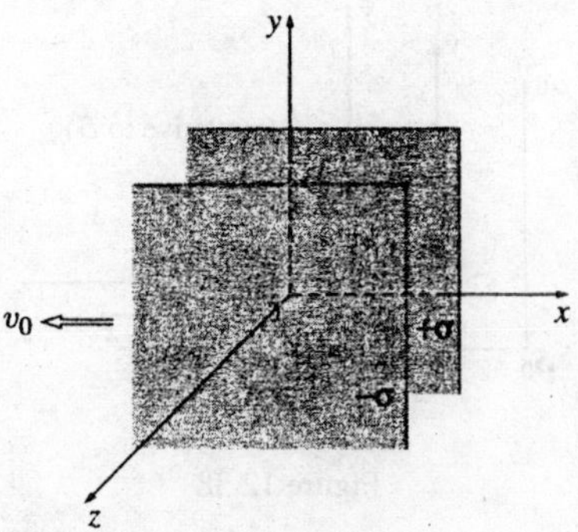

Figure 12.39

whereas

$$\bar{B}_z = -\gamma\left(1 + \frac{vv_0}{c^2}\right)\mu_0\sigma\left(\frac{v + v_0}{1 + vv_0/c^2}\right) = \gamma(B_z - \mu_0\epsilon_0 v E_y).$$

Or, since $\mu_0\epsilon_0 = 1/c^2$,

$$\left.\begin{aligned}\bar{E}_y &= \gamma(E_y - vB_z),\\ \bar{B}_z &= \gamma\left(B_z - \frac{v}{c^2}E_y\right).\end{aligned}\right\} \tag{12.102}$$

This tells us how E_y and B_z transform—to do E_z and B_y we simply align the same capacitor parallel to the xy plane instead of the xz plane (Fig. 12.39). The fields in $\mathcal{S}$ are then

$$E_z = \frac{\sigma}{\epsilon_0}, \quad B_y = \mu_0\sigma v_0.$$

(Use the right-hand rule to get the sign of B_y.) The rest of the argument is identical—everywhere we had E_y before, read E_z, and everywhere we had B_z, read $-B_y$:

$$\left.\begin{aligned}\bar{E}_z &= \gamma(E_z + vB_y),\\ \bar{B}_y &= \gamma\left(B_y + \frac{v}{c^2}E_z\right).\end{aligned}\right\} \tag{12.103}$$

As for the x components, we have already seen (by orienting the capacitor parallel to the yz plane) that

$$\bar{E}_x = E_x. \tag{12.104}$$

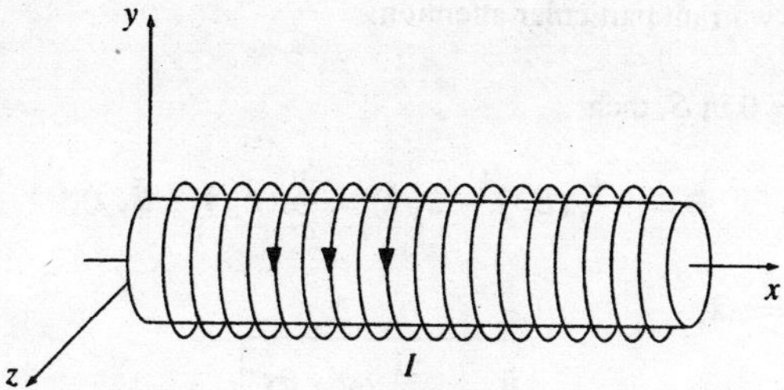

Figure 12.40

Since in this case there is no accompanying magnetic field, we cannot deduce the transformation rule for B_x. But another configuration will do the job: Imagine a long *solenoid* aligned parallel to the x axis (Fig. 12.40) and at rest in $\mathcal{S}$. The magnetic field within the coil is

$$B_x = \mu_0 n I, \tag{12.105}$$

where n is the number of turns per unit length, and I is the current. In system $\bar{\mathcal{S}}$, the length contracts, so n *increases:*

$$\bar{n} = \gamma n. \tag{12.106}$$

On the other hand, time *dilates:* The $\mathcal{S}$ clock, which rides along with the solenoid, runs slow, so the current (charge *per unit time*) in $\bar{\mathcal{S}}$ is given by

$$\bar{I} = \frac{1}{\gamma} I. \tag{12.107}$$

The two factors of γ exactly cancel, and we conclude that

$$\bar{B}_x = B_x.$$

Like **E**, the component of **B** *parallel* to the motion is unchanged.

Let's now collect together the complete set of transformation rules:

$$\boxed{\begin{array}{lll} \bar{E}_x = E_x, & \bar{E}_y = \gamma(E_y - vB_z), & \bar{E}_z = \gamma(E_z + vB_y), \\ \bar{B}_x = B_x, & \bar{B}_y = \gamma\left(B_y + \frac{v}{c^2}E_z\right), & \bar{B}_z = \gamma\left(B_z - \frac{v}{c^2}E_y\right). \end{array}} \tag{12.108}$$

Two special cases warrant particular attention:

1. If $\mathbf{B} = 0$ in $\mathcal{S}$, then

$$\bar{\mathbf{B}} = \gamma \frac{v}{c^2}(E_z \hat{\mathbf{y}} - E_y \hat{\mathbf{z}}) = \frac{v}{c^2}(\bar{E}_z \hat{\mathbf{y}} - \bar{E}_y \hat{\mathbf{z}}),$$

or, since $\mathbf{v} = v\,\hat{\mathbf{x}}$,

$$\boxed{\bar{\mathbf{B}} = -\frac{1}{c^2}(\mathbf{v} \times \bar{\mathbf{E}}).} \tag{12.109}$$

2. If $\mathbf{E} = 0$ in $\mathcal{S}$, then

$$\bar{\mathbf{E}} = -\gamma v(B_z \hat{\mathbf{y}} - B_y \hat{\mathbf{z}}) = -v(\bar{B}_z \hat{\mathbf{y}} - \bar{B}_y \hat{\mathbf{z}}),$$

or

$$\boxed{\bar{\mathbf{E}} = \mathbf{v} \times \bar{\mathbf{B}}.} \tag{12.110}$$

In other words, if either **E** or **B** is zero (at a particular point) in *one* system, then in any other system the fields (at that point) are very simply related by Eq. 12.109 or Eq. 12.110.

Example 12.14

Magnetic field of a point charge in uniform motion. Find the *magnetic* field of a point charge q moving at constant velocity **v**.

Solution: In the particle's *rest* frame ($\mathcal{S}_0$) the magnetic field is zero (everywhere), so in a system $\mathcal{S}$ moving to the right at speed v,

$$\mathbf{B} = -\frac{1}{c^2}(\mathbf{v} \times \mathbf{E}).$$

We calculated the *electric* field in Ex. 12.13. The magnetic field, then, is

$$\mathbf{B} = \frac{\mu_0}{4\pi} \frac{qv(1 - v^2/c^2)\sin\theta}{[1 - (v^2/c^2)\sin^2\theta]^{3/2}} \frac{\hat{\boldsymbol{\phi}}}{R^2}, \tag{12.111}$$

where $\hat{\boldsymbol{\phi}}$ aims counterclockwise as you face the oncoming charge. Incidentally, in the nonrelativistic limit ($v^2 \ll c^2$), Eq. 12.111 reduces to

$$\mathbf{B} = \frac{\mu_0}{4\pi} q \frac{\mathbf{v} \times \mathbf{R}}{R^2},$$

which is exactly what you would get by naïve application of the Biot-Savart law to a point charge (Eq. 5.40).

Problem 12.41 Why can't the electric field in Fig. 12.35b have a z component? After all, the *magnetic* field does.

Problem 12.42 A parallel-plate capacitor, at rest in $\mathcal{S}_0$ and tilted at a 45° angle to the x_0 axis, carries charge densities $\pm\sigma_0$ on the two plates (Fig. 12.41). System $\mathcal{S}$ is moving to the right at speed v relative to $\mathcal{S}_0$.

(a) Find $\mathbf{E}_0$, the field in $\mathcal{S}_0$.

(b) Find $\mathbf{E}$, the field in $\mathcal{S}$.

(c) What angle do the plates make with the x axis?

(d) Is the field perpendicular to the plates in $\mathcal{S}$?

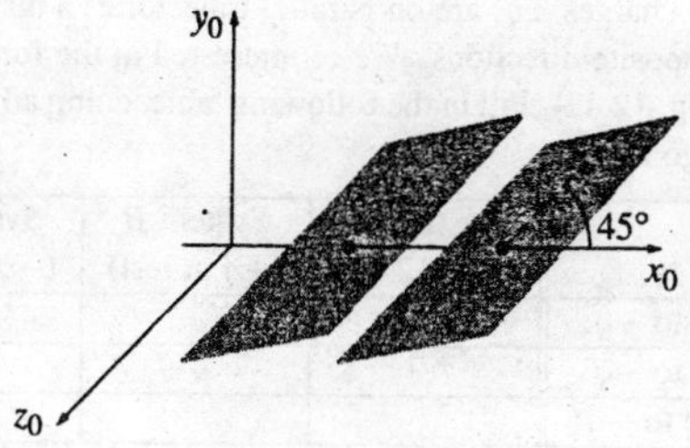

Figure 12.41

Problem 12.43

(a) Check that Gauss's law, $\oint \mathbf{E} \cdot d\mathbf{a} = (1/\epsilon_0)Q_{\text{enc}}$, is obeyed by the field of a point charge in uniform motion, by integrating over a sphere of radius R centered on the charge.

(b) Find the Poynting vector for a point charge in uniform motion. (Say the charge is going in the z direction at speed v, and calculate $\mathbf{S}$ at the instant q passes the origin.)

Problem 12.44

(a) Charge q_A is at rest at the origin in system $\mathcal{S}$; charge q_B flies by at speed v on a trajectory parallel to the x axis, but at $y = d$. What is the electromagnetic force on q_B as it crosses the y axis?

(b) Now study the same problem from system $\bar{\mathcal{S}}$, which moves to the right with speed v. What is the force on q_B when q_A passes the $\bar{y}$ axis? [Do it two ways: (i) by using your answer to (a) and transforming the force; (ii) by computing the fields in $\bar{\mathcal{S}}$ and using the Lorentz force law.]

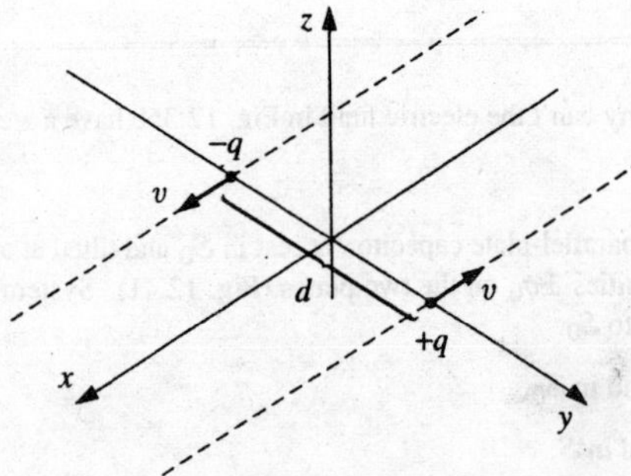

Figure 12.42

Problem 12.45 Two charges $\pm q$, are on parallel trajectories a distance d apart, moving with equal speeds v in opposite directions. We're interested in the force on $+q$ due to $-q$ at the instant they cross (Fig. 12.42). Fill in the following table, doing all the consistency checks you can think of as you go along.

	System A (Fig. 12.42)	System B ($+q$ at rest)	System C ($-q$ at rest)
E at $+q$ due to $-q$:			
B at $+q$ due to $-q$:			
F on $+q$ due to $-q$:			

Problem 12.46

(a) Show that $(\mathbf{E} \cdot \mathbf{B})$ is relativistically invariant.

(b) Show that $(E^2 - c^2 B^2)$ is relativistically invariant.

(c) Suppose that in one inertial system $\mathbf{B} = 0$ but $\mathbf{E} \neq 0$ (at some point P). Is it possible to find another system in which the *electric* field is zero at P?

Problem 12.47 An electromagnetic plane wave of (angular) frequency ω is traveling in the x direction through the vacuum. It is polarized in the y direction, and the amplitude of the electric field is E_0.

(a) Write down the electric and magnetic fields, $\mathbf{E}(x, y, z, t)$ and $\mathbf{B}(x, y, z, t)$. [Be sure to define any auxiliary quantities you introduce, in terms of ω, E_0, and the constants of nature.]

(b) This same wave is observed from an inertial system $\bar{S}$ moving in the x direction with speed v relative to the original system S. Find the electric and magnetic fields in $\bar{S}$, and express them in terms of the $\bar{S}$ coordinates: $\bar{\mathbf{E}}(\bar{x}, \bar{y}, \bar{z}, \bar{t})$ and $\bar{\mathbf{B}}(\bar{x}, \bar{y}, \bar{z}, \bar{t})$. [Again, be sure to define any auxiliary quantities you introduce.]

(c) What is the frequency $\bar{\omega}$ of the wave in $\bar{S}$? Interpret this result. What is the wavelength $\bar{\lambda}$ of the wave in $\bar{S}$? From $\bar{\omega}$ and $\bar{\lambda}$, determine the speed of the waves in $\bar{S}$. Is it what you expected?

(d) What is the ratio of the intensity in $\bar{S}$ to the intensity in S? As a youth, Einstein wondered what an electromagnetic wave would look like if you could run along beside it at the speed of light. What can you tell him about the amplitude, frequency, and intensity of the wave, as v approaches c?

12.3.3 The Field Tensor

As Eq. 12.108 indicates, **E** and **B** certainly do *not* transform like the spatial parts of the two 4-vectors—in fact, the components of **E** and **B** are stirred together when you go from one inertial system to another. What sort of an object is this, which has six components and transforms according to Eq. 12.108? *Answer*: It's an **antisymmetric, second-rank tensor**.

Remember that a 4-vector transforms by the rule

$$\bar{a}^{\mu} = \Lambda^{\mu}_{\nu} a^{\nu} \tag{12.112}$$

(summation over ν implied), where Λ is the Lorentz transformation matrix. If $\bar{S}$ is moving in the x direction at speed v, Λ has the form

$$\Lambda = \begin{pmatrix} \gamma & -\gamma\beta & 0 & 0 \\ -\gamma\beta & \gamma & 0 & 0 \\ 0 & 0 & 1 & 0 \\ 0 & 0 & 0 & 1 \end{pmatrix}, \tag{12.113}$$

and Λ^{μ}_{ν} is the entry in row μ, column ν. A (second-rank) tensor is an object with *two* indices, which transform with *two* factors of Λ (one for each index):

$$\bar{t}^{\mu\nu} = \Lambda^{\mu}_{\lambda} \Lambda^{\nu}_{\sigma} t^{\lambda\sigma}. \tag{12.114}$$

A tensor (in 4 dimensions) has $4 \times 4 = 16$ components, which we can display in a 4×4 array:

$$t^{\mu\nu} = \left\{ \begin{array}{cccc} t^{00} & t^{01} & t^{02} & t^{03} \\ t^{10} & t^{11} & t^{12} & t^{13} \\ t^{20} & t^{21} & t^{22} & t^{23} \\ t^{30} & t^{31} & t^{32} & t^{33} \end{array} \right\}.$$

However, the 16 elements need not all be different. For instance, a *symmetric* tensor has the property

$$t^{\mu\nu} = t^{\nu\mu} \quad \textbf{(symmetric tensor)}. \tag{12.115}$$

In this case there are 10 distinct components; 6 of the 16 are repeats ($t^{01} = t^{10}$, $t^{02} = t^{20}$, $t^{03} = t^{30}$, $t^{12} = t^{21}$, $t^{13} = t^{31}$, $t^{23} = t^{32}$). Similarly, an *anti*symmetric tensor obeys

$$t^{\mu\nu} = -t^{\nu\mu} \quad \textbf{(antisymmetric tensor)}. \tag{12.116}$$

Such an object has just 6 distinct elements—of the original 16, six are repeats (the same ones as before, only this time with a minus sign) and four are zero (t^{00}, t^{11}, t^{22}, and t^{33}). Thus, the general antisymmetric tensor has the form

$$t^{\mu\nu} = \begin{Bmatrix} 0 & t^{01} & t^{02} & t^{03} \\ -t^{01} & 0 & t^{12} & t^{13} \\ -t^{02} & -t^{12} & 0 & t^{23} \\ -t^{03} & -t^{13} & -t^{23} & 0 \end{Bmatrix}.$$

Let's see how the transformation rule 12.114 works, for the six distinct components of an antisymmetric tensor. Starting with $\bar{t}^{01}$, we have

$$\bar{t}^{01} = \Lambda^0_\lambda \Lambda^1_\sigma t^{\lambda\sigma},$$

but according to Eq. 12.113, $\Lambda^0_\lambda = 0$ unless $\lambda = 0$ or 1, and $\Lambda^1_\sigma = 0$ unless $\sigma = 0$ or 1. So there are four terms in the sum:

$$\bar{t}^{01} = \Lambda^0_0\Lambda^1_0 t^{00} + \Lambda^0_0\Lambda^1_1 t^{01} + \Lambda^0_1\Lambda^1_0 t^{10} + \Lambda^0_1\Lambda^1_1 t^{11}.$$

On the other hand, $t^{00} = t^{11} = 0$, while $t^{01} = -t^{10}$, so

$$\bar{t}^{01} = (\Lambda^0_0\Lambda^1_1 - \Lambda^0_1\Lambda^1_0)t^{01} = (\gamma^2 - (\gamma\beta)^2)t^{01} = t^{01}.$$

I'll let you work out the others—the complete set of transformation rules is

$$\left.\begin{array}{lll} \bar{t}^{01} = t^{01}, & \bar{t}^{02} = \gamma(t^{02} - \beta t^{12}), & \bar{t}^{03} = \gamma(t^{03} + \beta t^{31}), \\ \bar{t}^{23} = t^{23}, & \bar{t}^{31} = \gamma(t^{31} + \beta t^{03}), & \bar{t}^{12} = \gamma(t^{12} - \beta t^{02}). \end{array}\right\} \quad (12.117)$$

These are precisely the rules we derived on physical grounds for the electromagnetic fields (Eq. 12.108)—in fact, we can construct the **field tensor** $F^{\mu\nu}$ by direct comparison:[15]

$$F^{01} \equiv \frac{E_x}{c}, \quad F^{02} \equiv \frac{E_y}{c}, \quad F^{03} \equiv \frac{E_z}{c}, \quad F^{12} \equiv B_z, \quad F^{31} \equiv B_y, \quad F^{23} \equiv B_x.$$

Written as an array,

$$F^{\mu\nu} = \begin{Bmatrix} 0 & E_x/c & E_y/c & E_z/c \\ -E_x/c & 0 & B_z & -B_y \\ -E_y/c & -B_z & 0 & B_x \\ -E_z/c & B_y & -B_x & 0 \end{Bmatrix}. \quad (12.118)$$

Thus relativity completes and perfects the job begun by Oersted, combining the electric and magnetic fields into a single entity, $F^{\mu\nu}$.

If you followed that argument with exquisite care, you may have noticed that there was a *different* way of imbedding **E** and **B** in an antisymmetric tensor: instead of comparing

[15]Some authors prefer the convention $F^{01} \equiv E_x$, $F^{12} \equiv cB_z$, and so on, and some use the opposite signs. Accordingly, most of the equations from here on will look a little different, depending on the text.

the first line of Eq. 12.108 with the first line of Eq. 12.117, and the second with the second, we could relate the first line of Eq. 12.108 to the *second* line of Eq. 12.117, and vice versa. This leads to **dual tensor,** $G^{\mu\nu}$:

$$G^{\mu\nu} = \left\{ \begin{array}{cccc} 0 & B_x & B_y & B_z \\ -B_x & 0 & -E_z/c & E_y/c \\ -B_y & E_z/c & 0 & -E_x/c \\ -B_z & -E_y/c & E_x/c & 0 \end{array} \right\}. \tag{12.119}$$

$G^{\mu\nu}$ can be obtained directly from $F^{\mu\nu}$ by the substitution $\mathbf{E}/c \to \mathbf{B}$, $\mathbf{B} \to -\mathbf{E}/c$. Notice that this operation leaves Eq. 12.108 unchanged—that's why both tensors generate the correct transformation rules for **E** and **B.**

Problem 12.48 Work out the remaining five parts to Eq. 12.117.

Problem 12.49 Prove that the symmetry (or antisymmetry) of a tensor is preserved by Lorentz transformation (that is: if $t^{\mu\nu}$ is symmetric, show that $\bar{t}^{\mu\nu}$ is also symmetric, and likewise for antisymmetric).

Problem 12.50 Recall that a *covariant* 4-vector is obtained from a *contravariant* one by changing the sign of the zeroth component. The same goes for tensors: When you "lower an index" to make it covariant, you change the sign if that index is zero. Compute the tensor invariants

$$F^{\mu\nu}F_{\mu\nu},\quad G^{\mu\nu}G_{\mu\nu},\quad \text{and}\quad F^{\mu\nu}G_{\mu\nu},$$

in terms of **E** and **B.** Compare Prob. 12.46.

Problem 12.51 A straight wire along the z axis carries a charge density λ traveling in the $+z$ direction at speed v. Construct the field tensor and the dual tensor at the point $(x, 0, 0)$.

12.3.4 Electrodynamics in Tensor Notation

Now that we know how to represent the fields in relativistic notation, it is time to reformulate the laws of electrodynamics (Maxwell's equations and the Lorentz force law) in that language. To begin with, we must determine how the *sources* of the fields, ρ and **J**, transform. Imagine a cloud of charge drifting by; we concentrate on an infinitesimal volume V, which contains charge Q moving at velocity **u** (Fig. 12.43). The charge density is

$$\rho = \frac{Q}{V},$$

and the current density[16] is

$$\mathbf{J} = \rho\mathbf{u}.$$

---

[16]I'm assuming all the charge in V is of one sign, and it all goes at the same speed. If not, you have to treat the constituents separately: $\mathbf{J} = \rho_+\mathbf{u}_- + \rho_-\mathbf{u}_-$. But the argument is the same.

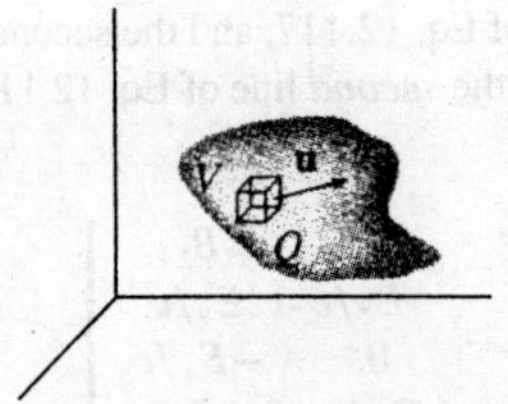

Figure 12.43

I would like to express these quantities in terms of the **proper charge density** ρ_0, the density *in the rest system of the charge:*

$$\rho_0 = \frac{Q}{V_0},$$

where V_0 is the rest volume of the chunk. Because one dimension (the one along the direction of motion) is Lorentz-contracted,

$$V = \sqrt{1 - u^2/c^2}\, V_0. \tag{12.120}$$

and hence

$$\rho = \rho_0 \frac{1}{\sqrt{1 - u^2/c^2}}, \quad \mathbf{J} = \rho_0 \frac{\mathbf{u}}{\sqrt{1 - u^2/c^2}}. \tag{12.121}$$

Comparing this with Eqs. 12.40 and 12.42, we recognize here the components of *proper velocity,* multiplied by the invariant ρ_0. Evidently charge density and current density go together to make a 4-vector:

$$J^\mu = \rho_0 \eta^\mu, \tag{12.122}$$

whose components are

$$\boxed{J^\mu = (c\rho, J_x, J_y, J_z).} \tag{12.123}$$

We'll call it the **current density 4-vector.**

The continuity equation (Eq. 5.29),

$$\nabla \cdot \mathbf{J} = -\frac{\partial \rho}{\partial t},$$

expressing the local conservation of charge, takes on a nice compact form when written in terms of J^μ. For

$$\nabla \cdot \mathbf{J} = \frac{\partial J_x}{\partial x} + \frac{\partial J_y}{\partial y} + \frac{\partial J_z}{\partial z} = \sum_{i=1}^{3} \frac{\partial J^i}{\partial x^i},$$

while

$$\frac{\partial \rho}{\partial t} = \frac{1}{c}\frac{\partial J^0}{\partial t} = \frac{\partial J^0}{\partial x^0}. \tag{12.124}$$

while

$$\frac{\partial \rho}{\partial t} = \frac{1}{c}\frac{\partial J^0}{\partial t} = \frac{\partial J^0}{\partial x^0}. \tag{12.124}$$

Thus, bringing $\partial\rho/\partial t$ over to the left side, we have:

$$\boxed{\frac{\partial J^\mu}{\partial x^\mu} = 0,} \tag{12.125}$$

with summation over μ implied. Incidentally, $\partial J^\mu/\partial x^\mu$ is the four-dimensional *divergence* of J^μ, so the continuity equation states that the current density 4-vector is divergenceless.

As for Maxwell's equations, they can be written

$$\boxed{\frac{\partial F^{\mu\nu}}{\partial x^\nu} = \mu_0 J^\mu, \quad \frac{\partial G^{\mu\nu}}{\partial x^\nu} = 0,} \tag{12.126}$$

with summation over ν implied. Each of these stands for four equations—one for every value of μ. If $\mu = 0$, the first equation reads

$$\begin{aligned}\frac{\partial F^{0\nu}}{\partial x^\nu} &= \frac{\partial F^{00}}{\partial x^0} + \frac{\partial F^{01}}{\partial x^1} + \frac{\partial F^{02}}{\partial x^2} + \frac{\partial F^{03}}{\partial x^3} \\ &= \frac{1}{c}\left(\frac{\partial E_x}{\partial x} + \frac{\partial E_y}{\partial y} + \frac{\partial E_z}{\partial z}\right) = \frac{1}{c}(\nabla \cdot \mathbf{E}) \\ &= \mu_0 J^0 = \mu_0 c\rho,\end{aligned}$$

or

$$\nabla \cdot \mathbf{E} = \frac{1}{\epsilon_0}\rho.$$

This, of course, is Gauss's law. If $\mu = 1$, we have

$$\begin{aligned}\frac{\partial F^{1\nu}}{\partial x^\nu} &= \frac{\partial F^{10}}{\partial x^0} + \frac{\partial F^{11}}{\partial x^1} + \frac{\partial F^{12}}{\partial x^2} + \frac{\partial F^{13}}{\partial x^3} \\ &= -\frac{1}{c^2}\frac{\partial E_x}{\partial t} + \frac{\partial B_z}{\partial y} - \frac{\partial B_y}{\partial z} = \left(-\frac{1}{c^2}\frac{\partial \mathbf{E}}{\partial t} + \nabla \times \mathbf{B}\right)_x \\ &= \mu_0 J^1 = \mu_0 J_x.\end{aligned}$$

Combining this with the corresponding results for $\mu = 2$ and $\mu = 3$ gives

$$\nabla \times \mathbf{B} = \mu_0 \mathbf{J} + \mu_0\epsilon_0\frac{\partial \mathbf{E}}{\partial t},$$

which is Ampère's law with Maxwell's correction.

Meanwhile, the second equation in 12.126, with $\mu = 0$, becomes

$$\begin{aligned}\frac{\partial G^{0\nu}}{\partial x^\nu} &= \frac{\partial G^{00}}{\partial x^0} + \frac{\partial G^{01}}{\partial x^1} + \frac{\partial G^{02}}{\partial x^2} + \frac{\partial G^{03}}{\partial x^3} \\ &= \frac{\partial B_x}{\partial x} + \frac{\partial B_y}{\partial y} + \frac{\partial B_z}{\partial z} = \nabla \cdot \mathbf{B} = 0\end{aligned}$$

(the third of Maxwell's equations), whereas $\mu = 1$ yields

$$\begin{aligned}\frac{\partial G^{1\nu}}{\partial x^\nu} &= \frac{\partial G^{10}}{\partial x^0} + \frac{\partial G^{11}}{\partial x^1} + \frac{\partial G^{12}}{\partial x^2} + \frac{\partial G^{13}}{\partial x^3} \\ &= -\frac{1}{c}\frac{\partial B_x}{\partial t} - \frac{1}{c}\frac{\partial E_z}{\partial y} + \frac{1}{c}\frac{\partial E_y}{\partial z} = -\frac{1}{c}\left(\frac{\partial \mathbf{B}}{\partial t} + \nabla \times \mathbf{E}\right)_x = 0.\end{aligned}$$

So, combining this with the corresponding results for $\mu = 2$ and $\mu = 3$,

$$\nabla \times \mathbf{E} = -\frac{\partial \mathbf{B}}{\partial t},$$

which is Faraday's law. In relativistic notation, then, Maxwell's four rather cumbersome equations reduce to two delightfully simple ones.

In terms of $F^{\mu\nu}$ and the proper velocity η^μ, the *Minkowski* force on a charge q is given by

$$\boxed{K^\mu = q\eta_\nu F^{\mu\nu}.} \tag{12.127}$$

For if $\mu = 1$, we have

$$\begin{aligned}K^1 &= q\eta_\nu F^{1\nu} = q(-\eta^0 F^{10} + \eta^1 F^{11} + \eta^2 F^{12} + \eta^3 F^{13}) \\ &= q\left[\frac{-c}{\sqrt{1-u^2/c^2}}\left(\frac{-E_x}{c}\right) + \frac{u_y}{\sqrt{1-u^2/c^2}}(B_z) + \frac{u_z}{\sqrt{1-u^2/c^2}}(-B_y)\right] \\ &= \frac{q}{\sqrt{1-u^2/c^2}}[\mathbf{E} + (\mathbf{u} \times \mathbf{B})]_x,\end{aligned}$$

with a similar formula for $\mu = 2$ and $\mu = 3$. Thus,

$$\mathbf{K} = \frac{q}{\sqrt{1-u^2/c^2}}[\mathbf{E} + (\mathbf{u} \times \mathbf{B})], \tag{12.128}$$

and therefore, referring back to Eq. 12.70,

$$\mathbf{F} = q[\mathbf{E} + (\mathbf{u} \times \mathbf{B})],$$

which is the Lorentz force law. Equation 12.127, then, represents the Lorentz force law in relativistic notation. I'll leave for you the interpretation of the zeroth component (Prob. 12.54).

Problem 12.52 Obtain the continuity equation (12.125) directly from Maxwell's equations (12.126).

Problem 12.53 Show that the second equation in (12.126) can be expressed in terms of the field tensor $F^{\mu\nu}$ as follows:

$$\frac{\partial F_{\mu\nu}}{\partial x^{\lambda}}+\frac{\partial F_{\nu\lambda}}{\partial x^{\mu}}+\frac{\partial F_{\lambda\mu}}{\partial x^{\nu}}=0. \tag{12.129}$$

Problem 12.54 Work out, and interpret physically, the $\mu=0$ component of the electromagnetic force law, Eq. 12.127.

12.3.5 Relativistic Potentials

From Chapter 10 we know that the electric and magnetic fields can be expressed in terms of a scalar potential V and a vector potential $\mathbf{A}$:

$$\mathbf{E}=-\nabla V-\frac{\partial \mathbf{A}}{\partial t},\qquad \mathbf{B}=\nabla\times\mathbf{A}. \tag{12.130}$$

As you might guess, V and $\mathbf{A}$ together constitute a 4-vector:

$$\boxed{A^{\mu}=(V/c, A_x, A_y, A_z).} \tag{12.131}$$

In terms of this **4-vector potential** the field tensor can be written

$$\boxed{F^{\mu\nu}=\frac{\partial A^{\nu}}{\partial x_{\mu}}-\frac{\partial A^{\mu}}{\partial x_{\nu}}.} \tag{12.132}$$

(Observe that the differentiation is with respect to the *covariant* vectors x_μ and x_ν; remember, that changes the sign of the zeroth component: $x_0=-x^0$. See Prob. 12.55.)

To check that Eq. 12.132 is equivalent to Eq. 12.130, let's evaluate a few terms explicitly. For $\mu=0$, $\nu=1$,

$$\begin{aligned}F^{01}&=\frac{\partial A^1}{\partial x_0}-\frac{\partial A^0}{\partial x_1}=-\frac{\partial A_x}{\partial(ct)}-\frac{1}{c}\frac{\partial V}{\partial x}\\&=-\frac{1}{c}\left(\frac{\partial \mathbf{A}}{\partial t}+\nabla V\right)_x=\frac{E_x}{c}.\end{aligned}$$

That (and its companions with $\nu=2$ and $\nu=3$) is the first equation in 12.130. For $\mu=1$, $\nu=2$, we get

$$F^{12}=\frac{\partial A^2}{\partial x_1}-\frac{\partial A^1}{\partial x_2}=\frac{\partial A_y}{\partial x}-\frac{\partial A_x}{\partial y}=(\nabla\times\mathbf{A})_z=B_z,$$

which (together with the corresponding results for F^{13} and F^{23}) is the second equation in 12.130.

The potential formulation automatically takes care of the homogeneous Maxwell equation ($\partial G^{\mu\nu}/\partial x^\nu = 0$). As for the inhomogeneous equation ($\partial F^{\mu\nu}/\partial x^\nu = \mu_0 J^\mu$), that becomes

$$\frac{\partial}{\partial x_\mu}\left(\frac{\partial A^\nu}{\partial x^\nu}\right) - \frac{\partial}{\partial x_\nu}\left(\frac{\partial A^\mu}{\partial x^\nu}\right) = \mu_0 J^\mu. \tag{12.133}$$

This is an intractable equation as it stands. However, you will recall that the potentials are not uniquely determined by the fields—in fact, it's clear from Eq. 12.132 that you could add to A^μ the gradient of any scalar function λ:

$$A^\mu \longrightarrow A^{\mu\prime} = A^\mu + \frac{\partial \lambda}{\partial x_\mu}, \tag{12.134}$$

without changing $F^{\mu\nu}$. This is precisely the **gauge invariance** we noted in Chapter 11; we can exploit it to simplify Eq. 12.133. In particular, the Lorentz gauge condition (Eq. 10.12)

$$\nabla \cdot \mathbf{A} = -\frac{1}{c^2}\frac{\partial V}{\partial t}$$

becomes, in relativistic notation,

$$\frac{\partial A^\mu}{\partial x^\mu} = 0. \tag{12.135}$$

In the Lorentz gauge, therefore, Eq. 12.133 reduces to

$$\boxed{\Box^2 A^\mu = -\mu_0 J^\mu,} \tag{12.136}$$

where $\Box^2$ is the **d'Alembertian,**

$$\Box^2 \equiv \frac{\partial}{\partial x_\nu}\frac{\partial}{\partial x^\nu} = \nabla^2 - \frac{1}{c^2}\frac{\partial^2}{\partial t^2}. \tag{12.137}$$

Equation 12.136 combines our previous results into a single 4-vector equation—it represents the most elegant (and the simplest) formulation of Maxwell's equations.[17]

[17] Incidentally, the *Coulomb* gauge is a *bad* one, from the point of view of relativity, because its defining condition, $\nabla \cdot \mathbf{A} = 0$, is destroyed by Lorentz transformation. To restore this condition, it is necessary to perform an appropriate gauge transformation every time you go to a new inertial system, in *addition* to the Lorentz transformation itself. In this sense A^μ is not a true 4-vector, in the Coulomb gauge.

Problem 12.55 You may have noticed that the **four-dimensional gradient** operator $\partial/\partial x^\mu$ functions like a *covariant* 4-vector—in fact, it is often written ∂_μ, for short. For instance, the continuity equation, $\partial_\mu J^\mu = 0$, has the form of an invariant product of two vectors. The corresponding *contravariant* gradient would be $\partial^\mu \equiv \partial x_\mu$. *Prove* that $\partial^\mu \phi$ is a (contravariant) 4-vector, if ϕ is a scalar function, by working out its transformation law, using the chain rule.

Problem 12.56 Show that the potential representation (Eq. 12.132) automatically satisfies $\partial G^{\mu\nu}/\partial x^\nu = 0$. [*Suggestion:* Use Prob. 12.53.]

More Problems on Chapter 12

Problem 12.57 Inertial system $\bar{S}$ moves at constant velocity $\mathbf{v} = \beta c(\cos\phi\,\hat{\mathbf{x}} + \sin\phi\,\hat{\mathbf{y}})$ with respect to S. Their axes are parallel to one another, and their origins coincide at $t = \bar{t} = 0$, as usual. Find the Lorentz transformation matrix Λ (Eq. 12.25).

$$\left[Answer: \begin{pmatrix} \gamma & -\gamma\beta\cos\phi & -\gamma\beta\sin\phi & 0 \\ -\gamma\beta\cos\phi & (\gamma\cos^2\phi + \sin^2\phi) & (\gamma - 1)\sin\phi\cos\phi & 0 \\ -\gamma\beta\sin\phi & (\gamma-1)\sin\phi\cos\phi & (\gamma\sin^2\phi + \cos^2\phi) & 0 \\ 0 & 0 & 0 & 1 \end{pmatrix} \right]$$

Problem 12.58 Calculate the **threshold** (minimum) momentum the pion must have in order for the process $\pi + p \to K + \Sigma$ to occur. The proton p is initially at rest. Use $m_\pi c^2 = 150$, $m_K c^2 = 500$, $m_p c^2 = 900$, $m_\Sigma c^2 = 1200$ (all in MeV). [*Hint:* To formulate the threshold condition, examine the collision in the center-of-momentum frame (Prob. 12.30). *Answer:* 1133 MeV/c]

Problem 12.59 A particle of mass m collides elastically with an identical particle at rest. *Classically,* the outgoing trajectories always make an angle of 90°. Calculate this angle *relativistically,* in terms of ϕ, the scattering angle, and v, the speed, in the center-of-momentum frame. [*Answer:* $\tan^{-1}(2c^2/v^2\gamma\sin\phi)$]

Problem 12.60 Find x as a function of t for motion starting from rest at the origin under the influence of a constant *Minkowski* force in the x direction. Leave your answer in implicit form (t as a function of x). [*Answer:* $2Kt/mc = z\sqrt{1+z^2} + \ln(z + \sqrt{1+z^2})$, where $z \equiv \sqrt{2Kx/mc^2}$]

! **Problem 12.61** An electric dipole consists of two point charges ($\pm q$), each of mass m, fixed to the ends of a (massless) rod of length d. (Do *not* assume d is small.)

(a) Find the net self-force on the dipole when it undergoes hyperbolic motion (Eq. 12.62) along a line perpendicular to its axis. [*Hint:* Start by appropriately modifying Eq. 11.90.]

(b) Notice that this self-force is *constant* (t drops out), and points in the direction of motion—just right to *produce* hyperbolic motion. Thus it is possible for the dipole to undergo *self-sustaining accelerated motion* with no external force at all![18] [Where do you suppose the energy comes from?] Determine the self-sustaining force, F, in terms of m, q, and d. [*Answer:* $(2mc^2/d)\sqrt{(\mu_0 q^2/8\pi m d)^{2/3} - 1}$]

[18] F. H. J. Cornish, *Am. J. Phys.* **54**, 166 (1986).

Problem 12.62 An ideal magnetic dipole moment **m** is located at the origin of an inertial system $\bar{S}$ that moves with speed v in the x direction with respect to inertial system S. In $\bar{S}$ the vector potential is

$$\bar{\mathbf{A}} = \frac{\mu_0}{4\pi}\frac{\bar{\mathbf{m}} \times \hat{\bar{\mathbf{r}}}}{\bar{r}^2},$$

(Eq. 5.83), and the electric potential $\bar{V}$ is zero.

(a) Find the scalar potential V in S. [*Answer:* $(1/4\pi\epsilon_0)(\hat{\mathbf{R}} \cdot (\mathbf{v} \times \mathbf{m})/c^2 R^2)(1 - v^2/c^2)/(1 - (v^2/c^2)\sin^2\theta)^{3/2}$]

(b) In the nonrelativistic limit, show that the scalar potential in S is that of an ideal *electric* dipole of magnitude

$$\mathbf{p} = \frac{\mathbf{v} \times \mathbf{m}}{c^2},$$

located at $\bar{\mathcal{O}}$.

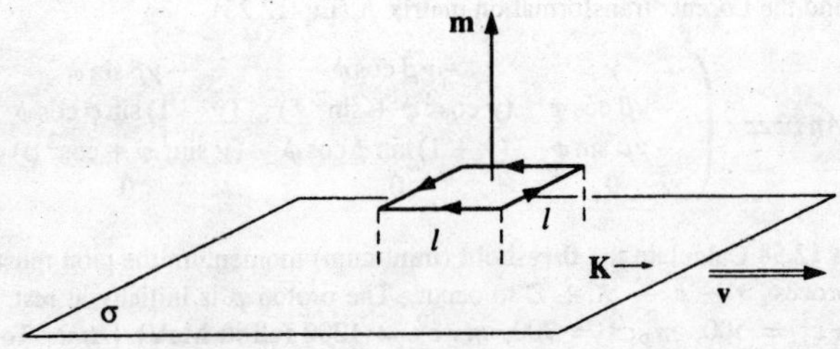

Figure 12.44

! **Problem 12.63** A stationary magnetic dipole, $\mathbf{m} = m\,\hat{\mathbf{z}}$, is situated above an infinite uniform surface current, $\mathbf{K} = K\,\hat{\mathbf{x}}$ (Fig. 12.44).

(a) Find the torque on the dipole, using Eq. 6.1.

(b) Suppose that the surface current consists of a uniform surface charge σ, moving at velocity $\mathbf{v} = v\,\hat{\mathbf{x}}$, so that $\mathbf{K} = \sigma\mathbf{v}$, and the magnetic dipole consists of a uniform line charge λ, circulating at speed v (same v) around a square loop of side l, as shown, so that $m = \lambda v l^2$. Examine the same configuration from the point of view of system $\bar{S}$, moving in the x direction at speed v. In $\bar{S}$ the surface charge is at *rest*, so it generates no magnetic field. Show that in this frame the current loop carries an *electric* dipole moment, and calculate the resulting torque, using Eq. 4.4.

Problem 12.64 In a certain inertial frame S, the electric field **E** and the magnetic field **B** are neither parallel nor perpendicular, at a particular space-time point. Show that in a different inertial system $\bar{S}$, moving relative to S with velocity **v** given by

$$\frac{\mathbf{v}}{1 + v^2/c^2} = \frac{\mathbf{E} \times \mathbf{B}}{B^2 + E^2/c^2},$$

the fields $\bar{\mathbf{E}}$ and $\bar{\mathbf{B}}$ are *parallel* at that point. Is there a frame in which the two are *perpendicular*?

Problem 12.65 Two charges $\pm q$ approach the origin at constant velocity from opposite directions along the x axis. They collide and stick together, forming a neutral particle at rest. Sketch the electric field before and shortly after the collision (remember that electromagnetic "news" travels at the speed of light). How would you interpret the field after the collision, physically?[19]

Problem 12.66 "Derive" the Lorentz force law, as follows: Let charge q be at rest in $\bar{S}$, so $\bar{\mathbf{F}} = q\bar{\mathbf{E}}$, and let $\bar{S}$ move with velocity $\mathbf{v} = v\,\hat{\mathbf{x}}$ with respect to S. Use the transformation rules (Eqs. 12.68 and 12.108) to rewrite $\bar{\mathbf{F}}$ in terms of $\mathbf{F}$, and $\bar{\mathbf{E}}$ in terms of $\mathbf{E}$ and $\mathbf{B}$. From these deduce the formula for $\mathbf{F}$ in terms of $\mathbf{E}$ and $\mathbf{B}$.

Problem 12.67 A charge q is released from rest at the origin, in the presence of a uniform electric field $\mathbf{E} = E_0\hat{\mathbf{z}}$ and a uniform magnetic field $\mathbf{B} = B_0\,\hat{\mathbf{x}}$. Determine the trajectory of the particle by transforming to a system in which $\mathbf{E} = 0$, finding the path in that system and then transforming back to the original system. Assume $E_0 < cB_0$. Compare your result with Ex. 5.2.

Problem 12.68

(a) Construct a tensor $D^{\mu\nu}$ (analogous to $F^{\mu\nu}$), out of $\mathbf{D}$ and $\mathbf{H}$. Use it to express Maxwell's equations inside matter in terms of the free current density J_f^{μ}. [*Answer:* $D^{01} \equiv cD_x$, $D^{12} \equiv H_z$, etc.; $\partial D^{\mu\nu}/\partial x^{\nu} = J_f^{\mu}$.]

(b) Construct the dual tensor $H^{\mu\nu}$ (analogous to $G^{\mu\nu}$). [*Answer:* $H^{01} \equiv H_x$, $H^{12} \equiv -cD_z$, etc.]

(c) Minkowski proposed the **relativistic constitutive relations** for linear media:

$$D^{\mu\nu}\eta_{\nu} = c^2\epsilon F^{\mu\nu}\eta_{\nu} \quad \text{and} \quad H^{\mu\nu}\eta_{\nu} = \frac{1}{\mu}G^{\mu\nu}\eta_{\nu},$$

where ϵ is the proper[20] permittivity, μ is the proper permeability, and η^{μ} is the 4-velocity of the material. Show that Minkowski's formulas reproduce Eqs. 4.32 and 6.31, when the material is at rest.

(d) Work out the formulas relating $\mathbf{D}$ and $\mathbf{H}$ to $\mathbf{E}$ and $\mathbf{B}$ for a medium moving with (ordinary) velocity $\mathbf{u}$.

! **Problem 12.69** Use the Larmor formula (Eq. 11.70) and special relativity to derive the Liénard formula (Eq. 11.73).

Problem 12.70 The natural relativistic generalization of the Abraham-Lorentz formula (Eq. 11.80) would seem to be

$$K^{\mu}_{\text{rad}} = \frac{\mu_0 q^2}{6\pi c}\frac{d\alpha^{\mu}}{d\tau}.$$

This is certainly a 4-vector, and it reduces to the Abraham-Lorentz formula in the non-relativistic limit $v \ll c$.

---

[19]See E. M. Purcell, *Electricity and Magnetism*, 2d ed. (New York: McGraw-Hill, 1985), Sect. 5.7 and Appendix B (in which Purcell obtains the Larmor formula by masterful analysis of a similar geometrical construction), and R. Y. Tsien, *Am. J. Phys.* **40**, 46 (1972).

[20]As always, "proper" means "in the rest frame of the material."

(a) Show, nevertheless, that this is not a possible Minkowski force. [*Hint:* See Prob. 12.38d.]

(b) Find a correction term that, when added to the right side, removes the objection you raised in (a), without affecting the 4-vector character of the formula or its nonrelativistic limit.[21]

Problem 12.71 Generalize the laws of relativistic electrodynamics (Eqs. 12.126 and 12.127) to include magnetic charge. [Refer to Sect. 7.3.4.]

---

[21] For interesting commentary on the relativistic radiation reaction, see F. Rohrlich, *Am. J. Phys.* **65**, 1051 (1997).

Appendix A

Vector Calculus in Curvilinear Coordinates

A.1 Introduction

In this Appendix I sketch proofs of the three fundamental theorems of vector calculus. My aim is to convey the *essence* of the argument, not to track down every epsilon and delta. A much more elegant, modern, and unified—but necessarily also much longer—treatment will be found in M. Spivak's book, *Calculus on Manifolds* (New York: Benjamin, 1965).

For the sake of generality, I shall use arbitrary (orthogonal) curvilinear coordinates (u, v, w), developing formulas for the gradient, divergence, curl, and Laplacian in any such system. You can then specialize them to Cartesian, spherical, or cylindrical coordinates, or any other system you might wish to use. If the generality bothers you on a first reading, and you'd rather stick to Cartesian coordinates, just read (x, y, z) wherever you see (u, v, w), and make the associated simplifications as you go along.

A.2 Notation

We identify a point in space by its three *coordinates*, u, v, and w, (in the Cartesian system, (x, y, z); in the spherical system, (r, θ, ϕ); in the cylindrical system, (s, ϕ, z)). I shall assume the system is *orthogonal*, in the sense that the three *unit vectors*, $\hat{\mathbf{u}}$, $\hat{\mathbf{v}}$, and $\hat{\mathbf{w}}$, pointing in the direction of the increase of the corresponding coordinates, are mutually perpendicular. Note that the unit vectors are *functions of position*, since their *directions* (except in the Cartesian case) vary from point to point. Any vector can be expressed in terms of $\hat{\mathbf{u}}$, $\hat{\mathbf{v}}$, and $\hat{\mathbf{w}}$—in particular, the infinitesimal displacement vector from (u, v, w) to $(u + du, v + dv, w + dw)$ can be written

$$d\mathbf{l} = f\,du\,\hat{\mathbf{u}} + g\,dv\,\hat{\mathbf{v}} + h\,dw\,\hat{\mathbf{w}}, \tag{A.1}$$

where f, g, and h are functions of position characteristic of the particular coordinate system (in Cartesian coordinates $f = g = h = 1$; in spherical coordinates $f = 1, g = r, h = r\sin\theta$; and in cylindrical coordinates $f = h = 1, g = s$). As you'll soon see, these three functions tell you everything you need to know about a coordinate system.

A.3 Gradient

If you move from point (u, v, w) to point $(u + du, v + dv, w + dw)$, a scalar function $t(u, v, w)$ changes by an amount

$$dt = \frac{\partial t}{\partial u}\,du + \frac{\partial t}{\partial v}\,dv + \frac{\partial t}{\partial w}\,dw; \tag{A.2}$$

this is a standard theorem on partial differentiation.[1] We can write it as a dot product,

$$dt = \nabla t \cdot d\mathbf{l} = (\nabla t)_u\, f\,du + (\nabla t)_v\, g\,dv + (\nabla t)_w\, h\,dw, \tag{A.3}$$

provided we define

$$(\nabla t)_u \equiv \frac{1}{f}\frac{\partial t}{\partial u}, \quad (\nabla t)_v \equiv \frac{1}{g}\frac{\partial t}{\partial v}, \quad (\nabla t)_w \equiv \frac{1}{h}\frac{\partial t}{\partial w}.$$

The **gradient** of t, then, is

$$\boxed{\nabla t \equiv \frac{1}{f}\frac{\partial t}{\partial u}\hat{\mathbf{u}} + \frac{1}{g}\frac{\partial t}{\partial v}\hat{\mathbf{v}} + \frac{1}{h}\frac{\partial t}{\partial w}\hat{\mathbf{w}}.} \tag{A.4}$$

If you now pick the appropriate expressions for f, g, and h from Table A.1, you can easily generate the formulas for ∇t in Cartesian, spherical, and cylindrical coordinates, as they appear in the front cover of the book.

System	u	v	w	f	g	h
Cartesian	x	y	z	1	1	1
Spherical	r	θ	ϕ	1	r	$r\sin\theta$
Cylindrical	s	ϕ	z	1	s	1

Table A.1

From Eq. A.3 it follows that the *total* change in t, as you go from point **a** to point **b** (Fig. A.1), is

$$t(\mathbf{b}) - t(\mathbf{a}) = \int_{\mathbf{a}}^{\mathbf{b}} dt = \int_{\mathbf{a}}^{\mathbf{b}} (\nabla t)\cdot d\mathbf{l}, \tag{A.5}$$

which is the **fundamental theorem for gradients** (not much to prove, really, in this case). Notice that the integral is independent of the path taken from **a** to **b**.

[1] M. Boas, *Mathematical Methods in the Physical Sciences*, 2nd ed., Chapter 4, Sect. 3 (New York: John Wiley, 1983).

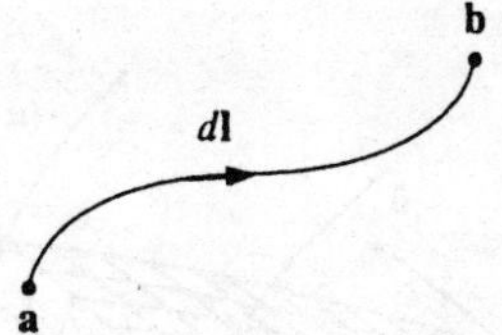

Figure A.1

A.4 Divergence

Suppose that we have a *vector* function,

$$\mathbf{A}(u, v, w) = A_u\,\hat{\mathbf{u}} + A_v\,\hat{\mathbf{v}} + A_w\,\hat{\mathbf{w}},$$

and we wish to evaluate the integral $\oint \mathbf{A}\cdot d\mathbf{a}$ over the surface of the infinitesimal volume generated by starting at the point (u, v, w) and increasing each of the coordinates in succession by an infinitesimal amount (Fig. A.2). Because the coordinates are orthogonal, this is (at least, in the infinitesimal limit) a rectangular solid, whose sides have lengths $dl_u = f\,du$, $dl_v = g\,dv$, and $dl_w = h\,dw$, and whose volume is therefore

$$d\tau = dl_u\,dl_v\,dl_w = (fgh)\,du\,dv\,dw. \tag{A.6}$$

(The sides are *not* just du, dv, dw—after all, v might be an *angle*, in which case dv doesn't even have the *dimensions* of length. The correct expressions follow from Eq. A.1.)

For the *front* surface,

$$d\mathbf{a} = -(gh)\,dv\,dw\,\hat{\mathbf{u}},$$

so that

$$\mathbf{A}\cdot d\mathbf{a} = -(ghA_u)\,dv\,dw.$$

The *back* surface is identical (except for the sign), *only this time the quantity* ghA_u *is to be evaluated at* $(u + du)$, *instead of* u. Since for any (differentiable) function $F(u)$,

$$F(u + du) - F(u) = \frac{dF}{du}\,du,$$

(in the limit), the front and back together amount to a contribution

$$\left[\frac{\partial}{\partial u}(ghA_u)\right] du\,dv\,dw = \frac{1}{fgh}\frac{\partial}{\partial u}(ghA_u)\,d\tau.$$

By the same token, the right and left sides yield

$$\frac{1}{fgh}\frac{\partial}{\partial v}(fhA_v)\,d\tau,$$

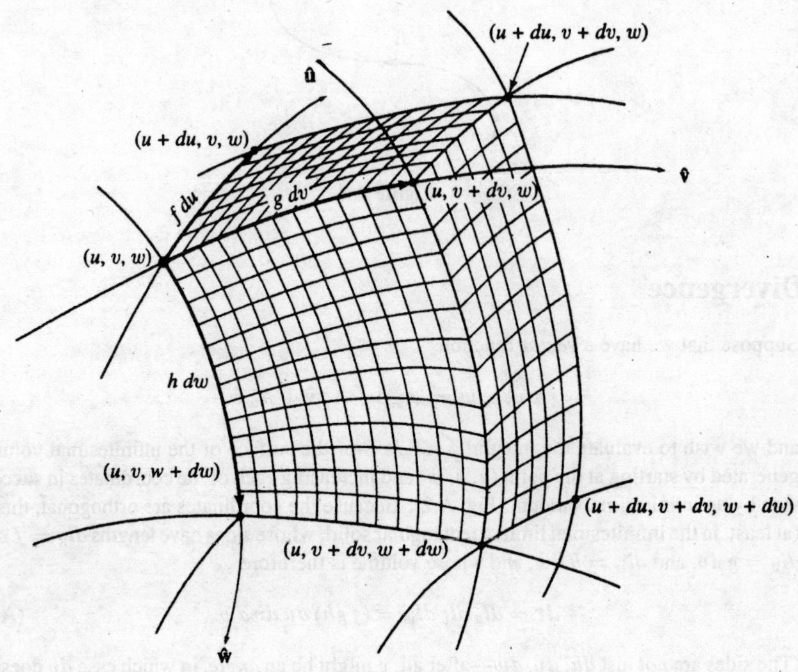

Figure A.2

and the top and bottom give

$$\frac{1}{fgh}\frac{\partial}{\partial w}(fgA_w)\,d\tau.$$

All told, then,

$$\oint \mathbf{A}\cdot d\mathbf{a} = \frac{1}{fgh}\left[\frac{\partial}{\partial u}(ghA_u) + \frac{\partial}{\partial v}(fhA_v) + \frac{\partial}{\partial w}(fgA_w)\right] d\tau. \tag{A.7}$$

The coefficient of $d\tau$ serves to define the **divergence** of **A** in curvilinear coordinates:

$$\boxed{\nabla\cdot\mathbf{A} \equiv \frac{1}{fgh}\left[\frac{\partial}{\partial u}(ghA_u) + \frac{\partial}{\partial v}(fhA_v) + \frac{\partial}{\partial w}(fgA_w)\right],} \tag{A.8}$$

and Eq. A.7 becomes

$$\oint \mathbf{A} \cdot d\mathbf{a} = (\nabla \cdot \mathbf{A})\, d\tau. \tag{A.9}$$

Using Table A.1, you can now derive the formulas for the divergence in Cartesian, spherical, and cylindrical coordinates, which appear in the front cover of the book.

As it stands, Eq. A.9 does not prove the divergence theorem, for it pertains only to *infinitesimal* volumes, and rather special infinitesimal volumes at that. Of course, a finite volume can be broken up into infinitesimal pieces, and Eq. A.9 can be applied to each one. The trouble is, when you then add up all the bits, the left-hand side is not just an integral over the *outer* surface, but over all those tiny *internal* surfaces as well. Luckily, however, these contributions cancel in pairs, for each internal surface occurs as the boundary of *two* adjacent infinitesimal volumes, and since $d\mathbf{a}$ always points *outward*, $\mathbf{A} \cdot d\mathbf{a}$ has the opposite sign for the two members of each pair (Fig. A.3). Only those surfaces that bound a *single* chunk—which is to say, only those at the outer boundary—survive when everything is added up. For *finite* regions, then,

$$\oint \mathbf{A} \cdot d\mathbf{a} = \int (\nabla \cdot \mathbf{A})\, d\tau, \tag{A.10}$$

and you need only integrate over the *external* surface.[2] This establishes the **divergence theorem.**

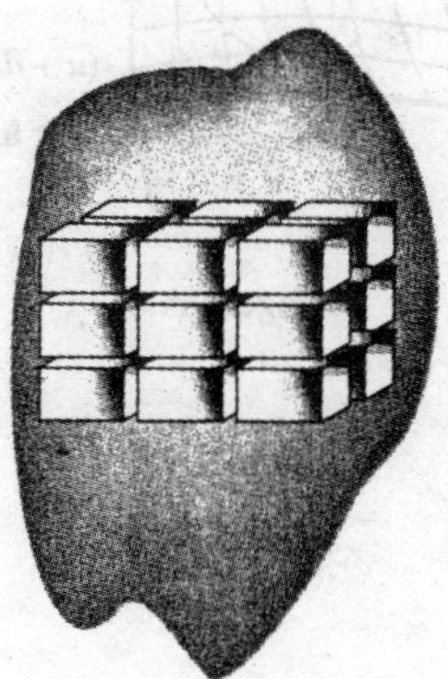

Figure A.3

[2]What about regions that cannot be fit perfectly by rectangular solids no matter *how* tiny they are—such as planes cut at an angle to the coordinate lines? It's not hard to dispose of this case; try thinking it out for yourself, or look at H. M. Schey's *Div, Grad, Curl and All That* (New York: W. W. Norton, 1973), starting with Prob. II-15.

A.5 Curl

To obtain the curl in curvilinear coordinates, we calculate the line integral,

$$\oint \mathbf{A} \cdot d\mathbf{l},$$

around the infinitesimal loop generated by starting at (u, v, w) and successively increasing u and v by infinitesimal amounts, holding w constant (Fig. A.4). The surface is a rectangle (at least, in the infinitesimal limit), of length $dl_u = f\,du$, width $dl_v = g\,dv$, and area

$$d\mathbf{a} = (fg)du\,dv\,\hat{\mathbf{w}}. \tag{A.11}$$

Assuming the coordinate system is right-handed, $\hat{\mathbf{w}}$ points out of the page in Fig. A.4. Having chosen this as the positive direction for $d\mathbf{a}$, we are obliged by the right-hand rule to run the line integral counterclockwise, as shown.

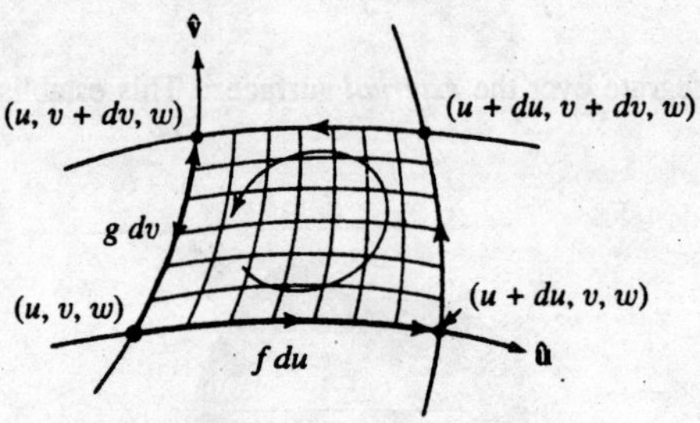

Figure A.4

Along the bottom segment,

$$d\mathbf{l} = f\,du\,\hat{\mathbf{u}},$$

so

$$\mathbf{A} \cdot d\mathbf{l} = (fA_u)\,du.$$

Along the top leg, the sign is reversed, and fA_u is evaluated at $(v + dv)$ rather than v. Taken together, these two edges give

$$\left[-(fA_u)\big|_{v+dv} + (fA_u)\big|_v\right] du = -\left[\frac{\partial}{\partial v}(fA_u)\right] du\,dv.$$

Similarly, the right and left sides yield

$$\left[\frac{\partial}{\partial u}(gA_v)\right] du\, dv,$$

so the total is

$$\begin{aligned}\oint \mathbf{A}\cdot d\mathbf{l} &= \left[\frac{\partial}{\partial u}(gA_v) - \frac{\partial}{\partial v}(fA_u)\right] du\, dv \\ &= \frac{1}{fg}\left[\frac{\partial}{\partial u}(gA_v) - \frac{\partial}{\partial v}(fA_u)\right] \hat{\mathbf{w}}\cdot d\mathbf{a}.\end{aligned} \tag{A.12}$$

The coefficient of $d\mathbf{a}$ on the right serves to define the w-component of the **curl**. Constructing the u and v components in the same way, we have

$$\boxed{\begin{aligned}\nabla\times\mathbf{A} \equiv\ & \frac{1}{gh}\left[\frac{\partial}{\partial v}(hA_w) - \frac{\partial}{\partial w}(gA_v)\right]\hat{\mathbf{u}} + \frac{1}{fh}\left[\frac{\partial}{\partial w}(fA_u) - \frac{\partial}{\partial u}(hA_w)\right]\hat{\mathbf{v}} \\ & + \frac{1}{fg}\left[\frac{\partial}{\partial u}(gA_v) - \frac{\partial}{\partial v}(fA_u)\right]\hat{\mathbf{w}},\end{aligned}} \tag{A.13}$$

and Eq. A.11 generalizes to

$$\oint \mathbf{A}\cdot d\mathbf{l} = (\nabla\times\mathbf{A})\cdot d\mathbf{a}. \tag{A.14}$$

Using Table A.1, you can now derive the formulas for the curl in Cartesian, spherical, and cylindrical coordinates.

Equation A.14 does not by itself prove Stokes' theorem, however, because at this point it pertains only to very special infinitesimal surfaces. Again, we can chop any *finite* surface into infinitesimal pieces and apply Eq. A.14 to each one (Fig. A.5). When we add them up, though, we obtain (on the left) not only a line integral around the outer boundary, but a lot of tiny line integrals around the internal loops as well. Fortunately, as before, the internal

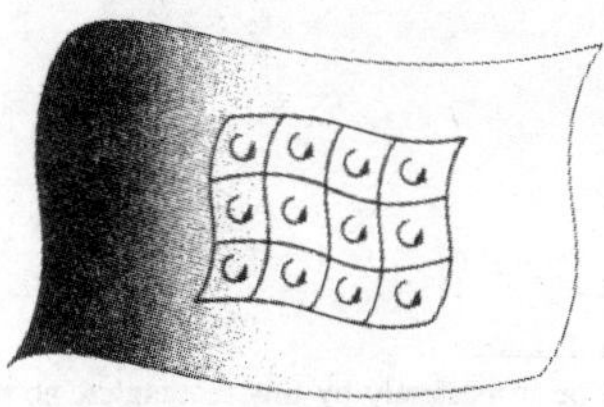

Figure A.5

contributions cancel in pairs, because every internal line is the edge of *two* adjacent loops running in opposite directions. Consequently, Eq. A.14 can be extended to finite surfaces,

$$\oint \mathbf{A} \cdot d\mathbf{l} = \int (\nabla \times \mathbf{A}) \cdot d\mathbf{a}, \tag{A.15}$$

and the line integral is to be taken over the external boundary only.[3] This establishes **Stokes' theorem.**

A.6 Laplacian

Since the **Laplacian** of a scalar is by definition the divergence of the gradient, we can read off from Eqs. A.4 and A.8 the general formula

$$\boxed{\nabla^2 t \equiv \frac{1}{fgh}\left[\frac{\partial}{\partial u}\left(\frac{gh}{f}\frac{\partial t}{\partial u}\right) + \frac{\partial}{\partial v}\left(\frac{fh}{g}\frac{\partial t}{\partial v}\right) + \frac{\partial}{\partial w}\left(\frac{fg}{h}\frac{\partial t}{\partial w}\right)\right].} \tag{A.16}$$

Once again, you are invited to use Table A.1 to derive the Laplacian in Cartesian, spherical, and cylindrical coordinates, and thus to confirm the formulas inside the front cover.

[3] What about surfaces that cannot be fit perfectly by tiny rectangles, no matter how small they are (such as triangles) or surfaces that do not correspond to holding one coordinate fixed? If such cases trouble you, and you cannot resolve them for yourself, look at H. M. Schey's *Div, Grad, Curl, and All That*, Prob. III-2 (New York: W. W. Norton, 1973).

Appendix B

The Helmholtz Theorem

Suppose we are told that the divergence of a vector function $\mathbf{F}(\mathbf{r})$ is a specified scalar function $D(\mathbf{r})$:

$$\nabla \cdot \mathbf{F} = D, \tag{B.1}$$

and the curl of $\mathbf{F}(\mathbf{r})$ is a specified vector function $\mathbf{C}(\mathbf{r})$:

$$\nabla \times \mathbf{F} = \mathbf{C}. \tag{B.2}$$

For consistency, $\mathbf{C}$ must be divergenceless,

$$\nabla \cdot \mathbf{C} = 0, \tag{B.3}$$

because the divergence of a curl is always zero. *Question:* can we, on the basis of this information, determine the function $\mathbf{F}$? If $D(\mathbf{r})$ and $\mathbf{C}(\mathbf{r})$ go to zero sufficiently rapidly at infinity, the answer is *yes*, as I will show by explicit construction.

I claim that

$$\mathbf{F} = -\nabla U + \nabla \times \mathbf{W}, \tag{B.4}$$

where

$$U(\mathbf{r}) \equiv \frac{1}{4\pi} \int \frac{D(\mathbf{r}')}{\mathscr{r}} \, d\tau', \tag{B.5}$$

and

$$\mathbf{W}(\mathbf{r}) \equiv \frac{1}{4\pi} \int \frac{\mathbf{C}(\mathbf{r}')}{\mathscr{r}} \, d\tau'; \tag{B.6}$$

the integrals are over all of space, and, as always, $\mathscr{r} = |\mathbf{r} - \mathbf{r}'|$. For if $\mathbf{F}$ is given by Eq. B.4, then its divergence (using Eq. 1.102) is

$$\nabla \cdot \mathbf{F} = -\nabla^2 U = -\frac{1}{4\pi} \int D \, \nabla^2 \left(\frac{1}{\mathscr{r}}\right) d\tau' = \int D(\mathbf{r}') \, \delta^3(\mathbf{r} - \mathbf{r}') \, d\tau' = D(\mathbf{r}).$$

(Remember that the divergence of a curl is zero, so the $\mathbf{W}$ term drops out, and note that the differentiation is with respect to $\mathbf{r}$, which is contained in $\mathscr{r}$.)

So the divergence is right; how about the curl?

$$\nabla \times \mathbf{F} = \nabla \times (\nabla \times \mathbf{W}) = -\nabla^2 \mathbf{W} + \nabla(\nabla \cdot \mathbf{W}). \tag{B.7}$$

(Since the curl of a gradient is zero, the U term drops out.) Now

$$-\nabla^2 \mathbf{W} = -\frac{1}{4\pi}\int \mathbf{C}\,\nabla^2\left(\frac{1}{\imath}\right) d\tau' = \int \mathbf{C}(\mathbf{r}')\delta^3(\mathbf{r}-\mathbf{r}')\,d\tau' = \mathbf{C}(\mathbf{r}),$$

which is perfect—I'll be done if I can just persuade you that the *second* term on the right side of Eq. B.7 vanishes. Using integration by parts (Eq. 1.59), and noting that derivatives of $\imath$ with respect to *primed* coordinates differ by a sign from those with respect to *unprimed* coordinates, we have

$$\begin{aligned} 4\pi \nabla \cdot \mathbf{W} &= \int \mathbf{C} \cdot \nabla\left(\frac{1}{\imath}\right) d\tau' = -\int \mathbf{C} \cdot \nabla'\left(\frac{1}{\imath}\right) d\tau' \\ &= \int \frac{1}{\imath} \nabla' \cdot \mathbf{C}\, d\tau - \oint \frac{1}{\imath} \mathbf{C} \cdot d\mathbf{a}. \end{aligned} \tag{B.8}$$

But the divergence of $\mathbf{C}$ is zero, by assumption (Eq. B.3), and the surface integral (way out at infinity) will vanish, as long as $\mathbf{C}$ goes to zero sufficiently rapidly.

Of course, that proof tacitly assumes that the integrals in Eqs. B.5 and B.6 *converge*—otherwise U and $\mathbf{W}$ don't exist at all. At the large r' limit, where $\imath \approx r'$, the integrals have the form

$$\int^{\infty} \frac{X(r')}{r'} r'^2\, dr' = \int^{\infty} r' X(r')\, dr'. \tag{B.9}$$

(Here X stands for D or $\mathbf{C}$, as the case may be). Obviously, $X(r')$ must go to zero at large r'—but that's not enough: if $X \sim 1/r'$, the integrand is constant, so the integral blows up, and even if $X \sim 1/r'^2$, the integral is a logarithm, which is still no good at $r' \to \infty$. Evidently the divergence and curl of $\mathbf{F}$ must go to zero *more rapidly than* $1/r^2$ for the proof to hold. (Incidentally, this is *more* than enough to ensure that the surface integral in Eq. B.8 vanishes.)

Now, assuming these conditions on $D(\mathbf{r})$ and $\mathbf{C}(\mathbf{r})$ are met, is the solution in Eq. B.4 *unique?* The answer is clearly *no,* for we can add to $\mathbf{F}$ any vector function whose divergence and curl both vanish, and the result still has divergence D and curl $\mathbf{C}$. However, it so happens that there is *no* function that has zero divergence and zero curl everywhere *and* goes to zero at infinity (see Sect. 3.1.5). So if we include a requirement that $\mathbf{F}(\mathbf{r})$ goes to zero as $r \to \infty$, then solution B.4 is unique.[1]

[1]Typically we *do* expect the electric and magnetic fields to go to zero at large distances from the charges and currents that produce them, so this is not an unreasonable stipulation. Occasionally one encounters artificial problems in which the charge or current distribution itself extends to infinity—infinite wires, for instance, or infinite planes. In such cases other means must be found to establish the existence and uniqueness of solutions to Maxwell's equations.

Now that all the cards are on the table, I can state the **Helmholtz theorem** more rigorously:

> If the divergence $D(\mathbf{r})$ and the curl $\mathbf{C}(\mathbf{r})$ of a vector function $\mathbf{F}(\mathbf{r})$ are specified, and if they both go to zero faster than $1/r^2$ as $r \to \infty$, and if $\mathbf{F}(\mathbf{r})$ goes to zero as $r \to \infty$, then $\mathbf{F}$ is given uniquely by Eq. B.4.

The Helmholtz theorem has an interesting **corollary**:

> Any (differentiable) vector function $\mathbf{F}(\mathbf{r})$ that goes to zero faster than $1/r$ as $r \to \infty$ can be expressed as the gradient of a scalar plus the curl of a vector:[2]

$$\mathbf{F}(\mathbf{r}) = \nabla\left(\frac{-1}{4\pi}\int \frac{\nabla' \cdot \mathbf{F}(\mathbf{r}')}{\mathscr{r}}\, d\tau'\right) + \nabla \times \left(\frac{1}{4\pi}\int \frac{\nabla' \times \mathbf{F}(\mathbf{r}')}{\mathscr{r}}\, d\tau'\right). \tag{B.10}$$

For example, in electrostatics $\nabla \cdot \mathbf{E} = \rho/\epsilon_0$ and $\nabla \times \mathbf{E} = 0$, so

$$\mathbf{E}(\mathbf{r}) = -\nabla\left(\frac{1}{4\pi\epsilon_0}\int \frac{\rho(\mathbf{r}')}{\mathscr{r}}\, d\tau'\right) = -\nabla V, \tag{B.11}$$

where V is the scalar potential, while in magnetostatics $\nabla \cdot \mathbf{B} = 0$ and $\nabla \times \mathbf{B} = \mu_0 \mathbf{J}$, so

$$\mathbf{B}(\mathbf{r}) = \nabla \times \left(\frac{\mu_0}{4\pi}\int \frac{\mathbf{J}(\mathbf{r}')}{\mathscr{r}}\, d\tau'\right) = \nabla \times \mathbf{A}, \tag{B.12}$$

where **A** is the vector potential.

---

[2] As a matter of fact, any differentiable vector function *whatever* (regardless of its behavior at infinity) can be written as a gradient plus a curl, but this more general result does not follow directly from the Helmholtz theorem, nor does Eq. B.10 supply the explicit construction, since the integrals, in general, diverge.

Appendix C

Units

In our units (the **Système International**) Coulomb's law reads

$$\mathbf{F} = \frac{1}{4\pi\epsilon_0}\frac{q_1 q_2}{\mathscr{r}^2}\hat{\mathscr{r}} \quad \text{(SI)}. \tag{C.1}$$

Mechanical quantities are measured in meters, kilograms, seconds, and charge is in **coulombs** (Table C.1). In the **Gaussian system**, the constant in front is, in effect, absorbed into the unit of charge, so that

$$\mathbf{F} = \frac{q_1 q_2}{\mathscr{r}^2}\hat{\mathscr{r}} \quad \text{(Gaussian)}. \tag{C.2}$$

Mechanical quantities are measured in centimeters, grams, seconds, and charge is in **electrostatic units** (or **esu**). For what it's worth, an esu is evidently a $(\text{dyne})^{1/2}$-centimeter. Converting electrostatic equations from SI to Gaussian units is not difficult: just set

$$\epsilon_0 \to \frac{1}{4\pi}.$$

For example, the energy stored in an electric field (Eq. 2.45),

$$U = \frac{\epsilon_0}{2}\int E^2\,d\tau \quad \text{(SI)},$$

becomes

$$U = \frac{1}{8\pi}\int E^2\,d\tau \quad \text{(Gaussian)}.$$

(Formulas pertaining to fields inside dielectrics are not so easy to translate, because of differing definitions of displacement, susceptibility, and so on; see Table C.2.)

Quantity	SI	Factor	Gaussian
Length	meter (m)	10^2	centimeter
Mass	kilogram (kg)	10^3	gram
Time	second (s)	1	second
Force	newton (N)	10^5	dyne
Energy	joule (J)	10^7	erg
Power	watt (W)	10^7	erg/second
Charge	coulomb (C)	3×10^9	esu (statcoulomb)
Current	ampere (A)	3×10^9	esu/second (statampere)
Electric field	volt/meter	$(1/3) \times 10^{-4}$	statvolt/centimeter
Potential	volt (V)	$1/300$	statvolt
Displacement	coulomb/meter^2	$12\pi \times 10^5$	statcoulomb/centimeter^2
Resistance	ohm (Ω)	$(1/9) \times 10^{-11}$	second/centimeter
Capacitance	farad (F)	9×10^{11}	centimeter
Magnetic field	tesla (T)	10^4	gauss
Magnetic flux	weber (Wb)	10^8	maxwell
H	ampere/meter	$4\pi \times 10^{-3}$	oersted
Inductance	henry (H)	$(1/9) \times 10^{-11}$	second^2/centimeter

Table C.1 **Conversion Factors.** [*Note:* Except in exponents, every "3" is short for $\alpha \equiv 2.99792458$ (the numerical value of the speed of light), "9" means α^2, and "12" is 4α.]

The Biot-Savart law, which for us reads

$$\mathbf{B} = \frac{\mu_0}{4\pi} I \int \frac{d\mathbf{l} \times \hat{\boldsymbol{\imath}}}{\imath^2} \quad \text{(SI)}, \tag{C.3}$$

becomes, in the Gaussian system,

$$\mathbf{B} = \frac{I}{c} \int \frac{d\mathbf{l} \times \hat{\boldsymbol{\imath}}}{\imath^2} \quad \text{(Gaussian)}, \tag{C.4}$$

where c is the speed of light, and current is measured in esu/s. The Gaussian unit of magnetic field (the **gauss**) is the one quantity from this system in everyday use: people speak of volts, amperes, henries, and so on (all SI units), but for some reason they tend to measure magnetic fields in gauss (the Gaussian unit); the correct SI unit is the **tesla** (10^4 gauss).

One major virtue of the Gaussian system is that electric and magnetic fields have the same dimensions (in principle, one could measure the electric fields in gauss too, though no one uses the term in this context). Thus the Lorentz force law, which we have written

$$\mathbf{F} = q(\mathbf{E} + \mathbf{v} \times \mathbf{B}) \quad \text{(SI)}, \tag{C.5}$$

	SI	Gaussian
Maxwell's equations		
In general:	$\nabla\cdot\mathbf{E}=\frac{1}{\epsilon_0}\rho$ $\nabla\times\mathbf{E}=-\partial\mathbf{B}/\partial t$ $\nabla\cdot\mathbf{B}=0$ $\nabla\times\mathbf{B}=\mu_0\mathbf{J}+\mu_0\epsilon_0\partial\mathbf{E}/\partial t$	$\nabla\cdot\mathbf{E}=4\pi\rho$ $\nabla\times\mathbf{E}=-\frac{1}{c}\partial\mathbf{B}/\partial t$ $\nabla\cdot\mathbf{B}=0$ $\nabla\times\mathbf{B}=\frac{4\pi}{c}\mathbf{J}+\frac{1}{c}\partial\mathbf{E}/\partial t$
In matter:	$\nabla\cdot\mathbf{D}=\rho_f$ $\nabla\times\mathbf{E}=-\partial\mathbf{B}/\partial t$ $\nabla\cdot\mathbf{B}=0$ $\nabla\times\mathbf{H}=\mathbf{J}_f+\partial\mathbf{D}/\partial t$	$\nabla\cdot\mathbf{D}=4\pi\rho_f$ $\nabla\times\mathbf{E}=-\frac{1}{c}\partial\mathbf{B}/\partial t$ $\nabla\cdot\mathbf{B}=0$ $\nabla\times\mathbf{H}=\frac{4\pi}{c}\mathbf{J}_f+\frac{1}{c}\partial\mathbf{D}/\partial t$
D and H		
Definitions:	$\mathbf{D}=\epsilon_0\mathbf{E}+\mathbf{P}$ $\mathbf{H}=\frac{1}{\mu_0}\mathbf{B}-\mathbf{M}$	$\mathbf{D}=\mathbf{E}+4\pi\mathbf{P}$ $\mathbf{H}=\mathbf{B}-4\pi\mathbf{M}$
Linear media:	$\mathbf{P}=\epsilon_0\chi_e\mathbf{E},\quad \mathbf{D}=\epsilon\mathbf{E}$ $\mathbf{M}=\chi_m\mathbf{H},\quad \mathbf{H}=\frac{1}{\mu}\mathbf{B}$	$\mathbf{P}=\chi_e\mathbf{E},\quad \mathbf{D}=\epsilon\mathbf{E}$ $\mathbf{M}=\chi_m\mathbf{H},\quad \mathbf{H}=\frac{1}{\mu}\mathbf{B}$
Lorentz force law	$\mathbf{F}=q(\mathbf{E}+\mathbf{v}\times\mathbf{B})$	$\mathbf{F}=q\left(\mathbf{E}+\frac{\mathbf{v}}{c}\times\mathbf{B}\right)$
Energy and power		
Energy:	$U=\frac{1}{2}\int(\epsilon_0E^2+\frac{1}{\mu_0}B^2)\,d\tau$	$U=\frac{1}{8\pi}\int\left(E^2+B^2\right)d\tau$
Poynting vector:	$\mathbf{S}=\frac{1}{\mu_0}(\mathbf{E}\times\mathbf{B})$	$\mathbf{S}=\frac{c}{4\pi}(\mathbf{E}\times\mathbf{B})$
Larmor formula:	$P=\frac{1}{4\pi\epsilon_0}\frac{2}{3}\frac{q^2a^2}{c^3}$	$P=\frac{2}{3}\frac{q^2a^2}{c^3}$

Table C.2 Fundamental Equations in SI and Gaussian Units.

(indicating that E/B has the dimensions of *velocity*), takes the form

$$\mathbf{F} = q\left(\mathbf{E} + \frac{\mathbf{v}}{c} \times \mathbf{B}\right) \quad \text{(Gaussian).} \tag{C.6}$$

In effect, the magnetic field is "scaled up" by a factor of c. This reveals more starkly the parallel structure of electricity and magnetism. For instance, the total energy stored in electromagnetic fields is

$$U = \frac{1}{8\pi}\int (E^2 + B^2)\, d\tau \quad \text{(Gaussian),} \tag{C.7}$$

eliminating the ϵ_0 and μ_0 that spoil the symmetry in the SI formula,

$$U = \frac{1}{2}\int \left(\epsilon_0 E^2 + \frac{1}{\mu_0}B^2\right) d\tau \quad \text{(SI).} \tag{C.8}$$

Table C.2 lists some of the basic formulas of electrodynamics in both systems. For equations not found here, and for Heaviside-Lorentz units, I refer you to the appendix of J. D. Jackson, *Classical Electrodynamics*, 3rd ed. (New York: John Wiley, 1999), where a more complete listing is to be found.[1]

[1] For an interesting "primer" on electrical SI units see N. M. Zimmerman, *Am. J. Phys.* **66**, 324 (1998).

Index

Abraham-Lorentz formula, 467, 469. 472, 545-546
Absorption, 392-398
Absorption coefficient, 402-403
Acausality, 421, 425, 467
Acceleration,
  field, 438, 460
  ordinary, 521
  proper, 521
Advanced potentials, 425
Advanced time, 425
Alfven's theorem, 341
Ampere (unit), 208, 216
Ampère, A. M., xiii
Ampère dipole, 258, 284
Ampère's law, 225, 232, 321-326, 539
  applications of, 225-232
  in matter, 269-271
  symmetry for, 229
Amperian loop, 225, 239
Angle,
  azimuthal, 38, 43
  of incidence, 388
  of reflection, 388
  of refraction, 388
  polar, 38
Angular momentum, 358-362
  density, 358
Anomalous dispersion, 403-404
Antisymmetric tensor, 535, 537
Atomic polarizability, 161, 200
Auxiliary fields:
  D, 175-182, 271, 273, 328, 545
  H, 269-277, 328-332, 545
Azimuthal angle, 38, 43
Azimuthal symmetry, 137

BAC-CAB rule, 8
Back emf, 314, 317
Bar electret, 170, 178
Bar magnet, 265, 274
Barn and ladder paradox, 491-492
Betatron, 336
Biot-Savart law, 215-220, 339, 532
Bohr atom,
  lifetime, 464-465
  polarizability, 163
Bohr magneton, 252
Bound charge, 166-173, 186, 328
Bound current, 263-268, 277, 328
Boundary conditions:
  for dielectrics, 178-179, 182, 186, 198, 331-333
  for electrodynamics, 53, 331-333
  for electromagnetic waves, 384, 387, 396
  for electrostatics, 87-90
  for Laplace's equation, 116-120
  for magnetic materials, 273-274, 283, 331-333
  for magnetostatics, 240-242
  for Maxwell's equations, 323, 556
  for waves on a string, 370-373
Boundary value problems, 121-145, 186
Bremsstrahlung, 464
Brewster's angle, 390-391

Capacitance, 104
Capacitor, 103-106
  charging, 105-106, 324-325

dielectric-filled, 183
discharging, 290-291
energy in, 105-106, 191
parallel-plate, 74, 104, 183, 231, 525
Cartesian coordinates, 4, 127, 547-548
Cauchy's formula, 404
Causality, 421, 425, 467, 506
Cavity:
in conductor, 99, 117
in dielectric, 177
in magnetic material, 272-273
resonant, 415
Center of momentum, 511
cgs units, xv, 327, 559-562
Charge:
bound, 166-173, 186, 328
conservation (*see* Conservation)
electric, xiv, 58-59
enclosed, 68
free, 160, 175, 393
induced, 97-101
invariance, 525
magnetic (*see* Monopole)
quantization, xiv, 362
Charge density:
line, 62
surface, 62, 102
volume, 62
Child-Langmuir law, 108
Circular polarization, 374
Clausius-Mossotti equation, 200
Coaxial cable, 75, 411-412
Coefficient of:
absorption, 402-403
dispersion, 404
reflection, 386, 391-392
refraction, 404
transmission, 386, 391-392
Colliding beam, 515-516
Collision:
classical, 483
elastic, 514
relativistic, 514-516
Completeness, 132
Complex notation, 369, 378, 382, 401
amplitude, 369
permittivity, 402
susceptibility, 401
wave number, 402
Component, 5, 39
Compton scattering, 514-515
Compton wavelength, 515
Conductivity, 285-286
Conductor, 96-103, 160, 285
"good" and "poor", 393
perfect, 285, 334, 341, 405
surface charge on, 123, 126, 288
Conservation laws:
global, xiv, 345
local (*see* Continuity equation)
relativistic, 510-516
Conservation, electrodynamic, 345-363
angular momentum, 358-361
charge, xiv, 214, 327, 354, 538
energy, 346-349, 386
momentum, 355-357, 442
Constitutive relation, 180, 275, 330, 545
Continuity equation, xiv, 214-215, 327, 345-346, 348, 356, 538-539
Contravariant vector, 501, 543
Coordinates:
Cartesian, 4, 548
curvilinear, 38, 547-554
cylindrical, 43-45, 548
inversion of, 12
rotation of, 10-12
spherical, 38-43, 548
translation of, 12
Cosines, law of, 3
Coulomb (unit), 59, 559
Coulomb gauge, 421, 541
Coulomb's law, xv, 59, 62
magnetic, 328
Covariant vector, 501, 543
Critical angle, 413
Cross product, 3, 6
Curie point, 281
Curl, 16, 19, 552-553

in Cartesian coordinates, 19
in curvilinear coordinates, 553
in cylindrical coordinates, 44
in spherical coordinates, 42
of **A**, 234, 416
of **B**, 221-225
of **D**, 178
of **E**, 65, 76, 302, 330
of **H**, 269, 330
Curl-less fields, 53, 77-79
Current, 208-214
bound, 263-268, 277
displacement, 323
enclosed, 222, 225, 269, 322
free, 269, 277
induced, 304
polarization, 329
steady, 215
Current density, 211-214
four-vector, 537-538
surface, 211-212
volume, 212-213
Curvilinear coordinates, 38, 547-554
Cutoff frequency, 409-411
Cycloid motion, 205-207, 545
Cyclotron motion, 205, 519-520
Cylindrical coordinates, 43-45, 548

D, (*see* Displacement, electric)
d'Alembertian, 422, 542
Del operator, 16
Delta function:
Dirac, 45-52, 157
Kronecker, 158, 352
Density of field lines, 65
Derivative, 13
normal, 90
Diamagnetism, 255, 260-263, 335, 337
Dielectric, 160
constant, 180
linear, 179-196
Diode, vacuum, 107
Dipole, electric, 66, 146, 149-155
energy of, in electric field, 165
energy of interaction of two, 165
field of:
static, 66, 153-155
oscillating, 447
force on, 164-165
induced, 160-163
moment, 149
perfect, 150, 154
permanent, 163
physical, 150, 154
potential of:
static, 146-147, 149
oscillating, 446
radiation, 444-450
runaway motion, 543
torque on, 164
Dipole, magnetic, 243-246
Ampère model, 258, 284
energy of, in magnetic field, 281
energy of interaction of two, 282
field of:
static, 246, 253-254
oscillating, 453
force on, 257-259, 282-283
Gilbert model, 258, 284, 454
moment, 244, 254
of electron, 252
moving, 544
perfect, 245-246
physical, 245-246
potential of:
static, 244, 246
oscillating, 453
radiation, 451-455, 459
Thomson, 362
torque on, 255-257, 259
Dirac, P. A. M., 362
Dirac delta function, 45-52, 157
Dirichlet's theorem, 130
Discharge of capacitor, 290-291
Discontinuity,
in **B**, 241, 274
in **E**, 88-89
Dispersion, 398-405

anomalous, 403-404
coefficient, 404
Displacement current, 323, 325, 330, 339
Displacement, electric, 175-179
Displacement vector:
finite, 1, 8-9
four-vector, 502
infinitesimal:
Cartesian, 9
curvilinear, 547
cylindrical, 44
spherical, 40
Divergence, 16, 17, 549-551
four-dimensional, 539
in Cartesian coordinates, 17
in curvilinear coordinates, 550
in cylindrical coordinates, 44
in spherical coordinates, 42
of **A**, 234
of **B**, 221-223, 330
of **D**, 329-330
of **E**, 65, 69
of **H**, 273
Divergenceless fields, 54, 240
Divergence theorem, 31, 551
Domain, 278-280
Dot product, 2, 5, 501
Drift velocity, 234, 289
Drude, P. K. I., 289
Dual tensor, 537, 545
Duality transformation, 342, 454
Dumbbell model, 469-470

Earnshaw's theorem, 115
Earth's magnetic field, 216
Eddy currents, 298-299, 305
Ehrenfest's paradox, 493
Einstein, A., 303, 478-479
Einstein summation convention, 501
Einstein velocity addition rule, 482-483, 497-498
Einstein's postulates, 477-482
Elastic collision, 514
Electret, 170, 178
Electric (*see* Charge, Current, Dipole, Displacement, Energy, Field, Force, Polarization, Potential, Susceptibility)
Electric field, 58, 61
average over a sphere, 156-157
curl of, 65
divergence of, 65
Electric field of:
dynamic configurations:
arbitrary charge distribution, 427, 457
oscillating electric dipole, 447
oscillating magnetic dipole, 453
parallel-plate capacitor, moving, 525-526, 533
point charge, arbitrary motion, 435-438
point charge, constant velocity, 439, 527-528
point charge moving in straight line, 441
rotating electric dipole, 450
static configurations:
bar electret, 170, 178
conducting sphere in dielectric medium, 199
conducting sphere in external field, 141-142
continuous charge distribution, 61
dielectric cylinder in external field, 190
dielectric sphere in external field, 186-188
dipole, 153-155, 157
disk, 64
finite line, 62-63
infinite cylinder, 72
infinite plane, 73
infinite line, 63, 75
line charge, 62
overlapping spheres, 75, 172
parallel-plate capacitor, 74
point charge distribution, 60

point charge near conducting plane, 121-122
point charge near dielectric plane, 188-190
polarized object, 166-170
ring, 64
sphere, 64, 70
spherical shell, 64, 75
surface charge distribution, 62
uniformly polarized cylinder, 173
uniformly polarized object, 283
uniformly polarized sphere, 168
volume charge distribution, 62
Electromagnetic
induction, 301-320
mass, 472
paradox, 472
radiation, xiv, 443
spectrum, 377
waves, 364-415
Electromotance, 293
Electromotive force, 285-300, 314
Electron,
dipole moment, 252
discovery, 208
radiative time constant, 467
spin, 252, 362
Electrostatics, 59, 215, 225, 232-233
Electrostatic pressure, 103
emf, 285-300, 314
Enclosed charge, 68
Enclosed current, 222, 225, 269, 322
Energy:
conservation, 346-349, 386, 510
flux, 347
in electric field, 346-348
in magnetic field, 317-321, 346
of capacitor, 106
of charge in static field, 90-91
of continuous charge distribution, 93-95
of dipole, 165, 281-282
of electromagnetic wave, 380-382
of inductor, 317
of linear dielectric, 191-193
of point charge distribution, 91
of point charge near conducting plane, 124
of spherical shell, 94-95
of static charge distribution, 90
Energy density:
electromagnetic, 348, 380
electrostatic, 93-96
in linear media, 348
magnetostatic, 318-319
of electromagnetic wave, 380-383
Energy-momentum four-vector, 510
Energy, relativistic, 510
kinetic, 510
rest, 510
Equipotential, 79, 97
Equivalence principle, 476
Ether, 479-481
drag, 480
wind, 479-481
Euler's formula, 369
Evanescent wave, 414
Event, 493

Farad (unit), 104
Faraday, M., xiii, 301
Faraday cage, 101
Faraday's law, 301-310, 321, 336, 378-379, 540
Ferromagnetic domain, 278-280
Ferromagnetism, 255, 278-282
Feynman disk paradox, 359-361
Field (*see also* Electric, Magnetic)
electric, 58
average over sphere, 156-157
curl of, 65
divergence of, 65
in conductor, 97, 285-286
induced, 302, 305-310
macroscopic, 173-175
microscopic, 173-175
line, 65-67
magnetic,

average over a sphere, 253
curl of, 221-225
divergence of, 221-223, 330
macroscopic, 268
microscopic, 268
point, 9, 60
tensor, 535-537, 541
theory, xiii, 52-55, 525
Flux,
electric, 65, 67
magnetic, 295, 300
Flux density, 271
energy, 347
momentum, 356
Flux integral, 24
Flux rule, 296-298, 302-303, 478-479
Force:
density, 351
electric:
between point charges, 59, 439
on conductor, 102-103
on dielectric, 193-196, 199
on electric dipole, 164-165
on point charge in field, 60, 204
on point charge near conducting plane, 123-124
on point charge near dielectric plane, 188-190
on surface charge, 102-103
electromagnetic:
between point charges, 439
Lorentz, 204, 209, 519
magnetic:
between current loops, 250
between monopoles, 328
between parallel currents, 202-204, 217, 220, 522-525
between parallel planes, 231
on current, 209, 211-212
on magnetic dipole, 257-259, 282
on magnetized material, 262
on point charge, 204
Minkowski, 518-519, 521, 540, 543
relativistic, 516
Fourier series, 130
Fourier transform, 370, 412
Fourier's trick, 130, 140
Four vector, 500-502
acceleration, 521
charge/current, 537-538
displacement, 502
energy/momentum, 510
gradient, 543
Minkowski force, 518, 521, 540, 543
position/time, 500
potential, 541-543
velocity, 507-508
Free charge, 175, 186, 393
Free current, 269, 277
Fresnel equations, 390-392
Fringing field, 194-195
Fundamental theorem of calculus, 28
for curls, 34
for divergences, 31, 551
for gradients, 29, 548
Future, 504

Galilean transformation, 494, 502
Galileo, 477
Galileo's principle of relativity, 477
Galileo's velocity addition rule, 482
Gauge:
Coulomb, 421, 541
invariance, 542
Lorentz, 421-422, 542
transformation, 419-420
Gauss (unit), 216, 560
Gaussian:
"pillbox", 72-73
surface, 70-73
units, xv, 327, 559-562
Gauss's law, 65, 67-69, 232, 321, 539
applications of, 70
inside matter, 175-177
symmetry for, 71
Gauss's theorem, 31
Gedanken experiment, 483
Generalized Coulomb field, 438

Generator, 294-300
Gilbert dipole, 258, 284, 454
Global conservation law, xiv, 345
Gradient, 13, 14, 548
- four-dimensional, 543
- in Cartesian coordinates, 13, 14
- in curvilinear coordinates, 548
- in cylindrical coordinates, 44
- in spherical coordinates, 42
- theorem, 29, 548

Green's identity, 56, 121
Green's reciprocity theorem, 157-158
Green's theorem, 31, 56
Ground, 118
Group velocity, 399, 410
Guided wave, 405-412
Gyromagnetic ratio, 252

H, 269-274
Hall effect, 247
Harmonic function, 111
Heaviside-Lorentz units, xv, 561
Helical motion, 205
Helmholtz coil, 249
Helmholtz theorem, 52-53, 555-557
Henry (unit), 313
Hertz, H., xiii, 323
Hidden momentum, 357, 361, 520-521
Homogeneous medium, 182
Horizon, 434-435
Hyperbolic
- geometry, 505-506
- motion, 434, 441, 476, 509, 516, 543

Hysteresis, 280-281

Images, method of, 121-125
- dipole and conducting plane, 165
- parallel cylinders, 127
- point charge and conducting plane, 121-124, 475
- point charge and conducting sphere, 124-126
- point charge and dielectric plane, 190

Incidence:
- angle of, 388
- plane of, 387-388

Incident wave, 370, 384
Index of refraction, 383, 388, 398, 403
Induced
- charge, 97-101, 123, 126
- current, 304
- dipole, 160-163
- electric field, 305-310
- emf, 302

Inductance, 310-316
- mutual, 310-312, 321
- self, 313-315

Induction, 271, 301-320
Inertial system, 477
Inhomogeneous wave equation, 422
Isotropic medium, 184
Insulator, 96, 160
Integration by parts, 37
Intensity, 381
Internal resistance, 293
Internal reflection, 413
Interval, spacetime, 502-503
- lightlike, 503, 505-506
- spacelike, 503, 505-506
- timelike, 503, 505-506

Invariance
- of charge, 525
- of mass, 511
- time-reversal, 425

Invariant, 501, 511, 534, 537
- interval, 501-503
- product, 501

Inversion, 12, 412
Irrotational field, 53, 77-79
Isotropic medium, 184

Jefimenko's equations, 427-429
Joule heating law, 290
Jumping ring, 304-305

Kinetic energy, 510
Kronecker delta, 158, 352

LC circuit, 316
Langevin equation, 200-201
Laplace's equation, 83, 110-114, 116
  in one dimension, 111-112
  in three dimensions, 114, 116
  in two dimensions, 112-114
Laplacian, 23
  in Cartesian coordinates, 22, 111
  in curvilinear coordinates, 235, 554
  in cylindrical coordinates, 44
  in spherical coordinates, 42
  of **A**, 235
  of a scalar, 23
  of V, 83, 87, 110
  of a vector, 23, 235
Larmor formula, 458, 462
Law of cosines, 3
Left-handed coordinates, 6
Legendre polynomials, 138, 148
Lenz's law, 303-304
Levi-Civita symbol, 283
Levitation, 335-336
Liénard formula, 463, 545
Liénard-Wiechert potentials, 429-434
Lifetime, 486, 488
Light, 364-415
  cone, 504
  speed of,
    linear medium, 383
    universal, 481
    vacuum, 376, 480-481
Lightlike interval, 503
Line charge, 62
Line current, 208-209
Line element:
  Cartesian, 9
  curvilinear, 547
  cylindrical, 44
  spherical, 40
Line integral, 24
Linear combination, 130, 369-370
Linear equation, 130, 367
Linear medium, 382-384
  electric, 179-196
  magnetic, 274-277
Linear polarization, 374
Local conservation (*see* Continuity)
Longitudinal wave, 373
Lorentz contraction, 481, 489-493, 497
  paradox, 490-491
Lorentz force law, 202-214, 232, 519, 540, 545
Lorentz gauge, 421-422, 426, 542
Lorentz transformation, 493, 500, 543
Lorentz, H. A., xiii, 469, 481
Lorentz-Lorenz equation, 200

Macroscopic field, 173-175, 268
Magnet, 265, 274
Magnetic (*see* Charge, Dipole, Energy, Field, Flux, Force, Magnetization, Potential, Susceptibility)
Magnetic field, 202-204, 271, 522
  average over a sphere, 253
  curl of, 221-225
  divergence of, 221-223, 330
  in superconductor, 325
  of earth, 216
Magnetic field of:
  dynamic configurations:
    arbitrary charge distribution, 428, 457
    charging capacitor, 324-325
    oscillating electric dipole, 447
    oscillating magnetic dipole, 453
    parallel-plate capacitor, moving, 528-530
    point charge, arbitrary motion, 219, 435-439
    point charge, constant velocity, 440, 532
    solenoid, moving, 530-531
  static configurations:
    bar magnet, 265, 274
    circular loop, 218
    dipole, 246, 253-254
    finite solenoid, 220
    finite straight line, 216

in cavity, 272-273
infinite plane, 226
infinite solenoid, 220, 227, 232, 249
infinite straight line, 217, 221, 226
magnetized object, 263-264, 268
solenoid filled with magnetic material, 275-276
sphere of linear material in external field, 277
spinning sphere, 237, 240, 253
toroidal coil, 229-230
uniformly magnetized cylinder, 265
uniformly magnetized object, 288
uniformly magnetized sphere, 264-265

Magnetic induction, 271, 301-320
Magnetic monopole, 232-233, 248, 258, 327-328, 342, 546
Magnetic susceptibility, 274-275
Magnetism, 522-525
Magnetization, 255-263, 328-330
Magnetomechanical ratio, 252
Magnetostatics, 215, 225, 232, 240, 339
Mass:
electromagnetic, 472
relativistic, 510
renormalization, 472
rest, 510

Massless particle, 512-515
Matrix:
Lorentz transformation, 500
rotation, 11

Maxwell, J. C., xiii, 321-323, 376
Maxwell's equations, 232, 321, 539, 542
in gaussian units, 561
inside matter, 328-330
in vacuum, 327
tensor form, 539
with magnetic monopoles, 327, 546

Maxwell stress tensor, 351-355
Meissner effect, 325
Merzbacher's puzzle, 340
Method (*see* Images, Relaxation, Separation of variables)
Michelson-Morley experiment, 480-481
Microscopic field, 173-175, 268
Minkowski, H., 504
Minkowski
constitutive relations, 545
diagram, 503
force, 518-519, 521-522, 540, 543

mks units, xv, 327, 559-562
Momentum:
angular, 358-361
conservation, 355-357, 442, 510
density, 355-356, 380
flux, 356
four-vector, 510
hidden, 357, 361, 520-521
in electromagnetic field, 349-357
in electromagnetic wave, 380-382
relativistic, 509-511

Monochromatic wave, 376-380
Monopole:
electric, 147, 149, 458
magnetic, 232-233, 243, 248, 327-328, 342, 362, 546

Motional emf, 294-300, 478-479
Multipole expansion:
of electrostatic potential, 146-152
of magnetostatic potential, 242
of radiation fields, 458

Mutual inductance, 310-312, 321

Neumann formula, 311
Newton's laws:
first, 477
second, 472, 516
third, 349-351, 442, 469, 517-518

Normal derivative, 90
Normal incidence, 384-386
Normal vector, 26, 89, 241, 332

Oblique incidence, 386-392
Observer, 484
Octopole, 147, 151, 159, 458

Oersted, C., xiii, 536
Ohm (unit), 287
Ohm's law, 285-291
Operator, 16
Ordinary
  acceleration, 521
  force, 516, 519
  velocity, 507-508
Orthogonal coordinates, 547
Orthogonal functions, 130, 132, 140
Orthogonality, 130, 132, 140

Paradox (*see* Barn and ladder, Ehrenfest, Electromagnetic mass, Feynman disk, Lorentz contraction, Merzbacher, Time dilation, Twin)
Parallel-plate capacitor, 74, 104-105, 183, 231, 525-526
Paramagnetism, 255-257, 263
Past, 504
Path independence, 24-25, 30, 53, 78
Path integral, 24
Perfect conductor, 285, 334, 341, 405
Permanent magnet, 265, 280
Permeability, 216, 274-275, 278, 545
  of free space, 216, 275
  relative, 275
Permittivity, 180, 545
  complex, 402
  of free space, 59, 180
  relative, 180
Phase, 367
  constant, 367, 395-396
  transition, 281
  velocity, 399, 410
Photon, 504, 513-515
Pill box, 72-73
Pinch effect, 247
Planck formula, 513
Plane:
  of incidence, 387-388
  of polarization, 386
  wave, 376-380
Plasma, 247, 341
Point charge (*see* Electric, Force, Magnetic, Monopole, Potential)
Poisson's equation, 83, 110, 235, 274
  for **A**, 235
  for V, 83, 87, 110
Polar angle, 38
Polar molecule, 163
Polarizability,
  atomic, 161
  tensor, 162-163
Polarization (of a medium), 161, 166
  current, 329
  electric, 161, 166, 328-330
  induced, 161
  magnetic (*see* Magnetization)
Polarization (of a wave), 373-375
  angle, 374
  circular, 374
  linear, 374
  vector, 374
Pole (magnetic), 232-233, 248, 258
Position vector, 8-9
Position-time four-vector, 500
Postulates, Einstein's, 477-482
Potential (*see also* Scalar, Vector), 557
  advanced, 425
  electric, 77-82
  four-vector, 541-543
  in electrodynamics, 416-422
  Liènard-Wiechert, 429-434
  magnetic scalar, 236, 239, 251
  magnetic vector, 234-246, 252
  retarded, 422-427
Potential energy, 79
  of a charge configuration, 93
  of a point charge, 91
Power:
  dissipated in resistor, 290, 348
  in electromagnetic wave, 381
  radiated by:
    arbitrary source, 457-458
    oscillating electric dipole, 448, 454
    oscillating magnetic dipole, 453-454

point charge, 460-465
Poynting's theorem, 346-349
Poynting vector, 347, 380-383
Preacceleration, 467, 469, 475
Present, 504
Pressure:
electromagnetic, 353
electrostatic, 103
radiation, 382
Principle (*see* Equivalence, Relativity, Superposition)
Product rules, 20
Propagation vector, 379
Proper
acceleration, 521
time, 507-509
velocity, 507-509
Pseudoscalar, 12
Pseudovector, 12, 204
Pulsar, 474

Quadrupole:
electric, 147, 150-151, 158, 459
magnetic, 243
moment, 158
radiation, 458-459
Quantization of charge, xiv, 362
Quasistatic, 308-309, 429
Quotient rules, 21

RC circuit, 290-291
RL circuit, 320
Radiation, 443-476
damping, 468
electromagnetic, xiv, 443-444, 460
field, 438, 460
pressure, 382
reaction, 465-473, 476
resistance, 450, 454
synchrotron, 465
zone, 446, 453, 456
Radiation by:
arbitrary source, 454-459
electric dipole, 444-450
electric quadrupole, 458-459
magnetic dipole, 451-454, 473-474
point charge, 460-465
in hyperbolic motion, 476
rotating electric dipole, 450
surface current, 474-475
Rapidity, 502
Reference point:
for electric dipole, 151-152
for magnetic dipole, 245
for potential, 78, 80, 82
Reflection, 384-392
angle of, 388
at conducting surface, 396-398
coefficient, 386, 391-392
internal, 413
law of, 388
waves on a string, 370-373
Refraction, 384-392
angle of, 388
coefficient of, 404
index of, 383, 398, 403
law of, 388
Relativistic,
constitutive relations, 545
dynamics, 516-522
electrodynamics, 522-543
energy, 509-511
kinematics, 511-516
mass, 510
mechanics, 507-522
momentum, 509-511
potentials, 541-543
Relativity:
of simultaneity, 483-484, 496-497
principle of, 477-483
special, xi, 477-546
Relaxation, method of, 113
Renormalization,
of charge, 183
of mass, 472
Resistance, 287
Resistivity, 285-286
Resistor, 286

Resonant cavity, 415
Rest energy, 510
Rest mass, 510
Retarded
 position, 429-432
 potentials, 422-427
 time, 423
Reversion of series, 471
Right hand rule, 3
Right-handed coordinates, 6
Rogrigues formula, 138, 144
Rotation, 10
Rotation matrix, 11
Runaway motion, 467, 469, 543

Saturation, 280
Scalar, 1
Scalar potential, 53, 416-442
 dynamic configurations:
  arbitrary charge distribution, 423, 456
  oscillating electric dipole, 446
  oscillating magnetic dipole, 451
  point charge, arbitrary motion, 432
  point charge, constant velocity, 433-434
 static configurations:
  average over a sphere, 114-115
  conducting sphere in external field, 141-142
  continuous charge distribution, 83-84
  disk, 86
  electric dipole, 149
  finite cylinder, 87
  infinite line, 85-86
  multipole expansion, 146-152
  point charges, 84
  polarized matter, 166-170
  ring, 86
  specified charge on surface of sphere, 142-143
  specified electric field, 78, 251
  specified potential on surface of sphere, 139-140
  spherical shell, 81, 85, 144
  surface charge, 85
  uniformly charged object, 283
  uniformly charged sphere, 82, 87
  uniformly polarized sphere, 168-169, 172
  volume charge, 84
Scalar potential, magnetic, 236, 239, 251, 274
Scalar product, 2, 5, 7, 501-502
Second derivative, 22-23
Second-rank tensor, 11-12, 535
Self-force, 469-472
Self-inductance, 312-315
Semiconductor, 286
Separation of variables, 127-145
 Cartesian coordinates, 127-136
 cylindrical coordinates, 145
 spherical coordinates, 137-145
Separation vector, ix, 9, 15, 59, 224
Shear, 353
Shielding, 183
Simultaneity, 483-484, 496-497
Sinusoidal waves, 367-370
SI units, xv, 327, 559-562
Skin depth, 394
Sky, blueness of, 449
Snell's law, 388
Solenoid, 220, 227-228
Solenoidal field, 54, 240
Source charge, 9, 58-59, 202
Source point, 9, 60
Space charge, 107
Spacelike interval, 503
Spacetime:
 diagram, 503-506
 interval, 502-503
 structure, 500-506
Special relativity, xi, 477-546
Spectrum, electromagnetic, 377
Speed,
 of charges in wire, 234, 289

of light in linear medium, 383
of light in vacuum, 376, 480-481
of waves on a string, 366
Spherical coordinates, 38-43
Spherical wave, 412
Standing wave, 367, 410
Stationary charge, 59, 215
Steady current, 215
Step function, 49
Stokes' theorem, 34, 552-554
Stress, 353
Stress tensor, 351-355
String, waves on, 364-374
Summation convention, 501-502
Sun, age of, 109
Sunset, redness of, 449
Superconductor, 335
Superluminal velocity, 399, 484-485
Superposition principle, 58, 81, 96
Surface charge, 62, 102-103, 288
Surface current, 211-212
Surface element, 26, 40
Surface integral, 24, 26
Susceptibility:
complex, 401
electric, 179, 200
magnetic, 274-275, 278
tensor, 184
Symmetric tensor, 535, 537
Symmetry:
azimuthal, 137
duality, 342
for Ampére's law, 229
for Gauss's law, 71
of **E**, **B**, **D**, and **H**, 283
of Maxwell's equation, 327-328
Synchronization, 484, 487-488, 496
Synchrotron radiation, 465

TE waves, 407-411
TEM waves, 407
Tensor, 11-12
antisymmetric, 535, 537
contravariant, 537
covariant, 537
dual, 537, 545
field, 535-537
polarizability, 162-163
second-rank, 11-12, 535
stress, 351-355
susceptibility, 184
symmetric, 535, 537
Terminal velocity, 300, 336
Tesla (unit), 216, 560
Test charge, 58-59, 202
Theta function, 49
Third law, 349-351, 442, 469, 517-518
Thompson-Lampard theorem, 159
Thomson dipole, 362
Three-dimensional wave equation, 376
Threshold, 543
Time,
advanced, 425
constant, 291, 315, 393, 467
dilation, 485-489, 496-497
paradox, 487
retarded, 423
reversal, 425
Timelike interval, 503
TM waves, 407
Toroidal coil, 229-230, 320
Torque:
on electric dipole, 164
on magnetic dipole, 255-257, 259
Total internal reflection, 413
Transformation:
duality, 342, 454
Galilean, 494, 502
gauge, 419-420
Lorentz, 493-498, 500-501, 543
of angles, 493, 499
of charge and current density, 538
of electromagnetic fields, 525-532
of forces, 518
of lengths, 489-493, 497
of momentum and energy, 510
of velocity, 508
Transformer, 338

Translation, 12
Transmission:
  coefficient, 386, 391-392
  line, 340, 411-412
  of waves on a string, 370-373
Transparency, 383
Transverse wave, 373-375, 378, 394
Triangle diagram:
  electrodynamics, 441
  electrostatics, 87
  magnetostatics, 240, 250-251
Triple product, 7
Tunneling, 414, 475-476
Twin paradox, 488, 499

Uniformly
  magnetized cylinder, 265
  magnetized object, 283, 288
  magnetized sphere, 264-265
  moving charge, 440, 532
  polarized cylinder, 173
  polarized object, 167, 283
  polarized sphere, 168-170
Uniqueness theorem, 116-120, 198, 252
Units, 559-562
  ampere, 208, 216
  coulomb, 59, 559
  esu (electrostatic unit), 559-560
  farad, 104
  gauss, 216, 560
  henry, 313
  ohm, 287
  tesla, 216 560
  volt, 81
Unit systems (*see* Gaussian, Heaviside-Lorentz, SI)
Unit vectors, ix, 3-4, 9, 39, 42
  Cartesian, 4
  curvilinear, 39, 547
  cylindrical, 43
  normal, 89
  spherical, 38, 42
Universal speed of light, 481

Vacuum polarization, 183
Vector, 1
  addition, 2, 5
  area, 57, 244
  component, 5, 39
  contravariant, 501
  covariant, 501
  displacement, 1, 8-9
  four-, 500-502
  magnitude, 1
  operator, 16
  polarization, 374
  position, 8
  product,
    cross, 3, 6
    dot, 2, 5
    scalar, 2, 5, 501-502
    vector, 3, 6
  propagation, 379
  pseudo-, 12, 204
  separation, ix, 9, 15, 59-60, 224
  subtraction, 2
  triple products, 7
  unit, (*see* Unit vectors)
Vector potential, 54, 234-246, 416-442
  direction of, 238
  dynamic configurations:
    arbitrary charge distribution, 423, 456
    oscillating electric dipole, 446,
    oscillating magnetic dipole, 453
    point charge, arbitrary motion, 433
    point charge, constant velocity, 433-434
  static configurations:
    arbitrary current configuration, 235-236
    finite line current, 239
    infinite line current, 239
    infinite plane current, 239
    infinite solenoid, 238
    magnetic dipole, 244-246
    magnetized material, 263-264
    multipole expansion, 242-246

specified magnetic field, 251
spinning sphere, 236-237, 253
uniform magnetic field, 239
Velocity:
addition rules, 482-483, 497-498
drift, 234, 289
field, 438, 460
four-, 507-509
group, 399
of light in linear medium, 383
of light in vacuum, 376, 480-481
ordinary, 507
phase, 399
proper, 507
waves on a string, 366
wave, 399
Visible range, 377
Volt (unit), 81
Voltmeter, 337
Volume:
charge, 62
current, 212
integral, 24, 27
Volume element,
Cartesian, 27
curvilinear, 549
cylindrical, 44
spherical, 40

Wave:
complex, 369
guide, 405, 408
length, 368
number, 368
vector, 379
velocity, 366, 376, 399
Wave equation, 364-367, 375-376
for **A**, 422
for **B**, 375-376
for **E**, 375-376
for V, 422
general solution, 367
homogeneous, 366, 376
inhomogeneous, 422
one-dimensional, 366
three-dimensional, 376
Waves:
dispersive, 398
electromagnetic, 364-415
evanescent, 414
guided, 405-412
in conductors, 392-398
in free space, 375-382
in linear media, 382-392
longitudinal, 373
monochromatic, 376
on a string, 364-367
plane, 376-380
sinusoidal, 367-370
spherical, 412
standing, 367, 410
transverse, 373-375, 378
water, 404
Work:
and emf, 295, 317
and potential, 90-91
-energy theorem, 516
relativistic, 516
Work done (*see also* Energy):
against back emf, 317
by magnetic force, 207, 210-211
in charging a capacitor, 105-106
in setting up a charge configuration, 91-93
in moving a charge, 90-91
in moving a dielectric, 194-196
in moving a wire loop, 294-296
in polarizing a dielectric, 191-193
World line, 503-504

VECTOR DERIVATIVES

Cartesian. $d\mathbf{l} = dx\,\hat{\mathbf{x}} + dy\,\hat{\mathbf{y}} + dz\,\hat{\mathbf{z}}$; $d\tau = dx\,dy\,dz$

$$\textit{Gradient}: \quad \nabla t = \frac{\partial t}{\partial x}\hat{\mathbf{x}} + \frac{\partial t}{\partial y}\hat{\mathbf{y}} + \frac{\partial t}{\partial z}\hat{\mathbf{z}}$$

$$\textit{Divergence}: \quad \nabla \cdot \mathbf{v} = \frac{\partial v_x}{\partial x} + \frac{\partial v_y}{\partial y} + \frac{\partial v_z}{\partial z}$$

$$\textit{Curl}: \quad \nabla \times \mathbf{v} = \left(\frac{\partial v_z}{\partial y} - \frac{\partial v_y}{\partial z}\right)\hat{\mathbf{x}} + \left(\frac{\partial v_x}{\partial z} - \frac{\partial v_z}{\partial x}\right)\hat{\mathbf{y}} + \left(\frac{\partial v_y}{\partial x} - \frac{\partial v_x}{\partial y}\right)\hat{\mathbf{z}}$$

$$\textit{Laplacian}: \quad \nabla^2 t = \frac{\partial^2 t}{\partial x^2} + \frac{\partial^2 t}{\partial y^2} + \frac{\partial^2 t}{\partial z^2}$$

Spherical. $d\mathbf{l} = dr\,\hat{\mathbf{r}} + r\,d\theta\,\hat{\boldsymbol{\theta}} + r\sin\theta\,d\phi\,\hat{\boldsymbol{\phi}}$; $d\tau = r^2\sin\theta\,dr\,d\theta\,d\phi$

$$\textit{Gradient}: \quad \nabla t = \frac{\partial t}{\partial r}\hat{\mathbf{r}} + \frac{1}{r}\frac{\partial t}{\partial \theta}\hat{\boldsymbol{\theta}} + \frac{1}{r\sin\theta}\frac{\partial t}{\partial \phi}\hat{\boldsymbol{\phi}}$$

$$\textit{Divergence}: \quad \nabla \cdot \mathbf{v} = \frac{1}{r^2}\frac{\partial}{\partial r}(r^2 v_r) + \frac{1}{r\sin\theta}\frac{\partial}{\partial \theta}(\sin\theta\, v_\theta) + \frac{1}{r\sin\theta}\frac{\partial v_\phi}{\partial \phi}$$

$$\textit{Curl}: \quad \nabla \times \mathbf{v} = \frac{1}{r\sin\theta}\left[\frac{\partial}{\partial \theta}(\sin\theta\, v_\phi) - \frac{\partial v_\theta}{\partial \phi}\right]\hat{\mathbf{r}} + \frac{1}{r}\left[\frac{1}{\sin\theta}\frac{\partial v_r}{\partial \phi} - \frac{\partial}{\partial r}(r v_\phi)\right]\hat{\boldsymbol{\theta}} + \frac{1}{r}\left[\frac{\partial}{\partial r}(r v_\theta) - \frac{\partial v_r}{\partial \theta}\right]\hat{\boldsymbol{\phi}}$$

$$\textit{Laplacian}: \quad \nabla^2 t = \frac{1}{r^2}\frac{\partial}{\partial r}\left(r^2\frac{\partial t}{\partial r}\right) + \frac{1}{r^2\sin\theta}\frac{\partial}{\partial \theta}\left(\sin\theta\frac{\partial t}{\partial \theta}\right) + \frac{1}{r^2\sin^2\theta}\frac{\partial^2 t}{\partial \phi^2}$$

Cylindrical. $d\mathbf{l} = ds\,\hat{\mathbf{s}} + s\,d\phi\,\hat{\boldsymbol{\phi}} + dz\,\hat{\mathbf{z}}$; $d\tau = s\,ds\,d\phi\,dz$

$$\textit{Gradient}: \quad \nabla t = \frac{\partial t}{\partial s}\hat{\mathbf{s}} + \frac{1}{s}\frac{\partial t}{\partial \phi}\hat{\boldsymbol{\phi}} + \frac{\partial t}{\partial z}\hat{\mathbf{z}}$$

$$\textit{Divergence}: \quad \nabla \cdot \mathbf{v} = \frac{1}{s}\frac{\partial}{\partial s}(s v_s) + \frac{1}{s}\frac{\partial v_\phi}{\partial \phi} + \frac{\partial v_z}{\partial z}$$

$$\textit{Curl}: \quad \nabla \times \mathbf{v} = \left[\frac{1}{s}\frac{\partial v_z}{\partial \phi} - \frac{\partial v_\phi}{\partial z}\right]\hat{\mathbf{s}} + \left[\frac{\partial v_s}{\partial z} - \frac{\partial v_z}{\partial s}\right]\hat{\boldsymbol{\phi}} + \frac{1}{s}\left[\frac{\partial}{\partial s}(s v_\phi) - \frac{\partial v_s}{\partial \phi}\right]\hat{\mathbf{z}}$$

$$\textit{Laplacian}: \quad \nabla^2 t = \frac{1}{s}\frac{\partial}{\partial s}\left(s\frac{\partial t}{\partial s}\right) + \frac{1}{s^2}\frac{\partial^2 t}{\partial \phi^2} + \frac{\partial^2 t}{\partial z^2}$$

VECTOR IDENTITIES

Triple Products

(1) $\mathbf{A}\cdot(\mathbf{B}\times\mathbf{C})=\mathbf{B}\cdot(\mathbf{C}\times\mathbf{A})=\mathbf{C}\cdot(\mathbf{A}\times\mathbf{B})$

(2) $\mathbf{A}\times(\mathbf{B}\times\mathbf{C})=\mathbf{B}(\mathbf{A}\cdot\mathbf{C})-\mathbf{C}(\mathbf{A}\cdot\mathbf{B})$

Product Rules

(3) $\nabla(fg)=f(\nabla g)+g(\nabla f)$

(4) $\nabla(\mathbf{A}\cdot\mathbf{B})=\mathbf{A}\times(\nabla\times\mathbf{B})+\mathbf{B}\times(\nabla\times\mathbf{A})+(\mathbf{A}\cdot\nabla)\mathbf{B}+(\mathbf{B}\cdot\nabla)\mathbf{A}$

(5) $\nabla\cdot(f\mathbf{A})=f(\nabla\cdot\mathbf{A})+\mathbf{A}\cdot(\nabla f)$

(6) $\nabla\cdot(\mathbf{A}\times\mathbf{B})=\mathbf{B}\cdot(\nabla\times\mathbf{A})-\mathbf{A}\cdot(\nabla\times\mathbf{B})$

(7) $\nabla\times(f\mathbf{A})=f(\nabla\times\mathbf{A})-\mathbf{A}\times(\nabla f)$

(8) $\nabla\times(\mathbf{A}\times\mathbf{B})=(\mathbf{B}\cdot\nabla)\mathbf{A}-(\mathbf{A}\cdot\nabla)\mathbf{B}+\mathbf{A}(\nabla\cdot\mathbf{B})-\mathbf{B}(\nabla\cdot\mathbf{A})$

Second Derivatives

(9) $\nabla\cdot(\nabla\times\mathbf{A})=0$

(10) $\nabla\times(\nabla f)=0$

(11) $\nabla\times(\nabla\times\mathbf{A})=\nabla(\nabla\cdot\mathbf{A})-\nabla^2\mathbf{A}$

FUNDAMENTAL THEOREMS

Gradient Theorem : $\int_{\mathbf{a}}^{\mathbf{b}}(\nabla f)\cdot d\mathbf{l}=f(\mathbf{b})-f(\mathbf{a})$

Divergence Theorem : $\int(\nabla\cdot\mathbf{A})\,d\tau=\oint\mathbf{A}\cdot d\mathbf{a}$

Curl Theorem : $\int(\nabla\times\mathbf{A})\cdot d\mathbf{a}=\oint\mathbf{A}\cdot d\mathbf{l}$

BASIC EQUATIONS OF ELECTRODYNAMICS

Maxwell's Equations

In general :

$$\begin{cases} \nabla \cdot \mathbf{E} = \frac{1}{\epsilon_0}\rho \\ \nabla \times \mathbf{E} = -\frac{\partial \mathbf{B}}{\partial t} \\ \nabla \cdot \mathbf{B} = 0 \\ \nabla \times \mathbf{B} = \mu_0 \mathbf{J} + \mu_0\epsilon_0 \frac{\partial \mathbf{E}}{\partial t} \end{cases}$$

In matter :

$$\begin{cases} \nabla \cdot \mathbf{D} = \rho_f \\ \nabla \times \mathbf{E} = -\frac{\partial \mathbf{B}}{\partial t} \\ \nabla \cdot \mathbf{B} = 0 \\ \nabla \times \mathbf{H} = \mathbf{J}_f + \frac{\partial \mathbf{D}}{\partial t} \end{cases}$$

Auxiliary Fields

Definitions :

$$\begin{cases} \mathbf{D} = \epsilon_0 \mathbf{E} + \mathbf{P} \\ \mathbf{H} = \frac{1}{\mu_0}\mathbf{B} - \mathbf{M} \end{cases}$$

Linear media :

$$\begin{cases} \mathbf{P} = \epsilon_0 \chi_e \mathbf{E}, \quad \mathbf{D} = \epsilon \mathbf{E} \\ \mathbf{M} = \chi_m \mathbf{H}, \quad \mathbf{H} = \frac{1}{\mu}\mathbf{B} \end{cases}$$

Potentials

$$\mathbf{E} = -\nabla V - \frac{\partial \mathbf{A}}{\partial t}, \qquad \mathbf{B} = \nabla \times \mathbf{A}$$

Lorentz force law

$$\mathbf{F} = q(\mathbf{E} + \mathbf{v} \times \mathbf{B})$$

Energy, Momentum, and Power

Energy : $U = \frac{1}{2}\int \left(\epsilon_0 E^2 + \frac{1}{\mu_0}B^2\right) d\tau$

Momentum : $\mathbf{P} = \epsilon_0 \int (\mathbf{E} \times \mathbf{B})\, d\tau$

Poynting vector : $\mathbf{S} = \frac{1}{\mu_0}(\mathbf{E} \times \mathbf{B})$

Larmor formula : $P = \frac{\mu_0}{6\pi c} q^2 a^2$

FUNDAMENTAL CONSTANTS

ϵ_0	$=$	$8.85 \times 10^{-12}\,\mathrm{C^2/Nm^2}$	(permittivity of free space)
μ_0	$=$	$4\pi \times 10^{-7}\,\mathrm{N/A^2}$	(permeability of free space)
c	$=$	$3.00 \times 10^{8}\,\mathrm{m/s}$	(speed of light)
e	$=$	$1.60 \times 10^{-19}\,\mathrm{C}$	(charge of the electron)
m	$=$	$9.11 \times 10^{-31}\,\mathrm{kg}$	(mass of the electron)

SPHERICAL AND CYLINDRICAL COORDINATES

Spherical

$$\begin{cases} x = r\sin\theta\cos\phi \\ y = r\sin\theta\sin\phi \\ z = r\cos\theta \end{cases} \qquad \begin{cases} \hat{\mathbf{x}} = \sin\theta\cos\phi\,\hat{\mathbf{r}} + \cos\theta\cos\phi\,\hat{\boldsymbol{\theta}} - \sin\phi\,\hat{\boldsymbol{\phi}} \\ \hat{\mathbf{y}} = \sin\theta\sin\phi\,\hat{\mathbf{r}} + \cos\theta\sin\phi\,\hat{\boldsymbol{\theta}} + \cos\phi\,\hat{\boldsymbol{\phi}} \\ \hat{\mathbf{z}} = \cos\theta\,\hat{\mathbf{r}} - \sin\theta\,\hat{\boldsymbol{\theta}} \end{cases}$$

$$\begin{cases} r = \sqrt{x^2+y^2+z^2} \\ \theta = \tan^{-1}(\sqrt{x^2+y^2}/z) \\ \phi = \tan^{-1}(y/x) \end{cases} \qquad \begin{cases} \hat{\mathbf{r}} = \sin\theta\cos\phi\,\hat{\mathbf{x}} + \sin\theta\sin\phi\,\hat{\mathbf{y}} + \cos\theta\,\hat{\mathbf{z}} \\ \hat{\boldsymbol{\theta}} = \cos\theta\cos\phi\,\hat{\mathbf{x}} + \cos\theta\sin\phi\,\hat{\mathbf{y}} - \sin\theta\,\hat{\mathbf{z}} \\ \hat{\boldsymbol{\phi}} = -\sin\phi\,\hat{\mathbf{x}} + \cos\phi\,\hat{\mathbf{y}} \end{cases}$$

Cylindrical

$$\begin{cases} x = s\cos\phi \\ y = s\sin\phi \\ z = z \end{cases} \qquad \begin{cases} \hat{\mathbf{x}} = \cos\phi\,\hat{\mathbf{s}} - \sin\phi\,\hat{\boldsymbol{\phi}} \\ \hat{\mathbf{y}} = \sin\phi\,\hat{\mathbf{s}} + \cos\phi\,\hat{\boldsymbol{\phi}} \\ \hat{\mathbf{z}} = \hat{\mathbf{z}} \end{cases}$$

$$\begin{cases} s = \sqrt{x^2+y^2} \\ \phi = \tan^{-1}(y/x) \\ z = z \end{cases} \qquad \begin{cases} \hat{\mathbf{s}} = \cos\phi\,\hat{\mathbf{x}} + \sin\phi\,\hat{\mathbf{y}} \\ \hat{\boldsymbol{\phi}} = -\sin\phi\,\hat{\mathbf{x}} + \cos\phi\,\hat{\mathbf{y}} \\ \hat{\mathbf{z}} = \hat{\mathbf{z}} \end{cases}$$

书　　名: Introduction to Electrodynamics 3rd ed.

作　　者: David J. Griffiths

中 译 名: 电动力学导论 第 3 版

责任编辑: 高蓉 刘慧

出 版 者: 世界图书出版公司北京公司

印 刷 者: 三河国英印务有限公司

发　　行: 世界图书出版公司北京公司 (北京朝内大街 137 号　100010)

联系电话: 010-64015659

电子信箱: kjsk@vip.sina.com

开　　本: 24 开

印　　张: 25

版　　次: 2011 年 9月

版权登记: 图字:01-2005-4319

书　　号: 978-7-5062-7289-6 / O・551　　定　　价: 39.00 元